Translational Medicine Research

Volume 1

Aims and Scope

In collaboration with National Infrastructures for Translational Medicine (Shanghai), the largest translational medicine research center in China, the book series "Translational Medicine Research" offers a state-of-the-art resource for physicians and researchers alike who are interested in the rapidly evolving field of translational medicine. It features original and observational investigations in the broad fields of laboratory, clinical and public health research, providing practical and up-to-date information on significant research from all subspecialties of medicine and broadening readers' horizons, from bench to bed and bed to bench.

With a focus on global interdisciplinary academic collaboration, the series aims to expedite the translation of scientific discovery into new or improved standards of management and health outcomes practice.

Series Description

Translational medicine converts promising laboratory discoveries into clinical applications and elucidates clinical questions with the use of bench work, aiming to facilitate the prediction, prevention, diagnosis and treatment of diseases. The development of translational medicine will accelerate disease control and the process of finding solutions to key health problems. It is a multidisciplinary endeavor that integrates research from the medical sciences, basic sciences and social sciences, with the aim of optimizing patient care and preventive measures that may extend beyond health care services. Therefore, close and international collaboration between all parties involved is essential to the advancement of translational medicine.

To enhance the aforementioned international collaboration as well as to provide a forum for communication and cross-pollination between basic, translational and clinical research practitioners from all relevant established and emerging disciplines, the book series "Translational Medicine Research" features original and observational investigations in the broad fields of laboratory, clinical and public health research, aiming to provide practical and up-to-date information on significant research from all subspecialties of medicine and to broaden readers' vision horizons, from bench to bed and bed to bench.

Produced in close collaboration with National Infrastructures for Translational Medicine (Shanghai), the largest translational medicine research center in China, the book series offers a state-of-the-art resource for physicians and researchers alike who are interested in the rapidly evolving field of translational medicine. Prof. Zhu Chen, the Editor-in-Chief of the series, is a hematologist at Shanghai Jiao Tong University, China's former Minister of Health, and chairman of the center's scientific advisory board.

More information about this series at http://www.springer.com/series/13024

Robert Chunhua Zhao
Editor

Stem Cells: Basics and Clinical Translation

上海交通大學出版社
SHANGHAI JIAO TONG UNIVERSITY PRESS

Springer

Editor
Robert Chunhua Zhao
Center of Excellence in Tissue Engineering
Institute of Basic Medical Sciences School
of Basic Medicine
Beijing
China

ISSN 2451-991X ISSN 2451-9928 (electronic)
Translational Medicine Research
ISBN 978-94-017-7272-3 ISBN 978-94-017-7273-0 (eBook)
DOI 10.1007/978-94-017-7273-0

Jointly published with Shanghai Jiao Tong University Press, Shanghai

Library of Congress Control Number: 2015943820

Springer Dordrecht Heidelberg New York London

Printed on acid-free paper

Springer Science+Business Media B.V. Dordrecht is part of Springer Science+Business Media
(www.springer.com)

Series Foreword

Over the years, a chasm between biomedical researchers and the patients who may benefit from their discoveries has been opened. On one hand, millions of patients with diseases such as cancer are anxiously waiting for new remedies to save their lives. On the other hand, many exciting basic science discoveries do not have opportunities to find practical applications. Recently emerging translational medicine aims to tie basic research to clinical results and optimize both patient care and preventive measures.

Translational medicine converts promising laboratory discoveries into clinical applications and elucidates clinical questions with the use of bench work, aiming to facilitate prediction, prevention, diagnosis, and treatment of diseases. With the ultimate goal to develop more effective preventive/therapeutic approaches and improve clinical outcomes and health levels, translational medicine is therefore a people (patients and the general population as a whole)-oriented medical practice.

The past three decades have witnessed tremendous advances in China in the development of living conditions, food and nutrition, and the health care system. However, while the economy grows and society rapidly transforms, the health care system faces multiple problems. China bears a complex disease spectrum: On one hand, communicable diseases frequently seen in developing countries remain a heavy burden; on the other hand, chronic diseases commonly found in developed countries are also the leading causes of death and disability in China. The situation shows that the health care system in China is facing great challenges, and a state effort is needed to meet these challenges. Therefore China is deepening its reform to improve its people's welfare. The development of translational medicine will accelerate disease control and finding solutions for health problems.

Translational medicine is a multidisciplinary program that integrates research from the medical sciences, basic sciences, and social sciences, with the aim of optimizing patient care and preventive measures that may extend beyond health care services. Therefore, close collaboration in an international scale among all the parties is essential to the development of translational medicine.

To enhance the aforementioned international collaboration as well as to provide a forum for communication and cross-fertilization among basic, translational, and clinical research practitioners, we launch the book series "Translational Medicine Research". It features original and observational investigations in the broad fields of laboratory, clinical, and public health research, aiming to provide practical up-to-date information in significant research from all subspecialties of medicine and to broaden the readers' vision and horizon from bench to bed and bed to bench.

In close collaboration with National Infrastructures for Translational Medicine (Shanghai), the book series "Translational Medicine Research" serves as a state-of-the-art resource for physicians and translational medical researchers alike who are interested in the rapidly evolving field of translational medicine. As the Editor-in-Chief, I welcome all the researchers in related areas to report the latest bench-to-bedside researches in this series, so that the series can promote human health by accelerating the knowledge dissemination in global community.

Shanghai, China Zhu Chen
May 2015

Preface

It is my great honor to be invited by Professor Zhu Chen as the chief editor of the *Stem Cell Fascicle of Published Engineering*. Chinese scientists have been devoting to the basic research and clinical application in the field of stem cell biology and have obtained a series of original achievement in the flow of exploration of fundamental theory, key scientific problem research, clinical trials of products, and analysis of mechanism back in the lab laboratory in the past decade.

The essence of life lies in stem cells. Stem cell research currently is the hottest and leading part in the field of life science research. Stem cell research acts as a great role in the development of life science providing new medical method for human health service and leading the development of new drugs and technological industrialization. As it were, stem cell research occupies currently the most influential field of scientific research and economy, which advances with times, broad and profound, covers and contains everything. The interdisciplinary research of basis and clinic attracts many senior and junior scientists to participate in, which blooms the field of stem cell.

In China, under the long-term influence of traditional medicine, which is quite different from the western, we have got a unique understanding of life. Thus, stem cells have been given a special definition and called flesh and blood of human body and best product for nourishing and nurturing. Integrating innovative thinking and life science technology into traditional Chinese medicine has promoted stem cell research to achieve in basic scientific findings and clinical applications. The discovery of stem cell is hailed as the model of technological innovation of oriental medicine to which the government has attached high importance by sponsoring considerably through related departments chronically. In the field of life science, China predominates the IPR of core technology of stem cell and has built a leading industry of stem cells all over the world especially in the parts of new drugs and clinical transportation from the same starting line with other countries. It will promote considerably the development of stem cell industry, lead the technological innovation, and make our country one of the developed countries in regenerative medicine of stem cells.

The research of stem cell involves nearly every field of life science and biomedicine and the application of stem cell therapy covers almost all of the clinical major diseases in present. In recent years, stem cell research has repeatedly been named as annual important scientific achievement by the international scientific community for the reason that stem cell and its technology holds great promises for human to cure refractory diseases for a long and healthy life. Therefore, in near future, the storage of stem cell will contribute to cure several major diseases for human and it is sure that the research of stem cell will involve in more fields of health service. The differentiation of multiple system and refinement of division of work will be a trend in the field of stem cell research. Inducing the stem cell to differentiate in specific direction by reprogramming technology makes the treatment of stem cell more targeted. For instance, blood type incompatibility or blood supply insufficiency can be excluded by self-renewal of blood cells.

The field of stem cell research is vibrant and has a bright future; time waits for no one and we race against it. We should develop new technologies constantly, accelerate the research of innovative drugs and establish professional, systemic platform for resource sharing of stem cell and biological treatment research to evaluate immune function of stem cell. We should integrate and increase the utilization efficiency of clinical resource and mobilize fully the enthusiasm of stem cell research in various clinical departments to take advantage of interdisciplinary superiority of stem cell. We should implement the research of clinical transportation and integrate production, education and research to push forward effectively the advance of life science and transformation medicine, drive the development of technological industry of life, and promote national health industry one more step forward by doing what we can do.

Finally I want to express gratitude to the team of authors, which is the authority in the field.

Professor Yanan Du develops and integrates micro/nano scale technologies and biomaterials to advance the understanding and applications of cell/tissue engineering and therapy for drug testing, pathology investigation, and regenerative medicine.

Professor Xiaosong Gu does excellent work in based tissue engineering for the treatment of neurological injuries based on marrow mesenchymal stem cells.

Professor Ying Han explores the clinical applications and mechanisms of stem cell therapy in liver diseases.

Professor Zongjin Li has used state-of-the-art molecular biology technology and developed molecular imaging assays for the study of intact biological systems, especially stem cell therapy.

Professor Naihe Jing mainly focuses on BMP signaling pathway and epigenetic regulation of central neural system development and neural differentiation of pluripotent stem cell.

Professor Xuetao Pei's main research fields are stem cell biology, stem cell therapy and regenerative medicine.

Professor Ji Wu is a pioneer in the study of female germline stem cells in mammals.

Professor Kaichun Wu has been interested in the stem cell research and application for human intestinal diseases including inflammatory bowel disease.

Dr. Rongwen Xi uses drosophila as a genetic model organism for the understanding of microenvironmental and intrinsic mechanism regulating self-renewal and differentiation of stem cell tissues.

Professor Xinhua Xiao's main research interests include the role of genetics and epigenetics in the pathogenesis and progression of type 2 diabetes and monogenic disorders of glucose metabolism.

Professor Limei Yu focuses on the study of mesenchymal stem cells and epithelial cells derived from amnion membrane.

Professor Qi Zhou is a pioneer in the study of iPSCs.

Professor Jianhong Zhu is pioneer in the study of neural stem cells.

Professor Lee is a chaser of regenerative medicine through stem cell based gene therapy.

Professor Nedime Serakinci has internationally proven extensive expertize in Telomere/telomerase biology as well as in stem cell biology and pioneer in cellular immortalization of primary stem cells.

Dr. Mahmut Cerkez Ergoren specialized on the fundamental processes like mutation, recombination and polymorphism that generate DNA diversity and genome instability and thus contribute to disease and drive evolution.

Professor Philip Pastides is an orthopaedic surgeon based in London with an interest in trauma and biomechanics.

Dr. Wasim Khan is an orthopaedic surgeon based in London with an interest in stem cell biology and musculoskeletal tissue engineering.

Professor Seok-Goo Cho focuses on researching the underlying mechanisms of mesenchymal stem cell-based immune modulation for various applications and ultimately aims to enhance the therapeutic potential of mesenchymal stem cells through novel approaches including combinatory cell-based therapy, gene-modification and polarization in immune-mediated disorders.

Dr. Nayoun Kim pursues basic and translational research aimed at employing mesenchymal stem cell-based therapy to regulate the immune system in various pathological conditions.

Dr. Robert Henning is the Director of the Center for Cardiovascular Research at the James A. Haley Hospital/University of South Florida College of Medicine and has a special research interest in human umbilical cord stem cells in the repair of hearts damaged from myocardial infarction and myocardial ischemia and injury.

Contents

Part I
Basic Concepts of Stem Cells

Chapter 1
Primordial Germ Cells and Germ Line Stem Cells

Ji Wu, Zhuxia Zheng, Hu Wang, Xingxing Mei, Xingbao Ding and Xiaoyong Li

Abstract Germ cells are a sexual reproductive cell type at any stage from primordial germ cells (PGCs) to mature gametes. Germ line stem cells are important for genetic transmission to future generations. In this review, we focus on female germ line stem cells (FGSCs), spermatogonial stem cells (SSCs), and PGCs. In addition, we summarize current research progress concerning PGC specification, migration, and development, SSC properties, their niche, and fate decisions, as well as the history and current research of FGSCs and their applications.

Keywords Primordial germ cells · Germ line stem cells · Spermatogonial stem cells

1.1 The Origin and Fate of Germ Cells in Mammals

In many organisms, a primary event during development is the segregation of germ cells from somatic cells. Germ cell development ensures the perpetuation of genetic information across the generations. In mammals, primordial germ cells (PGCs) are the first cell type established during embryogenesis and are the common precursors of both oocytes and spermatozoa.

1.1.1 PGC Specification in Mammals

In many invertebrates, PGCs are determined by the inheritance of maternal factors deposited in the egg, and only blastomeres containing germ cell determinants

J. Wu (✉) · Z. Zheng · H. Wang · X. Mei · X. Ding · X. Li
Key Laboratory for the Genetics of Developmental and Neuropsychiatric Disorders (Ministry of Education), Bio-X Institutes, Shanghai Jiao Tong University, No. 800, Dongchuan Road, Minhang District 200240, Shanghai, China
e-mail: jiwu@sjtu.edu.cn

R.C. Zhao (ed.), *Stem Cells: Basics and Clinical Translation*, Translational Medicine Research 1, DOI 10.1007/978-94-017-7273-0_1

develop into germ cells. However, in mammals, pluripotent epiblast cells acquire a germ cell fate in response to extrinsic and intrinsic signaling molecules. Prior to gastrulation, the mouse embryo consists of three distinct cell lineages: the epiblast, extraembryonic endoderm, and trophectoderm. PGCs are derived from the proximal fraction of the population of epiblast cells that will mainly give rise to the extraembryonic mesoderm. Transplantation experiments have demonstrated that signals from extraembryonic tissues are critical for PGC fate specification (Tam and Zhou 1996). In the mouse, PGCs are identified as an alkaline phosphatase-positive cell population in the developing allantois [embryonic day (E) 6.5–7.5].

Bone morphogenetic protein (BMP) signaling from extraembryonic tissues is essential for PGC specification in mouse embryos. *Bmp4*, which is expressed in the extraembryonic ectoderm prior to gastrulation and subsequently in the extraembryonic mesoderm, is required for the generation of PGCs. A previous study of *Bmp4* mutant embryos revealed a significant reduction in the number of PGCs in heterozygous mutant embryos, and no PGCs were detected in homozygous mutant embryos (Lawson et al. 1999). *Bmp8b*, which is expressed in the extraembryonic ectoderm in pregastrula and gastrula stage mouse embryos, is also required for PGC generation (Ying et al. 2000). *Bmp4* and *Bmp8b* may form heterodimers to induce the formation of PGCs (Ying et al. 2001). Targeted inactivation of the *Bmp2* gene, which is primarily expressed in the endoderm of pregastrula and gastrula stage mouse embryos, significantly reduces the number of PGCs (Ying and Zhao 2001). Moreover, WNT signaling in the epiblast plays a role in PGC formation. Wnt3 in the epiblast ensures its responsiveness to BMP4 for PGC differentiation (Ohinata et al. 2009). *Dullard* (also known as C-terminal domain nuclear envelope phosphatase 1; *Ctdnep1*) is a member of the serine/threonine phosphatase family of the C-terminal domain of eukaryotic RNA polymerase II. A recent study revealed that *Dullard* is essential for the formation of PGCs in the mouse embryo as a positive regulator of WNT signaling (Tanaka et al. 2013).

After induction by BMP and WNT signals, epiblast cells are regulated by PR domain proteins PRDM1 (also known as B lymphocyte induced maturation protein 1, *Blimp1*) and PRDM14. PRDM1, a potential transcriptional repressor of a histone methyltransferase subfamily, has a critical role in the foundation of the mouse germ cell lineage. PRDM1 promotes the expression of *Stella* (also known as *Dppa3*), a definitive PGC marker, and represses the expression of somatic cell genes, particularly members of the Hox gene family. In *Prdm1* knockout mouse embryos, PGC-like cells fail to repress the expression of somatic cell genes, *Hoxa1* and *Hoxb1*, and PGCs lacking PRDM1 do not properly migrate or proliferate (Ohinata et al. 2005). *Prdm14*, a PR domain-containing transcriptional regulator, has been found to be important for PGC specification in mice. Similar to *Prdm1*-knockout mice, PGCs are almost completely lost by E12.5 in *Prdm14* mutant embryos (Yamaji et al. 2008). Another study has demonstrated that a conserved mesodermal factor, T, which is induced by WNT3, is essential for the activation of *Prdm1* and *Prdm14* via binding to distinct regulatory elements in these genes for direct upregulation, thereby delineating the downstream PGC program (Aramaki et al. 2013). Recently, an in vitro study revealed that simultaneous overexpression of *Prdm1*, *Prdm14*, and *Tfap2c* (also known as *AP2γ*) rapidly and efficiently directs epiblastlike cells derived from embryonic stem cells (ESCs) or induced pluripotent stem cells or (iPSCs) into a PGC state (Nakaki et al. 2013). Another study demonstrated that, in principle, PRDM1, AP2γ, and PRDM14 are sufficient for PGC specification and the unprecedented resetting of the epigenome toward a basal state (Magnusdottir et al. 2013).

1.1.2 PGC Migration in Mammals

In the mouse, PGCs begin to migrate from the primitive streak to the endoderm (the future hindgut) at E7.5. They continue to migrate through the hindgut endoderm at E8.0 and then migrate bilaterally toward the dorsal body wall at E9.5 to finally colonize the gonadal ridges at E10.5.

There are several factors that play important roles during PGC migration in mammals. PGCs lacking β1 integrins fail to migrate normally to the gonads (Anderson et al. 1999). IFITM (interferon-induced transmembrane) proteins are cell surface proteins implicated in diverse cellular processes including cell adhesion. Knockdown of *Ifitm1* by RNA interference in the primitive streak leads to failure of PGC migration into the endoderm (Tanaka et al. 2005). During the later stage of migration, the interaction of stromal cell-derived factor 1 (expressed by the body wall mesenchyme and genital ridges) and its G-protein-coupled receptor, chemokine (CXC motif) receptor 4 (expressed by the migrating germ cells), is required for the colonization of the gonads by PGCs (Molyneaux et al. 2003). *Foxc1* encodes a forkhead/winged-helix transcription factor expressed in many embryonic tissues. Many PGCs fail to migrate normally to the gonadal ridge in *Foxc1* null mouse embryos, remaining trapped in the hindgut, although the germ cells are specified correctly (Mattiske et al. 2006).

1.1.3 PGC Development in Mammals

Following gonadal sex determination, germ cells in the testis initially proliferate and then undergo mitotic cell cycle arrest at G0/G1. The germ cells that differentiate from PGCs to type A spermatogonia, including spermatogonial stem cells (SSCs), are termed gonocytes. Gene expression patterns change dynamically during the transition from PGCs to gonocytes and SSCs (Culty 2009). After arriving at the genital ridge at approximately 10.5 days post-coitus (dpc), female germ cells are called oogonia and develop into clusters of cells called germ line cysts or oocyte nests. Subsequently, the oogonia enter meiosis and become oocytes. During fetal and neonatal development, germ line cysts break apart into single oocytes, which are intruded by pregranulosa cells to form primordial follicles (Pepling 2006, 2012; Pepling and Spradling 2001).

In mammals, meiotic initiation occurs at different time points in male and female germ cells. Female germ cells enter meiosis at around 13.5 dpc and arrest at the diplotene stage beginning at 17.5 dpc (Speed 1982), whereas male germ cells start meiosis at puberty. Retinoic acid (RA) is produced in the mesonephros of both sexes, which is postulated to diffuse or flow into the adjacent gonad. *Stra8* (stimulated by retinoic acid gene 8), which is induced by RA, is a premeiotic gene required for meiotic initiation. In the fetal ovary, high levels of RA induce germ cells to enter meiosis (Baltus et al. 2006; Bowles et al. 2006; Koubova et al. 2006; Vernet et al. 2006; Lin et al. 2008). However, meiosis is not triggered in the fetal testis because RA is degraded by the retinoid enzyme CYP26B1. In *Cyp26b1*-knockout male fetal gonads, germ cells enter meiosis (Bowles et al. 2006). A study of *Cyp26b1*-, *Fgf9* (fibroblast growth factor 9)-, and double-knockout embryos demonstrated that fibroblast growth factor (FGF) 9 produced in the fetal testis acts directly on germ cells to inhibit meiosis, making them less responsive to RA (Bowles et al. 2010). A recent study showed that PRC1 (polycomb repressive complex 1) has gene dosage effects on PGC development and coordinating the timing of sex differentiation of female PGCs by antagonizing extrinsic RA signaling to ensure proper timing of meiotic induction (Yokobayashi et al. 2013). In addition, *Dazl* (Lin et al. 2008), *Msx1/2* (Le Bouffant et al. 2011), *Dmrt1* (Matson et al. 2010; Krentz et al. 2011), *Nodal* (Souquet et al. 2012), and Notch pathways (Feng et al. 2014) regulate the initiation of meiosis. However, a study of *Raldh2* (retinaldehyde dehydrogenase-2)-knockout mice lacking RA synthesis and signaling in the mesonephros and adjacent gonad revealed that STRA8 expression in the fetal ovary does not require RA signaling (Kumar et al. 2011).

The conventional theory is that all germ cells in the fetal ovary enter meiosis, thereby committing to oogenesis. The number of germ cells is determined after birth. However, this view has been challenged. There are reports that female germ line stem cells (FGSCs) with the ability to produce functional oocytes still exist in neonatal and adult mouse ovaries (Zou et al. 2009). Subsequently, FGSCs have been discovered in the ovaries of reproductive-age woman (White et al. 2012) and rats (Zhou et al. 2014).

1.2 Female Germ Line Stem Cells

1.2.1 Introduction

FGSCs are a new class of germ cells in mammals. The recent identification and isolation of FGSCs from mouse and human ovaries have opened a new research area in stem cell biology, developmental biology, and reproductive biology as well as reproductive medicine. Although we know little about FGSCs and significant research needs to be performed at present, we believe that FGSCs might shed light on the preservation of fertility in reproductive-age women under the conditions of premature ovarian failure or chemotherapy. Recently, FGSCs were isolated and cultured from postnatal mammals, which allows us to study their biological characteristics and applications. In this section, we will discuss the progress of FGSC research.

1.2.2 History of FGSC Research

In the early 1950s, it was thought that postnatal germ line stem cells (GSCs) only existed in the male testis. However, in females, a fixed number of primordial follicles exist in the ovaries, and the defined follicle pool serves as the source of oogenesis over the life span of mammals (Zuckerman 1951; Rudkin and Griech 1962; Borum 1967; Peters and Crone 1967). From then on, although there have been different views from some researchers (Ying and Zhao 2001; Ohinata et al. 2005, 2009; Tanaka et al. 2013; Yamaji et al. 2008), the existence of a non-renewing follicle pool after birth in mammals has become a central dogma in classical reproductive biology.

Recently, Johnson et al. (2004) suggested that the female ovary may have regenerative activity in juvenile and adult mice in vivo by examining changes in follicle numbers from birth to adulthood. Subsequently, they showed that peripheral blood (PB) or bone marrow (BM) transplantation restores oocyte production in wild-type mice sterilized by chemotherapy and in ataxia telangiectasia-mutated gene-deficient mice (Johnson et al. 2004). Therefore, they concluded that BM and PB may be potential sources of female germ cells that can generate oocytes in adulthood. Unfortunately, this view spawned a wave of skepticism and controversy, as well as reports with contradictory findings, claiming there is no evidence for the formation of oocytes from BM cells in mice (Eggan et al. 2006; Gosden 2004).

In 2009, our laboratory successfully isolated FGSCs from neonatal and adult mouse ovaries by two enzymatic digestion steps and mouse vasa homolog (MVH)-magnetic bead sorting. Furthermore, a neonatal mouse FGSC line was established and cultured for more than 1 year, whereas the adult mouse FGSCs was cultured for more than 6 months. These long-term cultured FGSCs maintained a normal karyotype, high telomerase activity, and their capacity to differentiate into functional oocytes, and offspring were generated after transplantation into ovaries (Zou et al. 2009). Considering the low purification efficiency based on MVH-magnetic bead sorting, we screened other germ cell-specific markers and found that the germ line-specific protein Fragilis as a selection maker can remarkably improve the purification efficiency (Zou et al. 2011). Moreover, transgenic or gene knockdown mice were prepared by FGSC transplantation. The gene transfer efficiency was up to 29–37 % (Zhang et al. 2011). In addition, we isolated and cultured rat FGSCs with the abilities to produce fat-1 transgenic rats after transplantation in vivo and differentiate into oocytes in vitro (Zhou et al. 2014).

In 2012, White et al. extended our previously described protocol and culture system by isolating FGSCs from adult mice and reproductive-age

(20–30-year-old) women using MVH as the selection maker and fluorescence-activated cell sorting (FACS) (White et al. 2012; Woods and Tilly 2013).

In fact, FGSCs are found not only in rodents (mice and rats) and primates (humans), but also in other animals including vertebrate species such as fish including *zebra fish* (Wong et al. 2013) and *teleost medaka* (Nakamura et al. 2010). More importantly, using a retrospective phylogenetic-based method, a study showed preservation of the female germ line in both young and old mice (Reizel et al. 2012). Therefore, the existence of FGSCs has been demonstrated through cell biology and genetic analysis.

1.2.3 Current FGSC Research Progress

1.2.3.1 FGSC Origin and Their Location

In the mouse, PGCs arise within the proximal epiblast, begin to migrate along the hindgut at E8.5, and then arrive at the genital ridge at around E10.5. PGCs proliferate during their migration, thereby increasing their population. In the gonadal ridge, PGCs are considered as oogonia (Durcova-Hills et al. 2003). The oogonia divide mitotically in a short period. Subsequently, oogonia cease mitosis and enter meiosis I and arrest at this phase. Based on the current research of FGSCs, not all oogonia enter into meiosis, and a small number of GSCs exist during reproductive life (Bukovsky et al. 2008). However, the exact biological processes of differentiation from PGCs to FGSCs are unknown. Single-cell analysis and real-time, high-resolution imaging systems might facilitate future studies of these processes.

To investigate the location of FGSCs, bromodeoxyuridine (BrdU) has been injected into female mice followed by dual immunofluorescence staining of BrdU and MVH. The results indicated that FGSCs are located in the cortical surface of ovaries (Zou et al. 2009).

1.2.3.2 FGSC Isolation and Culture

Separation of FGSCs from ovaries requires knowledge of both the ovarian tissue structure and cell morphology. A schematic diagram of the major steps for FGSC isolation is shown in Fig. 1.1. Generally, there are two main methods to obtain pure FGSCs from a single cell suspension after two enzymatic (collagenase IV and trypsin) digestion steps, namely the differential plating method and the immunotargeted purification method [magnetic-activated cell sorting (MACS) and FACS]. Immunotargeting is largely based on a specific antibody targeting the surface markers on GSCs.

To establish FGSC lines, FGSC isolation and purification protocols have been developed based on techniques for isolation and in vitro expansion of SSCs. Such a method described in our online protocol is able to isolate FGSCs from ovarian tissue (Wang et al. 2013). It is important to note that the homogeneity of

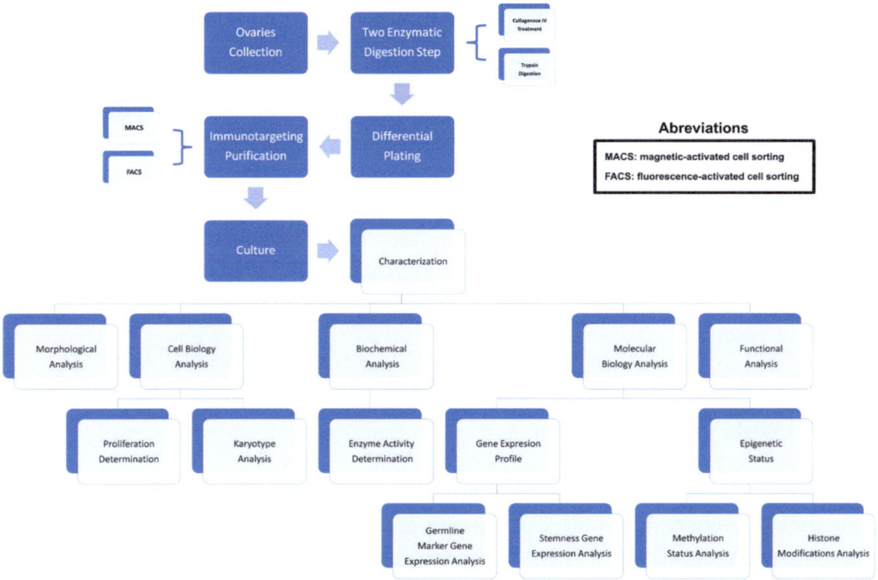

Fig. 1.1 Schematic diagram of major steps for FGSCs

the starting materials and standardization of the isolation protocol are key factors for obtaining desired cells. In present, there is no unique surface marker for GSCs (SSCs and FGSCs) purification. Therefore, the markers should be carefully selected for GSC isolation. Using germ line surface markers may obtain GSCs, whereas other pluripotency-related makers such as stage-specific embryonic antigen (SSEA)-1 may not be appropriate for GSC selection (Nakaki et al. 2013). In our opinion, regardless of the surface marker, probing the biological identity of the obtained cells is an issue of urgent priority.

In addition, a stable culture system is crucial to maintain the propagation and features of FGSCs in vitro. From our experience in stem cell culture, we believe that basic medium, a feeder layer, and growth factors play a major role in FGSC culture, although there is still some discrepancy between optimal culture conditions and the microenvironment of FGSCs in vivo. Growth factors, such as glial cell line-derived neurotrophic factor (GDNF), FGF2, epidermal growth factor (EGF), and leukemia inhibitory factor (LIF), are important for SSC and FGSC propagation (Wu et al. 2008; Xiong et al. 2011; Yuan et al. 2009). Among these factors for in vitro culture, GDNF is critical for GSC proliferation.

1.2.3.3 FGSC Characterization

FGSCs can be characterized based on SSC and other stem cell-related research by their morphology and gene expression profiles, as well as functional assays.

Interestingly, isolated FGSCs have a morphology common with freshly isolated SSCs, including a large cell body with little cytoplasm, helical nuclei with slight staining, a large nucleus/cytoplasm ratio, and a nuclear diameter of 12–20 μm. The growth pattern of FGSCs and SSCs is also similar. For example, most FGSCs grow with a typical grapelike morphology in primary culture. Both FGSCs and SSCs express germ cell-specific markers (MVH, Fragilis, Blimp-1, Dazl, and Stella) but not pluripotency-related proteins (Nanog, SSEA-1, and Sox2). Moreover, long-term cultured FGSCs maintain a normal karyotype (40, XX), alkaline phosphatase activity, and a female imprinting pattern (Zou et al. 2009). In addition, the most important functional analysis of FGSCs is through oogenesis in vitro or in vivo.

1.2.3.4 FGSC Transplantation

In the mouse, transplantation has been used as a functional assay to study the biological characteristics of GSCs. Although SSC transplantation is considered as a quite mature technology, FGSC transplantation research is still lacking. In fact, transplantation can be divided into two categories: direct in situ injection (Zou et al. 2009) and tissue grafting (White et al. 2012). Although the grafting has advantage that GSCs still remain in their microenvironment and interact with their neighboring or supporting cells, the direct injection strategy can meet the need of gametogenesis requirement (Zou et al. 2009). Recent studies have shown that the combination of organ culture and transplantation provides a new strategy for functional sperm preparation in vitro (Gohbara et al. 2010; Yokonishi et al. 2013; Sato et al. 2011a, b, 2012, 2013). However, whether functional oocytes can be generated from FGSCs using this platform is still unknown.

To eliminate effects from endogenous germ cells, recipient females of transplantation can be sterilized with cyclophosphamide and busulphan. Furthermore, non-endogenous germ cells generated by genetic modification would be more convenient for transplantation. To ensure a good outcome after transplantation, some critical points need consideration, such as carefully moving the intestines away from the inside of the abdominal cavity and not damaging the connective tissue or underlying structures of the ovaries. The details of transplantation have been described previously (Wang et al. 2013).

1.2.4 Applications of FGSCs

Stem cells have a great potential for use in regenerative medicine because of their self-renewal and multi-lineage differentiation abilities. From a clinical perspective, as a new type of adult stem cell, FGSCs may be applicable from the preservation of endangered species to ovarian aging therapy, as well as treating infertility caused by radiation and chemotherapy, even though embryo and oocyte cryopreservation are currently available to restore fertility. Moreover, FGSCs are an alternative source of mitochondria for ooplasmic transfer (Harvey et al. 2007; Barritt et al. 2001). From a basic research perspective, as a female germ cell precursor, FGSCs can be studied to understand the molecular mechanisms of oogenesis and folliculogenesis. Although numerous reports have shown that pluripotent stem cells including ESCs and iPSCs are able to differentiate into oocytes (Hubner et al. 2003; Hayashi et al. 2012), their direct differentiation is currently limited by low efficiency.

Since the discovery of pluripotent stem cells, they have created a new research area. Recent studies have reported that mouse SSCs can be converted into pluripotent stem cells under certain culture conditions (Ko et al. 2010; Conrad et al. 2008; Guan et al. 2006; Kanatsu-Shinohara et al. 2004, 2008; Seandel et al. 2007; Golestaneh et al. 2009; Kossack et al. 2009). Based on our previous research, we have found that FGSCs share common characters with SSCs, including their shape, growth pattern, and functions during gametogenesis (Wu et al. 2013). Therefore, we attempted FGSC conversion to pluripotent stem cells. As a result, similar to SSC conversion, we found that FGSCs can be converted into pluripotent stem cells under certain culture conditions (Wang et al. 2014). Consequently, the generation of patient-specific FGSC-derived pluripotent stem cells is feasible and provides a foundation for personalized regenerative applications. Moreover, SSCs can transdifferentiate into reproductive and non-reproductive cells and tissues in certain microenvironments (Zhang et al. 2013; Simon et al. 2009). Whether FGSCs can transdifferentiate into other types of cells is still unknown. If FGSC transdifferentiation occurs, FGSCs will become more widely applicable.

Although FGSCs have a wide range of applications, which is similar to that of SSCs (shown in Fig. 1.2), we must have a clear understanding of these cells. To reveal more aspects of FGSC biology, studies of SSCs in mice and FGSCs in *Caenorhabditis elegans* and *Drosophila* can offer us new insights for further exploration. More importantly, new technologies and equipment used by scientists with different backgrounds will be helpful to further FGSC research.

1.3 Spermatogonial Stem Cells

1.3.1 Introduction

Continual spermatogenesis lays the foundation for male fertility, which is highly dependent on SSCs, a very small population accounting for only about 0.02–0.03 % of the germ cell population (Tegelenbosch and de Rooij 1993). The existence of SSCs has been proposed since the 1950s, but the related research progress has been difficult and little has been clarified in this field (De Rooij and Russell 2000). Traditional studies of SSCs highly relied on morphology and not considered the deeper aspects of their molecular mechanisms. In 1994, Brinster and colleagues developed a transplantation technique to investigate SSC functionally (Brinster and Avarbock 1994).

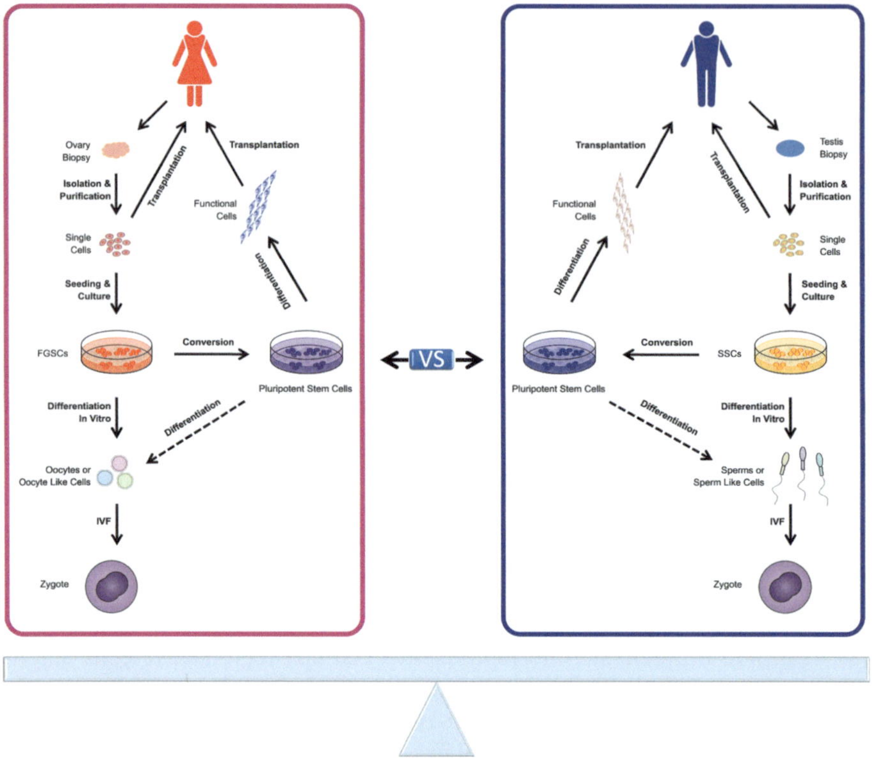

Fig. 1.2 FGSCs have a wide range of applications similar to that of SSCs

Briefly, donor testicular cells are dissociated and transplanted into the efferent duct of infertile recipient mice. After 6 weeks to 2 months, offspring are produced with the donor haplotype. This technique is of great importance because it allows relatively easy identification of SSCs and counting of SSC numbers by considering that each colony in the seminiferous tubules arises from a single SSC (Kanatsu-Shinohara et al. 2006). Another important milestone was the development of an in vitro SSC culture system in 2003 by Kanatsu-Shinohara et al. (2003). In the presence of GDNF, FGF2, LIF, EGF, and other cytokines, germ cells from neonatal mice are able to proliferate and form clusters of spermatogonia in long-term culture in vitro. Transplantation experiments have confirmed that SSC numbers are greatly increased in this system. In vitro culture systems are of great importance because they allow in vitro studies and the generation of large numbers of SSCs for molecular and biochemical studies (Kanatsu-Shinohara and Shinohara 2013). Owing to these two techniques and other traditional methods, SSC studies have advanced further to molecular mechanisms and signal transduction.

1.3.2 SSC Properties

1.3.2.1 SSC Classification

SSCs residue on the basement compartment of seminiferous tubules and are surrounded by a highly complex microenvironment called the niche that is responsible for sophisticated and orchestrated regulation of the balance between SSC self-renewal and differentiation (Kostereva and Hofmann 2008). In mice, SSCs are undifferentiated spermatogonial cells of As (A-single), Apr (A-paired), and Aal (A-aligned chains of 4, 8, and 16 cells, and 32 in rare cases) configurations based on their topographical arrangement (Ohinata et al. 2005; Yamaji et al. 2008). Undifferentiated as spermatogonial cells are thought to be the most primitive type of spermatogonial cell, which will divide into two Apr cells interconnected by intercellular bridges, further division produces intercellular interconnected Aal4, Aal8, and Aal16 cells. Aal spermatogonia convert to differentiating type A spermatogonia (A1–4) that further progress to In (intermediate) and B spermatogonia. Finally, type B spermatogonia divide into primary spermatocytes, and mitosis converts to meiosis to produce haploid spermatozoa (Phillips et al. 2010). A single SSC undergoes 11–12 divisions on average to eventually produce 2048 or 4096 spermatozoa (De Rooij and Russell 2000) (Fig. 1.3). Classically, Apr and Aal cells are thought to be the progenitors committed to differentiate. However, increasing evidence shows that some Apr and Aal cells are potential SSCs. Through lineage tracing in a transplantation assay, Nakagawa et al. (2007) found that transit-amplifying spermatogonia are also able to form colonies in a germ cell-depleted testis,

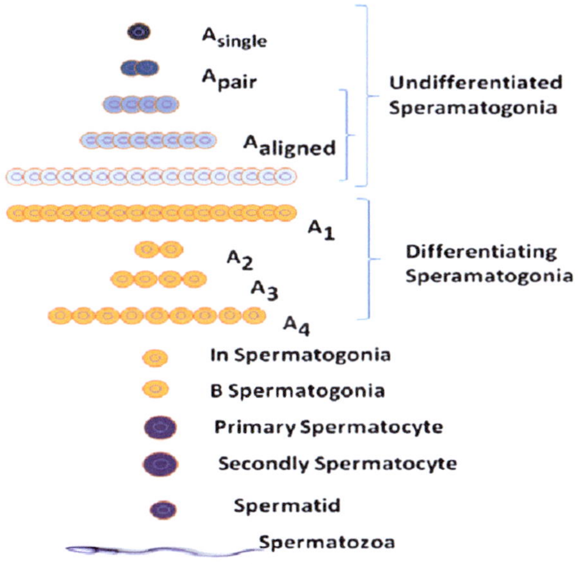

Fig. 1.3 The process of spermatogenesis

indicating their stem cell ability. Moreover, during tissue regeneration after testis injury, a significantly greater number of cells contribute to the stem cell pool than that under normal conditions, further confirming the progenitor cell potential.

Compared with rodents, the true identity of SSCs remains largely unknown in adult humans. Unlike propose model in mice previously mentioned, in a widely accepted model, there are two kinds of type A spermatogonia in human: a dark type A spermatogonia (A_{dark}) and pale type A spermatogonia (A_{pale}) according to their staining pattern and nuclear morphology. Both types of A spermatogonia are stem cells. A_{pale} spermatogonia are the active stem cells responsible for normal self-renewal and generating type B spermatogonia, which further divide into spermatozoa, while A_{dark} spermatogonia are the reserve stem cells with rare mitotic activity but are active under injury and disease states (Clermont 1963, 1966, 1972). However, this model has been challenged by Ehmcke and colleagues who proposed a revised model in which A_{pale} spermatogonia are self-renewing progenitors and A_{dark} spermatogonia are the true stem cells. In a non-human study, they showed that A_{pale} spermatogonia undergo higher mitosis than previously thought and their increase in number affects the total number of germ cells (Ehmcke and Schlatt 2006; Ehmcke et al. 2006). Nevertheless, a lack of evidence has limited our understanding of human SSCs and more studies are required.

1.3.2.2 SSC Characteristics

The lack of specific SSC markers has greatly hindered our understanding of SSCs. However, expression profiles are slowly being revealed spermatogonia stem (progenitor) cells (SPC), indicating exclusive expression of many genes. The strong adherence of SSCs to laminin, the main component of the extracellular matrix of basement membranes, led to the clarification of β1 and α6 integrins as surface markers of SSCs (Shinohara et al. 1999). Subsequently, more surface markers have been identified, such as thymus cell antigen-1 (Thy-1), Ep-CAM, CD9, GDNF receptors GFRα1 and c-Ret, and GPR125, some of which allow the enrichment of SSCs by FACS and MACS (Buageaw et al. 2005; Kanatsu-Shinohara et al. 2004; Tokuda et al. 2007; Kubota et al. 2003; Anderson et al. 1999; Naughton et al. 2006). GFRα1, a co-receptor of GNDF with c-Ret, tends to be expressed in As and Aal cells and appears to represent a relatively primitive proportion of spermatogonia. Combined with gravity sedimentation on a bovine serum albumin gradient, Hofmann et al. were able to isolate SSCs to 98 % purity using GFRα1 for antibody selection. However, purification with GFRα1 is only possible from pubescent mice but not adults, and the cell survival in culture is low (Hofmann et al. 2005; Ebata et al. 2005). Many studies have successfully enriched SSCs with an antibody against Thy-1 and realized their long-term cultivation. However, the cells are a mixture of spermatogonia at various stages, which is sufficient for most researchers (Kubota et al. 2004). c-kit is the receptor for stem cell factor (SCF), which was previously thought to be expressed by SSCs but later identified as a marker of differentiation (Shinohara et al. 1999). In multi-parameter cell sorting, negative selection with c-kit and positive selection with another surface marker will result in a higher percentage of SSCs. c-kit is expressed in late Aal spermatogonia to early spermatocytes, and its expression is often used to identify Aal cell conversion to differentiating spermatogonia A1 (Shinohara et al. 2000; Lennartsson and Rönnstrand 2012; Zhang et al. 2013).

Many transcription factors that promote self-renewal have been proposed as SSC markers. According to their response to GDNF, a key extracellular factor that promotes self-renewal, transcription factors can be divided into GDNF-dependent or GDNF-independent factors. GDNF is the most important extrinsic factor that regulates SSC self-renewal in a dose-dependent manner, and it is essential to culture SSCs in vitro (Meng et al. 2000). To define downstream effectors of GDNF signaling, Oatley and Brinster conducted microarray analysis of cultured germ cells. In their study, GDNF was removed and re-added to the cultured cells, and then, microarray analysis was performed at various time points to determine GDNF-inducible factors. Six genes responded most dramatically to GDNF, which were proposed to be the downstream effectors of GDNF signals, including Bcl6b, Etv5, Lhx1, Egr2, Egr3, and Tspan8 (Oatley et al. 2006). In vitro disruption of Bcl6b with siRNA significantly affects the proliferation of SSCs. Moreover, Bcl6b-null mice exhibit the same progressive defect as GDNF-null mice, further confirming that Bcl6b is a downstream effector of GDNF (Oatley et al. 2006). Subsequently, Etv5-knockout mice were generated and showed a similar phenotype (Chen et al. 2005). Inhibitor of DNA binding protein 4 (ID4) is another GDNF-inducible factor. However, ID4 is unique because it is exclusively expressed in As cells but not in Apr or Aal cells (Oatley et al. 2011). Recently, NONOS2, an RNA-binding protein that is preferentially expressed in As and Apr cells, was found to be a downstream effector of GDNF signaling. A lack of NANOS2 results in the same phenotype as that of GDNF-null mice, whereas NANOS2 overexpression will compensate for GFRα1 depletion in mice (Sada et al. 2012). GDNF signals through three pathways to downstream effectors for the promotion of SSC self-renewal, including PI3 K-AKT, SFK, and Ras/ERK1/2, which also cross talk with each other (Lee et al. 2007; Braydich-Stolle et al. 2007; He et al. 2008). Promyelocytic leukemia zink factor (PLZF), also known as ZFP145 and ZBTB16, is the first identified intrinsic factor that is exclusively expressed in undifferentiated spermatogonia in the testis. Disruption of PLZF leads to progressive germ cell loss, indicating the essential role of PLZF in SSC maintenance (Buaas et al. 2004; Costoya et al. 2004). The exact mechanism of PLZF has not been fully clarified in the maintenance of SSCs, although some details have been revealed, which will be discussed below. Oct4, another GDNF-independent maintenance factor, is also exclusively expressed in undifferentiated spermatogonia in the adult testis. Oct4 disruption in cultured GSCs notably reduces both their proliferation and survival rates, suggesting its indispensable role in SSC self-renewal (Dann et al. 2008). However, the downstream and upstream molecules of OCT4 signaling are almost unknown in SPCs, which require further study.

1.3.2.3 Reversibility and Heterogeneity

Recent studies have proposed that undifferentiated spermatogonia are not uniform as previously thought and that morphologically classified SPCs exhibit different molecular and biology characters among themselves. At all stages of undifferentiated SSCs (As, Apr, and Aal cells), NGN3 expression can be detected, but there are undifferentiated SSCs that are negative for NGN3, suggesting molecular heterogeneity among undifferentiated SSCs (Yoshida et al. 2007). Lineage tracing under the control of NGN3 expression revealed that most labeled cells are committed to differentiation, while a very small population are stem cells that account for only about 10 % of total SSCs (Nakagawa et al. 2007). Quantitative analysis revealed that about 10 % of As spermatogonia are GFRα1 negative, but transplantation has demonstrated their clonogenic ability. In contrast, cells selected using GFRα1 show almost no clonogenic ability. In addition, GFRα1 expression is not the same among Apr spermatogonia (Grisanti et al. 2009). Taking these evidences together, heterogeneity does exist among undifferentiated SSCs. Such heterogeneity raises the possibility that there may be more stages among the undifferentiated types of SSCs. Morphological classification may be not truly reflect the actual status of undifferentiated spermatogonia, and a better classification system combined with molecular characters should be developed.

Under normal conditions, the vast majority of NGN3-expressing spermatogonial cells are transit-amplifying cells committed to differentiate. However, in transplantation assays, a significantly larger number of NGN3-expressing spermatogonial cells are clonogenic and contribute to regeneration in lineage tracing experiments. This finding demonstrates that in addition to true stem cells, transit-amplifying cells can revert to SSCs (Nakagawa et al. 2007; Yoshida et al. 2007). Nakagawa et al. (2007) referred to this subpopulation as potential stem cells. Such a functionally distinct population of undifferentiated spermatogonia possesses the potential for self-renewal but do not show this ability in undisrupted testis. Similar results have been obtained in two studies showing that c-kit-positive spermatogonia both in vivo and in vitro, which are usually thought to be the differentiating subpopulation, are also able to regenerate recipient testis, although with a significantly lower ability compared with that of the c-kit-negative fraction (Barroca et al. 2009; Morimoto et al. 2009). Considering the existence of potential stem cells, transplantation assays may overestimate the number of true SSCs. However, estimations of total SSCs at <2000 per testis according to transplantation assays are similar to those in a study by Nakagawa et al. (2007) based on the boundary of one SSC territory, although this strategy is somehow confusing (Nakaki et al. 2013; Reizel et al. 2012). It is still unclear whether only some SPCs are reversible or all SPCs are able to convert to SSCs under certain conditions. The former possibility may indicate the complexity or heterogeneity of the undifferentiated spermatogonia population, while the latter may represent the phenotypic plasticity. If heterogeneity is important, a certain phase may mark the undirected differentiation. If there is a certain phase, it might be possible to characterize a molecular phenotype that marks the specific point, but no such marker has been revealed thus far. It is also possible that undifferentiated spermatogonia show plasticity and can be reversed under certain conditions, such as transplantation and tissue regeneration, which is important for the robustness of the spermatogenesis system.

1.3.3 SSC Niche

In sexually reproductive animals, the SSC niche is the specialized microenvironment that harbors the stem cells and precisely regulates their self-renewal and differentiation. In *Drosophila* and *C. elegans*, whose gonads are polarized, its localization is identified by a specialized somatic compartment that supports the stem cells. However, in mammals, this specialized microenvironment has not been proposed (Spradling et al. 2001; Yamashita and Fuller 2005). Seminiferous tubules are comprised of basal and adluminal compartments separated by tight junctions among Sertoli cells, and all spermatogonia lie in the basal compartment. Together with morphological research showing that SSCs are situated on the basal membrane, we can speculate that the SSC niche must be located somewhere adjoining the basal membrane in the basal compartment. Sertoli cells and the basal membrane composed of peritubular and extracellular matrix provide the structural basis for the SSC niche. However, no report has shown the functional difference of sertoli cells and the basal membrane, suggesting that these two components are not the main factors that dictate the location of the SSC niche (Wang et al. 2013). Another factor must maintain the SSC niche, which is probably derived from interstitial cells. Oatley et al. (2009) reported that cultured thy1+ germ cells are enriched for colony-stimulating factor 1 receptor (Csf1r), and the addition of colony-stimulating factor 1 (Csf1) greatly enhances the self-renewal of SSCs, but not the total germ cells, as confirmed by transplantation assays (Oatley et al. 2009). Csfr1 is expressed by Leydig cells that are not evenly to distributed in the interstitium of seminiferous tubules, suggesting that interstitial cells, such as Leydig cells, may contribute to the formation of the SSC niche. Yoshida et al. (2007) showed that undifferentiated spermatogonia are more likely to reside within the area that the vasculature goes through in the seminiferous interval using time-lapse imaging of green fluorescent protein. This result suggests that the SSC niche is located in this area because SSCs are a subpopulation of undifferentiated spermatogonia. Moreover, an alternate pattern of the vasculature system leads to rearrangement of the undifferentiated spermatogonia, which further confirms the vasculature-associated niche location (Yoshida et al. 2007). Despite the association of the vasculature and Leydig cells, the true location of the niche is still under debate owing to the lack of specific SSC markers (Fig. 1.4).

It is thought that the number of SSC niches decides the potential SSC number. Therefore, it is important to reveal which factors influence the number of niches. Ectopically expressed human GDNF in mouse sertoli cells results in a dramatic increase of SSCs in the testis, indicating that high GDNF levels may increase the number of SSC niches (Yomogida et al. 2003). Oatley et al. (2011) found that the

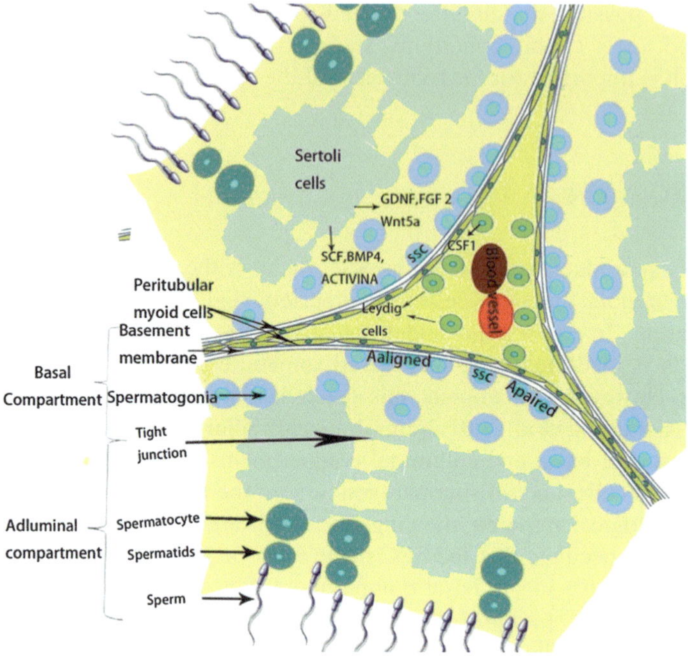

Fig. 1.4 A brief outlook of niche structure

number of SSCs is possibly dictated by the number of Sertoli cells that secrete GDNF. A threefold increase in the number of colony-forming cells in seminiferous tubules of recipient mice occurs after transplantation of SSCs from PTU-treated mice with an increased number of sertoli cells at puberty compared with that in normal mice. Furthermore, PTU-treated mice exhibit threefold more accessible niches than that in normal mice with normal sertoli cells. In addition, they found that expansion of the niche number is influenced by neither the vasculature nor the interstitial cell populations (Oatley et al. 2011). SSC numbers are strictly regulated by GDNF in a dose-dependent manner. Overexpression of GDNF results in accumulation of SSCs, and heterozygous mutants show depletion of germ cells including SSCs (Meng et al. 2000). Follicle-stimulating hormone (FSH) upregulates GDNF and during testis damage. GDNF expression is increased possibly through FSH to maintain a normal number of SSC niches (Tadokoro et al. 2002). Other factors, such as Sin3A and RA, may also affect the niche number. Germ cells transplanted into RA-deficient mice show less colony formation, whereas mice with sertoli cells lacking Sin3A show germ cell depletion (Payne et al. 2010; McLean et al. 2002).

The study of the niche has been difficult owing to the complicated three-dimensional structure of seminiferous tubules in vivo. Developing a three-dimensional culture system may provide a possible strategy to solve this problem. Using the testicular cells of infertile mice as feeder cells, Kanatsu-Shinohara et al. (2012)

reconstructed the SSC niche to some extent in a culture system, and the cultured SSCs migrated beneath the sertoli cells and formed cobblestone colonies. In this system, they found that the chemokine CXCL12 contributes to the SSC homing efficiency. Another study from Yokonishi and colleagues demonstrated that dissociated testicular cells are able to aggregate in suspension culture and form seminiferous tubules after transfer and culture on the surface of an agarose gel. When cultured GSCs were added to this autoreconstructed system, the GSCs differentiate before the meiotic phase (Yokonishi et al. 2013). This autoreconstruction of testicular cells is extraordinary and may be of great value for in vitro study of the niche.

1.3.4 Fate Decisions of SSCs

SSC self-renewal and differentiation must be sophistically balanced to maintain normal spermatogenesis and avoid tumorigenesis. When the balance shifts to self-renewal, there is accumulation of stem cells and decreases in the number of developed germ cells. In contrast, when the balance shifts to differentiation, maintenance defects will occur, eventually leading to infertility. This effect is best illustrated by paracrine regulation of GDNF secreted by sertoli cells. Meng et al. (2000) developed two transgenic mouse strains: one with overexpression of GDNF and the other with heterozygous ablation of GDNF. All transgenic mice showed decreased germ cell development and reduced fertility, and some even showed infertility. Mice overexpressing GDNF under the control of the promoter of testis-specific, human translation elongation factor showed larger clusters of spermatogonia, indicating accumulation of undifferentiated spermatogonia. Moreover, as the mice aged, these clusters grew larger and began to invade into the interstitium, and most of the mice generated non-metastatic tumors after 1 year of age. Through BrdU incorporation and apoptosis staining, they found no marked enhancement in the total proliferation rate. Thus, it was inhibition of differentiation rather than hyperproliferation that was responsible for the SSC accumulation. In $GDNF^{+/-}$ mice, although most were fertile, the depletion of germ cells increased with age and eventually resulted in only Sertoli cells in seminiferous tubules, indicating a maintenance defect of SSCs (Meng et al. 2000). This dose-dependent effect highlights the importance of precise regulation of GDNF and the role of GDNF in SSC fate decisions. GDNF is the ligand for co-receptors GFRα1/Ret, and its binding is able to activate several intrinsic signaling pathways such as PI3K/AKT/MEK and Src (He et al. 2008; Braydich-Stolle et al. 2007; Oatley et al. 2007). Many downstream transcription factors have been revealed, such as Bcl6b, Etv5, NANOS2, and ID4 (Oatley et al. 2006, 2011; Sada et al. 2009, 2012). Knockout or overexpression of these transcription factors have been performed in mice, confirming their indispensable roles in SSC maintenance. In addition to the most important extrinsic factor (GDNF), FGF2 and CSF1 play a role in SSC self-renewal, but not in the balance between self-renewal and differentiation (Oatley et al. 2009; Ishii et al. 2012).

An important intrinsic factor that regulates SSC fate is the transcriptional repressor Plzf. Ablation of Plzf leads to progressive germ cell loss due to exhaustion of the SSCs. One possible mechanism has been proposed by Filipponi et al. (2007). They revealed that Plzf directly binds to the promoter region of the kit gene, a character of spermatogonia differentiation, thus repressing the expression of kit, the SCF receptor (Filipponi et al. 2007). By inducing Redd1 expression, Plzf can oppose with mTORC1 whose hyperactivity leads to stem cell exhaustion. Through activation of PI3 K/AKT signaling, mTORC1 activates as a downstream effector. However, activation of mTORC1 suppresses the expression of the GDNF co-receptor GFRα1/Ret, which in turn inhibits GDNF signaling for SPC self-renewal. In the absence of Plzf, excessive amounts of GDNF are able to promote SPC self-renewal. Thus, the fate of SPCs is controlled by cross talk between Plzf-Redd1-mTORC1 and AKT/PI3K-mTORC1 signaling pathways in which mTORC1 plays the central role. The addition of rapamycin, a specific inhibitor of mTORC1, to Plzf$^{-/-}$ cultured GSCs restores expression of the receptor and partly rescues the GDNF signal (Hobbs et al. 2010). Another study from the same group revealed that Plzf–Sall4 antagonism decides the fate of SPCs. Sall4, a zing-finger transcription factor, restricts Plzf to non-cognate chromatin domains and induces the expression of the differentiation factor kit. In turn, Plzf opposes Sall4 functions and induces Sall1 expression. In vitro treatment of GSCs with RA, an indispensable extrinsic factor for initiation of differentiation, downregulates Plzf expression and increases the number of kit-positive spermatogonia. Though Sall4 transient upregulation in vivo, this change is accompanied by increasing numbers of kit+ spermatogonia (Hobbs et al. 2012). Considering the importance of Plzf in SSC fate decision, it is crucial to determine the regulatory mechanisms of Plzf expression. Identifying such factors will greatly enhance our understanding of the mechanisms of SSC fate decisions.

1.4 Perspectives

The expression profiles of undifferentiated spermatogonia involved in self-renewal and differentiation have been gradually revealed during the last few decades. However, there are no specific molecular markers that are unique to the subpopulation with both self-renewal and commitment abilities. Id4 may be a potential marker because it is the only identified molecule that is exclusively expressed by some As cells (Oatley et al. 2011). Identifying such molecules will greatly improve our understanding of fate decisions. It is still unclear whether there is a stage that marks the irreversibility of spermatogonia or whether it is the biological plasticity that leads to the heterogeneity among undifferentiated spermatogonia. Clarification of these aspects may change the current model of SSC development.

Development of a three-dimensional culture system for SSCs is another future challenge. In vivo studies have been difficult owing to the complex microenvironment. However, a three-dimensional SSC culture system will greatly facilitate niche studies. An in vitro culture system that supports the entire differentiation process and is able to produce haploid germ cells would be a useful tool. In vitro culture of SSCs has resolved the problem of rarity, but only establishment of an in vitro differentiation system will realize clinical applications. Dann et al. (2008) reported that expansion of cultured GSC with RA increases the number of germ cells positive for kit. However, they also noted premature meiosis and incomplete differentiation (Dann et al. 2008). Sato and Katagiri (2013) developed an organ culture system that supports all differentiation stages. Mature sperm can be obtained using this system, and in vitro fertilization has generated offspring (Sato et al. 2013). Nevertheless, this system still depends on testis organ fragments, thereby limiting its application to clinical use. Transplantation of rat or hamster SSCs into mouse testes results in exogenous spermatogenesis, which indicates conservation among species (Ogawa et al. 1999). Combining organ culture with xenogeneic testis culture is possible to fully support spermatogenesis, which needs further validation.

The signaling network that promotes SSC self-renewal and differentiation remains largely unknown. Microarray and transgenic mouse analyses have provided many potential genes that are important for SSC maintenance and differentiation, such as PHF13, SALL4, CDH1, and OCT4, but their exact roles have not been clarified (Tokuda et al. 2007; Gassei and Orwig 2013; Bordlein et al. 2011). As previously mentioned, the upstream molecules of Plzf remain unknown. Clarification of such networks will provide a better understanding of SSC self-renewal and differentiation to easily manipulate SSCs. Rodent SSC culture is efficient, but the culture of SSCs from other species is difficult (Kanatsu-Shinohara and Shinohara 2013). Understanding the molecular networks may improve the culture efficiency in other species.

References

Anderson R, Fassler R, Georges-Labouesse E, Hynes RO, Bader BL, Kreidberg JA, Schaible K, Heasman J, Wylie C. Mouse primordial germ cells lacking beta1 integrins enter the germline but fail to migrate normally to the gonads. Development. 1999a;126:1655–64.

Anderson R, Schaible K, Heasman J, Wylie C. Expression of the homophilic adhesion molecule, Ep-CAM, in the mammalian germ line. J Reprod Fertil. 1999b;116:379–84.

Aramaki S, Hayashi K, Kurimoto K, Ohta H, Yabuta Y, Iwanari H, Mochizuki Y, Hamakubo T, Kato Y, Shirahige K, Saitou M. A mesodermal factor, T, specifies mouse germ cell fate by directly activating germline determinants. Dev Cell. 2013;27:516–29.

Baltus AE, Menke DB, Hu YC, Goodheart ML, Carpenter AE, de Rooij DG, Page DC. In germ cells of mouse embryonic ovaries, the decision to enter meiosis precedes premeiotic DNA replication. Nat Genet. 2006;38:1430–4.

Barritt JA, Brenner CA, Malter HE, Cohen J. Mitochondria in human offspring derived from ooplasmic transplantation. Hum Reprod. 2001;16:513–6.

Barroca V, Lassalle B, Coureuil M, Louis JP, Le Page F, Testart J, Allemand I, Riou L, Fouchet P. Mouse differentiating spermatogonia can generate germinal stem cells in vivo. Nat Cell Biol. 2009;11:190–6.

Bordlein A, Scherthan H, Nelkenbrecher C, Molter T, Bosl MR, Dippold C, Birke K, Kinkley S, Staege H, Will H, Winterpacht A. SPOC1 (PHF13) is required for spermatogonial stem cell differentiation and sustained spermatogenesis. J Cell Sci. 2011;124:3137–48.

Borum K. Oogenesis in the mouse. A study of the origin of the mature ova. Exp Cell Res. 1967;45:39–47.

Bowles J, Knight D, Smith C, Wilhelm D, Richman J, Mamiya S, Yashiro K, Chawengsaksophak K, Wilson MJ, Rossant J, Hamada H, Koopman P. Retinoid signaling determines germ cell fate in mice. Science. 2006;312:596–600.

Bowles J, Feng CW, Spiller C, Davidson TL, Jackson A, Koopman P. FGF9 suppresses meiosis and promotes male germ cell fate in mice. Dev Cell. 2010;19:440–9.

Braydich-Stolle L, Kostereva N, Dym M, Hofmann MC. Role of Src family kinases and N-Myc in spermatogonial stem cell proliferation. Dev Biol. 2007;304:34–45.

Brinster RL, Avarbock MR. Germline transmission of donor haplotype following spermatogonial transplantation. Proc Natl Acad Sci. 1994;91:11303–7.

Buaas FW, Kirsh AL, Sharma M, McLean DJ, Morris JL, Griswold MD, de Rooij DG, Braun RE. Plzf is required in adult male germ cells for stem cell self-renewal. Nat Genet. 2004;36:647–52.

Buageaw A, Sukhwani M, Ben-Yehudah A, Ehmcke J, Rawe VY, Pholpramool C, Orwig KE, Schlatt S. GDNF family receptor alpha1 phenotype of spermatogonial stem cells in immature mouse testes. Biol Reprod. 2005;73:1011–6.

Bukovsky A, Gupta SK, Virant-Klun I, Upadhyaya NB, Copas P, Van Meter SE, Svetlikova M, Ayala ME, Dominguez R. Study origin of germ cells and formation of new primary follicles in adult human and rat ovaries. Methods Mol Biol. 2008;450:233–65.

Chen C, Ouyang W, Grigura V, Zhou Q, Carnes K, Lim H, Zhao GQ, Arber S, Kurpios N, Murphy TL, Cheng AM, Hassell JA, Chandrashekar V, Hofmann MC, Hess RA, Murphy KM. ERM is required for transcriptional control of the spermatogonial stem cell niche. Nature. 2005;436:1030–4.

Clermont Y. The cycle of the seminiferous epithelium in man. Am J Anat. 1963;112:35–51.

Clermont Y. Renewal of spermatogonia in man. Am J Anat. 1966;118:509–24.

Clermont Y. Kinetics of spermatogenesis in mammals: seminiferous epithelium cycle and spermatogonial renewal. Physiol Rev. 1972;52:198–236.

Conrad S, Renninger M, Hennenlotter J, Wiesner T, Just L, Bonin M, Aicher W, Buhring HJ, Mattheus U, Mack A, Wagner HJ, Minger S, Matzkies M, Reppel M, Hescheler J, Sievert KD, Stenzl A, Skutella T. Generation of pluripotent stem cells from adult human testis. Nature. 2008;456:344–9.

Costoya JA, Hobbs RM, Barna M, Cattoretti G, Manova K, Sukhwani M, Orwig KE, Wolgemuth DJ, Pandolfi PP. Essential role of Plzf in maintenance of spermatogonial stem cells. Nat Genet. 2004;36:653–9.

Culty M. Gonocytes, the forgotten cells of the germ cell lineage. Birth Defects Res C. 2009;87:1–26.

Dann CT, Alvarado AL, Molyneux LA, Denard BS, Garbers DL, Porteus MH. Spermatogonial stem cell self-renewal requires OCT4, a factor downregulated during retinoic acid-induced differentiation. Stem Cells. 2008;26:2928–37.

De Rooij DG, Russell LD. All you wanted to know about spermatogonia but were afraid to ask. J Androl. 2000;21:776–98.

Durcova-Hills G, Wianny F, Merriman J, Zernicka-Goetz M, McLaren A. Developmental fate of embryonic germ cells (EGCs), in vivo and in vitro. Differentiation. 2003;71:135–41.

Ebata KT, Zhang X, Nagano MC. Expression patterns of cell-surface molecules on male germ line stem cells during postnatal mouse development. Mol Reprod Dev. 2005;72:171–81.

Eggan K, Jurga S, Gosden R, Min IM, Wagers AJ. Ovulated oocytes in adult mice derive from non-circulating germ cells. Nature. 2006;441:1109–14.

Ehmcke J, Schlatt S. A revised model for spermatogonial expansion in man: lessons from non-human primates. Reproduction. 2006;132:673–80.

Ehmcke J, Wistuba J, Schlatt S. Spermatogonial stem cells: questions, models and perspectives. Hum Reprod Update. 2006;12:275–82.

Feng YM, Liang GJ, Pan B, Qin XS, Zhang XF, Chen CL, Li L, Cheng SF, De Felici M, Shen W. Notch pathway regulates female germ cell meiosis progression and early oogenesis events in fetal mouse. Cell Cycle. 2014;13:782–91.

Filipponi D, Hobbs RM, Ottolenghi S, Rossi P, Jannini EA, Pandolfi PP, Dolci S. Repression of kit expression by Plzf in germ cells. Mol Cell Biol. 2007;27:6770–81.

Gassei K, Orwig KE. SALL4 expression in gonocytes and spermatogonial clones of postnatal mouse testes. PLoS ONE. 2013;8:e53976.

Gohbara A, Katagiri K, Sato T, Kubota Y, Kagechika H, Araki Y, Araki Y, Ogawa T. In vitro murine spermatogenesis in an organ culture system. Biol Reprod. 2010;83:261–7.

Golestaneh N, Kokkinaki M, Pant D, Jiang J, DeStefano D, Fernandez-Bueno C, Rone JD, Haddad BR, Gallicano GI, Dym M. Pluripotent stem cells derived from adult human testes. Stem Cells Dev. 2009;18:1115–26.

Gosden RG. Germline stem cells in the postnatal ovary: is the ovary more like a testis? Hum Reprod Update. 2004;10:193–5.

Grisanti L, Falciatori I, Grasso M, Dovere L, Fera S, Muciaccia B, Fuso A, Berno V, Boitani C, Stefanini M, Vicini E. Identification of spermatogonial stem cell subsets by morphological analysis and prospective isolation. Stem Cells. 2009;27:3043–52.

Guan K, Nayernia K, Maier LS, Wagner S, Dressel R, Lee JH, Nolte J, Wolf F, Li M, Engel W, Hasenfuss G. Pluripotency of spermatogonial stem cells from adult mouse testis. Nature. 2006;440:1199–203.

Harvey AJ, Gibson TC, Quebedeaux TM, Brenner CA. Impact of assisted reproductive technologies: a mitochondrial perspective of cytoplasmic transplantation. Curr Top Dev Biol. 2007;77:229–49.

Hayashi K, Ogushi S, Kurimoto K, Shimamoto S, Ohta H, Saitou M. Offspring from oocytes derived from in vitro primordial germ cell-like cells in mice. Science. 2012;338:971–5.

He ZP, Jiang JJ, Kokkinaki M, Golestaneh N, Hofmann MC, Dym M. GDNF upregulates c-fos transcription via the Ras/ERK1/2 pathway to promote mouse spermatogonial stem cell proliferation. Stem Cells. 2008;26:266–78.

Hobbs RM, Seandel M, Falciatori I, Rafii S, Pandolfi PP. Plzf regulates germline progenitor self-renewal by opposing mTORC1. Cell. 2010;142:468–79.

Hobbs RM, Fagoonee S, Papa A, Webster K, Altruda F, Nishinakamura R, Chai L, Pandolfi PP. Functional antagonism between Sall4 and Plzf defines germline progenitors. Cell Stem Cell. 2012;10:284–98.

Hofmann MC, Braydich-Stolle L, Dym M. Isolation of male germ-line stem cells; influence of GDNF. Dev Biol. 2005;279:114–24.

Hubner K, Fuhrmann G, Christenson LK, Kehler J, Reinbold R, De La Fuente R, Wood J, Strauss JF 3rd, Boiani M, Scholer HR. Derivation of oocytes from mouse embryonic stem cells. Science. 2003;300:1251–6.

Ishii K, Kanatsu-Shinohara M, Toyokuni S, Shinohara T. FGF2 mediates mouse spermatogonial stem cell self-renewal via upregulation of Etv5 and Bcl6b through MAP2K1 activation. Development. 2012;139:1734–43.

Johnson J, Canning J, Kaneko T, Pru JK, Tilly JL. Germline stem cells and follicular renewal in the postnatal mammalian ovary. Nature. 2004;428:145–50.

Kanatsu-Shinohara M, Shinohara T. Spermatogonial stem cell self-renewal and development. Annu Rev Cell Dev Biol. 2013;29:163–87.

Kanatsu-Shinohara M, Ogonuki N, Inoue K, Miki H, Ogura A, Toyokuni S, Shinohara T. Long-term proliferation in culture and germline transmission of mouse male germline stem cells. Biol Reprod. 2003;69:612–6.

Kanatsu-Shinohara M, Inoue K, Lee J, Yoshimoto M, Ogonuki N, Miki H, Baba S, Kato T, Kazuki Y, Toyokuni S, Toyoshima M, Niwa O, Oshimura M, Heike T, Nakahata T, Ishino F, Ogura A, Shinohara T. Generation of pluripotent stem cells from neonatal mouse testis. Cell. 2004a;119:1001–12.

Kanatsu-Shinohara M, Toyokuni S, Shinohara T. CD9 is a surface marker on mouse and rat male germline stem cells. Biol Reprod. 2004b;70:70–5.
Kanatsu-Shinohara M, Inoue K, Miki H, Ogonuki N, Takehashi M, Morimoto T, Ogura A, Shinohara T. Clonal origin of germ cell colonies after spermatogonial transplantation in mice. Biol Reprod. 2006;75:68–74.
Kanatsu-Shinohara M, Lee J, Inoue K, Ogonuki N, Miki H, Toyokuni S, Ikawa M, Nakamura T, Ogura A, Shinohara T. Pluripotency of a single spermatogonial stem cell in mice. Biol Reprod. 2008;78:681–7.
Kanatsu-Shinohara M, Inoue K, Takashima S, Takehashi M, Ogonuki N, Morimoto H, Nagasawa T, Ogura A, Shinohara T. Reconstitution of mouse spermatogonial stem cell niches in culture. Cell Stem Cell. 2012;11:567–78.
Ko K, Arauzo-Bravo MJ, Kim J, Stehling M, Scholer HR. Conversion of adult mouse unipotent germline stem cells into pluripotent stem cells. Nat Protoc. 2010;5:921–8.
Kossack N, Meneses J, Shefi S, Nguyen HN, Chavez S, Nicholas C, Gromoll J, Turek PJ, Reijo-Pera RA. Isolation and characterization of pluripotent human spermatogonial stem cell-derived cells. Stem Cells. 2009;27:138–49.
Kostereva N, Hofmann MC. Regulation of the spermatogonial stem cell niche. Reprod Domest Anim. 2008;43(Suppl 2):386–92.
Koubova J, Menke DB, Zhou Q, Capel B, Griswold MD, Page DC. Retinoic acid regulates sex-specific timing of meiotic initiation in mice. Proc Natl Acad Sci U S A. 2006;103:2474–9.
Krentz AD, Murphy MW, Sarver AL, Griswold MD, Bardwell VJ, Zarkower D. DMRT1 promotes oogenesis by transcriptional activation of Stra8 in the mammalian fetal ovary. Dev Biol. 2011;356:63–70.
Kubota H, Avarbock MR, Brinster RL. Spermatogonial stem cells share some, but not all, phenotypic and functional characteristics with other stem cells. Proc Natl Acad Sci. 2003;100:6487–92.
Kubota H, Avarbock MR, Brinster RL. Culture conditions and single growth factors affect fate determination of mouse spermatogonial stem cells. Biol Reprod. 2004;71:722–31.
Kumar S, Chatzi C, Brade T, Cunningham TJ, Zhao XL, Duester G. Sex-specific timing of meiotic initiation is regulated by Cyp26b1 independent of retinoic acid signalling. Nat Commun 2011;2.
Lawson KA, Dunn NR, Roelen BAJ, Zeinstra LM, Davis AM, Wright CVE, Korving JPWFM, Hogan BLM. Bmp4 is required for the generation of primordial germ cells in the mouse embryo. Genes Dev. 1999;13:424–36.
Le Bouffant R, Souquet B, Duval N, Duquenne C, Herve R, Frydman N, Robert B, Habert R, Livera G. Msx1 and Msx2 promote meiosis initiation. Development. 2011;138:5393–402.
Lee J, Kanatsu-Shinohara M, Inoue K, Ogonuki N, Miki H, Toyokuni S, Kimura T, Nakano T, Ogura A, Shinohara T. Akt mediates self-renewal division of mouse spermatogonial stem cells. Development. 2007;134:1853–9.
Lennartsson J, Rönnstrand L. Stem cell factor receptor/c-Kit: from basic science to clinical implications. Physiol Rev. 2012;92:1619–49.
Lin Y, Gill ME, Koubova J, Page DC. Germ cell-intrinsic and -extrinsic factors govern meiotic initiation in mouse embryos. Science. 2008;322:1685–7.
Magnusdottir E, Dietmann S, Murakami K, Gunesdogan U, Tang FC, Bao SQ, Diamanti E, Lao KQ, Gottgens B, Surani MA. A tripartite transcription factor network regulates primordial germ cell specification in mice. Nat Cell Biol. 2013;15:905–U322.
Matson CK, Murphy MW, Griswold MD, Yoshida S, Bardwell VJ, Zarkower D. The mammalian doublesex homolog DMRT1 is a transcriptional gatekeeper that controls the mitosis versus meiosis decision in male germ cells. Dev Cell. 2010;19:612–24.
Mattiske D, Kume T, Hogan BLM. The mouse forkhead gene Foxc1 is required for primordial germ cell migration and antral follicle development. Dev Biol. 2006;290:447–58.
McLean DJ, Russell LD, Griswold MD. Biological activity and enrichment of spermatogonial stem cells in vitamin A-deficient and hyperthermia-exposed testes from mice based on colonization following germ cell transplantation. Biol Reprod. 2002;66:1374–9.

Meng X, Lindahl M, Hyvönen ME, Parvinen M, de Rooij DG, Hess MW, Raatikainen-Ahokas A, Sainio K, Rauvala H, Lakso M. Regulation of cell fate decision of undifferentiated spermatogonia by GDNF. Science. 2000;287:1489–93.

Molyneaux KA, Zinszner H, Kunwar PS, Schaible K, Stebler J, Sunshine MJ, O'Brien W, Raz E, Littman D, Wylie C, Lehmann R. The chemokine SDF1/CXCL12 and its receptor CXCR4 regulate mouse germ cell migration and survival. Development. 2003;130:4279–86.

Morimoto H, Kanatsu-Shinohara M, Takashima S, Chuma S, Nakatsuji N, Takehashi M, Shinohara T. Phenotypic plasticity of mouse spermatogonial stem cells. PLoS ONE. 2009;4:e7909.

Nakagawa T, Nabeshima Y-i, Yoshida S. Functional identification of the actual and potential stem cell compartments in mouse spermatogenesis. Dev Cell. 2007;12:195–206.

Nakaki F, Hayashi K, Ohta H, Kurimoto K, Yabuta Y, Saitou M. Induction of mouse germ-cell fate by transcription factors in vitro. Nature 2013;501:222–+.

Nakamura S, Kobayashi K, Nishimura T, Higashijima S, Tanaka M. Identification of germline stem cells in the ovary of the teleost medaka. Science. 2010;328:1561–3.

Naughton CK, Jain S, Strickland AM, Gupta A, Milbrandt J. Glial cell-line derived neurotrophic factor-mediated RET signaling regulates spermatogonial stem cell fate. Biol Reprod. 2006;74:314–21.

Oatley JM, Avarbock MR, Telaranta AI, Fearon DT, Brinster RL. Identifying genes important for spermatogonial stem cell self-renewal and survival. Proc Natl Acad Sci. 2006;103:9524–9.

Oatley JM, Avarbock MR, Brinster RL. Glial cell line-derived neurotrophic factor regulation of genes essential for self-renewal of mouse spermatogonial stem cells is dependent on Src family kinase signaling. J Biol Chem. 2007;282:25842–51.

Oatley JM, Oatley MJ, Avarbock MR, Tobias JW, Brinster RL. Colony stimulating factor 1 is an extrinsic stimulator of mouse spermatogonial stem cell self-renewal. Development. 2009;136:1191–9.

Oatley MJ, Kaucher AV, Racicot KE, Oatley JM. Inhibitor of DNA binding 4 is expressed selectively by single spermatogonia in the male germline and regulates the self-renewal of spermatogonial stem cells in mice. Biol Reprod. 2011a;85:347–56.

Oatley MJ, Racicot KE, Oatley JM. Sertoli cells dictate spermatogonial stem cell niches in the mouse testis. Biol Reprod. 2011b;84:639–45.

Ogawa T, Dobrinski I, Avarbock MR, Brinster RL. Xenogeneic spermatogenesis following transplantation of hamster germ cells to mouse testes. Biol Reprod. 1999;60:515–21.

Ohinata Y, Payer B, O'Carroll D, Ancelin K, Ono Y, Sano M, Barton SC, Obukhanych T, Nussenzweig M, Tarakhovsky A, Saitou M, Surani MA. Blimp1 is a critical determinant of the germ cell lineage in mice. Nature. 2005;436:207–13.

Ohinata Y, Ohta H, Shigeta M, Yamanaka K, Wakayama T, Saitou M. A signaling principle for the specification of the germ cell lineage in mice. Cell. 2009;137:571–84.

Payne CJ, Gallagher SJ, Foreman O, Dannenberg JH, Depinho RA, Braun RE. Sin3a is required by sertoli cells to establish a niche for undifferentiated spermatogonia, germ cell tumors, and spermatid elongation. Stem Cells. 2010;28:1424–34.

Pepling ME. From primordial germ cell to primordial follicle: mammalian female germ cell development. Genesis. 2006;44:622–32.

Pepling ME. Follicular assembly: mechanisms of action. Reproduction. 2012;143:139–49.

Pepling ME, Spradling AC. Mouse ovarian germ cell cysts undergo programmed breakdown to form primordial follicles. Dev Biol. 2001;234:339–51.

Peters H, Crone M. DNA synthesis in oocytes of mammals. Arch Anat Microsc Morphol Exp. 1967;56:160–70.

Phillips BT, Gassei K, Orwig KE. Spermatogonial stem cell regulation and spermatogenesis. Philos Trans R Soc Lond B Biol Sci. 2010;365:1663–78.

Reizel Y, Itzkovitz S, Adar R, Elbaz J, Jinich A, Chapal-Ilani N, Maruvka YE, Nevo N, Marx Z, Horovitz I, Wasserstrom A, Mayo A, Shur I, Benayahu D, Skorecki K, Segal E, Dekel N, Shapiro E. Cell lineage analysis of the mammalian female germline. PLoS Genet. 2012;8:e1002477.

Rudkin GT, Griech HA. On the persistence of oocyte nuclei from fetus to maturity in the laboratory mouse. J Cell Biol. 1962;12:169–75.

Sada A, Suzuki A, Suzuki H, Saga Y. The RNA-binding protein NANOS2 is required to maintain murine spermatogonial stem cells. Science. 2009;325:1394–8.

Sada A, Hasegawa K, Pin PH, Saga Y. NANOS2 acts downstream of glial cell line-derived neurotrophic factor signaling to suppress differentiation of spermatogonial stem cells. Stem Cells. 2012;30:280–91.

Sato T, Katagiri K, Yokonishi T, Kubota Y, Inoue K, Ogonuki N, Matoba S, Ogura A, Ogawa T. In vitro production of fertile sperm from murine spermatogonial stem cell lines. Nat Commun. 2011a;2:472.

Sato T, Katagiri K, Gohbara A, Inoue K, Ogonuki N, Ogura A, Kubota Y, Ogawa T. In vitro production of functional sperm in cultured neonatal mouse testes. Nature. 2011b;471:504–7.

Sato T, Katagiri K, Kubota Y, Ogawa T. In vitro sperm production from mouse spermatogonial stem cell lines using an organ culture method. Nature Protocols 2013;8:2098–2104

Sato T, Yokonishi T, Komeya M, Katagiri K, Kubota Y, Matoba S, Ogonuki N, Ogura A, Yoshida S, Ogawa T. Testis tissue explantation cures spermatogenic failure in c-Kit ligand mutant mice. Proc Natl Acad Sci U S A. 2012;109:16934–8.

Sato T, Katagiri K, Kubota Y, Ogawa T. In vitro sperm production from mouse spermatogonial stem cell lines using an organ culture method. Nat Protoc. 2013;8:2098–104.

Seandel M, James D, Shmelkov SV, Falciatori I, Kim J, Chavala S, Scherr DS, Zhang F, Torres R, Gale NW. Generation of functional multipotent adult stem cells from GPR125+ germline progenitors. Nature. 2007;449:346–50.

Shinohara T, Avarbock MR, Brinster RL. β1-and α6-integrin are surface markers on mouse spermatogonial stem cells. Proc Natl Acad Sci. 1999;96:5504–9.

Shinohara T, Orwig KE, Avarbock MR, Brinster RL. Spermatogonial stem cell enrichment by multiparameter selection of mouse testis cells. Proc Natl Acad Sci U S A. 2000;97:8346–51.

Simon L, Ekman GC, Kostereva N, Zhang Z, Hess RA, Hofmann MC, Cooke PS. Direct transdifferentiation of stem/progenitor spermatogonia into reproductive and nonreproductive tissues of all germ layers. Stem Cells. 2009;27:1666–75.

Souquet B, Tourpin S, Messiaen S, Moison D, Habert R, Livera G. Nodal signaling regulates the entry into meiosis in fetal germ cells. Endocrinology. 2012;153:2466–73.

Speed RM. Meiosis in the foetal mouse ovary. I. An analysis at the light microscope level using surface-spreading. Chromosoma. 1982;85:427–37.

Spradling A, Drummond-Barbosa D, Kai T. Stem cells find their niche. Nature. 2001;414:98–104.

Tadokoro Y, Yomogida K, Ohta H, Tohda A, Nishimune Y. Homeostatic regulation of germinal stem cell proliferation by the GDNF/FSH pathway. Mech Dev. 2002;113:29–39.

Tam PP, Zhou SX. The allocation of epiblast cells to ectodermal and germ-line lineages is influenced by the position of the cells in the gastrulating mouse embryo. Dev Biol. 1996;178:124–32.

Tanaka SS, Nakane A, Yamaguchi YL, Terabayashi T, Abe T, Nakao K, Asashima M, Steiner KA, Tam PPL, Nishinakamura R. Dullard/Ctdnep1 modulates WNT signalling activity for the formation of primordial germ cells in the mouse embryo. PLoS ONE. 2013;8.

Tanaka SS, Yamaguchi YL, Tsoi B, Lickert H, Tam PP. IFITM/Mil/fragilis family proteins IFITM1 and IFITM3 play distinct roles in mouse primordial germ cell homing and repulsion. Dev Cell. 2005;9:745–56.

Tegelenbosch RA, de Rooij DG. A quantitative study of spermatogonial multiplication and stem cell renewal in the C3H/101 F1 hybrid mouse. Mutat Res. 1993;290:193–200.

Tokuda M, Kadokawa Y, Kurahashi H, Marunouchi T. CDH1 is a specific marker for undifferentiated spermatogonia in mouse testes. Biol Reprod. 2007;76:130–41.

Vernet N, Dennefeld C, Rochette-Egly C, Oulad-Abdelghani M, Chambon P, Ghyselinck NB, Mark M. Retinoic acid metabolism and signaling pathways in the adult and developing mouse testis. Endocrinology. 2006;147:96–110.

Wang H, Shi L, Xiang J, Ding X, Luo H, Wang S, Wu J. Isolation, culture and transplantation of female germline stem cells from neonatal and prepubertal mice. Protoc Exch. 2013;. doi:10.1038/protex.2013.004.

Wang H, Jiang M, Bi H, Chen X, He L, Li X, Wu J. Conversion of female germline stem cells from neonatal and prepubertal mice into pluripotent stem cells. J Mol Cell Biol. 2014;6:164–71.

White YAR, Woods DC, Takai Y, Ishihara O, Seki H, Tilly JL. Oocyte formation by mitotically active germ cells purified from ovaries of reproductive-age women. Nat Med. 2012;18:413–U176.

Wong TT, Tesfamichael A, Collodi P. Production of zebrafish offspring from cultured female germline stem cells. PLoS ONE. 2013;8:e62660.

Woods DC, Tilly JL. Isolation, characterization and propagation of mitotically active germ cells from adult mouse and human ovaries. Nat Protoc. 2013;8:966–88.

Wu J, Zhang Y, Tian GG, Zou K, Lee CM, Yu Q, Yuan Z. Short-type PB-cadherin promotes self-renewal of spermatogonial stem cells via multiple signaling pathways. Cell Signal. 2008;20:1052–60.

Wu J, Luo H, Wang H. Germline stem cells. Curr Top Dev Biol. 2013;102:97–126.

Xiong J, Wang H, Guo G, Wang S, He L, Chen H, Wu J. Male germ cell apoptosis and epigenetic histone modification induced by Tripterygium wilfordii Hook F. PLoS ONE. 2011;6:e20751.

Yamaji M, Seki Y, Kurimoto K, Yabuta Y, Yuasa M, Shigeta M, Yamanaka K, Ohinata Y, Saitou M. Critical function of Prdm14 for the establishment of the germ cell lineage in mice. Nat Genet. 2008;40:1016–22.

Yamashita YM, Fuller MT. Asymmetric stem cell division and function of the niche in the Drosophila male germ line. Int J Hematol. 2005;82:377–80.

Ying Y, Zhao GQ. Cooperation of endoderm-derived BMP2 and extraembryonic ectoderm-derived BMP4 in primordial germ cell generation in the mouse. Dev Biol. 2001;232:484–92.

Ying Y, Liu XM, Marble A, Lawson KA, Zhao GQ. Requirement of Bmp8b for the generation of primordial germ cells in the mouse. Mol Endocrinol. 2000;14:1053–63.

Ying Y, Qi XX, Zhao GQ. Induction of primordial germ cells from murine epiblasts by synergistic action of BMP4 and BMP8B signaling pathways. Proc Natl Acad Sci U S A. 2001;98:7858–62.

Yokobayashi S, Liang CY, Kohler H, Nestorov P, Liu Z, Vidal M, van Lohuizen M, Roloff TC, Peters AH. PRC1 coordinates timing of sexual differentiation of female primordial germ cells. Nature. 2013;495:236–40.

Yokonishi T, Sato T, Katagiri K, Komeya M, Kubota Y, Ogawa T. In vitro reconstruction of mouse seminiferous tubules supporting germ cell differentiation. Biol Reprod. 2013;89:1–6.

Yokonishi T, Sato T, Katagiri K, Ogawa T. In vitro spermatogenesis using an organ culture technique. Methods Mol Biol. 2013;927:479–88.

Yomogida K, Yagura Y, Tadokoro Y, Nishimune Y. Dramatic expansion of germinal stem cells by ectopically expressed human glial cell line-derived neurotrophic factor in mouse Sertoli cells. Biol Reprod. 2003;69:1303–7.

Yoshida S, Nabeshima Y, Nakagawa T. Stem cell heterogeneity: actual and potential stem cell compartments in mouse spermatogenesis. Ann N Y Acad Sci. 2007a;1120:47–58.

Yoshida S, Sukeno M, Nabeshima Y. A vasculature-associated niche for undifferentiated spermatogonia in the mouse testis. Science. 2007b;317:1722–6.

Yuan Z, Hou R, Wu J. Generation of mice by transplantation of an adult spermatogonial cell line after cryopreservation. Cell Prolif. 2009;42:123–31.

Zhang Y, Yang Z, Yang Y, Wang S, Shi L, Xie W, Sun K, Zou K, Wang L, Xiong J, Xiang J, Wu J. Production of transgenic mice by random recombination of targeted genes in female germline stem cells. J Mol Cell Biol. 2011;3:132–41.

Zhang Z, Gong Y, Guo Y, Hai Y, Yang H, Yang S, Liu Y, Ma M, Liu L, Li Z, Gao WQ, He Z. Direct transdifferentiation of spermatogonial stem cells to morphological, phenotypic and functional hepatocyte-like cells via the ERK1/2 and Smad2/3 signaling pathways and the inactivation of cyclin A, cyclin B and cyclin E. Cell Commun Signal. 2013a;11:67.

Zhang L, Tang J, Haines CJ, Feng H, Lai L, Teng X, Han Y. c-kit expression profile and regulatory factors during spermatogonial stem cell differentiation. BMC Dev Biol. 2013b;13:38.

Zhou L, Wang L, Kang JX, Xie WH, Li XY, Wu CQ, Xu B, Wu J. Production of fat-1 transgenic rats using a post-natal female germline stem cell line. Mol Hum Reprod. 2014;20:271–81.

Zou K, Yuan Z, Yang ZJ, Luo HC, Sun KJ, Zhou L, Xiang J, Shi LJ, Yu QS, Zhang Y, Hou RY, Wu J. Production of offspring from a germline stem cell line derived from neonatal ovaries. Nat Cell Biol. 2009;11:631–U424.

Zou K, Hou L, Sun K, Xie W, Wu J. Improved efficiency of female germline stem cell purification using fragilis-based magnetic bead sorting. Stem Cells Dev. 2011;20:2197–204.

Zuckerman S. The number of oocytes in the mature ovary. Recent Prog Horm Res. 1951;6:63–109.

Chapter 2
Human Amniotic Fluid-Derived and Amniotic Membrane-Derived Stem Cells

Limei Yu

Abstract Application of amnion membrane with multiple bioactive biomaterial has over 100 years of history. Amnion membrane- and amniotic fluid-derived stem cells mainly included mesenchymal stem cells and epithelial cells. They have special morphology and express some of stem cell markers, different immunophenotype molecules, and germ layer original protein markers for identification. Amnion membrane-derived stem cells may be isolated and purified by using two digestive enzymes, with different adherence time and subculture. They may differentiate into kinds of function cells of three germ layers in in vitro and in vivo. Amniotic fluid-derived and amniotic mesenchymal stem cells not only have the power of proliferation and plasticity feature, but also have other functions, such as immunoregulation, angiogenic potential, and secretion. Amniotic epithelial cells seem to play a more effective role in neuronal damage. The immunoregulation of amniotic mesenchymal stem cells is emphasized on effects and the mechanism. The transplantation of amnion membrane- and amniotic fluid-derived stem cells, and engineered seed cells generate significant therapeutic actions on regeneration of tissue or organ injury and autoimmune diseases, etc. Although the safety and effectiveness still need further investigations, amnion membrane, amnion membrane- and amniotic fluid-derived stem cells have been shown a broad application prospect. The mesenchymal stem cells are considered as available sources of regenerative treatment. As adult mesenchymal stem cells, are generally derived from the mesoderm, such as bone marrow, umbilical cord blood, adipose, amnion, amniotic fluid, Wharton's jelly, and mobilizing peripheral blood. They have multipotent differentiation capacity and can be differentiated into various cell types, except for self-renewal. Many studies have demonstrated that the stem cells identified from amniotic membrane and amniotic fluid are shown to have advantages for many reasons, including the possibility of noninvasive isolation, low immunogenicity,

L. Yu (✉)
Key Laboratory of Cell Engineering in Guizhou Province, Zunyi Medical College,
Dalian Road, Zunyi, Guizhou 563003, People's Republic of China
e-mail: ylm720@sina.com

R.C. Zhao (ed.), *Stem Cells: Basics and Clinical Translation*, Translational
Medicine Research 1, DOI 10.1007/978-94-017-7273-0_2

abundant sources, anti-inflammatory and non-tumorigenicity properties, and minimal ethical problem (Ilancheran et al. in Placenta, 30:2–10, 2009; Wolbank et al. in Tissue Eng, 13:1173–83, 2007; Ilancheran et al. in Biol Reprod, 77:577–588, 2007; Wolbank et al. in Tissue Eng Part A, 15:1843–1854, 2009; Moodley et al. in Am J Respir Crit Care Med, 182:643–651, 2010).

Keywords Amnion membrane · Mesenchymal stem cell · Amniotic fluid · Stem cell

2.1 Amniotic Membrane and Amniotic Fluid

Amniotic membrane (AM) is a component of the placenta that originates in the extraembryonic tissue and has functions to protect the fetus during pregnancy with supplemental nutrients. AM is composed of three major layers (Sippel et al. 2001): a single epithelial layer, a thick basement membrane, and an avascular mesenchyme. Amniotic fluid (AF) contains a large of heterologous cell population from different tissues of all three germs, while mainly derived from AM, fetal skin, fetal digestive tract, respiratory tract, and urethra cast-off cells of the developing embryo, and so on. Currently, the treated human AM is widely used as biomaterial for clinical treatment (see Table 2.1) (Feng and Yu 2014). Because it has the ability

Table 2.1 Clinical application of amniotic membrane

Origin form	Application	Advantage	Defect
Fresh, cryopreserved, or freeze-dried AM, as surgical dressing	Eyelid and skin burn, wound, ulcer, skin coloboma	Powerful anti-inflammatory, anti-biosis, anti-infection, immunoregulation, and barrier function	Closely plying-up with tissue to replace with difficulty
Fresh AM, as tissue graft or surgical dressing	Corneal injury, tympanoplasty, vestibuloplasty, angioplasty, urinary tract reconstruction	Secrete active substance, keep long time for drug release	Activity is decreased after preservation
Fresh AM, as AM-derived cells or engineered tissue	Cell therapy (type 1 diabetes), engineered tissue as artificial skin, engineered osseous and blood vessel, spinal cord injury (preclinical animal studies)	Absorbability and plasticity, a host material of natural biological membrane structure is beneficial to cell adherence and growth	Potential ethical issue
Fresh or freeze-dried AM, as biomaterial, tissue graft, trestle, nerve conduit	Gastroschisis, post-natal sternal repair, myelomeningocele, Mayer-Rokitansky-Kuster-Hauser syndrome	Prevention of adhesion, abroad origin, low immunogenicity, no tumorigenicity for autograft or allograft	Need to eliminate pathogenic micro-organism infection and hereditary disease

to produce scarring reduction, and antiangiogenic and anti-inflammatory properties, the AM is generally known as tissue engineering material for therapy of serious burns, skin, and corneal transplantation (Luo et al. 2004; Fairbairn et al. 2014; Fan et al. 2006; McGhee and Patel 2011; Mi et al. 2012; Shimazaki et al. 2002). Many studies have demonstrated that a lot of amniotic mesenchymal stem cells (AMSCs) and amniotic epithelial cells (AECs) are derived from AM. They are what is called AM-derived MSCs and AM epithelial cells, and they have powerful self-renewal and pluripotency (Ochsenbein-Kölble et al. 2003; Miki and Strom 2006; In't Anker et al. 2003). AM also is an important source of stem cells.

AF exists in bag of waters. Along with fetal development, the bag of waters is gradually filled with AF. AF contains water, protein, carbohydrate, lipid, fetus urine, and electrolyte composition (Westgren et al. 1995). The volume and element all come up with unceasing changes following fetal development. Human AF was formed at 2 weeks after fertilization in the amniotic cavity of early gestation. During pregnancy, AF is secreted mainly as a result of active transport of sodium and chloride, which is accompanied by transport of water through the chorio-AM and embryo's skin, as well as some of protein molecules. The production of urine and respiratory fluid both contributes to the volume of AF. AF is important to keep the fetus safe, and it supports organ development. The average volume is increasing from 270 to 400 ml at week 16 and week 20 of pregnancy. From weeks 15 to 20 of pregnancy, AF cells are routinely used to evaluate karyotyping and they are genetic and molecular tested for prenatal diagnostic testing (Bocian 2007). The human AF has been proposed as a source of stem cells. One of the adherence and shuttle cells is termed human amniotic fluid–derived MSCs (AF-MSCs), which were more studied on biological characteristics and therapeutic uses. Furthermore, there are also multidirectional differentiation potential stem cells in AF. Many studies have identified that human AF has been proposed as a source of stem cells. The adherence and shuttle AF-MSCs are like bone marrow or AM-derived MSCs. AF-MSCs also are extensively studied on biological characteristic and therapeutic bases (Tsai et al. 2007; Zheng et al. 2008; Lovati et al. 2011; Zhou et al. 2014).

2.2 Isolation and Culture

AMSCs, AECs, and AF-derived stem cells (AFSCs) express some of stem cell protein makers, some of immune molecules and have biological function of stem cells. Many methods have been established that a lot of AMSCs, AECs, and AFSCs can be isolated and successfully cultured. These cells have stem cell characteristics and differential growth features.

2.2.1 AMSCs and AECs

AM usually was discarded after delivery. Human and the other animal AM collection can be obtained via normal delivery without the use of invasive methodologies. To prevent contamination and damage to the tissues, AM samples were collected immediately after parturition using sterilized surgical equipment. The collected placenta samples were stored at 4 °C and transported to the laboratory as quickly as possible. The AM is mechanically peeled away from placenta or allantois. The collected amniotic membrane is washed with 0.9 % normal saline three or four times under sterile conditions to remove debris and blood. After washing, the AM is minced with a surgical blade and scissors. The minced tissue was digested and gently shaken at 37 °C for approximately 3–4 h with collagenase type I or 20 min with 0.05 % trypsin–0.02 % EDTA-2Na for three times (Filioli Uranio et al. 2014; Miki et al. 2005). Amnion digests will be then filtered through a micron nylon mesh and centrifuged for the collection of AECs (see Fig. 2.1). These AECs will be washed and suspended in Dulbecco-modified essential medium (DMEM) with 10 % heat-inactivated fetal bovine serum (FBS). After digestion, the supernatant will be mixed with an equal volume of DMEM with 10 % FBS and centrifuged. The pellet will be resuspended with DMEM. Remaining amnion trypsin digests will be treated again. This process was repeated three times. AECs will be collected and pooled with the previous cell suspension. AECs will be cultured in low-glucose DMEM and then supplemented with 10 % heat-inactivated fetal bovine serum, 10 ng/ml human epidermal growth factor, 100 UI/ml penicillin, and 100 μg/ml streptomycin for culture. Additive may be included in nonessential amino acid, β-mercaptoethanol, and sodium pyruvate. Cell cultures are usually used at passages 2 and 5 for experiment (Zhao et al. 2012). Human AECs have a short life, but could be established the immortalized human AECs by introducing with viral oncogenes E6/E7 and with human telomerase reverse transcriptase.

The remaining fragments will be digested with 0.75–0.94 mg/ml collagenase II and 0.075 mg/ml or 20 mg/ml DNase I for 1–3 h at 37 °C with 200-rpm shaking (Yao et al. 2013a). Amnion fragments will be then removed, mobilized cells were passed through a 100–300-μm cell strainer and collected by centrifugation at 400 g for 10 min. As reported above, these cells are referred to as human AMSCs (see Fig. 2.1). AMSCs at different passages are more than freshly isolated cells that will be plated at a density of 5×10^5 cells/ml to 1×10^6 cells/ml. Upon reaching about 80 % confluence, adherent cells will be trypsinized and then subcultured at a density of 1×10^5 cells/ml until passages 10 (Han et al. 2008), at least passages 22–24. The cell samples will be washed in phosphate-buffered saline and centrifuged at 350 or 400 g. The cells will be cultured in L-DMEM containing 10 % fetal bovine serum or umbilical cord blood serum with basic fibroblast growth factor (bFGF) 4–10 ng/ml or umbilical cord blood serum supplemented

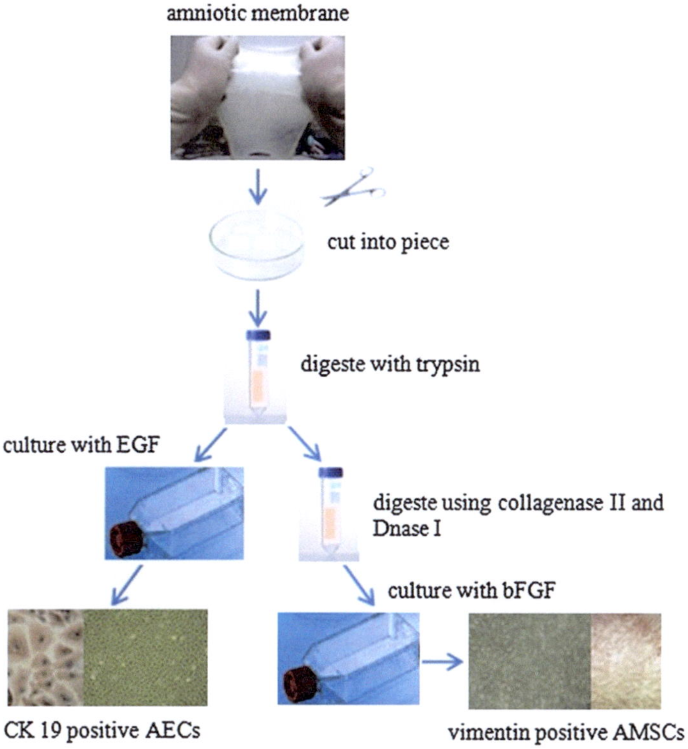

Fig. 2.1 The isolation, culture, and identification of AMSCs and AECs. The AM was peel off placenta, and then AMSCs and AECS were isolated by trypsin, collagenase II and DNAse I digestion. Two kinds of diAM-derived stem cells were identified by immunohistochemical staining, morphology, and flow cytometry. AMSCs are fusiformis and vimentin positive, but AECs are like flagstone and CK19 positive

with 1 % penicillin (100 UI/ml), streptomycin (100 μg/ml) or 0.25 mg/ml amphotericin B, with 2 mmol/L L-glutamine (Zhang et al. 2007). These cells cultured in a humidified atmosphere with 5 % CO_2. The culture medium will be changed every 3–5 days in primary culture and passaged after reaching 80 % confluency after trypsin digestion and centrifugation. The cell pellet was resuspended in complete culture medium. The cells will be cultured in a humidified atmosphere with 5 % CO_2. The basal culture medium was changed three times a week and passaged after reaching 80–90 % confluency. AMSCs were purified by differential adhesion and subculturing (Alviano et al. 2007). Literatures reported that AMSCs were cryopreserved after passage 3 in 50–90 % fetal bovine serum and dimethyl sulfoxide or protein-free cryopreservation for human mesenchymal stem cells (MSCs). These AMSCs were used for basic cytobiological and cell therapeutic researches.

2.2.2 AFSCs

Mesenchymal stem cells from amnion and AF are successfully isolated and cultured from human, as well as from the bovine. The isolation of AFSCs is a simpler process than isolation of AECs and AMSCs. After patients received detailed information, each participant gave her written informed consent. Two or three milliliters of amniotic fluid samples was obtained from 16- to 19-week pregnant women who underwent amniocentesis for fetal genetic determination in routine prenatal diagnosis (Tsai et al. 2004). Large numbers of AFSCs can be isolated by centrifuging the samples at 250–400 g for 10 min at room temperature and expanded in cultured condition. Cell samples will be used only when a normal karyotype was detected by the cytogenetic analysis. These cells proliferate rapidly with doubling times of 30–36 h and do not require supportive feeder layers for many passages, while maintaining chromosomal stability. AFSCs are cultured in serum-free culture medium or 10 % fetal bovine serum L-DMEM, or L-DMEM: F12 medium or α-minimal essential medium (α-MEM), supplemented 10 ng/ml EGF and 2 mmol/L L-glutamine in a humidified incubator at 37 °C with 5 % CO_2 (Ghaderi et al. 2011; Yang et al. 2013; Li et al. 2006). Culture medium was changed once every 3–5 days, suspension cells will be wiped off, and fibroblast-like colony will be scraped using cell scratcher. Suspension cells 4×10^4 cells/ml will be prepared and cultured by complete medium. Alternatively, AF-MSCs, a kind of shuttle cells of adherence, can be cultured in medium with 4–10 ng/ml bFGF until 70–80 % confluency in primary culture (Liu et al. 2009), and then, depurative cells will be continuously cultured by different adherence as passage 2 to 3. The AF-MSCs will be routinely subcultured every three days at 1:3 or 1:4 dilution and allowed to expand in complete medium. Cells usually will be maintained in culture for up to 4–8 passages and used for all the experiments. The pregnant metaphase AFSCs may be amplified at $1–5 \times 10^{12}$ cells/ml at the tenth generation from 20 to 40 ml AF. AF cell is not only used for prenatal diagnosis, but also as another source for stem cells of fetus. AFSCs also will be isolated and purified by immunomagnetic bead method and flow cytometry sorting. CD117-positive cells are one of sorting method by immunomagnetic bead method (Chen et al. 2009). MSCs from the amniochorionic membrane will be extracted using the markers $CD34^-$, $CD45^-$, $CD73^+$, $CD90^+$, $CD105^+$, and $CD29^+$ at the fluorescence-activated cell sorting analysis. The vitrification is a reliable and effective method for cryopreservation of human AF-MSCs (Moschidou et al. 2013).

Oxygen is a potent biochemical signaling molecule which exerts significant effects on the growth and development of mammalian cells. The state of hypoxia is cell-type dependent and affects critical cellular processes, such as proliferation, differentiation, adhesion, apoptosis, metabolism, extracellular matrix secretion, and growth factor expression. It has been demonstrated that hypoxic preconditioning of MSCs can reduce hypoxia-induced cell death, which is caused by the paracrine activity of MSCs. Hypoxia (1 or 5 % O_2) similarly not only increases the proliferation of AF-MSCs, but also maintains their constitutive characteristics (surface marker expression and differentiation potentials). Notably, more paracrine factors, vascular endothelial growth factor, and transforming growth factor beta 1, will be secreted into hypoxic conditioned medium from AF-MSCs than normoxic conditioned medium (Jun et al. 2014).

These cells may be cryopreserved, and the cell viability of revived cells is higher. The surface makers and multidirectional differentiation potential also are not affected. Therefore, ASCs, AECs, and AFSCs are suitable for large-scale culture and reservation (Janz Fde et al. 2012). Compared to other stem cells, amniotic cells and AFSCs can be easily collected during routine prenatal testing, and the AM can also be obtained during cesarean section after birth. These isolation methods are noninvasive progress without destroying human embryos and thus avoid ethical controversy the most.

2.3 Identification and Characterization

Among extrafetal tissues, recently, AM appeared to be an important stem cell source in different species, and AM-MSCs have been isolated and characterized in different species, including the human, horse, sheep, dog, rat, and cat (Marcus et al. 2008). AECs have been found only in humans, horses, and sheep. Human MSCs from AM or AF represent a population of multipotent adherent cells able to be differentiated into many lineages. The AMSCs can differentiate into all three germ layers for ectodermal, mesodermal, and endodermal lineage cells. They are positively expressed mesenchymal markers, such as CD44, CD73, CD29, CD105, and CD90, and negatively expressed hematopoietic markers, as CD34, CD45, CD11b, CD19, and human leukocyte antigen (HLA)-A, HLA-B, and DR antigens (Kim et al. 2007) (see Table 2.2 and Fig. 2.2). In addition, the AM expresses

Table 2.2 Comparison of biomarker proteins with AM- and AF-derived stem cells

Antigens	Expression		
	ASCs	AECs	AF-MSCs
Mesenchymal stem cell markers			
CD13	+		+
CD44	++	±	++
CD73	++	+	+
CD71		−	
CD90	++	++	+
CD105	++	++	±
Haematopoietic stem cell markers and immunoreactions moleculars			
CD34	−	−	−
CD45	−	−	−

(continued)

Table 2.2 (continued)

Antigens	Expression		
	ASCs	AECs	AF-MSCs
CD133			
CD14		±	−
CD19		−	−
CD11b	−		
CD117	−		
HLA-DR	−	−	±
HLA-A	±	+	±
HLA-B	±	+	±
HLA-C	±	+	±
HLA-G		++	
CD40	±	+	
CD40L	±	±	
CD80	±	±	
CD86	±	±	
Somatic stem cell markers			
CD24		+	
CD29	++	++	+
CD49d	+	−	
CD49f		++	
CD271		+	
CD166	+		+
CD146			++
Vimentin	++	−	
CK19	−	++	−
E-cadherin	±	++	
Nestin		++	
Stem cell markers			
Oct4	+	++	+
SOX-2	+	++	++
SSEA-3	+	++	−
SSEA-4	+	++	
Nanog	+	++	++
c-myc	+		+
Klf4		++	++
TRA1-60	+	+	−
TRA1-81	+	++	−

− negative expression; ± low expression; + moderate expression; ++ or +++ high expression

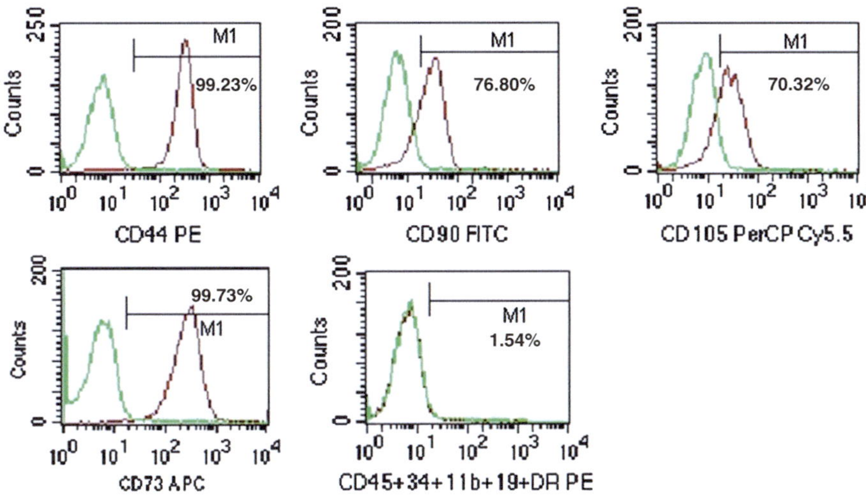

Fig. 2.2 Phenotypic analyses of human AMSCs by flow cytometry. Human AMSCs of passage 5 expressed positive CD44, CD90, CD150, and CD73, but negative CD45, CD34, CD11b, CD19, and HLA-DR by using human MSC identification kit from BD Biosciences

antiangiogenic and anti-inflammatory proteins. A lot of results show that the AMSCs are very important for advanced regenerative medicine, because inflammatory regulation and low immunogenicity remain indispensable factors, despite the pluripotent marker expression of AM-MSCs, such as Oct-4, Nanog, TRA-1-60, and TRA-1-81 (Miki et al. 2007; Yu et al. 2012). Human AECs have a low immunogenic profile and possess potent immunosuppressive properties and also have several characteristics similar to stem cells. They do not express CK-19 protein and do not form teratoma (Bilic et al. 2008). The cultured human AECs from P0 to P4 expressed and downregulated the stemness gene expression for Oct-4, Sox-2, Nanog3, FGF4, Rex-1, FZD-9, BST-1, and ABCG2. However, vimentin and nestin gene expression were upregulated (Simat et al. 2008).

Immunological rejection after therapy does not occur in the AM, and the cells are derived from AM and AF. For these reasons, AM and AM-derived stem cells might be useful sources for cell transplantation and tissue engineering for regenerative and autoimmune diseases with fewer ethical problems. After AMSCs are cryopreserved in 40 % FBS and 10 % DMSO with 50 % α-MEM culture medium, biological characteristics of revived AMSCs had no remarkable change, such as morphology, vitality, CD molecules, growth curve, cell cycles, and Oct 4 protein (Mann et al. 2013; Wang et al. 2012).

AF cells can be used as a source of fetal progenitor cells or otherwise discarded. Research results showed that AFSCs express embryonic stem cells-specific makers. AF- and AM-MSCs are same as bone marrow MSCs sharing similar morphological characteristics of the fibroblastoid shape. They possess the feature of adherence growth. After subculturing, the morphological change did not occur in fitting culture conditions. The growth curves showed the AM, and AF-MSCs had a similar proliferative capacity at passage 5 and passage 10. These cells have kept the length and activity of telomerase. The surface markers, karyotype, cell cycle, and apoptosis all do not change between passage 5 and passage 25 in subculture. The use of AFSCs could minimize the ethical objections as well. AF-MSCs have easy isolation, a high renewal activity, and maintenance stability. Oct-4, Nanog, SSEA-4, and SOX2 are all important regulation molecules on pluripotency and self-renewal. AFSCs also express MSC makers: CD29, CD44, CD58, CD73, CD90, CD105, CD117, and CD166. They are positive for major histocompatibility complex I class (MHC I) molecular antigen HLA-A, HLA-B, and HLA-C, but were negative, or mildly positive, for MHC class II antigen HLA-DR. And CD34, CD45, ABCG2, C-MET, SSEA-1, SSEA-3, TRA-1-60, and TRA-1-80 expressions are negative (see Table 2.2) (Rossi et al. 2014; Chen et al. 2011). The protein markers of AM- and AF-derived stem cells are analysed by flow cytometry, immunocytochemistry and immunofluorescence staining.

AMSCs express some of moderate stemness markers (FGF-2, LIF, Nanog, etc.), important signal molecules for stem cell maintenance and Wnt and Notch self-renewal genes, CD44-specific mesenchymal original markers and higher cell adhesion molecules and cell cycle regulators, and low differentiation markers after cultured passage 4–5, but did not express TERT gene by gene chip test, etc. (Terai et al. 2014). Furthermore, AF-MSCs enable to be use of autologous cells obtained from patients' tissues. Moreover, they maintain genetic stability and offer advantages of non-tumorigenicity and low immunogenic activity. These findings show that AF-MSCs are being considered as potential sources of treatment with diseases.

2.4 Function Features

2.4.1 Plasticity In Vivo

AM- and AF-derived stem cells show high proliferative capacity in culture and multilineage differentiation potential. This multipotential differentiation capability of these stem cells can be utilized for giving rise to a variety of differentiated cell types for tissue repair and regeneration. AMSCs, AECs, and AF-MSCs were seeded in special culture for differentiation studies (Miki and Strom 2006; Saito et al. 2012). When cultured in osteogenic medium, these stem cells displayed a significant increase in alkaline phosphatase activity and mRNA expression, Alizarin red S staining, and Runx2 mRNA expression (Kim et al. 2013). While maintaining in an adipogenic culture medium, these cells gave a time-dependent increase in PPARγ and FABP4 mRNA expression, glycerol-3-phosphate dehydrogenase activity, and positive lipid droplets to Oil Red Oil staining (Vidane et al. 2014). These results confirm that they can differentiate toward osteogenic and adipogenic phenotypes. Chondrogenic and neurogenic differentiations were assessed as previously reported. Chondrogenic was demonstrated by Alcian blue staining and neurogenic by conventional Nissl staining, which showed increasing ribosomes, and nerve cell shape was observed microscopically (Manochantr et al. 2010). Pleiotrophin is involved in the AEC-induced differentiation into dopaminergic neuron-like cells. Monolayer cultured human AMSC cell is differentiated into chondrocytes for the original cells of cartilage with transforming growth factor-β, dexamethasone, vitamin C, and insulin-transferrin-sodium selenite, which indicates that human amnion cells can be used as the seed cell of cartilage. Human AECs and AMCs have osteogenon characteristic when 3-glycerophosphate is added, which demonstrates that human amnion cells can be the resource of seed cells of bone tissue engineering.

The AM- and AF-MSCs were, respectively, seeded on plastic plates precoated with matrigel in L-DMEM containing FBS culture medium with 5-azacytidine. The differentiation capacities of AM- and AF-MSCs were detected to express GATA-4, cardiac troponin T, α-actin, Cx43, and Nkx2.5 genes, as myocardial genes after myocardial induction in vitro. Induced AMSCs and AECs all expressed desmin and α-actinin proteins after being treated with 5-azacytidine and bFGF or 5-azacytidine alone. Both human AMSCs and AECs possess the potential to differentiate into cardiomyocyte-like cells in vitro (Nagura et al. 2013; Bai et al. 2012; Han et al. 2011). And they might be candidate for cellular cardiomyoplasty for the treatment of heart failure caused by ischemic injury because adult cardiomyocytes do not regenerate. AF-MSCs also have the potential clinical application for myogenesis in cardiac regenerative therapy.

Combined approach of dexamethasone, hepatocyte growth factor, insulin-like growth factor, and other cytokines were used to induce the differentiation of human AECs and AF-MSCs into hepatocyte-like cells (Luo et al. 2011; Choi et al. 2013). The shuttle shape of AM- or AF-derived MSCs changed into polygon. The liver-like cells show changes on stem cell biomarker genes and liver cell special protein, the latter as the mRNA expression of alpha fetal protein, hepatocyte growth factor receptor the latter C-met, hepatocyte nuclear factor-3β, cytokeratin-18, expression of hepatic microsomal enzyme in vitro and in vivo. The differentiated cells also developed hepatocyte-specific functions, i.e., they secreted albumin, absorbed indocyanine green, and stored glycogen (Liu et al. 2011; Tamagawa et al. 2007; Miki et al. 2009). In the near future, coculture without contact of human AMSCs and normal human liver cell line also can successfully induce AMSC differentiation into liver-like cells.

After induction in vitro, AMSC is differentiated into neural stem-like cells that expressed higher levels of the neural stem cell markers, Nestin, Sox2, and Musashi. Interestingly, the neurotrophic factors, brain-derived neurotrophic factor, nerve growth factor, neurotrophin 3, glial cell-derived neurotrophic factor, and ciliary neurotrophic factor were markedly upregulated (Yan et al. 2013). AF-MAC-induced functional dopaminergic neuron-like cells in vitro showed increased activity in regeneration of dopaminergic neuron-like cells, increased migration distances, and improvement of animal behavior in the Parkinson's disease rat model (Liu et al.

2011). Following transplantation in a rat traumatic brain injury model, AECs showed significant improvements on neurological function and brain tissue morphology. Human AMSCs and AECs not only expressed neuron-specific enolase, neurofilament, glial fibrillary acidic protein, β-tubulin-III, microtubule-associated protein, and neuronal nuclei, but also the level of dopamine is rised at mean value in medium with all-trans retinoic acid (Chang et al. 2010). Human and sheep AECs can transform the neuron-like cell or neurons by coculture with traumatic brain tissue extracts or in conditional medium. Though AEC proliferation is not significant, the AECs cultured by low cell density survive more easily. Human AECs were induced to differentiate into neurocytes by using chemical inducer all-trans retinoic acid and astragalus, but astragalus induction has a higher cell survival rate, and the expression of Notch1 signal molecules is inhibited during the induction (Zhu et al. 2013; Chen and Wang 2012).

Human AMSCs and AECs may be induced to differentiate into insulin-secreting cells in nicotinamide and N_2 supplement medium. The induced ratio of insulin-positive cells or islet-like cells was above 70 %. The contents of insulin were 328.47 and 331.60 μIU/ml in the supernatant of cultured AMSCs and AECs, respectively (Peng et al. 2011; Zhao et al. 2012). This suggests that human-derived stem cells might become a new cell source of therapy for diabetes. The vascular endothelial cells were induced, when human AMSCs and AECs were cultured in DMEM with vascular endothelial growth factor and basic fibroblast growth factor. The CD34, CD54, and CD31 expressions of cultured AECs in induced medium with high and low glucoses showed no difference. However, the CD54 expression of AMSCs cultured in induced medium with high glucose was much lower than low-glucose DMEM. AMSCs have an angiogenic potential. These data suggest that human amniotic cells might become a seed cell of angiogenesis in tissue engineering and could become an alternative cell resource for repair of vascular injuries (Warrier et al. 2012). In addition, calcium-sensing receptor is a G-protein-coupled receptor able to bind calcium ions and plays a physiological role in regulating bone metabolism. Its agonist calcimimetics can prompt osteogenesis in AF-MSCs, perhaps being used in bone traumatic and degenerative damage (Di Tomo et al. 2013). AECs as an ideal stem cell resource for the cell replacement therapy, transplanted into the injured submandibular salivary gland in salivary gland dysfunction diseases. Notably, identification of GFP-labeled AFSCs and immunostaining with antihuman antigen-specific antibodies demonstrated that grafted human AFSCs survived and differentiated into granulosa cells during oocyte development in chemotherapy-induced sterility (Lai et al. 2013). Human AFSCs seem to be a good candidate for cell reprogramming in embryonic stem cell conditions with valproic acid administration, a transgene-free approach, and they are more efficiently reprogrammed to pluripotency than adult cells, as skin cells, except for differentiation into chondrocytes and lipoblast, etc. AF stem cell-induced pluripotent stem cells were able to form derivatives of the three embryonic germ layers, but also of the extraembryonic trophoblast lineage activating BMP signaling cascades and blocking of TGF-β/Activin/Nodal signaling (Moschidou et al. 2013; Li et al. 2009; Galende et al. 2010).

AECs possess a much greater ectodermal differentiation capacity, while AMSCs possess a much greater mesodermal differentiation capacity. Oct4, Nanog, and Sox2 are important transcriptional factors on stemness maintenance. But the canonical Wnt/ß-catenin signaling pathway appears to trigger human AF-MSCs osteoblastogenesis and adipogenic differentiation. The results unravel novel molecular determinants of AF-MSC commitment toward osteoblastogenesis, which may represent potential targets for improving their use in regenerative medicine. Although a plethora of molecules have been identified to have a role in modulating stem cell fate, the Wnt signaling is recognized as a key regulator of adult tissue homeostasis and remodeling through multiple so-called canonical and non-canonical pathways (D'Alimonte et al. 2013). Flt3, the receptor of Fms-related tyrosine kinase 3 ligand, is expressed in AMSCs. Fms-related tyrosine kinase 3 ligand is able to promote the proliferation of AMSCs effectively in vitro. However, the phenotype and ability of AMSCs to differentiate into mesenchymal lineages did not change (Li et al. 2014).

2.4.2 Immunomodulation

Human AECs, AMSCs, and AF-MSCs have a low immunogenic profile and possess potent immunosuppression. AM-derived stem cells have been shown to retain immunomodulatory properties and possessed strong inhibition of lymphocyte proliferation and survival when transplanted in immunocompetent animals without inducing any tumorigenic effect in vivo. It also inhibits cytokines interleukin (IL)-2 and interferon gamma IFN-γ production and suppresses the generation and maturation of monocyte-derived dendritic cells, as reported for MSC from other sources (Banas et al. 2008; Magatti et al. 2009; Xue et al. 2014). AF-MSCs are known to play a role in preventing rejection of the fetus and are thought to have low immunogenicity. As needed, AMSCs, AECs, and AF-MSCs could be stored and provided immediately for future autologous therapy. The autologous tissues made from patient-specific cells could be applied to non-rejected transplantation. AM-derived stem cells express human leukocyte antigen G (HLA-G) exposed to IFN-γ. The programmed cell death receptor 1 (PD-1, an inhibitory receptor that is normally expressed on activated T and B cells), is negative, but programmed death ligands 1 and 2 (PD-L1 and PD-L2) are typically upregulated on the AM stem cells. The two cells are also negative for the immunoglobulin-like transcript receptors 2, 3, and 4. There is some controversy about the expression of TRAIL, tumor necrosis factor alpha (TNF-α), and Fas ligand (Fas-L), all members of the TNF family for induction of apoptosis. However, in adaptive immunity, the level of IL-10 and TGF-β in the supernatant not only increased significantly in cocultures of AMSCs and peripheral blood mononuclear cells, but also inhibit proliferation. The level of IL-17 and IFN-γ also is lower in the presence of mitogens. Meanwhile, hepatic growth factor, indoleamine 2,3-dioxygenase (IDO), and cyclooxygenase 2 (COX-2) mRNA, were induced more in AMSCs. The inhibition of cyclooxygenase pathway partially reverted the antiproliferative effect of T cells. These factors previously documented to take part in the inhibitory effects of MSCs from other sources. The AM stem cells could be an interesting source of soluble factors in clinical application, without referring to rejection reaction (Insausti et al. 2014; Rossi et al. 2012). AMSCs can significantly suppress T lymphocyte proliferation, especially $CD8^{+}$ T cell. Otherwise, it can decrease activated Th1 and Tc1 percentages and slightly increase Th2 and Tc2 percentages. AMSCs have potential of alleviating acute graft-versus-host disease and maintaining graft versus leukemia. AMSCs and AECs that have HLA-G secretion feature may be involved in the suppression of the lytic activity of NK cells and B cells and modulate the maturation of dendritic cells (Insausti et al. 2014; Di Trapani et al. 2013; Fang et al. 2014).

Primary cultured AMSCs and AECs were treated with IL-1β as a model for acute chorioamnionitis and with CXCL10 for chronic chorioamnionitis. IL-33 and IL1RL1 (ST2) mRNA were not detected in AECs after incubation with IL-1β or CXCL10. IL-33 mRNA was expressed in AMSCs, and the level of expression has increased after incubation with IL-1β. IL1RL1 (ST2) mRNA expression has decreased in AMSCs after IL-1β treatment. However, IL-33 and IL1RL1 (ST2) mRNA expression in AMSCs did not change with CXCL10 treatment. The regulation of AMSCs and AECs is different from IL-1β- and CXCL10-induced inflammation and immune function (Kallapur et al. 2013).

2.4.3 Epithelial–Mesenchymal Transition (EMT)

EMT is a key cause of fibrosis disease and is also the pathological process in fibrosis. EMT of normal conjunctival tissues is a major reason for pterygium generation. An important maker is α-smooth muscle actin in epithelia transition into mesenchyma, but high expression of Oct3 and E-cadherin genes are makers on inhibition of EMT. Human amnion stem cells can significantly inhibit α-SMA expression and migration of human pterygium fibroblasts by coculture of amnion stem cells and pterygium fibroblasts in vitro. These results suggested that amnion stem cells have the potential to inhibit the generation and invasiveness of pterygium (Sha et al. 2014). In the culture medium with bFGF and the inclusion of an AM in the dermal matrix, most fibroblasts were α-SMA negative. The suppression of α-SMA expression enhanced epidermal differentiation and decreased TGF-β1 expression in the epidermis. The inhibition of TGF-β kinase completely suppressed α-SMA expression in the dermal matrix. The hyperproliferative epidermis expressed more TGF-β1, which is responsible for myofibroblast differentiation (Yang et al. 2011). AFSCs form epithelial tubules and cyst structures in 3D collagen gel. AFSCs continue to express MSC markers during cultivation in the gel. Thus, AFSCs may undergo epithelial–mesenchymal transition (Davydova et al. 2011). The cultured human AECs undergo EMT through the autocrine production of TGF-β. Multisubcultured AECs underwent morphological changes acquiring a mesenchymal shape. Epithelial cell markers E-cadherin and cytokeratins were lost, and typical mesenchymal markers, such as vimentin and α-SMA, appeared. The expression of SNAI1, MMP9, PAI1, or ACTA2 genes is associated with EMT increase. The expression of the transcription activators KLF4 or MTA3 was consistent with the downregulation of CDH1. The TGF-β receptor I (ALK5) inhibitor SB-431542 or TGF-β-neutralizing antibody can prevent EMT and preserve the AECs' epithelial phenotype (Alcaraz et al. 2013).

2.4.4 Angiogenic Potential

The AM is important in clinical applications as it is proangiogenic, antifibrotic, and antiscarring and has low immunogenicity (Koob et al. 2014). It has been recently reported that human AMSCs possess great angiogenic potential in vivo, except that AECs and AF-MSCs might be induced into vascular endothelial cells in condition medium with vascular endothelial growth factor and bFGF (Alviano et al. 2007). Vascular endothelial growth factor receptors 1 and 2 were expressed in induced human AMSCs and the expression of endothelium-specific markers such as FLT-1 KDR and ICAM-1. During the ameliorating peripheral neuropathy in sciatic nerve injury, AMSC injection promoted significant recovery of motor nerve conduction velocity and voltage amplitude, also augmented blood perfusion, and increased intraneural vascularity. Whole-mount fluorescent imaging analysis demonstrated that these MSCs exhibited higher engraftment and endothelial incorporation abilities in the sciatic nerve. The higher expression of proangiogenic factors was detected. Promoting angiogenesis is a therapeutic effect and mechanism of treating peripheral neuropathy, as same as repairing other tissue injury (Li et al. 2014; Warrier et al. 2012). AM-derived stem cells may potentially assist both bone and cartilage repair, due to their angiogenic potential, they may also pave the way for novel approaches in the development of tissue-engineered vascular grafts which are useful when vascularization of ischemic tissues is required (Petsche Connell et al. 2013). AF-MSCs supported vascular tubule formation in vivo more effectively than bone marrow MSCs, further enhancing their promise as vehicles for tissue repair and regeneration. There are differences in secreted angiogenic factors for angioinhibition, inflammatory response, migration, angiogenesis–vasculogenesis, tissue repair, and blood clotting between AF-MSCs and bone marrow MSCs using proteome arrays (Roubelakis et al. 2013).

2.4.5 Secretion Function

AEC is known to produce a "cocktail" of trophic factors, such as neurotrophin-3, nerve growth factor, fibroblast growth factor-2, IL-1, IL-4, IL-6, and so on (Venkatachalam et al. 2009). AECs are found to secrete some of the factors, brain-derived neurotrophic factor and ciliary neurotrophic factor for neuroprotective effect in rat retinal ganglion cells (Uchida et al. 2003). Amnion-derived cellular cytokine solution also accelerates the healing of skin burns. AF-MSCs express several specific neural stem/progenitor markers, such as nestin and connexin 43, and release amounts of brain-derived neurotrophic factor, as well as vascular endothelial growth factor. These factors can enhance cell recovery following neuronal damage through multiple rescue mechanisms and may provide a suitable stem cell therapeutic means for neurodegenerative disorders including Parkinson's disease (Payne et al. 2010). The use of human AF-MSCs as the feeder layer to establish human embryonic stem cell lines is promising, because of multiple biological active factor secretion of AFSCs. In addition, AMSCs and AECs quickly exert therapeutic effect on anti-inflammatory, promote angiogenesis, and reduce damage before these cells still have not found to be differentiated into specific tissue function cells in tissue or organ injury. The mechanism may be related to secretion or paracrine of AM- and AF-derived stem cells. Some of cytokines significantly enhanced in damage tissue, such as hepatocyte growth factor, bFGF, vascular endothelial growth factor, IL-4, IL-10, insulin-like growth factor-1, and granulocyte colony-stimulating factor. The action may be the same as bone marrow MSCs without direct and enough evidence on more cytokines' secretion (Carvajal et al. 2013; Yamahara et al. 2014).

2.5 The Potential Applications and Therapeutic Base

MSCs have shown therapeutic potential for repair and regeneration of tissues damaged by injury or diseases. Human adult stem cells are multipotent cells which are present in many tissues of the human body, including AM and AF, and present in bone, cartilage, muscle or fat, as well as a variety of other connective tissues by differentiating into various cells of other embryonic lineages, such as osteoblasts, chondrocytes, myoblasts, liver cells, islet cells, myocardial cells, neurocytes, and vascular endothelial cells or adipocytes. They are an important source for regenerative medicine, such as in osteogenesis imperfect, bone fracture, myocardial infarction, and liver, kidney, and spinal injury. These cells are also used to treat inflammation and immunopathy because they have anti-inflammatory and immunoregulation properties. AM- and AF-derived stem cells have low immunogenicity and have advantages on origin, ethic, amplification, and preservation in vitro, and AMSCs, AECs, and AFSCs have the extremely broad translation and application prospects in autologous and xenogenous treatments.

2.5.1 Cardiac Regeneration

After inducing cardiogenic differentiation of human AM- or AF-derived stem cells by cardiac lysis, growth differentiation factor-15, and bone morphogenetic protein-2, fibroblast growth factor 10, 5-azacytidine, respectively, the expressions of cardiac markers inward rectifier potassium channels 2.1, cardiac troponin T and myocyte enhancer factor 2. Nanog promoter-Cre plasmid and cytomegalovirus promoter-loxP-STOP-loxP-Red-puro(r) plasmid were cotransfected into immortalized human AMSCs (Otaka et al. 2013). These cells were treated with 5-azacytidine, trichostatin A, activin A, and bone morphogenetic protein-4, or cocultured with murine fetal cardiomyocytes. Then, expression of Nanog, Oct3/4, Sox2, and Klf4 was significantly higher. They expressed Nkx2.5, GATA-4, human atrial natriuretic peptide, cardiac troponin T, myosin light chain-2α, Mlc-2v, β-myosin heavy chain, hyperpolarization-activated cyclic nucleotide-gated channels, and inwardly rectifying potassium channels (Kir)-2.1. These induced cells could not contract (Nagura et al. 2013; Bai et al. 2012; Han et al. 2011; Shaw et al. 2011). Cocultured AF-MSCs spontaneously contracted in a synchronized manner and expressed the cardiac markers with neonatal cardiomyocytes (Guan et al. 2011). These results suggest that human AMSCs could be a useful cell source for cardiac regeneration therapy. In myocardial infarction rat, AMSCs and AECs may differentiate into myocardial cells, delay ventricular remodeling, and improve cardiac function in rats through epicardial and venous ejection fraction. Brdu-labeled positive human AM-derived stem cells were found in myocardial infarction region at 1, 4, and 6 weeks after human-derived stem cell transplantation. Cell engraftments expressed cardiac-specific protein connexin 43, α-actinin, and desmin. Ejection fraction and fractional shortening, diastolic anterior left ventricular wall, and systolic anterior left ventricular wall values of rats were all significantly higher in stem cell transplantation than in model rats (Fang et al. 2012; Wang et al. 2013a).

2.5.2 Neuronal Regeneration

Stem cell therapy is a potential treatment for spinal cord injury. Several studies have reported that AECs transplanted into the spinal cord transection rats can improve hind limb motor function and ameliorate the bladder function. Transplanted AECs survive well for a long time of 8 weeks and integrate well with the host. AECs survive in the transplanted environment, support the growth of host axons, prevent the formation of glial scar at the cut ends, and may prevent death in axotomized cells or attract the growth of new collateral sprouting (Sankar and Muthusamy 2003). Grafting AECs genetically modified to overexpress glial cell derived neurotrophic factor into spinal cord injury also rescue the axotomized rubrospinal neurons. Two weeks after spinal cord injury, human AECs were transplanted around the spinal cord lesion site of spinal cord injury-induced mechanical allodynia and thermal hyperalgesia. AECs significantly reduced mechanical allodynia, but have no effects on thermal hyperalgesia (Roh et al. 2013). This effect seems to be closely associated with the reduction of spinal cord microglial activity and NMDA receptor NR1 phosphorylation, microglial marker, and F4/80 expression of spinal cord, but not the increased expression of glial fibrillary acidic protein or induced nitric oxide synthetase. Human AECs displayed positive immunoreactivity to MAP-2, glial fibrillary acidic protein, and Nestin could secret the neurotrophic factors. bFGF can upregulate the TrkB receptor as a brain-derived neurotrophic factor preceding condition for activity. Neural stem cells and human umbilical cord blood mesenchymal stem cells grown either with AEC-conditioned medium or in transwells showed significantly improved survival and differentiation into dopaminergic neuron-like cells. AECs are a potential inducer to obtain DA neuron-like cells for an ethical and legal cell therapy for Parkinson's disease. Moreover, the neural differentiation and length of neurite were greater in exogenous FGF (Meng et al. 2007). This is related to secretion of brain-derived neurotrophic factor, etc. AECs may be regarded as a critical component of neural stem cells' niche. This microenvironment is the need for AEC–neural stem cells coculture and could potentially facilitate the production of neurons for future clinical applications. Transfected human bFGF to AECs was serving as neural stem cell-differentiated niche and be useful as a source of sustained trophic supported to improve neural stem cell differentiation toward neuron in vivo. AECs have beneficial effects by the neurotrophic factor secretion of AECs on rats with 6-OHDA-induced Parkinson's disease. AEC transplantation significantly ameliorated spatial memory deficits in double-transgenic mice of coexpressed APPswe- and PS1ΔE9-deleted genes, as well as increased acetylcholine levels and the number of hippocampal cholinergic neurites. This will be very beneficial for Alzheimer's disease therapy (Yang et al. 2010; Xue et al. 2012). Encouraging reports have revealed that human AECs can rescue injured brain tissue and improve functional recovery in experimental models of stroke or middle cerebral artery occlusion of rat (Liu et al. 2008). The potential stroke therapy may involve a reduction of local inflammation and modulation of the immune response, promotion of neural recovery, differentiation into neural tissue, re-innervation of lost connections, and secretion of necessary cytokines, growth factors, hormones, and/or neurotransmitters to restore cellular function. AECs cannot only survive in the cerebrum of rats with traumatic brain injury up to 4 weeks after transplantation, but also express the specific neuronal antigen MAP2 and improve the motor deficits of rats with traumatic brain injury. Human AECs can ameliorate behavioral dysfunction and reduce infarct size in the rat with middle cerebral artery occlusion or after stroke, due to neuronal differentiation and cytokine secretion by these cells (Liu et al. 2008). Transplanted glial cell-derived neurotrophic factor expressing AECs, as a transgene carrier in gene therapy, can protect against hippocampal neuronal death following traumatic brain injury. When it is transplanted into brain tissue as striatum of healthy or disease rats, AECs survived well for a long time, migrated for a distance, and did not induce immune rejection. The AECs labeled with 1,1′-dioctadecyl-3,3,3′,3′-tetrame thyllindocarbocyanine perchlorate can survive in the spinal cord of monkey for up to the maximum period of observation at 60 days. There is no evidence of immunological rejection probably due to the non-antigenic nature of the human AECs (Sankar and Muthusamy 2003).

Transplantation of AMSCs also can benefit to improve neurological function restoration of rats with spinal cord injury. These human nuclei of monoclonal antibody MABl281-positive cells survive in injury site and not express microtubule-associated protein and glial fibrillary acidic protein. Its mechanism might be related to upregulating NF-200 expression in the distal end of injured spinal cord. After spinal cord injury in rats, the combined treatment with methylprednisolone and AMSCs significantly reduces myeloperoxidase activity, the cell apoptosis, and the proinflammatory cytokines such as tumor necrosis factor-α, interleukin (IL)-1β, IL-6, and IL-17, and interferon-γ, but increases the levels of the anti-inflammatory cytokines IL-10, transforming growth factor-β1, and the survival rate of AMSCs in the injury site. That is to say anti-inflammatory and antiapoptotic effects are important mechanisms (Yu et al. 2012; Gao et al. 2014). AMSCs have been reported to be able to promote regeneration in central nervous tissue. AMSCs are induced to differentiate into motor neuron precursor cells. In these cells, neuron-specific enolase and synaptophysin expression levels are increased and glial fibrillary acidic protein expression is decreased. Human AMSC transplantation exhibits great potential for proliferation, is induced to differentiate into neuron-like cells and then significantly improves neurological symptoms following focal cerebral ischemia. Many of MSC-type cell studies have reported that MSCs have the ability to differentiate into neural-like cells or the neuron and neurotransmitter factors in vitro or in vivo. More and more researches have confirmed that the MSCs can be remyelinated in models of demyelination that do not involve inflammation (Hu et al. 2013; Li et al. 2012). AMSCs and AFSCs have been noted as new alternative sources that would be useful for clinical applications (Table 2.3). AM is a new composite matrix bridging both stumps of spinal cord transection in rats to promote recovery of motor function. Decellularized AM has been successfully applied as nerve conduit biomaterial to improve peripheral nerve regeneration in injury models. In the differentiation of human AM toward the Schwann cell with a sequential order of neuronal induction and growth factors, AM maintained high viability of brain derived neurotrophic factor and glial cell

Table 2.3 Potential clinical application of AFSCs

Target tissue	Disease	Application
Fetus	Fetal abnormality	Biochemical tests, prenatal diagnosis
Skin	Wound, burn	Tissue graft
Heart	Cardiac malformation	Autologous heart valve tissue engineering
β-cell	Type 1 diabetes	Preclinical animal studies
Central nervous system	Ischemic stroke	Cell transplantation, intravenous cell grafts, preclinical animal studies
	Alzheimer's disease	
	Parkinson's disease	Preclinical animal studies, cell injection
	Spinal cord injury	
Limb	Hind limb ischemia	Preclinical animal studies, cell injection, angiogenesis allograft
	Foot and ankle wounds	

derived neurotrophic factor secretion and p75 are upregulated and also exhibited a change in forming a multilayered epithelium with intense accumulations of the marker proteins (Banerjee et al. 2014; Liang et al. 2009).

After transplanting AF-MSCs into the striatum of ischemic rats, the grafted cells tended to survive and migrate toward injured brain regions. Immunohistochemical analysis showed that the cells had differentiated into neurons as well as astrocytes. This suggested that the AF-MSCs could be an alternative stem cell source on the ischemic brain tissue injury (Cipriani et al. 2007). The effectiveness of AF-MSCs has also been reported in the regeneration of the sciatic nerve. Glial cell line-derived neurotrophic factor-modified human AF-MSCs promoted nerve regeneration. More importantly, this factor expressed consecutively in the induced cells for up to four weeks. The combination of granulocyte colony-stimulating factor (50 μg/kg) administration by intraperitoneal injection and AF-MSC transplantation led to better outcomes. AF-MSCs can be recruited by expression of SDF-1α in muscle and nerve by intravenous administration after nerve crush injury. The increased deposition of AF-MSCs paralleled the expression profiles of SDF-1α and its receptor CXCR-4, regeneration markers in either muscle or nerve leading to improvements in neurobehavior (Yang et al. 2012).

2.5.3 Liver Disease

AM- and AF-derived stem cells were induced to liver-like cells or liver cells in vivo and vitro. The latter expressed albumin and cytokeratin 18, alpha fetal protein, oil drop, glycogen, and hepatocyte nuclear factor-3β. In carbon tetrachloride-induced acute hepatic injury mouse model, AMSC and AF-MSC transplantation shows obvious therapeutic effect on improved liver function and pathologic histological structure in situ and in intravenous injection. Transplanted cells are planted and survived in damaged livers as in D-galactosamine-induced hepatic failure rats or in carbon tetrachloride-induced acute liver injury mice (Gong et al. 2011; Zheng et al. 2012). Serum albumin level was significantly elevated, and serum alanine aminotransferase and aspartate aminotransferase level was decreased. Transplantation of human AMSCs and AF-MSCs can improve liver function in rat with carbon tetrachloride-induced hepatic cirrhosis yet. The extent of liver cirrhosis was obviously ameliorated. It has been reported that the transplantation of AMSCs significantly decreased hepatic fibrosis by regulating TGF-β signal transduction and decreasing hepatic stellate cell activity. AMSCs can be implanted and survived for 8 weeks in liver tissue of hepatic fibrosis rat (Zhang et al. 2011). In other words, the therapeutic action of AMSCs may be included in the inhibition of epithelial–mesenchymal transition and regulation of mesenchymal–epithelial transition. Human AEC culture medium treatment suppresses decreased TGF-β1 and collagen production in activated hepatic stellate cells, as well as inducing apoptosis and reducing proliferation. Human AEC culture medium treatment and secretion of AECs may be effective in ameliorating liver fibrosis (Hodge et al. 2014; Manuelpillai et al. 2010). The major acute-phase mediators associated with fulminant hepatic failure, including IL-1β, IL-6, and TNF-α, impair the regeneration of liver cells and stem cell grafts. AF-MSCs not only have the capacity to differentiate into hepatocytes, but also genetically modified to overexpress interleukin-1 receptor antagonist can improve liver function and increase survival rates in injured liver rats (Zheng et al. 2012). This may provide a novel therapeutic approach to the treatment of fulminant hepatic failure. Thereby, AMSC and AF-MSC transplantation provides a new approach for the treatment of fibrotic liver diseases, and they are also shown to survive and to achieve hepatocyte differentiation without apparent immunological rejection.

2.5.4 Kidney Injury

Acute kidney injury is emerging as a public health problem in developing and developed countries. Several pharmacological approaches to improve renal function and survival after an acute kidney injury episode have been largely unsuccessful in clinical practice. Adult stem cell therapy has provided new hopes of innovative intervention to enhance the limited capability of kidney regeneration. An important origin for cell therapy is MSCs, which are an attractive therapeutic tool by virtue of their unique biological properties, tropism for damaged tissues, and proregenerative capacity of tubular epithelial cells which after acute ischemic or toxic insults undergo dysfunction and lesion. The mechanisms were explicitly underlying the renoprotective effects of stem cell therapy of acute kidney injury. MSCs interact with damaged cells via the release of soluble factors and exosomes, improving microvesicles. Several biological effects, including antiapoptotic, promitogenic, immunomodulatory, and anti-inflammatory activities, have been analyzed in renal tissue (Herrera et al. 2004; Bruno et al. 2009). AM- and AF-derived stem cells transplantation are promising therapeutic tools already validated on prevent fibrosis, renal tubular epithelial cell regeneration, decrease ameliorate glomerulosclerosis and preserve renal function in a preclinical porcine of autotransplantation, mice and rat model of kidney injury, even of acute renal failure (Baulier et al. 2014; Chang et al. 2011; Lv et al. 2014; Perin et al. 2010).

Human AMSCs are distributed abundantly in kidney tubule mesenchyme and few in renal tubular epithelial cell of cisplatin-induced acute kidney injury through mice caudal vein transplantation. Some of hyperchromatic nuclei and larger volume of regenerative renal tubular epithelial cells were found. Renal function was obviously improved. The transplantation of AMSCs could promote the recovery of acute kidney injury. AMSC transplantation shows obvious therapeutic effect on acute kidney injury and not only alleviates tubular damage, but also may participate in the repair of damaged tissue (Yu et al. 2012). Researchers injected autologous AF-MSCs in the renal artery 6 days after renal transplantation. The AF-MSC injection improved glomerular and tubular functions, leading to full renal function recovery and abrogated fibrosis development at 3 months. The strong proof of concept generated by AF-MSC porcine model is a first step toward evaluation of MSC-based therapies in human kidney transplantation. It is an important mechanism that AF-MSCs secrete growth factors and anti-inflammatory cytokines, anti-oxidative stress, and immunomodulatory on the efficiency and the safety (Perin et al. 2010; Gosemann et al. 2012; Feng et al. 2013; Rota et al. 2012). Although the AMSCs homing and engraftment to sites of renal damage issue have also been reported, the effect of AM- and AF-derived stem cells are engrafted injured kidney predominantly exerted antiapoptotic effect, activated Akt, and stimulated proliferation of tubular cells, possibly via local release of factors, including IL-6, vascular endothelial growth factor, and stromal cell-derived factor-1. However, their actions are not completely known in the development of interstitial fibrosis, tubular atrophy, ischemia/reperfusion renal injuries, diabetic nephropathy, and lupus nephritis.

2.5.5 Skin Wounds and Burns

Extensive burns and full-thickness skin wounds can be devastating to patients, even when treated. The autologous split-thickness skin graft involves removing a piece of skin from a secondary surgical site for the patient, stretching the skin, and reapplying the graft on the wound or burn. Although skin autograft treatment yields a reasonable clinical outcome, if the wound is extensive, then the number and size of donor sites are limited if the wound is extensive. Allograft is an additional option, but they suffer from the need of immunosuppressive drugs to prevent immune rejection of the graft. Although such polymeric scaffold, biobrane, and new dermagraft result in improved wound healing over untreated controls, they are costly to produce and result in relatively poor cosmetic outcomes (Theoret 2009; Yannas et al. 2010; Rajangam and An 2013; Papanas et al. 2012). The results showed that bioprinting AF-MSCs could be an effective treatment and conquer the problems for large-scale wounds and burns. MSC treatment with acute and chronic wounds results in accelerated wound closure, increased epithelialization, formation of granulation tissue, and angiogenesis. MSCs have recently been shown to be also effective for improving in vivo skin expansion (Jadlowiec et al. 2012; Jiang et al. 2013; Skardal et al. 2012).

AFSCs are an attractive cell source for applications in skin regeneration unlike embryonic stem cells that form teratomas. Furthermore, AFSCs remain stable and show no signs of transformation in culture. The immunomodulatory and high proliferation properties of AFSCs suggest that they can be used as an "off-the-shelf" cell therapy product for wound healing (Yoon et al. 2010). AF-MSCs release some of the paracrine factors and their ability to accelerate the wound-healing process by stimulating proliferation and migration of dermal fibroblasts. These factors include various cytokines and chemokines that are known to be important in normal wound healing, as IL-8, IL-6, TGF-β, tumor necrosis factor receptor I, vascular endothelial growth factor, and EGF (Jun et al. 2014). The proteomic analysis showed that AFSCs secreted a number of growth factors at concentrations higher

in vivo. In parallel, AF-MSC-conditioned media induced endothelial cell migration in vitro. The increased wound closure rates and angiogenesis in wound site may be due to delivery of secreted trophic factors, rather than direct cell–cell interactions. AF-MSCs are resuspended in fibrin–collagen gel and "printed" over the wound site by bioprinting technology for the treatment of full-thickness skin wounds in nu/nu mice. The wound closure and re-epithelialization were significantly greater in wounds treated by fibrin–collagen gel only. Histological examination showed more increased microvessel density and capillary diameters in the AF-MSC treated wounds (Skardal et al. 2012).

AF-MSCs in hypoxic conditioned medium could enhance the proliferation and migration of human dermal fibroblasts in vitro and wound closure in a skin injury model, as compared to AF-MSCs in normoxic conditioned medium. However, the enhancement of fibroblast migration was inhibited by SB505124 and LY294002, inhibitors of TGF-β/Smad2 and PI3 K/AKT. Therefore, this enhanced wound healing is related to the increase in hypoxia-induced paracrine factors via activation of TGF-β/Smad2 and PI3 K/AKT pathways. Expression of TGF-β1 was more in albino rats with irradiated wounds than those injected intradermally with human AECs (Jun et al. 2014). The model groups showed severe inflammation, deficient healed dermis, and delayed re-epithelialization. SDF-1 expression was high, while CXCL-5 expression was high in AEC-transplanted rat causing accelerated wound healing. AECs showed a great effect on the quality of the dermis as well as bone marrow MSCs, while superiority of the epithelium and its appendages was achieved. Human AECs could be used safely in case of irradiated wounds (Mehanni et al. 2013).

AM is a biological dressing in the management of burns by rapid re-epithelialization and healing as it diminishes the oozing of plasma, bacterial count and fluid, protein, and heat loss. Dermal injection of freeze-dried AM extract also is a potential wound-healing substrate which can promote epidermal and dermal regeneration, while avoiding undesirable hyperproliferation of damaged tissue (Kang et al. 2013; Mohammadi et al. 2013).

2.5.6 Autoimmune Disease

MSCs have been shown to possess immunomodulatory properties, which suppress T cell proliferation, influence dendritic cell maturation and function, suppress B cell proliferation and terminal differentiation, and suppress immune modulation of other immune cells such as NK cells and macrophages. In terms of the clinical applications of MSCs, they are involved in four main areas: tissue regeneration for cartilage, bone, muscle, tendon, and neuronal cells; gene therapy vehicles; enhancement of hematopoietic stem cell engraftment; and treatment of immune diseases such as systemic lupus erythematosus, graft-versus-host disease, rheumatoid arthritis, autoimmune encephalomyelitis, acute pancreatitis, multiple sclerosis, and sepsis (Yi and Song 2012).

Rheumatoid arthritis is a chronic general autoimmune disease that is mediated by immunocompetent cell and multiple cytokines. Its main pathological feature is progressive joint damage. Human AMSCs may inhibit the development of collagen-induced arthritis by regulating the Foxp3+ Treg and Th17 cells. AMSC treatment caused lower arthritis index score, decreased volume of target joints, alleviated pathological damage of the joints, and decreased percentage of Th1, Th2, Th17, CTL1, CTL2, and NKT cells, but produced higher percentage of Foxp3+ Treg cells, compared with untreated arthritis rats and AMSC culture medium treatment (Xiao et al. 2013). The supernatant of cultured human AMSCs and AECs could inhibit lymphocyte proliferation. After mixed culturing with AMSCs, $CD4^+$ T cell subsets were enhanced, but $CD8^+$ T cell subsets were obviously suppressed. Th1 and Tc1 significantly decreased, but Th2 and Tc2 slightly increased in all experiments. AEC did not have significant difference compared with control. It was found that both human amniotic cells could secrete soluble cytokines to play immune suppression.

Human AMSCs (5×10^5 cells) were injected into the lateral cerebroventricle of rat with autoimmune encephalomyelitis, which was induced in guinea pigs by spinal cord homogenate, complete Freund's adjuvant, and Bordetella pertussis toxin. The behavior scores of human AMSC-treated rats were reduced gradually. After 3 weeks, AMSC-treated animals showed an improvement in inflammatory reaction in the brain and spinal cord, and the percentages of Treg, Th2, Tc1, and Tc2 lymphocyte subsets were increased obviously, whereas Th17 cells were decreased significantly. The concentrations of IFN-γ and IL-2 in the plasma were decreased, but concentrations of IL-4 and IL-10 were increased. AMSCs can improve neural function of rats with autoimmune encephalomyelitis and alleviate immunopathologic damage of neural tissues. Results suggested that the mechanism might relate to Foxp3+ Treg cell upregulation and Th17 cell downregulation (Fang et al. 2014). AEC is same as therapeutic effect on autoimmune encephalomyelitis and upregulation of Foxp3+ Treg cells and downregulation of Th17 cells (Li et al. 2013).

These immunomodulatory properties in vitro have generated enormous interest in the potential application of MSC in vivo as an immunosuppressive cellular therapy. Successful results have been obtained with the use of bone marrow MSC both for the prevention of graft-versus-host disease in solid organ transplantation and for the treatment of steroid-resistant acute graft-versus-host disease, arising after allogeneic hematopoietic cell transplantation. It is worth of further exploration that how many AMSCs and AF-MSCs are needed and how they affect the acute or chronic graft-versus-host disease of allogeneic hematopoietic cell transplantation.

2.5.7 Premature Ovarian Failure

Premature ovarian failure, a condition that causes amenorrhea and hypergonadotropic hypoestrogenism before the age of 40, affects 1 % of women in the general population. The occurrence of premature ovarian failure has increased in recent years. Premature ovarian failure usually can't be reversed and though currently available treatments yet. New treatment strategies are urgently required with regenerative medicine development. Stem cell transplantation has been reported to rescue ovarian function in a preclinical mouse model of chemotherapy-induced premature ovarian failure. AMSCs and AF-MSCs were injected into a cyclophosphamide or cis-platinum-induced premature ovarian failure mouse. These cells could be detected by fluorescence microscopy up to three and eight weeks after injection. Ovarian function was improved, and full recovery is by the regulation of local cytokines and perfect microenvironment of follicular development. The level of E2 was upregulated to reach the normal level, and the level of FSH was decreased as the same as normal mice. Follicular development and potentia generandi were not different compared with normal female mice. The red fluorescence protein-transduced $CD44^{+}/CD105^{+}$ human AF-MSCs could survive and proliferate in the ovary of long-term cyclophosphamide-induced premature ovarian failure mouse. The cells could be detected by fluorescence microscopy up to three weeks after transplanted into the ovaries. The ability of human AFSCs to differentiate into germ cell and oocyte-like cells has been previously documented. The function properties and long-term survival in vivo of AMSCs and AF-MSCs make them ideal seed cells for stem cell transplantation for premature ovarian failure treatment (Liu et al. 2012; Wang et al. 2013b). The grafted GFP-labeled AFSCs and immunostaining with antihuman antigen-specific antibodies demonstrated that they survived and differentiated into granulosa cells which directed oocyte maturation. Furthermore, labeling of ovarian tissue for anti-Müllerian hormone expression, a functional marker of folliculogenesis, was strong in injured ovaries but absent in negative controls. This result highlights the possibility of using AFSCs in regenerative medicine reproductive health (Lai et al. 2013). Intravenously injected AECs reached the ovaries of chemotherapy-treated mice and restored folliculogenesis for promoting reproductive health and improving the quality of life for female cancer survivors.

2.5.8 Hematopoietic Supporting and Improvement

It was reported that human AMSCs expressed multiple hematopoietic cytokines, including leukemia inhibitory factor, stem cell factor, macrophage colony-stimulating factor, granulocyte colony-stimulating factor, macrophage–granulocyte colony-stimulating factor, IL-3, IL-6, and IL-11. The number of suspension mononuclear cells of umbilical cord blood was enhanced in culture medium at the days 3–21. AMSC transplantation showed that it has a function to support hematopoiesis in vitro, as well as AFSCs (Ditadi et al. 2009). The stem cells derived from amnion and AF are cocultured with mononuclear cells from cryopreserved cord blood in a medium supplemented with cytokine stem cell factor, thrombopoietin, and granulocyte colony-stimulating factor. The hematopoietic stem and progenitor cells are distinctly amplified. The comparison of MSCs from bone marrow with MSCs from cord blood and AF showed no significant difference (Klein et al. 2013). High-dose chemotherapy often results in severe bone marrow damage. Beside drugs, stem cell transplantation has also been used as a strategy for the treatment of bone marrow damage. Human AMSC transplantation through the caudal vein, the body weight was higher than model mice after cisplatin-induced myelosuppression. The number of white blood cells, lymphocytes, mononuclear cells, red blood cells, and hemoglobin, hematocrit, and platelet of peripheral blood, and the number of bone marrow karyocytes were significantly higher than those in model mice. Bone marrow histopathological testing results showed that AMSC transplantation significantly improved femur bone marrow organizational structure compared with myelosuppression mice, especially megakaryocyte increases. The AMSCs with antihuman nuclei monoclonal antibody-fluorescein isothiocyanate were colonized in myeloid tissue (Mizokami et al. 2009; Yao et al. 2013a). It is clear that human AMSC transplantation can remarkably ameliorate hematopoietic function in myelosuppression condition.

2.5.9 Eye Injury

Limbal stem cell deficiency is a pathologic state that limbus of cornea or base material is damaged. It often leads to the corneal surface by invading conjunctival epithelium cells' ingrowth of fibrous tissue, stromal scarring, and neovascularization, which cause chronic pain and visual loss. AECs not only differentiate into corneal epithelial-like cells in vitro, but also provide a kind of niche that enhances the functional properties of human corneal endothelial cells via inhibiting P53-survivin-mitochondria axis (Yao et al. 2013b; Sha et al. 2013). Although there are differences on collagen type IV, V and laminin $\alpha 4$, BM40/SPARC, tenascin-C, amniotic and limbal epithelial basement membranes all show positive immunoreactivity for collagen type IV $\alpha 1$, $\alpha 2$, $\alpha 5$, and $\alpha 6$ chains; multiple type collagens, laminin, nidogen and fibulin; fibronectin, etc. AEC transplantation decreases corneal conjunctivization better than simple AM transplantation in limbal stem cell deficiency model through burning cornea by sodium hydroxide after cutting layers of limbus cornea (Dietrich-Ntoukas et al. 2012; Luo et al. 2013; Covre et al. 2011; Ricardo et al. 2009). The results indicate that transplantation of AECs represents an effective technique for ocular surface reconstruction in patients with severe limbal stem cell deficiency. Human AECs can be successfully reproduced in corneal stromal and formed 4–5 layer epithelial cells as the same as normal corneal epithelium. AECs also could be expected to reconstruct the corneal epithelium and by tissue engineering technology.

2.5.10 Other Tissue Injury

Human AFSCs were injected into the injured tibialis anterior muscles established by cardiotoxin and X-ray irradiation in Nod/Scid mice. The double-staining immunofluorescence showed that human-specific nuclear mitotic apparatus protein expressed in tibialis anterior muscles and no myogenic phenotype at 2 weeks after cell transplantation. The single-cell coexpressed hepatocyte growth factor receptor or myogenic regulatory factor at 4 weeks after cell transplantation. In some myofibers, human-specific nuclear mitotic apparatus proteins and laminin or desmin were also coexpressed (Ma et al. 2012). AF stem cells can participate in the regeneration of injured mouse muscle. After cell transplantation, transplanted cell also coexpressed NuMa and c-Met, Myf-5, and laminin or desmin at 4 weeks after cell transplantation (Ma et al. 2011). AF-MSCs improve survival and enhance repair of damaged intestine in necrotizing enterocolitis via a cyclooxygenase 2-dependent mechanism (Zani et al. 2014). Research demonstrated that AF-MSCs injected into an established model of necrotizing enterocolitis could improve survival, clinical status, gut structure, and function. Understanding the mechanism of this effect may help us to develop new cellular or pharmacological therapies for infants with necrotizing enterocolitis. Furthermore, human AM- and AF-derived stem cells are employed as a tool for basic research and studied in prevision of their use for cell-based therapies, although some irregularities in their epigenetic control are not dimness.

2.5.11 Potential Antitumor Therapeutic Strategies

Cancer therapeutic strategies principally include surgery, radiotherapy, chemotherapy and biotherapy. Despite the developments of therapy, cancer mortality rates are higher worldwide. Previous study reported that bone marrow MSC transplantation resulted in antitumor activity against non-Hodgkin's lymphoma (Secchiero et al. 2010). However, some bone marrow stem cells or precursor cells have been also shown to increase growth and metastasis of colon cancer, lymphoma, and melanoma cells in vivo (Audollent et al. 2011). Therefore, it is unclear whether MSCs promote or suppress tumor growth so far. Certain studies show that bone marrow MSCs have been promoted as an attractive option to use as cellular delivery vehicles to carry antitumor agents, owing to their ability to home into tumor sites and secrete cytokines (Hamada et al. 2005). However, MSC lineage to be used in the cell therapy needs to be carefully chosen to balance efficacy and safety for a particular tumor type. MSCs from AM and AF may be one of the tumor cell growth inhibitors or a new delivery vehicle for antitumor effects (Rolfo et al. 2014; Kang et al. 2012a, b). They inhibit proliferation of cancer cell lines of haematopoietic and non-haematopoietic origin by inducing cell cycle arrest, or induce C6 glioma apoptosis in vivo through the Bcl-2/caspase pathways (Magatti et al. 2012; Jiao et al. 2012). The two source stem cells also are capable of self-renewal and can generate differentiated progenies for organ development as well. They are considered as potential source for regenerative medicine and tissue replacement after disease. They are in an intermediate stage between pluripotent embryonic stem cells and lineage-restricted adult stem cells. Their non-tumorigenicity and no or low expression level of major histocompatibility complex antigens, and contribute to low immunogenicity and anti-inflammation. In non-engineered stem cell transplantation strategies, amnion-derived stem cells effectively target the tumor and suppressed the tumor growth by expressing cytotoxic cytokines or cancer suppressor gene. Additionally, they also have a potential as novel delivery vehicles transferring therapeutic genes to the cancer formation sites in gene-directed enzyme/prodrug combination therapy.

2.5.12 The Challenge of Clinical Translation

AM- and AF-derived stem cell-based therapeutic and seed cell-engineered strategies are showing huge potential in experimental studies. However, some problems of stem cells including safety and ethical issues have limited their clinical use. Thus, the AM and AF are considered as non-controversial sources because of the use of either heterologous embryonic stem cells or the less ethically disputed MSCs. A small amount of AF obtained by amniocentesis and amniotic membrane samples could produce enough MSCs for applications. However, some of the gene expressions of AFSCs were changed in different culture conditions. AMSCs even change into senescence as the same as other MSCs for times without number subculture. It is reported that AECs cause EMT after subculture after passage 5. How to preferably keep the biological characters and functions of AMSCs, AECs, and AFSCs in amplified culture system for more amplify and subculture in vitro? How to block or delay stem cells aging? How to effectively control EMT and mesenchymal–epithelial transition among AECs, AMSCs, or AF-MSCs in cultured conditions and induced directional differentiation? Whether are different therapeutic effects in the same tissue injury that is caused by different reasons? These are worth of further investigation. The epigenetics and regulation and control of AM- and AF-derived stem cells have not become directly involved for the biological mechanism, regeneration, restoration, and development. In tissue engineering, there are a number of problems that needed to be solved, such as interactions and histocompatibility. It is absolutely necessary that further studies on molecular mechanism of AM- and AF-derived stem cells demonstrate the prospects of potential therapeutic uses and safety for several diseases. Although we still need to do more investigation, even to break through the bottleneck, it is hoped translation that these beneficial effects of AMSCs, AECs, and AF-MSCs will gradually be developed into therapeutic outcomes for injury regeneration, autoimmune disease, and tissue engineering. Finally, we elaborate on the potential for these cells to promote regeneration of various tissue defects (Insausti et al. 2010; Murphy et al. 2010), including fetal tissue, the nervous system, heart, lungs, kidneys, bones, cartilage, and ovary.

References

Alcaraz A, Mrowiec A, Insausti CL, García-Vizcaíno EM, Ruiz-Canada C, López-Martínez MC, Moraleda JM, Nicolás FJ. Autocrine TGF-β induces epithelial to mesenchymal transition in human amniotic epithelialcells. Cell Transplant. 2013;22(8):1351–67.

Alviano F, Fossati V, Marchionni C, Arpinati M, Bonsi L, Franchina M, Lanzoni G, Cantoni S, Cavallini C, Bianchi F, Tazzari PL, Pasquinelli G, Foroni L, Ventura C, Grossi A, Bagnara GP. Term amniotic membrane is a high throughput source for multipotent mesenchymal stem cells with the ability to differentiate into endothelial cells in vitro. BMC Dev Biol. 2007;7:11.

Audollent R, Eveno C, Contreres JO, Hainaud P, Rampanou A, Dupuy E, Brouland JP, Pocard M. Bone marrow-derived endothelial and hematopoietic precursors cells enhance the metastasis of colon cancer in an orthotopic murine model. Int J Cancer. 2011;129(9):2304–5.

Bai J, Wang Y, Liu L, Chen J, Wang Y. Bio-characteristics of c-kit+ human amniotic fluid-derived mesenchymal stem cells and their differentiation into cardiomyocytes in vitro. Zhongguo Xiu Fu Chong Jian Wai Ke Za Zhi. 2012;26(2):152–7.

Banas RA, Trumpower C, Bentlejewski C, Marshall V, Sing G, Zeevi A. Immunogenicity and immunomodulatory effects of amnion-derived multipotent progenitor cells. Hum Immunol. 2008;69:321–8.

Banerjee A, Nürnberger S, Hennerbichler S, Riedl S, Schuh CM, Hacobian A, Teuschl A, Eibl J, Redl H, Wolbank S. In toto differentiation of human amniotic membrane towards the Schwann cell lineage. Cell Tissue Bank. 2014;15(2):227–39.

Baulier E, Favreau F, Le Corf A, Jayle C, Schneider F, Goujon JM, Feraud O, Bennaceur-Griscelli A, Hauet T, Turhan AG.). Amniotic fluid-derived mesenchymal stem cells prevent fibrosis and preserve renal function in a preclinicalporcine model of kidney transplantation. Stem Cells Transl Med. 2014;3(7):809–820.

Bilic G, Zeisberger SM, Mallik AS, Zimmermann R, Zisch AH. Comparative characterization of cultured human term amnion epithelial and mesenchymal stromal cells for application in cell therapy. Cell Transplant. 2008;17(8):955–68.

Bocian E. Future of prenatal cytogenetic studies: rapid aneuploidy testing or full karyotype. Ginekol Pol. 2007;78(11):881–7.

Bruno S, Grange C, Deregibus MC, Calogero RA, Saviozzi S, Collino F, Morando L, Busca A, Falda M, Bussolati B, Tetta C, Camussi G. Mesenchymal stem cell-derived microvesicles protect against acute tubular injury. J Am Soc Nephrol. 2009;20(5):1053–67.

Carvajal JA, Delpiano AM, Cuello MA, Poblete JA. Mechanical stretch increases brain natriuretic peptide production and secretion in the human fetal membranes. Reprod Sci. 2013;20(5):597–604.

Chang YJ, Hwang SM, Tseng CP, Cheng FC, Huang SH, Hsu LF, Hsu LW, Tsai MS. Isolation of mesenchymal stem cells with neurogenic potential from the mesoderm of the amniotic membrane. Cells Tissues Organs. 2010;192(2):93–105.

Chang JW, Hung SP, Wu HH, Wu WM, Yang AH, Tsai HL, Yang LY, Lee OK. Therapeutic effects of umbilical cord blood-derived mesenchymal stem cell transplantation in experimental lupus nephritis. Cell Transplant. 2011;20(2):245–57.

Chen XD, Wang JG. Astragalus induces human amniotic epithelial cells (WISH) to differentiate toward neurons, inhibits the expression of Notch1 and promotes cell survival. Sheng Li Xue Bao. 2012;64(6):617–24.

Chen WQ, Siegel N, Li L, Pollak A, Hengstschläger M, Lubec G. Variations of protein levels in human amniotic fluid stem cells CD117/2 over passages 5–25. J Proteome Res. 2009;8(11):5285–95.

Chen J, Lu Z, Cheng D, Peng S, Wang H. Isolation and characterization of porcine amniotic fluid-derived multipotent stem cells. PLoS ONE. 2011;6(5):e19964.

Choi SA, Choi HS, Kim KJ, Lee DS, Lee JH, Park JY, Kim EY, Li X, Oh HY, Lee DS, Kim MK. Isolation of canine mesenchymal stem cells from amniotic fluid and differentiation into hepatocyte-like cells. Vitro Cell Dev Biol Anim. 2013;49(1):42–51.

Cipriani S1, Bonini D, Marchina E, Balgkouranidou I, Caimi L, Grassi Zucconi G, Barlati S. Mesenchymal cells from human amniotic fluid survive and migrate after transplantation into adult rat brain. Cell Biol Int. 2007;31(8):845–850.

Covre JL, Loureiro RR, Cristovam PC, Ricardo JR, Haapalainen EF, Gomes JÁ. Comparison between amniotic membrane with and without epithelium as a support for human limbal epithelial cells cultured ex vivo. Arq Bras Oftalmol. 2011;74(2):114–7.

D'Alimonte I, Lannutti A, Pipino C, Di Tomo P, Pierdomenico L, Cianci E, Antonucci I, Marchisio M, Romano M, Stuppia L, Caciagli F, Pandolfi A, Ciccarelli R. Wnt signaling behaves as a "master regulator" in the osteogenic and adipogenic commitment of human amniotic fluid mesenchymal stem cells. Stem Cell Rev. 2013;9(5):642–54.

Davydova DA, Voroteliak EA, Bragina EE, Terskikh VV, Vasil'ev AV. Culture of human amniotic fluid stem cells in 3D collagen matrix. Tsitologiia. 2011;53(4):325–31.

Di Tomo P, Pipino C, Lanuti P, Morabito C, Pierdomenico L, Sirolli V, Bonomini M, Miscia S, Mariggiò MA, Marchisio M, Barboni B, Pandolfi A. Calcium sensing receptor expression in ovine amniotic fluid mesenchymal stem cells and the potential role of R-568 during osteogenic differentiation. PLoS One. 2013;8(9): e73816. doi:10.1371/journal. pone. 0073816. eCollection.

Di Trapani M1, Bassi G, Ricciardi M, Fontana E, Bifari F, Pacelli L, Giacomello L, Pozzobon M, Féron F, De Coppi P, Anversa P, Fumagalli G, Decimo I, Menard C, Tarte K, Krampera M. Comparative study of immune regulatory properties of stem cells derived from different tissues. Stem Cells Dev. 2013;22(22):2990–3002.

Dietrich-Ntoukas T, Hofmann-Rummelt C, Kruse FE, Schlötzer-Schrehardt U. Comparative analysis of the basement membrane composition of the human limbus epithelium and amniotic membrane epithelium. Cornea. 2012;31(5):564–9.

Ditadi A, de Coppi P, Picone O, Gautreau L, Smati R, Six E, Bonhomme D, Ezine S, Frydman R, Cavazzana-Calvo M, André-Schmutz I. Human and murine amniotic fluid c-Kit+ Lin- cells display hematopoietic activity. Blood. 2009;113(17):3953–60.

Fairbairn NG, Randolph MA, Redmond RW. The clinical applications of human amnion in plastic surgery. J Plast Reconstr Aesthet Surg. 2014;67(5):662–75.

Fan W, Yang Z, Luo J, Huang Y, Li X, Wang Z. A comparative study on acellular small intestinal submucosa and acellular amnion as dressings for traumatic skin defects. Zhongguo Xiu Fu Chong Jian Wai Ke Za Zhi. 2006;20:155–60.

Fang CH, Jin J, Joe JH, Song YS, So BI, Lim SM, Cheon GJ, Woo SK, Ra JC, Lee YY, Kim KS. In vivo differentiation of human amniotic epithelial cells into cardiomyocyte -like cells and cell transplantation effect on myocardial infarction in rats: comparison with cord blood and adipose tissue-derived mesenchymal stem cells. Cell Transplant. 2012;21(8):1687–96.

Fang N, Li D, Yu LM, Zhang T, Zhao CH, Chen DX. Therapeutic effects and immunological regulation of human amniotic mesenchymal stem cells on rats with experimental autoimmune encephalomyelitis. Mian Yi Xue Za Zhi. 2014a;30(1):6–12.

Fang N, Li D, Yu LM, Tao Zhang, Zhao CH, Chen DX. Therapeutic effect and immunological regulation of human aminiotic mesenchymal stem cells on rat with experimental autoimmune encephalomyelitis. Mian Yi Xue Za Zhi. 2014b;1:6–12.

Feng GW, Yu LM. Research progress of human amniotic membrane applications. Zhong Guo Sheng Wu Gong Cheng Za Zhi. 2014;31(4):930–4.

Feng J, Zhao L, Deng H, Wei M, Li J, Xu K. Immune tolerance of amniotic fluid stem cell-induced rat kidney graft and influences on oxidative stress. Transplant Proc. 2013;45(9):3394–401.

Filioli Uranio M, Dell'Aquila ME, Caira M, Guaricci AC, Ventura M, Catacchio CR, Martino NA, Valentini L. Characterization and in vitro differentiation potency of early-passage canine amnion- and umbilical cord-derivedmesenchymal stem cells as related to gestational age. Mol Reprod Dev. 2014;81(6):539–51.

Galende E, Karakikes I, Edelmann L, Desnick RJ, Kerenyi T, Khoueiry G, Lafferty J, McGinn JT, Brodman M, Fuster V, Hajjar RJ, Polgar K. Amniotic fluid cells are more efficiently reprogrammed to pluripotency than adult cells. Cell Reprogram. 2010;12(2):117–25.

Gao S, Ding J, Xiao HJ, Li ZQ, Chen Y, Zhou XS, Wang JE, Wu J, Shi WZ. Anti-inflammatory and anti-apoptotic effect of combined treatment with methylprednisolone and amniotic membrane mesenchymal stem cells after spinal cord injury in rats. Neurochem Res. 2014;39(8):1544–52.

Ghaderi S, Soheili ZS, Ahmadieh H, Davari M, Jahromi FS, Samie S, Rezaie-Kanavi M, Pakravesh J, Deezagi A. Human amniotic fluid promotes retinal pigmented epithelial cells' trans-differentiation into rod photoreceptors and retinal ganglion cells. Stem Cells Dev. 2011;20(9):1615–25.

Gong LM, Fang N, Chen DX, Wan EH, Zhang T, Liu ZL, Yu LM, Zhao CH. Zhong Guo Zu Zhi Gong Cheng Yan Jiu Yu Lin Chuang Kang Fu. 2011;15(36):6709–12.

Gosemann JH, Kuebler JF, Pozzobon M, Neunaber C, Hensel JH, Ghionzoli M, de Coppi P, Ure BM, Holze G. Activation of regulatory T cells during inflammatory response is not an exclusive property of stem cells. PLoS ONE. 2012;7(4):e35512.

Guan X, Delo DM, Atala A, Soker S. In vitro cardiomyogenic potential of human amniotic fluid stem cells. J Tissue Eng Regen Med. 2011;5(3):220–8.

Hamada H, Kobune M, Nakamura K, Kawano Y, Kato K, Honmou O, Houkin K, Matsunaga T, Niitsu Y. Mesenchymal stem cells (MSC) as therapeutic cytoreagents for gene therapy. Cancer Sci. 2005;96(3):149–56.

Han K, Lee JE, Kwon SJ, Park SY, Shim SH, Kim H, Moon JH, Suh CS, Lim HJ. Human amnion-derived mesenchymal stem cells are a potential source for uterine stem cell therapy. Cell Prolif. 2008;41:709–25.

Han X, Wan Q, Wu W, Zheng A, Li L, Liu X. Activin A and BMP-4 induce cardiomyocyte-like cells differentiation of human amniotic epithelial cells. Sheng Wu Yi Xue Gong Cheng Xue Za Zhi. 2011;28(6):1217–22.

Herrera MB, Bussolati B, Bruno S, Fonsato V, Romanazzi GM, Camussi G. Mesenchymal stem cells contribute to the renal repair of acute tubular epithelial injury. Int J Mol Med. 2004;14(6):1035–41.

Hodge A, Lourensz D, Vaghjiani V, Nguyen H, Tchongue J, Wang B, Murthi P, Sievert W, Manuelpillai U. Soluble factors derived from human amniotic epithelial cells suppress collagen production in human hepatic stellate cells. Cytotherapy. 2014;16(8):1132–44.

Hu W, Guan FX, Li Y, Tang YJ, Yang F, Yang B. New methods for inducing the differentiation of amniotic-derived mesenchymal stem cells into motor neuronprecursor cells. Tissue Cell. 2013;45(5):295–305.

Ilancheran S, Michalska A, Peh G, Wallace EM, Pera M, Manuelpillai U. Stem cells derived from human fetal membranes display multilineage differentiation potential. Biol Reprod. 2007;77(3):577–88.

Ilancheran S, Moodley Y, Manuelpillai U. Human fetal membranes: a source of stem cells for tissue regeneration and repair? Placenta. 2009;30:2–10.

In't Anker PS, Scherjon SA, Kleijburg-van der Keur C, Noort WA, Claas FH, Willemze R, Fibbe WE, Kanhai HH. Amniotic fluid as a novel source of mesenchymal stem cells for therapeutic transplantation. Blood. 2003;102:1548–1549.

Insausti CL, Blanquer M, Bleda P, Iniesta P, Majado MJ, Castellanos G, Moraleda JM. The amniotic membrane as a source of stem cells. Histol Histopathol. 2010;25:91–8.

Insausti CL, Blanquer M, García-Hernández AM, Castellanos G, Moraleda JM. Amniotic membrane-derived stem cells: immunomodulatory properties and potential clinical application. Stem Cells Cloning. 2014;7:53–63.

Jadlowiec C, Brenes RA, Li X, Lv W, Protack CD, Collins MJ, Dardik A. Stem cell therapy for critical limb ischemia: what can we learn from cell therapy for chronic wounds? Vascular. 2012;20(5):284–9.

Janz Fde L, Debes Ade A, Cavaglieri Rde C, Duarte SA, Romão CM, Morón AF, Zugaib M, Bydlowski SP. Evaluation of distinct freezing methods and cryoprotectants for human amniotic fluid stem cells cryopreservation. J Biomed Biotechnol. 2012;2012:649353.

Jiang D, Qi Y, Walker NG, Sindrilaru A, Hainzl A, Wlaschek M, MacNeil S, Scharffetter-Kochanek K. The effect of adipose tissue derived MSCs delivered by a chemically defined carrier on full-thickness cutaneous wound healing. Biomaterials. 2013;34(10):2501–15.

Jiao H, Guan F, Yang B, Li J, Song L, Hu X, Du Y. Human amniotic membrane derived-mesenchymal stem cells induce C6 glioma apoptosis in vivo through the Bcl-2/caspase pathways. Mol Biol Rep. 2012;39(1):467–73.

Jun EK, Zhang Q, Yoon BS, Moon JH, Lee G, Park G, Kang PJ, Lee JH, Kim A, You S. Hypoxic conditioned medium from human amniotic fluid-derived mesenchymal stem cells accelerates skin wound healing through TGF-β/SMAD2 and PI3 K/Akt pathways. Int J Mol Sci. 2014;15(1):605–628.

Jun EK, Zhang Q, Yoon BS, Moon JH, Lee G, Park G, Kang PJ, Lee JH, Kim A, You S. Hypoxic conditioned medium from human amniotic fluid-derived mesenchymal stem cells accelerates skin wound healing through TGF-β/SMAD2 and PI3 K/Akt pathways. Int J Mol Sci. 2014;15(1):605–28.

Kallapur SG, Presicce P, Senthamaraikannan P, Alvarez M, Tarantal AF, Miller LM, Jobe AH, Chougnet CA. Intra-amniotic IL-1β induces fetal inflammation in rhesus monkeys and alters the regulatory T cell/IL-17 balance. J Immunol. 2013;191(3):1102–9.

Kang NH, Hwang KA, Kim SU, Kim YB, Hyun SH, Jeung EB, Choi KC. Potential antitumor therapeutic strategies of human amniotic membrane and amniotic fluid-derived stem cells. Cancer Gene Ther. 2012a;19(8):517–22.

Kang NH, Yi BR, Lim SY, Hwang KA, Baek YS, Kang KS, Choi KC. Human amniotic membrane-derived epithelial stem cells display anticancer activity in BALB/c female nude mice bearing disseminated breast cancer xenografts. Int J Oncol. 2012b;40(6):2022–8.

Kang M, Choi S, Cho Lee AR. Effect of freeze dried bovine amniotic membrane extract on full thickness wound healing. Arch Pharm Res. 2013;36(4):472–8.

Kim J, Kang HM, Kim H, Kim MR, Kwon HC, Gye MC, Kang SG, Yang HS, You J. Ex vivo characteristics of human amniotic membrane-derived stem cells. Cloning Stem Cells. 2007;9(4):581–94.

Kim J, Jeong SY, Ju YM, Yoo JJ, Smith TL, Khang G, Lee SJ, Atala A. In vitro osteogenic differentiation of human amniotic fluid-derived stem cells on a poly(lactide-co-glycolide) (PLGA)-bladder submucosa matrix (BSM) composite scaffold for bone tissue engineering. Biomed Mater. 2013;8(1):014107.

Klein C, Strobel J, Zingsem J, Richter RH, Goecke TW, Beckmann MW, Eckstein R, Weisbach V. Ex vivo expansion of hematopoietic stem- and progenitor cells from cord blood in coculture with mesenchymal stroma cells from amnion, chorion, Wharton's jelly, amniotic fluid, cord blood, and bone marrow. Tissue Eng Part A. 2013;19(23–24):2577–85.

Koob Thomas J, Lim Jeremy J, Massee Michelle, Zabek Nicole, Rennert Robert, Gurtner Geoffrey, Li William W. Angiogenic properties of dehydrated human amnion/chorion allografts: therapeutic potential for soft tissue repair and regeneration. Vasc Cell. 2014;6:1–10.

Lai D, Wang F, Chen Y, Wang L, Wang Y, Cheng W. Human amniotic fluid stem cells hanve a potential to recover ovarian faunction in mice with chemotherapy-induced sterility. BMC Dev Biol. 2013;13:34. doi:10.1186/1471-213X-13-34.

Li XD, Geng HQ, Zhu Z, Jn Pan, Xie H, Chen F. Amniotic fluid: a novel source for mesenchymal stem cells. Zhong Hua Xiao Er Wai Ke Xue. 2006;27(9):452–99.

Li C, Zhou J, Shi G, Ma Y, Yang Y, Gu J, Yu H, Jin S, Wei Z, Chen F, Jin Y. Pluripotency can be rapidly and efficiently induced in human amniotic fluid-derived cells. Hum Mol Genet. 2009;18(22):4340–9.

Li F, Miao ZN, Xu YY, Zheng SY, Qin MD, Gu YZ, Zhang XG. Transplantation of human amniotic mesenchymal stem cells in the treatment of focal cerebral ischemia. Mol Med Rep. 2012;6(3):625–30.

Li D, Fang N, Chen DX, Yu LM, Zhang T, Zhao CH. Therapeutic effect and immunological regulation of human aminiotic epithelial cells on rat with experimental autoimmune encephalomyelitis. Zhong Guo Mian Yi Xue Za Zhi. 2013;29(10):1011–6.

Li F, Xu Y, Xu X, Xu B, Zhao J, Zhang X. Fms-related tyrosine kinase 3 ligand promotes proliferation of placenta amnion and chorion mesenchymal stem cells in vitro. Mol Med Rep. 2014a;10(1):322–8.

Li Y, Guo L, Ahn HS, Kim MH, Kim SW. Amniotic mesenchymal stem cells display neurovascular tropism and aid in the recovery of injured peripheral nerves. J Cell Mol Med. 2014b;18(6):1028–34.

Liang H, Liang P, Gao A, Qi Q, Liu E, Wu J, Xu X. A new composite matrix bridging both stumps of spinal cord transection in rats to promote recovery of motor function. Zhongguo Xiu Fu Chong Jian Wai Ke Za Zhi. 2009;23(11):1376–81.

Liu T, Wu J, Huang Q, Hou Y, Jiang Z, Zang S, Guo L. Human amniotic epithelial cells ameliorate behavioral dysfunction and reduce infarct size in the rat middle cerebral artery occlusion model. Shock. 2008;29(5):603–11.

Liu H, Liu DQ, Guan LD, Yan ZF, Wang J, He LJ, Lü Y, Nan X, Li YL, Pei XT. Effects of different culture conditions on isolation and expansion of stem cells from second-trimester amnioticfluids. Zhonghua Fu Chan Ke Za Zhi. 2009;44(4):241–5.

Liu H, Liu DQ, Li BW, Guan LD, Yan ZF, Li YL, Pei XT, Yue W, Wang M, Lu YP, Peng HM, Lv Y. Human amniotic fluid-derived stem cells can differentiate into hepatocyte-like cells in vitro and in vivo. Vitro Cell Dev Biol Anim. 2011a;47(9):601–8.

Liu T, Guo L, Liu Z, Huang Y, Cheng W. Induction of dopaminergic neuronal-like cells from $CD44^+$ human amniotic fluids that are ameliorative to behavioral recovery in a Parkinson's disease rat model. Int J Mol Med. 2011b;28(5):745–52.

Liu T, Huang Y, Guo L, Cheng W, Zou G. $CD44^+$/$CD105^+$ human amniotic fluid mesenchymal stem cells survive and proliferate in the ovary long-term in a mouse model of chemotherapy-induced premature ovarian failure. Int J Med Sci. 2012;9(7):592–602.

Lovati AB, Corradetti B, Lange Consiglio A, Recordati C, Bonacina E, Bizzaro D, Cremonesi F. Comparison of equine bone marrow, umbilical cord matrix and amniotic fluid-derived progenitor cells. Vet Res Commun. 2011;35(2):103–21.

Luo JC, Li XQ, Yang ZM. Preparation of human acellular amniotic membrane and its cytocompatibility and biocompatibility. Zhongguo Xiu Fu Chong Jian Wai Ke Za Zhi. 2004;18:108–11.

Luo H, Huang X, Huang F, Liu X. Transversal inducing differentiation of human amniotic epithelial cells into hepatocyte-like cells. Zhong Nan Da Xue Xue Bao Yi Xue Ban. 2011;36(6):525–31.

Luo H, Lu Y, Wu T, Zhang M, Zhang Y, Jin Y. Construction of tissue-engineered cornea composed of amniotic epithelial cells and acellular porcine cornea for treating corneal alkali burn. Biomaterials. 2013;34(28):6748–59.

Lv S, Cheng J, Sun A, Li J, Wang W, Guan G, Liu G, Su M. Mesenchymal stem cells transplantation ameliorates glomerular injury in streptozotocin-induced diabetic nephropathy in rats via inhibiting oxidative stress. Diabetes Res Clin Pract. 2014;104(1):143–54.

Ma X, Zhang S, Shang Y, Gao T, Wang X, Chen F, Zhou J. Involvement of human amniotic fluid colony derived stem cells in regeneration of mouse injured muscle. Zhongguo Xiu Fu Chong Jian Wai Ke Za Zhi. 2011;25(7):842–847.

Ma X, Zhang S, Zhou J, Chen B, Shang Y, Gao T, Wang X, Xie H, Chen F. Clone-derived human AF-amniotic fluid stem cells are capable of skeletal myogenic differentiation in vitro and in vivo. J Tissue Eng Regen Med. 2012;6(8):598–613.

Magatti M1, De Munari S, Vertua E, Parolini O. Amniotic membrane-derived cells inhibit proliferation of cancer cell lines by inducing cell cycle arrest.J Cell Mol Med. 2012;16(9):2208–1118.

Magatti M, De Munari S, Vertua E, Nassauto C, Albertini A, Wengler GS, Parolini O. Amniotic mesenchymal tissue cells inhibit dendritic cell differentiation of peripheral blood andamnion resident monocytes. Cell Transplant. 2009;18:899–914.

Mann A, Yadav RP, Singh J, Kumar D, Singh B, Yadav PS. Culture, characterization and differentiation of cells from buffalo (Bubalus bubalis) amnion. Cytotechnology. 2013;65(1):23–30.

Manochantr S, Tantrawatpan C, Kheolamai P, U-pratya Y, Supokawej A, Issaragrisil S. Isolation, characterization and neural differentiation potential of amnion derived mesenchymalstem cells. J Med Assoc Thai. 2010;93(Suppl 7):S183–91.

Manuelpillai U, Tchongue J, Lourensz D, Vaghjiani V, Samuel CS, Liu A, Williams ED, Sievert W. Transplantation of human amnion epithelial cells reduces hepatic fibrosis in immunocompetent CCl_4-treated mice. Cell Transplant. 2010;19(9):1157–1168.

Marcus AJ, Coyne TM, Rauch J, Woodbury D, Black IB. Isolation, characterization, and differentiation of stem cells derived from the rat amniotic membrane. Differentiation. 2008;76:130–44.

McGhee CN, Patel DV. Mooren's ulcer and amniotic membrane transplant: a simple surgical solution? Clin Experiment Ophthalmol. 2011;39(5):383–5.

Mehanni SS, Ibrahim NF, Hassan AR, Rashed LA. New approach of bone marrow-derived mesenchymal stem cells and human amniotic epithelial cells applications in accelerating wound healing of irradiated albino rats. Int J Stem Cells. 2013;6(1):45–54.

Meng XT, Chen D, Dong ZY, Liu JM. Enhanced neural differentiation of neural stem cells and neurite growth by amniotic epithelial cell co-culture. Cell Biol Int. 2007;31(7):691–8.

Mi S, David AL, Chowdhury B, Jones RR, Hamley IW, Squires AM, Connon CJ. Tissue engineering a fetal membrane. Tissue Eng Part A. 2012;18(3–4):373–81.

Miki T, Strom SC. Amnion-derived pluripotent/multipotent stem cells. Stem Cell Rev. 2006;2(2):133–42.

Miki T, Mitamura K, Ross MA, Stolz DB, Strom SC. Identification of stem cell marker-positive cells by immunofluorescence in term human amnion. J Reprod Immunol. 2007;75:91–96.

Miki T, Lehmann T, Cai H, Stolz DB, Strom SC. Stem cell characteristics of amniotic epithelial cells. Stem Cells. 2005;23:1549–59.

Miki T, Marongiu F, Ellis EC, Dorko K, Mitamura K, Ranade A, Gramignoli R, Davila J, Strom SC. Production of hepatocyte-like cells from human amnion. Methods Mol Biol. 2009;481:155–68.

Mizokami T, Hisha H, Okazaki S, Takaki T, Wang XL, Song CY, Li Q, Kato J, Hosaka N, Inaba M, Kanzaki H, Ikehara S. Preferential expansion of human umbilical cord blood-derived CD34-positive cells on major histocompatibility complex-matched amnion-derived mesenchymal stem cells. Haematologica. 2009;94(5):618–28.

Mohammadi AA, Johari HG, Eskandari S. Effect of amniotic membrane on graft take in extremity burns. Burns. 2013;39(6):1137–41.

Moodley Y, Ilancheran S, Samuel C, Vaghjiani V, Atienza D, Williams ED, Jenkin G, Wallace E, Trounson A, Manuelpillai U. Human amnion epithelial cell transplantation abrogates lung fibrosis and augments repair. Am J Respir Crit Care Med. 2010;182:643–51.

Moschidou D, Drews K, Eddaoudi A, Adjaye J, De Coppi P, Guillot PV. Molecular signature of human amniotic fluid stem cells during fetal development. Curr Stem Cell Res Ther. 2013;8(1):73–81.

Moschidou D, Mukherjee S, Blundell MP, Jones GN, Atala AJ, Thrasher AJ, Fisk NM, De Coppi P, Guillot PV. Human mid-trimester amniotic fluid stem cells cultured under embryonic stem cell conditions with valproic acidacquire pluripotent characteristics. Stem Cells Dev. 2013;22(3):444–58.

Murphy S, Rosli S, Acharya R, Mathias L, Lim R, Wallace E, Jenkin G. Amnion epithelial cell isolation and characterization for clinical use. Curr Protoc Stem Cell Biol. 2010 (Chapter 1:Unit 1E.6).

Nagura S, Otaka S, Koike C, Okabe M, Yoshida T, Fathy M, Fukahara K, Yoshimura N, Misaki T, Nikaido T. Effect of exogenous Oct4 overexpression on cardiomyocyte differentiation of human amniotic mesenchymal cells. Cell Reprogram. 2013;15(5):471–80.

Ochsenbein-Kölble N, Bilic G, Hall H, Huch R, Zimmermann R. Inducing proliferation of human amnion epithelial and mesenchymal cells for prospective engineering of membrane repair. J Perinat Med. 2003;31:287–94.

Otaka S, Nagura S, Koike C, Okabe M, Yoshida T, Fathy M, Yanagi K, Misaki T, Nikaido T. Selective isolation of nanog-positive human amniotic mesenchymal cells and differentiation into cardiomyocytes. Cell Reprogram. 2013;15(1):80–91.

Papanas N, Eleftheriadou I, Tentolouris N, Maltezos E. Advances in the topical treatment of diabetic foot ulcers. Curr Diabetes Rev. 2012;8(3):209–18.

Payne WG, Wachtel TL, Smith CA, Uberti MG, Ko F, Robson MC. Effect of amnion-derived cellular cytokine solution on healing of experimental partial-thickness burns. World J Surg. 2010;34(7):1663–1668.

Peng L, Wang J, Lu GX. Isolation of human amniotic mesenchymal cells and their differentiation potential into islet-like cellsin vitro. Nan Fang Yi Ke Da Xue Xue Bao. 2011;31(1):5–10.

Perin L, Sedrakyan S, Giuliani S, Da Sacco S, Carraro G, Shiri L, Lemley KV, Rosol M, Wu S, Atala A, Warburton D, De Filippo RE. Protective effect of human amniotic fluid stem cells in an immunodeficient mouse model of acute tubular necrosis. PLoS ONE. 2010;5(2):e9357.

Petsche Connell J, Camci-Unal G, Khademhosseini A, Jacot JG. Amniotic fluid-derived stem cells for cardiovascular tissue engineering applications. Tissue Eng Part B Rev. 2013;19(4):368–79.

Rajangam T, An SS. Fibrinogen and fibrin based micro and nano scaffolds incorporated with drugs, proteins, cells and genes for therapeutic biomedical applications. Int J Nanomedicine. 2013;8:3641–62.

Ricardo JR, Barros SL, Santos MS, Souza LB, Gomes JA. Amniotic membrane transplantation for severe acute cases of chemical ocular burn and Stevens-Johnson syndrome. Arq Bras Oftalmol. 2009;72:215–20.

Roh DH, Seo MS, Choi HS, Park SB, Han HJ, Beitz AJ, Kang KS, Lee JH. Transplantation of human umbilical cord blood or amniotic epithelial stem cells alleviates mechanical allodynia after spinal cord injury in rats. Cell Transplant. 2013;22(9):1577–90.

Rolfo A, Giuffrida D, Giuffrida MC, Todros T, Calogero AE. New perspectives for prostate cancer treatment: in vitro inhibition of LNCaP and PC3 cell proliferation by amnion-derived mesenchymal stromal cells conditioned media. Aging Male. 2014;17(2):94–101.

Rossi D, Pianta S, Magatti M, Sedlmayr P, Parolini O. Characterization of the conditioned medium from amniotic membrane cells: prostaglandins as key effectors of itsimmunomodulatory activity. PLoS ONE. 2012;7(10):e46956. doi:10.1371/journal.pone.0046956.

Rossi B, Merlo B, Colleoni S, Iacono E, Tazzari PL, Ricci F, Lazzari G, Galli C. Isolation and in vitro characterization of bovine amniotic fluid derived stem cells at different trimesters of pregnancy. Stem Cell Rev. 2014;10(5):712–24.

Rota C, Imberti B, Pozzobon M, Piccoli M, De Coppi P, Atala A, Gagliardini E, Xinaris C, Benedetti V, Fabricio AS, Squarcina E, Abbate M, Benigni A, Remuzzi G, Morigi M. Human amniotic fluid stem cell preconditioning improves their regenerative potential. Stem Cells Dev. 2012;21(11):1911–23.

Roubelakis MG, Tsaknakis G, Pappa KI, Anagnou NP, Watt SM. Spindle shaped human mesenchymal stem/stromal cells from amniotic fluid promote neovascularization. PLoS ONE. 2013;8(1):e54747.

Saito S, Lin YC, Murayama Y, Hashimoto K, Yokoyama KK. Human amnion- derived cells as a reliable source of stem cells. Curr Mol Med. 2012;12(10):1340–9.

Sankar V, Muthusamy R. Role of human amniotic epithelial cell transplantation in spinal cord injury repair research. Neuroscience. 2003;118(1):11–7.

Secchiero P, Zorzet S, Tripodo C, Corallini F, Melloni E, Caruso L, Bosco R, Ingrao S, Zavan B, Zauli G. Human bone marrow mesenchymal stem cells display anti-cancer activity in SCID mice bearing disseminated non-Hodgkin's lymphoma xenografts. PLoS ONE. 2010;5(6):e11140.

Sha X1, Wen Y, Liu Z, Song L, Peng J, Xie L. Inhibition of α-Smooth Muscle Actin Expression and Migration of Pterygium Fibroblasts by Coculture with Amniotic Mesenchymal Stem Cells. Curr Eye Res. 2014;39(11):1081–1089.

Sha X, Liu Z, Song L, Wang Z, Liang X. Human amniotic epithelial cell niche enhances the functional properties of human corneal endothelial cells via inhibiting P53-survivin-mitochondria axis. Exp Eye Res. 2013;116:36–46.

Shaw SW, David AL, De Coppi P. Clinical applications of prenatal and postnatal therapy using stem cells retrieved from amniotic fluid. Curr Opin Obstet Gynecol. 2011;23(2):109–16.

Shimazaki J, Aiba M, Goto E, Kato N, Shimmura S, Tsubota K. Transplantation of human limbal epithelium cultivated on amniotic membrane for the treatment of severe ocular surface disorders. Ophthalmology. 2002;109:1285–90.

Simat SF, Chua KH, Abdul Rahman H, Tan AE, Tan GC. The stemness gene expression of cultured human amniotic epithelial cells in serial passages. Med J Malaysia. 2008;63(Suppl A):53–54.

Sippel KC, Ma JJ, Foster CS. Amniotic membrane surgery. Curr Opin Ophthalmol. 2001;12:269–281.

Skardal A, Mack D, Kapetanovic E, Atala A, Jackson JD, Yoo J, Soker S. Bioprinted amniotic fluid-derived stem cells accelerate healing of large skin wounds. Stem Cells Transl Med. 2012;1(11):792–802.

Tamagawa T, Oi S, Ishiwata I, Ishikawa H, Nakamura Y. Differentiation of mesenchymal cells derived from human amniotic membranes into hepatocyte-likecells in vitro. Hum Cell. 2007;20:77–84.

Terai M, Izumiyama-Shimomura N, Aida J, Ishikawa N, Kuroiwa M, Arai T, Toyoda M, Nakamura KI, Takubo K. Arm-specific telomere dynamics of each individual chromosome in induced pluripotent stem cells revealed by quantitative fluorescence in situ hybridization. Tissue Cell. 2014;46(6):470–6.

Theoret C. Tissue engineering in wound repair: the three "R"s–repair, replace, regenerate. Vet Surg. 2009;38(8):905–13.

Tsai MS, Lee JL, Chang YJ, Hwang SM. Isolation of human multipotent mesenchymal stem cells from second-trimester amniotic fluid using a novel two-stage culture protocol. Hum Reprod. 2004;19(6):1450–6.

Tsai MS, Hwang SM, Chen KD, Lee YS, Hsu LW, Chang YJ, Wang CN, Peng HH, Chang YL, Chao AS, Chang SD, Lee KD, Wang TH, Wang HS, Soong YK. Functional network analysis of the transcriptomes of mesenchymal stem cells derived from amniotic fluid, amniotic membrane, cord blood, and bone marrow. Stem Cells. 2007;25:2511–23.

Uchida S, Suzuki Y, Araie M, Kashiwagi K, Otori Y, Sakuragawa N. Factors secreted by human amniotic epithelial cells promote the survival of rat retinal ganglion cells. Neurosci Lett. 2003;341(1):1–4.

Venkatachalam S, Palaniappan T, Jayapal PK, Neelamegan S, Rajan SS, Muthiah VP. Novel neurotrophic factor secreted by amniotic epithelial cells. Biocell. 2009;33(2):81–9.

Vidane AS, Souza AF, Sampaio RV, Bressan FF, Pieri NC, Martins DS, Meirelles FV, Miglino MA, Ambrósio CE. Cat amniotic membrane multipotent cells are nontumorigenic and are safe for use in cell transplantation. Stem Cells Cloning. 2014;7:71–78.

Wang Y, Bai J, Chen J, Liu LF, Wang Y. Comparative studies on different cryopreservation protocols of human amniotic fluid derived mesenchymal stem cells. Zhongguo Xiu Fu Chong Jian Wai Ke Za Zhi. 2012;26(2):141–5.

Wang YY, Fang N, Chen DX, Yu LM, Liu ZL, Wan X. Human amnion-derived mesenchymal stem cells engraftment and differentiate in rats with acute myocardial infarction. Zhong Guo Lao Nian Xue Za Zhi. 2013a;4(33):1583–6.

Wang F, Wang L, Yao X, Lai D, Guo L. Human amniotic epithelial cells can differentiate into granulosa cells and restore folliculogenesis in a mouse model of chemotherapy-induced premature ovarian failure. Stem Cell Res Ther. 2013b;4(5):124.

Warrier S, Haridas N, Bhonde R. Inherent propensity of amnion-derived mesenchymal stem cells towards endothelial lineage: vascularization from an avascular tissue. Placenta. 2012;33(10):850–8.

Westgren M, Ek S, Remberger M, Ringden O, Stangenberg M. Cytokines in fetal blood and amniotic fluid in Rh-immunized pregnancies. Obstet Gynecol. 1995;86:209–13.

Wolbank S, Peterbauer A, Fahrner M, Hennerbichler S, van Griensven M, Stadler G, Redl H, Gabriel C. Dose-dependent immunomodulatory effect of human stem cells from amniotic membrane: a comparison with human mesenchymal stem cells from adipose tissue. Tissue Eng. 2007;13:1173–83.

Wolbank S, Stadler G, Peterbauer A, Gillich A, Karbiener M, Streubel B, Wieser M, Katinger H, van Griensven M, Redl H, Gabriel C, Grillari J, Grillari-Voglauer R. Telomerase immortalized human amnion- and adipose-derived mesenchymal stem cells: maintenance of differentiation and immunomodulatory characteristics. Tissue Eng Part A. 2009;15:1843–54.

Xiao Y, Fang N, Chen DX, Yu LM, Zhang T, Zhao CH. Therapeutic effect and immunological regulation of human aminiotic mesenchymal stem cells on rat with collagen-induced arthristis. Man Yi Xue Za Zhi. 2013;29(6):461–6.

Xue S, Chen C, Dong W, Hui G, Liu T, Guo L. Therapeutic effects of human amniotic epithelial cell transplantation on double-transgenic mice co-expressing APPswe and PS1ΔE9-deleted genes. Sci China Life Sci. 2012;55(2):132–40.

Xue Y, Miao Z, Sun H. Effects of human amniotic mesenchymal stromal cells on rabbit T-cell responses in a xenolymphocyte reaction assay. Exp Clin Transplant. 2014;12(3):253–60.

Yamahara K, Harada K, Ohshima M, Ishikane S, Ohnishi S, Tsuda H, Otani K, Taguchi A, Soma T, Ogawa H, Katsuragi S, Yoshimatsu J, Harada-Shiba M, Kangawa K, Ikeda T. Comparison of angiogenic, cytoprotective, and immunosuppressive properties of human amnion- and chorion-derived mesenchymal stem cells. PLoS ONE. 2014;9(2):e88319.

Yan ZJ, Zhang P, Hu YQ, Zhang HT, Hong SQ, Zhou HL, Zhang MY and Xu RX. Neural stem-like cells derived from human amnion tissue are effective in treating traumatic brain injury in rat. Neurochem Res. 2013;38(5):1022–1033.

Yang DY, Sheu ML, Su HL, Cheng FC, Chen YJ, Chen CJ, Chiu WT, Yiin JJ, Sheehan J, Pan HC. Dual regeneration of muscle and nerve by intravenous administration of human amniotic fluid-derived mesenchymal stem cells regulated by stromal cell-derived factor-1α in a sciatic nerve injury model. J Neurosurg. 2012;116(6):1357–1367.

Yang X, Song L, Wu N, Liu Z, Xue S, Hui G. An experimental study on intracerebroventricular transplantation of human amniotic epithelial cells in a rat model of Parkinson's disease. Neurol Res. 2010;32(10):1054–9.

Yang L, Hashimoto K, Tohyama M, Okazaki H, Dai X, Hanakawa Y, Sayama K, Shirakata Y. Interactions between myofibroblast differentiation and epidermogenesis in constructing human living skin equivalents. J Dermatol Sci. 2011;65(1):50–7.

Yang JD, Choi DS, Cho YK, Kim TK, Lee JW, Choi KY, Chung HY, Cho BC, Byun JS. Arch Plast Surg. 2013;40(5):496–504.

Yannas IV, Tzeranis DS, Harley BA, So PT. Biologically active collagen-based scaffolds: advances in processing and characterization. Philos Trans A Math Phys Eng Sci. 2010;368(1917):2123–39.

Yao GP, Yu LM, Fan ZH, Ren F, Luo J, Zhang XY, Wang YY, Liu JW. Human aminion-derived mesenchymal stem cells transplantation improves hematopoietic function of myelosuppression mice. Zhong Guo Zu Zhi Gong Chengn Yan Jiu. 2013a;17(23):4243–55.

Yao M, Chen J, Yang XX, Zhang XL, Ji QS, Zhou Q, Xu JT. Differentiation of human amniotic epithelial cells into corneal epithelial-like cells in vitro. Int J Ophthalmol. 2013b;6(5):564–72.

Yi T, Song SU. Immunomodulatory properties of mesenchymal stem cells and their therapeutic applications. Arch Pharm Res. 2012;35(2):213–21.

Yoon BS, Moon JH, Jun EK, Kim J, Maeng I, Kim JS, Lee JH, Baik CS, Kim A, Cho KS, Lee JH, Lee HH, Whang KY, You S. Secretory profiles and wound healing effects of human amniotic fluid-derived mesenchymal stem cells. Stem Cells Dev. 2010;19(6):887–902.

Yu HF, Fang N, Che DX, YU LM. ZHang T, iu ZL, Wan WH, ZHao CH. Therapeutic effects of transplantation of human amnion - derived mesenchymal stem cells in rats with spinal cord injury. Zhong Hua Shen Jin Wai Ke Zha Zi. 2012;28(4): 411–414.

Yu H, Yao GP, Ren F, Fan ZH, Fang N, Wang YY, Yang J, Pu T, Yu LM. Human amnion-derived stem cells transplantation for the treatment of acute kidney injury. Zhongguo Zuzhi Gongcheng Yanjiu. 2012;16(49):9196–200.

Zani A, Cananzi M, Fascetti-Leon F, Lauriti G, Smith VV, Bollini S, Ghionzoli M, D'Arrigo A, Pozzobon M, Piccoli M, Hicks A, Wells J, Siow B, Sebire NJ, Bishop C, Leon A, Atala A, Lythgoe MF, Pierro A, Eaton S, De Coppi P. Amniotic fluid stem cells improve survival and enhance repair of damaged intestine in necrotising enterocolitis via a COX-2 dependent mechanism. Gut. 2014;63(2):300–9.

Zhang D, Jiang M, Miao D. Transplanted human amniotic membrane- derived mesenchymal stem cells ameliorate carbon tetrachloride-induced liver cirrhosis in mouse. PLoS One. 2011;6(2):e16789.

Zhang Lu, Ning Fang, Dai-xiong Chen, Zu-lin Liu, Wei-hong Wan, Jin-wei Liu, Tao Zhang. Differentiation of human amniotic mesenchymal cells into cardiomyocyte-like cells in vitro. Zhong Guo Sheng Wu Gong Chen Za Zhi. 2007;27(12):84–9.

Zhao YJ, Fang N, Chen DX, Yu LM, Yu HF, Zhao CH. Study on inducing differentiation of human amniotic epithelial cells into insulin secreting cells in vitro. Zhongguo Ying Yong Sheng Li Xue Za Zhi. 2012;28(2):139–43.

Zheng YB, Gao ZL, Xie C, Zhu HP, Peng L, Chen JH, Chong YT. Characterization and hepatogenic differentiation of mesenchymal stem cells from human amnioticfluid and human bone marrow: a comparative study. Cell Biol Int. 2008;32:1439–48.

Zheng YB, Zhang XH, Huang ZL, Lin CS, Lai J, Gu YR, Lin BL, Xie DY, Xie SB, Peng L, Gao ZL. Amniotic-fluid-derived mesenchymal stem cells overexpressing interleukin-1 receptor antagonist improve fulminant hepatic failure. PLoS One. 2012;7(7):e41392.

Zhou J, Wang D, Liang T, Guo Q, Zhang G. Amniotic fluid-derived mesenchymal stem cells: characteristics and therapeutic applications. Arch Gynecol Obstet. 2014;290(2):223–31.

Zhu X, Wang X, Cao G, Liu F, Yang Y, Li X, Zhang Y, Mi Y, Liu J, Zhang L. Stem cell properties and neural differentiation of sheep amniotic epithelial cells. Neural Regen Res. 2013;8(13):1210–9.

Chapter 3
Mesenchymal Stem Cells and Their Immunomodulatory Properties

Lianming Liao and Robert Chunhua Zhao

Abstract Mesenchymal stem cells (MSCs) are an important cell population that resides in a bone marrow microenvironment. In the past decade, MSCs have been discovered to have profound immunomodulatory functions both in vitro and in vivo. As MSCs can be expanded rapidly to numbers that are required for clinical application, clinical studies have been performed in immune diseases, bone marrow transplantation, and kidney transplantation. In this chapter, the mechanisms underlying MSCs' immunomodulating property and the potential clinical use of MSCs as a modulator of immune responses are reviewed.

Keywords MSC · Immunomodulatory functions · Interactions · T cells · Antigen-presenting cells

3.1 Introduction

The use of immunosuppressive agents has led to greatly improved organ graft survival rates and alleviation of autoimmune diseases. Nevertheless, side effects of immunosuppressive agents and patient morbidity due to lifelong immunosuppression remain substantial, especially high cancer incidence among the recipients. Active induction of tolerance allowing drug-free allograft acceptance with preserved immunocompetence has long been a goal for both immunologists and clinicians. This method is theoretically the only way to resolve rejection reaction of allogeneic transplantation and simultaneously keep the patients void of the side effects of immunosuppressive medication.

L. Liao (✉) · R.C. Zhao
Academy of Integrative Medicine, Fujian University of Traditional Chinese Medicine, Huatuo Road No. 1, Fuzhou 350112, Fujian, People's Republic of China
e-mail: llm@fjtcm.edu.cn

R.C. Zhao (ed.), *Stem Cells: Basics and Clinical Translation*, Translational Medicine Research 1, DOI 10.1007/978-94-017-7273-0_3

Stable chimerism is linked with permanent tolerance of donor organ or tissue transplants (Wekerle and Sykes 2001). Induction of mixed hematopoietic chimerism by bone marrow stem cell transplantation, which leads to stable donor-specific tolerance in allogeneic graft, has been reported (Sayegh et al. 1991; Helg et al. 1994; Sorof et al. 1995). However, the use of bone marrow transplantation is still largely restricted to patients with malignancies or severe immune diseases. The toxicity of the required host conditioning, the risk of engraftment failure, and the problem of graft-versus-host-disease (GVHD), especially when major histocompatibility complex (MHC) barriers are transgressed, hinder its clinical application. Embryonic-like stem cell transplantations have also been proved to be capable of establishing chimerism and inducing tolerance without supplementary host conditioning in animal model (Fandrich et al. 2002). However, there exists the well-known ethical obstacle for embryonic stem cells and tumorigenesis of embryonic-like stem cells transplantation in human need cautious evaluation. Therefore, development of a new cell population and procedure for donor-specific allograft tolerance induction is still a difficult task in organ transplantation.

In the past decade, bone-marrow-derived mesenchymal stem cells (MSCs) are garnering more and more attention for their immunoregulatory activity and regenerative ability. For example, administration of MSCs could prolong donor skin graft survival in mice (Han et al. 2003) and nonhuman primates (Bartholomew et al. 2002). The ability of MSCs to regulate immune responses could also be harnessed to reduce GVHD at the time of hematopoietic stem cell transplantation (Frassoni et al. 2002; Lazarus et al. 2000).

3.2 Characteristics of MSC

MSCs are multipotent cells that reside within the bone marrow and can be induced to differentiate into various components of the marrow microenvironment, such as bone, adipose, and stromal tissues under proper conditions (Fridenstein 1982; Pittenger et al. 1999; Deans and Moseley 2000). MSCs support the growth of hematopoietic progenitors by secreting a number of hematopoietic cytokines such as macrophage colony stimulating factor, interleukin (IL)-6, IL-7, IL-8, IL-11, IL-12, IL-14, IL-15, and leukemia inhibitory factor (Deans and Moseley 2000; Deans 2000). MSCs have been isolated in different species and are present in the bone marrow at low frequency (1 out of 104–105 mononuclear cells). Although MSCs are originally isolated from bone marrow, they can also be isolated from muscle (Asakura et al. 2001), pancreas (Hu et al. 2003), dermis (Young et al. 2001), adipose tissues (Zuk et al. 2001), lung (Noort et al. 2002), liver (Hu et al. 2001), and cord blood (Erices et al. 2000). The exact phenotype of MSCs in the tissue (i.e., the cell phenotype prior to culture) is still debated. Simmoms et al. described the first antibody (Stro-1) that targeted MSCs in the bone marrow (Simons and Torok-Storb 1991). Some typical markers of MSCs include CD105 (SH2 or endoglin), CD73 (SH3 or SH4), CD90, CD166, CD44, and CD29.

3.3 MSCs Exert Their Immunomodulation Function by Different Mechanisms

3.3.1 The Interaction Between MSCs and T Cells

The key orchestrators of the immune response in transplantation are T cells, which can react to alloantigen both directly, by recognizing intact foreign MHC molecules on donor antigen-presenting cells (APCs), and indirectly, as a result of interactions with processed donor antigens on self APCs (Sayegh and Turka 1998). In the thymus, double positive ($CD4^{+}CD8^{+}$) cells undergo positive and negative selections before leaving the thymus. Positive selection results in survival of T cells with antigen receptors that corecognize self-MHC molecules plus foreign peptides. T cells whose receptors do not detect self-MHC molecules die, presumably by failure to receive critical differentiating signals. Negative selection involves the removal of potentially autoreactive T cells that interact too well with self-MHC molecules plus self-peptides. The majority of cells with specificities for self-antigens are eliminated during development in the thymus (von Boehmer and Kisielow 1990). Otherwise, they will mature and migrate to the peripheral lymphoid organs. Cortical epithelial cells are essential for the process of positive selection because they display the self-peptide–MHC complexes that are recognized by $CD4^{+}CD8^{+}$ thymocytes and also provide essential differentiation factors (Anderson et al. 1993).

T cells can be physiologically silenced by a number of mechanisms, including deletion in the peripheral immune system; anergy, where they cannot adequately respond following restimulation with antigen; and suppression, which may be mediated by interactions with other cells or with soluble factors (Van Parijs and Abbas 1998; Charlton et al. 1994). The TH1 cell cytokines IL-2 and interferon gamma (IFN-γ) are considered detrimental to allografts, and the TH2 cell cytokine IL-4 is considered tolerogenic (Nickerson et al. 1994).The same mechanisms act in acquired transplantation tolerance (Li et al. 2001) and can be harnessed to achieve donor-specific tolerance by blunting the effects of alloreactive T cells.

Presently, although the relationship between MSCs and T cells is not well defined, several lines of evidence indicate that MSCs may modulate T cells by various mechanisms (Fig. 3.1). When added into a mixed lymphocyte reaction, either on day 0 or on day 3, baboon MSCs could inhibit an ongoing allogeneic response, leading to a greater than 50 % reduction in proliferative activity (Bartholomew et al. 2002). This effect could be maximized by escalating the dose of MSC and could be reduced with the addition of exogenous IL-2. The suppression of proliferative responses by MSCs did not appear to be dependent on the source of MSCs. MSCs were able to inhibit proliferation of T cells independent of whether they were of the same source of the responder, stimulator, or third party. Baboon MSCs could also inhibit proliferative response to potent T-cell mitogen ConA. Similarly, addition of MSCs to T cells stimulated by polyclonal activators (PHA or IL-2) resulted in suppression of proliferation (Di Nicola et al. 2002). MSC-inhibited

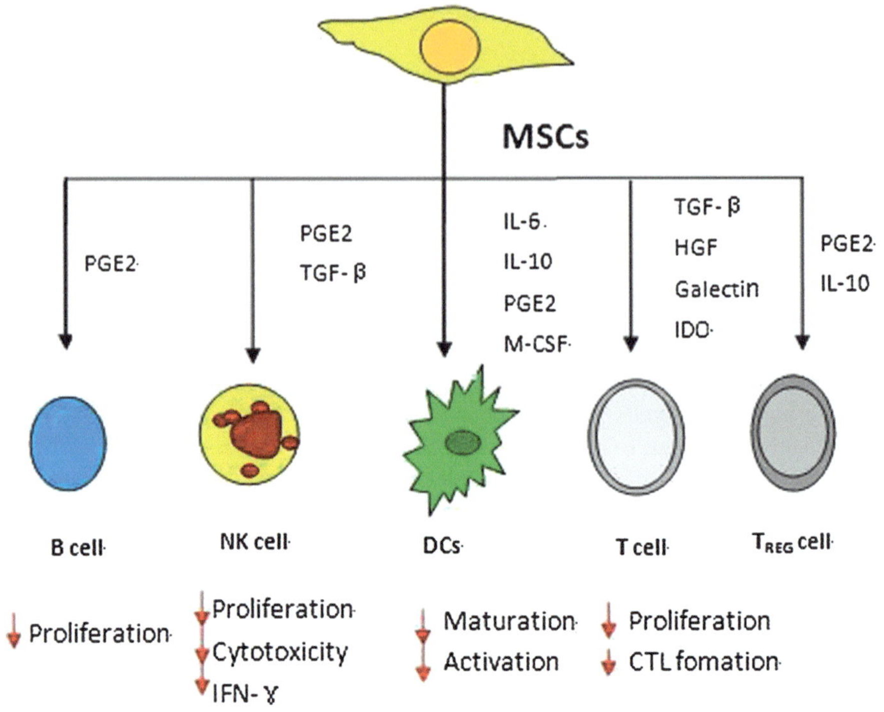

Fig. 3.1 MSCs inhibit the proliferation of NK cells, B cells, and T cells. This effect is mediated through the secretion of various factors, such as prostaglandin E2, human leukocyte antigen-G, and hepatocyte growth factor transforming growth factor-β. MSC also inhibits generation and maturation of DCs from monocytes. Abbreviations: *CTL* cytotoxic T cell; *DCs* dendritic cells; *HGF* hepatocyte growth factor; *IDO* indoleamine 2,3-dioxygenase; *MSCs* mesenchymal stem cells; *NK* natural killer; *PGE2* prostaglandin E2; T_{REG} *cells* regulatory T cells; and *TGF-β* transforming growth factor β

T lymphocytes were not apoptotic and efficiently proliferated on restimulation. MSCs significantly suppressed both $CD4^+$ and $CD8^+$ T cells (Chen et al. 2002). In a study on the immunogenicity and antigen-presenting ability of MSCs, Tse et al. (2003) demonstrated human MSCs failed to stimulate allogeneic peripheral blood mononuclear cells or T-cell proliferation in mixed cell cultures. MSCs actively suppressed proliferation of responder peripheral blood mononuclear cells stimulated by third-party allogeneic peripheral blood mononuclear cells as well as T cells stimulated by anti-CD3 and anti-CD28 antibodies. Similarly, marked inhibitory effects of allogeneic and autologous MSCs were also reported both in mixed lymphocyte reaction and after mitogenic lymphocyte stimulation by phytohaemagglutinin, concanavalin A, and protein A (Le Blanc et al. 2003). However, little, if any, inhibition occurred after stimulation with pokeweed mitogen. The inhibitory effect was also related to MSCs dose, as a minimum of MSCs was needed. Surprisingly, when the dosage was small, stimulatory effect was noted in some experiments. This phenomenon was also observed when MSCs were cocultured for 3 days with T cells isolated from cord blood and stimulated with phytohaemagglutinin for 60 h, in which high concentration of MSCs most often resulted in inhibition, while low concentration resulted in stimulation of T-cell proliferation (He et al. 2003). In another study, effect of MSCs on response of naive and memory T cells to their cognate antigenic epitopes was evaluated (Krampera et al. 2003). For this purpose, murine male transplantation antigens, HY, were selected to trigger immune response. C57BL/6 female mice immunized with male cells were the source of memory T cells, whereas C6 mice transgenic for HY-specific T-cell receptor provided naive T cells. Responder cells were stimulated in vitro with male spleen cells or HY peptides in the presence or absence of MSCs. MSCs inhibited HY-specific naive and memory T cells in a dose-dependent fashion and affected cell proliferation, cytotoxicity, and the number of interferon-gamma-producing HY peptide-specific T cells. However, MSC did not selectively target antigen-reactive T cells. The expression of MHC molecules and the presence in culture of antigen-presenting cells or of $CD4^{+}/CD25^{+}$ regulatory T cells were not required for MSCs to inhibit. This data demonstrate that autologous or allogeneic MSCs strongly suppress lymphocyte proliferation that is triggered by both cellular as well as nonspecific mitogenic stimuli in different in vitro models; this phenomenon has no immunologic restriction.

Suppression of lymphocyte proliferation by MSCs is likely due to the production of soluble factors. By using the transwell system, i.e., when MSCs were separated from T cells physically by a permeable membrane, T-cell proliferation was also significantly inhibited. After simultaneous addition of anti-transforming growth factor-β1 and anti-hepatocyte growth factor antibodies to bone marrow stromal cells-containing mixed lymphocyte reactions, T-cell proliferation was restored at values that were comparable to those detected in mixed lymphocyte reactions without bone marrow stromal cells, indicating transforming growth factor-β1 and hepatocyte growth factor were the mediators of marrow stromal cells' effects (Di Nicola et al. 2002). Chen et al. (2002) showed secretion of transforming growth factor-β1 by MSCs reached to 1 ng/ml in 72 h. Tse et al. demonstrated IL-10 secreted by MSCs also accounted for the suppressive activity by MSCs (Tse et al. 2003). However, the inhibitory activity was abrogated when MSCs were replaced by MSC culture supernatant (Krampera et al. 2003). This may suggest that pretreatment of MSCs with lymphocytes is necessary for MSCs to secret inhibitory factors.

In addition, the tryptophan catabolizing enzyme indoleamine 2,3-dioxygenase (IDO) has been suggested to play a role in the suppression of T-cell proliferation by MSCs (Meisel et al. 2004). Upon stimulation with IFN-γ, MSCs express IDO activity that degrades essential tryptophan and results in kynurenine breakdown products, resulting in reduced lymphocyte proliferation.

The families of human galectins are key players in the regulation of the innate and adaptive immune response (Yang et al. 2008). One family member, galectin-1, is a negative regulator of immune responses. Galectin-1 may inhibit proliferation and induce apoptosis of activated T cells (Blaser et al. 1998; Rabinovich et al. 2002; Perillo et al. 1995). Galectin-1 protein was detected intracellularly and on the cell surface of MSCs. It was reported that galectin-1 released into the cell culture supernatant by MSCs modulated the release of cytokines involved in GVHD and autoimmunity (e.g., tumor necrosis factor-α [TNF-α], IFN-γ, interleukin-2 [IL-2], and IL-10). Thus, galectin-1 may mediate the immunomodulatory effect of MSCs on allogeneic T cells (Gieseke et al. 2010).

MSC-induced T-cell anergy has been proposed as another potential mechanism of immune suppression. MSCs lack surface expression of costimulatory molecules, such as CD80 (B7-1) and CD86 (B7-2), and it is believed that MSCs can render T cells anergic (Glennie et al. 2005). Another level at which MSCs may modulate immune responses is through the induction of regulatory T cells (Treg). MSCs might induce formation of $CD4^+CD25^+$ regulatory T cells that were responsible for inhibition of allogeneic lymphocyte proliferation (Maccario et al. 2005; Tasso et al. 2012; Carrion et al. 2010). In the experimental autoimmune uveitis in mice, intraperitoneal injection of MSCs was able to significantly attenuate uveitis and that a significantly higher percentage of adaptive Treg was present in MSC-treated mice than in MSC-untreated animals. Furthermore, induction of antigen-specific Treg by MSCs was due at least in part to the secretion of TGF-β (Tasso et al. 2012).

3.3.2 The Interactions Between MSCs and Antigen-Presenting Cells

MSCs may also regulate the immune response through their interaction with dendritic cells (DCs) which play a key role in the induction of immunity (Fig. 3.1). MSCs may induce tolerance by inhibiting DC maturation and function, inhibit in vitro differentiation of DCs from monocytes and $CD34^+$ progenitors, and reduce secretion of pro-inflammatory cytokines (IL-12, IFN-γ, and TNF-α) and increase IL-10 secretion (Beyth et al. 2005). The production of IL-6, PGE2, IL-10, and M-CSF by MSCs may contribute to the inhibitory effect of MSCs on DC differentiation, although blocking studies indicate that these factors are not solely responsible for the inhibitory effect. DCs generated in the presence of MSCs were impaired in their response to maturation signals and exhibited no expression of CD83 or up-regulation of HLA-DR and costimulatory molecules (Jiang et al. 2005; Nauta et al. 2006; Zhang et al. 2004). Immature DCs generated in the presence of MSCs were strongly hampered in their ability to induce activation of T cells. These results suggest that MSCs suppress the differentiation of DCs, resulting in the formation of immature DCs that exhibit a suppressor or inhibitory phenotype.

3.3.3 The Interaction Between MSCs and B Cells

MSCs could inhibit the proliferation of B cells stimulated with anti-CD40L and IL-4, or with pokeweed mitogen (Augello et al. 2005). B cells were arrested in the G0/G1 phase of the cell cycle. Similar to T-cell suppression, the major mechanism of B-cell suppression is MSC production of soluble factors, as indicated by transwell experiments. In addition, B-cell differentiation was inhibited because IgM, IgG, and IgA production was significantly impaired. What is more, CXCR4, CXCR5, and CCR7 B-cell expressions, as well as chemotaxis to CXCL12, the CXCR4 ligand, and CXCL13, the CXCR5 ligand, were significantly down-regulated by MSCs, suggesting that these cells affect chemotactic properties of B cells (Corcione et al. 2006). Allogeneic MSCs have been shown to inhibit the proliferation, activation, and IgG secretion of B cells from BXSB mice that are used as an experimental model for human systemic lupus erythematosus (Fig. 3.1) (Deng et al. 2005).

3.3.4 Interaction Between MSCs and Natural Killer Cells

Natural killer (NK) cells exhibit spontaneous cytolytic activity that mainly targets cells that lack expression of HLA class I molecules. MSCs suppressed IL-2- or IL-15-driven NK-cell proliferation and IFN-γ production, and NK cells cultured for 4 to 5 days with IL-2 in the presence of MSCs have a reduced cytotoxic potential against K562 target cells (Krampera et al. 2006). Experiments with transwell culture systems indicated that MSCs suppressed the proliferation and cytokine production of IL-15 stimulated NK cells via soluble factors. At low NK-to-MSC ratios, MSCs alter the phenotype of NK cells and suppress proliferation, cytokine secretion, and cyto-toxicity against HLA-class I expressing targets. Some of these effects required cell-to-cell contact, whereas others were mediated by soluble factors, including PGE2 and transforming growth factor-β (Sotiropoulou et al. 2006). Indoleamine 2,3-dioxygenase also mediates MSC-induced inhibition of NK cells (Fig. 3.1) (Spaggiari et al. 2008).

MSCs secrete the soluble isoform HLA-G5. Blocking experiments using neutralizing anti-HLA-G antibody demonstrate that HLA-G5 contributed first to the suppression of allogeneic T-cell proliferation and then to the expansion of regulatory T cells. MSCs inhibited both NK cell-mediated cytolysis and interferon-gamma secretion through HLA-G5 (Selmani et al. 2008).

Taken together, numerous studies convincingly demonstrate that MSCs are able to modulate the function of different immune cells in vitro, particularly involving the suppression of T-cell proliferation. However, the mechanisms underlying the immunosuppressive effects of MSCS are still unclear.

3.3.5 MSCs Survive and Induce Immune Tolerance in the Host

It has been suggested that MSCs escape the immune system after they are infused to allogeneic recipient because they possess a cell surface phenotype that reflects poor recognition by T cells. For example, injection of genetically modified MSCs in baboon was not followed by their rejection because of the lack of immunogenicity of MSCs (Bartholomew et al. 2002; Devine et al. 2003). Indeed, the distinct immunophenotype profile of MSCs, i.e., no expression of costimulatory molecules B7-1, B7-2, CD40, and CD40L associated with the absence of MHC classes I and II expression, suggests that they may not be recognized by allogenic T cells and can escape host immune system' rejection. MSCs can easily migrate and reside in various tissues, which may result from their expression of a variety of adhesion molecule. In our study, when allogeneic murine MSCs were transplanted into lethally irradiated recipient mice 150 days before allogeneic skin transplantation, allogeneic donor skins were successfully transplanted and have survived for more than 100 days without any rejection reaction (Han et al. 2003). Immunohistochemistry staining showed donor MSCs could established long-term residency in gastrointestinal tissues, kidney, lung, liver, thymus, and skin (Deng et al. 2003). In a baboon model, following systemic infusion of GFP-marked MSC into an immunocompetent host, MSCs could be detected in a wide non-hematopoietic tissue distribution between 9 and 21 months later, including gastrointestinal tissues (colon, duodenum, jejunum, and ileum), kidney, skin, lung, thymus, and liver. Importantly, the results suggested that tissue distribution of MSC following systemic infusion was not affected by histocompatibility or prior conditioning. In the non-conditioned recipient, engraftment of MSC in these tissues was also achieved, although less abundant. When reinfused in non-human models, ex vivo–expanded human MSCs migrated to and became incorporated into several tissues of the recipient animals where MSCs were capable to elicit tissue-specific differentiation programs, indicating that MSCs have multiorgan homing capacity and an intrinsic degree of plasticity (Azizi et al. 1998; Toma et al. 2002; Saito et al. 2002; Liechty et al. 2000). Studies involving direct injection of MSCs into the rodent brain reported migration of cells within the brain and differentiation into glial populations. This approach has used xenogeneic transplant of human cells into the rat brain, as well as homologous mouse/mouse tracking studies (Azizi et al. 1998???). When purified human MSCs from adult bone marrow were injected into the left ventricle of CB17 SCID/beige adult mice, a limited number of cells survived and over time morphologically resembled the surrounding host cardiomyocytes. Immunohistochemistry revealed de novo expression of desmin, beta-myosin heavy chain, alpha-actinin, cardiac troponin T, and phospholamban at levels comparable to those of the host cardiomyocytes (Toma et al. 2002). In another xenograft model, bone marrow stromal cells were isolated from C57B1/6 mice and injected into immunocompetent adult Lewis rats. One week later, the recipient animals underwent coronary artery ligation and were sacrificed at various time points ranging from 1 day to 12 weeks after ligation. Labeled mice cells engrafted into the bone marrow cavities of the recipient rats for at least 13 weeks after transplantation without any immunosuppression. In the heart, some of these cells subsequently showed positive staining for cardiomyocyte-specific proteins, while other labeled cells participated in angiogenesis in the infarcted area. These findings indicate marrow stromal cells are adult stem cells with unique immunologic tolerance allowing their engraftment into a xenogeneic environment, while preserving their ability to be recruited to an injured myocardium to form a stable cardiac chimera (Saito et al. 2002). Similarly, human MSCs engrafted into fetal lamb could persist in multiple tissues for as long as 13 months after transplantation. Transplanted human cells underwent site-specific differentiation into chondrocytes, adipocytes, myocytes, cardiomyocytes, bone marrow stromal cells, and thymic stroma. Unexpectedly, there was a long-term engraftment even when cells were transplanted after the expected development of immunocompetence (Liechty et al. 2000). This is in contrast to the fact that fetal lambs develop the capacity to reject allogeneic skin grafts (Silverstein et al. 1964) and demonstrate allogeneic or xenogeneic hematopoietic engraftment failure (Zanjani et al. 1997) after 75 days of gestation. Thus, MSCs maintain their multipotential capacity after transplantation and seem to have unique immunologic characteristics that allow persistence in a xenogeneic environment. It is tempting to hypothesize that such wide distribution of infused MSCs in the host may induce peripheral tolerance.

Another finding that may account for MSCs' immunomodulation effect is that bone marrow stromal cells could migrate to the thymus after transplantation and thus may exert their immunomodulation effect there. This is first demonstrated by Li et al. (2000) who found donor-derived bone marrow stromal cells could migrate into the thymus and participate in the positive selection of T lymphocytes after bone marrow transplantation plus bone grafts. It therefore seems that bone marrow stromal cells may provide a scaffold for the adhesion of early T cells and, at least in culture, supply the appropriate stimuli for thymus precursor cell proliferation (BardaSaad et al. 1999). MSCs could secrete transforming growth factor-β1 in vitro and transforming growth factor-β1 is a potent inhibitor of T-cell proliferation, both in IL-2- and IL-4-derived response (Ruegemer et al. 1990; Ahuja et al. 1993; Fox et al. 1993). Transforming growth factor-β1 also significantly inhibited triple-negative ($CD3^{-}CD4^{-}CD8^{-}$) thymocytes in vitro (Mossalayi et al. 1995). Thus, MSCs migrating to thymus may inhibit proliferation of T cell by secretion of transforming growth factor-β1.

Preclinical animal studies demonstrated that MSCs can prolong allograft survival and alleviate autoimmune disease. When donor MSCs were intravenously administrated to MHC-mismatched recipient baboons prior to placement of autologous, donor, and third-party skin grafts, MSCs led to prolonged skin graft survival when compared to control animals (11.3 vs. 7 days) (Bartholomew et al. 2002). In a murine allograft system, we showed that allogeneic donor skins were successfully transplanted and have survived for more than 100 days without any rejection reaction with pre-infusion of donor MSCs (Han et al. 2003).

Allogeneic bone marrow transplantation associated with bone grafts was found to be efficient in the treatment of autoimmune disorders, such as in the MLR/lpr mouse model of lupus (Ishida et al. 1994). In these experiments, stromal cells have been assumed to play a critical role as compared to hematopoietic stem cells. In order to determine the real impact of MSCs in these experiments, the adherent cells were removed from the total bone marrow samples before transplantation. In this case, 75 % of the treated animals died within 90 days. In contrast, complementation of adherent cell-depleted bone marrow with stromal cells permitted the mice to survive 48 weeks and cured the autoimmune disease, suggesting that MSCs play a critical role in the complex immunoregulation of T- and B cells.

3.4 Applications of MSCs

3.4.1 GVHD

A potential application of MSCs in bone marrow transplantation is the prevention and treatment of steroid-resistant GVHD. Severe GVHD is a life-threatening complication after HSC transplantation. Unfortunately, steroids, the first-line treatment for GVHD, have a response rate of 30–50 %. In patients with severe steroid-resistant acute GVHD, the overall survival is low (Deeg 2007). In a case report, a 20-year-old woman with high-risk acute myelogenous leukemia was transplanted with granulocyte colony stimulating factor-mobilized peripheral blood $CD34^+$ hematopoietic stem cells and bone-marrow-derived MSCs from her HLA haplotype-mismatched father after myeloablative conditioning therapy. The patient engrafted rapidly and had no acute or chronic GVHD. Since transplantation, the patient has shown an enduring trilineage hematological complete response without any evidence of leukemia relapse at 31 months (Beyth et al. 2005). Several pilot studies have shown the efficiency of MSCs in treating steroid-resistant acute GVHD (Ringdén et al. 2006; Fang et al. 2007; Fang et al. 2006). These findings were confirmed by a phase II study (Le Blanc et al. 2008). Recently, Weng et al. (2010) reported that MSCs derived from HLA-identical sibling donors or HLA-disparate third-party donors were also effective as a salvage therapy for refractory chronic GVHD.

3.4.2 Organ Transplantation

MSCs may also offer therapeutic opportunities in organ transplantation by inhibiting T-cell proliferation, cytotoxic T-cell activity, B-cell activation, and differentiation and DC maturation and thereby blunting the effector arm of the alloresponse. In a baboon skin transplant model, a single intravenous administration of donor-type MSCs into MHC-mismatched recipients resulted in significant prolonged graft survival (Bartholomew et al. 2002). In a mouse transplant model, intraportal administration of MSCs extended heart allograft survival from 10 days in untreated controls to a median survival time of 40 days, with 33 % of MSC-treated recipients showing long-term tolerance (Casiraghi et al. 2008). Ding et al. (2009) showed that MSCs protected islet allografts from rejection. In the life-sustaining mouse islet allograft model, the allogeneic islets were rejected within 30 days. Surprisingly, administration of MSCs prevented rejection and led to long-term normoglycemia. In a recent report, 1 and 2 × 106 MSC/kg recipient body weights were infused at the time of renal transplantation and at two weeks of post–transplant, respectively. Preliminary results indicate that induction therapy with MSC appears to be equally effective as Basiliximab in the prevention of acute rejection and is associated with better clinical outcomes as far as early renal graft function and rate of infections (Tan et al. 2012). In an other pilot study, donor-derived bone marrow MSCs combined with a sparing dose of tacrolimus (50 % of standard dose) were shown to have a comparable effects with standard dose of tacrolimus in terms of acute rejection, graft function, and patient and graft survival within 12 months after kidney transplantation (Peng et al. 2013).

3.4.3 Chronic Inflammatory Autoimmune Diseases

MSCs have shown promise in exerting an anti-inflammatory immunomodulatory role in some autoimmune disease with little evidence of toxicity. They are effective for the treatment of autoimmune disease in various animal models, such as systemic lupus erythematosus (SLE), autoimmune enteropathy, autoimmune encephalomyelitis, autoimmune type 1 diabetes, and autoimmune rheumatic diseases (Parekkadan et al. 2008; Zappia et al. 2005; Fiorina et al. 2009; Sun et al. 2009). Clinical studies for refractory SLE patients using allogeneic MSCs demonstrated improvement in serological markers and renal function (Sun et al. 2009; Liang et al. 2010). Clinical trials with MSCs for diabetes and lupus nephritis are underway (http://clinicaltrials.gov/). Autologous BM-derived MSCs have been shown to be potently antiproliferative to stimulated T cells from both healthy donors and autoimmune patients (RA, systemic sclerosis, Sjoegren's, SLE) (Bocelli-Tyndall et al. 2007). Interestingly, it was reported that functional abnormalities existed in BM-derived MSCs from both patients with SLE and MRL/lpr mice (Sun et al. 2007), which suggests that abnormal MSCs may contribute to the development of SLE and allogeneic MSCs from healthy donors, may be superior to autologous ones in treating SLE. Recently, investigators also tried to treat Crohn's disease with MSCs. Crohn's disease is chronic inflammatory disorder of the gastrointestinal tract. Refractory patients do not respond to steroids, immunosuppressive agents (e.g., azathioprine, 6-mercaptopurine and methotrexate), or anti-TNF therapy and suffer from a poor quality of life. Duijvestein et al. reported that autologous BM-derived MSC therapy in patients with refractory Crohn's disease was promising. MSCs were infused intravenously at a target dose of 1-2 × 106 cells/kg body weight. In eight patients treated, Crohn's disease activity index scores improved in five patients, clinical response was seen in three patients at week 6 (Duijvestein et al. 2010). In another pilot study, 10 patients with fistulising Crohn's disease were treated with autologous BM-derived MSCs (Ciccocioppo et al. 2011). MSCs were injected both into the lumen and the wall of the fistula tracks. Twelve months afterward, the sustained complete closure (seven cases) or incomplete closure (three cases) of fistula tracks with a parallel reduction of Crohn's disease and perianal disease activity indexes ($p < 0.01$ for both), and rectal mucosal healing was achieved.

3.4.4 Conclusions and Future Directions

Ex vivo-generated MSCs might be useful in clinical situations in which engraftment failure is high, such as human leukocyte antigen-mismatched sibling, matched unrelated donor marrow, and umbilical cord blood transplantation, and may decrease GVHD and facilitate the engraftment and proliferation of hematopoietic progenitors. Reinfusion of MSCs aimed at exploiting immunoregulatory role might eventually be of relevance also in the setting of allografting with reduced conditioning regimens. The mechanism of its ability in immune treatment and its direct immunomodulatory therapeutic effect are not well understood and await further research. For example, although MSCs do not express MHC antigens at the time of in vitro culture, they are certain to express these antigens after they differentiate into committed cells in vivo. Why they still can stay in the host cannot be explained by their lacking of immunogenicity. As MSCs may be expanded as many as 40 generations and result in an increase of more than 10,000 fold in number while still maintain their multipotent mesenchymal lineages capability and phenotype, they are feasible for ex vivo implantation in clinical settings. Moreover, MSCs do not present alloantigen and do not require MHC expression to exert their inhibitory effect, suggesting that they can be derived from a donor irrespective of their MHC haplotype and be prepared as an "off-the-shelf" reagent for any patient.

References

Ahuja SS, Paliogianni F, Yarnada H, Balow JE, Boumpas DT. Effect of transforming growth factor-β on early and late activation events in human T cells. J Immunol. 1993;150:3109.

Anderson G, Jenkinson EJ, Moore NC, Owen JJ. MHC class II-positive epithelium and mesenchyme cells are both required for T-cell development in the thymus. Nature. 1993;362:70–3.

Asakura A, Komaki M, Rudnicki M. Muscle satellite cells are multipotential stem cells that exhibit myogenic, osteogenic, and adipogenic differentiation. Differentiation. 2001;68:245–53.

Augello A, Tasso R, Negrini SM, Amateis A, Indiveri F, Cancedda R, Pennesi G. Bone marrow mesenchymal progenitor cells inhibit lymphocyte proliferation by activation of the programmed death 1 pathway. Eur J Immunol. 2005;35:1482–90.

Azizi SA, Stokes D, Augelli BJ, DiGirolamo C, Prockop DJ. Engraftment and migration of human bone marrow stromal cells implanted in the brains of albino rats—similarities to astrocyte grafts. Proc Natl Acad Sci USA. 1998;95:3908–14.

BardaSaad M, Rozenszajn LA, Ashush H, Shav-Tal Y, Nun AB, Zipori D. Adhesion molecules involved in the interactions between early T cells and mesenchymal bone marrow stromal cells. Exp Hematol. 1999;27:834–44.

Bartholomew A, Sturgeon C, Siatskas M, Ferrer K, McIntosh K, Patil S, Hardy W, Devine S, Ucker D, Deans R, Moseley A, Hoffman R. Mesenchymal stem cells suppress lymphocyte proliferation in vitro and prolong skin graft survival in vivo. Exp Hematol. 2002;30:42–8.

Beyth S, Borovsky Z, Mevorach D, Liebergall M, Gazit Z, Aslan H, Galun E, Rachmilewitz J. Human mesenchymal stem cells alter antigen-presenting cell maturation and induce T-cell unresponsiveness. Blood. 2005;105:2214–9.

Blaser C, Kaufmann M, Müller C, Wells V, Mallucci L, Pircher H. Beta-galactoside-binding protein secreted by activated T cells inhibits antigen-induced proliferation of T cells. Eur J Immunol. 1998;28:2311–9.

Bocelli-Tyndall C, Bracci L, Spagnoli G, Braccini A, Bouchenaki M, Ceredig R, Pistoia V, Martin I, Tyndall A. Bone marrow mesenchymal stromal cells (BM-MSCs) from healthy donors and auto-immune disease patients reduce the proliferation of autologous and allogeneic-stimulated lymphocytes in vitro. Rheumatology. 2007;46:403–8.

Carrion F, Nova E, Ruiz C, Diaz F, Inostroza C, Rojo D, Mönckeberg G, Figueroa FE. Autologous mesenchymal stem cell treatment increased T regulatory cells with no effect on disease activity in two systemic lupus erythematosus patients. Lupus. 2010;19:317–22.

Casiraghi F, Azzollini N, Cassis P, Imberti B, Morigi M, Cugini D, Cavinato RA, Todeschini M, Solini S, Sonzogni A, Perico N, Remuzzi G, Noris M. Pretransplant infusion of mesenchymal stem cells prolongs the survival of a semiallogeneic heart transplant through the generation of regulatory T cells. J Immunol. 2008;181:3933–46.

Charlton B, Auchincloss H Jr, Fathman CG. Mechanisms of transplantation tolerance. Annu Rev Immunol. 1994;12:707–34.

Chen JL, Guo ZK, Xu C, Li YH, Hou CM, Mao N, Chen H. Mesenchymal stem cells suppress allogeneic T cell responses by secretion of TGF-beta1. Zhongguo Shi Yan Xue Ye Xue Za Zhi. 2002;10:285–8.

Ciccocioppo R, Bernardo ME, Sgarella A, Maccario R, Avanzini MA, Ubezio C, Minelli A, Alvisi C, Vanoli A, Calliada F, Dionigi P, Perotti C, Locatelli F, Corazza GR. Autologous bone marrow-derived mesenchymal stromal cells in the treatment of fistulising Crohn's disease. Gut. 2011;60:788–98.

Corcione A, Benvenuto F, Ferretti E, Giunti D, Cappiello V, Cazzanti F, Risso M, Gualandi F, Mancardi GL, Pistoia V, Uccelli A. Human mesenchymal stem cells modulate B-cell functions. Blood. 2006;107:367–72.

Deans RJ. Mesenchymal stem cells: cell and gene therapy applications. Eur Cytokine Networks. 2000;11:323–4.

Deans RJ, Moseley AB. Mesenchymal stem cells: biology and potential clinical uses. Exp Hematol. 2000;28:875–84.

Deeg HJ. How I treat refractory acute GVHD. Blood. 2007;109:4119–26.

Deng W, Han Q, Liao L. Long-term Distribution of Adult Stem Cells in Different Tissues after Transplantation. Stem Cell and Cellular Therapy. 2003;1:62–5.

Deng W, Han Q, Liao L, You S, Deng H, Zhao RC. Effects of allogeneic bone marrow-derived mesenchymal stem cells on T and B lymphocytes from BXSB mice. DNA Cell Biol. 2005;24:458–63.

Devine SM, Cobbs C, Jennings M, Bartholomew A, Hoffman R. Mesenchymal stem cells distribute to a wide range of tissues following systemic infusion into nonhuman primates. Blood. 2003;101:2999–3001.

Di Nicola M, Carlo-Stella C, Magni M, Milanesi Marco, Longoni PD, Matteucci P. Human bone marrow stromal cells suppress T-lymphocyte proliferation induced by cellular or nonspecific mitogenic stimuli. Blood. 2002;99:3838–43.

Ding Y, Xu D, Feng G, Bushell A, Muschel RJ, Wood KJ. Mesenchymal stem cells prevent the rejection of fully allogenic islet grafts by the immunosuppressive activity of matrix metalloproteinase-2 and -9. Diabetes. 2009;58:1797–806.

Duijvestein M, Vos AC, Roelofs H, Wildenberg ME, Wendrich BB, Verspaget HW, Kooy-Winkelaar EM, Koning F, Zwaginga JJ, Fidder HH, Verhaar AP, Fibbe WE, van den Brink GR, Hommes DW. Autologous bone marrow-derived mesenchymal stromal cell treatment for refractory luminal Crohn's disease: results of a phase I study. Gut. 2010;59:1662–9.

Erices A, Conget P, Minguell JJ. Mesenchymal progenitor cells in human umbilical cord blood. Br J Haematol. 2000;109:235–42.

Fandrich F, Lin X, Chai GX, Schulze M, Ganten D, Bader M, Holle J, Huang DS, Parwaresch R, Zavazava N, Binas B. Preimplantation-stage stem cells induce long-term allogeneic graft acceptance without supplementary host conditioning. Nat Med. 2002;8:171–8.

Fang B, Song YP, Liao LM, Han Q, Zhao RC. Treatment of severe therapy-resistant acute graft-versus-host disease with human adipose tissue-derived mesenchymal stem cells. Bone Marrow Transplant. 2006;38:389–90.

Fang B, Song Y, Liao L, Zhang Y, Zhao RC. Favorable response to human adipose tissue-derived mesenchymal stem cells in steroid-refractory acute graft-versus-host disease. Transplant Proc. 2007;39:3358–62.

Fiorina P, Jurewicz M, Augello A, Vergani A, Dada S, La Rosa S, Selig M, Godwin J, Law K, Placidi C, Smith RN, Capella C, Rodig S, Adra CN, Atkinson M, Sayegh MH, Abdi R. Immunomodulatory function of bone marrow-derived mesenchymal stem cells in experimental autoimmune type 1 diabetes. J Immunol. 2009;183:993–1004.

Fox FE, Ford HC, Douglas R, Cherian S, Nowell PC. Evidence that TGF-β can inhibit T-lymphocyte proliferation through paracrine and autocrine mechanisms. Cell Immunol. 1993;150:45–58.

Frassoni FLM, Bacigalupo A, Gluckman E. Expanded mesenchymal stem cells (MSC), coinfused with HLA identical hemopoietic stem cell transplants, reduce acute and chronic graft versus host disease: a matched pair analysis. Bone Marrow Transplant. 2002;29:75.

Fridenstein A. Stromal bone marrow cells and the hematopoietic microenvironment. Arkh Patol. 1982;44:3–11.

Gieseke F, Böhringer J, Bussolari R, Dominici M, Handgretinger R, Müller I. Human multipotent mesenchymal stromal cells use galectin-1 to inhibit immune effector cells. Blood. 2010;116:3770–9.

Glennie S, Soeiro I, Dyson PJ, Lam EW, Dazzi F. Bone marrow mesenchymal stem cells induce division arrest anergy of activated T cells. Blood. 2005;105:2821–7.

Han Q, Deng W, Li C, Zhao RC. Allogeneic adult stem cells establish long-term residence in recipient tissues and facilitate skin transplantation. Exp Hematol. 2003;31:158.

He J, Zhang Y, Jiang XX, Liu G, Liu YL, Li HL, Mao N. Effect of human bone marrow mesenchymal stem cell on cord blood T lymphocyte transformation. Zhongguo Shi Yan Xue Ye Xue Za Zhi. 2003;11:11–4.

Helg C, Chapuis B, Bolle JF, Morel P, Salomon D, Roux E, Antonioli V, Jeannet M, Leski M. Renal transplantation without immunosuppression in a host with tolerance induced by allogeneic bone marrow transplantation. Transplantation. 1994;58:1420–2.

Hu Y, Zhang LY, Ma GJ, Jiang XY, Zhao CH. Phenotypical and biological characteristics of human fetal marrow and liver mesenchymal stem cells. Zhongguo Shi Yan Xue Ye Xue Za Zhi. 2001;9:289–93.

Hu Y, Liao L, Wang Q, Ma L, Ma G, Jiang X, Zhao RC. Isolation and identification of mesenchymal stem cells from human fetal pancreas. J Lab Clin Med. 2003;141:342–9.

Ishida T, Inaba M, Hisha H, Sugiura K, Adachi Y, Nagata N, Ogawa R, Good RA, Ikehara S. Requirement of donor-derived stromal cells in the bone marrow for successful allogeneic bone marrow transplantation. Complete prevention of recurrence of autoimmune diseases in MRL/MP-Ipr/Ipr mice by transplantation of bone marrow plus bones (stromal cells) from the same donor. J Immunol. 1994;152:3119–27.

Jiang XX, Zhang Y, Liu B, Zhang SX, Wu Y, Yu XD, Mao N. Human mesenchymal stem cells inhibit differentiation and function of monocyte-derived dendritic cells. Blood. 2005;105:4120–6.

Krampera M, Glennie S, Dyson J, Scott D, Laylor R, Simpson E, Dazzi F. Bone marrow mesenchymal stem cells inhibit the response of naive and memory antigen-specific T cells to their cognate peptide. Blood. 2003;101:3722–9.

Krampera M, Cosmi L, Angeli R, Pasini A, Liotta F, Andreini A, Santarlasci V, Mazzinghi B, Pizzolo G, Vinante F, Romagnani P, Maggi E, Romagnani S, Annunziato F. Role for interferon-gamma in the immunomodulatory activity of human bone marrow mesenchymal stem cells. Stem Cells. 2006;24:386–98.

Lazarus H, Curtin P, Devine S. Role of mesenchymal stem cells in allogeneic transplantation: early phase 1 clincial results. Blood. 2000;96:392.

Le Blanc K, Tammik L, Sundberg B, Haynesworth SE, Ringden O. Mesenchymal stem cells inhibit and stimulate mixed lymphocyte cultures and mitogenic responses independently of the major histocompatibility complex. Scand J Immunol. 2003;57:11–20.

Le Blanc K, Frassoni F, Ball L, Locatelli F, Roelofs H, Lewis I, Lanino E, Sundberg B, Bernardo ME, Remberger M, Dini G, Egeler RM, Bacigalupo A, Fibbe W, Ringdén O. Developmental Committee of the European Group for Blood and Marrow Transplantation. Mesenchymal stem cells for treatment of steroid-resistant, severe, acute graft-versus-host disease: a phase II study. Lancet. 2008;371:1579–86.

Li Y, Hisha H, Inaba M, Lian Z, Yu C, Kawamura M, Yamamoto Y, Nishio N, Toki J, Fan H, Ikehara S. Evidence for migration of donor bone marrow stromal cells into recipient thymus after bone marrow transplantation plus bone grafts: a role of stromal cells in positive selection. Exp Hematol. 2000;28:950–60.

Li XC, Strom TB, Turka LA, Wells AD. T cell death and transplantation tolerance. Immunity. 2001;14:407–16.

Liang J, Zhang H, Hua B, Wang H, Lu L, Shi S, Hou Y, Zeng X, Gilkeson GS, Sun L. Allogenic mesenchymal stem cells transplantation in refractory systemic lupus erythematosus: a pilot clinical study. Ann Rheum Dis. 2010;69:1423–9.

Liechty KW, MacKenzie TC, Shaaban AF, Radu A, Moseley AM, Deans R, Marshak DR, Flake AW. Human mesenchymal stem cells engraft and demonstrate site-specific differentiation after in utero transplantation in sheep. Nat Med. 2000;6:1282–6.

Maccario R, Podesta M, Moretta A, Cometa A, Comoli P, Montagna D, Daudt L, Ibatici A, Piaggio G, Pozzi S, Frassoni F, Locatelli F. Interaction of human mesenchymal stem cells with cells involved in alloantigen-specific immune response favors the differentiation of CD4+ T-cell subsets expressing a regulatory/suppressive phenotype. Haematologica. 2005;90:516–25.

Meisel R, Zibert A, Laryea M, Göbel U, Däubener W, Dilloo D. Human bone marrow stromal cells inhibit allogeneic T-cell responses by indoleamine 2,3-dioxygenase-mediated tryptophan degradation. Blood. 2004;103:4619–21.

Mossalayi MD, Mentz F, Ouaaz F, Dalloul AH, Blanc C, Debre P, Ruscetti FW. Early human thymocyte proliferation is regulated by an externally controlled autocrine transforming growth factor-β1 mechanism. Blood. 1995;85:3594–601.

Nauta AJ, Kruisselbrink AB, Lurvink E, Willemze R, Fibbe WE. Mesenchymal stem cells inhibit generation and function of both CD34+-derived and monocyte-derived dendritic cells. J Immunol. 2006;177:2080–7.

Nickerson P, Steurer W, Steiger J, Zheng X, Steele AW, Strom TB. Cytokines and the Th1/Th2 paradigm in transplantation. Curr Opin Immunol. 1994;6:757–64.

Noort WA, Kruisselbrink AB, in't Anker PS, Kruger M, van Bezooijen RL, de Paus RA, Heemskerk MH, Löwik CW, Falkenburg JH, Willemze R, Fibbe WE. Mesenchymal stem cells promote engraftment of human umbilical cord blood-derived CD34+ cells in NOD/SCID mice. Exp Hematol. 2002;30:870–8.

Parekkadan B, Tilles AW, Yarmush ML. Bone marrow-derived mesenchymal stem cells ameliorate autoimmune enteropathy independently of regulatory T cells. Stem Cells. 2008;26:1913–9.

Peng Y, Ke M, Xu L, Liu L, Chen X, Xia W, Li X, Chen Z, Ma J, Liao D, Li G, Fang J, Pan G, Xiang AP. donor-derived mesenchymal stem cells combined with low-dose tacrolimus prevent acute rejection after renal transplantation: a clinical pilot study. Transplantation. 2013;95:161–8.

Perillo NL, Pace KE, Seilhamer JJ, Baum LG. Apoptosis of T cells mediated by galectin-1. Nature. 1995;378:736–9.

Pittenger MF, Mackay AM, Beck SC, Jaiswal RK, Douglas R, Mosca JD, Moorman MA, Simonetti DW, Craig S, Marshak DR. Multi-lineage potential of adult human mesenchymal stem cells. Science. 1999;284:143–7.

Rabinovich GA, Ramhorst RE, Rubinstein N, Corigliano A, Daroqui MC, Kier-Joffé EB, Fainboim L. Induction of allogenic T-cell hyporesponsiveness by galectin-1-mediated apoptotic and non-apoptotic mechanisms. Cell Death Differ. 2002;9:661–70.

Ringdén O, Uzunel M, Rasmusson I, Remberger M, Sundberg B, Lönnies H, Marschall HU, Dlugosz A, Szakos A, Hassan Z, Omazic B, Aschan J, Barkholt L, Le Blanc K. Mesenchymal stem cells for treatment of therapy-resistant graft-versus-host disease. Transplantation. 2006;81:1390–7.

Ruegemer JJ, Ho SN, Augustine JA, Schlager JW, Bell MP, McKean DJ, Abraham RT. Regulatory effects of transforming growth factor-β on IL-2- and IL-4-dependent T cell cycle progression. J Immunol. 1990;144:1769–76.

Saito T, Kuang JQ, Bittira B, Al-Khaldi A, Chiu RC. Xenotransplant cardiac chimera: immune tolerance of adult stem cells. Ann Thorac Surg. 2002;74:19–24.

Sayegh MH, Turka LA. The role of T-cell costimulatory acti-vation pathways in transplant rejection. N Engl J Med. 1998;338:1813–21.

Sayegh MH, Fine NA, Smith JL, Rennke HG, Milford EL, Tilney NL. Immunologic tolerance to renal allografts after bone marrow transplants from the same donors. Ann Intern Med. 1991;114:954–5.

Selmani Z, Naji A, Zidi I, Favier B, Gaiffe E, Obert L, Borg C, Saas P, Tiberghien P, Rouas-Freiss N, Carosella ED, Deschaseaux F. Human leukocyte antigen-G5 secretion by human mesenchymal stem cells is required to suppress T lymphocyte and natural killer function and to induce CD4+CD25highFOXP3+ regulatory T cells. Stem Cells. 2008;26:212–22.

Silverstein AM, Prendergast RA, Kraner KL. Fetal response to antigenic stimulus IV rejection of skin homografts by the fetal lamb. J Exp Med. 1964;119:955–64.

Simons PG, Torok-Storb B. Identification pf stromal cell precursors in human bone marrow by a novel monoclonal antibody STRO-1. Blood. 1991;78:55–62.

Sorof JM, Koerper MA, Portale AA, Potter D, DeSantes K, Cowan M. Renal transplantation without chronic immunosuppression after T cell-depleted, HLA-mismatched bone marrow transplantation. Transplantation. 1995;59:1633–5.

Sotiropoulou PA, Perez SA, Gritzapis AD, Baxevanis CN, Papamichail M. Interactions between human mesenchymal stem cells and natural killer cells. Stem Cells. 2006;24:74–85.

Spaggiari GM, Capobianco A, Abdelrazik H, Becchetti F, Mingari MC, Moretta L. Mesenchymal stem cells inhibit natural killer-cell proliferation, cytotoxicity, and cytokine production: role of indoleamine 2,3-dioxygenase and prostaglandin E2. Blood. 2008;111:1327–33.

Sun LY, Zhang HY, Feng XB, Hou YY, Lu LW, Fan LM. Abnormality of bone marrow-derived mesenchymal stem cells in patients with systemic lupus erythematosus. Lupus. 2007;16:121–8.

Sun L, Akiyama K, Zhang H, Yamaza T, Hou Y, Zhao S, Xu T, Le A, Shi S. Mesenchymal stem cell transplantation reverses multiorgan dysfunction in systemic lupus erythematosus mice and humans. Stem Cells. 2009;27:1421–32.

Tan J, Wu W, Xu X, Liao L, Zheng F, Messinger S, Sun X, Chen J, Yang S, Cai J, Gao X, Pileggi A, Ricordi C. Induction therapy with autologous mesenchymal stem cells in living-related kidney transplants: a randomized controlled trial. JAMA. 2012;307:1169–77.

Tasso R, Ilengo C, Quarto R, Cancedda R, Caspi RR, Pennesi G. Mesenchymal stem cells induce functionally active T-regulatory lymphocytes in a paracrine fashion and ameliorate experimental autoimmune uveitis. Invest Ophthalmol Vis Sci. 2012;53:786–93.

Toma C, Pittenger MF, Cahill KS, Byrne BJ, Kessler PD. Human mesenchymal stem cells differentiate to a cardiomyocyte phenotype in the adult murine heart. Circulation. 2002;105:93–8.

Tse WT, Pendleton JD, Beyer WM, Egalka MC, Guinan EC. Suppression of allogeneic T-cell proliferation by human marrow stromal cells: implications in transplantation. Transplantation. 2003;75:389–97.

Van Parijs L, Abbas AK. Homeostasis and self-tolerance in the immune system: turning lymphocytes off. Science. 1998;280:243–8.

von Boehmer H, Kisielow P. Self-nonself discrimination by T cells. Science. 1990;248:1369–73.

Wekerle T, Sykes M. Mixed chimerism and transplantation tolerance. Annu Rev Med. 2001;52:353–70.

Weng JY, Du X, Geng SX, Peng YW, Wang Z, Lu ZS, Wu SJ, Luo CW, Guo R, Ling W, Deng CX, Liao PJ, Xiang AP. Mesenchymal stem cell as salvage treatment for refractory chronic GVHD. Bone Marrow Transplant. 2010;45:1732–40.

Yang RY, Rabinovich GA, Liu FT. Galectins: structure, function and therapeutic potential. Expert Rev Mol Med. 2008;10:e17.

Young HE, Steele TA, Bray RA, Hudson J, Floyd JA, Hawkins K, Thomas K, Austin T, Edwards C, Cuzzourt J, Duenzl M, Lucas PA, Black AC Jr. Human reserve pluripotent mesenchymal stem cells are present in the connective tissues of skeletal muscle and dermis derived from fetal, adult, and geriatric donors. Anat Rec. 2001;264:51–62.

Zanjani E, Almeida-Porada G, Ascensao J, Mackintosh F, Flake A. Transplantation of hematopoietic stem cells in utero. Stem cell. 1997;15:79–93.

Zappia E, Casazza S, Pedemonte E, Benvenuto F, Bonanni I, Gerdoni E, Giunti D, Ceravolo A, Cazzanti F, Frassoni F, Mancardi G, Uccelli A. Mesenchymal stem cells ameliorate experimental autoimmune encephalomyelitis inducing T-cell anergy. Blood. 2005;106:1755–61.

Zhang W, Ge W, Li C, You S, Liao L, Han Q, Deng W, Zhao RC. Effects of mesenchymal stem cells on differentiation, maturation, and function of human monocyte-derived dendritic cells. Stem Cells Dev. 2004;13:263–71.

Zuk PA, Zhu M, Mizuno H, Huang J, Futrell JW, Katz AJ, Benhaim P, Lorenz HP, Hedrick MH. Multilineage cells from human adipose tissue: implications for cell-based therapies. Tissue Eng. 2001;7:211–28.

Chapter 4
Induced Pluripotent Stem Cell, a Rising Star in Regenerative Medicine

Yihuan Mao, Yukai Wang, Libin Wang and Qi Zhou

Abstract The generation of induced pluripotent stem cells (iPSCs) is a milestone in biomedical researches. With the iPS technology, mouse somatic cells could be reprogrammed into a pluripotent state similar to that of embryonic stem cells (ESCs) by overexpression of only several defined transcription factors. This technology also enables the successful generation of human iPSCs from various human somatic cells and therefore avoids the ethical conflicts of ESCs that are derived from human embryos. The iPSCs generated from somatic cells of patients are generally considered to be free of immune rejection in autogenous transplantation therapy, thus they hold great therapeutic potential for producing personalized regenerative medicine. Besides, the autologous iPSCs from patients and the subsequent differentiation can be directly used for the disease modeling and drug screening in dishes. Though holding invaluable therapeutic potential, the iPS technology also faces some challenges such as immunogenicity, oncogenicity, in vitro differentiation propensities, and genome or epigenome integrities, which must be carefully evaluated before clinical translation. In this chapter, we briefly summarize the history and progress of the iPSC researches and the application of iPSCs in regenerative medicine and disease modeling.

Keywords Stem cell · Induced pluripotent stem cell · Regenerative medicine

4.1 Introduction

ESCs are cells derived from the inner cell mass (ICM) in blastocyst, which can maintain self-renewal in certain condition in vitro and have the pluripotency to differentiate into cells of all the three germ layers (Yamanaka 2012). The mouse

---

Y. Mao · Y. Wang · L. Wang · Q. Zhou (✉)
State Key Laboratory of Reproductive Biology, Institute of Zoology,
Chinese Academy of Sciences, 1 Beichen West Road,
Chaoyang District, Beijing 100101, People's Republic of China
e-mail: qzhou@ioz.ac.cn

R.C. Zhao (ed.), *Stem Cells: Basics and Clinical Translation*, Translational
Medicine Research 1, DOI 10.1007/978-94-017-7273-0_4

embryonic stem cells (ESCs) were generated by Evans MJ and Martin GR in 1981 (Evans and Kaufman 1981; Martin 1981).

Cell reprogramming refers to the conversion of the somatic cell type to pluripotent cell type, the study of which originated from a theory proposed by Spemann that a differentiated nucleus could still be totipotent. In his book, the concept of nuclear transfer (NT) was first proposed (Speman 1938). In 1952, Briggs and King accomplished NT and first produced normal swimming tadpoles of *Rana pipiens* by transplanting the nuclei of blastula embryos into enucleated eggs of frogs (Briggs and King 1952). However, they found that transplantation of the nuclei of gastrula embryos which were only slightly older than blastula ones failed to produce normal embryos and concluded that there probably existed irreversible process inside the nuclei during cellular differentiation (Briggs and King 1957). However, in 1962, the British scientist John B. Gurdon successfully got cloned frog from fully differentiated cells, demonstrating that the somatic nuclei still held the developmental potency as embryonic nuclei. The first successful somatic cell nuclear transfer (SCNT) in the world initiated the reprogramming research, which was later explained by the discovery that the differentiation process only involved changes in the nuclear gene expression rather than the genome integrity. Another milestone of SCNT was the birth of Dolly in 1996 (Campbell et al. 1996), which was the first clone mammal obtained by transferring the nuclei of cultured mammary gland cells from an adult sheep to enucleated eggs from another sheep by I. Wilmut et al. This event first confirmed the pluripotency of mammalian nuclei. Then, in 2007, Byrne JA and his colleagues derived rhesus macaque embryonic stem cell lines from adult skin fibroblasts by SCNT (Byrne et al. 2007). All of the above events have given people the hope that one day we could clone ourselves; however, this idea could not be carried out for ethical issues. Many efforts have failed in generating human ntES cells. Eventually, in 2013, the group of Shoukhrat Mitalipov succeeded in producing human ntES cell lines by optimizing SCNT protocols. The efficiency of obtaining specific differentiated types of cells is important when it comes to therapeutic application of SCNT.

Besides SCNT, many other kinds of reprogramming procedures emerged, including cell fusion (Harris 1967; Pomerantz and Blau 2004), induction of pluripotency by defined transcription factors, and direct reprogramming (Weintraub et al. 1989; Xie et al. 2004; Orkin and Zon 2008). Cell fusion is another way to reprogram cells. During this process, two somatic cells are fused to be a multinuclear cell or heterokaryons, in which the two nuclei are separated by a cell division inhibitor. In the heterokaryons, the more dominant nuclear can reprogram the other nuclear by imposing its own pattern of gene expression on the other one. As early as 1967, Harris H. introduced a mature hen erythrocyte into a Hela cell and this resulted in nuclear reactivation (Harris 1967). Several years later, Pomerantz and Blau (2004) fused a human liver cell with a multinucleated muscle cell. Besides, some scientists fuse various somatic cells with embryonic carcinoma cells (Solter 2006), embryonic germ cells (EG) (Tada et al. 1997, 2003), or ES cells (Tada et al. 2001, 2003; Cowan et al. 2005; Yu et al. 2006) and these resulted in reprogramming the somatic cell nuclei to a pluripotent state.

Here, we refer to direct reprogramming. As early as 1989, Weintraub H et al. turned pigment, nerve, fat, liver, and fibroblast cells into muscular cells by forced expression of the *MyoD* gene, a master regulatory gene for myogenesis (Weintraub et al. 1989). Xie et al. (2004) succeeded in switching B cells to macrophages by forced expression of C/EBP α and C/EBP β. In 2006, Shinya Yamanaka made a groundbreaking discovery that won him the Nobel Prize 6 years later. He turned mouse embryonic fibroblasts (MEF) into stem cells named induced pluripotent stem cells (iPSC) by the overexpression of 4 transcription factors, *Oct4*, *Sox2*, *Klf4*, and *c-Myc* (Takahashi and Yamanaka 2006). To detect transcriptional factors that can reprogram somatic cells into pluripotent cells like ESCs, Yamanaka (Tokuzawa et al. 2003) started a screen for factors within a pool of 24 pluripotency-associated embryonic factors. When introduced all these 24 factors, MEF turned out to be ESCs-like. And then, after rounds of elimination of individual factors, finally Yamanaka got iPSCs from mouse fibroblasts by overexpression of only four defined transcription factors (Takahashi and Yamanaka 2006). Their pluripotency of iPSCs has been confirmed by chimeric experiment (Okita et al. 2007) and tetraploid complementation experiment (Zhao et al. 2009; Boland et al. 2009). Subsequently, iPSCs have also been generated from rat (Buehr et al. 2008), monkey (Liu et al. 2008), pig (Esteban et al. 2009), dog (Koh et al. 2010), rabbit (Honda et al. 2010), and human (Takahashi et al. 2007; Yu et al. 2007) by similar approaches.

The iPS technology has great therapeutic potential as a regenerative source of patient-specific cells. For one thing, we can use iPSCs to achieve custom-tailored cell therapy, in which defective cells are being replaced by normal cells derived from other types of cells, and finally derive replacement brain, heart, pancreas, or other types of cells from more accessible organs such as the skin and urine of the patient while avoiding immunological rejection. For another, we can use these cells to build disease models for disease research and drug screening. In 2012, John B. Gurdon and Shinya Yamanaka shared the Nobel Prize for physiology or medicine for their discovery that mature cells can be reprogrammed to stem cells.

Many researchers have carried out the preclinical studies of iPS technology and overcome various obstacles. First is the integration of iPSCs. Human iPSCs were generated by viral vectors such as lentivirus and retrovirus, which led to the integration of the exotic transcript factors into the genome and therefore may cause tumors and genome instability. Then, several non-integration techniques have emerged to solve this problem, such as using plasmids or adenovirus vectors, and using proteins or mRNAs encoded by the reprogramming factors, but they turned out to be inefficient. Second is immunogenicity of iPSCs. As was reported by Zhao et al., some of iPSC-derived differentiated cells can give rise to T cell-dependent immune response in allogeneic and even syngeneic recipients. Third is the insufficient reprogramming of iPSCs as was reported by several studies of gene expression, epigenetic modification, and differentiation. This part will be discussed in the section below.

Therefore, before clinical translation, the genome or epigenome integrities, immunogenicity, oncogenicity, and in vitro differentiation propensities of iPSCs must be evaluated objectively. And a specific safety assessment criteria system must be proposed for this new style of therapy. Below, we will review the characteristics of iPSCs, the development of iPS technology and theory, and discuss recent advances in the potential therapeutic application of iPSCs in mammalian systems including mouse, rat, monkey, and human.

4.2 Characteristics of iPSCs

iPSCs show morphology and proliferation characteristics similar to those of ESCs in ES medium. They can permanently proliferate, self-renew, and have potential to differentiate into cells of all the three germ layers. The properties of mouse iPSCs are shown below.

- Morphology
  iPSCs have morphology indistinguishable from ESCs. The iPS clones are domed and have large nucleoli and scant cytoplasm. All cells inside the clone come from one single cell. The surfaces of the clones are soft, and it is hard to distinguish one single cell from another inside the clone.
- Permanently proliferation and self-renewal
- iPSCs can maintain a pluripotent and self-renewal state by LIF/STAT3-dependent signaling, which is similar to ESCs. iPSCs can be cultured in ES mediums such as serum-free medium of ES, N_2B_{27} medium with the addition of leukemia inhibitory factor (LIF), and two inhibitors (2i): PD0325901 (MEK inhibitor) and CHIR99021 (GSK3β inhibitor).
- Expression of pluripotent markers
  iPSCs express ESC markers, including *Oct4, Nanog, Sox2, Fgf4, Dppa3, and SSEA-1*. Besides, they are positive for alkaline phosphatase and have high telomerase activity. The epigenetic characteristics of iPSCs are shown below. The promoters of *Oct4* and *Nanog* show increased acetylation of histone H3 and decreased dimethylation of lysine 9 of histone H3, which means that the *Oct4* and *Nanog* are both activated.
- In vitro and in vivo differentiate into all three germ layers
  iPSCs have pluripotency to differentiate into cells of all the three germ layers both in vitro and in vivo. In the teratoma formation experiment, iPSCs can generate teratomas that contain all three germ layers including neural tissues, cartilage, and columnar epithelium when subcutaneously injected into nude mice. Then, in the in vitro differentiation experiment, firstly the iPS clones become embryoid bodies (EBs) in non-coated plastic dishes, and then, the EBs are grown in tissue culture dishes and they attach to the bottom and differentiate. Several days later, markers of all three germ layers can be detected.
- Contribution to germ line in chimera organisms

A more convincing experiment to test pluripotency of stem cells is chimera formation. Being equivalent to mouse ESC, iPSC microinjected into blastocysts has the ability to contribute to germ line of chimeric mice (Okita et al. 2007).

- Tetraploid complementation
  The most stringent assay to test pluripotency is the tetraploid complementation experiment. In this assay, 4N blastocysts are obtained by electric fusion of 2-cell embryo and can only develop into placentas. Then, the iPSCs are injected into the tetraploid blastocysts. Therefore, any viable mouse obtained totally come from the injected iPSCs. In 2009, group of Qi Zhou and Kristen Baldwin conducted the tetraploid complementation experiment of mouse iPSCs and succeeded in obtaining viable mice (Zhao et al. 2009; Boland et al. 2009). The result confirmed the sincere pluripotency of mouse iPSCs for the first time in the world and provided a solid theoretical foundation for the application of iPS technology.
  And as previously reported by Liu et al. in 2010, the activation state of the imprinted *Dlk1-Dio3* region correlates with pluripotent levels of mouse ESCs and iPSCs (Liu et al. 2010). This region starts with Dlk1 and ends with Dio3, which is activated in fully pluripotent mouse ESCs or iPSCs while repressed in partially pluripotent ones.

Equivalently, the iPS reprogramming process was shown to work in human cells, and human iPSCs were derived from adult human dermal fibroblasts (HDF) in 2007 using human OSKM transcription factors (Takahashi et al. 2007). Human iPSCs are similar to human ESCs in morphology, proliferation, pluripotent markers, epigenetic status of pluripotent cell-specific genes, gene expression, and telomerase activity. Besides, they can differentiate into all three germ layers in vitro and in teratomas. Because of the ethical issues, chimera formation and tetraploid compensation assays cannot be conducted.

However, as reported, human iPSCs were not identical to human ES cells. DNA microarray analyses have detected differences between them. Further studies are essential to answer the question whether human iPSCs represent is an alternative source in medical applications.

4.3 Development of iPS Technology and Theory

In 2006, Takahashi and Yamanaka got iPSCs from mouse fibroblasts by overexpression of four defined transcription factors (*Oct4, Klf4, Sox2, c-Myc*) (Takahashi and Yamanaka 2006). Subsequently, iPSCs have been derived from other animals and humans (Takahashi et al. 2007; Yu et al. 2007) by similar approaches. Particularly, the human iPSCs can be generated by introducing OSKM (*Oct4, Klf4, Sox2, c-Myc*) (Takahashi et al. 2007) or *Oct4*, *Sox2*, *Nanog*, and *Lin28* into somatic cells (Yu et al. 2007).

However, there are lots of challenges before the clinical translation of iPSCs. Yamanaka et al. reported that the mouse iPSCs were not safe enough, and about 20 % of chimeric mice derived from them developed tumors. The same question also existed in human iPSCs. In October 2007, Yamanaka et al. reported that they generated human iPSCs from adult HDF and other somatic cells using the same approach as in mice. They found that each iPS clone contained 3–6 retroviral integration sites for each factor. That is to say, each clone had more than 20 retroviral integration sites in total, which may increase the risk of tumorigenesis. The tumors may be caused by, at least in part, the reactivation of the oncogene *c-Myc* retrovirus. This problem and the low reprogramming efficiency limit the application of iPSCs in regenerative medicine. Hence, researchers have been modifying reprogramming protocols or creating other methods to avoid integrations into host genome and enhance the induction efficiency. Reprogramming methods have been rapidly improving over the last few years. Now, there are several ways to deliver or express the reprogramming factors, including non-integration virus vectors, dox-inducible expression systems, protein, mRNA, miRNA, and molecule compounds that minimize or eliminate the risk of insertion into the host cell genome. Factor delivery methods for iPSC derivation are shown below.

Exactly the same day, Yamanaka et al. removed the oncogene *c-Myc* and finally got iPSCs from mouse and human fibroblasts (Takahashi et al. 2007). In this modified protocol, non-iPS background cells were fewer, and the iPSCs were consistently of high quality. Importantly, mice derived from these iPSCs developed no tumor during the research. Besides, they could efficiently isolate iPS cells without drug selection. Furthermore, by this approach, they generated human iPSCs. However, the efficiency of OSK is lower than that of OSKM both in mice and in human, and in human it was less than 0.001 %. Wernig et al. (2008) further demonstrated that *c-Myc* was not indispensable during the induction of pluripotency from mouse fibroblasts. Their data showed that the three-factor-induced iPSCs developed fewer tumors than the four-factor ones. They also found that the reprogramming process was largely delayed in the absence of *c-Myc* and the efficiency was similarly low as it was in human. The presence of *c-Myc* can promote the establishment of three endogenous pluripotency genes *Oct4, Nanog,* and *Sox2*. As reported in 2007, Marius Wernig et al. suggested that exogenous *c-Myc* increased the fraction of reprogrammed cells and quickened the overall process probably by accelerating stochastic events leading to the generation of iPSCs. One major function of *c-Myc* is enhancing proliferation; hence, other non-oncogenic reprogramming factors possessing similar function or growth factors maybe can replace *c-Myc*.

Researchers also have been sparing no effort in enhancing efficiency. On July 2008, Danwei Huangfu et al. discovered small-molecule compounds to improve reprogramming efficiency during the induction of pluripotent stem cells from mouse fibroblasts (Huangfu et al. 2008). They used several DNA methyltransferase and histone deacetylase (HDAC) inhibitors which could improve the reprogramming efficiency of SCNT by 2–5-fold. Particularly, valproic acid (VPA), an HDAC inhibitor, improved the efficiency by at least 100-fold when using Oct4-GFP reporter, and was up to 11.8 ± 2.2 %. It also allowed reprogramming without c-Myc and increased the efficiency by 50-fold. Then, soon after, this group generated between 1 and 5 human iPS cell lines out of every 100,000 BJ or NHDF cells by only *Oct4*, *Sox2* and VPA, the efficiency of which was similar to the previous result of human iPSCs transfected by three factors (*Oct4*, *Sox2* and *Klf4*). In the same year, Yan Shi et al. set out a phenotypic screening for compounds from a library of about 2000 known small molecules and finally generated mouse iPSCs by two small molecules (BIX-01294 and BayK8644) and the transfection of *Oct4* and *Klf4* (Shi et al. 2008). The BIX-01294 (BIX) is a G9a histone methyltransferase (G9a HMTase) inhibitor, and the BayK8644 is an L-channel calcium agonist. The former causes epigenetic modifications directly while the latter indirectly. In 2010, Xu Yuan et al. reported a new molecule, a protein arginine methyltransferase inhibitor AMI-5, combining with a transforming growth factor (TGF)-β inhibitor A-83-01 and the transfection of *Oct4* succeeded in getting mouse iPSCs (Esteban et al. 2010). And such iPSCs produced viable pups through tetraploid complementation assays. In 2011, Jai-Hee Moon et al. showed that *Bmi1*, which leads MEFs to transdifferentiate into NSC-like cells, combining with *Oct4* can reprogram MEFs into iPSCs (Moon et al. 2011).

These discoveries demonstrated that the epigenetic modification may be a significant process during the generation of iPSCs and encouraged researchers to screen small molecules to increase the reprogramming efficiency and even replace one to several reprogramming factors. Surprisingly, Deng's group used only 7 small-molecule compounds and finally generated mouse iPSCs, the efficiency of which was up to 0.2 %. They used Forskolin, 2-methyl-5-hydroxytryptamine, and D4476 as a replacement for *Oct4* and another combination called "VC6T" [VPA, CHIR99021, 616452, tranylcypromine] to finish the induction process. The effect of small molecules was reversible and could be turned by concentrations change. Moreover, they are cell permeable, non-immunogenic, and more cost-effective; and easily synthesized, preserved, and standardized.

In 2008, group of Stadtfeld (2008) and group of Shinya Yamanaka (Okita et al. 2008) generated non-integrating mouse iPSCs. Matthias Stadtfeld introduced non-integrating adenoviruses which transiently expressed *OSKM* into mouse fibroblasts or liver cells and obtained iPSCs. The replication-incompetent adenovirus vectors consisted of the cDNA of *OSKM* and the hCMV IE (human cytomegalovirus immediate early) promoter. The same day, Shinya Yamanaka et al. introduced two expression plasmid vectors repeatedly into mouse fibroblasts and also got non-integrating iPSCs. They constructed two plasmids: One is the pCX-OKS-2A plasmid containing the cDNA of OKS (in order), CAG constitutively active promoter (Niwa et al. 1991), and 2A self-cleaving peptide from the foot-and-mouth disease virus (Hasegawa et al. 2007; Hsiao et al. 2008); the other is the pCX-M plasmid containing the cDNA of c-Myc and CAG constitutively active promoter. PCR assay and Southern blot analysis showed no integration into host genome in iPSCs from both two groups. However, the induction efficiency of transient expression was relatively low. The major problem of the two transient expression systems above, however, was lower induction efficiency caused by the reduction in reprogramming factors expression after cell proliferation.

Subsequently, *piggybac* (PB) transposon and Cre/Loxp were applied to delete reprogramming factors. PB transposon is derived from moth and is highly active in mice and mammal cells and had been applied to gene mutation and delivery (Ding et al. 2005; Wu et al. 2006; Cadiñanos and Bradley 2007). Kosuke Yusa et al. used this system for two main reasons: (1) DNA fragments as huge as 10 kb can be transposed while keeping the same transposition efficiency; (2) it causes no mutations upon removal from the host genome. In 2009, they use the PB transposon to deliver the OSKM and reprogrammed mouse fibroblasts into iPSCs with efficiency equal to retrovirus (Yusa et al. 2009). The PB transposon vector contained PB transposon, constitutively active CAG promoter, OSKM, or OSKML linked to 2A from foot-and-mouth disease virus and Thosea asigna virus. Besides, in this experiment, they found that VPA could not enhance reprogramming efficiency, unlike previous result reported by Huangfu et al. (2008), while *Lin28* could enhance, which was in accord with previous report. Soon after that, Knut Woltjen et al. demonstrated efficient generation of mice and human iPSCs using doxycycline-inducible transcription factors delivered by PB transposition (Woltjen et al. 2009). Otherwise, Frank Soldner et al. use Cre/Loxp system to excise TFs and succeeded in derivation of human iPSCs from five Parkinson's disease (PD) patients (Soldner et al. 2009).

In 2009, Robert L Judson et al. used miRNA combination of miR-291-3p, miR-294, and miR-295 and the transfection of OSK increased the reprogramming efficiency. These miRNAs are downstream effectors of *c-Myc* and can replace *c-Myc* during reprogramming. Better than cMyc, however, the miRNAs induced a homogeneous population of iPS cell colonies. In 2011, Frederick Anokye-Danso et al. only used miR302/367 cluster to reprogram mouse and human somatic cells to iPSCs without transfection of TFs and its efficiency showed two orders of magnitude over standard OSKM protocols (Anokye-Danso et al. 2011). The effect of miRNAs is higher than that of TFs because they do not need protein translation which causes a faster response on protein expression by suppression of mRNA translation and its stability. Besides, miRNAs generally target scores or hundreds of mRNAs that coordinate expression of many different proteins, which can rapidly impose a dominant phenotypic change in cell identity.

The same day, in 2009, two groups generated unintegrated iPSCs by a different approach. Yu et al. 2009 obtained human iPS cells using non-integrating episomal vectors. Zhou et al. (2009) generated mouse iPSCs using recombinant cell-penetrating reprogramming proteins. Furthermore, Hyun-Jai Cho et al. got mouse iPSCs reprogrammed by recombinant proteins (Kim et al. 2009). Soon, Dohoon Kim generated human iPSCs by direct delivery of reprogramming proteins (Cho et al. 2010).

In 2009, Noemi Fusaki et al. efficiently got transgene-free human iPSCs using Sendai virus (SeV) (Fusaki et al. 2009), an RNA virus that does not alter the host genome. SeV-derived transgenes were diluted after cell division. Besides, SeV can be erased easily by antibody-mediated negative selection using cell surface marker HN expressing on SeV-infected cells. Subsequently, in 2010, Yakubov et al. got

Table 4.1 Methods to generate iPSCs and their efficiencies

Vector	Factors	Approximate efficiency (%)
Retrovirus	*OSKM, OSK, OSK + VPA, or OS +* VPA	0.001–1
Lentivirus	*OSKM or miR302/367 cluster +* VPA	0.1–1.1
Inducible lentivirus	*OSKM or OSKMN*	0.1–2
loxP-flanked lentivirus	*OSK*	0.1–1
Adenovirus	*OSKM*	0.001
Sendai virus	*OSKM*	1
Transposon	*OSKM*	0.1
Episomal plasmid	*OSNL*	0.001
Protein	*OS*	0.001
Modified mRNA	*OSKM or OSKML +* VPA	1–4.4
MicroRNA	miR-200c, miR-302s or miR-369s	0.1

Note the meanings of acronyms are showed here: *O OCT4, S SOX2, K KLF4, M c-MYC, L LIN28, N NANOG*, and *VPA* valproic acid

human iPSCs from human foreskin fibroblasts by mRNAs of OSKM (Yakubov et al. 2010). This RNA-induced protocol provided a new way to reprogramming without DNA vectors. Recently, these two approaches are widely used for their high reprogramming efficiency.

The above development of iPS field demonstrates that genome integration or the continued presence of exogenous reprogramming factors is dispensable for reprogramming human somatic cells to pluripotent state (Table 4.1). The progress in this field paves the way for human iPSCs to clinical application.

4.4 Potential Application of iPSCs in Disease Treatment

One of the ultimate goals of regenerative medicine is to obtain autologous organs in vitro and make them work when transplanted into patients and therefore cure the diseases. The idea of using stem cells for cell replacement therapy and disease models dates back to the generation of human ESCs in 1998 (Thomson et al. 1998) because of their potency to produce cells of all the three germ layers. While the human induced pluripotent stem cells have potential for clinical application in the same way. However, challenges must be addressed before the clinical application of human iPSCs. The application has two aspects: cell-based disease models and cell replacement therapy. Here, we will review recent progress in these two fields.

Cell-based disease models are cell lines in vitro which mimic human diseases using cells. We use these models to better understand, diagnose, and treat human disease, as well as did preclinical analysis of drug candidates. Utilizing human iPSCs to simulate diseases has two steps: deriving disease-specific iPSCs from accessible tissues of patients and in vitro differentiation into disease-specific cells.

Various human iPSCs have been generated from accessible tissues of patients with specific diseases. In 2008, In-Hyun Park and his colleagues generated disease-specific iPSCs from patients with genetic diseases, including PD, Down syndrome (DS)/trisomy 21, Huntington disease (HD), and juvenile-onset, type 1 diabetes mellitus (JDM) (Park et al. 2008). Besides, many reports about animal iPSCs demonstrated the therapeutic potential of iPSCs. Particularly, mouse iPSCs can differentiate into hematopoietic (Hanna et al. 2007; Schenke-Layland et al. 2008; Xu et al. 2009), cardiovascular (Schenke-Layland et al. 2008; Narazaki et al. 2008; Kuzmenkin et al. 2009), neural (Wernig et al. 2008), hepatic progenitor cells (Cantz et al. 2008), and so on. What is more, iPSCs have been used for restoration of physiological function of disease tissues in vivo. The applications of iPSCs in various diseases are reviewed as follows. Figure 4.1 shows cell therapy using patient-specific iPSCs in the future. First, the somatic cells of patients are reprogrammed to iPSCs, the gene of which will be repaired by gene editing technologies. And the healthy iPSCs are differentiated into functional somatic cells of certain type which can rescue the diseases after transplantation.

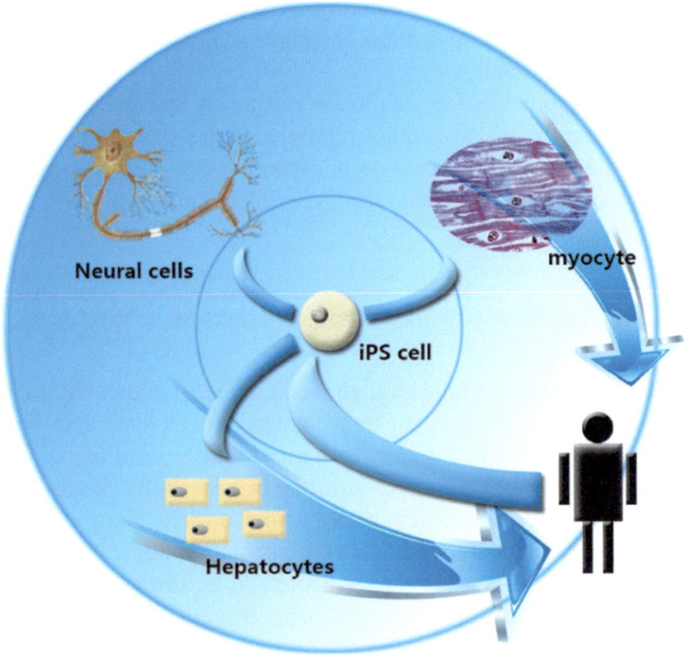

Fig. 4.1 Schematic overview of iPSC-based cell therapy. Patient autologous iPSCs are derived and directly differentiated into neural cells, hepatocytes, or myocytes, etc. and are further transplanted back into the patient

4.4.1 iPSCs and Hematological Diseases

4.4.1.1 Sickle Cell Anemia

In 2007, Hanna et al. grafted autologous iPSC-derived hematopoietic progenitors into a humanized sickle cell anemia mouse model and finally rescued the mice (Hanna et al. 2007). They first reprogrammed patient fibroblasts into iPSCs by retrovirus transfection of OSKM. Then, homologous recombination technology was used to correct the sickle hemoglobin gene of iPSCs. The iPSCs were differentiated into hematopoietic progenitors in vitro. Subsequently, they graft the autologous corrected iPSCs into patients after irradiation.

In 2011, several groups corrected patient-specific iPSCs by gene targeting modification technologies. Zou et al. corrected the sickle cell disease mutation with 2 mutant β-globin alleles of patient-specific iPSCs by homologous recombination using ZFN. Another group of Vittorio Sebastiano achieved in situ correction of the sickle cell anemia mutation in patient iPSCs using ZFN (Sebastiano et al. 2011).

4.4.1.2 Hemophilia A

Hemophilia A is a congenital bleeding disease and is caused by genetic deficiency in clotting factor VIII. In 2009, endothelial cells derived from mouse iPSCs were injected directly into the liver of irradiated hemophilia. Mice extended their survival for more than 3 months and rescued depleted plasma FVIII levels (Xu et al. 2009). In 2014, Jia et al. used patient-specific iPSCs from urine cells and modeled the hemophilia A (Jia et al. 2014). They produced iPSCs from urine cells of HA patients by integration-free reprogramming approach using episomal vectors containing *Oct4, Sox2, Klf4* and *SV40LT*. Then, the iPSCs were differentiated into hepatocyte-like cells (HLCs) which functioned normally but produced no FVIII. Such cells recapitulated the phenotype of hemophilia A. This report used non-integration reprogramming approach and derived iPSCs from urine cells, a very accessible resource. Future study will focus on gene correction and cell replacement therapy.

4.4.1.3 Fanconi Anemia

Fanconi anemia (FA) is an autosomal recessive blood disorder with a frequency of 1/350,000 births, which is caused by a genetic defect in DNA repair-associated protein clusters (Moustacchi 2003). FA is characterized by progressive aplastic anemia, pancytopenia, and various congenital malformations. All of these symptoms are noted predispositions to acute myeloid leukemia.

In 2009, Raya et al. obtained genetic-corrected FA-specific iPSCs from patients (Raya et al. 2009). The primary dermal fibroblasts or skin cells from several patients were reprogrammed either with or without genetic correction with lentiviral vectors of *FAND2* or *FANCA*. In their improved induction protocols, somatic cells underwent two-round infection with retrovirus encoding OSKM based on mouse stem cell virus. And they demonstrated that the iPSCs that were reprogrammed from somatic cells after gene correction could differentiate into haematopoietic progenitors of the myeloid and erythroid lineages. And such cells have normal phenotypes similar to healthy ones. However, the iPSCs were not integration free and failed to recapitulate the FA phenotypes at cell level.

Most recently, in 2014, Liu et al. succeeded in generating integration-free FA-specific iPSCs from fibroblasts of a patient (Liu et al. 2014). The patient had a biallelic truncating mutation in the *FANCA* gene. They reprogrammed the somatic cells by episomal vectors encoding *Oct4, Klf4, Sox2, L-Myc, Lin28,* and *p53* shRNA. This group either in situ corrected the patient-specific iPSCs or generated isogenic FA-deficient human ESCs. Furthermore, such iPSCs and ESCs and their derivatives could recapitulate the FA phenotypes at cell level. Such model can apply for drug screening.

4.4.1.4 Engineered Blood Vessels

Generating functional tissues is the ultimate goal of regenerative medicine. In 2013, Samuel et al. successfully obtained functionally competent and durable artificial blood vessels using human iPSCs (Samuel et al. 2013). In their experiments, they used three markers (CD34, neuropilin 1, and human kinase insert domain-containing receptor) to select endothelial precursor cells from human iPSCs of healthy donors. And they subsequently used an efficient 2D culture system to facilitate the cell expansion of the endothelial precursor cells. Such endothelial precursor cells formed functional blood vessels and stayed in mice for as long as 280 days. Besides, they generated mesenchymal precursor cells from the same human iPS cell line and also got functional blood vessels using the iPSC-derived mesenchymal precursor cells. The stability of these engineered blood vessels showed application potential in long-time generation of vascular for tissue engineering and therapy of vascular diseases. Besides, this group also generated endothelial precursor cells and mesenchymal precursor cells from type 1 diabetic-specific iPSCs and obtained blood vessels in vivo.

4.4.2 iPSCs and Cardiovascular Diseases

4.4.2.1 Long QT Syndrome

Long QT syndrome (LQTS) is a rare genetic cardiovascular disease. The hearts of LQTS patients have delayed repolarization after a heartbeat, which creates prolongation of the QT interval on an electrocardiogram, which increased the risk of torsades de pointes (TDP). This disease is fatal for it may cause fainting, palpitations, and even sudden death.

About 30–35 % of the disease cases are of LQTS type 1, which are caused by the mutation of *Kcnq1* gene encoding the repolarizing potassium channel. The gene mediates the delayed rectifier IKs current. In 2010, Alessandra Moretti et al. derived patient-specific iPSCs from patients with LQTS type 1, who have autosomal dominant missense mutation in the *Kcnq1* gene, using OSKM retroviral vector infection method (Kim et al. 2009). Then, they differentiated the iPSCs directly into functional cardiac myocytes. Such LQTS-specific cardiac myocytes demonstrated "ventricular," "nodal," or "atrial" phenotype. They discovered significant prolongation of the duration of the action potential and proneness to catecholamine-induced tachyarrhythmia in "ventricular" and "atrial" cells. Moreover, they reported that β-blockade could treat the symptom of catecholamine-induced tachyarrhythmia.

LQTS type 2 is the second commonest type and makes up approximately 25–30 % of all the LQTS cases and is caused by the mutation of *human ether-a-go-go-related* gene (*hERG*, also called *Kcnh2*). Subsequently, in 2010, Cho et al. generated patient-specific iPSCs from a patient of LQTS type 2 who was possessed of A614V missense mutation in the *hERG* gene (Cho et al. 2010). And they differentiated the iPSCs into cardiomyocytes and similarly discovered significant prolongation of the duration of the action potential which stemmed from a marked reduction of the cardiac potassium current IKr. Furthermore, the group applied such cardiomyocytes to assess existing and potential drugs.

4.4.3 iPSCs and Liver Diseases

Liver diseases are the 16th leading cause of death worldwide; and, liver transplantation is necessary for liver failure. However, alternative tissue sources such as hepatic cell lines and fetal hepatocytes are hard to proliferate and do not show all hepatocyte phenotypes and functions. Large-scale-generation functional hepatocytes from iPSCs hold large promise for research on liver development, drug metabolism and toxicity, host–pathogen interactions, and even cell replacement therapy for liver diseases.

iPSCs can be differentiated efficiently into HLCs which have phenotypic and morphologic features similar to hepatocytes by a stepwise and reproducible approach (Zhang et al. 2014). Furthermore, to establish in vitro conditions for iPSC-derived HLCs, which can model the diseases more precisely, some groups have been trying to recreate spatial organization and cell–cell interaction similar to the in vivo conditions. Schwartz et al. have optimized a microscale structure which can maintain some hepatocyte phenotypic functions for as long as several weeks (Schwartz et al. 2014). Besides, three-dimensional coculture of adult hepatocytes with liver sinusoidal endothelial cells can maintain their phenotypes simultaneously.

Gene targeting modification technology can correct the mutant genes of the patient-specific iPSCs. In 2011, Yusa et al. achieved correction of α-antitrypsin deficiency (*A1at* or *Serpina1*) in human iPSCs (Yusa et al. 2011). They use a combination of zinc finger nucleases and piggybac technology to correct a biallelic point mutation in the α-antitrypsin. This report provided the potential of genetic correction in patient-specific iPSCs for autologous transplantation.

4.4.4 iPSCs and Spinal Cord Injury

Spinal cord injury (SCI) means any injury to the spinal cord caused not by disease but by trauma. In 2010, a study by Tsuji et al. suggested the potential application of mouse iPSCs for SCI (Tsuji et al. 2010). They directly differentiated mouse iPSCs into neural spheres and pre-evaluated the tumorigenic activities of the cells by transplantation into brain of NOD or SCID mouse to screen the safe cells. The safe neural spheres can be differentiated in vitro into electrophysiologically functional neurons, oligodendrocytes, and astrocytes. When transplanted into the spinal cord nine days after contusive injury, the spheres produced all three neural lineages without tumor formation. Notably, the unsafe neural spheres during the pre-evaluation exhibited intense tumor formation activities and suddenly lost locomotor function after functional rescue in the mouse model. Furthermore, Satoshi Nori et al. transplanted human iPSC-derived neural spheres into SCI mouse model and recovered the motor function which persisted through the experiment period of 112 d. What is more, the mice developed no tumor.

4.4.5 iPSCs and Diabetes and Related Complications

Diabetes mellitus is a metabolic disease mainly defined by high blood sugar levels for a long time. The symptoms are frequent urination, increased appetite, and thirst. It can cause many complications. There are three types of diabetes. Type 1 diabetes is a major type of this disease, which is caused by the dysfunction of insulin-secreting pancreas cells. Type 2 diabetes is caused by insulin resistance, and type 3 diabetes is called gestational diabetes, which means pregnant women have high blood sugar levels. Until now, type 1 diabetes and type 2 diabetes cannot be truly cured. Upon the generation of patient-specific iPSCs, scientists have been trying to cure the type 1 diabetes using iPSCs.

In 2007, Jiang and his colleagues obtained insulin-secreting isletlike cell clusters from human ESCs (Jiang et al. 2007). One year after the generation of human iPSCs, Keisuke Tateishi et al. succeeded in generating insulin-secreting isletlike cells from human iPSCs (Tateishi et al. 2008). The iPSCs were reprogrammed from skin fibroblasts by retroviral transfection of OSKM and then differentiated into insulin-secreting isletlike cell clusters through definitive and pancreatic endoderm. The cell clusters contained glucagon-positive and C-peptide-positive cells and secreted C-peptide in response to glucose stimulation. In 2009, Zhang et al.

improved the differentiation efficiency of human ESCs and iPSCs into mature insulin-secreting cells (Zhang et al. 2009). The same year, Maehr et al. generated iPSCs from patients with type 1 diabetes (Maehr et al. 2009). In 2011, Deng HK group first got pancreatic insulin-secreting cells from iPSCs of rhesus monkey (Zhu et al. 2011).

In 2010, Toshihiro Kobayashi et al. generated rat pancreas in mouse by injecting rat wild-type iPSCs into mouse $Pdx1^{-/-}$ (pancreatogenesis-disabled) blastocysts (Kobayashi et al. 2010). The above $Pdx1^{-/-}$ mice have functional pancreas, did not develop pancreatic insufficiency, and grew to adult.

4.4.6 iPSCs and Brain Diseases

4.4.6.1 Rett's Syndrome

Rett's syndrome (RTT), originally known as cerebroatrophic hyperammonemia, is a rare genetic progressive neurological disorder relative to the gray matter of the brain. This disease is characterized by small hands and feet, decreased rate of head growth (some patients suffer from microcephaly), impaired motor function such as repeated hand movements, degradation of developmental skills, seizures, and autistic disorder (Amir et al. 1999). It is caused by mutations in the X-linked gene *MeCP2* which encodes methyl-CpG-binding protein 2 (Amir et al. 1999). In 2010, Maria C.N. Marchetto and his colleagues generated iPSCs from RTT patients and derived functional neurons from these iPS cell lines (Marchetto et al. 2010). Such RTT-specific neurons recapitulated early stage of RTT, which showed decreased spine density, fewer synapses, smaller soma size, impaired electrophysiology, and altered calcium signaling process. What is more, they also used such neurons to screen drugs which could treat synaptic defects. And they confirmed that the mutation of MECP2 recapitulated the neuronal disease. This study was valuable because it suggested that presymptomatic defects may possibly serve as new markers of RTT, treatment in the early stage of development may be useful, and it facilitated drug screening.

4.4.6.2 Dravet Syndrome and Mild Febrile Seizures

70 % of Dravet syndrome, also known as severe myoclonic epilepsy of infancy (SMEI), mild febrile seizures or generalized epilepsy with febrile seizures plus, are caused by the mutation of *SCN1A* (Claes et al. 2001; Scheffer et al. 2009). This gene encodes the α1-subunit of the Nav1.1 voltage-gated sodium channel. The phenotypes of mild febrile seizures vary from classical mild febrile seizures which do not need treatment and SMEI which is the most rare and severe one. The SMEI is characterized by febrile seizures since the first year of an infant and the subsequent epilepsy, damaged psychomotor development and ataxia. What

is more, the disorder is always resistant to pharmacotherapy and its mechanism was not good to understand because of the difficulty in obtaining disease-specific neurons.

In 2013, Gao's group succeeded in modeling them by the iPS technology and finally got patient-specific neurons (Jiao et al. 2013). They obtained such iPSCs from a patient with SMEI and another with mild febrile seizures. Subsequently, they generated functional glutamatergic neurons by differentiating the iPSCs of patient or direct reprogramming of fibroblasts into neurons, of which the function was confirmed by electrophysiological assays. Such neurons showed a hyperexcitable state. Significantly, such disease-specific iPSC-derived neurons could recapitulate the pathophysiology of the diseases and have response to an antiepileptic drug.

4.4.6.3 Parkinson's Disease

iPSCs were also used for PD models. PD (also known as idiopathic or primary parkinsonism, hypokinetic rigid syndrome (HRS) or paralysis agitans) is one kind of degenerative disorders of the central nervous system. The death of dopamine neurons in the substantia nigra, a region of the midbrain, causes the motor symptoms of PD. An important obstacle to the research on this disease was the inaccessibility of diseased tissue. Thus, researchers in the stem cell field have been trying to figure out the disease mechanism and rescue the symptoms using iPSCs.

In 2008, it was reported by Wernig et al. that mouse iPSCs generated from fibroblasts using OSKM could be efficiently differentiated into neural precursor cells which can generate neuronal and glial cell types in vitro (Wernig et al. 2008). And after grafted into mouse fetal brain, the neural precursor cells could migrate to different parts of the brain and be differentiated into neuronal and glial cells, which have right morphology and electrophysiological function. Further assays were carried out to test whether such neural precursor cells could rescue PD. In an adult rat model of PD, midbrain dopamine (DA) neurons derived from iPSCs were transplanted into the brains and finally improved the behavior of the rats. Then, in 2009, Frank Solder and his colleagues produced PD-specific hiPSCs, which had no integration into host genome using Cre recombinase to excise external DNA (Soldner et al. 2009). This report demonstrated that the remaining transgene expression could change cellular molecular characteristics. Therefore, integration-free hiPSCs are more suitable for modeling human diseases.

In the other hand, researchers have investigated genes involved in the PD using iPSCs. The α-synuclein dysfunction is critically related to PD, dementia with Lewy bodies, and multiple system atrophy. Besides, it was reported that *leucine-rich repeat kinase 2* (Lrrk2) regulated the progression of neuropathology induced by PD-related mutant α-synuclein, and Pink1 encoded a mitochondrial kinase which was related to the regulation of mitochondrial degradation. With the improvement of gene modification technology, specific gene mutant iPSCs were generated. In 2011, DA neurons derived from Lrrk2 mutant iPSCs were reported to be more susceptible to oxidative stress (Nguyen et al. 2011). In detail, these

DA neurons derived from LRRK2 mutant iPSCs demonstrated higher expression of key oxidative stress-response genes and α-synuclein protein. Such mutant iPSCs can be used to build sporadic PD model and screen drugs. Soon later, Philip Seibler et al. generated iPSCs from human skin fibroblasts of three PD patients who have nonsense or missense mutation of *PINK1* gene (Seibler et al. 2011). DA neurons derived from the iPSCs exhibited increased mitochondrial copy number, upregulation of PGC-1, an important regulator of mitochondrial biogenesis, and decreased recruitment of lentivirally expressed Parkin to mitochondria during mitochondrial depolarization. Significantly, these symptoms were rescued by lentiviral expression of wild-type *PINK1*. In the same year, another group generated PD-specific iPSCs with triplication of *SNCA*, gene encoding the α-synuclein (Devine et al. 2011). In their report, iPSC-derived DA neurons were generated from a patient with triplication of *SNCA* or an unaffected first-degree relative. The expression of α-synuclein from the patient was double that of the unaffected relative, which summed the cause of PD in these individuals.

More recently, Chee Yeun Chung et al. generated cortical neurons from iPSCs of patients with mutant α-synuclein, who were probable to develop PD dementia, and identified and rescued α-synuclein toxicity in these neurons (Chung et al. 2013). Finally, they found that the small molecule NAB2, and the ubiquitin ligase Nedd4 it affects, can rescue PD phenotypes in these neurons. Besides human iPSCs, human ESCs have been also used in the PD. Human ES-derived DA neurons were efficiently grafted in animal models of PD.

4.4.6.4 Alzheimer's Disease

Alzheimer's disease (AD) is the commonest type of age-related dementia, the main symptoms of which are cognitive disturbance and progressive memory loss. They are familial AD (FAD) and sporadic AD (SAD). The mutations of *presenilin* (*PS*) 1 and 2 can give rise to autosomal dominant early-onset FAD. In 2011, iPSC-derived neurons were generated from the fibroblasts of FAD patients with mutation in *PS1* and *PS2* (Yagi et al. 2011). These neurons exhibited higher amyloid β42 secretion, which recapitulated the molecular pathogenesis of mutant PS family genes. Moreover, secretion of amyloid β42 from these neurons violently responded to γ-secretase inhibitors and modulators, which indicated the potential for identification and validation of drugs. In 2012, the iPS technology made it possible to probe both familial and SAD (Israel et al. 2012). iPSC-derived neurons were generated from the primary fibroblasts of two patients with FAD caused by the duplication of the amyloid-β precursor protein (APP) gene, two patients with SAD, and two persons without AD for negative control. The results showed a direct relationship between APP proteolytic processing, but not amyloid-β, in tau phosphorylation and GSK-3β activation in neurons. Besides, neurons with the genome of one SAD patient exhibited the phenotypes seen in FAD ones. These results demonstrated that the iPS technology could be applied to observe phenotypes of AD.

4.4.6.5 Familial Dysautonomia

Familial dysautonomia (FD), also known as hereditary sensory and autonomic neuropathy type III or Riley–Day syndrome, is an autosomal recessive disorder of autonomic nervous system. It's a severe disease that causes many deaths in infancy and childhood. The disease is caused by a single point mutation in *IKBKAP* gene (Slaugenhaupt et al. 2001) related to transcriptional elongation (Close et al. 2006). It is characterized by hypotonia, weak suck reflex, and hypothermia in newborn; retarded physical development, poor motor in coordination, and poor temperature in early childhood; and other symptoms including relative insensitivity to pain, reduction or absence of tears, absence of corneal reflex, depression of deep tendon reflexes, and postural hypotension (Axelrod 2002).

In 2009, Lee et al. first generated iPSCs from a FD patient by lentiviral vectors encoding *OKSM* and differentiated them into peripheral neurons (Lee et al. 2009). They discovered defects in the *IKBKAP* splicing, neurogenesis, and migration of neural crest precursors derived from the iPSCs above. Such neural crest precursors served as a useful model to explore the pathogenesis of FD. However, larger sets of FD-specific iPSCs are necessary to further confirm the phenotype of FD.

However, several challenges should be addressed in the future research. They are described as follows. First, a high yield of iPSC-derived neurons still need complicated differentiation procedures and prolonged culture period. Second, there still exists heterogeneity of differentiated neuronal cell types depending on the clonal variability and culture conditions in current protocols. Therefore, developing faster neuronal differentiation methods with minimal clonal variation is necessary for future drug screening.

4.4.7 iPSCs and Eye Diseases

As is known, many kinds of blindness are caused by the dysfunction of retinal photoreceptors. Induced stem cells have potential to cure damaged retina following injury or diseases.

As early as 2009, Hiramia et al. generated retinal cells from mouse and human iPSCs (Hiramia et al. 2009). In 2011, Zhou et al. achieved the differentiation of swine iPSCs into rod photoreceptors, which could integrate into the outer nuclear layer of the retina where photoreceptors resided (Zhou et al. 2011). Moreover, in 2014, Zhong et al. generated 3D retinal tissue consisting of functional photoreceptors derived from human iPSCs (Zhong et al. 2014). The iPSCs they generated could spatiotemporally recapitulate every main step during the retinal development process. Besides, they could form 3D retinal cups containing all the main retinal cell types which were properly arranged in their own layers. Notably, the photoreceptors were advanced mature and demonstrated the start of outer-segment disk formation and photosensitivity. In 2014, Riazifar et al. generated retinal ganglion cells from both human ESCs and iPSCs using chemically induced specification approach (Riazifar et al. 2014).

Amazingly, on June 26, 2013, world's first human clinical application of iPSCs was approved by Japan government. Masayo Takahashi and his group at the Riken Center for Developmental Biology in Kobe had applied to a screening panel in February to go ahead with the clinical research. The study involved surgical transplantations of patient-specific iPSC-derived retina cells that were planned to treat age-related macular degeneration, an intractable disease of the eyes in the summer of 2014 in Kobe.

4.4.8 iPSCs and Potential Cancer Therapy

Cancer, also called malignant tumor, is a group of fatal diseases. It led to about 14.6 % or 8.2 million deaths of all human deaths, according to the World Cancer Report 2014 from World Health Organization (WHO).

How can human iPSCs treat cancer? The potential value of this technology in cancer treatment may be shown in three aspects.

First is drug screening. We can generate human iPSCs from the cancer cells and use such iPSCs to screen agents targeting cancer cells. Second is immune therapy (Yang and Baltimore 2005). It is reported that human iPSCs derived from T lymphocytes maintained the pre-rearranged T cell receptor gene (Brown et al. 2010; Loh et al. 2010). Thus, suppose we extract large numbers of T cells that carry specificity against certain tumor antigens from cancer patients and generate iPSCs; furthermore, we differentiate these iPSCs into T cells with active function and transplant such autologous T cells into the patients. However, one vital problem of this method is safety. Third is tissue replacement. We may use certain tissues derived from iPSCs of the cancer patients in vitro to replace the tissues impaired by chemotherapy, radiation or the surgical treatment during the process of cancer therapy. As most of the mutations occurred to the cancer patients are not inherited but acquired, we can derive healthy iPSCs from healthy tissues of patients and conduct transplantation. However, one vital problem of this method is the engraftment of the iPSCs. So far, just a few human iPSC-derived cell types have succeeded in being engrafted in animal models.

4.4.9 iPSCs and Infertility

Infertility refers to the inability to conceive child, which includes female infertility and male infertility. In 2014, Cyril Ramathal et al. generated iPSCs from azoospermic men (Ramathal et al. 2014). The patients have deletions which encompassed three Y chromosome azoospermia factor regions. Therefore, the patients produced few or even no sperm and developed infertility. The patient-specific iPSCs have difficulty in differentiating into germ cells in vitro. When transplanted into mouse seminiferous tubules, undifferentiated iPSCs robustly differentiated into germ cell-like cells. The germ cell-like cells localized near the basement membrane showed morphology similar to fetal germ cells and expressed markers of primordial germ cells. But all the iPSCs that exited seminiferous tubules formed tumors. Especially, iPSCs with more severe deficiency in azoospermia factor deletions produced much fewer germ cell-like cells, which indicated that the differentiation process induced by interspecific transplantation of human iPSCs was depend on donor genetic status. Further investigations should be done before iPS technology finally used in human infertility therapy.

4.5 Concluding Note

The iPS technology has initiated a new era of regenerative medicine. This technology holds the potential to provide large quantity of various functional human cells, which can be applied to establish disease models for studying disease mechanism, screening drugs, and cell therapy. Many progresses have been made in this field in recent years. However, there are still lots of challenges before practical application. One is the safety issue, including oncogenicity, genome or epigenome integrities, and immunogenicity. Also, there are still limited reports about iPSCs showing phenotypes of complex diseases in vitro. Though the iPSCs can model monogenic diseases, it is difficult to model complex diseases with iPSCs. In spite of the above challenges, lots of developments have been made to move iPSC research forward, such as the recently emerged genome editing technology including ZFN, TALEN, and CRISPR-Cas9 system, which can greatly facilitate the gene correction in iPSCs of patient. Of note, it's uplifting that the first clinical trial of human iPSCs was conducted in Japan for patients of age-related macular degeneration in 2014 (World Health Organization 2012). We hope that the technical obstacles will be solved before long and the iPS technology will benefit patients and industries in the near future.

References

Amir RE, Van den Veyver IB, Wan M, Tran CQ, Francke U, Zoghbi HY. Rett syndrome is caused by mutations in X-linked MECP2, encoding methyl-CpG-binding protein 2. Nat Genet. 1999;23:185–8.

Anokye-Danso F, Trivedi CM, Juhr D, Gupta M, Cui Z, Tian Y, Zhang Y, Yang W, Gruber PJ, Epstein JA, Morrisey EE. Highly efficient miRNA-mediated reprogramming of mouse and human somatic cells to pluripotency. Cell Stem Cell. 2011;8:376–88.

Axelrod FB. Hereditary sensory and autonomic neuropathies. Clin Auton Res. 2002;12.

Boland MJ, Hazen JL, Nazor KL, Rodriguez AR, Gifford W, Martin G, Kupriyanov S, Baldwin KK. Adult mice generated from induced pluripotent stem cells. Nature. 2009;461:91–4.

Briggs R, King TJ. Changes in the nuclei of differentiating endoderm cells as revealed by nuclear transplantation. J Morphol. 1957;100:269–311.

Briggs R, King TJ. Transplantation of living nuclei from blastula cells into enucleated frog eggs. Proc Natl Acad Sci USA. 1952;38:455–463.

Brown ME, Rondon E, Rajesh D, Mack A, Lewis R, Feng X, Zitur LJ, Learish RD, Nuwaysir EF. Derivation of induced pluripotent stem cells from human peripheral blood T lymphocytes. PLoS ONE. 2010;5:e11373.

Buehr M, Meek S, Blair K, Yang J, Ure J, Silva J, McLay R, Hall J, Ying Q-L, Smith A. Capture of authentic embryonic stem cells from rat blastocysts. Cell. 2008;135:1287–98.

Byrne JA, Pedersen DA, Clepper LL, Nelson M, Sanger WG, Gokhale S, Wolf DP, Mitalipov SM. Producing primate embryonic stem cells by somatic cell nuclear transfer. Nature. 2007;450:497–502.

Cadiñanos J, Bradley A. Generation of an inducible and optimized piggyBac transposon system. Nucleic Acids Res. 2007;35:e87.

Campbell KHS, McWhir J, Ritchie WA, Wilmut I. Sheep cloned by nuclear transfer from a cultured cell line. Nature. 1996;380:64–6.

Cantz T, Bleidissel M, Stehling M, Scholer HR. In vitro differentiation of reprogrammed murine somatic cells into hepatic precursor cells. Biol Chem. 2008;389:889–96.

Cho HJ, Lee CS, Kwon YW, Paek JS, Lee SH, Hur J, Lee EJ, Roh TY, Chu IS, Leem SH, Kim Y, Kang HJ, Park YB, Kim HS. Induction of pluripotent stem cells from adult somatic cells by protein-based reprogramming without genetic manipulation. Blood. 2010;116:386–95.

Chung CY, Khurana V, Auluck PK, Tardiff DF, Mazzulli JR, Soldner F, Baru V, Lou Y, Freyzon Y, Cho S, Mungenast AE, Muffat J, Mitalipova M, Pluth MD, Jui NT, Schule B, Lippard SJ, Tsai LH, Krainc D, Buchwald SL, Jaenisch R, Lindquist S. Identification and rescue of alpha-synuclein toxicity in Parkinson patient-derived neurons. Science. 2013;342:983–7.

Claes L, Del-Favero J, Ceulemans B, Lagae L, Van Broeckhoven C, De Jonghe P. De Novo mutations in the sodium-channel gene SCN1A cause severe myoclonic epilepsy of infancy. Am J Hum Genet. 2001;68:1327–32.

Close P, Hawkes N, Cornez I, Creppe C, Lambert CA, Rogister B, Siebenlist U, Merville M-P, Slaugenhaupt SA, Bours V. Transcription impairment and cell migration defects in elongator-depleted cells: implication for familial dysautonomia. Mol Cell. 2006;22:521–31.

Cowan CA, Atienza J, Melton DA, Eggan K. Nuclear reprogramming of somatic cells after fusion with human embryonic stem cells. Science. 2005;309:1369–73.

Devine MJ, Ryten M, Vodicka P, Thomson AJ, Burdon T, Houlden H, Cavaleri F, Nagano M, Drummond NJ, Taanman JW, Schapira AH, Gwinn K, Hardy J, Lewis PA, Kunath T. Parkinson's disease induced pluripotent stem cells with triplication of the alpha-synuclein locus. Nat Commun. 2011;2:440.

Ding S, Wu X, Li G, Han M, Zhuang Y, Xu T. Efficient transposition of the piggyBac (PB) transposon in mammalian cells and mice. Cell. 2005;122:473–83.

Esteban MA, Xu J, Yang J, Peng M, Qin D, Li W, Jiang Z, Chen J, Deng K, Zhong M. Generation of induced pluripotent stem cell lines from Tibetan miniature pig. J Biol Chem. 2009;284:17634–40.

Esteban MA, Wang T, Qin B, Yang J, Qin D, Cai J, Li W, Weng Z, Chen J, Ni S, Chen K, Li Y, Liu X, Xu J, Zhang S, Li F, He W, Labuda K, Song Y, Peterbauer A, Wolbank S, Redl H, Zhong M, Cai D, Zeng L, Pei D. Vitamin C enhances the generation of mouse and human induced pluripotent stem cells. Cell Stem Cell. 2010;6:71–9.

Evans MJ, Kaufman MH. Establishment in culture of pluripotential cells from mouse embryos. Nature. 1981;292:154–156.

Fusaki N, Ban H, Nishiyama A, Saeki K, Hasegawa M. Efficient induction of transgene-free human pluripotent stem cells using a vector based on sendai virus, an RNA virus that does not integrate into the host genome. Proc Jpn Acad Ser B Phys Biol Sci. 2009;85:348–62.

Hanna J, Wernig M, Markoulaki S, Sun CW, Meissner A, Cassady JP, Beard C, Brambrink T, Wu LC, Townes TM, Jaenisch R. Treatment of sickle cell anemia mouse model with iPS cells generated from autologous skin. Science. 2007;318:1920–3.

Harris H. The reactivation of the red cell nucleus. J Cell Sci. 1967;2:23–32.
Hasegawa K, Cowan AB, Nakatsuji N, Suemori H. Efficient multicistronic expression of a transgene in human embryonic stem cells. Stem Cells. 2007;25:1707–12.
Hirami Y, Osakada F, Takahashi K, Okita K, Yamanaka S, Ikeda H, Yoshimura N, Takahashi M. Generation of retinal cells from mouse and human induced pluripotent stem cells. Neurosci Lett. 2009;458:126–31.
Honda A, Hirose M, Hatori M, Matoba S, Miyoshi H, Inoue K, Ogura A. Generation of induced pluripotent stem cells in rabbits potential experimental models for human regenerative medicine. J Biol Chem. 2010;285:31362–9.
Hsiao EC, Yoshinaga Y, Nguyen TD, Musone SL, Kim JE, Swinton P, Espineda I, Manalac C, Conklin BR. Marking embryonic stem cells with a 2A self-cleaving peptide: a NKX2-5 emerald GFP BAC reporter. PLoS ONE. 2008;3:e2532.
Huangfu D, Maehr R, Guo W, Eijkelenboom A, Snitow M, Chen AE, Melton DA. Induction of pluripotent stem cells by defined factors is greatly improved by small-molecule compounds. Nat Biotechnol. 2008;26:795–7.
Israel MA, Yuan SH, Bardy C, Reyna SM, Mu Y, Herrera C, Hefferan MP, Van Gorp S, Nazor KL, Boscolo FS, Carson CT, Laurent LC, Marsala M, Gage FH, Remes AM, Koo EH, Goldstein LS. Probing sporadic and familial Alzheimer's disease using induced pluripotent stem cells. Nature. 2012;482:216–20.
Jia B, Chen S, Zhao Z, Liu P, Cai J, Qin D, Du J, Wu C, Chen Q, Cai X. Modeling of hemophilia a using patient-specific induced pluripotent stem cells derived from urine cells. Life Sci. 2014;108:22.
Jiang J, Au M, Lu K, Eshpeter A, Korbutt G, Fisk G, Majumdar AS. Generation of insulin-producing islet-like clusters from human embryonic stem cells. Stem Cells. 2007;25:1940–53.
Jiao J, Yang Y, Shi Y, Chen J, Gao R, Fan Y, Yao H, Liao W, Sun XF, Gao S. Modeling dravet syndrome using induced pluripotent stem cells (iPSCs)and directly converted neurons. Hum Mol Genet. 2013;22:4251.
Kim D, Kim CH, Moon JI, Chung YG, Chang MY, Han BS, Ko S, Yang E, Cha KY, Lanza R, Kim KS. Generation of human induced pluripotent stem cells by direct delivery of reprogramming proteins. Cell Stem Cell. 2009;4:472–6.
Kobayashi T, Yamaguchi T, Hamanaka S, Kato-Itoh M, Yamazaki Y, Ibata M, Sato H, Lee YS, Usui J, Knisely AS, Hirabayashi M, Nakauchi H. Generation of rat pancreas in mouse by interspecific blastocyst injection of pluripotent stem cells. Cell. 2010;142:787–99.
Koh S, Tsai S, Bischoff S, Olby NJ, Piedrahita J. Generation of putative induced pluripotent stem cells from adult canine fibroblast by three factor reprogramming (Oct3/4, Klf4, and Sox2). In: Biology of reproduction. Society for the Study of Reproduction. pp. 198–199.
Kuzmenkin A, Liang H, Xu G, Pfannkuche K, Eichhorn H, Fatima A, Luo H, Šarić T, Wernig M, Jaenisch R. Functional characterization of cardiomyocytes derived from murine induced pluripotent stem cells in vitro. FASEB J. 2009;23:4168–80.
Lee G, Papapetrou EP, Kim H, Chambers SM, Tomishima MJ, Fasano CA, Ganat YM, Menon J, Shimizu F, Viale A, Tabar V, Sadelain M, Studer L. Modelling pathogenesis and treatment of familial dysautonomia using patient-specific iPSCs. Nature. 2009;461:402–6.
Liu L, Luo G-Z, Yang W, Zhao Z, Zheng Q, Lv Z, Li W, Wu H-J, Wang L, Wang X-J, Zhou Q. Activation of the imprinted Dlk1-Dio3 region correlates with pluripotency levels of mouse stem cells. J Biol Chem. 2010;285:19483–19490.
Liu H, Zhu F, Yong J, Zhang P, Hou P, Li H, Jiang W, Cai J, Liu M, Cui K. Generation of induced pluripotent stem cells from adult rhesus monkey fibroblasts. Cell Stem Cell. 2008;3:587–90.
Liu G-H, Suzuki K, Li M, Qu J, Montserrat N, Tarantino C, Gu Y, Yi F, Xu X, Zhang W, Ruiz S, Plongthongkum N, Zhang K, Masuda S, Nivet E, Tsunekawa Y, Soligalla RD, Goebl A, Aizawa E, Kim NY, Kim J, Dubova I, Li Y, Ren R, Benner C, del Sol A, Bueren J, Trujillo JP, Surralles J, Cappelli E, Dufour C, Esteban CR, Belmonte JCI. Modelling fanconi anemia pathogenesis and therapeutics using integration-free patient-derived iPSCs. Nat Commun. 2014;5:4330.

Loh YH, Hartung O, Li H, Guo C, Sahalie JM, Manos PD, Urbach A, Heffner GC, Grskovic M, Vigneault F, Lensch MW, Park IH, Agarwal S, Church GM, Collins JJ, Irion S, Daley GQ. Reprogramming of T cells from human peripheral blood. Cell Stem Cell. 2010;7:15–9.

Maehr R, Chen S, Snitow M, Ludwig T, Yagasaki L, Goland R, Leibel RL, Melton DA. Generation of pluripotent stem cells from patients with type 1 diabetes. Proc Natl Acad Sci U S A. 2009;106:15768–73.

Marchetto MC, Carromeu C, Acab A, Yu D, Yeo GW, Mu Y, Chen G, Gage FH, Muotri AR. A model for neural development and treatment of Rett syndrome using human induced pluripotent stem cells. Cell. 2010;143:527–39.

Martin GR. Isolation of a pluripotent cell line from early mouse embryos cultured in medium conditioned by teratocarcinoma stem cells. Proc Natl Acad Sci. 1981;78:7634–8.

Moon JH, Heo JS, Kim JS, Jun EK, Lee JH, Kim A, Kim J, Whang KY, Kang YK, Yeo S, Lim HJ, Han DW, Kim DW, Oh S, Yoon BS, Scholer HR, You S. Reprogramming fibroblasts into induced pluripotent stem cells with Bmi1. Cell Res. 2011;21:1305–15.

Moustacchi E. Fanconi's anemia. Orphanet encyclopedia October 2003.

Narazaki G, Uosaki H, Teranishi M, Okita K, Kim B, Matsuoka S, Yamanaka S, Yamashita JK. Directed and systematic differentiation of cardiovascular cells from mouse induced pluripotent stem cells. Circulation. 2008;118:498–506.

Nguyen HN, Byers B, Cord B, Shcheglovitov A, Byrne J, Gujar P, Kee K, Schule B, Dolmetsch RE, Langston W, Palmer TD, Pera RR. LRRK2 mutant iPSC-derived DA neurons demonstrate increased susceptibility to oxidative stress. Cell Stem Cell. 2011;8:267–80.

Niwa H, Yamamura K, Miyazaki J. Efficient selection for high-expression transfectants with a novel eukaryotic vector. Gene. 1991;108:193–9.

Okita K, Ichisaka T, Yamanaka S. Generation of germline-competent induced pluripotent stem cells. Nature. 2007;448:313–7.

Okita K, Nakagawa M, Hyenjong H, Ichisaka T, Yamanaka S. Generation of mouse induced pluripotent stem cells without viral vectors. Science. 2008;322:949–53.

Orkin SH, Zon LI. Hematopoiesis: an evolving paradigm for stem cell biology. Cell. 2008;132:631–44.

Park IH, Arora N, Huo H, Maherali N, Ahfeldt T, Shimamura A, Lensch MW, Cowan C, Hochedlinger K, Daley GQ. Disease-specific induced pluripotent stem cells. Cell. 2008;134:877–86.

Pomerantz J, Blau HM. Nuclear reprogramming: a key to stem cell function in regenerative medicine. Nat Cell Biol. 2004;6:810–6.

Ramathal C, Durruthy-Durruthy J, Sukhwani M, Arakaki JE, Turek PJ, Orwig KE, Pera RAR. Fate of iPSCs derived from azoospermic and fertile men following xenotransplantation to murine seminiferoustubules. Cell Rep. 2014;7:1284.

Raya A, Rodriguez-Piza I, Guenechea G, Vassena R, Navarro S, Barrero MJ, Consiglio A, Castella M, Rio P, Sleep E, Gonzalez F, Tiscornia G, Garreta E, Aasen T, Veiga A, Verma IM, Surralles J, Bueren J, Izpisua Belmonte JC. Disease-corrected haematopoietic progenitors from fanconi anaemia induced pluripotent stem cells. Nature. 2009;460:53–59.

Riazifar H, Jia Y, Chen J, Lynch G, Huang T. Chemically induced specification of retinal ganglion cells from human embryonic and induced pluripotent stem cells. Stem Cells Transl Med. 2014;3:424–32.

Samuel R, Daheron L, Liao S, Vardam T, Kamoun WS, Batista A, Buecker C, Schäfer R, Han X, Au P. Generation of functionally competent and durable engineered blood vessels from human induced pluripotent stem cells. Proc Natl Acad Sci. 2013;110:12774–9.

Scheffer IE, Zhang Y-H, Jansen FE, Dibbens L. Dravet syndrome or genetic (generalized) epilepsy with febrile seizures plus? Brain Dev. 2009;31:394–400.

Schenke-Layland K, Rhodes KE, Angelis E, Butylkova Y, Heydarkhan-Hagvall S, Gekas C, Zhang R, Goldhaber JI, Mikkola HK, Plath K. Reprogrammed mouse fibroblasts differentiate into cells of the cardiovascular and hematopoietic lineages. Stem Cells. 2008;26:1537–46.

Schwartz RE, Fleming HE, Khetani SR, Bhatia SN. Pluripotent stem cell-derived hepatocyte-like cells. Biotechnol Adv. 2014;32:504–13.

Sebastiano V, Maeder ML, Angstman JF, Haddad B, Khayter C, Yeo DT, Goodwin MJ, Hawkins JS, Ramirez CL, Batista LF, Artandi SE, Wernig M, Joung JK. In situ genetic correction of the sickle cell anemia mutation in human induced pluripotent stem cells using engineered zinc finger nucleases. Stem Cells. 2011;29:1717–26.

Seibler P, Graziotto J, Jeong H, Simunovic F, Klein C, Krainc D. Mitochondrial Parkin recruitment is impaired in neurons derived from mutant PINK1 induced pluripotent stem cells. J Neurosci. 2011;31:5970–6.

Shi Y, Desponts C, Do JT, Hahm HS, Scholer HR, Ding S. Induction of pluripotent stem cells from mouse embryonic fibroblasts by Oct4 and Klf4 with small-molecule compounds. Cell Stem Cell. 2008;3:568–74.

Slaugenhaupt SA, Blumenfeld A, Gill SP, Leyne M, Mull J, Cuajungco MP, Liebert CB, Chadwick B, Idelson M, Reznik L. Tissue-specific expression of a splicing mutation in the IKBKAP gene causes familial dysautonomia. Am J Hum Genet. 2001;68:598–605.

Soldner F, Hockemeyer D, Beard C, Gao Q, Bell GW, Cook EG, Hargus G, Blak A, Cooper O, Mitalipova M, Isacson O, Jaenisch R. Parkinson's disease patient-derived induced pluripotent stem cells free of viral reprogramming factors. Cell. 2009;136:964–77.

Solter D. From teratocarcinomas to embryonic stem cells and beyond: a history of embryonic stem cell research. Nat Rev Genet. 2006;7:319–27.

Speman H. Embryonic development and induction. Am J Med Sci. 1938;196:738.

Stadtfeld M, Nagaya M, Utikal J, Weir G, Hochedlinger K. Induced pluripotent stem cells generated without viral integration. Science. 2008;322:945–949.

Tada M, Tada T, Lefebvre L, Barton SC, Surani MA. Embryonic germ cells induce epigenetic reprogramming of somatic nucleus in hybrid cells. EMBO J. 1997;16:6510–20.

Tada M, Takahama Y, Abe K, Nakatsuji N, Tada T. Nuclear reprogramming of somatic cells by in vitro hybridization with ES cells. Curr Biol. 2001;11:1553–8.

Tada M, Morizane A, Kimura H, Kawasaki H, Ainscough JF, Sasai Y, Nakatsuji N, Tada T. Pluripotency of reprogrammed somatic genomes in embryonic stem hybrid cells. Dev Dyn. 2003;227:504–10.

Takahashi K, Yamanaka S. Induction of pluripotent stem cells from mouse embryonic and adult fibroblast cultures by defined factors. Cell. 2006;126:663–76.

Takahashi K, Tanabe K, Ohnuki M, Narita M, Ichisaka T, Tomoda K, Yamanaka S. Induction of pluripotent stem cells from adult human fibroblasts by defined factors. Cell. 2007;131:861–72.

Tateishi K, He J, Taranova O, Liang G, D'Alessio AC, Zhang Y. Generation of insulin-secreting islet-like clusters from human skin fibroblasts. J Biol Chem. 2008;283:31601–7.

Thomson JA, Itskovitz-Eldor J, Shapiro SS, Waknitz MA, Swiergiel JJ, Marshall VS, Jones JM. Embryonic stem cell lines derived from human blastocysts. Science. 1998;282:1145–7.

Tokuzawa Y, Kaiho E, Maruyama M, Takahashi K, Mitsui K, Maeda M, Niwa H, Yamanaka S. Fbx15 is a novel target of Oct3/4 but is dispensable for embryonic stem cell self-renewal and mouse development. Mol Cell Biol. 2003;23:2699–708.

Tsuji O, Miura K, Okada Y, Fujiyoshi K, Mukaino M, Nagoshi N, Kitamura K, Kumagai G, Nishino M, Tomisato S. Therapeutic potential of appropriately evaluated safe-induced pluripotent stem cells for spinal cord injury. Proc Natl Acad Sci. 2010;107:12704–9.

Weintraub H, Tapscott SJ, Davis RL, Thayer MJ, Adam MA, Lassar AB, Miller AD. Activation of muscle-specific genes in pigment, nerve, fat, liver, and fibroblast cell lines by forced expression of MyoD. Proc Natl Acad Sci USA. 1989;86:5434–8.

Wernig M, Meissner A, Cassady JP, Jaenisch R. c-Myc is dispensable for direct reprogramming of mouse fibroblasts. Cell Stem Cell. 2008a;2:10–2.

Wernig M, Zhao JP, Pruszak J, Hedlund E, Fu D, Soldner F, Broccoli V, Constantine-Paton M, Isacson O, Jaenisch R. Neurons derived from reprogrammed fibroblasts functionally integrate into the fetal brain and improve symptoms of rats with Parkinson's disease. Proc Natl Acad Sci USA. 2008b;105:5856–61.

Woltjen K, Michael IP, Mohseni P, Desai R, Mileikovsky M, Hamalainen R, Cowling R, Wang W, Liu P, Gertsenstein M, Kaji K, Sung HK, Nagy A. piggyBac transposition reprograms fibroblasts to induced pluripotent stem cells. Nature. 2009;458:766–70.

World Health Organization. Global data on visual impairments 2010. 2012. URL: http://www.who.int/blindness/GLOBALDATAFINALforweb.pdf. Accessed 28 Feb 2013[WebCite Cache].

Wu SC-Y, Meir Y-JJ, Coates CJ, Handler AM, Pelczar P, Moisyadi S, Kaminski JM. piggyBac is a flexible and highly active transposon as compared to sleeping beauty, Tol2, and Mos1 in mammalian cells. Proc Natl Acad Sci. 2006;103:15008–13.

Xie H, Ye M, Feng R, Graf T. Stepwise reprogramming of B cells into macrophages. Cell. 2004;117:663–76.

Xu D, Alipio Z, Fink LM, Adcock DM, Yang J, Ward DC, Ma Y. Phenotypic correction of murine hemophilia a using an iPS cell-based therapy. Proc Natl Acad Sci. 2009;106:808–13.

Yagi T,Ito D, Akamatsu W, Nihei Y, Yoshizaki T, Hideyuki O, Norihiro S, Shinya Y. Modeling familial Alzheimer's disease with induced pluripotent stem cells. Hum Mol Genet. 2011;20:4530–4539.

Yakubov E, Rechavi G, Rozenblatt S, Givol D. Reprogramming of human fibroblasts to pluripotent stem cells using mRNA of four transcription factors. Biochem Biophys Res Commun. 2010;394:189–93.

Yamanaka S. Induced pluripotent stem cells: past, present, and future. Cell Stem Cell. 2012;10:678–84.

Yang L, Baltimore D. Long-term in vivo provision of antigen-specific T cell immunity by programming hematopoietic stem cells. Proc Natl Acad Sci USA. 2005;102:4518–23.

Yu J, Vodyanik MA, He P, Slukvin II, Thomson JA. Human embryonic stem cells reprogram myeloid precursors following cell-cell fusion. Stem Cells. 2006;24:168–76.

Yu J, Vodyanik MA, Smuga-Otto K, Antosiewicz-Bourget J, Frane JL, Tian S, Nie J, Jonsdottir GA, Ruotti V, Stewart R, Slukvin II, Thomson JA. Induced pluripotent stem cell lines derived from human somatic cells. Science. 2007;318:1917–20.

Yu J, Hu K, Smuga-Otto K, Tian S, Stewart R, Slukvin II, Thomson JA. Human induced pluripotent stem cells free of vector and transgene sequences. Science. 2009;324:797–801.

Yusa K, Rad R, Takeda J, Bradley A. Generation of transgene-free induced pluripotent mouse stem cells by the piggyBac transposon. Nat Methods. 2009;6:363–9.

Yusa K, Rashid ST, Strick-Marchand H, Varela I, Liu PQ, Paschon DE, Miranda E, Ordonez A, Hannan NR, Rouhani FJ, Darche S, Alexander G, Marciniak SJ, Fusaki N, Hasegawa M, Holmes MC, Di Santo JP, Lomas DA, Bradley A, Vallier L. Targeted gene correction of alpha1-antitrypsin deficiency in induced pluripotent stem cells. Nature. 2011;478:391–4.

Zhang D, Jiang W, Liu M, Sui X, Yin X, Chen S, Shi Y, Deng H. Highly efficient differentiation of human ES cells and iPS cells into mature pancreatic insulin-producing cells. Cell Res. 2009;19:429–38.

Zhang R, Takebe T, Sekine K, Koike H, Zheng Y, Taniguchi H. Identification of proliferating human hepatic cells from human induced pluripotent stem cells. Transplant Proc. 2014;46:1201–4.

Zhao XY, Li W, Lv Z, Liu L, Tong M, Hai T, Hao J, Guo CL, Ma QW, Wang L, Zeng F, Zhou Q. iPS cells produce viable mice through tetraploid complementation. Nature. 2009;461:86–90.

Zhong X, Gutierrez C, Xue T, Hampton C, Vergara MN, Cao LH, Peters A, Park TS, Zambidis ET, Meyer JS, Gamm DM, Yau KW, Canto-Soler MV. Generation of three-dimensional retinal tissue with functional photoreceptors from human iPSCs. Nat Commun. 2014;5:4047.

Zhou H, Wu S, Joo JY, Zhu S, Han DW, Lin T, Trauger S, Bien G, Yao S, Zhu Y, Siuzdak G, Scholer HR, Duan L, Ding S. Generation of induced pluripotent stem cells using recombinant proteins. Cell Stem Cell. 2009;4:381–4.

Zhou L, Wang W, Liu Y, Fernandez J, de Castro T, Ezashi BP Telugu, Roberts RM, Kaplan HJ, Dean DC. Differentiation of induced pluripotent stem cells of swine into rod photoreceptors and their integration into the retina. Stem Cells. 2011;29:972–80.

Zhu FF, Zhang PB, Zhang DH, Sui X, Yin M, Xiang TT, Shi Y, Ding MX, Deng H. Generation of pancreatic insulin-producing cells from rhesus monkey induced pluripotent stem cells. Diabetologia. 2011;54:2325–36.

Chapter 5
Stem Cells and Hematopoietic Cell Engineering

Xiaoyan Xie, Yanhua Li, Yanxun Sun, Jin Zhang, Fang Fang, Wen Yue and Xuetao Pei

Abstract Hematopoietic stem cells (HSCs) and progenitor cells possess the potential to develop all type of blood cells in the circulatory system. The generation of these blood cells as well as HSCs/HPCs ex vivo might meet the deficiency in blood cell transfusion and blaze new trails in immunotherapy, hematological disease treatment, and cancer therapy. In this chapter, the potential clinical application of hematopoietic cell–engineering products is summarized. The obstacles, current strategies, and progress in hematopoietic cell engineering are discussed as well.

Keywords Stem cells · Hematopoietic stem cell · Engineering

5.1 The History and Current Situation of Blood Cell Therapy

The application of blood has a nearly thousand-year of history, which can be generally divided into three main stages: the ancient era which included blood baths and the drinking of blood; the modern era, which included whole-blood transfusion; and an even more modern era of component blood transfusion. The term "narrow blood transfusion" refers to whole-blood infusion and generalized transfusion including whole blood, whereas the transfusion of all kinds of tangible or intangible components of blood, strictly speaking, hematopoietic stem cells

X. Xie · Y. Li · Y. Sun · J. Zhang · F. Fang · W. Yue · X. Pei (✉)
Stem Cell and Regenerative Medicine Lab, Beijing Institute of Transfusion Medicine, Beijing 100850, China
e-mail: peixt@nic.bmi.ac.cn

X. Xie · Y. Li · Y. Sun · J. Zhang · F. Fang · W. Yue · X. Pei
South China Research Center for Stem Cell and Regenerative Medicine, AMMS, Guangzhou 510005, China

R.C. Zhao (ed.), *Stem Cells: Basics and Clinical Translation*, Translational Medicine Research 1, DOI 10.1007/978-94-017-7273-0_5

(HSCs) (bone marrow (BM) or peripheral blood), is a special kind of blood transfusion. In 1656, blood was successfully transfused into blood vessels. In 1667, a French aristocrat transfused 280 ml of calf blood to a "tramp" with mental disorders in an attempt to treat his mental problems. Although he had a severe immune response, the man's body eventually accepted the input of animal blood, and he miraculously survived, and his health was maintained for a period of time. After this event, blood transfusion therapy was accepted by some innovative doctors. The process of blood transfusion has had a long process of development, approximately 300 years. From 1817 to 1818, blood transfusion was applied directly from one human to another, and approximately three of eight cases of such were successful. However, in general, by this time transfusion therapy was still in the exploratory stage because there was no relevant knowledge (such as blood type) to further the practice of blood transfusion. Although blood transfusion caused many deaths, some doctors found that blood transfusion really can save lives. In 1901, Viennese pathologist Karl Landsteiner first discovered the human ABO blood group and aggregation rule, which provided a solid pathophysiological foundation for modern blood transfusion. In 1912, French doctor Alexis Carrel won the Nobel Prize in Physiology for inventing the technique of vascular anastomosis for blood transfusion, thus promoting greater acceptance of transfusion therapy. In the 10 years that followed, other doctors gradually established blood anticoagulation and cross-matching technology. These epoch-making discoveries made blood transfusion safe and effective, and blood transfusion became the clinical "rescue" treatment for many severe diseases. Landsteiner also won the Nobel Prize in Physiology or Medicine in 1930. Because blood in vitro clotted quickly, physicians began to explore methods of anticoagulation. In 1914, sodium citrate anticoagulation was discovered. In 1943, the discovery of the ACD solution prolonged the period of blood preservation up to 21 days. In 1959, component blood transfusion was put forward by Gibson, but it was not truly developed until the early 1970s and to the late 1980 s. The amount of component blood applied in developed countries now accounts for 80 to 90 % of all blood used.

The percentage of component blood transfusion is one of the most important markers by which to measure the level of medical technology in a country or region. At present, the international proportion of component blood accounts for >90 % of total component blood applied and for <10 % of whole blood; developed countries account for >95 %. This progress has laid a solid foundation for all countries in the world to establish a blood bank, thus turning the dream of blood transfusion therapy into a solid reality. Clinical blood transfusion treatment is currently the most widely used, the most maturely developed, and the most massive means of cell therapy used in surgery and trauma, chronic anemia, hematologic malignancies, tumor radiation and chemotherapy, and the treatment of many other diseases. Through the continuous development and innovation in the clinical application of and research on blood transfusion, from which the first mature cell therapy method was developed, blood transfusion therapy has gradually developed into one of the most dynamic research fields as well as one of the important means of treatment in various hospital departments settings.

In recent years, with the rapid development of medical technology and people's health needs, the demand for blood in medical institutions has shown a rapid trend in growth. According to statistics, presently >99 % of blood in clinical use comes from voluntary blood donation. However, participation in unpaid blood donation is relatively weak in many countries worldwide. For example, in China, the citizens' blood donation rate is approximately 8.7 %. Compared with the proportion of the middle-income countries in the world, there is a large gap. In addition, although the total volume has increased from 800 tons/year 10 years ago to ≥3600 tons/y currently, the China's blood supply is still very low, and the gap between blood supply and blood demand is very conspicuous. It is estimated that the annual gap in China's blood supply gap is approximately 10 to 15 %. Furthermore, there is a seasonal blood shortage in many large cities, which has drawn national attention to solving the problem of insufficient blood supply. In addition, solving the problem of various types of pathogen contamination in the blood supply is a great challenge. Clinical practice has proven that a variety of viruses can be spread by blood transfusion, thus causing viral spread—especially viruses of the lipid envelope such as hepatitis B virus (HBV), hepatitis C virus (HCV), and the AIDS virus (HIV)—as well as serious consequences after blood transfusion. Transfusion of blood that has gone through strict testing still carries the risk of iatrogenic infection, and the main reasons for this are as follows:

- The existence of a certain proportion of false-negative results by reagent testing means that virus-carrying blood will inadvertently be used in clinical blood therapy.
- Blood donors are in a "window" of infection when antibodies have not been produced or are at low concentrations.

In large cities, such as Beijing, all collected blood is tested using the latest and most sensitive detection technology—i.e., nucleic acid detection technology—to detect viruses before blood is transfused, but this technology still cannot detect pathogenic viruses of the blood during the window of infection. The emergency blood demand during wartime and natural disasters is of course larger, and because blood security requirements are greater at these times, the attrition rate of the country's blood supply concomitantly increases. With prompt administration, blood transfusion can save the lives of wounded individuals who are losing blood. Blood has been an important "drug" since the early treatment of war and trauma injuries, but it is difficult to fully meet the blood requirements under the current system.

5.2 Various Sources of Stem Cells for Ex Vivo Hematopoietic Cell Generation

The transplantation of HSCs has been used to treat various diseases including malignant and nonmalignant hematologic conditions, immunodeficiencies, and metabolic disorders. Infused HSCs can replace and reconstitute the hematopoietic and immune systems, thus sparing patients from the effects of chemotherapy or radiotherapy. Therefore, harvesting sufficient HSCs is important for autologous or allogeneic stem cell transplantation. For hematopoietic support or immune therapy, large numbers of HSCs can produce sufficient progeny cells, such as erythrocytes, megakaryocytes, and immune cells, ex vivo. These downstream blood cell products also provide potential new strategies in clinical multiple-disease therapy. Thus, harvesting or manufacturing sufficient numbers of HSCs is the first step for the treatment of both hematological and nonhematogical diseases.

5.2.1 The Sources of HSCs in Clinical Application

The sources of HSCs in clinical application are mainly BM and mobilized peripheral blood. In the last decade, umbilical cord blood (UCB) has become a new resource of HSCs for allogeneic transplantation (Barker et al. 2005). Due to the low number of collectable stem cells from a single UCB unit, their initial use was limited to pediatric therapies. Clinical application of UCB HSCs in adults would become feasible if there were a culture method that can effectively expand HSCs while maintaining their self-renewal capacity. In recent years, numerous attempts have been made to expand human UCB HSCs in vitro. Most ex vivo cultures of human HSCs use cytokine mixtures in serum-free medium that includes SCF, TPO, IL-3, IL-6, and Flt3-L (Murray et al. 1999). The manipulation of some signaling pathways, such as Notch and Wnt, has also shown effectiveness for the ex vivo expansion of HSC availability (Delaney et al. 2010; Perry et al. 2011). Several laboratories have explored expanding the numbers of HSCs with aryl hydrocarbon receptors, copper chelators, stromal support, and automated continuous perfusion of culture systems or "bioreactors" (Boitano et al. 2010; Ferreira et al. 2012; Peled et al. 2004; da Silva et al. 2005). Overexpression of transcription factors, such as SALL4 and HOXB4, can also increase the number of HSCs in vitro (Yang et al. 2011; Amsellem et al. 2003). Several clinical trials have been initiated with ex vivo expanded human $CD34^+$ CB progenitors (Oran and Shpall 2012). Although the in vitro culture method for the expansion of HSC availability has been improved, it still needs to be optimized to obtain more transplantable HSCs.

5.2.2 Pluripotent Stem Cell–Derived HSCs

Currently HSCs used during clinical transplantation are usually derived from cord blood, adult BM, or peripheral blood. Given the shortage of donors and the limited supply of HSCs, human ESCs have become an alternative source of HSCs and mature blood cells for therapeutic purposes. ESCs have the capacity for nearly unlimited proliferation and can differentiate into cells/tissues of all three germ layers. In recent years, induced pluripotent stem cells (iPSCs) with the characteristics of ESCs have been obtained using adult somatic cell reprogramming, which involves the overexpression of key transcription factors (Takahashi et al. 2007) or culturing with combinations of exogenous small molecules (Hou et al. 2013). The development of iPSC technology has revolutionized the possibilities of personalized cell therapy. These iPSCs could serve as an embryo-free source of patient-specific stem cells with decreased human leukocyte antigen (HLA) mismatching. Taylor suggested that a tissue bank from 150 selected homozygous HLA-typed volunteers could match 93 % of the United Kingdom's population with a minimal requirement for immunosuppression (Taylor et al. 2012). The use of iPSC technology could also be applied to modify mutated genes from patients with inherited disorders such as sickle cell anemia (Zou et al. 2011) and α-thalassemia (Chang and Bouhassira 2012). Several groups have generated gene-corrected β-thalassemia iPSCs from patients; these could be induced to differentiate into hematopoietic progenitor cells (HPCs) and then into erythroblasts expressing normal β-globin (Wang et al. 2012; Fan et al. 2012; Ma et al. 2013).

It has been reported that pluripotent stem cells (PSCs) can be induced to differentiate into hematopoietic cells using three different culture methods: embryoid body formation, stromal cell–based coculturing, and monolayer culturing (Dang et al. 2002; Vodyanik et al. 2005; Niwa et al. 2011). These different induction strategies often engage several sets of cytokines to provide a simulated microenvironment for hematopoiesis. PSCs undergo several stages when differentiating into hematopoietic cells. Bone morphogenetic protein-4 (BMP-4) and basic fibroblast growth factor (bFGF) have been employed to induce the differentiation of mesoderm progenitors (Bernardo et al. 2011). Vascular endothelial growth factor (VEGF) and bFGF have been used to promote hemangioblast specification (Wang et al. 2012). Hematopoietic cytokines—such as stem cell factor (SCF), Fms-like tyrosine kinase 3 (Flt-3), interleukin-3 (IL-3), IL-6, thrombopoietin (TPO), and erythropoietin (EPO)—have been used to promote the formation of hematopoietic cells. Until now, the generation efficiency of $CD34^+$ hematopoietic cells from PSCs has required improvement. McKinney-Freeman et al. applied a network biology-based analysis to reconstruct gene regulatory networks during sequential stages of HSC development (McKinney-Freeman et al. 2012). They found that HSCs from in vitro–differentiated embryonic stem cells (ESCs) closely resembled definitive HSCs; however, they lacked a Notch signaling signature. Their results suggested that an exogenous Notch ligand could be added to the induction system and thus facilitate hematopoietic commitment (McKinney-Freeman et al. 2012). Notch ligand has also been used to promote the expansion of cord blood HSCs (Delaney et al. 2010). Although in vitro culture methods for the differentiation of PSCs into hematopoietic cells have been improved, optimization is required to obtain HSCs that are more suitable for transplantation. To efficiently generate engrafted hematopoietic cells, the forced expression of certain transcription factors, such as HOXB4 and RUNX1a, is an alternative approach (Bowles et al. 2006; Ran et al. 2013). Recently, Daley's laboratory reported a strategy to respecify $CD34^+CD45^+$ myeloid precursors from PSCs into multilineage progenitors that can be expanded in vitro and engrafted in vivo using five transcription factors (Doulatov et al. 2013). A greater understanding of hematopoiesis mechanisms might help us develop more efficient induction strategies for hematopoietic cells.

5.2.3 Lineage Reprogramming of Somatic Cells into HSCs

The generation of HSCs from PSCs is still limited by multiple step induction and low differentiation efficiency. Direct conversion of human autologous somatic cells into engraftable HSCs offers tremendous clinical application potential. Several methods have been reported to reprogram somatic cells into hematopoietic cells. Cellular fusion technique has been used to directly convert somatic cells into fetal HSCs (Sandler et al. 2011). The enforced expression of transcription factors, such as OCT-4–activated hematopoietic transcription factor, together with specific cytokine treatment has allowed a generation of multilineage blood progenitors to be produced from human fibroblasts (Szabo et al. 2010). Introduction of a combination of transcription factors has also been used to reprogram endothelial-like precursor cells or endothelial cells into hematopoietic cells (Pereira et al. 2013; Sandler et al. 2014). Direct conversion between different blood cell types is also feasible; for example, B cells can be reprogrammed into macrophages (Xie et al. 2004). The progress in these findings and techniques offers a potential pathway to new autologous cell therapies for the treatment of hematological and nonhematological diseases.

5.2.4 Conclusion

BM and mobilized peripheral blood are the main sources of clinical HSC transplantation. In addition, UCB is now an accepted source of allogeneic HSC transplantation. Single UCB has shown a benefit in children with hematological malignancies (Locatelli et al. 1999). Double UCB also presents wider applicability in adults (Brunstein et al. 2011). Human placenta and chorion are potential additional sources of HSCs for transplantation (Bárcena et al. 2011). To obtain large numbers of autologous HSCs, new approaches have been in development to generate HSCs from PSCs or somatic cells using microenvironmental induction and/or reprogramming strategies. Prospective large-scale production of functional HSCs with multiple approaches might be used in clinical HSC transplantation or blood cell-based therapy.

5.3 Hematopoietic Cell Engineering for Blood Cell Generation

Stem cells have the potential of self-renewal, proliferation, and differentiation into all cell types of the body. In recent years, with the development of stem cell–related technologies, it is possible to directly induce stem cells toward the cells of many tissues we need in vitro. This possibility would greatly facilitate the development of clinical application.

The hematopoietic system is a highly active metabolism system. Because of the differentiation potential of HSCs and progenitor cells into multiple lineages of blood cells, HSCs and HPCs are promising cell sources for the treatment of prevalent and currently incurable diseases (Lim et al. 2011). Hematopoiesis and its regulation depend on a variety of hematopoietic growth factors, stromal cells, and extracellular matrix as well as the interaction among and balance of these factors. The complex process of hematopoiesis involves cell proliferation, differentiation, development, maturation, migration, integration, aging, apoptosis, and cancerization.

By using prevalent methods, we cannot obtain enough cells, including stem cells and their descendents, to satisfy the need for clinical use. Tremendous efforts have been made to develop new efficient methods of differentiating stem cells into target cells to overcome the current limitations. However, scientists soon learned that it is extremely difficult to control the fate and differentiation of PSCs (Lim et al. 2011). New methods should be found to solve this formidable problem. A new therapeutic technology, called "hematopoietic cell engineering," which is the third generation of cell therapy, has attracted increasing attention.

Hematopoietic cell engineering, a promising strategy for the treatment of numerous diseases, is the use of the high proliferative capacity and multilineage-differentiation potential of HSCs/progenitor cells to simulate or partially mimic hematopoiesis in vitro, including the regulation of stromal cells/hematopoietic growth factors, by cell-engineering technique.

Hematopoietic cell engineering focuses on the separation, purification, amplification, directional differentiation, and gene modification of HSCs/progenitor cells in vitro. By amplifying and differentiating early hematopoietic progenitor/precursor cells, the terminal aim of hematopoietic cell engineering is to obtain large amounts of blood cells, including erythrocytes, granulocytes/macrophages, megakaryocytes/platelets, dendritic cells (DCs), NK cells, T/B lymphocytes, and all of the other blood components.

Hematopoietic cell engineering, including cell regulation and the modification of defective genes, will not only meet the needs of the basic research, it will eventually become more widely and effectively accepted in the fields of stem cell transplantation, biological immune therapy, gene therapy, etc. Below are the possible benefits of hematopoietic cell engineering.

1. Decreasing the incidence of transfusion-transmitted infections: Blood transfusion is now playing an indispensable role in rescuing wounded patients in urgent situations. However, blood transfusion can also cause adverse patient reactions and serious complications. The main concern of patients receiving a blood transfusion is the risk of acquiring a transfusion-transmitted infection (Schmidt et al. 2014) including viruses, bacteria, spirochetes, and protozoa. Regarding viruses, for example, the rates of transfusion-related HIV1/2, HBV, and HCV infections are increasing. In addition, blood transfusion can also cause some patient immune reactions. Red blood cells (RBCs), platelets, and leukocytes can become triggers of an immune response. Therefore, the safety of blood transfusion is always a concern. Through the process of hematopoietic cell engineering, we may avoid the problem of transfusion-transmitted infection via gene modification and thus pave the way for the safety of transfusion in the future.
2. Amplification of RBCs as blood reserves: The differentiation and amplification of HSCs has attractive prospects in the field of transfusion. HSCs can directional differentiate and amplified into erythroblasts, RBCs, and neutrophils in vitro. The infusion of mixed cells can partially replace the traditional source of blood. It not only alleviates the shortage of blood to a certain extent, it also has advantages regarding the safety of blood transfusion and may decrease or even prevent the incidence of blood-borne infections.
3. Hematopoietic support therapy after HSC transplantation and high-dose chemotherapy: The reduction of neutrophils and platelets is a major limitation of malignant tumor chemotherapy. The strategies of HSC transplantation plus high-dose chemoradiotherapy can improve the therapeutic effect on malignant tumors and prolong patient survival. One aspect of the significance of HSC engineering lies in the ability to obtain a sufficient number of HSCs for transplantation. Another significance of HSC engineering is the ability to produce large amounts of young blood cells for transfusion after the treatment. The time lapse of $CD34^+$ cells to differentiate into mature neutrophils and platelets is approximately 10 days. The infusion of partially differentiated neutrophils and/or megakaryocyte progenitor cells cultured in vitro at a certain time may render those cells mature soon after infusion and play a physiological role. Thus, hematopoietic cell engineering can play important role in hematopoietic support in terms of infection prevention and treatment of bleeding and anemia, compatability with high-dose chemotherapy, etc.
4. Immune therapy associated with hematopoietic cells. Immune therapy mediated by lymphokine-activated killer cells, cytokine-induced killer (CIK) cells, cytotoxic T lymphocytes (CTL), and DCs is gradually becoming an important treatment strategy for malignant tumors and other diseases. In addition, HSCs are one of the ideal target cells for gene therapy. On this basis, gene therapy based on hematopoietic cell engineering holds bright prospects in the treatment of severe immunodeficiency, genetic diseases, malignant tumor, and AIDS (Caplan 2000).

5.4 Regulation Mechanisms in Stem Cell-Blood Cell Development

Hematopoiesis is a continuous process in which stem/progenitor cells develop into mature blood cellular components. As hematopoietic cells differentiate from stem cells to the mature lineages, they gradually become more committed to their ultimate lineage and lose their multipotentiality and self-renewal ability and in turn gain more specialized functionalities. During maturation, HSCs and HPCs differentiate into all blood-forming elements including myeloid (monocytes, macrophages, neutrophils, basophils, eosinophils, erythrocytes, megakaryocytes/platelets, and DCs) and lymphoid lineages (T cells, B cells, and NK cells) (Kaufman 2009). Decisions regarding the fate of blood cells in the hematpoietic system are dictated by a complex regulatory mechanism involving hematopoietic cytokines, cytokine-mediated transcriptional regulators, and manifold intercellular signaling pathways.

5.4.1 Cytokine and Cytokine-Induced Transcriptional Regulation of HSC Differentiation

Cytokines are small soluble proteins that can either be secreted or membrane-bound. They are produced by a variety of different cell types and can act both locally within the microenvironment of HSCs/progenitor cells and systemically by way of the bloodstream or lymphatic vessels. Two general models for the role of cytokines in hematopoietic differentiation have been proposed: the instructive model and the stochastic model. In the instructive model, cytokines transmit specific signals to multipotential hematopoietic cells, thus directing their lineage commitment and differentiation. In the stochastic model, lineage commitment and terminal determination are intrinsically determined by cytokines providing permissive growth and survival signals (D'Andrea 1994). Cytokines of the hematopoietic system include interleukins (ILs), colony-stimulating factors (CSFs), interferons, EPO, and TPO. Both gain-of-function and loss-of-function experiments in vivo have confirmed the cytokine-dependent regulation of lineage-specific blood cell formation (Fig. 5.1) (Metcalf 2008; Laiosa et al. 2006).

5.4.1.1 Interleukin-6

Inflammatory cytokines, particularly interleukin 6 (IL-6), play a critical role in myeloid cells differentiation IL-6 is a pleiotropic cytokine secreted by numerous immune and nonimmune cells. In a state of emergency granulopoiesis, IL-6

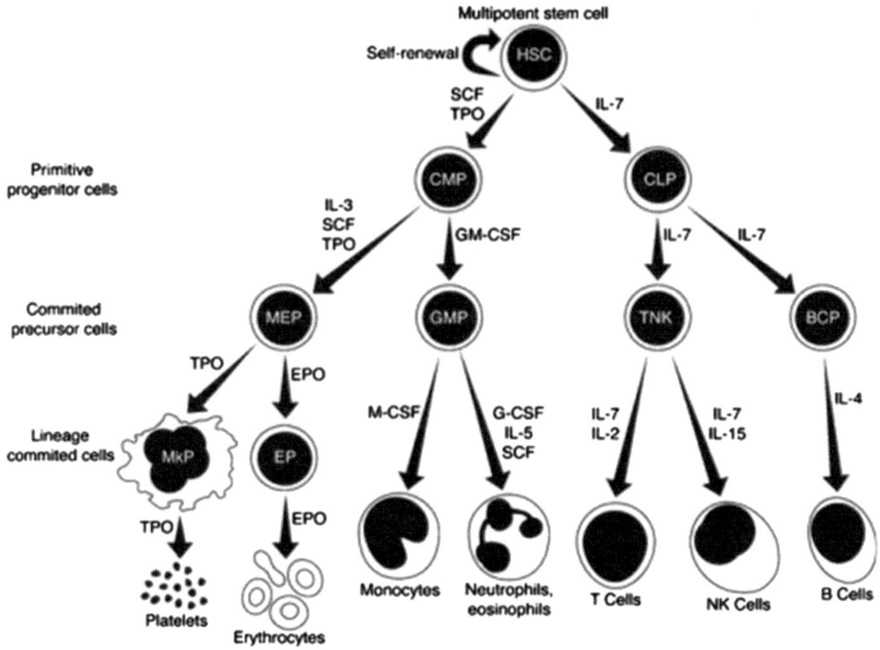

Fig. 5.1 Hematopoiesis and the role of cytokines in vivo. Cytokines act on both multipotential progenitors and their committed offspring (Robb 2007)

mediates hematopoietic progenitor differentiation toward myeloid lineage through the expression of the Id1 transcription factor (Maeda et al. 2009; Newburger 2008). IL-6 interacts with the CAAT/enhancer binding protein–beta (C/EBP-β) protein, which is a leucine zipper transcription factor (Poli et al. 1990). C/EBP-β has been observed during myeloid differentiation, which is a key transcription factor regulating monocytic gene expression (Liu et al. 2009; Ramji and Foka 2002). Furthermore, C/EBP-β expression is induced dramatically during macrophage differentiation (Natsuka et al. 1992; Scott et al. 1992); for example, C/EBP-β–deficient mice lacked emergent neutrophil production in response to cytokines and/or infection (Hirai et al. 2006).

5.4.1.2 GM-CSF

Granulocyte-macrophage colony–stimulating factor (GM-CSF) is a critical hematopoietic growth factor that controls the differentiation of myeloid progenitors into granulocytes, eosinophils, monocytes, megakaryocytes, and erythrocytes (Metcalf 2008). GM-CSF was the first cytokine shown to efficiently promote DC development in vitro and has been used to induce DC differentiation from human monocytes as well as human and mouse HPCs (Inaba et al. 1992; Sallusto and Lanzavecchia 1994; Caux et al. 1996; Caux et al. 1992).

5.4.1.3 M-CSF

Macrophage colony–stimulating factor (M-CSF) is an important inducer of monocytic development. M-CSF signals through its corresponding tyrosine kinase receptor, which is widely expressed on the common myeloid progenitor (CMP) and granulocyte-macrophage progenitor (GMP) myeloid compartments. (Auffray et al. 2009) M-CSF is regulated by the transcription factor MafB to restrict lineage commitment toward myelopoiesis. (Hegde et al. 1999) Hematopoietic cells with myeloid and lymphoid potential exhibit a high sensitivity to M-CSF and undergo differentiation toward myeloid cells at high M-CSF concentrations (Himes et al. 2001). In fact, high M-CSF concentrations inhibit proper myeloid differentiation and promote the differentiation of DCs into macrophage-like cells (Menetrier-Caux et al. 1998). Furthermore, combined treatment with IL-3 and M-CSF efficiently differentiates embryonic stem cell lines toward homogenous monocytic cells (Karlsson et al. 2008).

5.4.1.4 EPO

Signaling by EPO through its receptor EpoR is necessary for the production of definitive erythrocytes in murine fetal liver. EPO has recently been shown to increase erythroid and decrease granulocyte/macrophage (GM) output from hematopoietic stem and progenitor cells by the suppression of nonerythroid fates at a transcriptional level, thus suggesting an instructive role for EPO on hematopoietic stem and progenitor cells in vivo (Grover et al. 2014).

5.4.1.5 TPO

TPO is the chief cytokine that regulates megakaryocyte production by signaling through its receptor Mpl. Megakaryocytes are the only hematopoietic cell lineage affected by the loss of TPO signaling, e.g., there is less than half in $TPO^{-/-}$ and $Mpl^{-/-}$ mice (Alexander et al. 1996; Carver-Moore et al. 1996).

5.4.2 Transcription Regulators that Control the Differentiation of HSC

The cell-fate specification of HSC is intricately controlled by intercellular and intracellular signaling mechanisms (Kaushansky 2006; Mikkola and Orkin 2006). These mechanisms commonly target transcriptional regulators, which in turn establish complex transcriptional networks. Each commitment step requires the activation of lineage-specific genes, whereas unnecessary and conflicting genes

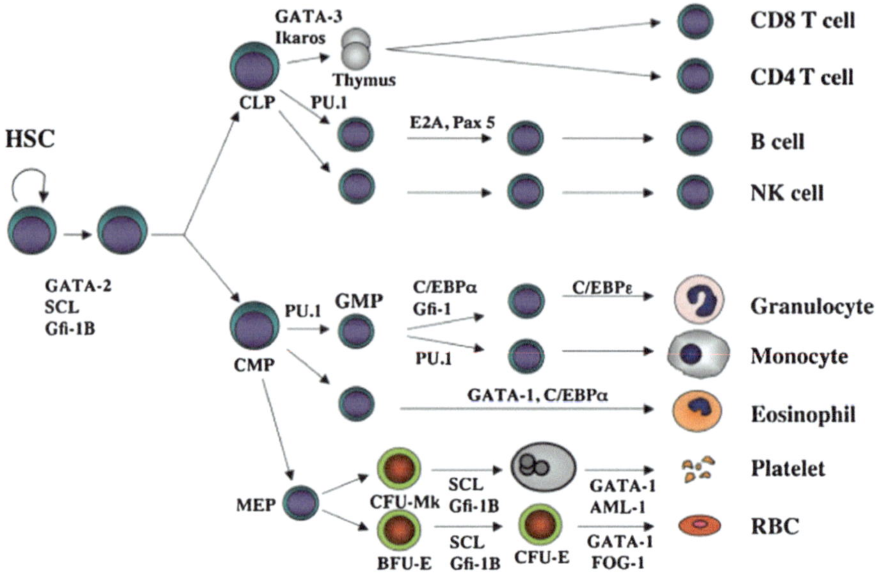

Fig. 5.2 Transcription factors regulating hematopoietic differentiation (Nakajima 2011)

are concomitantly repressed. This balance involves the concerted actions of multiple transcriptional activators, repressors, and epigenetic modifiers (Teitell and Mikkola 2006). Several transcriptional factors—such as PU.1, C/EBPα, GATA1/2, Pax5, and SCL genes—have been identified as essential for the development of definitive hematopoietic cells during embryogenesis (Fig. 5.2).

5.4.2.1 Purine-Rich Box 1 (PU.1)

PU.1 is an ETS-family transcription factor. This transcription factor is an ideal marker of early cell fate change because it is both required and sufficient to drive myeloid fate in early multipotent stem and progenitor cell populations (Iwasaki and Akashi 2007). As HSCs and multipotential progenitors differentiate, PU.1 is maintained or upregulated in some lineages and downregulated in others. For example, PU.1 is expressed in CMP; however, the level of its expression increases in granulocyte monocyte progenitor (GMP) and their progeny (granulocytes and monocytes), but it decreases in megakaryocyte–erythrocyte progenitors and their derivatives (megakaryocytes and erythroid cells), which suggests that the effect of PU.1 is dose-dependent (Hromas et al. 1993; Chen et al. 1995; Akashi et al. 2000; Miyamoto et al. 2002). Disruption of PU.1 from hematopoiesis has shown that PU.1 fails to initiate a commitment to both myeloid and lymphoid lineages, such as CMPs, GMPs, and CLPs, in adult hematopoiesis but not for MEP development. These observations indicate that PU.1 is not required for the specification of monocytic precursors but is important for their further functions (Iwasaki et al. 2005).

5.4.2.2 C/EBPα

There are six members of the CCAAT/enhancer binding protein (C/EBP) family: C/EBPα, β, γ, δ, ε, and ξ) (Ramji and Foka 2002). C/EBP-family proteins play a relevant role in the process of hematopoiesis and (myelo)monocytic differentiation, especially that of C/EBPα. C/EBPα is the predominant isoform in immature granulocytes (Hohaus et al. 1995; Scott et al. 1992). C/EBPα expression is upregulated during early differentiation stages toward GMP in humans and mice (Friedman 2007). C/EBPα knock-out mice showed inhibited differentiation of CMP toward GMP resulting in a decreased number of GMP, MDP, monocytes, and macrophages as well as granulocytes in combination with an accumulation of immature myeloid progenitors (Zhang et al. 1997; Zhang et al. 2004; Heath et al. 2004). C/EBPα also caused a lineage switch toward the myeloid lineage in megakaryocyte-erythroid progenitors (MEPs). This supports the assumption that C/EBPα suppresses erythroid lineage formation but promotes determination of the myeloid lineage (Suh et al. 2006; Fukuchi et al. 2006). C/EBPα is a critical factor for the myeloblast to undergo promyelocyte transition. The conditional expression of C/EBPα in a bipotential hematopoietic cell line induces granulocytic differentiation and G-CSF receptor expression and blocks monocytic differentiation (Radomska et al. 1998).

5.4.2.3 GATA1

GATA-1 is a key erythroid transcription factor required for the development of normal erythroid and megakaryocytic lineages. Expression of GATA-1 in Myb-Ets–transformed myeloblasts reprograms them into eosinophils or thromboblasts while at the same time suppressing myelomonocytic differentiation (Kulessa et al. 1995). GATA1 was also found to repress the expression and function of PU.1. GATA-1 and PU.1 functionally antagonize each other through direct physical interaction. In Xenopus embryos, ectopic expression of PU.1 blocks erythropoiesis during normal development. Introduction of exogenous GATA-1 can trigger Xenopus embryos to resume differentiation and undergo terminal cell division to lose their tumorigenicity (Rekhtman et al. 1999). Recent work in zebrafish shows that transcription intermediate factor-1γ (tif-1ϒ) modulates the myeloid-erythroid fate decision controlled by the GATA1–PU.1 balance in various teleost hematopoietic populations (Kulessa et al. 1995).

5.4.2.4 GATA-2

GATA2 is an another transcriptional factor that belongs to the GATA family, and it is important for the development of HSCs. GATA-2 is expressed in HSCs (HSCs), multipotent hematopoietic progenitors, erythroid precursors, megakaryocytes, eosinophils, and mast cells (Monteiro et al. 2011; Tsai et al. 1994; Tsai and Orkin

1997; Minegishi et al. 1999). GATA-2 regulates early hematopoiesis by controlling the genesis and/or survival of HSCs and/or multipotent progenitors (Tsai et al. 1994; Tsai and Orkin 1997; Ling et al. 2007). GATA-2 is crucial for mast cell development, and GATA-2 knock-out hematopoietic precursors are competent to undergo erythroid and myeloid terminal differentiation (Tsai and Orkin 1997). Overexpression of GATA-2 in a human leukemia cell line (K562) inhibited erythroid and promoted megakaryocyte differentiation (Ikonomi et al. 2000).

5.4.2.5 Pax5

Pax5 is a multifunctional transcriptional regulator expressed throughout the B-cell lineage from the pro-B cell stage until its downregulation in plasma cells (Busslinger 2004; Singh et al. 2005). Pax5 regulates B-cell lineage commitment by increasing the expression of the characteristic B-cell genes CD19 and Igα as well as suppressing myeloid genes such as c-fms and PD-1. Conditional inactivation of Pax5 in mature B cells resulted in de-differentiation to lymphoid progenitors, which gave rise to functional T cells that can develop into T lymphocytes, natural killer (NK) cells, and macrophages (Rolink et al. 1999).

5.4.2.6 SCL/Tal-1

Stem cell leukemia/T-cell acute lymphocytic leukemia-1 (SCL/tal1) is a member of the basic helix–loop–helix family of transcription factors, which has important functions to stimulate the generation of hemangioblasts, which differentiate into both blood and endothelial cells (Chung et al. 2002; Ema et al. 2003; Gering et al. 2003; Patterson et al. 2007). Mice lacking SCL had severe defects in blood formation and lack of any hematopoietic lineages (Porcher et al. 1996). In contrast to the critical role that SCL plays in HSC generation, it is not required for HSC function in mice. However, in the absence of SCL, megakaryocytic and erythroid differentiation are severely hampered (Mikkola et al. 2003).

5.4.3 Signaling Pathways Regulating HSC Differentiation

Signal transduction is the process by which cells receive and respond to information from the outside environment, which often occurs through secreted proteins binding receptors on the cell's surface, which alter intracellular mediators to affect gene expression. Many signaling pathways—such as TGFβ/BMP, Wnt, and Notch, which controls hematopoiesis in the embryo—are involved in the differentiation of HSC.

5.4.3.1 TGF-β/BMP Signalling Pathway

TGFβ is the foundational member of the TGF-β superfamily and has several mammalian isoforms (TGFβ1–TGFβ5). The binding of TGFβ to its receptor (I, II, or III) activates receptor serine/threonine kinase, which in turn activates several intracellular signaling pathways. Through detailed studies of hematopoietic progenitors in vitro, (Fortunel et al. 2000) TGF- β has been characterized as a well-known regulator of hematopoiesis. TGF- β can enhance the differentiation of myeloid progenitors into granulocyte or monocytes (Ottmann and Pelus 1988; Keller et al. 1991; Keller et al. 1994). TGF-β1 inhibits the proliferation of both human and murine hematopoietic stem and primitive progenitor cells in vitro, whereas further differentiated progenitors are more resistant to TGF-β inhibition (Batard et al. 2000; Sitnicka et al. 1996; Keller et al. 1990; Jacobsen et al. 1991). Bone morphogenetic proteins (BMPs) are also members of the TGF-β superfamily. They bind as dimers to cell surface receptors consisting of type I and type II serine/threonine kinase subunits. Bmp4 signalling has been shown to be directly required to trigger HSC specification from the hemogenic endothelium of the dorsal aorta in zebrafish (Wilkinson et al. 2009). BMP4 has also been reported to induce hematopoietic differentiation in both murine and human ESC in vitro (Johansson and Wiles 1995; Chadwick et al. 2003). At high concentration, BMP4-mediated signaling acts to maintain the proliferation of HSCs derived from human UCB rather than promoting their differentiation (Bhatia et al. 1999).

5.4.3.2 The Notch Pathway

In mammals, the Notch pathway comprises different proteins including four different Notch receptors (Notch 1 through 4), two Jagged ligands (Jag 1 and 2), three Delta-like ligands (Dll 1, 3, and 4), nuclear transcription factor RBPj, and specific cofactors such as Mastermind (Kopan and Ilagan 2009). Notch is a highly conserved signaling pathway involved in the regulation of cell-fate acquisition and differentiation in several systems, and its role in both adult and embryonic hematopoiesis has been carefully studied (Fortini 2009). Notch signalling is essential for T-cell specification and differentiation in the thymus as well as for splenic marginal zone B-cell development (Maillard et al. 2003; Radtke et al. 2004). Deletion of RBPjκ/CSL results in increased B-cell differentiation and blockage of T-cell development (Han et al. 2002). Several investigators have reported that alterations in Notch signaling has minimal effects on myelopoiesis. Enforced expression of the Notch intracellular domain (NICD) or Notch target genes in BM progenitors abrogates B-cell development and promotes myeloid differentiation in a non–cell autonomous manner (Schroeder et al. 2003; Kawamata et al. 2002). Mice deficient in downstream Notch effectors exhibit defective B-cell and myeloid development (Kawamata et al. 1998). Early studies using myeloid cell lines showed that forced Notch 1 or Notch 2 activation could inhibit myeloid differentiation in a cytokine-dependent manner (Milner and Bigas 1999; Bigas et al. 1998). These observations indicate that Notch-mediated alterations in lymphocyte development can affect myelopoiesis.

5.4.3.3 Wnt Signaling Pathway

Wnt/β-catenin is a highly evolutionary conserved signaling pathway required for the proper development of most tissues and organisms. There are two main intracellular signal transduction pathways, the β-catenin/TCF–dependent pathway and the β-catenin–independent pathway, which are usually referred to as the "canonical pathway" and the "noncanonical pathway," respectively. Wnt proteins play a critical role in the development of multiple hematopoietic lineages. However, the effect of Wnt/β -catenin signaling on the regulation of HSC/HPC function is complex. Some investigators have shown that the constitutive activation of β-catenin made lymphoid or myeloid progenitor lineages unstable and allowed the expansion of multipotential cells in culture (Baba et al. 2005, 2006). Others investigators found that the expression of constitutively active β-cateninin hematopoietic cells in vivo leads to loss and blockade of their differentiation and maturation (Kirstetter et al. 2006; Scheller et al. 2006). In vitro stimulation of HSC with recombinant Wnt3a lead to the de-differentiation of committed B cells to more stem cell–like cells, but the noncanonical Wnt5a protein had opposite effects (Malhotra et al. 2008).

Emerging evidence indicates that multiple signaling pathways are activated during HSC differentiation and that these pathways may play critical roles in facilitating the differentiation process. Several other signals have been implicated as regulators of HSC specification, although their roles have not been studied as well as those of BMP, Notch, and Wnt. These pathways include JAK-STAT, Ras, Raf/MEK/ERK cascade, p38 MAPK, and JNK/SAPK, PI3 K-Akt pathway.

5.4.4 Summary

Decisions of blood cell fate in the hematpoietic system are made by way of multiple mechanisms. Cytokine-activated signaling pathways can regulate the expression or activity of key proteins. Activated proteins may promote the cell-cycle arrest necessary for differentiation to occur, whereas other signaling molecules may act primarily to sustain the survival of cells undergoing differentiation. Still others may act to drive the phenotypic changes associated with the differentiation process. Furthermore, investigators studying HSC differentiation will also need to consider specific transcription factors and the microenvironmental cross-talk between stem cells and multiple niche compartments in BMs. Such cross-talk may occur in a stage- or lineage-specific manner. In addition, nonproteinaceous regulators of HSC differentiation should also be investigated. Recent studies have shown that micro-RNAs and epigenetic regulators play critical roles in the regulation of HSCs differentiation (Undi et al. 2013; Oh and

Humphries 2012). More focused research—such as how multicomponent signaling networks are functionally integrated into long-term decisions of cell fate and how identical pathways can produce different specific responses in different target cells—is required to understand the mechanism of HSC differentiation .The understanding of decisions of blood cell fate in HSCs is of great fundamental interest and will be critical for the successful development of therapeutic applications.

5.5 Progress and the Potential Clinical Application of Hematopoietic Cell–Engineering Products

5.5.1 Definite Hematopoietic Cell Induction

HSCs have the capacity to develop into all kind of blood cells in the body. Currently, investigators are reporting the numbers of blood cells—including RBCs, megakaryocytes and platelets, neutrophils, dendric cells (DCs), and T cells—that could be induced in vitro from HSCs or PSCs (ESCs or iPSCs, respectively). These "manmade" blood cells have shown clinical application potential. The progresses of in vitro blood cell manufacture and their applications are summarized below.

5.5.2 Red Blood Cells

The transfusion of RBCs is indicated to achieve a fast increase in the supply of oxygen to the tissues, e.g., when the concentration of hemoglobin (Hb) is low and/or the oxygen carrying capacity is reduced, in the presence of inadequate physiological mechanisms of compensation (Liumbruno et al. 2009). Major surgery and trauma require the delivery of >16 million RBC transfusions every year in the United States. In addition, RBC transfusion is also commonly used in the treatment of anemia and complications from radiotherapy and chemotherapy. However, insufficient blood donation, especially for rare blood types, and potential infectious risks call for alternative sources of RBCs to be developed urgently. RBCs generated in vitro from human hematopoietic stem/progenitor cells (HSCs) derived from cord and adult blood or human PSCs may represent an important resource for providing blood to patients with rare blood types as well as supplementing the general blood supply during emergencies. Furthermore, the semifinished products, erythroid precursors, may ultimately serve as a novel cell-based therapy providing a renewable source of RBCs (Migliaccio et al. 2009).

Typically, the development of HSCs into RBCs involves different "players" including CMPs, MEPs, burst forming unit–erythroids (BFU-Es), colony forming unit–erythroids (CFU-Es), proerythroblasts (ProE), erythroblasts/basophilic normoblasts (BasoE), polychromatophilic normoblasts (PolyE), orthochromatic normoblasts (OrthoE), reticulocytes (Retic), and mature RBCs. The maturation of RBCs usually takes place in the BM because this microenvironment provides a niche for HSCs. The in vitro production of mature RBCs could be fulfilled by simulating the HSC niche in vivo. Using optimized media, combinations of cytokines, and stromal cells, researchers have successfully expanded and induced erythrocytes from HSCs. Neildez-Nquyen et al. reported that sequential application of specific growth factor combinations in serum-free culture medium resulted in erythroid precursors that could fully mature when transplanted into immunodeficient mice (Neildez-Nguyen et al. 2002). These immature blood cells could be used for the treatment of chronic anemia and in gene therapy; however, they are not fully functional immediately after transfusion. Giarratana et al. (2005) showed the application of cytokines and coculture on stromal cells for the large-scale ex vivo production of mature human RBCs from HSCs of diverse origins. Enucleated RBCs meet the demand of functional blood cells during acute hemorrhage; however, the dependency on stromal cells hampers the application of this method. Miharada et al. (2006) attempted to develop an in vitro stromal cell–independent model for producing RBCs. Despite this progress, the generation of large-scale, fully mature, and clinically applicable RBCs remains an obstacle to researchers.

Pluoripotent stem cells (PSCs), including ESCs and induced pluoripotent stem cells (iPSCs), possess the potential to differentiate into all derivatives of the three primary germ layers as well as blood cells. Considering the unlimited expansion and differentiation ability of PSCs, they have become an alternative resource for RBC manufacture. A single PSC line with universal O and rhesus (RhD)-negative blood type might produce sufficient universal RBCs for transfusion. PSCs have been shown to differentiate into an erythroid population (Qiu et al. 2008). Series reports came in 2008 when Lu, Ma, and Qiu successfully developed RBCs from human ESCs separately (Qiu et al. 2008; Lu et al. 2008; Ma et al. 2008). These erythroid cells began to express β-globin, and approximately 40 % of the RBCs became enucleated. These differentiated erythroid cells showed similar functions as RBCs derived from cord blood. However, the scalable expansion and generation of mature erythrocytes from PSCs remains a problem. Recently, Hirose reported that the immortalization of erythroblasts by overexpression of c-MYC and BCL-XL enabled large-scale erythrocyte production from human PSCs (Hirose et al. 2013). Another group showed the large-scale transformation of human ESCs into functional erythrocytes using a sequential four-step procedure (Lu et al. 2008). They reported that 10^{11} to 10^{12} nucleated erythroid cells can be produced from approximately 10^7 human ESCs; their differentiated RBCs had functions comparable with normal adult RBCs. However, full maturation and efficient enucleation of erythroid cells from PSCs have yet to be accomplished. A deeper understanding of the process of erythrocyte maturation would be helpful in directing the ex vivo generation of functionally mature RBCs from stem cells.

In terms of developmental stage, HSCs are in the intermediate phase between PSCs and RBCs. The in vitro induction and expansion of RBCs from HSCs can be divided into three steps: erythroid lineage determination, expansion of erythroid progenitors, and RBC maturation (Zeuner et al. 2012). In some instances, an extra step for HSC amplification is necessary beforehand. Cytokines and supplements applied to promote erythroid progenitor proliferation include lipids, SCF, IL-3, EPO, and/or a glucocorticoid receptor (GR) agonist. Reports trying to improve the expansion and differentiation of erythroid progenitors have shown that chromatin-modifying agents (such as valproic acid), hypoxia, agents that stabilize the transcription factor HIF-1α, and steroid hormones have a role during these processes (Migliaccio et al. 2010; Chaurasia et al. 2011; Leberbauer et al. 2005; Narla 2011; Flygare et al. 2011). For the maturation of erythrocytes, EPO and IGF-I are required. However, regulators for erythrocyte enucleation remain to be identified. It was suggested that no cytokines other than stromal cells or matrix are necessary for enucleation (Keerthivasan et al. 2011). The establishment of massive in vitro–expansion methods for human RBCs is a prerequisite for the clinical application of these cells; therefore, greater effort should be applied to achieve greater proliferation capacities as well as greater specificity of fully mature RBCs.

Although RBCs generated ex vivo are not in as much demand for transfusion, appropriate animal models and proof-of-principle human transfusion studies have confirmed their function in vivo. Immunodeficient mice were first used to evaluate the in vivo function of cultured RBCs. Hu and coworkers further suggested that by depleting macrophages, human RBCs could develop and function better in immunodeficient mice (Hu et al. 2011). Primates are also excellent animal models based on their close relationship to humans. In 2011, Giarratana and colleagues produced the first transplantable RBCs ex vivo (Giarratana et al. 2011). In their study, 10^{10} RBCs were generated from peripheral $CD34^+$ HSCs and transplanted into a health volunteer. These RBCs survived more than 26 days in the circulatory system, thus verifying their quality and function in vivo.

Progress has been made during the last 10 years in the ex vivo generation of a large number of RBCs from stem cells. However, several issues remain to be resolved. During the next 5 to 10 years, much work needs to be performed regarding the efficient differentiation of erythroid cells from stem cells, the switch to adult β-globin, and the enucleation of erythrocytes. Given the limited number of HSCs derived from cord blood or adult sources, large-scale manufacturing of RBCs from PSCs will be required. However, challenges remain in the production of a large number of RBCs that can be used clinically. Appropriate bioreactor design and culture will be required to realize RBC manufacturing for clinical applications. The cost of RBC products is another consideration. Biosafety is particularly important for the application of RBCs generated ex vivo; good manufacturing practices are needed to produce stem cell-derived RBCs. This should encompass the choice of stem cells, the source of reagents, and the use of defined component-free animal culture environments as well as quality-control and quality-assurance procedures. Novel RBC products from stem cells are highly suitable as an alternative to blood transfusion. If the barrier of obtaining enough RBCs from stem cells is resolved, these cells might become the first stem cell–derived products to be reliably used in a clinical setting.

5.5.3 *Megakaryocytes/Platelets*

Platelets are small cell structures in blood that are essential to the process of blood clotting (hemostasis). They work with clotting factors in plasma to help prevent bleeding. Thrombocytopenia is a condition in which the platelet count is lower than the normal level. Currently platelet transfusion has steadily grown to become an essential part of the treatment of thrombocytopenia caused by cancer, hematological malignancies, marrow failure, HSC transplantation, and surgery (Stroncek and Rebulla 2007). The demand for platelet transfusion has increased dramatically in recent decades. In the United States, total platelet transfusion comprises >2.17 million apheresis-equivalent units/y. These platelets are derived entirely from human donors. Considering the clinically significant immunogenicity, the associated risk of sepsis, the blood inventory shortages due to high demand, and the 5-day blood shelf life, the production of mature blood cells from stem cells by way of large-scale manufacture is an alternative way to meet transfusion demands (Thon 2014).

Platelets come from special cells in the BM called "megakaryocytes." The development of platelets from HSCs involves CMPs, common MK-erythroid progenitors (MEPs), megakaryoblasts, megakaryocyte progenitors, and megakaryocytes. Subsequently, megakaryocytes become polyploid through endomitosis. The cells increase in size, become full of platelet-specific granules, expand their cytoplasmic content of cytoskeletal proteins, and develop a highly tortuous invaginated membrane system. The branches of mature magakaryocytes, which are called "proplatelets," extend into the sinusoidal blood vessels of the BM and finally form platelets under blood shear stress (Machlus and Italiano 2013). In case of nonemergency conditions, megakaryocytes are an alternative selection for platelet transfusion because the infused megakaryocytes can shed platelets gradually in vivo.

The strategy used for megakaryocyte/platelet manufacture from HSCs involves stepwise induction, which mimics the four phases of MK/platelet development in vivo: (1) stem cell self-renewal; (2) MK progenitor amplification; (3) MK differentiation/maturation; and (4) platelet release. In a three-step strategy, the first step was to amplify the progenitors with TPO, SCF, Flt3L, IL-6, Notch-ligands, SR1, and IGFBP2; the second step was to support MK differentiation by TPO, SCF, IL-6, and IL-9 combination; and the third step was to promote platelet biogenesis using TPO, SCF, IL-9, SDF-1, and FGF-4 (Reems et al. 2010; Lee et al. 2014). Although cytokine cocktails varied in different studies, TPO and SCF were the two most important cytokines for megakaryocyte expansion and differentiation. The results of a representative study were published in 2006 when Matsunaga et al. reported the large-scale generation of human platelets from cord blood $CD34^+$ cells using a three-phase culture system: (1) $CD34^+$ cells were cultured on stromal cells in serum-free medium supplemented with SCF, FL, and TPO for 14 days; (2) the cells were transferred onto fresh stromal cells with medium containing IL-11 in addition to the original cytokine cocktail; and (3) the cells were cultured stromal free in a medium containing SCF, FL, TPO, and IL-11 for another 5 days to recover platelet fractions from the supernatant. Approximately 2.5 to 3.4 U of platelets could be generated by this method (Matsunaga et al. 2006).

In addition to HSCs, PSCs were proven to be the origin of megakaryocytes and platelets as well. Gaur et al. established an OP9 stromal cell coculture system to generate megakaryocytes from human embryonic stem cells (hESCs) (Gaur et al. 2006). Later on, Takayama et al. (2008) optimized the protocol by taking stromal cells together with VEGF to induce sac-like structures from hESCs. The spherical cells inside the sac could be induced to mature megakaryocytes in the presence of TPO. Feeder-free induction protocols were subsequently developed. Yu et al. provided a three-stage chemical defined method to sequentially induce mesendoderm cells (with BMP4 and Wnt3a), hemato-vascular precursors (with VEGF and bFGF), and hematopoietic progenitors (with retinoic acid) from hESCs (Yu et al. 2010). Human-induced pluripotent stem cells (iPSCs) were also proven to produce platelets under appropriate induction. By overexpressing BMI1 and BCL-XL, respectively, to suppress senescence and apoptosis and the constraining overexpression of c-MYC to promote proliferation, Eto's group generated immortalized megakaryocyte progenitor cell lines (imMKCLs) from iPSC-derived hematopoietic progenitors. These imMKCLs were capable of producing sufficient functional platelets in vitro (Nakamura et al. 2014). Megakaryon Corporation was built based on this achievement, and related clinical trials will be launched in 2015. In addition to PSCs, a recent article also showed that fibroblasts could be transdifferentiated into megakaryocytes by introducing three factors: p45NF-E2, Maf G, and Maf K. This exciting discovery provided a convenience for personalized cell therapy with cells other than iPSCs (Masuda et al. 2013).

The manufacture of platelets for clinical transfusion purposes requires a series of technological advancements that will foster a development of bioreactors to maximize yields at each of the four biological phases of in vivo MK/platelet development. In an antecedent study to increase platelet yield from hESCs or iPSC-derived MKs, Nakagawa created a two-dimensional flow culture system that mimics the blood vessel system. Platelets from iPSC-derived MKs with the system were generated in numbers sufficient for transfusion therapy (Nakagawa et al. 2013).

Although induced platelets derived from stem cells provide promising therapeutic potential, and although above-mentioned reports suggested that platelets could also be produced on a clinical scale, obstacles still existing before we can use these cells in clinical applications. One problem is that platelets, especially platelets generated in vitro, can easily be activated by the environment. These platelets become functionless after being activated in vitro. Owing to this and inadequate platelet production, currently there is no evidence for direct transfusion of stem cell–derived platelets. However, proof-of-principle studies have confirmed the applicability of stem cell–derived megakaryocytes in the therapy of thrombocytopenia. A study from our group has confirmed the in vivo safety of megakaryocytic progenitors derived from cord blood mononuclear cells in phase I clinical trials (Xi et al. 2013). Data from that study also suggested the effectiveness of these cells. Such stem cell–based cell therapy will soon come into clinical application.

For the in vitro generation of platelets from stem cells, the problems remaining to be solved include the following:

- The number of platelets produced in vitro per MK are orders of magnitude lower than the number produced in vivo.
- The quality of the generated platelets is unclear, and the safety of producing platelets on an industrial scale by way of biogenesis remains unknown.
- To boost platelet generation in vitro demands further understanding and better imitation of the physiological development of megakaryocytes and platelets.

Nonetheless, precedent studies provide proof-of-principle that developing in vitro strategies to manufacture platelets is feasible.

5.5.4 Immune Cells

The immune system can be divided into the innate immune system and the adaptive immune system. Monocytes, neutrophils, eosinophils, basophils, macrophages, DCs, and NK cells comprise the innate immune system, which responses attack microbes indiscriminately. In the adaptive immune response, DCs, helper T cells, NK cells, and antigen-specific CTL participate, respectively, in antigen presentation, immune response orchestration, and killing infected cells. Cellular immunity is an important immune response when the body encounters viruses, bacteria, and cancers; it is also plays an important role in tissue transplantation. Adoptive cell transfer after host preconditioning by lymphodepletion represents an important advance in cancer immunotherapy. Immune effector cells—such as lymphocytes, macrophages, DCs, NK cells, CTLs, etc.—work together to defend the body against cancer by targeting abnormal antigens expressed on the surface of the tumor due to mutation. Stem cell–derived immune cells have now shown great feasibility in adjuvant cancer immunotherapy (Gattinoni et al. 2006).

5.5.5 DCs

Antigen-presenting cells (APCs) have been shown to play a crucial role in the induction of tumor-protective immune responses by generating tumor-specific T cells. DCs are the most potent APCs among others like monocytes, macrophages, and B cells. Currently DCs have increasingly become the adjuvant of choice in new approaches to cancer immunotherapy. DC trials have shown encouraging results in treating various types of cancer such as lung and breast. Owing to limited autologous DC numbers, large donor-to-donor variability, and frequent DC defects in cancer patients, alternative DC sources are appealing.

DC can be generated from $CD34^+$ cells using GM-CSF and TNF-α or $CD14^+$ monocytes derived from peripheral blood using GM-CSF and IL-4. Because DC therapy requires a cell number of 10^8/patient, scale-up DC generation from PSCs, including hESCs and hiPSCs, was recently developed. Given the unique proliferative capacity, the hPSC-based process has the potential for mass production of DCs at a scale $>10^{10}$ cells. The HSC is the intermediate stage of hPSC-DC induction. Similar to deriving HSCs, there are two methods for DC differentiation from hPSCs: (1) through the formation of EBs in suspension culture; and (2) by coculture with mouse OP9 stromal cells. To fulfill the potential of hPSC-derived DCs, large-scale production in bioreactors is a critical step toward clinical application. The use of bioreactors—including spinner flasks, rotating wall vessels, and perfusion bioreactors, which use the advantage of shear stress—has been shown to be possible in scale-up DC generation (Li et al. 2014; Reinhard et al. 2002).

5.5.6 NK Cells

NK cells are part of the innate immune system that are derived from $CD34^+$ HSCs and undergo differentiation by way of precursor NK cells in BM through sequential acquisition of functional surface receptors. NK cells express their own repertoire of receptors, including activating and inhibitory receptors, that bind to major histocompatibility complex (MHC) class I– or class I–related molecules. The balance between activating and inhibitory receptors determines the function of NK cells to kill targets (Suk Ran et al. 2007) NK cells are an alluring option for immunotherapy due to their ability to kill infected cells or cancer cells without previous sensitization.

Throughout the past 20 years, different groups have been able to reproduce NK cell development in vitro, and NK cell ontogeny studies have provided the basis for the establishment of protocols to produce NK cells in vitro for immunotherapy. Notably, the use of NK cells for immunotherapy relies on the availability of a great number of NK cells (1×10^7 to 2×10^7 cells/kg) with optimal cytotoxic activity, whereas autologous NK cells in peripheral blood is limited. (Luevano et al. 2012). Derived NK cells have become a superior choice to induce stem cells for cancer immunotherapy. Regarding HSC–to–K cell differentiation, Miller et al. suggested that intimate contact with stromal cells is needed for the most primitive progenitors to differentiate into NK cells, but it is not longer required after the first step when commitment happens. Stromal cells are also required for the large-scale generation of NK cells (Miller et al. 1994) Willams et al. (1998) worked out a step-wise NK cell–induction protocol in 1998: In phase 1, IL-7, SCF, and FLT3 ligand (Flt3L) acted on the NK lineage commitment with the expression of IL-2Rβ; and in phase 2, IL-15 was used to further the expansion and maturation of NK cells. Clinical trials were carried out based on these strategies. In

2011, Spanholtz et al. used a closed-system culture process to produce large-scale, highly active, and functional NK cells for a phase I dose-finding trial in elderly AML patients (Spanholtz et al. 2011).

Human ESC–derived hematopoietic precursors have also proven be able to produce functional NK cells. Recently, Woll et al. efficiently generated functional NK cells from H9 hESCs by a two-step culture method. These NK cells expressed maturation markers, including KIRs, natural cytotoxic receptors, and CD16, and possessed the ability to lyse tumor cells by direct cell-mediated cytotoxicity and ADCC. Furthermore, these hESC-derived NK cells could mediate an effective antitumor response in an in vivo xenogeneic mouse model, which was more effective compared with UCB-derived NK cells. The NK cell products enhanced cell-based therapies with the potential to serve as a "universal" source of antitumor lymphocytes (Knorr and Kaufman 2010; Woll et al. 2009).

5.5.7 CIK Cells

CIK cells are polyclonal T-effector cells generated when cultured under cytokine stimulation. CIK cells exhibit potent, non–MHC-restricted cytolytic activities against susceptible tumor cells of both autologous and allogeneic origins. During the past 20 years, CIK cells have evolved from experimental observations into early clinical studies with encouraging preliminary efficacy toward susceptible autologous and allogeneic tumor cells in both therapeutic and adjuvant settings. CIK cells can be generated successfully from healthy donors as well as from patients treated with chemotherapy for various malignancies and patients undergoing peripheral blood progenitor cell (PBSC) leukapheresis. Stem cells have again become an alternative source of CIK cells. The feasibility of large-scale expansion was reported for cord blood and even for washout of leftover mononuclear cells from cord blood unit bags (Introna et al. 2006; Linn and Hui 2010). The therapeutic potential of cord blood–derived CIK cells was proven later on by Niu et al. who showed that CB-CIK, together with chemotherapy, had a significant effect on the treatment of solid malignancies (Niu et al. 2011). These correlated observations open up the possibility of stem cell-based cancer immunotherapy in near future.

5.6 Gene Therapy with Hematopoietic Stem Cells

Cell-based therapies are fast-growing forms of personalized medicine that make use of the steady advances in stem cell manipulation and gene-transfer technologies. Ex vivo cell therapies are based on the ability to isolate stem, progenitor, or differentiated cells from a patient or a normal donor, expand them ex vivo with or without genetic modification, and administer them to the patient to establish a transient or, more often, a stable graft of the infused cells and their progeny.

Because the best developed and most successful cell therapy is HSC transplantation (HCT), ex vivo gene-transfer procedures can be integrated into standard HCT protocols and achieve rapid clinical translation. HSC gene therapy represents an emerging therapeutic option for several monogenic diseases of the blood and the immune system as well as for storage disorders, and it may become a first-line treatment option for selected disease conditions.

The current protocol for ex vivo HSC gene therapy includes HSC isolation, ex vivo expansion, retrovirus infection with functional gene, and modified HSC infusion after a few days of manipulation. HSC gene therapy has been applied to the treatment of severe combined immuno-deficiency (SCID). The 10-year long-term follow-up of the clinical studies gives exciting outcomes: Most treated patients have had their immunodeficiency corrected. However, negative results were also observed, in which vector-induced hematological malignancy occurred. Better understanding of the mechanisms of vector insertional mutagenesis and the exploration of novel vectors might finally realize the gene therapy procedure (Hacein-Bey-Abina et al. 2010; Naldini 2011).

5.7 Conclusion

The generation of RBCs from stem cells in vitro has been called "blood pharming" by the Defense Advanced Research Projects Agency (DARPA). Accordingly, the manufacture of blood cells with other functions, such as megakaryocytes/platelets, DCs, T cells, etc., can be called blood pharming as well based on their great therapeutic significance. To fulfill hematopoietic cell generation at a clinical scale may eventually provide new opportunities for human health.

References

Akashi K, Traver D, Miyamoto T, Weissman IL. A clonogenic common myeloid progenitor that gives rise to all myeloid lineages. Nature. 2000;404:193–7.

Alexander WS, Roberts AW, Nicola NA, Li R, Metcalf D. Deficiencies in progenitor cells of multiple hematopoietic lineages and defective megakaryocytopoiesis in mice lacking the thrombopoietic receptor c-Mpl. Blood. 1996;87:2162–70.

Amsellem S, Pflumio F, Bardinet D, Izac B, Charneau P, Romeo PH, Dubart-Kupperschmitt A, Fichelson S. Ex vivo expansion of human hematopoietic stem cells by direct delivery of the hoxb4 homeoprotein. Nat Med. 2003;9:1423–7.

Auffray C, Sieweke MH, Geissmann F. Blood monocytes: development, heterogeneity, and relationship with dendritic cells. Annu Rev Immunol. 2009;27:669–92.

Baba Y, Garrett KP, Kincade PW. Constitutively active beta-catenin confers multilineage differentiation potential on lymphoid and myeloid progenitors. Immunity. 2005;23:599–609.

Baba Y, Yokota T, Spits H, Garrett KP, Hayashi S, Kincade PW. Constitutively active beta-catenin promotes expansion of multipotent hematopoietic progenitors in culture. J Immunol. 2006;177:2294–303.

Bárcena A, Muench MO, Kapidzic M, Gormley M, Goldfien GA, Fisher SJ. Human placenta and chorion: potential additional sources of hematopoietic stem cells for transplantation. Transfusion. 2011;51(Suppl 4):94S–105S.

Barker JN, Weisdorf DJ, DeFor TE, Blazar BR, McGlave PB, MillerJS Verfaillie CM, Wagner JE. Transplantation of 2 partially HLA-matched umbilical cord blood units to enhance engraftment in adults with hematologic malignancy. Blood. 2005;105:1343–7.

Batard P, Monier MN, Fortunel N, Ducos K, Sansilvestri-Morel P, Phan T, Hatzfeld A, Hatzfeld JA. TGF-(beta)1 maintains hematopoietic immaturity by a reversible negative control of cell cycle and induces CD34 antigen up-modulation. J Cell Sci. 2000;113:383–90.

Bernardo AS, Faial T, Gardner L, Niakan KK, Ortmann D, Senner CE, Callery EM, Trotter MW, Hemberger M, Smith JC, Bardwell L, Moffett A, Pedersen RA. BRACHYURY and CDX2 mediate BMP-induced differentiation of human and mouse pluripotent stem cells into embryonic and extraembryonic lineages. Cell Stem Cell. 2011;9:144–55.

Bhatia M, Bonnet D, Wu D, Murdoch B, Wrana J, Gallacher L, Dick JE. Bone morphogenetic proteins regulate the developmental program of human hematopoietic stem cells. J Exp Med. 1999;189:1139–48.

Bigas A, Martin DI, Milner LA. Notch1 and Notch2 inhibit myeloid differentiation in response to different cytokines. Mol Cell Biol. 1998;18:2324–33.

Boitano AE, Wang J, Romeo R, Bouchez LC, Parker AE, Sutton SE, Walker JR, Flaveny CA, Perdew GH, Denison MS, Schultz PG, Cooke MP. Aryl hydrocarbon receptor antagonists promote the expansion of human hematopoietic stem cells. Science. 2010;329:1345–8.

Bowles KM, Vallier L, Smith JR, Alexander MR, Pedersen RA. HOXB4 overexpression promotes hematopoietic development by human embryonic stem cells. Stem Cells. 2006;24:1359–69.

Brunstein CG, Fuchs EJ, Carter SL, Karanes C, Costa LJ, Wu J, Devine SM, Wingard JR, Aljitawi OS, Cutler CS, Jagasia MH, Ballen KK, Eapen M, O'Donnell PV. Alternative donor transplantation after reduced intensity conditioning: results of parallel phase 2 trials using partially HLA-mismatched related bone marrow or unrelated double umbilical cord blood grafts. Blood. 2011;118:282–8.

Busslinger M. Transcriptional control of early B cell development. Annu Rev Immunol. 2004;22:55–79.

Caplan AI. Mesenchymal stem cells and gene therapy. Clin Orthop Relat Res. 2000;379(Suppl):S67–70.

Carver-Moore K, Broxmeyer HE, Luoh SM, Cooper S, Peng J, Burstein SA, Moore MW, de Sauvage FJ. Low levels of erythroid and myeloid progenitors in thrombopoietin-and c-mpl-deficient mice. Blood. 1996;88:803–8.

Caux C, Dezutter-Dambuyant C, Schmitt D, Banchereau J. GM-CSF and TNF-alpha cooperate in the generation of dendritic langerhans cells. Nature. 1992;360:258–61.

Caux C, Vanbervliet B, Massacrier C, Dezutter-Dambuyant C, de Saint-Vis B, Jacquet C, Yoneda K, Imamura S, Schmitt D, Banchereau J. CD34+ hematopoietic progenitors from human cord blood differentiate along two independent dendritic cell pathways in response to GM-CSF+TNF alpha. J Exp Med. 1996;184:695–706.

Chadwick K, Wang L, Li L, Menendez P, Murdoch B, Rouleau A, Bhatia M. Cytokines and BMP-4 promote hematopoietic differentiation of human embryonic stem cells. Blood. 2003;102:906–15.

Chang CJ, Bouhassira EE. Zinc-finger nuclease-mediated correction of α-thalassemia in iPS cells. Blood. 2012;120:3906–14.

Chaurasia P, Berenzon D, Hoffman R. Chromatin-modifying agents promote the ex vivo production of functional human erythroid progenitor cells. Blood. 2011;117:4632.

Chen H, Ray-Gallet D, Zhang P, Hetherington CJ, Gonzalez DA, Zhang DE, Moreau-Gachelin F, Tenen DG. PU.1 (Spi-1) autoregulates its expression in myeloid cells. Oncogene. 1995;11:1549–60.

Chung YS, Zhang WJ, Arentson E, Kingsley PD, Palis J, Choi K. Lineage analysis of the hemangioblast as defined by FLK1 and SCL expression. Development. 2002;129:5511–20.

da Silva CL, Gonçalves R, Crapnell KB, Cabral JM, Zanjani ED, Almeida-Porada G. A human stromal-based serum-free culture system supports the ex vivo expansion/maintenance of bone marrow and cord blood hematopoietic stem/progenitor cells. Exp Hematol. 2005;33:828–35.

D'Andrea AD. Hematopoietic growth factors and the regulation of differentiative decisions. Curr Opin Cell Biol. 1994;6:804–8.

Dang SM, Kyba M, Perlingeiro R, Daley GQ, Zandstra PW. Efficiency of embryoid body formation and hematopoietic development from embryonic stem cells in different culture systems. Biotechnol Bioeng. 2002;78:442–53.

Delaney C, Heimfeld S, Brashem-Stein C, Voorhies H, Manger RL, Bernstein ID. Notch-mediated expansion of human cord blood progenitor cells capable of rapid myeloid reconstitution. Nat Med. 2010;16:232–6.

Doulatov S, Vo LT, Chou SS, Kim PG, Arora N, Li H, Hadland BK, Bernstein ID, Collins JJ, Zon LI, Daley GQ. Induction of multipotential hematopoietic progenitors from human pluripotent stem cells via respecification of lineage-restricted precursors. Cell Stem Cell. 2013;13:459–70.

Ema M, Faloon P, Zhang WJ, Hirashima M, Reid T, Stanford WL, Orkin S, Choi K, Rossant J. Combinatorial effects of Flk1 and Tal1 on vascular and hematopoietic development in the mouse. Genes Dev. 2003;17:380–93.

Fan Y, Luo Y, Chen X, Li Q, Sun X. Generation of human β-thalassemia induced pluripotent stem cells from amniotic fluid cells using a single excisable lentiviral stem cell cassette. J Reprod Dev. 2012;58:404–9.

Ferreira MS, Jahnen-Dechent W, Labude N, Bovi M, Hieronymus T, Zenke M, Schneider RK, Neuss S. Cord blood-hematopoietic stem cell expansion in 3d fibrin scaffolds with stromal support. Biomaterials. 2012;33:6987–97.

Flygare J, Estrada VR, Shin C, Gupta S, Lodish HF. HIF1α synergizes with glucocorticoids to promote BFU-E progenitor self-renewal. Blood. 2011;117:3435.

Fortini ME. Notch signaling: the core pathway and its posttranslational regulation. Dev Cell. 2009;16:633–47.

Fortunel NO, Hatzfeld A, Hatzfeld JA. Transforming growth factor-beta: pleiotropic role in the regulation of hematopoiesis. Blood. 2000;96:2022–36.

Friedman AD. Transcriptional control of granulocyte and monocyte development. Oncogene. 2007;26:6816–28.

Fukuchi Y, Shibata F, Ito M, Goto-Koshino Y, Sotomaru Y, Ito M, Kitamura T, Nakajima H. Comprehensive analysis of myeloid lineage conversion using mice expressing an inducible form of C/EBP alpha. EMBO J. 2006;25:3398–410.

Gattinoni L, Powell DJ, Rosenberg SA, Restifo NP. Adoptive immunotherapy for cancer: building on success. Nat Rev Immunol. 2006;6:383–93.

Gaur M, Kamata T, Wang S, Moran B, Shattil SJ, Leavitt AD. Megakaryocytes derived from human embryonic stem cells: a genetically tractable system to study megakaryocytopoiesis and integrin function. J Thromb Haemost. 2006;4:436–42.

Gering M, Yamada Y, Rabbitts TH, Patient RK. Lmo2 and Scl/Tal1 convert non-axial mesoderm into haemangioblasts which differentiate into endothelial cells in the absence of Gata1. Development. 2003;130:6187–99.

Giarratana M-C, Rouard H, Dumont A, Kiger L, Safeukui I, Le Pennec P-Y, François S, Trugnan G, Peyrard T, Marie T, Jolly S, Hebert N, Mazurier C, Mario N, Harmand L, Lapillonne H, Devaux J-Y, Douay L. Proof of principle for transfusion of in vitro–generated red blood cells. Blood. 2011;118:5071.

Giarratana M-C, Kobari L, Lapillonne H, Chalmers D, Kiger L, Cynober T, Marden MC, Wajcman H, Douay L. Ex vivo generation of fully mature human red blood cells from hematopoietic stem cells. Nat Biotech. 2005;23:69–74.

Grover A, Mancini E, Moore S, Mead AJ, Atkinson D, Rasmussen KD, O'Carroll D, Jacobsen SE, Nerlov C. Erythropoietin guides multipotent hematopoietic progenitor cells toward an erythroid fate. J Exp Med. 2014;211:181–8.

Hacein-Bey-Abina S, Hauer J, Lim A, Picard C, Wang GP, Berry CC, Martinache C, Rieux-Laucat F, Latour S, Belohradsky BH, Leiva L, Sorensen R, Debré M, Casanova JL, Blanche S, Durandy A, Bushman FD, Fischer A, Cavazzana-Calvo M. Efficacy of gene therapy for X-linked severe combined immunodeficiency. N Engl J Med. 2010;363:355–64.

Han H, Tanigaki K, Yamamoto N, Kuroda K, Yoshimoto M, Nakahata T, Ikuta K, Honjo T. Inducible gene knockout of transcription factor recombination signal binding protein-J reveals its essential role in T versus B lineage decision. Int Immunol. 2002;14:637–45.

Heath V, Suh HC, Holman M, Renn K, Gooya JM, Parkin S, Klarmann KD, Ortiz M, Johnson P, Keller J. C/EBPalpha deficiency results in hyperproliferation of hematopoietic progenitor cells and disrupts macrophage development in vitro and in vivo. Blood. 2004;104:1639–47.

Hegde SP, Zhao J, Ashmun RA, Shapiro LH. c-Maf induces monocytic differentiation and apoptosis in bipotent myeloid progenitors. Blood. 1999;94:1578–89.

Himes SR, Tagoh H, Goonetilleke N, Sasmono T, Oceandy D, Clark R, Bonifer C, Hume DA. A highly conserved c-fms gene intronic element controls macrophage-specific and regulated expression. J Leukoc Biol. 2001;70:812–20.

Hirai H, Zhang P, Dayaram T, Hetherington CJ, Mizuno S, Imanishi J, Akashi K, Tenen DG. C/EBPbeta is required for 'emergency' granulopoiesis. Nat Immunol. 2006;7:732–9.

Hirose S, Takayama N, Nakamura S, Nagasawa K, Ochi K, Hirata S, Yamazaki S, Yamaguchi T, Otsu M, Sano S, Takahashi N, Sawaguchi A, Ito M, Kato T, Nakauchi H, Eto K. Immortalization of erythroblasts by c-MYC and BCL-XL enables large-scale erythrocyte production from human pluripotent stem cells. Stem Cell Reports. 2013;1:499–508.

Hohaus S, Petrovick MS, Voso MT, Sun Z, Zhang DE, Tenen DG. PU.1 (Spi-1) and C/EBP alpha regulate expression of the granulocyte-macrophage colony-stimulating factor receptor alpha gene. Mol Cell Biol. 1995;15:5830–45.

Hou P, Li Y, Zhang X, Liu C, Guan J, Li H, Zhao T, Ye J, Yang W, Liu K, Ge J, Xu J, Zhang Q, Zhao Y, Deng H. Pluripotent stem cells induced from mouse somatic cells by small-molecule compounds. Science. 2013;341:651–4.

Hromas R, Orazi A, Neiman RS, Maki R, Van Beveran C, Moore J, Klemsz M. Hematopoietic lineage- and stage-restricted expression of the ETS oncogene family member PU.1. Blood. 1993;82:2998–3004.

Hu Z, Van Rooijen, Yang Y-G. Macrophages prevent human red blood cell reconstitution in immunodeficient mice. Blood. 2011;118:5938.

Ikonomi P, Rivera CE, Riordan M, Washington G, Schechter AN, Noguchi CT. Overexpression of GATA-2 inhibits erythroid and promotes megakaryocyte differentiation. Exp Hematol. 2000;28:1423–31.

Inaba K, Inaba M, Romani N, Aya H, Deguchi M, Ikehara S, Muramatsu S, Steinman RM. Generation of large numbers of dendritic cells from mouse bone marrow cultures supplemented with granulocyte/macrophage colony-stimulating factor. J Exp Med. 1992;176:1693–702.

Introna M, Franceschetti M, Ciocca A, Borleri G, Conti E, Golay J, Rambaldi A. Rapid and massive expansion of cord blood-derived cytokine-induced killer cells: an innovative proposal for the treatment of leukemia relapse after cord blood transplantation. Bone Marrow Transplant. 2006;38:621–7.

Iwasaki H, Akashi K. Hematopoietic developmental pathways: on cellular basis. Oncogene. 2007;26:6687–96.

Iwasaki H, Somoza C, Shigematsu H, Duprez EA, Iwasaki-Arai J, Mizuno S, Arinobu Y, Geary K, Zhang P, Dayaram T, Fenyus ML, Elf S, Chan S, Kastner P, Huettner CS, Murray R, Tenen DG, Akashi K. Distinctive and indispensable roles of PU.1 in maintenance of hematopoietic stem cells and their differentiation. Blood. 2005;106:1590–600.

Jacobsen SE, Keller JR, Ruscetti FW, Kondaiah P, Roberts AB, Falk LA. Bidirectional effects of transforming growth factor beta (TGF-beta) on colony-stimulating factor-induced human myelopoiesis in vitro: differential effects of distinct TGF-beta isoforms. Blood. 1991;78:2239–47.

Johansson BM, Wiles MV. Evidence for involvement of activin A and bone morphogenetic protein 4 in mammalian mesoderm and hematopoietic development. Mol Cell Biol. 1995;15:141–51.
Karlsson KR, Cowley S, Martinez FO, Shaw M, Minger SL, James W. Homogeneous monocytes and macrophages from human embryonic stem cells following coculture-free differentiation in M-CSF and IL-3. Exp Hematol. 2008;36:1167–75.
Kaufman DS. Toward clinical therapies using hematopoietic cells derived from human pluripotent stem cells. Blood. 2009;114:3513–23.
Kaushansky K. Lineage-specific hematopoietic growth factors. N Engl J Med. 2006;354:2034–45.
Kawamata S, Sakaida H, Hori T, Maeda M, Uchiyama T. The upregulation of p27Kip1 by rapamycin results in G1 arrest in exponentially growing T-cell lines. Blood. 1998;91:561–9.
Kawamata S, Du C, Li K, Lavau C. Notch1 perturbation of hemopoiesis involves non-cell-autonomous modifications. J Immunol. 2002;168:1738–45.
Keerthivasan G, Wickrema A, Crispino JD. Erythroblast enucleation. Stem Cells Int. 2011;2011:9.
Keller JR, Bartelmez SH, Sitnicka E, Ruscetti FW, Ortiz M, Gooya JM, Jacobsen SE. Distinct and overlapping direct effects of macrophage inflammatory protein-1 alpha and transforming growth factor beta on hematopoietic progenitor/stem cell growth. Blood. 1994;84(7):2175–2181.
Keller JR, Mcniece IK, Sill KT, Ellingsworth LR, Quesenberry PJ, Sing GK, Ruscetti FW. Transforming growth factor beta directly regulates primitive murine hematopoietic cell proliferation. Blood. 1990;75:596–602.
Keller JR, Jacobsen SE, Sill KT, Ellingsworth LR, Ruscetti FW. Stimulation of granulopoiesis by transforming growth factor beta: synergy with granulocyte/ macrophage-colony-stimulating factor. Proc Natl Acad Sci USA. 1991;88:7190–4.
Kirstetter P, Anderson K, Porse BT, Jacobsen SE, Nerlov C. Activation of the canonical Wnt pathway leads to loss of hematopoietic stem cell repopulation and multilineage differentiation block. Nat Immunol. 2006;7:1048–56.
Knorr DA, Kaufman DS. Pluripotent stem cell-derived natural killer cells for cancer therapy. Transl Res J Lab Clin Med. 2010;156:147–54.
Kopan R, Ilagan MX. The canonical Notch signaling pathway: unfolding the activation mechanism. Cell. 2009;137:216–33.
Kulessa H, Frampton J, Graf T. GATA-1 reprograms avian myelomonocytic cell lines into eosinophils, thromboblasts, and erythroblasts. Genes Dev. 1995;9:1250–62.
Laiosa CV, Stadtfeld M, Graf T. Determinants of lymphoid-myeloid lineage diversification. Annu Rev Immunol. 2006;24:705–38.
Leberbauer C, Boulmé F, Unfried G, Huber J, Beug H, Müllner EW. Different steroids co-regulate long-term expansion versus terminal differentiation in primary human erythroid progenitors. Blood. 2005;105:85.
Lee E-J, Godara P, Haylock D. Biomanufacture of human platelets for transfusion: rationale and approaches. Exp Hematol. 2014;42:332–346.
Li Y, Liu M, Yang ST. Dendritic cells derived from pluripotent stem cells: potential of large scale production. World J Stem Cells. 2014;6:1–10.
Lim JM, Lee M, Lee EJ, Gong SP, Lee ST. Stem cell engineering: limitation, alternatives, and insight. Ann N Y Acad Sci. 2011;1229:89–98. doi:10.1111/j.1749-6632.2011.06093.x.
Ling KW, Ottersbach K, van Hamburg JP, Oziemlak A, Tsai FY, Orkin SH, Ploemacher R, Hendriks RW, Dzierzak E. GATA-2 plays two functionally distinct roles during the ontogeny of hematopoietic stem cells. J Exp Med. 2007;200:871–82.
Linn YC, Hui KM. Cytokine-induced NK-like T cells: from bench to bedside. J Biomed Biotechnol. 2010;2010:435745.
Liu Y, Nonnemacher MR, Wigdahl B. CCAAT/enhancer-binding proteins and the pathogenesis of retrovirus infection. Future Microbiol. 2009;4:299–321.

Liumbruno G, Bennardello F, Lattanzio A, Piccoli P, Rossetti G. Recommendations for the transfusion of red blood cells. Blood Transfus. 2009;7:49–64.

Locatelli F, Rocha V, Chastang C, Arcese W, Michel G, Abecasis M, Messina C, Ortega J, Badell-Serra I, Plouvier E, Souillet G, Jouet JP, Pasquini R, Ferreira E, Garnier F, Gluckman E. Factors associated without come after cord blood transplantation in children with acute leukemia. Eurocord-Cord Blood Transplant Group Blood. 1999;93:3662–71.

Lu S-J, Feng Q, Park JS, Vida L, Lee B-S, Strausbauch M, Wettstein PJ, Honig GR, Lanza R. Biologic properties and enucleation of red blood cells from human embryonic stem cells. Blood. 2008;112:4475.

Lu SJ, Feng Q, Park JS, Vida L, Lee BS, Strausbauch M, Wettstein PJ, Honig GR, Lanza R. Biologic properties and enucleation of red blood cells from human embryonic stem cells. Blood. 2008;112:4475–84.

Luevano M, Madrigal A, Saudemont A. Generation of natural killer cells from hematopoietic stem cells in vitro for immunotherapy. Cell Mol Immunol. 2012;9:310–20.

Ma F, Ebihara Y, Umeda K, Sakai H, Hanada S, Zhang H, Zaike Y, Tsuchida E, Nakahata T, Nakauchi H, Tsuji K. Generation of functional erythrocytes from human embryonic stem cell-derived definitive hematopoiesis. Proc Natl Acad Sci USA. 2008;105:13087–92.

Ma N, Liao B, Zhang H, Wang L, Shan Y, Xue Y, Huang K, Chen S, Zhou X, Chen Y, Pei D, Pan G. Transcription activator-like effector nuclease (TALEN)-mediated gene correction in integration-free β-thalassemia induced pluripotent stem cells. J Biol Chem. 2013;288:34671–9.

Machlus KR, Italiano JE. The incredible journey: from megakaryocyte development to platelet formation. J Cell Biol. 2013;201:785–96.

Maeda K, Malykhin A, Teague-Weber BN, Sun XH, Farris AD, Coggeshall KM. Interleukin-6 aborts lymphopoiesis and elevates production of myeloid cells in systemic lupus erythematosus-prone B6.Sle1.Yaa animals. Blood. 2009;113:4534–40.

Maillard I, He Y, Pear WS. From the yolk sac to the spleen: New roles for Notch in regulating hematopoiesis. Immunity. 2003;18:587–9.

Malhotra S, Baba Y, Garrett KP, Staal FJ, Gerstein R, Kincade PW. Contrasting responses of lymphoid progenitors to canonical and noncanonical Wnt signals. J Immunol. 2008;181:3955–64.

Masuda S, Li M, Belmonte JCI. In vitro generation of platelets through direct conversion: first report in my knowledge (iMK). Cell Res. 2013;23:176–8.

Matsunaga T, Tanaka I, Kobune M, Kawano Y, Tanaka M, Kuribayashi K, Iyama S, Sato T, Sato Y, Takimoto R, Takayama T, Kato J, Ninomiya T, Hamada H, Niitsu Y. Ex vivo large-scale generation of human platelets from cord blood CD34+ Cells. Stem Cells. 2006;24:2877–87.

McKinney-Freeman S, Cahan P, Li H, Lacadie SA, Huang HT, Curran M, Loewer S, Naveiras O, Kathrein KL, Konantz M, Langdon EM, Lengerke C, Zon LI, Collins JJ, Daley GQ. The transcriptional landscape of hematopoietic stem cell ontogeny. Cell Stem Cell. 2012;11:701–14.

Menetrier-Caux C, Montmain G, Dieu MC, Bain C, Favrot MC, Caux C, Blay JY. Inhibition of the differentiation of dendritic cells from CD34(+) progenitors by tumor cells: role of interleukin-6 and macrophage colony-stimulating factor. Blood. 1998;92:4778–91.

Metcalf D. Hematopoietic cytokines. Blood. 2008;111:485–91.

Migliaccio AR, Whitsett C, Migliaccio G. Erythroid cells in vitro: from developmental biology to blood transfusion products. Curr Opin Hematol. 2009;16:259–68.

Migliaccio G, Sanchez M, Masiello F, Tirelli V, Varricchio L, Whitsett C, Migliaccio AR. Humanized culture medium for clinical expansion of human erythroblasts. Cell Transplant. 2010;19:453–69.

Miharada K, Hiroyama T, Sudo K, Nagasawa T, Nakamura Y. Efficient enucleation of erythroblasts differentiated in vitro from hematopoietic stem and progenitor cells. Nat Biotech. 2006;24:1255–6.

Mikkola HK, Orkin SH. The journey of developing hematopoietic stem cells. Development. 2006;133:3733–44.

Mikkola HK, Klintman J, Yang H, Hock H, Schlaeger TM, Fujiwara Y, Orkin SH. Haematopoietic stem cells retain long-term repopulating activity and multipotency in the absence of stem-cell leukaemia SCL/tal-1 gene. Nature. 2003;421:547–51.

Miller J, Alley K, McGlave P. Differentiation of natural killer (NK) cells from human primitive marrow progenitors in a stroma-based long-term culture system: identification of a CD34+7+ NK progenitor. Blood. 1994;83:2594.

Milner LA, Bigas A. Notch as a mediator of cell fate determination in hematopoiesis: evidence and speculation. Blood. 1999;93:2431–48.

Minegishi N, Ohta J, Yamagiwa H, Suzuki N, Kawauchi S, Zhou Y, Takahashi S, Hayashi N, Engel JD, Yamamoto M. The mouse GATA-2 gene is expressed in the para-aortic splanchnopleura and aorta-gonads and mesonephros region. Blood. 1999;93:4196–207.

Miyamoto T, Iwasaki H, Reizis B, Ye M, Graf T, Weissman IL, Akashi K. Myeloid or lymphoid promiscuity as a critical step in hematopoietic lineage commitment. Dev Cell. 2002;3:137–47.

Monteiro R, Pouget C, Patient R. The gata1/pu.1 lineage fate paradigm varies between blood populations and is modulated by tif1γ. EMBO J. 2011;30:1093–103.

Murray LJ, Young JC, Osborne LJ, Luens KM, Scollay R, Hill BL. Thrombopoietin, flt3, and kit ligands together suppress apoptosis of human mobilized $CD34^{+}$ cells and recruit primitive $CD34^{+}$ Thy-1^{+} cells into rapid division. Exp Hematol. 1999;27:1019–28.

Nakagawa Y, Nakamura S, Nakajima M, Endo H, Dohda T, Takayama N, Nakauchi H, Arai F, Fukuda T, Eto K. Two differential flows in a bioreactor promoted platelet generation from human pluripotent stem cell–derived megakaryocytes. Exp Hematol. 2013;41:742–8.

Nakajima H. Role of transcription factors in differentiation and reprogramming of hematopoietic cells. Keio J Med. 2011;60:47–55.

Nakamura S, Takayama N, Hirata S, Seo H, Endo H, Ochi K, K-i Fujita T, Koike K-i Harimoto, Dohda T, Watanabe A, Okita K, Takahashi N, Sawaguchi A, Yamanaka S, Nakauchi H, Nishimura S, Eto K. Expandable megakaryocyte cell lines enable clinically applicable generation of platelets from human induced pluripotent stem cells. Cell Stem Cell. 2014;14:535–48.

Naldini L. Ex vivo gene transfer and correction for cell-based therapies. Nat Rev Genet. 2011;12:301–15.

Narla A, Dutt S, McAuley JR, Al-Shahrour F, Hurst S, McConkey M, Neuberg D, Ebert BL. Dexamethasone and lenalidomide have distinct functional effects on erythropoiesis. Blood. 2011;118:2296.

Natsuka S, Akira S, Nishio Y, Hashimoto S, Sugita T, Isshiki H, Kishimoto T. Macrophage differentiation-specific expression of NF-IL6, a transcription factor for interleukin-6. Blood. 1992;79:460–6.

Neildez-Nguyen TMA, Wajcman H, Marden MC, Bensidhoum M, Moncollin V, Giarratana M-C, Kobari L, Thierry D, Douay L. Human erythroid cells produced ex vivo at large scale differentiate into red blood cells in vivo. Nat Biotech. 2002;20:467–72.

Newburger PE. IL6 to the rescue. Blood. 2008;111:3914–5.

Niu Q, Wang W, Li Y, Qin S, Wang Y, Wan G, Guan J, Zhu W. Cord blood-derived cytokine-induced killer cells biotherapy combined with second-line chemotherapy in the treatment of advanced solid malignancies. Int Immunopharmacol. 2011;11:449–56.

Niwa A, Heike T, Umeda K, Oshima K, Kato I, Sakai H, Suemori H, Nakahata T, Saito MK. A novel serum-free monolayer culture for orderly hematopoietic differentiation of human pluripotent cells via mesodermal progenitors. PLoS ONE. 2011;6:e22261.

Oh IH, Humphries RK. Concise review: multidimensional regulation of the hematopoietic stem cell state. Stem Cells. 2012;30:82–8.

Oran B, Shpall E. Umbilical cord blood transplantation: a maturing technology. Hematology Am Soc Hematol Educ Program. 2012;2012:215–22.

Ottmann OG, Pelus LM. Differential proliferative effects of transforming growth factor-beta on human hematopoietic progenitor cells. J Immunol. 1988;140:2661–5.

Patterson LJ, Gering M, Eckfeldt CE, Green AR, Verfaillie CM, Ekker SC, Patient R. The transcription factors Scl and Lmo2 act together during development of the hemangioblast in zebrafish. Blood. 2007;109:2389–98.
Peled T, Landau E, Mandel J, Glukhman E, Goudsmid NR, Nagler A, Fibach E. Linear polyamine copper chelatortetraethylenepentamine augments long-term ex vivo expansion of cord blood-derived $CD34^{+}$ cells and increases their engraftment potential in nod/scid mice. Exp Hematol. 2004;32:547–55.
Pereira CF, Chang B, Qiu J, Niu X, Papatsenko D, Hendry CE, Clark NR, Nomura-Kitabayashi A, Kovacic JC, Ma'ayan A, Schaniel C, Lemischka IR, Moore K. Induction of a hemogenic program in mouse fibroblasts. Cell Stem Cell. 2013;13:205–18.
Perry JM, He XC, Sugimura R, Grindley JC, Haug JS, Ding S, Li L. Cooperation between both Wnt/beta-catenin and PTEN/PI3 K/AKT signaling promotes primitive hematopoietic stem cell self-renewal and expansion. Genes Dev. 2011;25:1928–42.
Poli V, Mancini FP, Cortese R. IL-6DBP, a nuclear protein involved in interleukin-6 signal transduction, defines a new family of leucine zipper proteins related to C/EBP. Cell. 1990;63:643–53.
Porcher C, Swat W, Rockwell K, Fujiwara Y, Alt FW, Orkin SH. The T cell leukemia oncoprotein SCL/tal-1 is essential for development of all hematopoietic lineages. Cell. 1996;86:47–57.
Qiu C, Olivier EN, Velho M, Bouhassira EE. Globin switches in yolk sac–like primitive and fetal-like definitive red blood cells produced from human embryonic stem cells. Blood 2008.
Qiu C, Olivier EN, Velho M, Bouhassira EE. Globin switches in yolk sac-like primitive and fetal-like definitive red blood cells produced from human embryonic stem cells. Blood. 2008;111:2400–8.
Radomska HS, Huettner CS, Zhang P, Cheng T, Scadden DT, Tenen DG. CCAAT/enhancer binding protein alpha is a regulatory switch sufficient for induction of granulocytic development from bipotential myeloid progenitors. Mol Cell Biol. 1998;18:4301–14.
Radtke F, Wilson A, Mancini SJ, MacDonald HR. Notch regulation of lymphocyte development and function. Nat Immunol. 2004;5:247–53.
Ramji DP, Foka P. CCAAT/enhancer-binding proteins: structure, function and regulation. Biochem J. 2002;365:561–75.
Ran D, Shia WJ, Lo MC, Fan JB, Knorr DA, Ferrell PI, Ye Z, Yan M, Cheng L, Kaufman DS, Zhang DE. RUNX1a enhances hematopoietic lineage commitment from human embryonic stem cells and inducible pluripotent stem cells. Blood. 2013;121:2882–90.
Reems J-A, Pineault N, Sun S. In vitro megakaryocyte production and platelet biogenesis: state of the art. Transfus Med Rev. 2010;24:33–43.
Reinhard G, Marten A, Kiske SM, Feil F, Bieber T, Schmidt-Wolf IGH. Generation of dendritic cell-based vaccines for cancer therapy. Br J Cancer. 2002;86:1529–33.
Rekhtman N, Radparvar F, Evans T, Skoultchi AI. Direct interaction of hematopoietic transcription factors PU.1 and GATA-1: functional antagonism in erythroid cells. Genes Dev. 1999;13:1398–411.
Robb L. Cytokine receptors and hematopoietic differentiation. Oncogene. 2007;26:6715–23.
Rolink AG, Nutt SL, Melchers F, Busslinger M. Long-term in vivo reconstitution of T-cell development by Pax5-deficient B-cell progenitors. Nature. 1999;401:603–6.
Sallusto F, Lanzavecchia A. Efficient presentation of soluble antigen by cultured human dendritic cells is maintained by granulocyte/macrophage colony-stimulating factor plus interleukin 4 and downregulated by tumor necrosis factor alpha. J Exp Med. 1994;179:1109–18.
Sandler VM, Lailler N, Bouhassira EE. Reprogramming of embryonic human fibroblasts into fetal hematopoietic progenitors by fusion with human fetal liver CD34+ cells. PLoS ONE. 2011;6:e18265.
Sandler VM, Lis R, Liu Y, Kedem A, James D, Elemento O, Butler JM, Scandura JM, Rafii S. Reprogramming human endothelial cells to haematopoietic cells requires vascular induction. Nature. 2014;511:312–8.
Scheller M, Huelsken J, Rosenbauer F, Taketo MM, Birchmeier W, Tenen DG, Leutz A. Hematopoietic stem cell and multilineage defects generated by constitutive beta-catenin activation. Nat Immunol. 2006;7:1037–47.

Schmidt Michael, Geilenkeuser Wolf-Jochen, Sireis Walid, Seifried Erhard, Transfus Kai Hourfar. Emerging pathogens—how safe is blood? Med Hemother. 2014;41:10–7. doi:10.1159/000358017.

Schroeder T, Kohlhof H, Rieber N, Just U. Notch signaling induces multilineage myeloid differentiation and up-regulates PU.1 expression. J Immunol. 2003;170:5538–48.

Scott LM, Civin CI, Rorth P, Friedman AD. A novel temporal expression pattern of three C/EBP family members in differentiating myelomonocytic cells. Blood. 1992;80:1725–35.

Singh H, Medina KL, Pongubala JM. Contingent gene regulatory networks and B cell fate specification. Proc Natl Acad Sci USA. 2005;102:4949–53.

Sitnicka E, Ruscetti FW, Priestley GV, Wolf NS, Bartelmez SH. Transforming growth factor beta 1 directly and reversibly inhibits the initial cell divisions of long-term repopulating hematopoietic stem cells. Blood. 1996;88:82–8.

Spanholtz J, Preijers F, Tordoir M, Trilsbeek C, Paardekooper J, de Witte T, Schaap N, Dolstra H. Clinical-grade generation of active NK cells from cord blood hematopoietic progenitor cells for immunotherapy using a closed-system culture process. PLoS ONE. 2011;6:e20740.

Stroncek DF, Rebulla P. Platelet transfusions. Lancet. 2007;370:427–38.

Suh HC, Gooya J, Renn K, Friedman AD, Johnson PF, Keller JR. C/EBPalpha determines hematopoietic cell fate in multipotential progenitor cells by inhibiting erythroid differentiation and inducing myeloid differentiation. Blood. 2006;107:4308–16.

Suk Ran Y, Jin Woong C, Inpyo C. Development of natural killer cells from hematopoietic stem cells. Mol Cells. 2007;24:1–8.

Szabo E, Rampalli S, Risueño RM, Schnerch A, Mitchell R, Fiebig-Comyn A, Levadoux-Martin M, Bhatia M. Direct conversion of human fibroblasts to multilineage blood progenitors. Nature. 2010;468:521–6.

Takahashi K, Tanabe K, Ohnuki M, Narita M, Ichisak T, Tomoda K, Yamanaka S. Induction of pluripotent stem cells from adult human fibroblasts by defined factors. Cell. 2007;131:861–72.

Takayama N, Nishikii H, Usui J, Tsukui H, Sawaguchi A, Hiroyama T, Eto K, Nakauchi H. Generation of functional platelets from human embryonic stem cells in vitro via ES-sacs, VEGF-promoted structures that concentrate hematopoietic progenitors. Blood. 2008;111:5298–306.

Taylor CJ, Peacock S, Chaudhry AN, Bradley JA, Bolton EM. Generating an iPSC bank for HLA-matched tissue transplantation based on known donor and recipient HLA types. Cell Stem Cell. 2012;11:147–52.

Teitell MA, Mikkola HKA. Transcriptional activators, repressors, and epigenetic modifiers controlling hematopoietic stem cell development. Pediatr Res. 2006;59:33R–9R.

Thon JN, Mazutis L, Wu S, Sylman JL, Ehrlicher A, Machlus KR, Feng Q, Lu S, Lanza R, Neeves KB, Weitz DA, Italiano JE. Platelet bioreactor-on-a-chip. Blood. 2014;124:1857.

Tsai FY, Orkin SH. Transcription factor GATA-2 is required for proliferation/survival of early hematopoietic cells and mast cell formation, but not for erythroid and myeloid terminal differentiation. Blood. 1997;89:3636–43.

Tsai FY, Keller G, Kuo FC, Weiss M, Chen J, Rosenblatt M, Alt FW, Orkin SH. An early haematopoietic defect in mice lacking the transcription factor GATA-2. Nature. 1994;371:221–6.

Undi RB, Kandi R, Gutti RK. MicroRNAs as haematopoiesis regulators. Adv Hematol. 2013;2013:695754.

Vodyanik MA, Bork JA, Thomson JA, Slukvin II. Human embryonic stem cell-derived $CD34^+$ cells: efficient production in the coculture with OP9 stromal cells and analysis of lymphohematopoietic potential. Blood. 2005;105:617–26.

Wang Y, Zheng CG, Jiang Y, Zhang J, Chen J, Yao C, Zhao Q, Liu S, Chen K, Du J, Yang Z, Gao S. Genetic correction of β-thalassemia patient-specific iPS cells and its use in improving hemoglobin production in irradiated SCID mice. Cell Res. 2012a;22:637–48.

Wang C, Tang X, Sun X, Miao Z, Lv Y, Yang Y, Zhang H, Zhang P, Liu Y, Du L, Gao Y, Yin M, Ding M, Deng H. TGFβ inhibition enhances the generation of hematopoietic progenitors from human ES cell-derived hemogenic endothelial cells using a stepwise strategy. Cell Res. 2012b;22:194–207.

Wilkinson RN, Pouget C, Gering M, Russell AJ, Davies SG, Kimelman D, Patient R. Hedgehog and Bmp polarize hematopoietic stem cell emergence in the zebrafish dorsal aorta. Dev Cell. 2009;16:909–16.

Williams NS, Klem J, Puzanov IJ, Sivakumar PV, Schatzle JD, Bennett M, Kumar V. Natural killer cell differentiation: insights from knockout and transgenic mouse models and in vitro systems. Immunol Rev. 1998;165:47–61.

Woll PS, Grzywacz B, Tian X, Marcus RK, Knorr DA, Verneris MR, Kaufman DS. Human embryonic stem cells differentiate into a homogeneous population of natural killer cells with potent in vivo antitumor activity. Blood. 2009;113:6094.

Xi J, Zhu H, Liu D, Nan X, Zheng W, Liu K, Shi W, Chen L, Lv Y, Yan F, Li Y, Xie X, Wang Y, Yue W, Xu X, Wei X, Zhu J, Huang X, Pei X. Infusion of megakaryocytic progenitor products generated from cord blood hematopoietic stem/progenitor cells: results of the phase 1 study. PLoS ONE. 2013;8:e54941.

Xie H, Ye M, Feng R, Graf T. Stepwise reprogramming of B cells into macrophages. Cell. 2004;117:663–76.

Yang J, Aguila JR, Alipio Z, Lai R, Fink LM, Ma Y. Enhanced self-renewal of hematopoietic stem/progenitor cells mediated by the stem cell gene Sall4. J Hematol Oncol. 2011;4:38.

Yu C, Liu Y, Miao Z, Yin M, Lu W, Lv Y, Ding M, Deng H. Retinoic acid enhances the generation of hematopoietic progenitors from human embryonic stem cell–derived hemato-vascular precursors. Blood. 2010;116:4786.

Zeuner A, Martelli F, Vaglio S, Federici G, Whitsett C, Migliaccio AR. Concise review: stem cell-derived erythrocytes as upcoming players in blood transfusion. Stem Cells. 2012;30:1587–96.

Zhang DE, Zhang P, Wang ND, Hetherington CJ, Darlington GJ, Tenen DG. Absence of granulocyte colony-stimulating factor signaling and neutrophil development in CCAAT enhancer binding protein alpha-deficient mice. Proc Natl Acad Sci USA. 1997;94:569–74.

Zhang P, Iwasaki-Arai J, Iwasaki H, Fenyus ML, Dayaram T, Owens BM, Shigematsu H, Levantini E, Huettner CS, Lekstrom-Himes JA, Akashi K, Tenen DG. Enhancement of hematopoietic stem cell repopulating capacity and self-renewal in the absence of the transcription factor C/EBP alpha. Immunity. 2004;21:853–63.

Zou J, Mali P, Huang X, Dowey SN, Cheng L. Site-specific gene correction of a point mutation in human ips cells derived from an adult patient with sickle cell disease. Blood. 2011;118:4599–608.

Chapter 6
Signaling Pathways Regulating Stem Cells

Xingting Guo, Jun Chen, Zhaohui Li and Rongwen Xi

Abstract Stem cells have a unique ability to self-renewal and a potential to differentiate into one or more types of specialized cells. These properties in vivo are commonly maintained and regulated by signaling pathways involved in cell-to-cell communications. Signaling communications occur between stem cells and the niche supporting cells, as well as between stem cells and their differentiated daughter cells. Different types of stem cells within the same tissue are also coordinately regulated by signaling pathways to maintain tissue architecture and function. Signaling pathways are also critical in mediating stem cell activation in response to tissue damage for accelerated regeneration. This chapter will review signaling mechanisms in controlling various behaviors of several well-characterized tissue stem cells, including self-renewal, differentiation, and regenerative activation of stem cells.

Keywords Stem cell · Signaling pathway · Mechanisms

6.1 Introduction

Adult stem cells or tissue stem cells have been a research of focus for many years, due to their prominent roles in a variety of biological processes, including tissue homeostasis, regeneration, organogenesis, and tumorigenesis. Stem cells are a population of undifferentiated cells that can self-renew via mitosis and differentiate into specialized progenies. There are many types of adult stem cells, such as hematopoietic stem cells (HSCs), neuronal stem cells (NSCs), intestinal stem cells (ISCs), and germline stem cells (GSCs) (Li and Xie 2005; Gancz and Gilboa 2013). Tightly controlled proliferation and differentiation of these stem cells

X. Guo · J. Chen · Z. Li · R. Xi (✉)
National Institute of Biological Sciences, Beijing 102206, China
e-mail: xirongwen@nibs.ac.cn

R.C. Zhao (ed.), *Stem Cells: Basics and Clinical Translation*, Translational Medicine Research 1, DOI 10.1007/978-94-017-7273-0_6

throughout adulthood is critical for normal tissue homeostasis and damage repair. Once such regulation is disrupted, uncontrolled stem cells may lead to tumor initiation or stem cell depletion and consequently degenerative diseases (Voog and Jones 2010).

Adult stem cells commonly reside in a specialized tissue microenvironment or niche, and secreted signals from the niche have important roles in regulating stem cell maintenance, proliferation, and differentiation (Schofield 1977; Morrison and Spradling 2008). These secreted niche signals usually directly activate surface receptors on the cell membrane of stem cells, followed by the activation of signal cascades in the cytoplasm and eventually the activation of transcription factors and gene expression programs, which controls the proliferation, survival, or differentiation of stem cells. Signaling pathways are also commonly involved in mediating the response of stem cells to external stimuli (nutrient, tissue damage, and bacterial infection, etc.) to coordinate stem cell activity with demands. Recent advances in cell biology and biochemistry have largely illustrated details of these signaling pathways, and genetic analysis on model organisms such as *Drosophila melanogaster* and *Mus musculus* has provided insights into their functions in stem cell regulation. In the following parts, we will provide an overview of several commonly utilized signaling pathways and their roles in the regulation of adult stem cells.

6.2 Wnt Pathway

The Wnt signaling is highly conserved from invertebrates to vertebrates (Klaus and Birchmeier 2008). Since the discovery of the *Drosophila* segment polarity gene Wingless and its homologue Int-1 in the murine (Nusse and Varmus 1982; Rijsewijk et al. 1987), more mutations identified in *Drosophila* with similar phenotypes with $wg^{-/-}$ have led to the identification of downstream signaling components, which constitute the canonical Wnt signaling cascade (Nusse et al. 1991). Activation of the pathway requires the binding of secreted Wnt ligands to Frizzled receptors and LRP5–LRP6 co-receptors in the plasma membrane (Finch et al. 1997). The central player during Wnt signal transduction is a cytosolic protein called β-catenin. Under quiescent state, newly synthesized β-catenin binds to a destructive complex consisting of two scaffold proteins—the tumor suppressor adenomatous polyposis coli (APC) and Axin—and two kinases in this complex, CKI and GSK3, and sequentially phosphorylates a set of conserved Ser and Thr residues in the N-terminus of β-catenin afterward (Peters et al. 1999; Amit et al. 2002). The phosphorylated signal can be recognized by a b-TrCP-containing E3 ligase and lead to proteasomal degradation of β-catenin (Liu et al. 2002; Aberle et al. 1997). Meanwhile, with little β-catenin in the nucleus, transcription factors such as TCF and LEF interact with c-repressor Groucho or CtBP to repress Wnt-specific target genes (Cavallo et al. 1998; Fang et al. 2006).

Upon activation by Wnt binding, LRP5–LRP6 phosphorylation and Dishevelled aggression serve as mediator and lead to the translocation of Axin to the plasma membrane and thus inactivates the destructive complex (Bilić et al. 2007). Cytosolic β-catenin could accumulate and enter the nucleus afterward, and the mechanism of the translocation still remains unclear. In the nucleus, it forms a transcriptionally active complex with LEF and TCF transcription factors by displacing Groucho and interacting with other co-activators such as BCl9, Pygopus, CBP (CREB-binding protein), or Hyrax, promoting activation of Wnt target genes (Hecht et al. 2000; Mosimann et al. 2006) (Fig. 6.1).

In addition to canonical Wnt pathway described above, some Wnt ligands and Frizzled receptors, and the Dishevelleds, are capable of activating a β-catenin independent, non-canonical Wnt signaling cascade, including the planar cell polarity (PCP) pathway and the Ca^{2+}-dependent Wnt signaling pathway (Seifert and Mlodzik 2007; Veeman et al. 2003). They may function in regulating polarization of cells and directed cell motility, as well as transforming capacity in cell culture. However, mutations in non-canonical Wnt pathway components have not been reported to be associated with tumorigenesis.

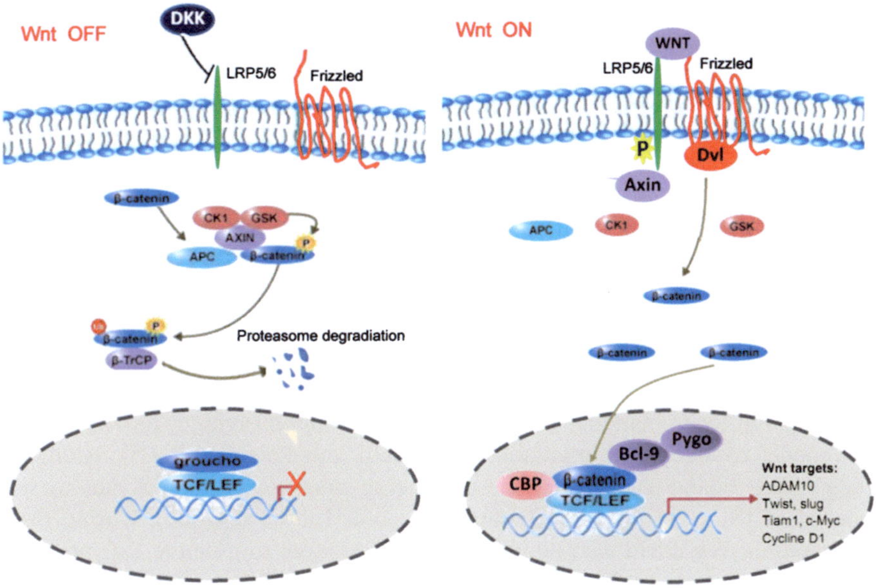

Fig. 6.1 The canonical Wnt signaling pathway. Under quiescent state, newly synthesized β-catenin binds to and is phosphorylated by a destructive complex consisting of two scaffold proteins—APC and axin—and two kinases, CKI and GSK3 (Peters et al. 1999; Amit et al. 2002). The phosphorylated β-catenin can be recognized by a b-TrCP-containing E3 ligase, which leads to proteasomal degradation of β-catenin (Liu et al. 2002; Aberle et al. 1997). Wnt binding to the receptors leads to inactivation of the destructive complex (Bilić et al. 2007). Cytosolic β-catenin then accumulates and enters the nucleus. In the nucleus, β-catenin forms a transcriptionally active complex with LEF and TCF transcription factors and other co-activators such as BCl9, Pygopus, CBP (CREB-binding protein), or Hyrax to displace Groucho and promote activation of Wnt target genes (Hecht et al. 2000; Mosimann et al. 2006)

Decades of studies on model organisms have revealed that a variety of stem cells are regulated by the Wnt signaling pathway. In vitro-cultured mouse ESCs were laid on a layer of fibroblasts in order to obtain the necessary factors, among which Wnt signal was found to be necessary for mESC self-renewal and pluripotency maintenance (Young 2011). Other approaches that activate this pathway are also effective, including overexpression of an active form of β-catenin or treatment with GSK3 inhibitors (Sato et al. 2003). However, in terms of human ESC, whether Wnt/β-catenin signaling maintains them in an undifferentiated and self-renewing state, or whether it promotes differentiation, remains controversial (Day et al. 2005). During the process of gastrulation, Wnt pathway is required for generation of primitive streak (PS), which will ultimately differentiate into mesoderm or definitive endoderm, whereas ectoderm formation is blocked by this pathway (Lindsley et al. 2006; Aubert et al. 2002). Further differentiation after germ layer formation is also regulated by Wnt/β-catenin signaling; for example, it inhibits cardiac differentiation and may redirect the cells to alternate mesodermal fates like hematopoietic lineage (Murry and Keller 2008; Trompouki et al. 2011). Reports also show that non-canonical Wnt/PCP and Wnt/β-catenin signaling cooperates to regulate the cell-fate choice in asymmetrically dividing cells in Xenopus embryo by restricting Lrp6 to the basolateral part of the stem cell (Glinka et al. 2011).

During adulthood, Wnt/β-catenin pathway also regulates proliferation and differentiation of adult stem cells, among which the best well characterized might be ISC. In mammalian small intestine, ISCs are located in the bottom of the crypt, while their progenies move upward to the villi. Paneth cells surrounding ISCs secrete WNT3 to maintain ISCs via induced expression of Wnt target genes such as *cMyc* and *cyclin D* (Li and Xie 2005). Other targets of β-catenin/Tcf signaling include ephrin receptors EphB2 and EphB3, which function in establishing crypt–villus boundaries and positioning of paneth cells at the crypt bottom (Batlle et al. 2002). Besides, additional Wnt signals could promote differentiation of paneth cells by the activation of genes specific to paneth cells, such as *cryptidin* (van Es et al. 2005).

Compared to its mammalian counterparts, *Drosophila* midgut provides a simpler model to study ISC regulation (Micchelli and Perrimon 2005). Along the midgut, Wg ligand secreted by the underlying muscle cells contributes to self-renewal of ISCs to maintain the stem cell pool (Lin et al. 2008). During tissue regeneration, Wg could also be induced in progenitors to promote ISC proliferation (Cordero et al. 2012). Hair follicle stem cells are also regulated by the Wnt/β-catenin signaling pathway. In comparison with ISCs, bulge stem cells remain quiescent when undergoing a resting phase and reside in a Wnt-restricted environment (DasGupta and Fuchs 1999). TCF3 with no β-catenin association may function to maintain skin stem cells in an undifferentiated state through repression of specific TCF target genes. Once entering regenerative phase, Wnt signaling pathway is activated and β-catenin stabilized, which will relieve TCF3 repression. Meanwhile, β-catenin interacts with another LEF/TCF member LEF1, which will activate bulge stem cells and generate new hair follicles (Lowry et al. 2005). As the stem cells proliferate and undergo a differentiated lineage, β-catenin binds to other TFs to promote hair shaft cell differentiation, while β-catenin deficiency leads to the genesis of epidermal cell, indicating that Wnt signaling also participates in fate choice (Lowry et al. 2005). These studies indicate that by binding to different partners, Wnt/β-catenin pathway could coordinately regulate self-renewal, proliferation as well as terminal differentiation of hair follicle stem cells (Blanpain et al. 2007; Choi et al. 2013).

HSC residing in bone marrow is the origin of blood cells and immune cells. Wnt ligands, secreted by HSCs themselves as well as by the microenvironment, are responsible for self-renewal of HSC and progenitor cells, as well as maintaining them in an undifferentiated state (Rattis et al. 2004). Deletion of β-catenin in mouse models blocks long-term growth and maintenance of HSC and reduced possibility to develop BCRABL-induced chronic myelogenous leukemia (Austin et al. 1997; Jamieson et al. 2004).

In the nervous system, both central and peripheral, astrocytes generate Wnt3a to promote proliferation and neuronal fate commitment of neural precursors via canonical Wnt/β-catenin pathway (Lee et al. 2004). Wnt also promotes differentiation of NSCs into neuronal and astrocyte lineages in a time- and location-dependent manner (Toledo et al. 2008). In other systems, including mammary stem cells (MaSC) and airway stem cells, canonical Wnt pathway regulates their maintenance and self-renewal in a manner similar to that described above (Zeng and Nusse 2010; Zhang et al. 2008).

The common requirement in various types of stem cells makes the Wnt signaling cascade critical in both organ development and tumorigenesis. Depletion of Wnt/β-catenin activities will result in a series of morphological and functional defects, including the absence of intestinal crypts or hair follicles (Pinto et al. 2003; Andl et al. 2002). On the contrary, aberrant activation of Wnt pathway will lead to hyper-accumulation of stem and progenitor cells, which further might induce tumorigenesis. Leukemia, breast cancer, and the majority of familial and sporadic colon tumors are, to some extent, associated with mutations that lead to constant activation of Wnt signaling pathway (Jamieson et al. 2004; Korkaya et al. 2009; Phelps et al. 2009).

6.3 Notch Pathway

The first Notch mutant was identified in *Drosophila* a century ago by Morgan (1917). The Notch pathway is evolutionarily conserved from invertebrates to vertebrates and plays fundamental roles in a broad range of development processes.

In the canonical Notch pathway of *Drosophila*, the receptor Notch, a single-pass transmembrane receptor consisting of a large extracellular region with epidermal growth factor-like repeats and an intracellular region, interacts with the membrane-bound ligands Delta(Dl) or Serrate(Ser) (Delta-like and Jagged in mammalian) from the neighboring cells. A cascade of proteolytic events is triggered after the ligand–receptor interaction, resulting in the release of the intracellular domain (NICD) of Notch to the cytoplasm. The NICD then translocates into the nucleus to form a transcriptional complex with the co-activator Mastermind (Mam) and the DNA-binding protein Suppressor of Hairless [Su(H)] (CSL, CBF1/RBPJK in mammalian) to regulate gene expression (Bray 2006; Kopan and Ilagan 2009) (Fig. 6.2). There is also increasing evidence for a ligand- or transcription-independent non-canonical Notch pathway that exerts important biological functions, which we will not discuss here but have been reviewed elsewhere by Heitzler (2010), Andersen et al. (2012).

Notch signaling has been implicated to function critically in many kinds of stem cell lineages, such as stem cells in skin, nervous system, hematopoietic system, muscle, liver, kidney, and intestine. As mentioned above, Notch signaling is a very simple signaling cascade even without a second messenger, but its function is highly context dependent and can be modulated at multiple levels, allowing it to play a variety of biological roles from stem cell maintenance to multiple cell lineage differentiation. Here, we mainly focus on its roles in embryonic stem cells (ESCs) and adult stem cells in hematopoietic system and intestine.

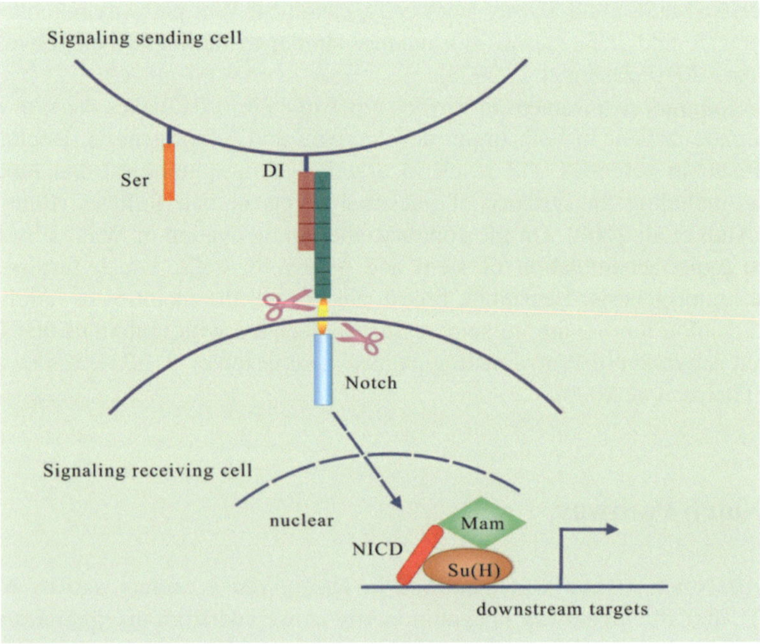

Fig. 6.2 The canonical Notch signaling pathway in *Drosophila*. The single-pass transmembrane receptor Notch can be activated by the membrane-bound ligands, Delta (*Dl*) or Serrate (*Ser*), from neighboring signal-sending cells. The ligand–receptor interaction induces a cascade of proteolytic events, which leads to the release of Notch intracellular domain (*NICD*) from the cell membrane. The NICD translocates into the nucleus, where it forms a transcriptional complex with the co-activator Mastermind and the DNA-binding protein Suppressor of Hairless (Su(H)) to regulate its target gene expression

There is no detectable Notch activity in human embryonic stem cells (hESCs), and Notch is not required for the hESC maintenance (Noggle et al. 2006). But Notch plays important roles in cell-fate decision during ESC differentiation. Activation of Notch signaling in ESCs under differentiation condition promotes the neural commitment of ESCs, resulting in differentiation into neuroectodermal progenitor cells (Lowell et al. 2006). Conversely, loss or downregulation of the Notch activity leads to cardiac mesodermal differentiation (Schroeder et al. 2003; Nemir et al. 2006; Jang et al. 2008). Therefore, Notch is not required for ESC maintenance but modulates the outcome during differentiation.

The hematopoietic system is relatively complex, which consists of multiple cell lineages. Notch has involved in many aspects of hematopoiesis. Notch signaling is required for the generation of HSCs in the aorta-gonad-mesonephros (AGM) region during embryogenesis and in the long-term definitive hematopoiesis past the early fetal liver stage (Kumano et al. 2003; Robert-Moreno et al. 2008; Hadland et al. 2004). It also functions in HSC maintenance in the marrow. Many Notch ligands are expressed by the hematopoietic niche (Karanu et al. 2001; Fernandez et al. 2008), and increased Notch activity in HSCs promotes expansion of HSCs and hematopoietic progenitor cells in vivo and in vitro (Calvi et al. 2003; Butler et al. 2010; Stier et al. 2002; Varnum-Finney et al. 2003; Karanu et al. 2000). In contrast, inhibition of Notch activity in HSCs displays no detectable effect on HSC maintenance (Mancini et al. 2005; Maillard et al. 2008). Therefore, Notch signaling is dispensable for adult HSC maintenance, but its activation is sufficient to promote HSC proliferation. In addition, Notch signaling is also essential for the differentiation of HSCs and other hematopoietic progenitor cells. Enforced Notch activity in the hematopoietic progenitor cells promotes T cell commitment (de La Coste et al. 2005; Radtke et al. 2004), and loss of the activity leads to T cell deficiency (Radtke et al. 1999). Notch also plays a role in cell-fate decision among myeloid progenitors, and activation of Notch both in vivo and in vitro induces megakaryocyte development (Mercher et al. 2008). Therefore, the role of Notch signaling in hematogenesis is important for lineage commitment at multiple branch points of hematopoiesis. Further investigation into the mechanism of the Notch function in different lineages may facilitate the understanding of the complexity of Notch function in hematopoietic system.

Compared to the mammalian system, the *Drosophila* midgut is a much simpler model system for studying signaling regulation of stem cells. The midgut ISCs usually asymmetrically divide into a new ISC and an intermediate enteroblast (EB), and EB can further differentiate into either an enterocyte (EC) or an enteroendocrine (ee) cell (Micchelli and Perrimon 2005; Ohlstein and Spradling 2005, 2007). Notch signaling plays crucial roles in the binary cell-fate decision and terminal differentiation of EB. The Notch ligand Dl is specifically expressed in ISCs and activates Notch activity in their immediate daughter EBs. Overexpression of the NICD in progenitor cells invariably induces EC differentiation, while loss of Notch leads to the expansion of ISC-like cells and ee cells. Therefore, it has been proposed that EBs that receive high levels of Notch activity will differentiate into ECs, whereas EBs that receive low or no Notch activity will differentiate into ee cells (Micchelli and Perrimon 2005; Ohlstein and Spradling 2005, 2007). In mammalian intestine, Notch regulates both the maintenance and differentiation of ISC. In contrast to that in *Drosophila* gut, ISCs in mammals are also signal-receiving cells where the receptors Notch1 and Notch2 are expressed (Fre et al. 2011; Pellegrinet et al. 2011). Activation of Notch in ISCs promotes their amplification, and activation of Notch in progenitor cells favors absorptive cell differentiation over secretory cell differentiation (Fre et al. 2005; Ueo et al. 2012). Conversely, loss of Notch activity leads to the loss of ISCs and increased production of the secretory goblet cells (Ueo et al. 2012; Riccio et al. 2008; Milano et al. 2004). In addition to HSCs and ISCs, Notch is involved in the regulation of many other types of stem cells, which will not be discussed here but are summarized by Liu and Carolina (Liu et al. 2010; Perdigoto and Bardin 2013).

Taken together, Notch signaling is involved in many aspects of stem cell behavior, including maintenance, cell-fate decision, and terminal differentiation. Immediate questions remaining to be answered include how the specificity of Notch function in a different context is achieved and how the pathway is regulated to meet the needs during tissue homeostasis and in response to environmental changes.

6.4 EGFR Pathway

As a central element for a variety of cellular response and signaling transduction network, the epidermal growth factor receptor (EGFR) family has fundamental roles in the development of multicellular organisms. EGFR signaling pathway has been reported to regulate many cellular functions including the cell survival, motility, proliferation, and cell-fate decision. Disruption of the signaling pathway is frequently implicated in the development of human tumors, proposing EGFR as a prognostic marker or target in cancer therapy. In this part, we will focus on the critical role of EGFR signaling pathway as regulators of stem cell properties including maintenance and differentiation.

The role of EGFR in regulating stem cell proliferation and self-renewal has been well described in many tissues, including neural system, intestine, and mammalian epidermis.

Proliferation of stem cells is crucial for tissue homeostasis and regeneration during wound repairing, especially in tissues with high turnover. A typical system is *Drosophila* intestine, where EGFR promotes the proliferative capacity of ISCs to maintain gut homeostasis and regeneration after damage (Jiang et al. 2011). The EGFR ligand, Vein, is specifically expressed in visceral muscle surrounding the midgut epithelium as a proliferating niche signal. And two additional EGFR ligands, Spitz and Keren, serve as autocrine signals to redundantly promote ISC proliferation and maintenance (Biteau and Jasper 2011; Xu et al. 2011). Damage in midgut epithelium induces multiple EGFR ligands to activate EGFR/Ras/MAPK signaling pathway in ISCs, which is required in ISC proliferation and tissue regeneration (Jiang et al. 2011). The synergetic cooperation of EGFR with other signaling pathways, including JAK/STAT and Wingless signaling, is essential for ISC maintenance (Jiang et al. 2011; Xu et al. 2011; Buchon et al. 2010).

Similar in *Drosophila* intestine, interaction of EGFR with other signaling pathways is also crucial in the regulation of stem cell maintenance in nervous system. In the adult brain, the subventricular zone (SVZ) and the dentate gyrus are the niches that maintain neural stem cells (NSCs) and neural progenitor cells (NPCs), and balance of sizes between these two populations is critical for brain homeostasis (Alvarez-Buylla and Lim 2004). In the SVZ, Notch signaling is required to maintain NSCs, while EGFR is responsible for the development of NPCs (Hitoshi et al. 2002; Alexson et al. 2006; Lillien and Raphael 2000). Through direct interaction between NSCs and NPCs, the cooperation of EGFR signaling and Notch signaling occurs to maintain the balance between these two populations (Aguirre et al. 2010).

Along with its well-described role in regulating cell survival and proliferation, EGFR signaling also functions in the differentiation of certain types of stem cells or progenitor cells. In *Drosophila* eye disc, EGFR activation triggers differentiation of all retinal cell types (Freeman 1996). The generation of distinct cell fates depends on the combinatorial effect of EGFR signaling with other signal responses (Flores et al. 2000). Combination of EGFR and Notch signaling regulates cone cell specification, while specification of R7 cells requires the Sevenless receptor tyrosine kinase besides EGFR and Notch signaling (Flores et al. 2000; Cooper and Bray 2000). In other species like freshwater planarians, EGFR signaling is also essential in the process of differentiation and morphogenesis. Silencing planarian EGFR gene using RNAi results in the abnormal differentiation of certain cell types and various tissue defects (Fraguas et al. 2011).

EGFR also regulates the mobility of stem cells. In various tissues, EGFR signaling is linked to a more mobile phenotype in both immature progenitors and committed cells. In transplanted embryonic progenitor cells, continuous EGF signaling stimulates their proliferation and migration (Fricker-Gates et al. 2000). In the case of mature tissue system, active EGFR signaling also enhances the migration of NSCs and HSCs, providing a possible pharmacological strategy for cellular transplantation in disease therapy (Ryan et al. 2010; Boockvar et al. 2003; Ayuso-Sacido et al. 2010).

EGFR signaling pathway plays a central role in a variety of fundamental cellular functions including cell growth, proliferation, transformation, and mobilization. The ability that the single receptors function in such diverse processes attributes to the cell types that receive the EGF stimulation and combinatorial effect of multiple signaling pathways. Due to the complexity of the EGFR transduction network, the mechanism regulating the interconnected network and the resulting responses still remains incompletely clear. Since EGFR signaling is frequently implicated in hyper-proliferative diseases, more mechanistic studies of its involvement in cellular responses under both normal and pathological conditions are needed for the understanding of disease mechanisms and the development of therapies.

6.5 BMP Pathway

BMP signaling has diverse functions in multicellular organism development and recently has been reported to function as an essential regulator of stem cell maintenance and cell-fate decision. In this section, we will discuss how BMP signaling regulates stem cell properties and its potential role in cancer development.

BMPs belong to TGF-β superfamily and signal through receptor-mediated intercellular pathway to regulate expression of target genes. There are two types of receptors, one with type I receptor (Bmpr II) and three with type II receptors (Alk2, Alk3, and Alk6). Activation of the BMP pathway involves the formation of heteromeric complex of type I receptor and type II receptor upon ligand binding, which mediates the phosphorylation of type I receptor. Then, downstream intercellular messengers are three classes of Smad proteins: receptor-mediated Smad1/5/8 (R-Smad), the common mediator Smad4 (Co-Smad), and the inhibitory Smad6/7 (I-Smad). Activated type I receptor mediates R-Smad phosphorylation, which induces the formation of R-Smad/Co-Smad complex. The heteromeric Smad complex then translocates to the nucleus and regulates target gene expression in cooperation with other transcription factors. I-Smad functions to negatively regulate the Smad signaling pathway. A downstream pathway in parallel with the canonical BMP pathway is TAK1/MAPK pathway. TAK1 is a MAPKKK tyrosine kinase, which is activated by linking to the receptor mediated by X-linked inhibitor of apoptosis (XIAP). Notably, TAK1 also participates in JNK and NF-kB pathway, providing a possible means of cross talk between BMP and other signaling pathways.

The BMP pathway has a role in maintaining mammalian embryonic stem cell self-renewal. ESCs are stem cells with widest developmental capacity, which can contribute to all three germ layers: the ectoderm, mesoderm, and endoderm (Chambers and Smith 2004). ESCs are derived from the inner cell mass (ICM) of the blastocyst embryos and can be stably maintained in in vitro culture under a proper condition, thus providing a widely used system to study self-renewal and commitment of stem cells.

A series of studies in mouse ES cells have shown that BMP pathway is able to maintain cultured mES cells in undifferentiated state (Ying et al. 2003). mESCs can be cultured with a layer of mouse embryonic fibroblast (MEF) cells to produce supporting factors including leukemia inhibitory factor (LIF). LIF effectively supports mESCs' self-renewal in culture conditions containing serum, but in the absence of feeder cells or serum, LIF alone cannot maintain pluripotency but induce neural differentiation of mESCs (Ying et al. 2003; Ying and Smith 2003). However, treatment in combination with LIF and BMP4 suppressed neural differentiation and is sufficient to maintain pluripotency of mESCs without feeder cells or serum (Ying et al. 2003). It is well known that in contrast to LIF, which favors neural differentiation through Stat3, BMP signaling inhibits neural differentiation (Ying et al. 2003; Tropepe et al. 2001). Therefore, coordination of LIF and BMP signaling and their balanced mutual inhibition are crucial for mESC maintenance.

By contrast, BMPs in human ESCs promote differentiation (Pera et al. 2004). Unlike mESCs, hESCs need basic fibroblast factors (bFGF) rather than LIF to support self-renewal and pluripotency when cultured with a feeder layer or fibroblast-conditioned media. High level of BMP signaling was found in unconditionally cultured hESCs. Moreover, in the absence of feeder layer or conditioned media but in the presence of bFGF, cultured hESCs require exogenous BMP antagonist Noggin to maintain pluripotency (Wang et al. 2005; Xu et al. 2005). This divergent response to BMP signaling may be due to the fundamental distinctions between mESCs and hESCs, as they are probably at different pluripotent states (Pera and Trounson 2004).

BMP signaling is also one of the key regulators of cell-fate commitment in stem cell differentiation. Here, we take neural crest stem cells (NCSCs) as an example. In vertebrates, the neural crest originates from dorsal neural tube during early development and will later migrate and generate multiple cell types including melanocytes, smooth muscle, and neurons and glia of peripheral neural system. Among the neural crest are subsets of pluripotent NCSCs, which can self-renew and differentiate into diverse cell types. In rat NCSCs, BMPs can induce their differentiation into autonomic precursor cells expressing Mash1, while continuous BMP signaling contributes to neuronal commitment (Shah et al. 1996). Therefore, cell-fate determination directed by BMPs is a multistep process.

The BMP pathway has a central role in the maintenance of GSCs in *Drosophila*. GSCs and their surrounding niche cells are located at the germarium in *Drosophila* ovary. The cap cells directly contacting GSCs, together with the filament cells, form the supporting niche for GSCs (Lin 2002; Xie and Spradling 2000). BMPs, including Dpp and Gbb, are key molecules that are secreted by cap cells in GSC maintenance and self-renewal. Dpp/Gbb signaling promotes GSC expansion and disruption of their expression resulting in GSC loss (Lin 2002; Xie and Spradling 2000). Inhibition of the differentiation-promoting gene, bag of marbles (bam), is critical for Dpp/Gbb to regulate GSC self-renewal (Chen and McKearin 2003; Song et al. 2004). This machinery also functions in maintaining GSCs in *Drosophila* testis, where hub cell plays the role of niche to support GSCs. But in testis, other pathways including JAK/STAT signaling are also essential for GSC maintenance (Kiger et al. 2001; Tulina and Matunis 2001).

In summary, BMP signaling pathway has diverse functions in different organisms at different developmental stages. Much evidence has shown that BMP signaling plays an essential role in promoting self-renewal of many types of stem cells, such as mammalian ESC, *Drosophila* GSC and various somatic stem cells. BMP signaling is also involved in the regulation of cell-fate determination, which is well studied in the development of NCSCs. In most cases, BMP functions in coordination with other signaling pathways to regulate stem cell properties. Therefore, balanced control of stem cell activity involves delicate mechanisms, which usually requires the cooperation of multiple signaling pathways.

6.6 JAK-STAT Pathway

The Janus kinase/signal transducers and activators of transcription (JAK-STAT) pathway, initially discovered in mammalian about two decades ago (Firmbach-Kraft et al. 1990; Wilks et al. 1991; Shuai et al. 1993; Muller et al. 1993; Watling et al. 1993; Shuai et al. 1993), are conserved between *Drosophila* and mammals. Subsequent studies in *Drosophila* and mammals have uncovered its role in regulating diverse biological processes.

The JAK-STAT pathway consists of three main components: the receptors, JAKs, and STATs. The two JAKs, which are associated with the closed receptor, are brought together after ligand–receptor binding and phosphorylate each other. The phosphorylated JAK can recruit and phosphorylate STAT in the cytoplasm. Then, two phosphorylated STATs dimerize and translocate into the nucleus and bind to DNA to promote transcription. The JAK-STAT signaling pathway can be activated by several kinds of ligands, such as interferons, interleukins, and growth factors. It can also be regulated by multiple modulators at several levels. For example, protein phosphatases can inhibit active JAK by removing its phosphates, and protein inhibitors of activated STAT (PIAS) can prevent binding of active STAT to DNA (Ungureanu et al. 2003, 2005; Rakesh and Agrawal 2005).

A role for the JAK-STAT pathway in stem cell maintenance was first implicated in *Drosophila* germline stem cell niche (Tulina and Matunis 2001; Kiger et al. 2001), and now, it has become clear that it plays important roles in a variety of stem cells. The JAK-STAT pathway plays an important role in the maintenance of ESC in mouse and swine and in facilitating the reprogramming (Onishi et al. 2012; Wu et al. 2014; Tang et al. 2012; Tang and Tian 2013; Ernst et al. 1999; Hao et al. 2006). But its function between mouse and human is a little controversial (Dreesen and Brivanlou 2007; Humphrey et al. 2004). Here, we mainly review its role in regulating tissue-specific stem cells and cancers.

In the *Drosophila* testis, the JAK/STAT signaling ligand Upd is specifically secreted from the hub, the niche of the GSCs, and somatic cyst stem cells (CySCs), which activates the JAK-STAT signaling pathway in GSCs and CySCs (Tulina and Matunis 2001; Kiger et al. 2001). JAK-STAT activity in CySCs is sufficient to induce self-renewal of both GSCs and CySCs, while its activity in GSCs is required for DE-cadherin-mediated attachment of GSCs to the hub (Leatherman and DiNardo 2008, 2010). There is evidence that STAT3 can regulate spermatogonial stem cell differentiation in mouse (Oatley et al. 2010). In addition, the differentiated GSC induced by the depletion of JAK-STAT signal can undergo dedifferentiation when the JAK-STAT signal is restored to its normal level (Brawley and Matunis 2004).

Similar to the stem cells in testis, the *Drosophila* prohemocyte progenitor cells are supported by a niche named posterior signaling center (PSC). Signals from PSC lead to JAK-STAT activation in the prohemocytes for prohemocyte maintenance (Krzemień et al. 2007). But unlike in testis, JAK-STAT activation in prohemocytes is not likely induced by Upd, and other signals from PSC are needed

(Makki et al. 2010). JAK-STAT signaling is also required for hematocyte–lamellocyte transition after infestation (Stofanko et al. 2010). The JAK-STAT pathway also functions in hematopoietic stem cell maintenance and differentiation in mammalian (Bradley et al. 2004; Snow et al. 2002; Kato et al. 2005; Wang et al. 2009). For example, the loss of STAT5 activity in mice leads to the deficiency of multipotent hematopoietic progenitors and consequently several hematopoietic cell lineages, which causes inefficient repopulation upon irradiation (Bradley et al. 2004; Snow et al. 2002). The mechanism of JAK-STAT regulation in mammalian hematopoiesis is rather complex and is extensively reviewed by Stine and Matunis (2013).

Together with other signaling pathways, the JAK-STAT pathway regulates both proliferation of ISCs and differentiation of intestinal EBs to maintain homeostasis of the fly midgut. Loss of the JAK-STAT activity in ISCs decreases the proliferation rate of ISC and vice versa (Beebe et al. 2010; Lin et al. 2010; Liu et al. 2010). JAK-STAT activity is also essential for EB differentiation, as mutation in the pathway blocks the progenitor cell differentiation at EB stage (Beebe et al. 2010; Lin et al. 2010). The JAK-STAT pathway can also induce rapid proliferation and differentiation of the progenitor cells in response to stress or bacterial infection (Jiang et al. 2009; Buchon et al. 2009).

The JAK-STAT signaling pathway is important in controlling the balance between stem cell self-renewal and differentiation in *Drosophila* neural system. It is required for neuroepithelial (NE) stem cell maintenance, and the loss of JAK-STAT activity leads to the loss of NE due to precocious differentiation (Yasugi et al. 2008; Ngo et al. 2010; Wang et al. 2011). The JAK-STAT pathway also plays important roles during neurogenesis in mammals to ensure the appropriate generation of the right cell type at different developmental stages of the neural system. For example, the elevated JAK-STAT activity is required for neurogenesis–gliogenesis transition (Barnabé-Heider et al. 2005; He et al. 2005).

The JAK-STAT pathway may also function in cancer stem cells (CSCs). The upregulation of JAK-STAT signaling activity has been found in many CSCs (Zhou et al. 2007; Birnie et al. 2008; Cook et al. 2014), and the activity is required for the maintenance of CSCs (Zhou et al. 2007; Cook et al. 2014; Sherry et al. 2009). Consistent with its role in CSCs, many inhibitors that target the JAK-STAT pathway could be useful in the treatment of certain types of cancers (Hart et al. 2011; Pardanani et al. 2013; Harrison et al. 2012; Verstovsek et al. 2012; Mascarenhas et al. 2014).

In conclusion, the JAK-STAT pathway plays an important role in regulating tissue homeostasis by regulating the balance between stem cell self-renewal and differentiation. It usually runs in parallel or interacts with many other signaling pathways to control the stem cell maintenance and differentiation. Many studies of the JAK-STAT pathway have been done in *Drosophila* because of less redundancy and complexity of this pathway. More investigations are needed in mammals to determine whether the lessons learned from *Drosophila* can be applied to mammalian stem cells and to study its relationship with other signaling pathways in the regulation of tissue stem cells and cancer.

6.7 Hedgehog Pathway

The Hedgehog (Hh) family proteins are key morphogens that direct cell patterning of embryonic tissues and tissue homeostasis throughout animal development. The Hh signaling pathway regulates diverse cellular responses including cell survival, proliferation, and fate determination. Disruption of Hh signaling is often involved in developmental disorders and tumorigenesis. Here, we will discuss the role of Hh signaling in regulating stem cell maintenance.

The *Hh* gene was first identified in *Drosophila*. Vertebrate Hh counterparts, including the Desert Hedgehog (Dhh), Indian Hedgehog (Ihh), and Sonic Hedgehog (Shh), were found shortly thereafter, and the developmental function of Hh molecules is evolutionarily conserved in *Drosophila* and vertebrates. Moreover, main components of Hh signal transduction pathway are also evolutionarily conserved. In both *Drosophila* and mammals, ligand-free Patched (Ptc) protein restrains the activation of the transmembrane protein Smoothened (Smo) by triggering Smo degradation and blocking membrane localization of Smo. Binding of Hh blocks Ptc activity and liberates Smo to translocate to the membrane. The downstream intercellular cascade resulting from Smo activation ended with the translocation of Ci/Gli family into nucleus to direct the expression of target genes. Significant differences still exist in the pathway components between invertebrates and mammals, especially the transduction machinery from the receptor to the Ci/Gli transcription factors.

In addition to its well-established function as developmental morphogen, Hh signaling also regulates stem cell self-renewal and tissue homeostasis. In mammals, the expression of several stem cell-related genes, including genes encoding MYC, BMI1, Cyclin D1, Nanog, and insulin-like growth factor 2 (IGF2), is promoted by Hh signaling (Davidson et al. 2012; Briscoe and Therond 2013).

Hh is required for the maintenance of stem cells in a variety of adult tissues. In neural system, blockage of Hh signaling by inhibiting Smo decreases the proliferative capacity of neural stem cell both in vitro and in vivo (Machold et al. 2003; Lai et al. 2003). In the case of HSC, treatment of Shh increases the expansion of HSCs in vitro and in vivo (Bhardwaj et al. 2001). Similar phenomena are found in *Drosophila* ovary system, where Hh is a major factor controlling proliferation and maintenance of ovary somatic stem cells (Zhang and Kalderon 2001).

Consistent with its crucial role in controlling cell patterning and stem cell self-renewal, Hh is implicated in the regeneration and damage repair of a variety of tissues. In newt lens and limb, the Hh proteins are often expressed in the injury-induced dedifferentiated cells, and tissue regeneration is blocked by treatment with cyclopamine, a specific inhibitor of Hh pathway (Tsonis et al. 2004; Imokawa and Yoshizato 1997; Roy and Gardiner 2002). Blocking Hh pathway by cyclopamine also disrupts tissue repair in other species, such as the fin of zebrafish and the facial nerves of mouse (Laforest et al. 1998; Akazawa et al. 2004).

However, Hh families do not always function to promote cell proliferation during tissue regeneration. Different tissue types within an organ system may have different responses to a single signaling molecule, depending on the type of responding cells and the coordination of multiple signaling pathways. A good example is the Hh signaling network in gastrointestinal tract, where epithelial proliferation is regulated by multiple signaling pathways including Hh signaling and Wnt signaling. The proliferation of epithelium in the esophagus, stomach, and pancreas is promoted by Hh signaling, but proliferation in the intestine is suppressed by Hh signaling through its negative effects on Wnt signaling pathway (Katoh and Katoh 2006).

In summary, Hh is an evolutionarily conserved signaling pathway that has various and critical functions in the process of embryonic morphogenesis and adult tissue homeostasis. Diverse cellular responses, including cell survival, growth, proliferation, and cell-fate specification, are regulated by Hh signaling pathway network. Moreover, Hh signaling pathway has been implicated in various human tumors, proposing Hh signaling components as potential targets in cancer therapy. However, because of its complexity, the machinery of Hh is not completely understood yet. Therefore, many questions remain to be studied in exploring the details of Hh signaling network as well as the mechanisms of this process in regulating tumor progression for developing therapeutic strategies.

6.8 Hippo Pathway

The Hippo pathway is a newly characterized, evolutionarily conserved signaling cascade. The first component of this pathway was first identified by Wan Yu's and Peter J. Bryant's group in 1995 through mosaic clonal screens for genes involved in tissue growth control in *Drosophila* (Xu et al. 1995; Justice et al. 1995). The mutation in this pathway leads to increased organ size through increased cell proliferation and decreased cell death. The number and the activity of the stem cells in organs may play an important role for the organ size (Stanger et al. 2007). Therefore, the Hippo pathway may control the organ size by regulating the tissue stem cells (Camargo et al. 2007; Song et al. 2010; Lee et al. 2010; Jansson and Larsson 2012; Nejigane et al. 2013).

The core components of this pathway have been well characterized (Xu et al. 1995; Justice et al. 1995; Tapon et al. 2002; Wu et al. 2003; Udan et al. 2003; Harvey et al. 2003; Jia et al. 2003; Pantalacci et al. 2003; Dong et al. 2007), which consists of a highly conserved kinase cascade (Ste20-like kinase Hippo and NDR family kinase Warts in *Drosophila*, MST1/2 and LATS1/2 in mammalian) and downstream transcription co-activators (Yorkie in *Drosophila*, YAP/TAZ in mammals). In *Drosophila*, Hippo (Hpo) forms an active complex with the scaffolding protein Salvador (Sav) to directly phosphorylate and activate Warts (Wts) and its regulatory protein Mob (Tapon et al. 2002; Lai et al. 2005). Then, the active Wts/Mob complex phosphorylates the Yorkie (Yki) to promote its binding to 14-3-3 protein, which inhibits the translocation of Yki into the nucleus, where it act as the co-activator for the TEAD/TEF family transcription factor Scalloped (Sd) to promote gene expression, thereby facilitating cell proliferation and survival (Dong et al. 2007; Vassilev et al. 2001; Huang et al. 2005; Mahoney et al. 2005; Oh and Irvine 2008; Ren et al. 2010; Staley and Irvine 2012). The signal transduction in mammals is similar to that in flies (Wu et al. 2003; Lai et al. 2005; Huang et al. 2005; Tao et al. 1999). Although the core components in this pathway are well characterized, its upstream regulators are not well defined. It has been shown that apical–basal polarity proteins, cellular junction proteins, and some extracellular hormones can regulate the Hippo pathway, a subject that has been comprehensively reviewed by Jung-Soon Mo (Ramos and Camargo 2012).

The Hippo pathway was firstly characterized to control organ size by inhibiting proliferation and promoting apoptosis, and the current studies show that Hippo pathway can also regulate stem cell self-renewal and has important roles in tissue regeneration (Lian et al. 2010; Varclas ct al. 2008; von Gisc ct al. 2012; Yimlamai et al. 2014). The Hippo pathway plays a critical role in regulating ESCs, adult progenitor cells, and CSCs. The phosphorylation of LATS together with YAP can suppress TEAD4 activity in the inside cell of preimplantation mouse embryo to distinguish mouse ICM from trophectoderm (Nishioka et al. 2009). YAP is also required for mouse ESC pluripotency, directly binds to a large number of pluripotency-related genes, and enhances reprogramming efficiency of mouse iPSs (Lian et al. 2010). TAZ is required for maintaining pluripotent gene expression in human ESC, and knockdown of TAZ results in differentiation (Varelas et al. 2008). YAP and TEAD2 are activated by LIF in mouse ESCs and are downregulated during differentiation (Tamm et al. 2011). In addition to ESC, the Hippo pathway also functions in tissue-specific progenitor cells. YAP activation in postnatal liver leads to dramatic increase in the liver size (Camargo et al. 2007). Mst1/2 mutation also results in liver overgrowth (Song et al. 2010). More recently, the Hippo pathway has been demonstrated to be essential for the maintenance of the differentiated state of hepatocyte. Its inactivation in vivo is sufficient for hepatocytes to dedifferentiate into progenitors (Yimlamai et al. 2014). The intestinal epithelium is also regulated by the Hippo pathway, and YAP overexpression results in the expansion of the intestinal progenitor cells (Zhou et al. 2011). Similarly, YAP activation in the skin leads to skin hyperplasia, which is driven by the excessive proliferation of the interfollicular stem cells (Schlegelmilch et al. 2011). But the overgrowth of the heart induced by Sav1 knockout results from the expansion of the cardiomyocytes, not the cardiac progenitors (Heallen et al. 2011), and TAZ overexpression leads to the myogenic differentiation (Jeong et al. 2010). In the nervous tissues, YAP co-localizes with neural progenitor maker Sox2, while activation of YAP or inactivation of MST1/2 results in neural progenitor expansion (Cao et al. 2008; Gee et al. 2011). More information about functions of the Hippo pathway in tissue-specific progenitor cells has been summarized (Ramos and Camargo 2012) elsewhere (Ramos and Camargo 2012). Abnormal activity of Hippo signaling has also implicated in various cancers. TAZ is essential for the self-renewal of breast CSCs and tumor progression (Cordenonsi et al. 2011). Moreover, upregulation of YAP1 acts as a determinant for maintaining esophageal CSC properties (Song et al. 2014). TAZ and YAP are also highly expressed and activated in a variety of human cancers (Liang et al. 2014; Perra et al. 2014; Steinhardt et al. 2008; Yue et al. 2014).

The role of the Hippo signaling pathway in stem cell regulation stimulates a new line of research in cancer- and degeneration-related diseases. Although it is evident that YAP activation can mediate Wnt or Notch signaling pathway in some tissues (Yimlamai et al. 2014; Heallen et al. 2011; Xin et al. 2011) and LIF (Tamm et al. 2011) and TGF-β or BMP signaling pathway (Varelas et al. 2008; Alarcón et al. 2009; Beyer et al. 2013) in others, the immediate upstream regulators of the Hippo pathway and the mechanisms that turn on and off the pathway are still not well understood. How the Hippo pathway integrates the inputs from these multiple signals to generate the correct outputs for context-dependent function? What are the target genes that drive the appropriate cellular response? In addition, the Hippo pathway could regulate stem cell expansion and tumorigenesis through different mechanisms, and loss of different components in this pathway sometimes leads to diverse phenotypes. Therefore, further studies are needed to fully elucidate the exact role of each component of the Hippo pathway in regulating stem cells. Answers to these questions will ultimately contribute to our understanding of tissue homeostasis control, regeneration, and tumorigenesis.

6.9 Insulin Pathway

Endocrine system plays an important role in coordinately regulating the growth of multiple organs. These systemic signals, along with short-range niche signals, can function together to regulate tissue stem cells (Gancz and Gilboa 2013).

Among these hormones, insulin is a well-characterized signal that is conserved in various organisms. After binding to insulin receptor (InR), it activates a downstream cascade that ultimately affects Forkhead Group O (Foxo) and tuberous sclerosis complex 2 (TSC2), two important nodes in metabolism and energy control. Thus, insulin pathway could serve as a link between nutrient state and stem cell activity (Grewal 2009). For example, as the most energy-consuming process in female *Drosophila*, oogenesis is tightly regulated by this pathway in multiple ways. For the regulation of GSCs, insulin binds to InR on the surface of GSCs and autonomously regulates its division. Besides, it can indirectly regulate GSC activity by regulating Notch signaling in the cap cells and DE-cadherin between cap cells and GSCs (Hsu et al. 2008; Hsu and Drummond-Barbosa 2009, 2011).

Another tissue that closely relates to food consumption and metabolism is the intestine. Either in *Drosophila* or in mice, intestine changes its size according to the abundance of food. This may attribute to the altered rate of ISC proliferation and number of ECs caused by the altered production of insulin (O'Brien et al. 2011). Locally secreted *Drosophila* insulin-like peptide 3 (DILP3) by visceral muscle cells underneath the ISCs is the major player, while systemic DILPs that originated from brain IPCs may also participate in this process. Elevated level of DILPs will lead to a switch from asymmetric to symmetric division of ISCs and expand the pool of stem cells. Besides, in response to tissue damage in the *Drosophila* midgut, systemic DILP2 secreted by brain cells participates in promoting ISC division and consequently epithelial regeneration (Amcheslavsky et al. 2009). The *Drosophila* ISCs show declined proliferation in response to caloric restriction. By contrast, ISCs in mouse small intestine increase their proliferation and number. ISCs' closest neighbors, the paneth cells, sense the caloric restriction via mTORC1 pathway. Repression of this pathway in paneth cells promotes ISC proliferation via a secreted enzyme that generates cyclic ADP ribose (cADPR) (Yilmaz et al. 2012). However, the activity of transient amplifying cells is reduced in response to caloric restriction. Therefore, different responses to insulin signaling between ISCs and transient amplifying cells may provide a mechanism to protect stem cell population under starvation conditions and at the same time to limit the production of stem cell progenies for energy reservation.

The insulin/Tor pathway also regulates homeostasis of HSCs. Cell autonomous activation of insulin/Tor pathway disrupts quiescence of HSC and finally lead to exhaustion of stem cells due to reduced self-renewal capacity. On the other hand, insulin/Tor pathway activation in the niche supports HSC self-renewal and preservation of the stem cell population (Chen et al. 2008; Kharas et al. 2010; Kobayashi et al. 2010). In addition to insulin, other hormones have also been found to regulate stem cells. Ecdysone is one example, which plays multiple roles in orchestrating development and homeostasis of GSCs in *Drosophila* (Ables and Drummond-Barbosa 2010; König et al. 2011; Morris and Spradling 2012). This hormone functions autonomously to promote GSC maintenance and self-renewal by Nurf-dependent activation of BMP activity, which otherwise will lead to differentiation. The major source of ecdysone has been located to maturing egg chambers, whose survival is dependent on food supply (Ables and Drummond-Barbosa 2010). Via this regulatory loop, the oogenesis could be modulated based on the nutrient status.

6.10 Conclusion Remarks

Tissue stem cells are important for maintaining tissue homeostasis due to their unlimited or prolonged ability of self-renewal and their cellular multipotency. These abilities are intrinsic properties but can be maintained and regulated by signaling pathways. As reviewed above, in both invertebrate and vertebrate models, self-renewal and differentiation of stem cells are commonly regulated by local signals as well as systemic signals. The local microenvironment ensures long-term maintenance of stem cells and proper division activity and differentiation potential required for tissue homeostasis. The systemic signals usually serve to coordinate stem cell activity with tissue/organ growth and with needs, such as nutrient availability. Although many signaling pathways are frequently utilized to control stem cells, there is no common signaling circuitry that controls stemness of all or most types of tissue stem cells. Instead, each type of tissue stem cells is usually regulated by a distinct set of signaling pathways. Some of them are critical for self-renewal and proliferation, whereas some of them are more important for the differentiation of progenitor cells. Several examples of stem cells and their regulation are summarized in Table 6.1. Because cells in different tissues have distinct transcriptional programs required for their specific functions, distinct regulatory mechanisms underlying self-renewal of different types of tissue stem cells may

Table 6.1 Signaling pathways regulating self-renewal, proliferation, and differentiation of stem cells

	Signaling pathways regulating self-renewal and proliferation	Signaling pathways regulating differentiation
Embryonic stem cell (mouse)	Wnt* (Young 2011), EGFR (Fricker-Gates et al. 2000), BMP* (Ying et al. 2003)	Notch (Lowell et al. 2006)
Embryonic stem cell (human)	EGFR (Fricker-Gates et al. 2000), TGF-β (James et al. 2005), bFGF (Wang et al. 2005)	Wnt* (Davidson et al. 2012; Kielman et al. 2002), BMP* (Pera et al. 2004)
Intestinal stem cell (*Drosophila*)	Wnt (Li and Xie 2005; Lin et al. 2008), EGFR (Jiang et al. 2011), JAK-STAT (Beebe et al. 2010), Hippo (Zhou et al. 2011), Integrin (O'Brien et al. 2011)	Notch (Ohlstein and Spradling 2007; Fre et al. 2005), JAK-STAT (Beebe et al. 2010), BMP (Auclair et al. 2007)
Intestinal stem cell (mammalian)	Wnt (Li and Xie 2005; Lin et al. 2008), Notch (mammalian)** (Pellegrinet et al. 2011)	Notch (Ohlstein and Spradling 2007; Fre et al. 2005), BMP (Auclair et al. 2007)
Hair follicle stem cell	Wnt (low) (Lowry et al. 2005), Hedgehog (Blanpain and Fuchs 2006)	Wnt (high) (Lowry et al. 2005), BMP, Notch (Blanpain and Fuchs 2006; Brack et al. 2008)
Hematopoietic stem cell	Wnt (Rattis et al. 2004), Notch (Butler et al. 2010)	Notch (Mercher et al. 2008)
Germline stem cell (*Drosophila* testis)	BMP (Kiger et al. 2001), JAK-STAT (Tulina and Matunis 2001), insulin (Ueishi et al. 2009)	
Cyst stem cell (*Drosophila* testis)	JAK-STAT (Stine and Matunis 2013), Hedgehog (Michel et al. 2012)	
Germline stem cell (*Drosophila* ovary)	BMP (Xie and Spradling 2000), Ecdysone (Ables and Drummond-Barbosa 2010), insulin (Ueishi et al. 2009)	
Follicle stem cell (*Drosophila* ovary)	Hedgehog (Zhang and Kalderon 2001), JAK-STAT (Vied et al. 2012), Wnt (Song and Xie 2003)	Notch (Adam and Montell 2004), Hippo (Chen et al. 2011)

(continued)

Table 6.1 (continued)

	Signaling pathways regulating self-renewal and proliferation	Signaling pathways regulating differentiation
Neural stem cell	Notch (Alvarez-Buylla and Lim 2004), EGFR (Aguirre et al. 2010), Hedgehog (Klein et al. 2001), JAK-STAT (Wang et al. 2011)	BMP (Shah et al. 1996)
Satellite muscle cell	Wnt7a (non-canonical) (Le Grand et al. 2009), Notch (Kuang et al. 2007)	Wnt (canonical) (Polesskaya et al. 2003)
Mammary stem cell	Wnt (Zeng and Nusse 2010), Hedgehog (Liu et al. 2006)	Notch (Dontu et al. 2004)
Airway stem cell	Wnt (Zhang et al. 2008), Notch (Paul et al. 2014)	Hippo (Zhao et al. 2014), Notch (Rock et al. 2011)

*Wnt and BMP have distinct functions in mouse and human ESCs; **Notch activation promotes ISC self-renewal in mammals, while inhibits ISC self-renewal in Drosophila

facilitate the execution of unique transcriptional programs required for cell-type-specific functions during cell lineage differentiation. Therefore, coupling stem cell identity with tissue identity could be an efficient and effective mechanism in controlling the self-renewal and differentiation of tissue stem cells.

Although signaling pathways involved in regulating various types of tissue stem cells have been elucidated and characterized, potential signaling cross talk or coordination for balanced self-renewal and differentiation is much less well understood. The downstream transcriptional factors that mediate signaling pathway-induced various stem cell behaviors are also in general less well understood. As signaling pathways have a central role in controlling stem cell properties and have been implicated in various diseases, further dissecting out their function and regulation in stem cells will not only contribute to our understanding of disease mechanisms, but also pave the way for regenerative medicine and drug discovery.

References

Aberle H, Bauer A, Stappert J, Kispert A, Kemler R. β-catenin is a target for the ubiquitin–proteasome pathway. EMBO J. 1997;16:3797–804.

Ables ET, Drummond-Barbosa D. The steroid hormone ecdysone functions with intrinsic chromatin remodeling factors to control female germline stem cells in *Drosophila*. Cell Stem Cell. 2010;7:581–92.

Adam JC, Montell DJ. A role for extra macrochaetae downstream of notch in follicle cell differentiation. Development. 2004;131:5971–80.

Aguirre A, Rubio ME, Gallo V. Notch and EGFR pathway interaction regulates neural stem cell number and self-renewal. Nature. 2010;467:323–7.

Akazawa C, Tsuzuki H, Nakamura Y, Sasaki Y, Ohsaki K, Nakamura S, Arakawa Y, Kohsaka S. The upregulated expression of sonic hedgehog in motor neurons after rat facial nerve axotomy. J Neurosci. 2004;24:7923–30.

Alarcón C, Zaromytidou A-I, Xi Q, Gao S, Yu J, Fujisawa S, Barlas A, Miller AN, Manova-Todorova K, Macias MJ. Nuclear CDKs drive smad transcriptional activation and turnover in BMP and TGF-β pathways. Cell. 2009;139:757–69.

Alexson TO, Hitoshi S, Coles BL, Bernstein A, van der Kooy D. Notch signaling is required to maintain all neural stem cell populations–irrespective of spatial or temporal niche. Dev Neurosci. 2006;28:34–48.

Alvarez-Buylla A, Lim DA. For the long run: maintaining germinal niches in the adult brain. Neuron. 2004;41:683–6.

Amcheslavsky A, Jiang J, Ip YT. Tissue damage-induced intestinal stem cell division in *Drosophila*. Cell Stem Cell. 2009; 4:49–61.

Amit S, Hatzubai A, Birman Y, Andersen JS, Ben-Shushan E, Mann M, Ben-Neriah Y, Alkalay I. Axin-mediated CKI phosphorylation of β-catenin at Ser 45: a molecular switch for the Wnt pathway. Genes Dev. 2002;16:1066–76.

Andersen P, Uosaki H, Shenje LT, Kwon C. Non-canonical notch signaling: emerging role and mechanism. Trends Cell Biol. 2012;22:257–65.

Andl T, Reddy ST, Gaddapara T, Millar SE. WNT signals are required for the initiation of hair follicle development. Dev Cell. 2002;2:643–53.

Aubert J, Dunstan H, Chambers I, Smith A. Functional gene screening in embryonic stem cells implicates Wnt antagonism in neural differentiation. Nat Biotechnol. 2002;20:1240–5.

Auclair BA, Benoit YD, Rivard N, Mishina Y, Perreault N. Bone morphogenetic protein signaling is essential for terminal differentiation of the intestinal secretory cell lineage. Gastroenterology. 2007;133:887–96.

Austin TW, Solar GP, Ziegler FC, Liem L, Matthews W. A role for the Wnt gene family in hematopoiesis: expansion of multilineage progenitor cells. Blood. 1997;89:3624–35.

Ayuso-Sacido A, Moliterno JA, Kratovac S, Kapoor GS, O'Rourke DM, Holland EC, Garcia-Verdugo JM, Roy NS, Boockvar JA. Activated EGFR signaling increases proliferation, survival, and migration and blocks neuronal differentiation in post-natal neural stem cells. J Neurooncol. 2010;97:323–37.

Barnabé-Heider F, Wasylnka JA, Fernandes KJ, Porsche C, Sendtner M, Kaplan DR, Miller FD. Evidence that embryonic neurons regulate the onset of cortical gliogenesis via cardiotrophin-1. Neuron. 2005;48:253–65.

Batlle E, Henderson JT, Beghtel H, van den Born MM, Sancho E, Huls G, Meeldijk J, Robertson J, van de Wetering M, Pawson T, Clevers H. Beta-catenin and TCF mediate cell positioning in the intestinal epithelium by controlling the expression of EphB/ephrinB. Cell. 2002;111:251–63.

Beebe K, Lee W-C, Micchelli CA. JAK/STAT signaling coordinates stem cell proliferation and multilineage differentiation in the *Drosophila* intestinal stem cell lineage. Dev Biol. 2010;338:28–37.

Beyer TA, Weiss A, Khomchuk Y, Huang K, Ogunjimi AA, Varelas X, Wrana JL. Switch enhancers interpret TGF-β and hippo signaling to control cell fate in human embryonic stem cells. Cell Rep. 2013;5:1611–24.

Bhardwaj G, Murdoch B, Wu D, Baker DP, Williams KP, Chadwick K, Ling LE, Karanu FN, Bhatia M. Sonic hedgehog induces the proliferation of primitive human hematopoietic cells via BMP regulation. Nat Immunol. 2001;2:172–80.

Bilić J, Huang Y-L, Davidson G, Zimmermann T, Cruciat C-M, Bienz M, Niehrs C. Wnt induces LRP6 signalosomes and promotes dishevelled-dependent LRP6 phosphorylation. Science. 2007;316:1619–22.

Birnie R, Bryce SD, Roome C, Dussupt V, Droop A, Lang SH, Berry PA, Hyde CF, Lewis JL, Stower MJ. Gene expression profiling of human prostate cancer stem cells reveals a pro-inflammatory phenotype and the importance of extracellular matrix interactions. Genome Biol. 2008;9:R83.

Biteau B, Jasper H. EGF signaling regulates the proliferation of intestinal stem cells in *Drosophila*. Development. 2011;138:1045–55.

Blanpain C, Fuchs E. Epidermal stem cells of the skin. Annu Rev Cell Dev Biol. 2006;22:339.
Blanpain C, Horsley V, Fuchs E. Epithelial stem cells: turning over new leaves. Cell. 2007;128:445–58.
Boockvar JA, Kapitonov D, Kapoor G, Schouten J, Counelis GJ, Bogler O, Snyder EY, McIntosh TK, O'Rourke DM. Constitutive EGFR signaling confers a motile phenotype to neural stem cells. Mol Cell Neurosci. 2003;24:1116–30.
Brack AS, Conboy IM, Conboy MJ, Shen J, Rando TA. A temporal switch from notch to Wnt signaling in muscle stem cells is necessary for normal adult myogenesis. Cell Stem Cell. 2008;2:50–9.
Bradley HL, Couldrey C, Bunting KD. Hematopoietic-repopulating defects from STAT5-deficient bone marrow are not fully accounted for by loss of thrombopoietin responsiveness. Blood. 2004;103:2965–72.
Brawley C, Matunis E. Regeneration of male germline stem cells by spermatogonial dedifferentiation in vivo. Science. 2004;304:1331–4.
Bray SJ. Notch signalling: a simple pathway becomes complex. Nat Rev Mol Cell Biol. 2006;7:678–89.
Briscoe J, Therond PP. The mechanisms of hedgehog signalling and its roles in development and disease. Nat Rev Mol Cell Biol. 2013;14:416–429.
Buchon N, Broderick NA, Chakrabarti S, Lemaitre B. Invasive and indigenous microbiota impact intestinal stem cell activity through multiple pathways in *Drosophila*. Genes Dev. 2009;23:2333–44.
Buchon N, Broderick NA, Kuraishi T, Lemaitre B. *Drosophila* EGFR pathway coordinates stem cell proliferation and gut remodeling following infection. BMC Biol. 2010;8:152.
Butler JM, Nolan DJ, Vertes EL, Varnum-Finney B, Kobayashi H, Hooper AT, Seandel M, Shido K, White IA, Kobayashi M. Endothelial cells are essential for the self-renewal and repopulation of notch-dependent hematopoietic stem cells. Cell Stem Cell. 2010;6:251–64.
Calvi L, Adams G, Weibrecht K, Weber J, Olson D, Knight M, Martin R, Schipani E, Divieti P, Bringhurst F. Osteoblastic cells regulate the haematopoietic stem cell niche. Nature. 2003;425:841–6.
Camargo FD, Gokhale S, Johnnidis JB, Fu D, Bell GW, Jaenisch R, Brummelkamp TR. YAP1 increases organ size and expands undifferentiated progenitor cells. Curr Biol. 2007;17:2054–60.
Cao X, Pfaff SL, Gage FH. YAP regulates neural progenitor cell number via the TEA domain transcription factor. Genes Dev. 2008;22:3320–34.
Cavallo RA, Cox RT, Moline MM, Roose J, Polevoy GA, Clevers H, Peifer M, Bejsovec A. *Drosophila* Tcf and groucho interact to repress wingless signalling activity. Nature. 1998;395:604–8.
Chambers I, Smith A. Self-renewal of teratocarcinoma and embryonic stem cells. Oncogene. 2004;23:7150–60.
Chen D, McKearin D. Dpp signaling silences bam transcription directly to establish asymmetric divisions of germline stem cells. Curr Biol. 2003;13:1786–91.
Chen C, Liu Y, Liu R, Ikenoue T, Guan K-L, Liu Y, Zheng P. TSC–mTOR maintains quiescence and function of hematopoietic stem cells by repressing mitochondrial biogenesis and reactive oxygen species. J Exp Med. 2008;205:2397–408.
Chen H, Wang C, Wang T, Liaw G, Hsu T, Lin T, Yu J. The Hippo pathway controls polar cell fate through Notch signaling during *Drosophila* oogenesis. Developmental biology. 2011;357:370.
Choi YS, Zhang Y, Xu M, Yang Y, Ito M, Peng T, Cui Z, Nagy A, Hadjantonakis AK, Lang RA, Cotsarelis G, Andl T, Morrisey EE, Millar SE. Distinct functions for Wnt/beta-catenin in hair follicle stem cell proliferation and survival and interfollicular epidermal homeostasis. Cell Stem Cell. 2013;13:720–33.
Cook AM, Li L, Ho Y, Lin A, Li L, Stein A, Forman S, Perrotti D, Jove R, Bhatia R. Role of altered growth factor receptor-mediated JAK2 signaling in growth and maintenance of human acute myeloid leukemia stem cells. Blood. 2014;123:2826–37.

Cooper MT, Bray SJ. R7 photoreceptor specification requires notch activity. Curr Biol. 2000;10:1507–10.

Cordenonsi M, Zanconato F, Azzolin L, Forcato M, Rosato A, Frasson C, Inui M, Montagner M, Parenti AR, Poletti A. The hippo transducer TAZ confers cancer stem cell-related traits on breast cancer cells. Cell. 2011;147:759–72.

Cordero JB, Stefanatos RK, Scopelliti A, Vidal M, Sansom OJ. Inducible progenitor-derived wingless regulates adult midgut regeneration in Drosophila. EMBO J. 2012;31:3901–17.

DasGupta R, Fuchs E. Multiple roles for activated LEF/TCF transcription complexes during hair follicle development and differentiation. Development. 1999;126:4557–68.

Davidson KC, Adams AM, Goodson JM, McDonald CE, Potter JC, Berndt JD, Biechele TL, Taylor RJ, Moon RT. Wnt/β-catenin signaling promotes differentiation, not self-renewal, of human embryonic stem cells and is repressed by Oct4. Proc Natl Acad Sci. 2012;109:4485–90.

Day TF, Guo X, Garrett-Beal L, Yang Y. Wnt/β-catenin signaling in mesenchymal progenitors controls osteoblast and chondrocyte differentiation during vertebrate skeletogenesis. Dev Cell. 2005;8:739–50.

de La Coste A, Six E, Fazilleau N, Mascarell L, Legrand N, Mailhé M-P, Cumano A, Laâbi Y, Freitas AA. In vivo and in absence of a thymus, the enforced expression of the notch ligands delta-1 or delta-4 promotes T cell development with specific unique effects. J Immunol. 2005;174:2730–7.

Dong J, Feldmann G, Huang J, Wu S, Zhang N, Comerford SA, Gayyed MF, Anders RA, Maitra A, Pan D. Elucidation of a universal size-control mechanism in *Drosophila* and Mammals. Cell. 2007;130:1120–33.

Dontu G, Jackson KW, McNicholas E, Kawamura MJ, Abdallah WM, Wicha MS. Role of Notch signaling in cell-fate determination of human mammary stem/progenitor cells. Breast Cancer Res. 2004;6:R605–15.

Dreesen O, Brivanlou AH. Signaling pathways in cancer and embryonic stem cells. Stem cell Rev. 2007;3:7–17.

Ernst M, Novak U, Nicholson SE, Layton JE, Dunn AR. The carboxyl-terminal domains of gp130-related cytokine receptors are necessary for suppressing embryonic stem cell differentiation involvement of stat3. J Biol Chem. 1999;274:9729–37.

Fang M, Li J, Blauwkamp T, Bhambhani C, Campbell N, Cadigan KM. C-terminal-binding protein directly activates and represses Wnt transcriptional targets in Drosophila. EMBO J. 2006;25:2735–45.

Fernandez L, Rodriguez S, Huang H, Chora A, Fernandes J, Mumaw C, Cruz E, Pollok K, Cristina F, Price JE, Ferkowicz MJ, Scadden DT, Clauss M, Cardoso AA, Carlesso N. Tumor necrosis factor-alpha and endothelial cells modulate notch signaling in the bone marrow microenvironment during inflammation. Exp Hematol. 2008;36:545–58.

Finch PW, He X, Kelley MJ, Uren A, Schaudies RP, Popescu NC, Rudikoff S, Aaronson SA, Varmus HE, Rubin JS. Purification and molecular cloning of a secreted, frizzled-related antagonist of Wnt action. Proc Natl Acad Sci USA. 1997;94:6770–5.

Firmbach-Kraft I, Byers M, Shows T, Dalla-Favera R, Krolewski J. tyk2, prototype of a novel class of non-receptor tyrosine kinase genes. Oncogene. 1990;5:1329–36.

Flores GV, Duan H, Yan H, Nagaraj R, Fu W, Zou Y, Noll M, Banerjee U. Combinatorial signaling in the specification of unique cell fates. Cell. 2000;103:75–85.

Fraguas S, Barberan S, Cebria F. EGFR signaling regulates cell proliferation, differentiation and morphogenesis during planarian regeneration and homeostasis. Dev Biol. 2011;354:87–101.

Fre S, Huyghe M, Mourikis P, Robine S, Louvard D, Artavanis-Tsakonas S. Notch signals control the fate of immature progenitor cells in the intestine. Nature. 2005;435:964–8.

Fre S, Hannezo E, Sale S, Huyghe M, Lafkas D, Kissel H, Louvi A, Greve J, Louvard D, Artavanis-Tsakonas S. Notch lineages and activity in intestinal stem cells determined by a new set of knock-in mice. PLoS ONE. 2011;6:e25785.

Freeman M. Reiterative use of the EGF receptor triggers differentiation of all cell types in the *Drosophila* eye. Cell. 1996;87:651–60.

Fricker-Gates RA, Winkler C, Kirik D, Rosenblad C, Carpenter MK, Bjorklund A. EGF infusion stimulates the proliferation and migration of embryonic progenitor cells transplanted in the adult rat striatum. Exp Neurol. 2000;165:237–47.

Gancz D, Gilboa L. Hormonal control of stem cell systems. Ann Rev Cell Dev Biol. 2013;29:137–62.

Gee ST, Milgram SL, Kramer KL, Conlon FL, Moody SA. Yes-associated protein 65 (YAP) expands neural progenitors and regulates Pax3 expression in the neural plate border zone. PLoS ONE. 2011;6:e20309.

Glinka A, Dolde C, Kirsch N, Huang YL, Kazanskaya O, Ingelfinger D, Boutros M, Cruciat CM, Niehrs C. LGR4 and LGR5 are R-spondin receptors mediating Wnt/β-catenin and Wnt/PCP signalling. EMBO Rep. 2011;12:1055–61.

Grewal SS. Insulin/TOR signaling in growth and homeostasis: a view from the fly world. Int J Biochem Cell Biol. 2009;41:1006–10.

Hadland BK, Huppert SS, Kanungo J, Xue Y, Jiang R, Gridley T, Conlon RA, Cheng AM, Kopan R, Longmore GD. A requirement for notch1 distinguishes 2 phases of definitive hematopoiesis during development. Blood. 2004;104:3097–105.

Hao J, Li T-G, Qi X, Zhao D-F, Zhao G-Q. WNT/β-catenin pathway up-regulates < i > Stat3 </i > and converges on LIF to prevent differentiation of mouse embryonic stem cells. Dev Biol. 2006;290:81–91.

Harrison C, Kiladjian J-J, Al-Ali HK, Gisslinger H, Waltzman R, Stalbovskaya V, McQuitty M, Hunter DS, Levy R, Knoops L. JAK inhibition with ruxolitinib versus best available therapy for myelofibrosis. N Engl J Med. 2012;366:787–98.

Hart S, Goh K, Novotny-Diermayr V, Hu C, Hentze H, Tan Y, Madan B, Amalini C, Loh Y, Ong L. SB1518, a novel macrocyclic pyrimidine-based JAK2 inhibitor for the treatment of myeloid and lymphoid malignancies. Leukemia. 2011;25:1751–9.

Harvey KF, Pfleger CM, Hariharan IK. The *Drosophila* mst ortholog, hippo restricts growth and cell proliferation and promotes apoptosis. Cell. 2003;114:457–67.

He F, Ge W, Martinowich K, Becker-Catania S, Coskun V, Zhu W, Wu H, Castro D, Guillemot F, Fan G. A positive autoregulatory loop of Jak-STAT signaling controls the onset of astrogliogenesis. Nat Neurosci. 2005;8:616–25.

Heallen T, Zhang M, Wang J, Bonilla-Claudio M, Klysik E, Johnson RL, Martin JF. Hippo pathway inhibits Wnt signaling to restrain cardiomyocyte proliferation and heart size. Science. 2011;332:458–61.

Hecht A, Vleminckx K, Stemmler MP, Van Roy F, Kemler R. The p300/CBP acetyltransferases function as transcriptional coactivators of β-catenin in vertebrates. EMBO J. 2000;19:1839–50.

Heitzler P. Chapter fourteen-biodiversity and noncanonical notch signaling. Curr Top Dev Biol. 2010;92:457–81.

Hitoshi S, Alexson T, Tropepe V, Donoviel D, Elia AJ, Nye JS, Conlon RA, Mak TW, Bernstein A, van der Kooy D. Notch pathway molecules are essential for the maintenance, but not the generation, of mammalian neural stem cells. Genes Dev. 2002;16:846–58.

Hsu H-J, Drummond-Barbosa D. Insulin levels control female germline stem cell maintenance via the niche in *Drosophila*. Proc Natl Acad Sci. 2009;106:1117–21.

Hsu H, Drummond-Barbosa D. Insulin signals control the competence of the *Drosophila* female germline stem cell niche to respond to notch ligands. Dev Biol. 2011;350:290.

Hsu H, LaFever L, Drummond-Barbosa D. Diet controls normal and tumorous germline stem cells via insulin-dependent and-independent mechanisms in *Drosophila*. Dev Biol. 2008;313:700–12.

Huang J, Wu S, Barrera J, Matthews K, Pan D. The hippo signaling pathway coordinately regulates cell proliferation and apoptosis by inactivating yorkie, the *Drosophila* homolog of YAP. Cell. 2005;122:421–34.

Humphrey RK, Beattie GM, Lopez AD, Bucay N, King CC, Firpo MT, Rose-John S, Hayek A. Maintenance of pluripotency in human embryonic stem cells is STAT3 independent. Stem Cells. 2004;22:522–30.

Imokawa Y, Yoshizato K. Expression of Sonic hedgehog gene in regenerating newt limb blastemas recapitulates that in developing limb buds. Proc Natl Acad Sci USA. 1997;94:9159–64.

James D, Levine AJ, Besser D, Hemmati-Brivanlou A. TGFbeta/activin/nodal signaling is necessary for the maintenance of pluripotency in human embryonic stem cells. Development. 2005;132:1273–82.

Jamieson CH, Ailles LE, Dylla SJ, Muijtjens M, Jones C, Zehnder JL, Gotlib J, Li K, Manz MG, Keating A. Granulocyte–macrophage progenitors as candidate leukemic stem cells in blast-crisis CML. N Engl J Med. 2004;351:657–67.

Jang J, Ku SY, Kim JE, Choi K, Kim YY, Kim HS, Oh SK, Lee EJ, Cho HJ, Song YH. Notch inhibition promotes human embryonic stem cell-derived cardiac mesoderm differentiation. Stem Cells. 2008;26:2782–90.

Jansson L, Larsson J. Normal hematopoietic stem cell function in mice with enforced expression of the hippo signaling effector YAP1. PLoS ONE. 2012;7:e32013.

Jeong H, Bae S, An SY, Byun MR, Hwang J-H, Yaffe MB, Hong J-H, Hwang ES. TAZ as a novel enhancer of MyoD-mediated myogenic differentiation. FASEB J. 2010;24:3310–20.

Jia J, Zhang W, Wang B, Trinko R, Jiang J. The *Drosophila* Ste20 family kinase dMST functions as a tumor suppressor by restricting cell proliferation and promoting apoptosis. Genes Dev. 2003;17:2514–9.

Jiang H, Patel PH, Kohlmaier A, Grenley MO, McEwen DG, Edgar BA. Cytokine/Jak/Stat signaling mediates regeneration and homeostasis in the *Drosophila* midgut. Cell. 2009;137:1343–1355.

Jiang H, Grenley MO, Bravo MJ, Blumhagen RZ, Edgar BA. EGFR/Ras/MAPK signaling mediates adult midgut epithelial homeostasis and regeneration in *Drosophila*. Cell Stem Cell. 2011;8:84–95.

Justice RW, Zilian O, Woods DF, Noll M, Bryant PJ. The *Drosophila* tumor suppressor gene warts encodes a homolog of human myotonic dystrophy kinase and is required for the control of cell shape and proliferation. Genes Dev. 1995;9:534–46.

Karanu FN, Murdoch B, Gallacher L, Wu DM, Koremoto M, Sakano S, Bhatia M. The notch ligand jagged-1 represents a novel growth factor of human hematopoietic stem cells. J Exp Med. 2000;192:1365–72.

Karanu FN, Murdoch B, Miyabayashi T, Ohno M, Koremoto M, Gallacher L, Wu D, Itoh A, Sakano S, Bhatia M. Human homologues of delta-1 and delta-4 function as mitogenic regulators of primitive human hematopoietic cells. Blood. 2001;97:1960–7.

Kato Y, Iwama A, Tadokoro Y, Shimoda K, Minoguchi M, Akira S, Tanaka M, Miyajima A, Kitamura T, Nakauchi H. Selective activation of STAT5 unveils its role in stem cell self-renewal in normal and leukemic hematopoiesis. J Exp Med. 2005;202:169–79.

Katoh Y, Katoh M. Hedgehog signaling pathway and gastrointestinal stem cell signaling network (review). Int J Mol Med. 2006;18:1019–23.

Kharas MG, Okabe R, Ganis JJ, Gozo M, Khandan T, Paktinat M, Gilliland DG, Gritsman K. Constitutively active AKT depletes hematopoietic stem cells and induces leukemia in mice. Blood. 2010;115:1406–15.

Kielman MF, Rindapää M, Gaspar C, van Poppel N, Breukel C, van Leeuwen S, Taketo MM, Roberts S, Smits R, Fodde R. Apc modulates embryonic stem-cell differentiation by controlling the dosage of β-catenin signaling. Nat Genet. 2002;32:594–605.

Kiger AA, Jones DL, Schulz C, Rogers MB, Fuller MT. Stem cell self-renewal specified by JAK-STAT activation in response to a support cell cue. Science. 2001;294:2542–5.

Klaus A, Birchmeier W. Wnt signalling and its impact on development and cancer. Nat Rev Cancer. 2008;8:387–98.

Klein RS, Rubin JB, Gibson HD, DeHaan EN, Alvarez-Hernandez X, Segal RA, Luster AD. SDF-1α induces chemotaxis and enhances Sonic hedgehog-induced proliferation of cerebellar granule cells. Development. 2001;128:1971–81.

Kobayashi H, Butler JM, O'Donnell R, Kobayashi M, Ding B-S, Bonner B, Chiu VK, Nolan DJ, Shido K, Benjamin L. Angiocrine factors from Akt-activated endothelial cells balance self-renewal and differentiation of haematopoietic stem cells. Nat Cell Biol. 2010;12:1046–56.

König A, Yatsenko AS, Weiss M, Shcherbata HR. Ecdysteroids affect *Drosophila* ovarian stem cell niche formation and early germline differentiation. EMBO J. 2011;30:1549–62.
Kopan R, Ilagan MXG. The canonical notch signaling pathway: unfolding the activation mechanism. Cell. 2009;137:216–33.
Korkaya H, Paulson A, Charafe-Jauffret E, Ginestier C, Brown M, Dutcher J, Clouthier SG, Wicha MS. Regulation of mammary stem/progenitor cells by PTEN/Akt/beta-catenin signaling. PLoS Biol. 2009;7:e1000121.
Krzemień J, Dubois L, Makki R, Meister M, Vincent A, Crozatier M. Control of blood cell homeostasis in *Drosophila* larvae by the posterior signalling centre. Nature. 2007;446:325–8.
Kuang S, Kuroda K, Le Grand F, Rudnicki MA. Asymmetric self-renewal and commitment of satellite stem cells in muscle. Cell. 2007;129:999–1010.
Kumano K, Chiba S, Kunisato A, Sata M, Saito T, Nakagami-Yamaguchi E, Yamaguchi T, Masuda S, Shimizu K, Takahashi T. Notch1 but not notch2 is essential for generating hematopoietic stem cells from endothelial cells. Immunity. 2003;18:699–711.
Laforest L, Brown CW, Poleo G, Geraudie J, Tada M, Ekker M, Akimenko MA. Involvement of the sonic hedgehog, patched 1 and bmp2 genes in patterning of the zebrafish dermal fin rays. Development. 1998;125:4175–84.
Lai K, Kaspar BK, Gage FH, Schaffer DV. Sonic hedgehog regulates adult neural progenitor proliferation in vitro and in vivo. Nat Neurosci. 2003;6:21–7.
Lai Z-C, Wei X, Shimizu T, Ramos E, Rohrbaugh M, Nikolaidis N, Ho L-L, Li Y. Control of cell proliferation and apoptosis by mob as tumor suppressor, mats. Cell. 2005;120:675–85.
Le Grand F, Jones AE, Seale V, Scimè A, Rudnicki MA. Wnt7a activates the planar cell polarity pathway to drive the symmetric expansion of satellite stem cells. Cell Stem Cell. 2009;4:535–47.
Leatherman JL, DiNardo S. Germline self-renewal requires cyst stem cells and stat regulates niche adhesion in *Drosophila* testes. Nat Cell Biol. 2010;12:806–11.
Leatherman JL, DiNardo S. Zfh-1 controls somatic stem cell self-renewal in the *Drosophila* testis and nonautonomously influences germline stem cell self-renewal. Cell stem cell. 2008;3:44–54.
Lee H-Y, Kléber M, Hari L, Brault V, Suter U, Taketo MM, Kemler R, Sommer L. Instructive role of Wnt/ß-catenin in sensory fate specification in neural crest stem cells. Science. 2004;303:1020–3.
Lee K-P, Lee J-H, Kim T-S, Kim T-H, Park H-D, Byun J-S, Kim M-C, Jeong W-I, Calvisi DF, Kim J-M. The hippo-salvador pathway restrains hepatic oval cell proliferation, liver size, and liver tumorigenesis. Proc Natl Acad Sci. 2010;107:8248–53.
Li L, Xie T. Stem cell niche: structure and function. Annu Rev Cell Dev Biol. 2005;21:605–31.
Lian I, Kim J, Okazawa H, Zhao J, Zhao B, Yu J, Chinnaiyan A, Israel MA, Goldstein LS, Abujarour R. The role of YAP transcription coactivator in regulating stem cell self-renewal and differentiation. Genes Dev. 2010;24:1106–18.
Liang K, Zhou G, Zhang Q, Li J, Zhang C. Expression of hippo pathway in colorectal cancer. Saudi J Gastroenterol. 2014;20:188.
Lillien L, Raphael H. BMP and FGF regulate the development of EGF-responsive neural progenitor cells. Development. 2000;127:4993–5005.
Lin H. The stem-cell niche theory: lessons from flies. Nat Rev Genet. 2002;3:931–40.
Lin G, Xu N, Xi R. Paracrine wingless signalling controls self-renewal of *Drosophila* intestinal stem cells. Nature. 2008;455:1119–23.
Lin G, Xu N, Xi R. Paracrine unpaired signaling through the JAK/STAT pathway controls self-renewal and lineage differentiation of *Drosophila* intestinal stem cells. J Mol Cell Biol. 2010;2:37–49.
Lindsley RC, Gill JG, Kyba M, Murphy TL, Murphy KM. Canonical Wnt signaling is required for development of embryonic stem cell-derived mesoderm. Development. 2006;133:3787–96.
Liu C, Li Y, Semenov M, Han C, Baeg G-H, Tan Y, Zhang Z, Lin X, He X. Control of β-catenin phosphorylation/degradation by a dual-kinase mechanism. Cell. 2002;108:837–47.

Liu S, Dontu G, Mantle ID, Patel S, Ahn N-S, Jackson KW, Suri P, Wicha MS. Hedgehog signaling and Bmi-1 regulate self-renewal of normal and malignant human mammary stem cells. Cancer Res. 2006;66:6063–71.

Liu J, Sato C, Cerletti M, Wagers A. Notch signaling in the regulation of stem cell self-renewal and differentiation. Curr Top Dev Biol. 2010a;92:367–409.

Liu W, Singh SR, Hou SX. JAK–STAT is restrained by notch to control cell proliferation of the *Drosophila* intestinal stem cells. J Cell Biochem. 2010b;109:992–9.

Lowell S, Benchoua A, Heavey B, Smith AG. Notch promotes neural lineage entry by pluripotent embryonic stem cells. PLoS Biol. 2006;4:e121.

Lowry WE, Blanpain C, Nowak JA, Guasch G, Lewis L, Fuchs E. Defining the impact of beta-catenin/Tcf transactivation on epithelial stem cells. Genes Dev. 2005;19:1596–611.

Machold R, Hayashi S, Rutlin M, Muzumdar MD, Nery S, Corbin JG, Gritli-Linde A, Dellovade T, Porter JA, Rubin LL, Dudek H, McMahon AP, Fishell G. Sonic hedgehog is required for progenitor cell maintenance in telencephalic stem cell niches. Neuron. 2003;39:937–50.

Mahoney W, Hong J, Yaffe M, Farrance I. The transcriptional co-activator TAZ interacts differentially with transcriptional enhancer factor-1 (TEF-1) family members. Biochem J. 2005;388:217–25.

Maillard I, Koch U, Dumortier A, Shestova O, Xu L, Sai H, Pross SE, Aster JC, Bhandoola A, Radtke F. Canonical notch signaling is dispensable for the maintenance of adult hematopoietic stem cells. Cell Stem Cell. 2008;2:356–66.

Makki R, Meister M, Pennetier D, Ubeda J-M, Braun A, Daburon V, Krzemień J, Bourbon H-M, Zhou R, Vincent A. A short receptor downregulates JAK/STAT signalling to control the *Drosophila* cellular immune response. PLoS Biol. 2010;8:e1000441.

Mancini SJ, Mantei N, Dumortier A, Suter U, MacDonald HR, Radtke F. Jagged1-dependent notch signaling is dispensable for hematopoietic stem cell self-renewal and differentiation. Blood. 2005;105:2340–2.

Mascarenhas JO, Cross NC, Mesa RA. The future of JAK inhibition in myelofibrosis and beyond. Blood Rev. 2014;28:189.

Mercher T, Cornejo MG, Sears C, Kindler T, Moore SA, Maillard I, Pear WS, Aster JC, Gilliland DG. Notch signaling specifies megakaryocyte development from hematopoietic stem cells. Cell Stem Cell. 2008;3:314–26.

Micchelli CA, Perrimon N. Evidence that stem cells reside in the adult *Drosophila* midgut epithelium. Nature. 2005;439:475–9.

Michel M, Kupinski AP, Raabe I, Bökel C. Hh signalling is essential for somatic stem cell maintenance in the *Drosophila* testis niche. Development. 2012;139:2663–9.

Milano J, McKay J, Dagenais C, Foster-Brown L, Pognan F, Gadient R, Jacobs RT, Zacco A, Greenberg B, Ciaccio PJ. Modulation of notch processing by γ-secretase inhibitors causes intestinal goblet cell metaplasia and induction of genes known to specify gut secretory lineage differentiation. Toxicol Sci. 2004;82:341–58.

Morgan TH. The theory of the gene. Am Nat. 1917;51:513–44.

Morris LX, Spradling AC. Steroid signaling within *Drosophila* ovarian epithelial cells sex-specifically modulates early germ cell development and meiotic entry. PLoS ONE. 2012;7:e46109.

Morrison SJ, Spradling AC. Stem cells and niches: mechanisms that promote stem cell maintenance throughout life. Cell. 2008;132:598–611.

Mosimann C, Hausmann G, Basler K. Parafibromin/Hyrax activates Wnt/Wg target gene transcription by direct association with β-catenin/Armadillo. Cell. 2006;125:327–41.

Muller M, Briscoe J, Laxton C, Guschin D, Ziemiecki A, Silvennoinen O, Harpur AG, Barbieri G, Witthuhn BA, Schindler C, et al. The protein tyrosine kinase JAK1 complements defects in interferon-alpha/beta and -gamma signal transduction. Nature. 1993;366:129–35.

Murry CE, Keller G. Differentiation of embryonic stem cells to clinically relevant populations: lessons from embryonic development. Cell. 2008;132:661–80.

Nejigane S, Takahashi S, Haramoto Y, Michiue T, Asashima M. Hippo signaling components, Mst1 and Mst2, act as a switch between self-renewal and differentiation in xenopus hematopoietic and endothelial progenitors. Int J Dev Biol. 2013;57:407–14.

Nemir M, Croquelois A, Pedrazzini T, Radtke F. Induction of cardiogenesis in embryonic stem cells via downregulation of notch1 signaling. Circ Res. 2006;98:1471–8.

Ngo KT, Wang J, Junker M, Kriz S, Vo G, Asem B, Olson JM, Banerjee U, Hartenstein V. Concomitant requirement for Notch and Jak/Stat signaling during neuro-epithelial differentiation in the *Drosophila* optic lobe. Dev Biol. 2010;346:284–295.

Nishioka N, Inoue K-I, Adachi K, Kiyonari H, Ota M, Ralston A, Yabuta N, Hirahara S, Stephenson RO, Ogonuki N. The hippo signaling pathway components lats and yap pattern tead4 activity to distinguish mouse trophectoderm from inner cell mass. Dev Cell. 2009;16:398–410.

Noggle SA, Weiler D, Condie BG. Notch signaling is inactive but inducible in human embryonic stem cells. Stem Cells. 2006;24:1646–53.

Nusse R, Varmus HE. Many tumors induced by the mouse mammary tumor virus contain a provirus integrated in the same region of the host genome. Cell. 1982;31:99–109.

Nusse R, Brown A, Papkoff J, Scambler P, Shackleford G, McMahon A, Moon R, Varmus H. A new nomenclature for int-1 and related genes: the Wnt gene family. Cell. 1991;64:231.

Oatley JM, Kaucher AV, Avarbock MR, Brinster RL. Regulation of mouse spermatogonial stem cell differentiation by STAT3 signaling. Biol Reprod. 2010;83:427–33.

O'Brien LE, Soliman SS, Li X, Bilder D. Altered modes of stem cell division drive adaptive intestinal growth. Cell. 2011;147:603–14.

Oh H, Irvine KD. In vivo regulation of yorkie phosphorylation and localization. Development. 2008;135:1081–8.

Ohlstein B, Spradling A. The adult *Drosophila* posterior midgut is maintained by pluripotent stem cells. Nature. 2005;439:470–4.

Ohlstein B, Spradling A. Multipotent *Drosophila* intestinal stem cells specify daughter cell fates by differential notch signaling. Science. 2007;315:988–92.

Onishi K, Tonge PD, Nagy A, Zandstra PW. Microenvironment-mediated reversion of epiblast stem cells by reactivation of repressed JAK–STAT signaling. Integr Biol. 2012;4:1367–76.

Pantalacci S, Tapon N, Léopold P. The salvador partner hippo promotes apoptosis and cell-cycle exit in *Drosophila*. Nat Cell Biol. 2003;5:921–7.

Pardanani A, Laborde R, Lasho T, Finke C, Begna K, Al-Kali A, Hogan W, Litzow M, Leontovich A, Kowalski M. Safety and efficacy of CYT387, a JAK1 and JAK2 inhibitor, in myelofibrosis. Leukemia. 2013;27:1322–7.

Paul MK, Bisht B, Darmawan DO, Chiou R, Ha VL, Wallace WD, Chon AT, Hegab AE, Grogan T, Elashoff DA, Alva-Ornelas JA, Gomperts BN. Dynamic changes in intracellular ROS levels regulate airway basal stem cell homeostasis through Nrf2-dependent notch signaling. Cell Stem Cell. 2014;15:199–214.

Pellegrinet L, Rodilla V, Liu Z, Chen S, Koch U, Espinosa L, Kaestner KH, Kopan R, Lewis J, Radtke F. Dll1-and dll4-mediated notch signaling are required for homeostasis of intestinal stem cells. Gastroenterology. 2011;140:1230–1240. e1237.

Pera MF, Trounson AO. Human embryonic stem cells: prospects for development. Development. 2004;131:5515–25.

Pera MF, Andrade J, Houssami S, Reubinoff B, Trounson A, Stanley EG, Ward-van Oostwaard D, Mummery C. Regulation of human embryonic stem cell differentiation by BMP-2 and its antagonist noggin. J Cell Sci. 2004;117:1269–80.

Perdigoto CN, Bardin AJ. Sending the right signal: Notch and stem cells. Biochim Biophys Acta (BBA)-Gen Subj. 2013;1830:2307–2322.

Perra A, Kowalik MA, Ghiso E, Ledda-Columbano GM, Di Tommaso L, Angioni MM, Raschioni C, Testore E, Roncalli M, Giordano S. YAP activation is an early event and a potential therapeutic target in liver cancer development. J Hepatol. 2014;61:1088.

Peters JM, McKay RM, McKay JP, Graff JM. Casein kinase I transduces Wnt signals. Nature. 1999;401:345–50.

Phelps RA, Chidester S, Dehghanizadeh S, Phelps J, Sandoval IT, Rai K, Broadbent T, Sarkar S, Burt RW, Jones DA. A two-step model for colon adenoma initiation and progression caused by APC loss. Cell. 2009;137:623–34.

Pinto D, Gregorieff A, Begthel H, Clevers H. Canonical Wnt signals are essential for homeostasis of the intestinal epithelium. Genes Dev. 2003;17:1709–13.

Polesskaya A, Seale P, Rudnicki MA. Wnt signaling induces the myogenic specification of resident CD45 + adult stem cells during muscle regeneration. Cell. 2003;113:841–52.

Radtke F, Wilson A, Stark G, Bauer M, van Meerwijk J, MacDonald HR, Aguet M. Deficient T cell fate specification in mice with an induced inactivation of notch1. Immunity. 1999;10:547–58.

Radtke F, Wilson A, Mancini SJ, MacDonald HR. Notch regulation of lymphocyte development and function. Nat Immunol. 2004;5:247–53.

Rakesh K, Agrawal DK. Controlling cytokine signaling by constitutive inhibitors. Biochem Pharmacol. 2005;70:649–57.

Ramos A, Camargo FD. The hippo signaling pathway and stem cell biology. Trends Cell Biol. 2012;22:339–46.

Rattis FM, Voermans C, Reya T. Wnt signaling in the stem cell niche. Curr Opin Hematol. 2004;11:88–94.

Ren F, Zhang L, Jiang J. Hippo signaling regulates yorkie nuclear localization and activity through 14-3-3 dependent and independent mechanisms. Dev Biol. 2010;337:303–12.

Riccio O, van Gijn ME, Bezdek AC, Pellegrinet L, van Es JH, Zimber-Strobl U, Strobl LJ, Honjo T, Clevers H, Radtke F. Loss of intestinal crypt progenitor cells owing to inactivation of both Notch1 and Notch2 is accompanied by derepression of CDK inhibitors p27Kip1 and p57Kip2. EMBO Rep. 2008;9:377–83.

Rijsewijk F, Schuermann M, Wagenaar E, Parren P, Weigel D, Nusse R. The *Drosophila* homolog of the mouse mammary oncogene int-1 is identical to the segment polarity gene wingless. Cell. 1987;50:649–57.

Robert-Moreno À, Guiu J, Ruiz-Herguido C, López ME, Inglés-Esteve J, Riera L, Tipping A, Enver T, Dzierzak E, Gridley T. Impaired embryonic haematopoiesis yet normal arterial development in the absence of the Notch ligand Jagged1. EMBO J. 2008;27:1886–95.

Rock JR, Gao X, Xue Y, Randell SH, Kong Y-Y, Hogan BL. Notch-dependent differentiation of adult airway basal stem cells. Cell Stem Cell. 2011;8:639–48.

Roy S, Gardiner DM. Cyclopamine induces digit loss in regenerating axolotl limbs. J Exp Zool. 2002;293:186–90.

Ryan MA, Nattamai KJ, Xing E, Schleimer D, Daria D, Sengupta A, Kohler A, Liu W, Gunzer M, Jansen M, Ratner N, Le Cras TD, Waterstrat A, Van Zant G, Cancelas JA, Zheng Y, Geiger H. Pharmacological inhibition of EGFR signaling enhances G-CSF-induced hematopoietic stem cell mobilization. Nat Med. 2010;16:1141–6.

Sato N, Meijer L, Skaltsounis L, Greengard P, Brivanlou AH. Maintenance of pluripotency in human and mouse embryonic stem cells through activation of Wnt signaling by a pharmacological GSK-3-specific inhibitor. Nat Med. 2003;10:55–63.

Schlegelmilch K, Mohseni M, Kirak O, Pruszak J, Rodriguez JR, Zhou D, Kreger BT, Vasioukhin V, Avruch J, Brummelkamp TR. Yap1 acts downstream of α-catenin to control epidermal proliferation. Cell. 2011;144:782–95.

Schofield R. The relationship between the spleen colony-forming cell and the haemopoietic stem cell. Blood Cells. 1977;4:7–25.

Schroeder T, Fraser ST, Ogawa M, Nishikawa S, Oka C, Bornkamm GW, Nishikawa S-I, Honjo T, Just U. Recombination signal sequence-binding protein Jκ alters mesodermal cell fate decisions by suppressing cardiomyogenesis. Proc Natl Acad Sci. 2003;100:4018–23.

Seifert JR, Mlodzik M. Frizzled/PCP signalling: a conserved mechanism regulating cell polarity and directed motility. Nat Rev Genet. 2007;8:126–38.

Shah NM, Groves AK, Anderson DJ. Alternative neural crest cell fates are instructively promoted by TGFbeta superfamily members. Cell. 1996;85:331–43.

Sherry MM, Reeves A, Wu JK, Cochran BH. STAT3 is required for proliferation and maintenance of multipotency in glioblastoma stem cells. Stem Cells. 2009;27:2383–92.

Shuai K, Ziemiecki A, Wilks AF, Harpur AG, Sadowski HB, Gilman MZ, Darnell JE. Polypeptide signalling to the nucleus through tyrosine phosphorylation of jak and stat proteins. Nature. 1993;366:580.

Shuai K, Stark GR, Kerr IM, Darnell J. A single phosphotyrosine residue of stat91 required for gene activation by interferon-gamma. Science. 1993;261:1744–6.

Snow JW, Abraham N, Ma MC, Abbey NW, Herndier B, Goldsmith MA. STAT5 promotes multilineage hematolymphoid development in vivo through effects on early hematopoietic progenitor cells. Blood. 2002;99:95–101.

Song X, Xie T. Wingless signaling regulates the maintenance of ovarian somatic stem cells in *Drosophila*. Development. 2003;130:3259–68.

Song X, Wong MD, Kawase E, Xi R, Ding BC, McCarthy JJ, Xie T. Bmp signals from niche cells directly repress transcription of a differentiation-promoting gene, bag of marbles, in germline stem cells in the *Drosophila* ovary. Development. 2004;131:1353–64.

Song H, Mak KK, Topol L, Yun K, Hu J, Garrett L, Chen Y, Park O, Chang J, Simpson RM. Mammalian Mst1 and Mst2 kinases play essential roles in organ size control and tumor suppression. Proc Natl Acad Sci. 2010;107:1431–6.

Song S, Ajani JA, Honjo S, Maru DM, Chen Q, Scott AW, Heallen TR, Xiao L, Hofstetter WL, Weston B. Hippo coactivator YAP1 upregulates SOX9 and endows stem-like properties to esophageal cancer cells. Cancer Res. 2014;13:canres. 3569.2013.

Staley BK, Irvine KD. Hippo signaling in *Drosophila*: recent advances and insights. Dev Dyn. 2012;241:3–15.

Stanger BZ, Tanaka AJ, Melton DA. Organ size is limited by the number of embryonic progenitor cells in the pancreas but not the liver. Nature. 2007;445:886–91.

Steinhardt AA, Gayyed MF, Klein AP, Dong J, Maitra A, Pan D, Montgomery EA, Anders RA. Expression of yes-associated protein in common solid tumors. Hum Pathol. 2008;39:1582–9.

Stier S, Cheng T, Dombkowski D, Carlesso N, Scadden DT. Notch1 activation increases hematopoietic stem cell self-renewal in vivo and favors lymphoid over myeloid lineage outcome. Blood. 2002;99:2369–78.

Stine RR, Matunis EL. JAK-STAT signaling in stem cells. In: Transcriptional and translational regulation of stem cells. The Netherlands: Springer; 2013. pp. 247–267.

Stofanko M, Kwon SY, Badenhorst P. Lineage tracing of lamellocytes demonstrates *Drosophila* macrophage plasticity. PLoS ONE. 2010;5:e14051.

Tamm C, Böwer N, Annerén C. Regulation of mouse embryonic stem cell self-renewal by a Yes–YAP–TEAD2 signaling pathway downstream of LIF. J Cell Sci. 2011;124:1136–44.

Tang Y, Tian XC. JAK-STAT3 and somatic cell reprogramming. JAK-STAT. 2013;2:e24935–e24935.

Tang Y, Luo Y, Jiang Z, Ma Y, Lin CJ, Kim C, Carter MG, Amano T, Park J, Kish S. Jak/Stat3 signaling promotes somatic cell reprogramming by epigenetic regulation. Stem Cells. 2012;30:2645–56.

Tao W, Zhang S, Turenchalk GS, Stewart RA, St MA, Chen W, Xu T. Human homologue of the *Drosophila* melanogaster lats tumour suppressor modulates CDC2 activity. Nat Genet. 1999;21:177–81.

Tapon N, Harvey KF, Bell DW, Wahrer DC, Schiripo TA, Haber DA, Hariharan IK. Salvador promotes both cell cycle exit and apoptosis in *Drosophila* and is mutated in human cancer cell lines. Cell. 2002;110:467–478.

Toledo EM, Colombres M, Inestrosa NC. Wnt signaling in neuroprotection and stem cell differentiation. Prog Neurobiol. 2008;86:281–96.

Trompouki E, Bowman TV, Lawton LN, Fan ZP, Wu D-C, DiBiase A, Martin CS, Cech JN, Sessa AK, Leblanc JL. Lineage regulators direct BMP and Wnt pathways to cell-specific programs during differentiation and regeneration. Cell. 2011;147:577–89.

Tropepe V, Hitoshi S, Sirard C, Mak TW, Rossant J, van der Kooy D. Direct neural fate specification from embryonic stem cells: a primitive mammalian neural stem cell stage acquired through a default mechanism. Neuron. 2001;30:65–78.

Tsonis PA, Vergara MN, Spence JR, Madhavan M, Kramer EL, Call MK, Santiago WG, Vallance JE, Robbins DJ, Del Rio-Tsonis K. A novel role of the hedgehog pathway in lens regeneration. Dev Biol. 2004;267:450–61.
Tulina N, Matunis E. Control of stem cell self-renewal in *Drosophila* spermatogenesis by JAK-STAT signaling. Science. 2001;294:2546–9.
Udan RS, Kango-Singh M, Nolo R, Tao C, Halder G. Hippo promotes proliferation arrest and apoptosis in the salvador/warts pathway. Nat Cell Biol. 2003;5:914–20.
Ueishi S, Shimizu H, Inoue YH. Male germline stem cell division and spermatocyte growth require insulin signaling in *Drosophila*. Cell Struct Funct. 2009;34:61–9.
Ueo T, Imayoshi I, Kobayashi T, Ohtsuka T, Seno H, Nakase H, Chiba T, Kageyama R. The role of Hes genes in intestinal development, homeostasis and tumor formation. Development. 2012;139:1071–82.
Ungureanu D, Vanhatupa S, Kotaja N, Yang J, Aittomäki S, Jänne OA, Palvimo JJ, Silvennoinen O. PIAS proteins promote SUMO-1 conjugation to STAT1. Blood. 2003;102:3311–3.
Ungureanu D, Vanhatupa S, Grönholm J, Palvimo JJ, Silvennoinen O. SUMO-1 conjugation selectively modulates STAT1-mediated gene responses. Blood. 2005;106:224–6.
van Es JH, Jay P, Gregorieff A, van Gijn ME, Jonkheer S, Hatzis P, Thiele A, van den Born M, Begthel H, Brabletz T. Wnt signalling induces maturation of paneth cells in intestinal crypts. Nat Cell Biol. 2005;7:381–6.
Varelas X, Sakuma R, Samavarchi-Tehrani P, Peerani R, Rao BM, Dembowy J, Yaffe MB, Zandstra PW, Wrana JL. TAZ controls Smad nucleocytoplasmic shuttling and regulates human embryonic stem-cell self-renewal. Nat Cell Biol. 2008;10:837–48.
Varnum-Finney B, Brashem-Stein C, Bernstein ID. Combined effects of notch signaling and cytokines induce a multiple log increase in precursors with lymphoid and myeloid reconstituting ability. Blood. 2003;101:1784–9.
Vassilev A, Kaneko KJ, Shu H, Zhao Y, DePamphilis ML. TEAD/TEF transcription factors utilize the activation domain of YAP65, a Src/Yes-associated protein localized in the cytoplasm. Genes Dev. 2001;15:1229–41.
Veeman MT, Axelrod JD, Moon RT. A second canon: functions and mechanisms of β-catenin-independent Wnt signaling. Dev Cell. 2003;5:367–77.
Verstovsek S, Mesa RA, Gotlib J, Levy RS, Gupta V, DiPersio JF, Catalano JV, Deininger M, Miller C, Silver RT. A double-blind, placebo-controlled trial of ruxolitinib for myelofibrosis. N Engl J Med. 2012;366:799–807.
Vied C, Reilein A, Field NS, Kalderon D. Regulation of stem cells by intersecting gradients of long-range niche signals. Dev Cell. 2012;23:836–48.
von Gise A, Lin Z, Schlegelmilch K, Honor LB, Pan GM, Buck JN, Ma Q, Ishiwata T, Zhou B, Camargo FD. YAP1, the nuclear target of Hippo signaling, stimulates heart growth through cardiomyocyte proliferation but not hypertrophy. Proc Natl Acad Sci. 2012;109:2394–9.
Voog J, Jones DL. Stem cells and the niche: a dynamic duo. Cell Stem Cell. 2010;6:103–15.
Wang W, Li Y, Zhou L, Yue H, Luo H. Role of JAK/STAT signaling in neuroepithelial stem cell maintenance and proliferation in the *Drosophila* optic lobe. Biochem Biophys Res Commun.2011;410:714–720.
Wang G, Zhang H, Zhao Y, Li J, Cai J, Wang P, Meng S, Feng J, Miao C, Ding M, Li D, Deng H. Noggin and bFGF cooperate to maintain the pluripotency of human embryonic stem cells in the absence of feeder layers. Biochem Biophys Res Commun. 2005;330:934–42.
Wang Z, Li G, Tse W, Bunting KD. Conditional deletion of STAT5 in adult mouse hematopoietic stem cells causes loss of quiescence and permits efficient nonablative stem cell replacement. Blood. 2009;113:4856–65.
Watling D, Guschin D, Müller M, Silvennoinen O, Witthuhn BA, Quelle FW, Rogers NC, Schindler C, Stark GR, Ihle JN. Complementation by the protein tyrosine kinase JAK2 of a mutant cell line defective in the interferon-& gamma; signal transduction pathway. Nature. 1993;366:166.
Wilks A, Harpur A, Kurban R, Ralph S, Zürcher G, Ziemiecki A. Two novel protein-tyrosine kinases, each with a second phosphotransferase-related catalytic domain, define a new class of protein kinase. Mol Cell Biol. 1991;11:2057–65.

Wu S, Huang J, Dong J, Pan D. hippo encodes a ste-20 family protein kinase that restricts cell proliferation and promotes apoptosis in conjunction with salvador and warts. Cell. 2003;114:445–56.

Wu H, Wu Y, Ai Z, Yang L, Gao Y, Du J, Guo Z, Zhang Y. Vitamin C enhances nanog expression via activation of the JAK/STAT signaling pathway. Stem Cells. 2014;32:166–76.

Xie T, Spradling AC. A niche maintaining germ line stem cells in the *Drosophila* ovary. Science. 2000;290:328–30.

Xin M, Kim Y, Sutherland LB, Qi X, McAnally J, Schwartz RJ, Richardson JA, Bassel-Duby R, Olson EN. Regulation of insulin-like growth factor signaling by yap governs cardiomyocyte proliferation and embryonic heart size. Sci Signal. 2011; 4:ra70.

Xu T, Wang W, Zhang S, Stewart RA, Yu W. Identifying tumor suppressors in genetic mosaics: the *Drosophila* lats gene encodes a putative protein kinase. Development. 1995;121:1053–63.

Xu RH, Peck RM, Li DS, Feng X, Ludwig T, Thomson JA. Basic FGF and suppression of BMP signaling sustain undifferentiated proliferation of human ES cells. Nat Methods. 2005;2:185–90.

Xu N, Wang SQ, Tan D, Gao Y, Lin G, Xi R. EGFR, wingless and JAK/STAT signaling cooperatively maintain *Drosophila* intestinal stem cells. Dev Biol. 2011;354:31–43.

Yasugi T, Umetsu D, Murakami S, Sato M, Tabata T. *Drosophila* optic lobe neuroblasts triggered by a wave of proneural gene expression that is negatively regulated by JAK/STAT. Development. 2008;135:1471–80.

Yilmaz ÖH, Katajisto P, Lamming DW, Gültekin Y, Bauer-Rowe KE, Sengupta S, Birsoy K, Dursun A, Yilmaz VO, Selig M. mTORC1 in the Paneth cell niche couples intestinal stem-cell function to calorie intake. Nature. 2012;486:490.

Yimlamai D, Christodoulou C, Galli GG, Yanger K, Pepe-Mooney B, Gurung B, Shrestha K, Cahan P, Stanger BZ, Camargo FD. Hippo pathway activity influences liver cell fate. Cell. 2014;157:1324–38.

Ying QL, Smith AG. Defined conditions for neural commitment and differentiation. Methods Enzymol. 2003;365:327–41.

Ying QL, Nichols J, Chambers I, Smith A. BMP induction of Id proteins suppresses differentiation and sustains embryonic stem cell self-renewal in collaboration with STAT3. Cell. 2003a;115:281–92.

Ying QL, Stavridis M, Griffiths D, Li M, Smith A. Conversion of embryonic stem cells into neuroectodermal precursors in adherent monoculture. Nat Biotechnol. 2003b;21:183–6.

Young RA. Control of the embryonic stem cell state. Cell. 2011;144:940–54.

Yue G, Sun X, Gimenez-Capitan A, Shen J, Yu L, Teixido C, Guan W, Rosell R, Liu B, Wei J. TAZ is highly expressed in gastric signet ring cell carcinoma. BioMed Res Int. 2014; 2014:6.

Zeng YA, Nusse R. Wnt proteins are self-renewal factors for mammary stem cells and promote their long-term expansion in culture. Cell Stem Cell. 2010;6:568–77.

Zhang Y, Kalderon D. Hedgehog acts as a somatic stem cell factor in the *Drosophila* ovary. Nature. 2001;410:599–604.

Zhang Y, Goss AM, Cohen ED, Kadzik R, Lepore JJ, Muthukumaraswamy K, Yang J, DeMayo FJ, Whitsett JA, Parmacek MS. A Gata6-Wnt pathway required for epithelial stem cell development and airway regeneration. Nat Genet. 2008;40:862–70.

Zhao R, Fallon TR, Saladi SV, Pardo-Saganta A, Villoria J, Mou H, Vinarsky V, Gonzalez-Celeiro M, Nunna N, Hariri LP, Camargo F, Ellisen LW, Rajagopal J. Yap tunes airway epithelial size and architecture by regulating the identity, maintenance, and self-renewal of stem cells. Dev Cell. 2014;30:151–65.

Zhou J, Wulfkuhle J, Zhang H, Gu P, Yang Y, Deng J, Margolick JB, Liotta LA, Petricoin E, Zhang Y. Activation of the PTEN/mTOR/STAT3 pathway in breast cancer stem-like cells is required for viability and maintenance. Proc Natl Acad Sci. 2007;104:16158–63.

Zhou D, Zhang Y, Wu H, Barry E, Yin Y, Lawrence E, Dawson D, Willis JE, Markowitz SD, Camargo FD. Mst1 and Mst2 protein kinases restrain intestinal stem cell proliferation and colonic tumorigenesis by inhibition of Yes-associated protein (Yap) overabundance. Proc Natl Acad Sci. 2011;108:E1312–20.

Author Biography

Dr. Rongwen Xi is currently an associate investigator at the National Institute of Biological Sciences, Beijing. His laboratory uses *Drosophila* as a genetic model organism for the understanding of microenvironmental and intrinsic mechanisms regulating self-renewal and differentiation of tissue stem cells. His recent findings include the characterization of the niche regulation of *Drosophila* midgut stem cells, the identification of a tumor suppressive role for polycomb group genes in stem cells, and the discovery of tubular sclerosis complex genes in stem cell maintenance.

Chapter 7
The Concept of Mesenchymal Stem Cell System: Bring More Insights into Functional Research of MSCs

Shihua Wang and Robert Chunhua Zhao

Abstract Mesenchymal stem cells have generated great interest among researchers and physicians due to their unique biological characteristics and potential clinical applications. Here, we propose for the first time the concept of mesenchymal stem cell system which is composed of all mesenchymal stem cells derived from different stages of embryonic development, from post-embryonic subtotipotent stem cells to MSC progenitors. Post-embryonic subtotipotent stem cells are leftover cells during embryonic development and are on the top of the system. MSC system is a combination of cells that possess different differentiation potential and ultimately give rise to cells that share a similar set of phenotypic markers. It is a more comprehensive view of MSCs and could better explain the heterogeneity of MSCs in differentiation potential and immunomodulatory functions. In a word, this concept constitutes an important part of the biological properties of MSC and will help researchers gain better insight into MSC.

Keywords Mesenchymal stem cells · Biological properties · System

7.1 Introduction

MSCs were first discovered in 1968 by Friedenstein as an adherent fibroblast-like population in the bone marrow capable of differentiating into adipocytes, chondrocytes, and osteocytes, both in vitro (Friedenstein et al. 1968) and in vivo

S. Wang · R.C. Zhao
Institute of Basic Medical Sciences Chinese Academy of Medical Sciences,
Center of Excellence in Tissue Engineering,
School of Basic Medicine Peking Union Medical College, Beijing 100005, China

R.C. Zhao (✉)
Peking Union Medical College Hospital, Beijing 100032, China
e-mail: zhaochunhua@vip.163.com

R.C. Zhao (ed.), *Stem Cells: Basics and Clinical Translation*, Translational Medicine Research 1, DOI 10.1007/978-94-017-7273-0_7

(Friedenstein et al. 1974). Caplan (1991) demonstrated that bone and cartilage turnover was mediated by MSCs, and the surrounding conditions were critical to inducing MSC differentiation. They termed these cells "mesenchymal stem cells," and the term "MSC" became popular after the work of Caplan in 1991. Later, the multilineage differentiation capability of MSCs was definitively demonstrated by Pittenger (1999). During the late 1990s, Kopen et al. (1999) then described the capacity of MSCs to transdifferentiate into ectoderm-derived tissues.

The defining characteristics of MSCs are inconsistent among investigators. Many laboratories have developed methods to isolate and expand MSCs, which invariably have subtle, and occasionally quite significant, differences. To address this problem, in 2006, the Mesenchymal and Tissue Stem Cell Committee of International Society for Cellular Therapy (ISCT) proposed a set of standards to define human MSCs for both laboratory-based scientific investigations and for pre-clinical studies. First, MSCs must be plastic-adherent when maintained in standard culture conditions using tissue culture flasks. Second, $\geq$95 % of the MSC population must express CD105, CD73, and CD90, as measured by flow cytometry. Additionally, these cells must lack the expression ($\leq$2 % positive) of CD45, CD34, CD14 or CD11b, CD79a or CD19, and HLA class II. Third, the cells must be able to differentiate into osteoblasts, adipocytes, and chondroblasts under standard in vitro differentiating conditions (Dominici et al. 2006).

MSCs have been identified in almost every tissue type, including placenta, umbilical cord blood, amniotic fluid, bone marrow, adipose tissue, and the liver. Most of the adult sources, including large volumes of normal bone marrow, are relatively difficult to access as a tissue source for the isolation of MSCs. In contrast, birth-associated tissues, including placenta, are readily and widely available. However, bone marrow remains the principal source of MSCs for most preclinical and clinical studies. It is estimated that MSCs represent only between approximately 0.01 and 0.001 % of the total nucleated cells within isolated bone marrow aspirates (Pittenger et al. 1999; Salem and Thiemermann 2010). Despite this low number, there remains a great enthusiasm in these cells, as they can be isolated easily from a small aspirate and culture expanded in vitro to significant numbers in approximately 8–10 weeks. MSCs from different sources have been studied, and each type has been reported to vary in their proliferative and multilineage differentiation potential (Salem and Thiemermann 2010). MSCs hold great promise for the treatment of a variety of diseases (Liao and Zhao 2008; Liu et al. 2006). In this review, we raise the concept of MSC system, which could provide better understanding of the heterogeneity of MSCs.

7.2 The Concept of MSC System

Although the biological characteristics and therapeutic potential of MSCs have been extensively studied, the in vivo behavior and developmental origin of these cells remain largely unknown. During embryonic development, MSCs arise from two major sources: neural crest and mesoderm. The adult MSCs are commonly considered to be of mesodermal origin, whereas embryonic MSCs derive mainly from the neural crest. The neural crest is a transient embryonic tissue that originates at the neural folds during vertebrate development. Morikawa et al. (2009) found that the development of MSCs partially originates from the neural crest. Takashima et al. (2007) showed that the earliest wave of MSCs in the embryonic trunk is generated from Sox1+ neuroepithelium, and they provided evidence that Sox1+ neuroepithelium gives rise to MSCs in part through a neural crest intermediate stage. The mesoderm is considered to be another major source of mesenchymal cells giving rise to skeletal and connective tissues (Dennis and Charbord 2002). Using hESCs directed toward mesendodermal differentiation, Vodyanik MA et al. showed that mesoderm-derived MSCs arise from a common endothelial and mesenchymal cell precursor, the mesenchymoangioblast, which is a transient population of cells within the APLNR+ mesodermal subset. Recently, the Olsen group revealed that vascular endothelial cells can transform into MSCs by an ALK2 receptor-dependent mechanism. Expressing mutant ALK2 in human endothelial cells causes an endothelial–mesenchymal transition (endMT) and the acquisition of a multipotent stem cell-like phenotype (Medici et al. 2010). This result indicates that endothelial cells could be an important source of MSCs in postnatal life. Conversely, the transition from MSCs to endothelial cells has also been described in several studies. These studies suggest a cycle of cell-fate transition from endothelium to MSCs and back to endothelium. Because multiple parallels could be drawn between the endMT described in adult tissues and that during hESC differentiation, one may wonder whether bipotential cells with endothelial and MSC potential similar to embryonic mesenchymoangioblasts are present and constitute an important element of the EndMT circuit in adults (Slukvin and Vodyanik 2011). The number of MSCs of neuroepithelial origin in the adult bone marrow decreases rapidly, which suggests that in postnatal life, the relative importance of MSCs derived from other developmental lineages decreases due to the increasing importance of mesodermal MSCs. We isolated $Flk1^{+}CD31^{-}CD34^{-}$ stem cells, which are MSCs from human fetal bone marrow, and found that they could differentiate into cells of the three germ layers, such as endothelial, hepatocyte-like, neural, and pancreatic progenitor cells (Fang et al. 2003, 2004; Feng et al. 2013; Li et al. 2013; Liu et al. 2007; Huang et al. 2012; Yang et al. 2011). Based on these results, we hypothesized that post-embryonic subtotipotent stem cells exist, which are left-over cells during embryonic development and are on the top of the MSC hierarchy.

Here, for the first time, we propose the concept of MSC system, which is composed of all mesenchymal stem cells from post-embryonic subtotipotent stem cells to MSCs progenitors. The MSC system is a combination of cells that are derived from different stages of embryonic development, possess different differentiation potential, and ultimately give rise to cells that share a similar set of phenotypic markers (Figs. 7.1 and 7.2). The MSC system could better explain the heterogeneity of MSCs in differentiation potential and immunomodulatory functions.

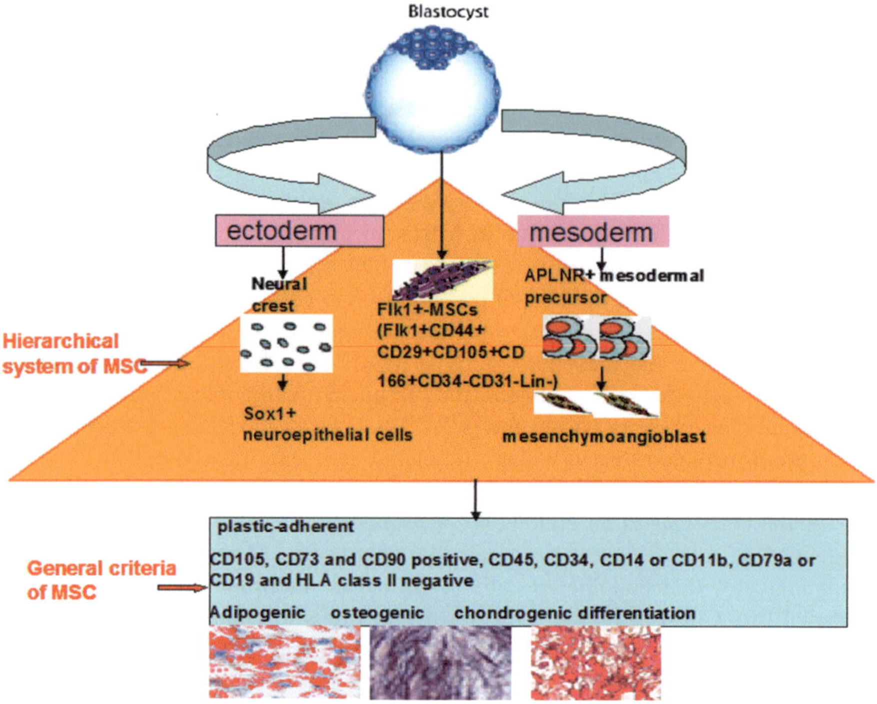

Fig. 7.1 The MSC system is a combination of cells that are derived from different stages of embryonic development

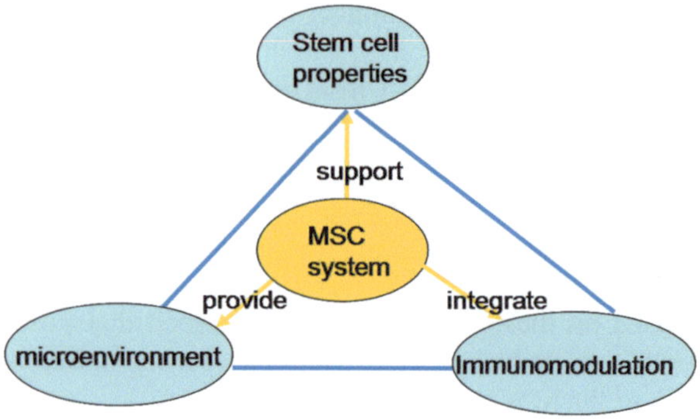

Fig. 7.2 The MSC system could better explain the heterogeneity of MSCs in differentiation potential and immunomodulatory functions

7.3 MSCs Have Different Differentiation Potential

As previously demonstrated, MSCs can differentiate into cells of mesenchymal lineages, such as osteoblasts, chondrocytes, and adipocytes, under culture conditions containing specific growth factors and chemical agents. In addition to the above-mentioned mesenchymal lineages, MSCs have been reported to give rise to cells of other lineages. Kopen et al. (1999) were the first researchers to demonstrate that bone marrow MSCs injected into the central nervous systems of newborn mice migrate throughout the brain and adopt morphological and phenotypic characteristics of astrocytes and neurons. Slukvin and Vodyanik (2011) reported that coculture with heat-shocked small airway epithelial cells induced human MSCs to differentiate into epithelial-like cells, as evidenced by their expression of keratins 17, 18, and 19, the Clara cell marker CC26, and the formation of adherens junctions with neighboring epithelial cells (Slukvin and Vodyanik 2011). These results give us a hint that we are the first group to demonstrate that Flk1+-MSCs (Flk1+CD44+CD29+CD105+CD166+CD34−CD31−Lin−) can give rise to multilineage cells of the three germ layers at the clone level.

7.4 MSCs Have Immunomodulatory Effects

MSCs lack immunogenicity because they express low levels of major histocompatibility complex-I (MHC-I) molecules and do not express MHC-II molecules or costimulatory molecules such as CD80, CD86, or CD40 (Guo et al. 2009; Sun et al. 2011). Another important property of MSCs is that MSCs could act on both innate and adaptive immune systems by interacting with a variety of immune cells such as T cells, B cells, NK cells, and DC cells. The immunomodulatory properties of MSCs were initially reported in T-cell proliferation assays using one of a variety of stimuli, including mitogens and alloantigens; these are settings in which the ability of MSCs to suppress T-cell proliferation can readily be determined (Abdi et al. 2008; Bartholomew et al. 2002; Le Blanc et al. 2003). MSCs regulate the proliferation, activation, and maturation of B lymphocytes in vitro in a dose-dependent and time-limited manner (Aggarwal and Pittenger 2005), and they can facilitate the immunosuppressive effect of cyclosporin A on T lymphocytes through Jagged-1-mediated inhibition of NF-kB signaling (Shi et al. 2011). We first reported that MSCs could inhibit the upregulation of CD1a, CD40, CD80, CD86, and HLA-DR during DC differentiation and prevent an increase of CD40, CD86, and CD83 expression during DC maturation (Zhang et al. 2004). We also demonstrated that in the presence of MSCs, the percentage of cells with a cDC phenotype is significantly reduced, whereas the percentage of pDC phenotypes increases, further suggesting that MSCs can significantly influence DC development (Chen et al. 2007). MSCs could drive maDCs to differentiate into a novel Jagged-2-dependent regulatory DC population and escape their apoptotic fate (Zhang et al. 2009).

MSCs express various receptors for growth factors and inflammatory cytokines, whose activation by external signals could lead to profound cellular responses (Liotta et al. 2008). One important type of receptors is Toll-like receptors (TLRs) which when bound by TLR ligands could induce the release of a list of inflammatory mediators. MSCs could polarize into either pro-inflammatory or anti-inflammatory phenotype through activation of different TLRs under specific

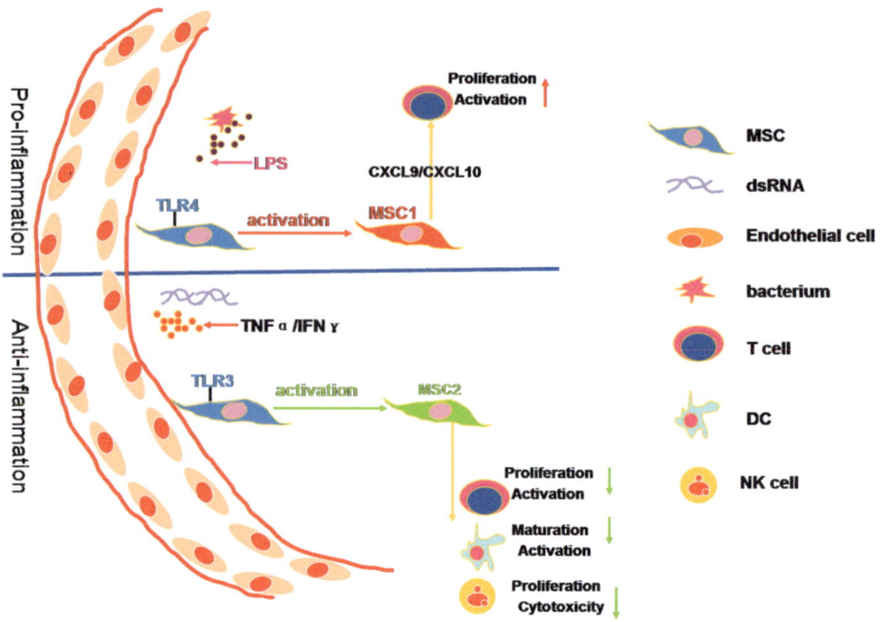

Fig. 7.3 TLR4 could prime MSCs into MSC1, while TLR3 or IFN-γ or TNF-α could induce MSCs into MSC2

microenvironments (Waterman et al. 2010) (Fig. 7.3). TLR4 could prime MSCs into MSC1, while TLR3 or IFN-γ or TNF-α could induce MSCs into MSC2. MSC1 expresses pro-inflammatory mediators such as interleukin(IL)-6 and IL-8, while MSC2 expresses mostly immunosuppressive ones such as IL-10, indoleamine 2,3-dioxygenase (IDO), and prostaglandin E2 (PGE2) (Waterman et al. 2010). The concept of MSC system integrates the apparently contradictory roles of MSCs in inflammation.

Tissue regeneration is the process of repair, replacement, or regeneration of cells in the tissues to restore impaired function resulting from causes such as congenital defects, diseases, and trauma. Tissue injury is always associated with the activation of immune/inflammatory cells both in the innate and adaptive immune responses. Local MSCs or circulating MSCs could be recruited into the damaged tissues by the inflammatory signals and promote the process of tissue repair through several modes of action such as differentiation, secreting bioactive molecules, and immunomodulation (Guo et al. 2013; Zhang et al. 2013; Li et al. 2010). These studies underscore the importance of MSCs' immunomodulatory capacity in regenerating injured tissues.

7.5 Conclusions

The MSC system is a concept embracing both vertical and horizontal properties of MSC. Vertically, it is composed of cells of different tissue origin and from different stages of embryonic development. Cells in the MSC system possess different differentiation potential and ultimately give rise to cells that share a similar set of phenotypic markers. These cells could serve as the seed cells for tissue regeneration. Horizontally, MSC system contains three important biological characteristics of MSC: stem cell properties of MSCs, MSCs as components of tissue microenvironment, and immunomodulatory functions of MSCs.

References

Abdi R, Fiorina P, Adra CN, Atkinson M, Sayegh MH. Immunomodulation by mesenchymal stem cells: a potential therapeutic strategy for type 1 diabetes. Diabetes. 2008;57:1759–67.

Aggarwal S, Pittenger MF. Human mesenchymal stem cells modulate allogeneic immune cell responses. Blood. 2005;105:1815–22.

Bartholomew A, Sturgeon C, Siatskas M, Ferrer K, McIntosh K, Patil S, Hardy W, Devine S, Ucker D, Deans R, Moseley A, Hoffman R. Mesenchymal stem cells suppress lymphocyte proliferation in vitro and prolong skin graft survival in vivo. Exp Hematol. 2002;30:42–8.

Caplan AI. Mesenchymal stem cells. J Orthop Res. 1991;9:641–50.

Chen L, Zhang W, Yue H, Han Q, Chen B, Shi M, Li J, Li B, You S, Shi Y, Zhao RC. Effects of human mesenchymal stem cells on the differentiation of dendritic cells from CD34 + cells. Stem Cells Dev. 2007;16:719–31.

Dennis JE, Charbord P. Origin and differentiation of human and murine stroma. Stem Cells. 2002;20:205–14.

Dominici M, Le Blanc K, Mueller I, Slaper-Cortenbach I, Marini F, Krause D, Deans R, Keating A, Prockop D, Horwitz E. Minimal criteria for defining multipotent mesenchymal stromal cells. The international society for cellular therapy position statement. Cytotherapy. 2006;8:315–7.

Fang B, Shi M, Liao L, Yang S, Liu Y, Zhao RC. Multiorgan engraftment and multilineage differentiation by human fetal bone marrow Flk1+/CD31−/CD34− Progenitors. J Hematother Stem Cell Res. 2003;12:603–13.

Fang B, Liao L, Shi M, Yang S, Zhao RC. Multipotency of Flk1CD34 progenitors derived from human fetal bone marrow. J Lab Clin Med. 2004;143:230–40.

Feng N, Han Q, Li J, Wang S, Li H, Yao X, Zhao RC. Generation of highly purified neural stem cells from human adipose-derived mesenchymal stem cells by Sox1 activation. Stem Cells Dev. (2013).

Friedenstein AJ, Petrakova KV, Kurolesova AI, Frolova GP. Heterotopic of bone marrow. Analysis of precursor cells for osteogenic and hematopoietic tissues. Transplantation. 1968;6:230–47.

Friedenstein AJ, Chailakhyan RK, Latsinik NV, Panasyuk AF, Keiliss-Borok IV. Stromal cells responsible for transferring the microenvironment of the hemopoietic tissues. Cloning in vitro and retransplantation in vivo. Transplantation. 1974;17:331–40.

Guo L, Ge J, Zhou Y, Wang S, Zhao RC, Wu Y. 3D Spheroid cultured mesenchymal stem cells devoid of embolism attenuate brain stroke injury after intra-arterial injection. Stem Cells Dev. (2013).

Guo M, Sun Z, Sun QY, Han Q, Yu CL, Wang DH, Qiao JH, Chen B, Sun WJ, Hu KX, Liu GX, Liu B, Zhao RC, Ai H. A modified haploidentical nonmyeloablative transplantation without T cell depletion for high-risk acute leukemia: successful engraftment and mild GVHD. Biol Blood Marrow Transplant. 2009;15:930–7.

Huang S, Wang S, Bian C, Yang Z, Zhou H, Zeng Y, Li H, Han Q, Zhao RC. Upregulation of miR-22 promotes osteogenic differentiation and inhibits adipogenic differentiation of human adipose tissue-derived mesenchymal stem cells by repressing HDAC6 protein expression. Stem Cells Dev. 2012;21:2531–40.

Kopen GC, Prockop DJ, Phinney DG. Marrow stromal cells migrate throughout forebrain and cerebellum, and they differentiate into astrocytes after injection into neonatal mouse brains. Proc Natl Acad Sci U S A. 1999;96:10711–6.

Le Blanc K, Tammik L, Sundberg B, Haynesworth SE, Ringden O. Mesenchymal stem cells inhibit and stimulate mixed lymphocyte cultures and mitogenic responses independently of the major histocompatibility complex. Scand J Immunol. 2003;57:11–20.

Li K, Han Q, Yan X, Liao L, Zhao RC. Not a process of simple vicariousness, the differentiation of human adipose-derived mesenchymal stem cells to renal tubular epithelial cells plays an important role in acute kidney injury repairing. Stem Cells Dev. 2010;19:1267–75.

Li J, Zhu L, Qu X, Lin R, Liao L, Wang J, Wang S, Xu Q, Zhao RC. Stepwise differentiation of human adipose-derived mesenchymal stem cells toward definitive endoderm and pancreatic progenitor cells by mimicking pancreatic development in vivo. Stem Cells Dev. 2013;22:1576–87.

Liao L, Zhao RC. An overview of stem cell-based clinical trials in China. Stem Cells Dev. 2008;17:613–8.

Liotta F, Angeli R, Cosmi L, Fili L, Manuelli C, Frosali F, Mazzinghi B, Maggi L, Pasini A, Lisi V, Santarlasci V, Consoloni L, Angelotti ML, Romagnani P, Parronchi P, Krampera M, Maggi E, Romagnani S, Annunziato F. Toll-like receptors 3 and 4 are expressed by human bone marrow-derived mesenchymal stem cells and can inhibit their T-cell modulatory activity by impairing notch signaling. Stem Cells. 2008;26:279–89.

Liu L, Sun Z, Chen B, Han Q, Liao L, Jia M, Cao Y, Ma J, Sun Q, Guo M, Liu Z, Ai H, Zhao RC. Ex vivo expansion and in vivo infusion of bone marrow-derived Flk-1+CD31−CD34− mesenchymal stem cells: feasibility and safety from monkey to human. Stem Cells Dev. 2006;15:349–57.

Liu Y, Yan X, Sun Z, Chen B, Han Q, Li J, Zhao RC. Flk-1+ adipose-derived mesenchymal stem cells differentiate into skeletal muscle satellite cells and ameliorate muscular dystrophy in mdx mice. Stem Cells Dev. 2007;16:695–706.

Medici D, Shore EM, Lounev VY, Kaplan FS, Kalluri R, Olsen BR. Conversion of vascular endothelial cells into multipotent stem-like cells. Nat Med. 2010;16:1400–6.

Morikawa S, Mabuchi Y, Niibe K, Suzuki S, Nagoshi N, Sunabori T, Shimmura S, Nagai Y, Nakagawa T, Okano H, Matsuzaki Y. Development of mesenchymal stem cells partially originate from the neural crest. Biochem Biophys Res Commun. 2009;379:1114–9.

Pittenger MF, Mackay AM, Beck SC, Jaiswal RK, Douglas R, Mosca JD, Moorman MA, Simonetti DW, Craig S, Marshak DR. Multilineage potential of adult human mesenchymal stem cells. Science. 1999;284:143–7.

Salem HK, Thiemermann C. Mesenchymal stromal cells: current understanding and clinical status. Stem Cells. 2010;28:585–96.

Shi D, Liao L, Zhang B, Liu R, Dou X, Li J, Zhu X, Yu L, Chen D, Zhao RC. Human adipose tissue-derived mesenchymal stem cells facilitate the immunosuppressive effect of cyclosporin A on T lymphocytes through Jagged-1-mediated inhibition of NF-kappaB signaling. Exp Hematol. 2011;39(214–224):e1.

Slukvin II, Vodyanik M. Endothelial origin of mesenchymal stem cells. Cell Cycle. 2011;10:1370–3.

Sun Z, Han Q, Zhu Y, Li Z, Chen B, Liao L, Bian C, Li J, Shao C, Zhao RC. NANOG has a role in mesenchymal stem cells' immunomodulatory effect. Stem Cells Dev. 2011;20:1521–8.

Takashima Y, Era T, Nakao K, Kondo S, Kasuga M, Smith AG, Nishikawa S. Neuroepithelial cells supply an initial transient wave of MSC differentiation. Cell. 2007;129:1377–88.
Waterman RS, Tomchuck SL, Henkle SL, Betancourt AM. A new mesenchymal stem cell (MSC) paradigm: polarization into a pro-inflammatory MSC1 or an immunosuppressive MSC2 phenotype. PLoS ONE. 2010;5:e10088.
Yang Z, Bian C, Zhou H, Huang S, Wang S, Liao L, Zhao RC. MicroRNA hsa-miR-138 inhibits adipogenic differentiation of human adipose tissue-derived mesenchymal stem cells through adenovirus EID-1. Stem Cells Dev. 2011;20:259–67.
Zhang W, Ge W, Li C, You S, Liao L, Han Q, Deng W, Zhao RC. Effects of mesenchymal stem cells on differentiation, maturation, and function of human monocyte-derived dendritic cells. Stem Cells Dev. 2004;13:263–71.
Zhang B, Liu R, Shi D, Liu X, Chen Y, Dou X, Zhu X, Lu C, Liang W, Liao L, Zenke M, Zhao RC. Mesenchymal stem cells induce mature dendritic cells into a novel Jagged-2-dependent regulatory dendritic cell population. Blood. 2009;113:46–57.
Zhang L, Li K, Liu X, Li D, Luo C, Fu B, Cui S, Zhu F, Zhao RC, Chen X. Repeated systemic administration of human adipose-derived stem cells attenuates overt diabetic nephropathy in rats. Stem Cells Dev. 2013;22:3074–86.

Chapter 8
Human Neural Development and Human Embryonic Stem Cell Neural Differentiation

Yunbo Qiao and Naihe Jing

Abstract Human neurodevelopment is a key process that is precisely controlled by signaling pathways, intrinsic factors and epigenetic modifications. The previous knowledge of human neurodevelopment has mainly archived from human structural development, which is interrelated with cognitive and behavioral development. The establishment of in vitro model of human embryonic stem cell (HESC) and induced human pluripotent stem cell (iHPSC) differentiation makes us available to investigate the underlying mechanisms of human neural development and to model human neuropsychiatric disorders. Although there are gaps between in vitro models and in vivo status, the progress in this field will bring the two worlds together.

Keywords Human neural development · Human embryonic stem cell · Neural differentiation

8.1 Human Neural Development

8.1.1 The Overview of Human Nervous System

Human neurodevelopment is a complex and precisely regulated process that occurs over the entire human life span from an egg. Human-specific features of neurodevelopment are coordinated with signaling pathways, genetic information, and epigenetic modifications, which are the important factors in the evolution of human specializations. The physical structures of human central nervous system (CNS) are described with three axes: the rostral–caudal, dorsal–ventral and medial–lateral axes. The human CNS is mainly composed of the spinal cord and brain. The brain

Y. Qiao · N. Jing (✉)
State Key Laboratory of Cell Biology, Institute of Biochemistry and Cell Biology,
Shanghai Institutes for Biological Sciences, Chinese Academy of Sciences,
320 Yue Yang Road, Shanghai 200031, China
e-mail: njing@sibcb.ac.cn

R.C. Zhao (ed.), *Stem Cells: Basics and Clinical Translation*, Translational Medicine Research 1, DOI 10.1007/978-94-017-7273-0_8

is the central organ responsible for human mental prowess, allowing us to exhibit highly sophisticated abilities, such as language, symbolic thinking, self-awareness, and cultural learning. The mechanisms underlying human neural development are a complex network, requiring understanding on genetic and epigenetic levels.

The spinal cord is the most caudal part of the central nervous system, which extends from the base of the skull to the first lumbar vertebra. The spinal cord receives sensory information from the skin, joints, and muscles of the trunk and limbs, and contains the motor neurons responsible for both voluntary and reflex movements. The spinal cord is divided into gray matter and white matter. The gray matter is composed of dorsal and ventral horns and the white matter is made up of rostral–caudal ascending and descending tracts of myelinated axons. The nerve fibers that link the spinal cord with muscles and sensory receptors are bundled in 31 pairs of spinal nerves, each of which has a dorsal sensory and a ventral motor division. Different classes of axons in dorsal or ventral roots convey distinct kinds of input information. The brain lays rostral to the spinal cord and is divided into six subregions, including the medulla, pons, midbrain, cerebellum, diencephalon, and telencephalon. The medulla, pons, and midbrain are collectively termed brain stem, which is continuous with the spinal cord and contains distinct nerve cell clusters that contribute to a variety of sensory and motor systems. Each of the six divisions is further divided into distinct areas according to the differential axes and anatomical or functional structures.

A general rule for central nervous system development is that cells are generated in sites different from those in which they will later reside. The basic columnar organization favors radial migration of cells in the early stages of neurodevelopment. Somas of later-generated neural cells migrate to the distant cerebral cortex and take positions external to somas of their predecessors. The final position along the radial vector may be influenced by afferent axons. Cell relationships in the developing cerebellar cortex are mainly established by the key migration of granule neurons in the reverse direction to the external surface inward past Purkinje dendrites and somas.

In the past decades, the understanding about neurodevelopment has been largely updated by numerous studies in various kinds of animal models. As for the experimental and ethical limitations, the knowledge of human neural development has mainly archived by the transcriptome analysis of human neural tissues and the in vitro model of human embryonic stem cell differentiation, which will be reviewed in the following text.

8.1.2 Early Human Neurulation (Referred to Moore et al. 1982)

After fertilization completes, the zygote passes along the uterine tube and undergoes cleavage into blastomeres. A ball of 12–16 blastomeres, called morula, enters the uterus and forms a blastocyst consisting of an inner cell mass (ICM), a blastocyst cavity, and the trophoblast. Then, the zona pellucida disappears and the blastocyst attaches to the endometrial epithelium by the end of the first week.

During the second week, rapid proliferation and differentiation of the trophoblast occur as follows: (1) the inner layer cytotrophoblast and the outer layer syncytiotrophoblast form; (2) lacunae develop in the syncytiotrophoblast and soon fuse to form lacunar networks; (3) the trophoblast erodes maternal sinusoids and endometrial glands; (4) a primitive uteroplacental circulation is established as blood seeps into the lacunar networks; (5) primary chorionic villi form on the external surface of the chorionic sac; (6) implantation is completed as the conceptus is wholly embedded within endometrium, and the defect in the surface epithelium is healed. Concurrent with these decidual reaction and yolk sac development, the amniotic cavity appears as a slit-like space between the cytotrophoblast and the inner cell mass, and the inner cell mass differentiates into a bilaminar embryonic disk consisting of epiblast and hypoblast. Moreover, the prochordal plate develops as a localized thickening of hypoblast, indicating the future cranial region of the embryo and the site of the future mouth.

The rapid development of human embryo during the third week occurs as the bilaminar embryonic disk is converted into a trilaminar disk, composed of three germ layers, called gastrulation. Formation of the primitive streak and the notochord is the important process occurring during gastrulation (days 14–19).

The primitive streak appears as a midline thickening of the embryonic epiblast. It gives rise to cells that migrate ventrally, laterally, and cranially between the epiblast and the hypoblast. As soon as the primitive streak has begun to produce these mesenchymal cells, the epiblast layer is known as the embryonic ectoderm, and the hypoblast is known as the embryonic endoderm. The cells produced by the primitive streak soon organize into the intraembryonic mesoderm.

Subsequently, the primitive knot gives rise to the cells that form the notochordal process. Cells from the primitive streak migrate to the edges of the embryonic disk, where they join the extraembryonic mesoderm covering the amnion and the yolk sac. By the end of the third week, mesoderm exists between the ectoderm and the endoderm everywhere except at the oropharyngeal membrane, in the midline occupied by the notochord (a derivative of the notochordal process), and at the cloacal membrane. The primitive pit extends into the notochordal process to form the notochordal canal. Openings develop in the floor of the notochordal canal that soon coalesce, leaving the notochordal plate. The notochordal plate infolds to form the notochord.

The formation of the neural plate, the neural folds, and their closure to form the neural tube is called neurulation, which ends at about 26 days. As the notochord develops, the embryonic ectoderm over it thickens to form the neural plate, which is induced by the developing notochord and the paraxial mesoderm on each side of it. The ectoderm of the neural plate, called neuroectoderm, gives rise to the CNS consisting of the brain and the spinal cord. The neural plate first appears cranial to the primitive knot, and dorsal to the notochordal process and the mesoderm adjacent to it. As the notochord forms and elongates, the neural plate broadens and extends cranially as far as the oropharyngeal membrane. On about day 18, the neural plate invaginates along its central axis to form a neural groove with neural folds on each side. By the end of the third week, the neural folds have begun to move together and fuse, converting the neural plate into neural tube. The neural tube separates itself from the surface ectoderm, and the free edges of the ectoderm fuse so that this layer becomes continuous over the back of the embryo. Subsequently, the surface ectoderm differentiates into the epidermis of the skin.

As the neural folds fuse, some ectodermal cells lying along the crest of each neural fold lost their epithelial affinities and attachments to the neighboring cells. As the neural tube separates from the surface ectoderm, these mesoblastic neural crest cells migrate inwardly and invade the mesoblast on each side of the neural tube. The neural crest, which locates between the neural tube and the overlying surface ectoderm, gives rise to the spinal ganglia and the ganglia of the autonomic nervous system.

8.1.3 Structural Brain Development (Referred to Kandel et al. 2012)

After the early neurulation of human nervous system, the anterior segment of the neural tube forms the three main parts of the brain: the forebrain, midbrain, and the hindbrain, and these brain regions further divide into subregions. The structural basis of human brain is essential for the neural activities and cognitive changes in Homo sapiens.

Structural brain development essentially includes neurogenesis, development of dendrites and axons, synaptogenesis, and myelination. Overall, brain development begins within the weeks of conception and continues through the adolescent period into young adulthood (Epstein 1986). While the basic architecture is assembled during the first two trimesters of fetal life, changes in neuronal connectivity and function occur in the last trimester and continue postnatally into adulthood (Dobbing and Sands 1973).

The cortical gray matter, which is composed of neural cell bodies and glial cells, non-neural cells to provide protection and support, increases in volume during childhood, peaking at around 4 years of age, but significantly decreases after 12 years of age, at different times in different brain regions (Pfefferbaum et al. 1994; Reiss et al. 1996; Giedd et al. 1996). The cerebral white matter increases throughout childhood and well into adulthood in a roughly linear pattern with some regional differences (Gogtay et al. 2004). Therefore, overall brain size increase and its pattern and timing result from different processes in different regions at different times.

Neurogenesis takes place before birth beginning in the fifth prenatal week and peaks between the third and forth prenatal months. While neurogenesis occurs primarily prenatally, the production of axons and dendrites continues for a considerable time postnatally. The proliferation and overproduction of axons, dendrites, and synapses are followed by a process of retracting synaptic connections. This synaptic pruning is dependent on the communication among neurons: more active synapses tend to be strengthened and less active ones tend to be eliminated.

Myelination of the cerebral cortex begins prenatally and continues in the frontal lobe into adulthood beyond the second decade of life. Myelin acts as a form of insulation and thereby increases conduction velocity. Myelination of axons after birth inhibits plasticity because myelinated axons lose their ability to branch out and connect with other neurons. Therefore, the regional pattern of myelination seems to be related to the functional maturation of the brain.

To summarize, developmental changes in brain anatomy occur before and after birth, well into young adulthood, even after adult brain size has been attained at around 6 years of age. Neurogenesis, growth of axons and dendrites, synaptogenesis, and myelination all show localized changes at different times. The formation and perpetuation of these internal structures relate to experience and usage of these structures as required by the environment. Therefore, the morphological development of the brain and cognitive development are tightly interrelated in a complex way.

8.1.4 Global Transcriptome of Human Brain

The pattern of hierarchical maturation of different brain regions is paralleled by a shift from diffuse to more focal recruitment of cortical regions with learning and cognitive development, which is presumably an experience-driven maturational process. The maturation of connecting fiber tracts between diverse brain regions is also correlated with the development of cognitive abilities. The development of human brain circuitry depends on the diversity and precise spatiotemporal regulation of its transcriptome. The formation of molecularly distinct and intricate neural circuits may have increased our susceptibility to certain psychiatric and neurological disorders. Recently, a comprehensive analysis of the spatiotemporal dynamics of gene expression and transcript variants was reported, which might enhance our understanding about human neurodevelopment.

The combination of histology-guided fine neuroanatomical molecular profiling and mapping of gene expression data into MRI coordinate space produced an anatomically accurate quantitative map of transcript distribution across the entire human brain (Hawrylycz et al. 2012). Michael et al. describe the generation of a transcriptional atlas of the adult human brain, comprising histological analysis and comprehensive microarray profiling of 900 neuroanatomically precise subdivisions from two individuals. Interestingly, transcriptional regulation varies enormously by anatomical location and constituent cell types, displaying robust molecular signatures, while the transcriptional profiling is highly conserved between individuals. Differential gene expression analysis demonstrates that brain-wide variation reflects the distributions of major cell types such as neurons, astrocytes, oligodendrocytes, and microglia. Local neighborhood relationships between fine anatomical subdivisions are correlated with discrete neuronal subtypes and genes involved with synaptic transmission. The neocortex displays a relatively homogeneous transcriptional pattern, but with distinct features associated selectively with primary sensorimotor cortices and with enriched frontal lobe expression. Intriguingly, the spatial topography of the neocortex is strongly reflected in its molecular topography—the closer two cortical regions, the more similar their transcriptomes. Finally, these data allow comparisons between humans and other animals, with particular relevance for investigating human neurological and neuropsychiatric disease.

Human neurodevelopment is a complex and precisely regulated process that occurs over a protracted period of time, and it is important to understand the complexity and dynamics of the transcriptome of the human brain. In the recent study, transcriptomes of 16 regions comprising the cerebellar cortex, mediodorsal nucleus of the thalamus, striatum, amygdala, hippocampus and 11 areas of the neocortex generated collected from 57 developing and adult postmortem brains of clinically unremarkable donors representing males and females of multiple ethnicities were constructed (Kang et al. 2011). The authors found that 86 % of the genes analyzed were expressed and that 90 % of these were differentially regulated at the whole-transcript or exon level across brain regions and/or time. The majority of these spatiotemporal differences were detected before birth, with subsequent increases in the similarity among regional transcriptomes. The transcriptome is organized into distinct co-expression networks and shows sex-biased gene expression and exon usage. Some of the inter-individual variations in the regional and developmental transcriptomes are associated with specific SNP genotypes, which may have altered expression-regulating elements. This study links the specific patterns of expression to specialized biological processes and uncovers deeper insights into the transcriptional foundations of human brain development and evolution.

8.2 Human Embryonic Stem Cell (HESC) Neural Differentiation

8.2.1 Directed Neural Specification of HESCs

The most detailed accounts of human neural development come from anatomical studies that outline a complex process by which neuroepithelial tissue forms the major CNS structures during neurogenesis (Kang et al. 2011), but this work does not provide insights into functional neuronal development and maturation. Therefore, human embryonic stem cells (HESCs), capable of differentiating into all cell types (Thomson et al. 1998; Reubinoff et al. 2000), allow the systematic functional evaluation of neural development under highly reproducible conditions. HESC is a powerful tool for the study of human development and disease and for applications in regenerative medicine. The use of HESCs differentiated toward CNS lineages has been of particular interest given the lack of effective therapies for many neurodegenerative and neuropsychiatric disorders and the availability of protocols for efficiently directing neuronal specification in vitro.

Interestingly, the early stages of neural development from HESCs correspond to early anatomical brain development (Zhang 2006), which facilitates the mechanistic studies of human neurodevelopment. Neural progenitor cells can be efficiently generated from HESCs and patterned to regionalized neural progenitors (Yan et al. 2005; Zhang et al. 2001). To explore the neuronal development into subtypes, multiple methods have been established to direct HESCs differentiation into motor neurons, dopamine neurons, and other subtypes (Perrier et al. 2004; Li et al. 2005; Maroof et al. 2013). Investigators have also developed methods for direct conversion of HESCs into induced functional neurons (Zhang et al. 2013), which can be obtained with nearly 100 % yield and purity in less than 2 weeks by forced expression of a single transcription factor-NGN2. Surprisingly, biologists have successfully developed a human pluripotent stem cell-derived cerebral organoids, which develop various discrete brain regions and recapitulate features of human cortical development (Lancaster et al. 2013). This three-dimensional organoid culture system provides an in vitro model to recapitulate human brain development and disease. Taken together, this cellular model of neural differentiation from HESCs forms a foundation to explore functional neuronal development and decipher how these physiological traits are regulated.

8.2.2 HESC Neural Differentiation and Disease Models

Given the lack of alternative sources, a major effort has been directed toward the development of differentiation protocols that convert pluripotent stem cells into neurons to allow examination of healthy human neurons and of neurons derived from patients with a variety of neurological diseases. The generation of HESCs and induced human pluripotent stem cells (iHPSCs) and their in vitro differentiation into potentially any desired cell type hold great promise and may revolutionize the study of human disease.

In this approach, fibroblasts from patients with poorly understood neurodevelopmental diseases are converted into iHPSCs that are then differentiated into neurons to study the pathogenesis of these diseases. Elegant studies have described differentiation protocols that produce distinct types of neurons in vitro. Early studies using iHPSCs have been primarily generated toward neurodegenerative disorders, which are known to affect specific neuron types such as midbrain dopamine neurons in Parkinson's disease or motor neurons in amyotrophic lateral sclerosis and spinal muscular atrophy (Kriks et al. 2011; Dimos et al. 2008; Ebert et al. 2009). Moreover, there has been considerable progress in establishing protocols for the derivation of human cortical projection neurons from HESCs and iHPSCs (Maroof et al. 2013; Espuny-Camacho et al. 2013; Shi et al. 2012; Marchetto and Cage 2012). Some studies also tried to model diseases, such as Alzheimer's disease and Rett syndrome, with iHPSCs to test the physiological characteristics and drug responses (Marchetto and Gage 2012; Kondo et al. 2013; Zhou and Tripathi 2012; Marchetto et al. 2010). More importantly, scientists have performed genetic correction of human-induced pluripotent stem cells from patients with neurodevelopmental diseases, such as spinal muscular atrophy (Corti et al. 2012; Maetzel et al. 2014).

Overall, these studies suggest that derivation of neurons from human pluripotent stem cells may allow scientists to examine specific subtypes of neurons, to generate human neurons for regenerative medicine, and to investigate changes in human neurons in neuropsychiatric disorders (Han et al. 2011). In addition, challenges will be the development of a variety of tools that most sensitively and reliably measure functional phenotypes that are relevant to disease, finally leading to the marriage between basic biology and translational medicine.

8.3 Lessons from HESCs for Human Neurodevelopment

8.3.1 The Superiority of HESC Neural Differentiation to Study Human Neurodevelopment

Current knowledge about cellular phenotypes in neurodevelopment and neurodegenerative diseases in humans was gathered from studies in postmortem brain tissues, which often represent the end-stage of the disease and are not always a fair representation of how the disease developed. Therefore, the use of HESCs and iHPSCs to model neurodegenerative and neurodevelopmental diseases has been broadly performed to observe the pathological phenotypes. Studying the in vitro phenotypic consequences of cell disease model can help to identify a molecular mechanism responsible for subtle alterations in the nervous system, which can be further checked in human patient tissues. A future challenge for neurodevelopmental disorders is the contribution of genetic background and environmental clues. Novel gene-targeting techniques in human pluripotent stem cells such as homologous recombination, zinc-finger nucleases, and CRISPR-Cas9 (Ran et al. 2013; Urnov et al. 2005) may help to eliminate background noise and individual variability. Effective gene targeting in HESCs could disrupt a specific disease-related gene, and the resulting neuronal behavior could be compared with the patient's neuron containing the mutation.

Once a consistent abnormal disease-related phenotype is identified, screening platforms can be developed to test compounds (proteins, small molecules, and small hairpin RNAs) that revert or protect the cellular phenotype. After rigorous testing, therapeutic compounds will emerge from the screenings that could potentially benefit a large cohort of patients. In the future, the direct conversion of astrocytes to motor neurons in spinal cord trauma patients or other neurodegenerative diseases in vivo might be an effective way for therapy of these diseases. Besides, the recapitulation of all stages of neural development from iHPSCs is a powerful tool to depict the exact moment of the disease onset and to optimize therapeutical interventions.

8.3.2 *The Gap Between Human Pluripotent Stem Cell Neural Differentiation and Human Neurodevelopment*

Scientists are now using the powerful iHPSC technology to investigate early stages of human development and to model diseases, providing an innovative way to understand disease pathology, whereas there is a gap between HPSC neural differentiation and human neurodevelopment.

The available lines of HESCs are notoriously variable with regard to epigenetic marks, expression profile, and differentiation propensity, while recent reports suggest that significant intrinsic variability remains in the generated iHPSC lines (Pick et al. 2009). The abnormal expression of imprinted genes in a significant number of iHPSC lines was identified. The genetic and epigenetic differences have been generally attributed to the introduction of reprogramming factors and donor gene expression (Ghosh et al. 2010). Unexpectedly, the expression profile analysis of integration-free iHPSC has shown a different expression signature with the original population and standard HESCs (Marchetto et al. 2009). Furthermore, the neuronal differentiation competence of different iHPSC lines was highly variable when compared with HESC differentiation (Hu et al. 2010). Determining the variability levels between different cell lines will help to screen a suitable cell model to elucidate more robust phenotypes on cells derived from diseased iHPSCs. In view of the variability between different iHPSC lines in undifferentiated and differentiated states, it is hard to assign the best controls for diseased iHPSC lines. Recently, it has been reported that pluripotent stem cells could be induced from mouse somatic cells by small-molecule compounds (Hou et al. 2013). If it could be achieved in iHPSCs, it may enormously narrow the gap between human pluripotent stem cells and human neurodevelopment and diseases. Along with the scientific progress, any obstacles between in vitro cell model and in vivo study could be solved, and a large cohort of patients could potentially benefit from the scientists' efforts.

8.4 Summary

Here, we have reviewed human brain growth and structural development in a comparative framework. Humans have a unique developmental phase directly after birth that contributes to the typical globular shape of the human brain. It is widely accepted that ontogenetic structural changes of the brain are tightly interrelated with cognitive and behavioral development. We have also reviewed the recent progress on the human brain transcriptome, which provides a huge set of information about human neurodevelopment. Future work should attempt to more tightly integrate our knowledge about brain development on structural development, cognitive development, and transcriptome in humans.

HESCs and iHPSCs, which could be efficiently differentiated into human neural progenitor cells or mature neurons, have been widely used to model human neurodevelopment and neuropsychiatric disorders. Although there are some distance between in vitro cultured neural cells and in vivo neural tissues, the great progress in this field will reduce this gap and bring the in vitro and in vivo world together.

References

Corti S, et al. Genetic correction of human induced pluripotent stem cells from patients with spinal muscular atrophy. Sci Transl Med. 2012;4:165ra162.
Dimos JT, et al. Induced pluripotent stem cells generated from patients with ALS can be differentiated into motor neurons. Science. 2008;321(5893):1218–21.
Dobbing J, Sands J. Quantitative growth and development of human brain. Arch Dis Child. 1973;48(10):757–67.
Ebert AD, et al. Induced pluripotent stem cells from a spinal muscular atrophy patient. Nature. 2009;457(7227):277–80.
Epstein HT. Stages in human brain development. Brain Res. 1986;395(1):114–9.
Espuny-Camacho I, et al. Pyramidal neurons derived from human pluripotent stem cells integrate efficiently into mouse brain circuits in vivo. Neuron. 2013;77(3):440–56.
Ghosh Z, et al. Persistent donor cell gene expression among human induced pluripotent stem cells contributes to differences with human embryonic stem cells. PLoS ONE. 2010;5(2):e8975.
Giedd JN, et al. Quantitative magnetic resonance imaging of human brain development: ages 4–18. Cereb Cortex. 1996;6(4):551–60.
Gogtay N, et al. Dynamic mapping of human cortical development during childhood through early adulthood. Proc Natl Acad Sci U S A. 2004;101(21):8174–9.
Han SS, Williams LA, Eggan KC. Constructing and deconstructing stem cell models of neurological disease. Neuron. 2011;70(4):626–44.
Hawrylycz MJ, et al. An anatomically comprehensive atlas of the adult human brain transcriptome. Nature. 2012;489(7416):391–9.
Hou P, et al. Pluripotent stem cells induced from mouse somatic cells by small-molecule compounds. Science. 2013;341(6146):651–4.
Hu BY, et al. Neural differentiation of human induced pluripotent stem cells follows developmental principles but with variable potency. Proc Natl Acad Sci U S A. 2010;107(9):4335–40.
Kang HJ, et al. Spatio-temporal transcriptome of the human brain. Nature. 2011;478(7370):483–9.
Kandel ER, et al. Principles of neural science. 5th ed. New York: McGraw-Hill; 2012.
Kondo T, et al. Modeling Alzheimer's disease with iPSCs reveals stress phenotypes associated with intracellular Abeta and differential drug responsiveness. Cell Stem Cell. 2013;12(4):487–96.
Kriks S, et al. Dopamine neurons derived from human ES cells efficiently engraft in animal models of Parkinson's disease. Nature. 2011;480(7378):547–51.
Lancaster MA, et al. Cerebral organoids model human brain development and microcephaly. Nature. 2013;501(7467):373–9.
Li XJ, et al. Specification of motoneurons from human embryonic stem cells. Nat Biotechnol. 2005;23(2):215–21.
Maetzel D, et al. Genetic and chemical correction of cholesterol accumulation and impaired autophagy in hepatic and neural cells derived from Niemann-Pick type C patient-specific iPS cells. Stem Cell Rep. 2014;2(6):866–80.

Marchetto MC, Gage FH. Modeling brain disease in a dish: really? Cell Stem Cell. 2012;10(6):642–5.
Marchetto MC, et al. Transcriptional signature and memory retention of human-induced pluripotent stem cells. PLoS ONE. 2009;4(9):e7076.
Marchetto MC, et al. A model for neural development and treatment of Rett syndrome using human induced pluripotent stem cells. Cell. 2010;143(4):527–39.
Maroof AM, et al. Directed differentiation and functional maturation of cortical interneurons from human embryonic stem cells. Cell Stem Cell. 2013;12(5):559–72.
Moore KL, et al. The developing human. 3rd ed. Philadelphia: Saunders; 1982.
Perrier AL, et al. Derivation of midbrain dopamine neurons from human embryonic stem cells. Proc Natl Acad Sci U S A. 2004;101(34):12543–8.
Pfefferbaum A, et al. A quantitative magnetic resonance imaging study of changes in brain morphology from infancy to late adulthood. Arch Neurol. 1994;51(9):874–87.
Pick M, et al. Clone- and gene-specific aberrations of parental imprinting in human induced pluripotent stem cells. Stem Cells. 2009;27(11):2686–90.
Ran FA, et al. Double nicking by RNA-guided CRISPR Cas9 for enhanced genome editing specificity. Cell. 2013;154(6):1380–9.
Reiss AL, et al. Brain development, gender and IQ in children volumetric imaging study. Brain. 1996;119(Pt 5):1763–74.
Reubinoff BE, et al. Embryonic stem cell lines from human blastocysts: somatic differentiation in vitro. Nat Biotechnol. 2000;18(4):399–404.
Shi Y, et al. Human cerebral cortex development from pluripotent stem cells to functional excitatory synapses. Nat Neurosci. 2012;15(3):477–486, S1.
Thomson JA, et al. Embryonic stem cell lines derived from human blastocysts. Science. 1998;282(5391):1145–7.
Urnov FD, et al. Highly efficient endogenous human gene correction using designed zinc-finger nucleases. Nature. 2005;435(7042):646–51.
Yan Y, et al. Directed differentiation of dopaminergic neuronal subtypes from human embryonic stem cells. Stem Cells. 2005;23(6):781–90.
Zhang SC. Neural subtype specification from embryonic stem cells. Brain Pathol. 2006;16(2):132–42.
Zhang SC, et al. In vitro differentiation of transplantable neural precursors from human embryonic stem cells. Nat Biotechnol. 2001;19(12):1129–33.
Zhang Y, et al. Rapid single-step induction of functional neurons from human pluripotent stem cells. Neuron. 2013;78(5):785–98.
Zhou Q, Tripathi P. How to remake a fibroblast into a neural stem cell. Cell Stem Cell. 2012;10(4):347–8.

Chapter 9
Molecular Imaging: The Key to Advancing Stem Cell Therapy

Ningning He and Zongjin Li

Abstract With their capacity for self-renewal and pluripotency, stem cells are a promising option for the treatment of degeneration diseases or injury, such as type I diabetes mellitus, Parkinson's disease, Huntington's disease, myocardial infraction, muscle damage, and many others. However, several issues should be addressed before clinical translation, such as (1) What is the optimal cell type, cell dosage, and delivery route? (2) When is the ideal timing for cell transplantation? (3) How long do these cells survive and do they integrate, proliferate, and differentiate? and (4) Can these physiologic processes be monitored in vivo? Molecular imaging offers the potential for monitoring the location, engraftment, and survival of the transplanted cell. In this chapter, we will describe the major imaging strategies and their applications for stem cell therapy. Furthermore, the perspective on the future role of molecular imaging in defining safety and efficacy for clinical implementation of stem cell therapies will be discussed.

Keywords Molecular imaging · Stem cell · Therapy · Reporter gene · Labeling

9.1 Introduction

Stem cell therapy has shown great promise for regenerative repair of injured or diseased tissues. However, several issues in stem cell transplantation such as monitoring the migration, differentiation, and distribution of transplanted cells should be addressed (Himmelreich and Dresselaers 2009). Therefore, the ability to track cell migration, cell homing, and cellular fate noninvasively in vivo is important

---

N. He · Z. Li (✉)
Department of Pathophysiology, Nankai University School of Medicine, 94 Weijin Road, Tianjin 300071, People's Republic of China
e-mail: zongjinli@nankai.edu.cn

R.C. Zhao (ed.), *Stem Cells: Basics and Clinical Translation*, Translational Medicine Research 1, DOI 10.1007/978-94-017-7273-0_9

in understanding the roles of transplanted cells in preclinical trials. Consequently, novel molecular imaging techniques are necessary for exploring the behaviors and the feasibility of cell transplantation therapy of stem cells (Li et al. 2009; Tong et al. 2013).

Along with the rapid development of sensitive, noninvasive technologies, several molecular imaging approaches have been implicated to track the fate of stem cells in vivo (Li et al. 2008; Gu et al. 2012). Stem cell-based therapies may be improved by the application of imaging technologies that allow investigators to track the location, engraftment, and survival of the transplanted cells. Over the past decade, investigations have produced promising clinical data regarding stem cell therapy, but design of trials and evaluation of treatments stand to be improved by emerging insight from imaging studies (Ransohoff and Wu 2012).

9.2 Stem Cell Therapy

A wide variety of stem or progenitor cells, including adult bone marrow stem cells, endothelial progenitor cells, mesenchymal stem cells (MSCs), resident cardiac stem cells, and embryonic stem cells, have been shown to have positive effects in preclinical studies and therefore hold promise for treating and curing degenerative diseases. There is increasing evidence that adult human tissues harbor stem cells and progenitor cells that can be used for stem cell therapy. Bone marrow contains pluripotent hematopoietic stem cells (HSCs) that continuously produce blood cells, which have the ability to self-renew and differentiate into progenitor cells, leading to all hematopoietic lineages. HSC-based gene therapy has been an established cellular therapy for patients with hematological diseases, and it has shown great potential for the emerging field of regenerative medicine (Lee et al. 2008). In addition to HSCs, bone marrow also contains progenitor cells that can differentiate into multiple mesenchymal phenotypes (Haynesworth et al. 1992; Pittenger et al. 1999). It has been proposed that the mesenchymal progenitors of bone marrow are part of a mesenchymal lineage analogous to that described for hematopoiesis and that the multipotential progenitors in bone marrow may be mesenchymal stem cells (MSCs) or multipotent bone marrow stromal cells. Moreover, MSCs have been derived from a variety of other tissues or organs, including adipose, placenta, and umbilical cord. Embryonic stem (ES) cells are capable of differentiation into any somatic cell type of the human body and have the potential for unlimited self-renewal (Thomson 1998). Furthermore, recent progresses on induced pluripotent stem cells (iPSCs) provide alternative source for stem cell therapy (Li et al. 2011). However, stem cell biology remains incompletely understood despite significant advances in the field. Inefficient stem cell differentiation, difficulty in verifying successful delivery to the target organ, and problems with engraftment all hamper the transition from laboratory animal studies to human clinical trials (Gu et al. 2012). To realize the full therapeutic potential of stem cell technology, it will be necessary to develop novel assessments that can be used easily to determine the exact cellular state of the transplanted cells, which can be achieved by the application of molecular imaging.

9.3 Molecular Imaging

Recently, noninvasive, imaging-based monitoring methods have been developed to track stem cell transplants by labeling the cells. The goal is to track the disposition, distribution, and migration of stem cells once introduced into the model organism (Lee et al. 2008). Advanced imaging technology can provide anatomic and functional assessment (e.g., conventional anatomic and functional imaging), as well as visualization of biological processes at the cellular and molecular level (e.g., molecular imaging). Both disciplines use various imaging technologies such as radionuclide, magnetic resonance, optical, and computed tomographic. How to select the most effective imaging strategy requires a determination of whether the imaging system can meet the necessary requirements for spatial and temporal resolution, sensitivity, and penetration depth for visualization of the imaging target.

Although conventional anatomic and functional imaging can provide an overview for cell delivery and an assessment of the effects of stem cell therapy, molecular imaging can be used to track stem cells in vivo and to study their potential mechanistic benefits. In molecular imaging, imaging probes are used to target the biological process of interest (Nguyen et al. 2011).

Moreover, the use of stem cells has been hampered by lack of understanding of their mechanisms of action and poor retention rates after delivery (Rodriguez-Porcel et al. 2012; Das et al. 2009). Recent research on stem cell therapy illustrated the potential of molecular imaging in monitoring not only the survival of progenitor cells but also their effect on the target tissue (Ale et al. 2013). The main strength of molecular imaging is the capacity of not only detecting organ dysfunction, but also providing insight on the mechanism that led to such dysfunction, opening the door for the understanding of the disease at the molecular level (Chen and Wu 2011). However, there is no single imaging modality that can provide all the answers, and the choice of imaging modalities should be selected depending on the specific question or need in mind.

9.4 Approaches for Noninvasive in Vivo Imaging of Stem Cell Therapy

Stem cell therapy is an exciting area of research that promises future treatment of many diseases (Chiu 2003). Molecular imaging techniques provide a means for noninvasive, repeated, and quantitative tracking of stem cell implant or transplant. However, to fully understand the beneficial effects of stem cell therapy, investigators must be able to track the biology and physiology of transplanted cells in living subjects over time (Li et al. 2007). The development of sensitive, noninvasive technologies to monitor this fundamental engraftment will greatly aid clinical implementation of cell therapy (Li et al. 2008).

9.4.1 Methodologies for Labeling Stem Cells

There are two main classes of molecular imaging techniques: direct stem cell labeling and reporter gene imaging. The former employs contrast agents such as magnetic particles, luminescent nanoparticles, or radionuclides to directly label the cell, whereas the latter genetically alters the cell to transcribe and translate a reporter protein (Fig. 9.1) (Gu et al. 2012). Direct labeling allows high spatial resolution (MRI) and high sensitivity (SPECT or PET) imaging and is best used to track the in vivo localization of cells in the hours to days following delivery. However, a common drawback these methods share is their reliance on physical labels. Superparamagnetic iron oxide (SPIO) and radionuclide probes are diluted with cell division and are not capable of tracking cell proliferation, especially when cells misbehave as in the case of ES cell-derived teratoma formation (Li et al. 2008). SPIO agents further suffer from the unique problem of being taken up by macrophages

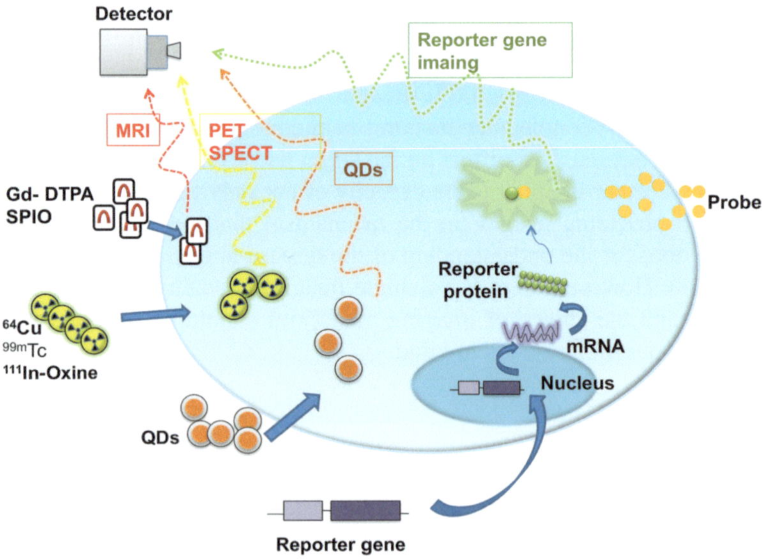

Fig. 9.1 Conceptual basis for direct imaging and indirect imaging. *Left*, Schema of direct imaging shows MRI or PET tracers; *Right*, The indirect reporter gene imaging that needs exogenous substrate, reporter protein, and reporter gene. Abbreviations: Gd-DTPA, gadolinium-diethylenetriamine pentaacetic acid; SPIO, superparamagnetic iron oxide; ^{99m}Tc, ^{99m}Tc-hexamethylpropylene amine oxime; ^{111}In-oxine, ^{111}In-oxyquinoline. (Reproduced from Tong et al. 2013)

after donor cell death (which may continue to produce signal even after cell death) and hence cannot be used to accurately monitor long-term cell survival and behavior (Bulte and Kraitchman 2004). By comparison, SPECT or PET tracers lose signal due to radioisotope decay. A typical PET radioisotope such as F-18 has a half-life of only 110 min and can only be used to image cells in the hours immediately following cell delivery (Hofmann et al. 2005). Although reporter gene imaging requires genomic manipulation and poses potential safety issues, it is the preferred labeling strategy because signal generation is dependent on cell viability. In this type of imaging, a gene coding for the synthesis of a detectable protein is introduced into a target cell line or tissue via viral or non-viral vectors. Reporter genes are only expressed by live cells, and the signal is propagated by daughter cells (Li et al. 2009). Therefore, in reporter gene imaging, stem cells can be genetically engineered to express various reporter genes before transplantation. The reporter genes can be detected by sensitive imaging devices such as the optical charge-coupled device (CCD), single photon emission computed tomography (SPECT), or positron emission tomography (PET) (Cao et al. 2006). Magnetic resonance (MR) imaging of stem cells is also an emerging application for monitoring cell engraftment (Li et al. 2008). Examples of commonly used reporter genes include firefly luciferase (Fluc) and herpes simplex virus thymidine kinase (HSV-tk), which can be detected by bioluminescence imaging (BLI) and PET, respectively (Cao et al. 2006).

9.4.2 Radionuclide Imaging

Radionuclide imaging is the sole direct labeling technique used thus far in human studies, involving both autologous bone marrow-derived stem cells and peripheral hematopoietic stem cells (Schachinger et al. 2008; Dedobbeleer et al. 2009). There are two main techniques for radionuclide imaging: positron emission tomography (PET) and single photon emission computed tomography (SPECT). Both PET and SPECT are highly sensitive tools for investigating in vivo bio-distribution of cells. Radionuclide imaging can be used in two ways: direct labeling and reporter gene labeling. Direct labeling requires isotopes such as ^{18}O, ^{13}N, ^{11}C, and ^{18}F, and genetic labeling can be performed with reporter genes such as herpes simplex virus thymidine kinase (HSV-tk) (Wu et al. 2003) or human sodium iodide symporter (hNIS) (Terrovitis et al. 2008).

9.4.3 Magnetic Resonance Imaging

While less sensitive than PET, SPECT, and optical imaging (BLI and fluorescence), MRI has higher spatial resolution and is useful for real-time tracking and monitoring cell survival. MRI provides valuable simultaneous anatomical and physiological data. Novel molecular agents are being developed to increase the sensitivity of this imaging modality. The ideal MRI contrast agent should be able to increase signal intensity without decreasing target cell activity, detect a small number of cells, and be retained by cells long enough and stably enough to permit longitudinal imaging (Lau et al. 2010). As with radionuclide imaging, MRI can employ both physical and genetic (such as β-galactosidase and the transferrin receptor) labels. The two most widely used physical labels are gadolinium chelates and superparamagnetic iron oxide (SPIO) nanoparticles. MRI protocols require that cells should be labeled prior to transplantation. The first MRI reporter gene encoded a creatine kinase, an enzyme that can generate phosphocreatine detectable by ^{31}P magnetic resonance spectroscopy (MRS) in the rat liver (Koretsky et al. 1990).

A number of studies have shown the feasibility of longitudinal noninvasive monitoring of transplanted cells in preclinical models using MRI. This approach could potentially be translated into clinical practice for evaluating stem cell survival and for monitoring therapeutic intervention during tissue rejection. MRI is well suited for stem cell tracking because it can provide both whole-body and detailed information on host organs with near microscopic anatomical resolution and excellent soft-tissue contrast (Mathiasen and Kastrup 2013).

9.4.4 Bioluminescence Imaging

Bioluminescence imaging (BLI) requires incorporation of an optical reporter gene, most commonly the firefly luciferase gene (Fluc, from the firefly Photinus pyralis). Photons are emitted when the optical probe, D-luciferin, administered intraperitoneally or intravenously, is oxidized by the firefly luciferase enzyme, which is encoded by the Fluc gene. Unlike fluorescence imaging, BLI requires no excitatory light source. The amount of background auto-fluorescence in vivo is low, so images are highly sensitive (Fig. 9.2). Other types of bioluminescence reporter gene and reporter probe [e.g., Renilla luciferase from sea pansy (Renilla reniformis) and coelenterazine] are also available (Bhaumik and Gambhir 2002; Leng et al. 2014). BLI is a highly sensitive method for tracking cell survival, especially compared to physical labels used in MR or to short-lived PET radionuclide tracers.

BLI has been used extensively in small animal studies to track and compare the survival, engraftment, and migration of a range of cell populations, including bone marrow stem cells, skeletal myoblasts, and MSCs in the infarcted myocardium (van der Bogt et al. 2008). In a hindlimb ischemia model, mouse embryonic stem cell-derived endothelial cells preferentially localized to the ischemic hindlimb and incorporated into the microvasculature, and their localization (as assessed by BLI) was associated with improvement in limb perfusion (as assessed by laser Doppler) (Huang et al. 2010). Furthermore, BLI has been used to evaluate various immunosuppressive regimens for inducing long-term tolerance to xenogenic and allogeneic ES cells and induced pluripotent stem cell (iPSC) transplantation (Pearl et al. 2011). Finally, insertion of the Fluc gene downstream of a tissue-specific promoter can lead to reporter activity upon cellular differentiation (Kammili et al. 2010).

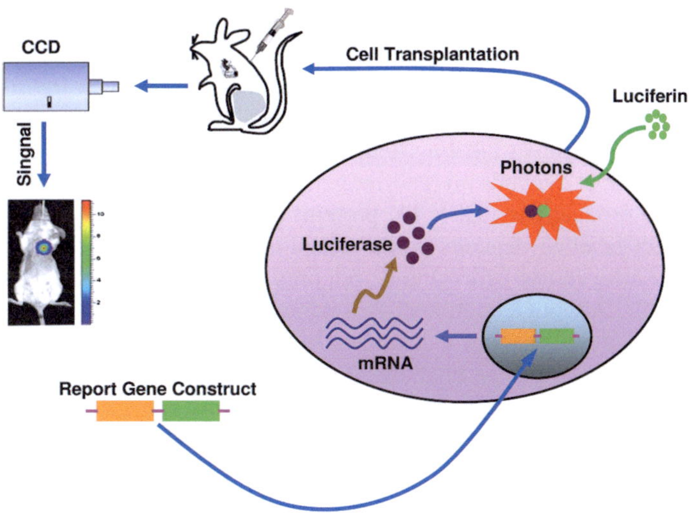

Fig. 9.2 Conceptual basis of tracking the survival of transplanted stem cells in living animals with bioluminescence. Stem cells are first stably transduced with a reporter gene such as firefly luciferase. Then, cells are transplanted into animal model. In the presence of reporter probe such as D-luciferin, the cells can emit photons that can be detected by a highly sensitive charge-coupled device (CCD) camera. Cell survival and migration can then be tracked longitudinally within the same animal. (Reproduced from Li et al. 2009)

9.4.5 Fluorescence Imaging

Fluorescent imaging has been used for decades, first of all excitatory visible light is needed, and the resultant shifts in wavelength are then recorded. Fluorescent reporter proteins include green fluorescent protein (GFP) which derived from the jellyfish Aequorea victoria, and excitatory violet light leads to GFP emission of green (509 nm) light (Massoud 2003). To date, there are many fluorescent proteins with point mutations that make different fluorescent properties (Kang and Chung 2008).

9.5 The Benefits and Drawbacks of Major Imaging Modalities

9.5.1 Radionuclide Imaging

The advantages of radioscintigraphic techniques include their picomolar sensitivity, good tissue penetration, and translation to clinical applications (Welling et al. 2011). Preclinical studies of PET or SPECT reporter genes have been used to track stem cell survival longitudinally. Stem cells are either transiently or stably transduced with a reporter gene (e.g., HSV-tk or hNIS) in vitro prior to transplantation. This technique has been used to track MSCs stably transduced with HSV-tk (via lentiviral vector) (Gyongyosi et al. 2008) or MSCs transiently transduced with HSV-tk (via adenoviral vector) (Willmann et al. 2009). The major advantage of reporter genes over direct labeling is that the signal directly reflects cell viability, because the readout depends on reporter gene expression and the interaction of the reporter gene product (i.e., HSV-TK protein) and the reporter probe. In contrast, because direct labeling depends on the decay of the radioisotope, positive results do not necessarily equate cell viability, but only denote probe presence (Ransohoff and Wu 2010). Although radionuclide imaging offer great sensitivity, there are several disadvantages to both techniques, including the leakage of radionuclides into nontarget cells (Stojanov et al. 2012), limited time window for imaging due to half-life decay, lower spatial resolution as compared to MRI, and the emission of ionizing radiation that may impair stem cell proliferation and survival.

9.5.2 Magnetic Resonance Imaging

The advantages of MRI are high resolution, good contrast, and label persistence. Though MRI has a high spatial resolution (μm), the sensitivity of MRI is low, in the micromolar range with gadolinium chelates and in the millimolar range with iodine-based contrast agents. This is because the percentage of dipoles that align correctly in the magnetic pulse is very low, so large amounts of contrast must be used. In studies of small and large animals, the lowest number of cells that could be detected was ~100, 000. Another problem with MRI is that physical labeling with SPIO does not distinguish viable from non-viable cells and cannot provide information about cell proliferation, since the initial number of iron particles used to label the parent cell can remain after cell death (Fig. 9.3) (Li et al. 2008). Moreover, MRI is not widely used in patients with implantable devices such as pacemakers or implantable cardioverter-defibrillators because of the potential interference between the device and the magnetic field (Sierra and Machado 2008). Finally, while genetic modification to express MR reporter genes is highly useful and sensitive to both cell proliferation and viability, hurdles involving detection sensitivity and the effects of modification on cell characteristics have yet to be resolved (Zhang and Wu 2007).

9.5.3 Bioluminescence Imaging

Though BLI is a highly sensitive and versatile imaging tool, it has several disadvantages. First, light transmission through an opaque animal is dependent on tissue type and depth. In addition, photon scatter and signal loss are nonlinear as a function of depth (Ntziachristos et al. 2005) and BLI only can be used for small

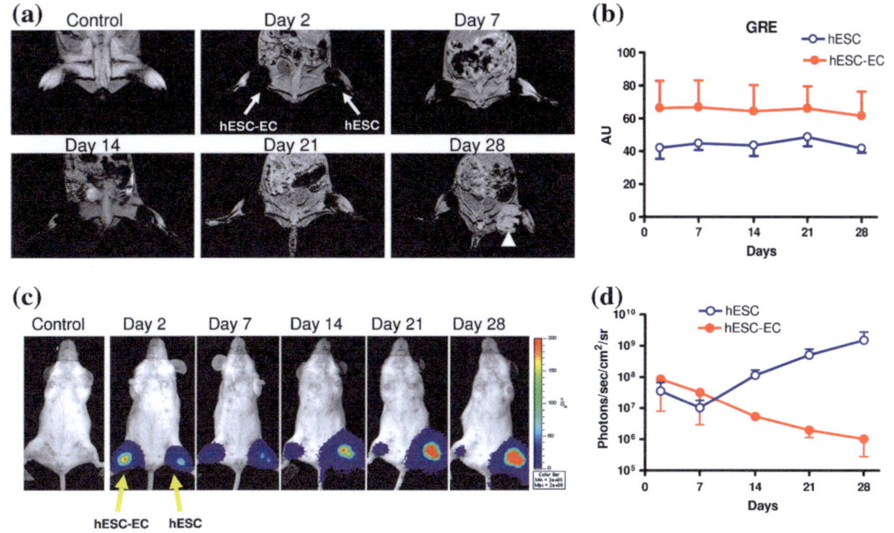

Fig. 9.3 Direct comparison of reporter gene imaging (genetic labeling) versus iron particle imaging (physical labeling) for tracking stem cells. **a** These predifferentiated human ESC-derived endothelial cells (hESC-ECs) and undifferentiated ESCs were SPIO-labeled (with Feridex) and then injected into mouse hindlimbs. MR images of one representative animal show the cells at days 2, 7, 14, 21, and 28. **b** MR does not show survival differences between the two groups, as the signal is steady throughout all imaging time points, with a higher signal in the hESC-EC group through day 28. **c** These same ESCs were transduced with the human ubiquitin promoter driving firefly luciferase (Fluc) and enhanced green fluorescence protein (eGFP). These cells were then cultured as in **a** prior to transplantation into the hindlimb of a mouse. **d** BLI showed divergent survival profiles for the two groups, with proliferation of ESC and acute donor cell death of predifferentiated hESC-ECs. This study demonstrated that MRI provided detailed information on the anatomical location of cells, but not on cell viability. Reporter gene imaging is a better indicator of cell viability and proliferation. (Reproduced from Li et al. 2008)

animal. A final problem is one shared by all tracking methods involving genetic manipulation. Reporter gene expression such as Fluc can decrease over time due to epigenetic silencing, especially when a viral promoter (e.g., cytomegalovirus promoter) is used (Krishnan et al. 2006).

9.5.4 Fluorescence Imaging

Fluorescence imaging has a high background due to significant auto-fluorescence and is also limited by scattered and shallow tissue penetration (Sutton et al. 2008). Another drawback of fluorescence imaging is its leakage to neighboring cells, a problem shared by SPIO-based MRI. Leakage to nearby cells following cell death could lead to inaccurate measurements and false positives, since signal would no longer correlate with cell viability (Ransohoff and Wu 2010).

9.6 Molecular Imaging of Stem Cell Therapy

Stem cell-based therapies can offer the treatment of many diseases and injuries, including cardiovascular disease, stroke, peripheral arterial disease, neurodegenerative diseases, and cancer. However, stem cell biology remains incompletely understood in spite of significant advances in this field. Low efficiency of stem cell differentiation, difficulty in detecting successful delivery to the target organ, and problems with engraftment all prevent the transition from laboratory animal studies to human clinical trials. There are still many unsolved questions about the ideal cell type for different patient populations in addition to dose, timing, and optimal delivery route. To take into consideration these factors and to optimize the cell therapies, it is important to perform noninvasively imaging technology to monitor the presence of transplanted cells and the metabolic biology of transplanted cells over time and to integrate this with the clinical effects seen in patients (Mathiasen and Kastrup 2013). Over the past decade, investigations have produced promising clinical data regarding cell therapy, but design of trials and evaluation of treatments stand to be improved by emerging insight from imaging studies (Ransohoff and Wu 2012).

9.6.1 Imaging Stem Cell Therapy for the Treatment of Cardiovascular Disease

Coronary artery disease (CAD) is the most common cause of death in Western countries (Yusuf et al. 2001a, b). Despite some advances in treatment of CAD, a large number of patients cannot be treated successfully. Several elegant studies demonstrate that endothelial progenitor cells harvested from the bone marrow (Kocher et al. 2001; Takahashi et al. 1999) or circulating peripheral blood (Vaughan and O'Brien 2012) can contribute to angiogenesis and functional regeneration of ischemic or infarcted myocardium (Li et al. 2007).

Noninvasive imaging could provide a comprehensive overview of cell therapy, ranging from the basic biology of cells to preclinical and clinical translation. Stem cell transplantation has been studied widely for the past decade as a novel therapy to reverse or minimize myocardial injury, with the goal of improving cardiac function (Cheng et al. 2012; Khan et al. 2012; Welt et al. 2013). Moreover, combined different stem cells may enhance therapeutic effect; for instance, human cardiac stem cells and MSCs altogether accelerate cardiac function improvement after myocardial infarction (Williams et al. 2013).

Direct labeling iron oxide nanoparticles, a well-defined agent, show an excellent safety performance with little impact on cellular function (Arbab et al. 2005; Kostura et al. 2004). Labeling of cells with fluorinated compound is also applied in stem/progenitor cells (Partlow et al. 2007). Reporter gene imaging of stem cells is another form used in investigating cardiac disease therapy, which can overcome some of the limitations caused by direct cell labeling. For instance, firefly

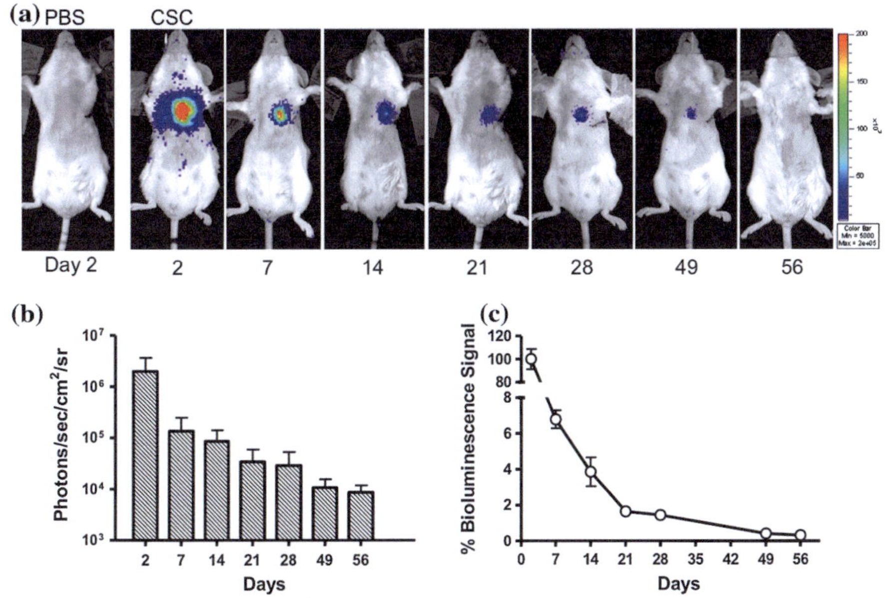

Fig. 9.4 Reporter gene imaging of CSC fate *after* transplantation. **a** A representative animal injected with 5×10^5 cardiac stem cells (CSCs) shows significant bioluminescence activity at day 2, which decreases progressively over the following 8 weeks. **b** Detailed quantitative analysis of signals from all animals transplanted with CSCs. Signal activity is expressed as photons/s/cm^2/sr. **c** Estimation of percent donor cell survival plotted as % signal activity (normalized to day 2) over the 8-week period after transplantation. (Reproduced from Li et al. 2009)

luciferase reporter gene labeled cardiac resident stem cells engraftment can be noninvasively tracked for 8 weeks (Fig. 9.4). Furthermore, the key biological interactions between transplanted cells and their host environment can also be obtained by reporter gene imaging in living subjects. So imaging of these molecules provides a better understanding to the changes produced by transplantation of stem cells in the development of atherosclerotic vascular disease, such as imaging of integrin $\alpha_v\beta_3$, exposed phosphatidylserine, and glucose metabolism. Recent study revealed that reporter gene imaging allows assessment of changes in the oxidative status of MSCs after delivery to ischemic myocardium in a rat model of myocardial ischemia/reperfusion (Psaltis et al. 2013).

9.6.2 Imaging Stem Cell Therapy for the Treatment of Brain Injury

Brain injury causes brain damage in the fetus and newborn infants and represents a major cause of cerebral palsy, learning disability, cognitive impairment, and epilepsy (Vannucci 2000). Early imaging studies in patients with stroke (Chollet et al.

1991; Ward 2004) and microstimulation in experimental models of stroke (Nudo et al. 1996; Brown et al. 2007) reported that in response to ischemic injury, the brain undergoes limited compensatory changes in an effort to recover from structural and functional loss (Cramer and Chopp 2000).

Neural stem cell-based therapy offers the prospect to rescue damaged tissue, to replace lost cells, and to restore neurological function after brain injury (Daadi et al. 2010). However, they cannot differentiate into multiple functional neural cell types and the quantity of cells generated cannot meet the global demand for autologous cell transplantation therapy. Human ES cell-derived neuron stem cells are amenable to large-scale manufacturing under a quality assurance program and have a broad potential for cell differentiation (Daadi et al. 2009). To understand the behavioral and functional properties of the transplanted cells, it is necessary to perform noninvasive imaging modalities. Previous studies have reported superparamagnetic iron oxide (SPIO)-based MRI of grafted cells in the course of migration in stroke experimental models (Hoehn et al. 2002; Guzman et al. 2007). Moreover, reporter gene-based molecular imaging techniques such as bioluminescence imaging (BLI) have been used to efficiently track the survival of hES cell-derived neuron stem cells (Daadi et al. 2009, 2010).

9.6.3 Imaging Stem Cell Therapy for the Treatment of Cancers

Endothelial progenitor cells (EPCs)/MSCs have shown tropism toward primary tumors or metastases and are thus potential vehicles for targeting tumor therapy (Leng et al. 2014; Su et al. 2013). A research has reported that human embryonic stem cell-derived endothelial cells (hESC-ECs) can be used as cellular delivery vehicles for therapy of metastatic breast cancer (Su et al. 2013). And optical bioluminescence imaging of luciferase reporter genes has been used to monitor tumor growth and regression (Goldman et al. 2011; Sun et al. 2010), to visualize the kinetics of tumor cell clearance by chemotherapeutics, and to track gene expression (Luker et al. 2012; Griesenbach et al. 2011). With their capacity of tumor-specific tropism, mesenchymal stem cells (MSCs) have been considered to be attractive vehicles for delivering therapeutic agents toward tumor sites (Yong et al. 2009; Kidd et al. 2010). Human umbilical cord-derived MSCs (hUC-MSCs) armed with a triple fusion (TF) gene containing the herpes simplex virus truncated thymidine kinase (HSV-ttk), Renilla luciferase (Rluc), and red fluorescent protein (RFP) were injected into pre-established GFP-Fluc double fusion MDA-MB-231 breast cancer model. Bioluminescence imaging of Fluc and Rluc provided the real-time monitor of tumor cells and hUC-MSCs simultaneously. Near-infrared (NIR) imaging was employed for further demonstrating the effect of hUC-MSCs on tumor cells in vivo, and the results showed that hUC-MSCs could inhibit tumor angiogenesis and increased apoptosis to a certain degree. Those results indicate that molecular imaging is an invaluable tool in tracking cell delivery and tumor response to stem cell therapies as well as cellular and molecular processes in tumor (Leng et al. 2014).

9.7 Multimodalities Molecular Imaging of Stem Cell Therapy

Multimodality medical imaging takes advantage of the strengths of different imaging modalities to provide a more complete picture of the anatomy under investigation (Wang et al. 2013). The idea of combining imaging technologies moved to the mainstream with the advent of the first successful commercial fused instrument is

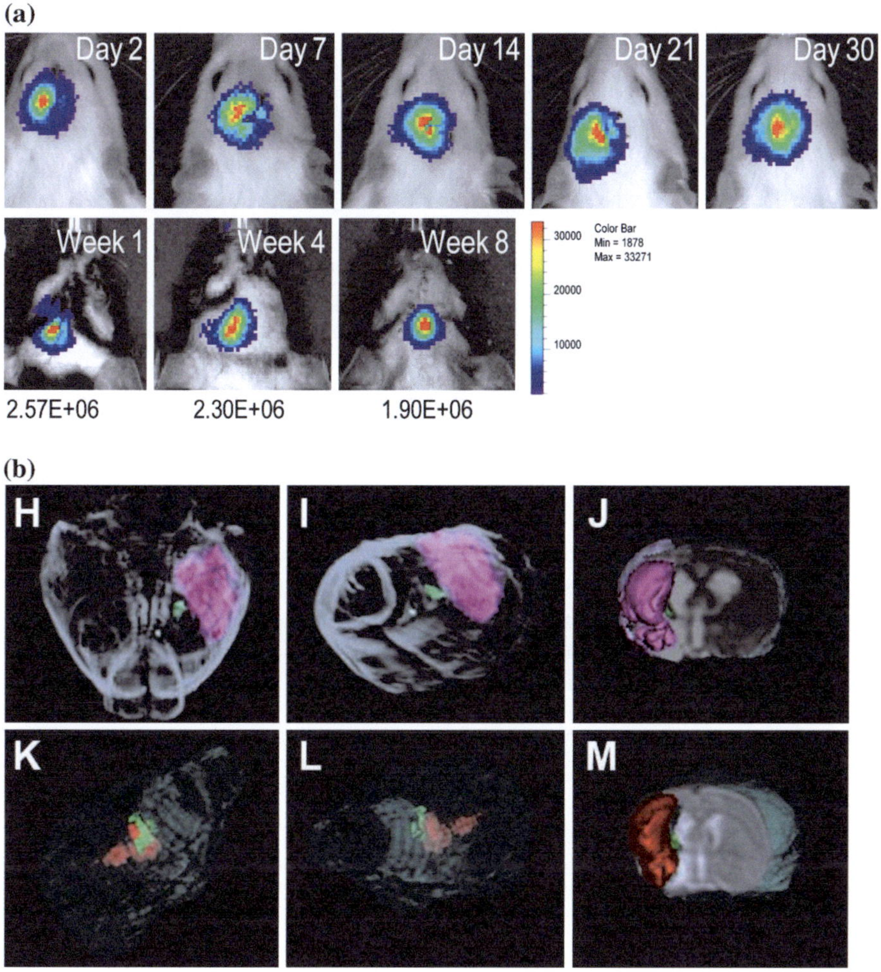

Fig. 9.5 MRI and firefly luciferase (Fluc) imaging analysis of the neural stem cell grafts in experimental stroke model. **a** Fluc activity in these animals shows a stable bioluminescence signal, which suggests the survival of the grafts and the non-proliferative property of the transplanted stem cells. **b** Three-dimensional surface rendering reconstruction of grafted rat brain from high-resolution T2-MRI illustrates the grafts (*green*) and stroke (*pink, red*) in a representative animal from the (**h**–**j**) low-dose and (**k**–**m**) intermediate-dose group. (Reproduced from Daadi et al. 2009)

PET-CT. PET-CT, an imaging technique that combines PET and an X-ray computed tomography (CT), gives chance to obtain better images having higher resolution by fusing both functional and anatomical images in the same imaging modality at the same time. Moreover, many other multimodality imaging technologies, including SPECT-CT and PET-MR, are commonly used in clinic. Real-time imaging of transplanted stem cells in stroke-damaged rat brain with bioluminescence imaging (BLI) and magnetic resonance imaging (MRI) revealed that cell survival can be tracked noninvasively by MRI and BLI for 2 months after transplantation and confirmed histologically (Daadi et al. 2009). However, reporter gene is a better marker for monitoring cell viability, whereas iron particle labeling is a better marker for high-resolution detection of cell location by MR (Fig. 9.5) (Li et al. 2008; Daadi et al. 2009).

9.8 Conclusion

A wide variety of stem or progenitor cells, including adult bone marrow stem cells, endothelial progenitor cells, MSCs, resident cardiac stem cells, and ES cells, have been shown to have positive effects in preclinical studies and therefore hold promise for treating and curing debilitating and deadly diseases. In addition to establishing the efficacy of stem cell therapy, there are some safety concerns that should be addressed in the routine clinical implementation of cell-based treatment, such as cell migration and tumorigenicity. Noninvasive molecular imaging could provide a comprehensive overview of cell therapy, which gives integral and comprehensive guidance ranging from the basic biology of cells to preclinical and clinical translation for the investigation of stem cell therapy. Although a variety of cell imaging techniques have been validated in preclinical models, translation of cell-tracking technologies into clinical settings will require a substantial amount of preparation and perseverance. With the efforts of researchers, molecular imaging is becoming a clinically feasible tool for the interrogation of the variability in cell survival, engraftment, and differentiation after stem cell transplantation.

Acknowledgments This work was partially supported by grants from the National Basic Research Program of China (2011CB964903) and National Natural Science Foundation of China (81371620).

References

Ale A, Siebenhaar F, Kosanke K, Aichler M, Radrich K, Heydrich S, Schiemann M, Bielicki I, Noel PB, Braren R, Maurer M, Walch AK, Rummeny EJ, Ntziachristos V, Wildgruber M. Cardioprotective C-kit(+) bone marrow cells attenuate apoptosis after acute myocardial infarction in mice—in-vivo assessment with fluorescence molecular imaging. Theranostics. 2013;3:903–13.

Arbab AS, Yocum GT, Rad AM, Khakoo AY, Fellowes V, Read EJ, Frank JA. Labeling of cells with ferumoxides–protamine sulfate complexes does not inhibit function or differentiation capacity of hematopoietic or mesenchymal stem cells. NMR Biomed. 2005;18:553–9.

Bhaumik S, Gambhir SS. Optical imaging of Renilla luciferase reporter gene expression in living mice. Proc Natl Acad Sci USA. 2002;99:377–82.

Brown CE, Li P, Boyd JD, Delaney KR, Murphy TH. Extensive turnover of dendritic spines and vascular remodeling in cortical tissues recovering from stroke. J Neurosci. 2007;27:4101–9.

Bulte JW, Kraitchman DL. Monitoring cell therapy using iron oxide MR contrast agents. Curr Pharm Biotechnol. 2004;5:567–84.

Cao F, Lin S, Xie X, Ray P, Patel M, Zhang X, Drukker M, Dylla SJ, Connolly AJ, Chen X, Weissman IL, Gambhir SS, Wu JC. In vivo visualization of embryonic stem cell survival, proliferation, and migration after cardiac delivery. Circulation. 2006;113:1005–14.

Chen IY, Wu JC. Cardiovascular molecular imaging: focus on clinical translation. Circulation. 2011;123:425–43.

Cheng K, Malliaras K, Li T-S, Sun B, Houde C, Galang G, Smith J, Matsushita N, Marban E. Magnetic enhancement of cell retention, engraftment, and functional benefit after intracoronary delivery of cardiac-derived stem cells in a rat model of ischemia/reperfusion. Cell Transplant. 2012;21:1121–35.

Chiu RC. Bone-marrow stem cells as a source for cell therapy. Heart Fail Rev. 2003;8:247–51.

Chollet F, DiPiero V, Wise RJ, Brooks DJ, Dolan RJ, Frackowiak RS. The functional anatomy of motor recovery after stroke in humans: a study with positron emission tomography. Ann Neurol. 1991;29:63–71.

Cramer SC, Chopp M. Recovery recapitulates ontogeny. Trends Neurosci. 2000;23:265–71.

Daadi MM, Li Z, Arac A, Grueter BA, Sofilos M, Malenka RC, Wu JC, Steinberg GK. Molecular and magnetic resonance imaging of human embryonic stem cell-derived neural stem cell grafts in ischemic rat brain. Mol Ther. 2009;17:1282–91.

Daadi MM, Davis AS, Arac A, Li Z, Maag AL, Bhatnagar R, Jiang K, Sun G, Wu JC, Steinberg GK. Human neural stem cell grafts modify microglial response and enhance axonal sprouting in neonatal hypoxic-ischemic brain injury. Stroke. 2010;41:516–23.

Das S, Girard L, Green T, Weitzman L, Lewis-Bowen A, Clark T. Building biomedical web communities using a semantically aware content management system. Brief Bioinform. 2009;10:129–38.

Dedobbeleer C, Blocklet D, Toungouz M, Lambermont M, Unger P, Degaute JP, Goldman S, Berkenboom G. Myocardial homing and coronary endothelial function after autologous blood CD34+ progenitor cells intracoronary injection in the chronic phase of myocardial infarction. J Cardiovasc Pharmacol. 2009;53:480–5.

Goldman SJ, Chen E, Taylor R, Zhang S, Petrosky W, Reiss M, Jin S. Use of the ODD-luciferase transgene for the non-invasive imaging of spontaneous tumors in mice. PLoS ONE. 2011;6:e18269.

Griesenbach U, Vicente CC, Roberts MJ, Meng C, Soussi S, Xenariou S, Tennant P, Baker A, Baker E, Gordon C, Vrettou C, McCormick D, Coles R, Green AM, Lawton AE, Sumner-Jones SG, Cheng SH, Scheule RK, Hyde SC, Gill DR, Collie DD, McLachlan G, Alton EW. Secreted Gaussia luciferase as a sensitive reporter gene for in vivo and ex vivo studies of airway gene transfer. Biomaterials. 2011;32:2614–24.

Gu E, Chen WY, Gu J, Burridge P, Wu JC. Molecular imaging of stem cells: tracking survival, biodistribution, tumorigenicity, and immunogenicity. Theranostics. 2012;2:335–45.

Guzman R, Uchida N, Bliss TM, He D, Christopherson KK, Stellwagen D, Capela A, Greve J, Malenka RC, Moseley ME, Palmer TD, Steinberg GK. Long-term monitoring of transplanted human neural stem cells in developmental and pathological contexts with MRI. Proc Natl Acad Sci USA. 2007;104:10211–6.

Gyongyosi M, Blanco J, Marian T, Tron L, Petnehazy O, Petrasi Z, Hemetsberger R, Rodriguez J, Font G, Pavo IJ, Kertesz I, Balkay L, Pavo N, Posa A, Emri M, Galuska L, Kraitchman DL, Wojta J, Huber K, Glogar D. Serial noninvasive in vivo positron emission tomographic tracking of percutaneously intramyocardially injected autologous porcine mesenchymal stem cells modified for transgene reporter gene expression. Circ Cardiovasc Imaging. 2008;1:94–103.

Haynesworth SE, Goshima J, Goldberg VM, Caplan AI. Characterization of cells with osteogenic potential from human marrow. Bone. 1992;13:81–8.

Himmelreich U, Dresselaers T. Cell labeling and tracking for experimental models using magnetic resonance imaging. Methods. 2009;48:112–24.

Hoehn M, Kustermann E, Blunk J, Wiedermann D, Trapp T, Wecker S, Focking M, Arnold H, Hescheler J, Fleischmann BK, Schwindt W, Buhrle C. Monitoring of implanted stem cell migration in vivo: a highly resolved in vivo magnetic resonance imaging investigation of experimental stroke in rat. Proc Natl Acad Sci USA. 2002;99:16267–72.

Hofmann M, Wollert KC, Meyer GP, Menke A, Arseniev L, Hertenstein B, Ganser A, Knapp WH, Drexler H. Monitoring of bone marrow cell homing into the infarcted human myocardium. Circulation. 2005;111:2198–202.

Huang NF, Niiyama H, Peter C, De A, Natkunam Y, Fleissner F, Li Z, Rollins MD, Wu JC, Gambhir SS, Cooke JP. Embryonic stem cell-derived endothelial cells engraft into the ischemic hindlimb and restore perfusion. Arterioscler Thromb Vasc Biol. 2010;30:984–91.

Kammili RK, Taylor DG, Xia J, Osuala K, Thompson K, Menick DR, Ebert SN. Generation of novel reporter stem cells and their application for molecular imaging of cardiac-differentiated stem cells in vivo. Stem Cells Dev. 2010;19:1437–48.

Kang JH, Chung JK. Molecular-genetic imaging based on reporter gene expression. J Nucl Med. 2008;49(Suppl 2):164S–79S.

Khan M, Meduru S, Gogna R, Madan E, Citro L, Kuppusamy ML, Sayyid M, Mostafa M, Hamlin RL, Kuppusamy P. Oxygen cycling in conjunction with stem cell transplantation induces NOS3 expression leading to attenuation of fibrosis and improved cardiac function. Cardiovasc Res. 2012;93:89–99.

Kidd S, Caldwell L, Dietrich M, Samudio I, Spaeth EL, Watson K, Shi Y, Abbruzzese J, Konopleva M, Andreeff M, Marini FC. Mesenchymal stromal cells alone or expressing interferon-beta suppress pancreatic tumors in vivo, an effect countered by anti-inflammatory treatment. Cytotherapy. 2010;12:615–25.

Kocher AA, Schuster MD, Szabolcs MJ, Takuma S, Burkhoff D, Wang J, Homma S, Edwards NM, Itescu S. Neovascularization of ischemic myocardium by human bone-marrow-derived angioblasts prevents cardiomyocyte apoptosis, reduces remodeling and improves cardiac function. Nat Med. 2001;7:430–6.

Koretsky AP, Brosnan MJ, Chen LH, Chen JD, Van Dyke T. NMR detection of creatine kinase expressed in liver of transgenic mice: determination of free ADP levels. Proc Natl Acad Sci USA. 1990;87:3112–6.

Kostura L, Kraitchman DL, Mackay AM, Pittenger MF, Bulte JW. Feridex labeling of mesenchymal stem cells inhibits chondrogenesis but not adipogenesis or osteogenesis. NMR Biomed. 2004;17:513–7.

Krishnan M, Park JM, Cao F, Wang D, Paulmurugan R, Tseng JR, Gonzalgo ML, Gambhir SS, Wu JC. Effects of epigenetic modulation on reporter gene expression: implications for stem cell imaging. FASEB J. 2006;20:106–8.

Lau JF, Anderson SA, Adler E, Frank JA. Imaging approaches for the study of cell-based cardiac therapies. Nat Rev Cardiol. 2010;7:97–105.

Lee Z, Dennis JE, Gerson SL. Imaging stem cell implant for cellular-based therapies. Exp Biol Med (Maywood). 2008;233:930–40.

Leng L, Wang Y, He N, Wang D, Zhao Q, Feng G, Su W, Xu Y, Han Z, Kong D, Cheng Z, Xiang R, Li Z. Molecular imaging for assessment of mesenchymal stem cells mediated breast cancer therapy. Biomaterials. 2014;35:5162–70.

Li Z, Wu JC, Sheikh AY, Kraft D, Cao F, Xie X, Patel M, Gambhir SS, Robbins RC, Cooke JP, Wu JC. Differentiation, survival, and function of embryonic stem cell derived endothelial cells for ischemic heart disease. Circulation. 2007;116:I46–54.

Li Z, Suzuki Y, Huang M, Cao F, Xie X, Connolly AJ, Yang PC, Wu JC. Comparison of reporter gene and iron particle labeling for tracking fate of human embryonic stem cells and differentiated endothelial cells in living subjects. Stem Cells. 2008;26:864–73.

Li Z, Han Z, Wu JC. Transplantation of human embryonic stem cell-derived endothelial cells for vascular diseases. J Cell Biochem. 2009;106:194–9.
Li Z, Hu S, Ghosh Z, Han Z, Wu JC. Functional characterization and expression profiling of human induced pluripotent stem cell- and embryonic stem cell-derived endothelial cells. Stem Cells Dev. 2011;20:1701–10.
Luker KE, Mihalko LA, Schmidt BT, Lewin SA, Ray P, Shcherbo D, Chudakov DM, Luker GD. In vivo imaging of ligand receptor binding with Gaussia luciferase complementation. Nat Med. 2012;18:172–7.
Massoud TFGS. Molecular imaging in living subjects: seeing funda-mental biological processes in a new light. Genes Dev. 2003;17:545–80.
Mathiasen AB, Kastrup J. Non-invasive in-vivo imaging of stem cells after transplantation in cardiovascular tissue. Theranostics. 2013;3:561–72.
Nguyen PK, Lan F, Wang Y, Wu JC. Imaging: guiding the clinical translation of cardiac stem cell therapy. Circ Res. 2011;109:962–79.
Ntziachristos V, Ripoll J, Wang LV, Weissleder R. Looking and listening to light: the evolution of whole-body photonic imaging. Nat Biotechnol. 2005;23:313–20.
Nudo RJ, Wise BM, SiFuentes F, Milliken GW. Neural substrates for the effects of rehabilitative training on motor recovery after ischemic infarct. Science. 1996;272:1791–4.
Partlow KC, Chen J, Brant JA, Neubauer AM, Meyerrose TE, Creer MH, Nolta JA, Caruthers SD, Lanza GM, Wickline SA. 19F magnetic resonance imaging for stem/progenitor cell tracking with multiple unique perfluorocarbon nanobeacons. FASEB J. 2007;21:1647–54.
Pearl JI, Lee AS, Leveson-Gower DB, Sun N, Ghosh Z, Lan F, Ransohoff J, Negrin RS, Davis MM, Wu JC. Short-term immunosuppression promotes engraftment of embryonic and induced pluripotent stem cells. Cell Stem Cell. 2011;8:309–17.
Pittenger MF, Mackay AM, Beck SC, Jaiswal RK, Douglas R, Mosca JD, Moorman MA, Simonetti DW, Craig S, Marshak DR. Multilineage potential of adult human mesenchymal stem cells. Science. 1999;284:143–7.
Psaltis PJ, Peterson KM, Xu R, Franchi F, Witt T, Chen IY, Lerman A, Simari RD, Gambhir SS, Rodriguez-Porcel M. Noninvasive monitoring of oxidative stress in transplanted mesenchymal stromal cells. JACC Cardiovasc Imaging. 2013;6(7):795–802.
Ransohoff KJ, Wu JC. Advances in cardiovascular molecular imaging for tracking stem cell therapy. Thromb Haemost. 2010;104:13–22.
Ransohoff JD, Wu JC. Imaging stem cell therapy for the treatment of peripheral arterial disease. Curr Vasc Pharmacol. 2012;10:361–73.
Rodriguez-Porcel M, Kronenberg MW, Henry TD, Traverse JH, Pepine CJ, Ellis SG, Willerson JT, Moye LA, Simari RD. Cell tracking and the development of cell-based therapies: a view from the cardiovascular cell therapy research network. JACC Cardiovasc Imaging. 2012;5:559–65.
Schachinger V, Aicher A, Dobert N, Rover R, Diener J, Fichtlscherer S, Assmus B, Seeger FH, Menzel C, Brenner W, Dimmeler S, Zeiher AM. Pilot trial on determinants of progenitor cell recruitment to the infarcted human myocardium. Circulation. 2008;118:1425–32.
Sierra M, Machado C. Magnetic resonance imaging in patients with implantable cardiac devices. Rev Cardiovasc Med. 2008;9:232–8.
Stojanov K, de Vries EF, Hoekstra D, van Waarde A, Dierckx RA, Zuhorn IS. [18F]FDG labeling of neural stem cells for in vivo cell tracking with positron emission tomography: inhibition of tracer release by phloretin. Mol Imaging. 2012;11:1–12.
Su W, Wang L, Zhou M, Liu Z, Hu S, Tong L, Liu Y, Fan Y, Kong D, Zheng Y, Han Z, Wu JC, Xiang R, Li Z. Human embryonic stem cell-derived endothelial cells as cellular delivery vehicles for treatment of metastatic breast cancer. Cell Transplant. 2013;22:2079–90.
Sun A, Hou L, Prugpichailers T, Dunkel J, Kalani MA, Chen X, Kalani MY, Tse V. Firefly luciferase-based dynamic bioluminescence imaging: a noninvasive technique to assess tumor angiogenesis. Neurosurgery. 2010;66:751–7.

Sutton EJ, Henning TD, Pichler BJ, Bremer C, Daldrup-Link HE. Cell tracking with optical imaging. Eur Radiol. 2008;18:2021–32.

Takahashi T, Kalka C, Masuda H, Chen D, Silver M, Kearney M, Magner M, Isner JM, Asahara T. Ischemia- and cytokine-induced mobilization of bone marrow-derived endothelial progenitor cells for neovascularization. Nat Med. 1999;5:434–8.

Terrovitis J, Kwok KF, Lautamaki R, Engles JM, Barth AS, Kizana E, Miake J, Leppo MK, Fox J, Seidel J, Pomper M, Wahl RL, Tsui B, Bengel F, Marban E, Abraham MR. Ectopic expression of the sodium-iodide symporter enables imaging of transplanted cardiac stem cells in vivo by single-photon emission computed tomography or positron emission tomography. J Am Coll Cardiol. 2008;52:1652–60.

Thomson JA. Embryonic stem cell lines derived from human blastocysts. Science. 1998;282:1145–7.

Tong L, Zhao H, He Z, Li Z. Current perspectives on molecular imaging for tracking stem cell therapy. Europe: InTech; 2013.

van der Bogt KE, Sheikh AY, Schrepfer S, Hoyt G, Cao F, Ransohoff KJ, Swijnenburg RJ, Pearl J, Lee A, Fischbein M, Contag CH, Robbins RC, Wu JC. Comparison of different adult stem cell types for treatment of myocardial ischemia. Circulation. 2008;118:S121–9.

Vannucci RC. Hypoxic-ischemic encephalopathy. Am J Perinatol. 2000;17:113–20.

Vaughan EE, O'Brien T. Isolation of circulating angiogenic cells. Methods Mol Biol. 2012;916:351–6.

Wang L, Wang Y, Li Z. Nanoparticle-based tumor theranostics with molecular imaging. Curr Pharm Biotechnol. 2013;14:683–92.

Ward NS. Functional reorganization of the cerebral motor system after stroke. Curr Opin Neurol. 2004;17:725–30.

Welling MM, Duijvestein M, Signore A, van der Weerd L. In vivo biodistribution of stem cells using molecular nuclear medicine imaging. J Cell Physiol. 2011;226:1444–52.

Welt FG, Gallegos R, Connell J, Kajstura J, D'Amario D, Kwong RY, Coelho-Filho O, Shah R, Mitchell R, Leri A. Effect of cardiac stem cells on left-ventricular remodeling in a canine model of chronic myocardial infarction. Circ Heart Fail. 2013;6:99–106.

Williams AR, Hatzistergos KE, Addicott B, McCall F, Carvalho D, Suncion V, Morales AR, Da Silva J, Sussman MA, Heldman AW, Hare JM. Enhanced effect of combining human cardiac stem cells and bone marrow mesenchymal stem cells to reduce infarct size and to restore cardiac function after myocardial infarction. Circulation. 2013;127:213–23.

Willmann JK, Paulmurugan R, Rodriguez-Porcel M, Stein W, Brinton TJ, Connolly AJ, Nielsen CH, Lutz AM, Lyons J, Ikeno F, Suzuki Y, Rosenberg J, Chen IY, Wu JC, Yeung AC, Yock P, Robbins RC, Gambhir SS. Imaging gene expression in human mesenchymal stem cells: from small to large animals. Radiology. 2009;252:117–27.

Wu JC, Chen IY, Sundaresan G, Min JJ, De A, Qiao JH, Fishbein MC, Gambhir SS. Molecular imaging of cardiac cell transplantation in living animals using optical bioluminescence and positron emission tomography. Circulation. 2003;108:1302–5.

Yong RL, Shinojima N, Fueyo J, Gumin J, Vecil GG, Marini FC, Bogler O, Andreeff M, Lang FF. Human bone marrow-derived mesenchymal stem cells for intravascular delivery of oncolytic adenovirus Delta24-RGD to human gliomas. Cancer Res. 2009;69:8932–40.

Yusuf S, Reddy S, Ounpuu S, Anand S. Global burden of cardiovascular diseases: part I: general considerations, the epidemiologic transition, risk factors, and impact of urbanization. Circulation. 2001a;104:2746–53.

Yusuf S, Reddy S, Ounpuu S, Anand S. Global burden of cardiovascular diseases: part II: variations in cardiovascular disease by specific ethnic groups and geographic regions and prevention strategies. Circulation. 2001b;104:2855–64.

Zhang SJ, Wu JC. Comparison of imaging techniques for tracking cardiac stem cell therapy. J Nucl Med. 2007;48:1916–9.

Chapter 10
Stem Cell and Peripheral Nerve Regeneration

Chengbin Xue, Sheng Yi and Xiaosong Gu

Abstract The application of stem cells has always attracted great interest in the field of peripheral nerve regeneration. In recent years, the rapid development of neural tissue engineering makes it possible to use stem cell transplantation to repair peripheral nerve injury. Seed/support cell or cellular source from stem cell has been known as one of the components for neural tissue engineering. The tissue-engineered nerve grafts (TENGs) support the regeneration of longer peripheral nerve gaps than scaffold alone. A number of TENGs have been used experimentally to bridge long peripheral nerve gaps in various animal models, where the desired outcome is peripheral nerve regeneration and functional recovery. Stem cells may improve the local microenvironment in nerve injury sites, providing necessary conditions for axonal regeneration. Nowadays, the types of stem cells and their application tend to diversify. Stem cells are more effective in providing necessary factors that promote peripheral nerve regeneration. So far, the application of stem cells for peripheral nerve regeneration is limited mainly because of the low survival rate of transplanted stem cells due to host immune rejection and changes in the local microenvironment. Here, we summarize the latest research progress and application strategies of stem cells in peripheral nerve regeneration. To push the translation of stem cell application for peripheral nerve regeneration into the clinic, we anticipate that a TENG with a close proximity to the regenerative microenvironment of the peripheral nervous system (PNS) will be developed.

Keywords Stem cells · Neural tissue engineering · Peripheral nerve regeneration

C. Xue · S. Yi · X. Gu (✉)
Jiangsu Key Laboratory of Neuroregeneration, Co-Innovation Center of Neuroregeneration, Nantong University, Nantong, China
e-mail: nervegu@ntu.edu.cn

R.C. Zhao (ed.), *Stem Cells: Basics and Clinical Translation*, Translational Medicine Research 1, DOI 10.1007/978-94-017-7273-0_10

10.1 Introduction

Peripheral nerve injury is a global clinical problem and causes a devastating impact on patients' quality of life (Noble et al. 1998; Robinson 2000; Taylor et al. 2008; Asplund et al. 2009). In the USA, about 360,000 people suffer from upper extremity paralytic syndromes annually, and 44,000 upper extremity procedures involved the nervous system during the period of 1989–1991 (Kelsey et al. 1997); over 300,000 cases of peripheral nerve injury occur annually in Europe (Mohanna et al. 2003). Although the peripheral nervous system (PNS) has a greater capacity of axonal regeneration than the central nervous system (CNS) after injury, spontaneous repair of peripheral nerve is nearly always unsatisfied with poor functional recovery. Various types of medical therapy have been carried out for several hundred years with the intention of improving outcomes (Artico et al. 1996; Battiston et al. 2009). Peripheral nerve repair consequently represents a unique challenge and opportunity to clinical and translational neurosciences. Various grafts between the nerve stumps are required to bridge the gap and support axonal regeneration in a substantial nerve gap of peripheral nerve injury. An autologous nerve graft, which is usually a functionally less important nerve segment from another site of the body (Johnson and Soucacos 2008), is accepted as the gold standard therapy for peripheral nerve injury. However, there are inherent disadvantages, including the limited supply of donor nerves, a second surgery, donor site morbidity, and a mismatch between the donor nerve and the recipient site (Ortiguela et al. 1987; Mackinnon and Hudson 1992). These collectively have encouraged the development of alternatives to autologous nerve grafts.

As is known, the microenvironment surrounding an injury site in PNS is often more permissive to axonal regeneration as compared to that in the CNS (Gu et al. 2011). Although peripheral nerve regeneration is ultimately determined by quality and speed of axonal outgrowth (Malin et al. 2009), Schwann cells (SCs) also play a critical role in the establishment of the regenerative microenvironment. It is the special importance of SCs in PNS that is responsible for the effectiveness of using SCs as support cells for the generation of TENGs (Johnson et al. 2005). Myelination of axons seems to be the most basic function of SCs because myelin sheath, as a unique component of the nervous system, can increase axonal conduction, especially salutatory conduction, thus allowing fast and efficient salutatory propagation of action potentials along the nerve (Honkanen et al. 2007; Salzer et al. 2008; Rumsey et al. 2009). On the other hand, the development of myelinated nerve fibers in PNS depends on complex interactions between SCs and axons. Axons can in turn promote the deposition of the basal lamina by SCs, which is required for the ensheathment of axons and the subsequent differentiation of SCs, and the maturation of fully myelinating SCs depends on contact and signaling from axons (Muir 2010; Wanner et al. 2006). Based on the insights into the interactions between SCs and axons, in this study, a unique scheme for incorporating biochemical cues into neural scaffold was adopted to establish an optimal regenerative microenvironment.

With progress in regenerative medicine, especially when tissue engineering and a subfield of neural tissue engineering has emerged, various biological and artificial nerve grafts, which are generally named tissue-engineered nerve grafts (TENGs), have been produced in attempts to supplement or even substitute for autologous nerve grafts. The typical TENGs involve both physiochemical and biological cues, which are provided by a biomaterial-based structure, as well as a multitude of cellular or molecular components. In recent years, many excellent review articles discuss the clinical applications and future directions of TENGs (Johnson and Soucacos 2008; Gu et al. 2011; Schmidt and Leach 2003; Chalfoun et al. 2006; Johnson et al. 2008; Seidlits et al. 2008; Deumens et al. 2010; Jiang et al. 2010; Khaing and Schmidt 2012; Rajaram et al. 2012; Zochodne 2012).

The application of various stem cells has always attracted much interest in the field of peripheral nerve regeneration. In recent years, the rapid development of neural tissue engineering makes it possible for the use of stem cell transplantation to repair peripheral nerve injury. Seed/support cell which usually is, or source from stem cell has been known as one of the components for neural tissue engineering. The TENGs can support longer peripheral nerve regeneration than the scaffold alone. A number of TENGs have been used experimentally to bridge long peripheral nerve gaps in various animal models, where the desired outcome is peripheral nerve regeneration and functional recovery. The stem cell may improve the local microenvironment in nerve injury sites, providing necessary conditions for axonal regeneration. Nowadays, the type of stem cell and their application tend to diversify. Stem cells are more effective to provide the necessary factors that promote peripheral nerve regeneration. So far, the application of stem cell for peripheral nerve regeneration is limited mainly because of the less survival rate of transplanted stem cell due to host immune rejection and changes in the local microenvironment. This chapter summarizes the latest research progress and application strategy of stem cells for peripheral nerve regeneration. To push the translation of stem cell application for peripheral nerve regeneration into the clinic, we anticipate that a TENG with a close proximity to the regenerative microenvironment of the PNS will be developed.

10.1.1 Types and Sources of Stem Cell for Peripheral Nerve Regeneration

10.1.1.1 Mesenchymal Stem Cells (MSCs)

The application of stem cells from different sources in the field of neural tissue engineering has attracted much interest, the bone marrow mesenchymal stem cells (also named bone marrow stromal cells, BMSCs) being undoubtedly the most important. MSCs are featured by their abilities to: (1) differentiate into mature cells and populate the resident tissue, having a therapeutic potential for regenerative medicine; (2) secrete growth factors or other soluble mediators; and (3) serve as a vehicle for protein drug delivery, namely perform gene therapy (Horwitz and Dominici 2008). They localize in the stromal compartment of the bone marrow, where they support hematopoiesis and differentiate into mesenchymal lineages (Johnson and Dorshkind 1986; Deryugina and Muller-Sieburg 1993; Bianco et al. 2001; Abdallah and Kassem 2008; Phinney and Prockop 2007; Franchi et al. 2012). Unorthodox plasticity of MSCs, however, has recently been described in that they have ability to cross oligolineage boundaries, in other words, to transdifferentiate into nonmesenchymal cells. Several in vitro studies have reported that MSCs can be induced to differentiate into neural lineages including neurons, astrocytes, oligodendrocytes, microglia, and SCs-like cells (Chen et al. 2006; Lu et al. 2008; Munoz-Elias et al. 2003; Sanchez-Ramos et al. 2000; Suzuki et al. 2004; Wislet-Gendebien et al. 2005; Woodbury et al. 2000), and specific approaches to induce transdifferentiation of rodent or human MSCs toward neural lineage cells have been established (Dezawa et al. 2001, 2004; Mimura et al. 2004). On the other hand, in vivo studies have also reported that after implanted into the brain, MSCs generate neural phenotypes specific to the injury site (Kocsis et al. 2002; Kopen et al. 1999; Lu et al. 2006). These results motivate explorations into the possibility of using MSCs as an alternative to SCs for peripheral nerve repair (Dezawa 2006). The MSCs are easily obtained through the aspiration of the bone marrow and expanded in a large scale by in vitro culture; BMSCs have found increasing applications in cell-based therapies for various diseases, including neural injury and disorders (Horwitz et al. 2002; Fickert et al. 2003; Ortiz et al. 2003; Kunter et al. 2006; Lee et al. 2006; Minguell and Erices 2006; Ringden et al. 2006). Despite the indispensable value of SCs for the construction of TENGs, autologous SCs are difficult to obtain in large number, and allogeneic SCs are involved in immunological rejections. Therefore, BMSCs have become a promising alternative to SCs for using as support cells within TENGs, showing considerable success in experimental studies (Hu et al. 2013; Ding et al. 2010; Yang et al. 2011).

MSCs pre-labeled with bromodeoxyuridine (BrdU) were injected into the distal stump of transected rat sciatic nerves. Dual immunofluorescence labeling showed that BrdU-reactive MSCs survived in the injected area for at least 33 days after implantation, and almost 5 % of BrdU cells exhibited Schwann cell-like phenotype (S-100 immunoreactivity). Walking track test at 18 and 33 days after implantation indicated that MSC implantation promoted functional recovery of injured nerves (Cuevas et al. 2002). GFP-labeled undifferentiated MSCs were seeded in a Matrigel-containing chitosan NGC to bridge a 5-mm rat sciatic nerve gap. After 6 weeks, the growth and myelination of axons and the sciatic functional index were significantly improved as compared to MSC-free NGC. Moreover, confocal microscopy confirmed that implanted MSCs adopted the Schwann cell-like phenotype (Zhang et al. 2005). Transdifferentiated MSCs, which had been transformed into Schwann cell-like cells under induction of a cocktail of cytokines, were implanted into devitalized muscle conduits for bridging a 20-mm rat sciatic nerve gap. The examinations at 6 weeks after implantation indicated that the transplanted MSCs were able to support peripheral nerve regeneration to a certain extent. Although the enhancement of nerve regeneration by MSC-containing NGCs was inferior to that by Schwann cell-containing NGCs or by autologous nerve grafts, such deficit in regenerative outcomes could be compensated by combining MSCs with muscle-derived allografts (Keilhoff et al. 2006a, b). Differentiated MSCs were suspended in Matrigel and transferred into hollow fibers, which were implanted across a 10-mm rat sciatic nerve gap. After 6 months, significant improvements in motor nerve conduction velocity and sciatic functional index were observed in the differentiated MSC-implanted animals, and GFP labeling clearly showed the trace of implanted MSCs within the regenerated nerves (Mimura et al. 2004). In a rhesus monkey model, an acellular allogeneic nerve graft seeded with autologous MSCs was implanted into a 10-mm radial nerve gap. After 8 weeks, nerve regeneration and functional restoration were observed, and the implanted MSCs exhibited Schwann cell-like phenotype as evidenced by double immunostaining for S-100 and BrdU (Wang et al. 2008).

The aforementioned observations lead to a transdifferentiation mechanism of MSCs, which assumes that the actions of MSCs on nerve regeneration, at least in part, stem from their ability to replace damaged neural cells via cellular differentiation; however, the mechanisms are also believed to be due to spontaneous fusion of MSCs with host cells rather than real transdifferentiation (Weimann et al. 2003a, b). The exact mechanisms behind the enhanced nerve regeneration in the presence of undifferentiated MSCs remain to be elucidated. The undifferentiated MSCs may contribute to nerve regeneration possibly by secreting growth factors and depositing basal lamina components (Chen et al. 2007; Wang et al. 2009). The production of neurotrophic molecules by MSCs can delay cell death and restore neural tissues (Borlongan et al. 2004; Chopp and Li 2002; Chopp et al. 2000; Crigler et al. 2006; Gu et al. 2010; Hofstetter et al. 2002; Munoz et al. 2005). Searching for undescribed mediators generated by MSCs will probably reveal a new array of important signaling secreted molecules (Horwitz and Dominici 2008). These results, together with the finding that local differentiation of MSCs is an uncommon event, suggest a new general paradigm for the neural activities of MSCs. Systemically infused MSCs exert a therapeutic effect primarily through the release of growth factors or other soluble mediators, which act on local even distant target tissues. Rather than serving as stem cells to replace neural cells and repair nerve tissues, MSCs behave as a small molecular factory secreting growth factors or other bioactive molecules to stimulate the reconstruction of nerve tissues or to establish a favorable microenvironment for nerve regeneration (Chopp and Li 2002; Caplan and Dennis 2006; Chen et al. 2002; Liu and Hwang 2005; Neuhuber et al. 2005). Some researchers propose that MSCs promote peripheral nerve regeneration not only via their direct release of growth factors, but also through indirect modulation of cellular behaviors of SCs (Wang et al. 2009). To sum up, the molecular mechanisms that are responsible for the favorable effects of MSCs on peripheral nerve regeneration seem to involve many aspects, including cell replacement, growth factor production, ECM molecule synthesis, microenvironment construction, and immune modulation, and they need further exploration.

Adipose tissue has also been identified as a niche for multipotent stem cells that have a phenotypic profile comparable to that of BMSCs and can differentiate into a myelinating Schwann cell-like phenotype in culture with lineage-specific stimuli (Kingham et al. 2007; Xu et al. 2008). In consequence, adipose-derived mesenchymal stem cells (AMSCs), also named adipose-derived stem cells (ADSCs), are potentially valuable because of their capability of multilineage differentiation in a manner resembling that of BMSCs. Importantly, AMSCs are superior to BMSCs in some aspects, such as the convenient harvesting of AMSCs through liposuction, a much less invasive method than bone marrow aspiration, and the greater availability of adipose tissue than bone marrow (Rider et al. 2008). To apply AMSCs for neural tissue engineering, many experimental studies in diverse animal models have been accomplished, in which different neural scaffolds containing either undifferentiated or differentiated AMSCs have bridged peripheral nerve gaps of different lengths (Yang et al. 2011; Erba et al. 2010; Zhang et al. 2010; di Summa et al. 2011; Scholz et al. 2011; Sun et al. 2011; Wei et al. 2011; Gu et al. 2012; Orbay et al. 2012; Shen et al. 2012; Tomita et al. 2012, 2013; Carriel et al. 2013; Mohammadi et al. 2013; Suganuma et al. 2013). All these studies indicate the favorable effects of AMSCs on peripheral nerve reconstruction and open a new approach for the use of support cells for constructing TENGs.

10.1.1.2 Embryonic Stem Cells (ESCs)

After peripheral nerve injury, many neurons die of insufficient nutrition. Because the number of neurons cannot be easily expanded by in vitro culture, there is a big difficulty to apply primary cultured neurons in nerve tissue engineering. ESCs have a great potential to proliferate unlimitedly and differentiate into neurons under various protocols; hence, they become a good candidate of support cells in cell-based therapies for neural injuries and disorders.

ESCs are undifferentiated, pluripotent cells derived from the inner cell mass of blastocyst-stage embryos and possess a nearly unlimited capacity for self-renewal and an ability to virtually differentiate into any kind of cell type in the body (Jakob 1984). The differentiation of ESCs can be modulated by the stimulation of growth factors (Schuldiner et al. 2000). Retinoic acid (RA) and nerve growth factor have been found to be potent enhancers of neuronal differentiation, eliciting extensive outgrowth of processes and expression of neuron-specific molecules (Schuldiner et al. 2001).

Extensive research has focused on the implantation of ESCs for treating the CNS disorders, while the potential of ESC-based therapy for the PNS injuries is largely unknown. Based on the assumption that implantation of neuronal cells derived from ESCs into denervated muscle would replace lost neurons or non-neuronal cells and prevent muscle atrophy in a peripheral nerve injury model, ESCs were differentiated into cholinergic motor neuron progenitors and labeled with florescence, followed by injection into gastrocnemius muscle of rats after denervation by ipilateral sciatic nerve transection. The observation showed that motor neuron progenitors prevented muscle atrophy after denervation for a brief time (Craff et al. 2007). In another study, ESC-derived neural progenitor cells were implanted into a 10-mm rat sciatic nerve gap, resulting in substantial axonal regrowth and nerve repair. The implanted cells survived until 3 months and differentiated into myelinating cells. Nerve stumps showed nearly normal diameter with longitudinally oriented, densely packed Schwann cell-like cell arrangement. Electrophysiological recordings confirmed that functional activity recovered across the nerve gap (Cui et al. 2008). As reported by Yohn et al. (2008), ESC-derived motor neurons could form functional synapses with denervated host muscle after implantation into transected tibial nerves, thereby attenuating the denervation-induced muscle atrophy.

Furthermore, it is believed that neuronal cells derived from ESCs provide new choices of support cells for incorporation to neural scaffolds, promoting peripheral nerve regeneration.

10.1.1.3 Neural Stem Cells (NSCs)

NSCs are multipotent cells that reside within paramedian generative zones present along the entire neuraxis throughout all stages of neural development and also during adult life (Gokhan and Mehler 2001). NSCs have the potential to differentiate into three major cellular elements of the nervous system, including neurons, astrocytes, and oligodendrocytes, and they can proliferate unlimitedly and undergo rapid cellular expansion in response to nerve injuries. The properties of NSCs, including multipotential differentiation, strong plasticity, high immigration ability, easily isolation and culture in vitro, and low immunogenicity, make NSCs an attractive source of support cells for the construction of TENGs (Alessandri et al. 2004). Chitosan NGCs seeded with NSCs were used as a tissue-engineered nerve graft to bridge 10-mm facial nerve gaps in rabbits. Nerve regeneration at 12 weeks after implantation was similar to that by autologous nerve grafting (Guo and Dong 2009). By GFP labeling, Hsu et al. (2009) found that 85 % of seeded NSCs were successfully aligned on the micropatterned poly (D,L-lactic acid) (PLA) NGC within 72 h, and the cells expressed the genes that are related to the production of neurotrophic factors and thus facilitated nerve repair and functional recovery in a 10-mm rat sciatic nerve injury model during a period of 6 weeks. Genetically modified NSCs can also serve as a source of neurotrophic factors. NSCs engineered to overexpress glial cell line-derived neurotrophic factor, which is known to protect motoneurons, were implanted to chronically denervated distal tibial nerve. There was a better regeneration of peroneal axons into tibial nerve with a reduction of chondroitin sulfate proteoglycan immunoreactivity in the ECM (Heine et al. 2004). Implantation of NSC overexpressing glial cell line-derived neurotrophic factor could significantly increase the nerve action potential amplitude, axonal area, and axonal number, as well as the labeling for S-100, NF, and beta III tubulin (Shi et al. 2009). After NSCs transfected with neurotrophin-3 or its receptor were incorporated into a PLGA NGC followed by 14-day culture, the viable NSCs were widely distributed within the NGC. This construct permitted the NSCs to differentiate toward neurons and to exhibit synaptic activities (Xiong et al. 2009).

10.1.1.4 Olfactory Ensheathing Cells (OECs)

Olfactory ensheathing cells (OECs) are neural crest cells which allow growth and regrowth of the primary olfactory neurons. Indeed, the primary olfactory system is characterized by its ability to give rise to new neurons even in adult animals. This particular ability is partly due to the presence of OECs which create a favorable microenvironment for neurogenesis (Guerout et al. 2014).

OECs are a unique type of glial cells that wrap olfactory axons and support their continual regeneration from the olfactory epithelium to the bulb (Su et al. 2013). OECs develop from a peripheral origin, the olfactory placode, and retain the ability to self-renew and differentiate, and are considered as peripheral nerve progenitor cells (Tohill and Terenghi 2004; Fairless and Barnett 2005). OECs that share both Schwann cell and astrocytic characteristics have been shown to promote axonal regeneration after transplantation. The tissue-engineered poly (lactic-co-glycolic acid) (PLGA) seeded with OECs was verified to improve peripheral nerve regeneration in a long sciatic nerve defect (Tan et al. 2013).

OECs have been studied in the context of enhancing repair of peripheral nerve by direct transplantation in different peripheral nerve lesion models for enhancement of axonal nerve regeneration by providing a scaffold for the regenerating axons as well as trophic factors and directional cues (Deumens et al. 2006). OECs are known to provide trophic factors conducive to axonal regeneration and survival. They may promote endogenous SCs mobilization possibly by a trophic influence (Cao et al. 2007; Au et al. 2007). Experimental studies performed in rodents show that transplantation of OECs into injured nerve or implantation of OEC-seeded conduits leads to an enhancement in axonal regeneration and improved functional outcome under some experimental conditions. Axonal dieback of the proximal nerve stump is reduced in the OEC transplanted nerves suggesting that the OECs provided early trophic support leading to earlier onset of regeneration. This could be critical for allowing the regenerating axons to navigate across the injury site before impeding scar tissue develops. However, OECs share many properties with SCs such as their production of neurotrophic factors and extracellular matrix (ECM) molecules as well as their ability to form peripheral myelin. Transplanted identified eGFP-expressing OECs integrate into the nerve injury site and remyelinate the regenerated axons, suggesting direct participation of OECs in the repair process (Radtke and Kocsis 2012).

OECs-containing silicone tubes were noted to support an improved axonal regeneration in 50 or 79 % of rats with a 15- or 12-mm sciatic nerve injury gap (Verdu et al. 1999). Another case showed that bridging of 15-mm nerve gap in rat sciatic nerve injury model with muscle-stuffed vein seeded with OECs as a substitute for autologous nerve graft. Neurophysiological recovery, as assessed by electrophysiological analysis, was faster in the constructed biological nerve conduit compared to that of autologous nerve graft (Lokanathan et al. 2014). Although the therapeutic potential of OECs in peripheral nerve repair is yet far from conclusive, there have been later studies reporting on the treatment of peripheral nerve injury by direct injection of OECs to the injured site (Andrews and Stelzner 2004; Dombrowski et al. 2006; Radtke et al. 2009; Guerout et al. 2011).

10.1.1.5 Induced Pluripotent Stem (iPS) Cells

Cell source is a major issue for tissue engineering and regenerative medicine. An exciting breakthrough in stem cell biology is that adult somatic cells (e.g., skin fibroblasts) can be reprogrammed into induced pluripotent stem cells (iPSCs) by the activation of a limited number of genes (transgenes) such as Oct3/4, Sox2, c-Myc, and KLF4 (Takahashi et al. 2007; Park et al. 2008) or Oct3/4, Sox2, Nanog, and Lin28 (Yu et al. 2007). The iPSCs derived from somatic cells make it possible for patient-specific cell therapies, which bypass immune rejection issue and ethical concerns of deriving and using ESCs as a cell source. The unlimited expansion potential of iPSCs also makes them a valuable cell source for tissue engineering. However, to use iPSCs as a cell source, many important issues remain to be addressed, such as the differences among various iPSC lines in differentiation and expansion and the appropriate differentiation stage of the cells for specific tissue engineering applications. In general, most of iPSC lines, as ESCs, can differentiate into neural crest stem cells (NCSCs), although the differentiation efficiency was different. NCSCs can differentiate into cell types of all three germ layers and represent a valuable model system to investigate the differentiation and therapeutic potential of stem cells (Rao and Anderson 1997; Morrison et al. 1999; Crane and Trainor 2006; Lee et al. 2007; Sauka-Spengler and Bronner-Fraser 2008). The adult cell sources are limited by the number of cells that can be obtained and complicated by the need to sacrifice additional nerves and tissues. Moreover, there is a lack of efficiency and consistency in cell isolation and expansion, which causes variability in therapeutic efficacy. In contrast, NCSCs derived from iPSCs can be immune compatible, expandable, and well characterized as a valuable cell source for the regeneration of peripheral nerve and other tissues (Wang et al. 2011).

Stem cells from sources other than the bone marrow are now getting more attention. The gliogenic secondary neurospheres derived from iPS cells have the ability to differentiate into SCs. The iPS cells were added to a PLC-based NGC, followed by implantation across a sciatic nerve gap in mice, showing regeneration of peripheral nerves and functional recovery (Uemura et al. 2012). iPSCs and their derivatives are valuable cell sources for tissue engineering.

10.1.1.6 Skin-Derived Precursors (SKPs)

Skin-derived precursors (SKPs) are stem cells found in the dermis (Fernandes et al. 2004; Blazejewska et al. 2009). These stem cells persist into adulthood, as they can be isolated from adult skin (Toma et al. 2001). In their endogenous environment, the dermis, they instruct hair follicle growth and contribute to maintaining the dermis and repairing it after injury. One niche where SKPs are found is at the base of hair follicles, in the dermal papilla (DP) and surrounding the follicle in the dermal sheath (DS) (Fernandes et al. 2004). The DP is the control center for hair growth (Jahoda et al. 1984), as such SKPs found in the DP instruct hair growth. SKPs can also migrate out of the DP. Upon injury to the dermis, SKPs migrate to the site of injury and differentiate into dermal fibroblasts, thereby replenishing the interfollicular dermis.

The skin dermis contains neural crest-related precursor cells, and the SKPs can be cultured to differentiate into neural crest cell types with the characteristics of neurons and SCs in the PNS (Fernandes et al. 2004; McKenzie et al. 2006). Both rodent and human SKPs are differentiated into SCs when transplanted into the brains of shiverer mice (McKenzie et al. 2006), which have a genetic deficiency in MBP and hypomyelination of the CNS (Dupouey et al. 1979). Both whisker pad SKPs and dorsal back SKPs differentiate into SCs with similar efficiency (Jinno et al. 2010). Of note, dorsal trunk SKPs are somite derived but can still differentiate into SCs, which are neural crest derived during development (Jinno et al. 2010).

SCs differentiated from SKPs express typical Schwann cell markers and can myelinate DRG axons in vitro. When transplanted into rodent models of peripheral nerve injury, SKP-SCs are able to myelinate axons in vivo and aid in injury repair (McKenzie et al. 2006; Walsh et al. 2009, 2010). The first experiment using SKP-SCs transplantation into the PNS used a crush model. Here, the sciatic nerve of a mouse was crushed with forceps (but not transected), and SKP-SCs were immediately transplanted distal to the crush site. SKP-SCs were able to myelinate the regenerating axons (McKenzie et al. 2006). SKP-SCs can also be used to repair transected nerves (Shakhbazau et al. 2014). In an acute model of peripheral injury, SKP-SCs were transplanted into an acellular nerve graft and used to bridge a 12-mm gap in the sciatic nerve. SKP-SCs promoted axon regeneration, myelination, and electrophysiological recovery (Walsh et al. 2009).

SKP-SCs were also able to regenerate chronic peripheral nerve injury. More motor neurons regenerated into the chronically denervated nerve with SKP-SCs transplantation compared with media controls, and these regenerated axons were larger. The compound muscle action potential (CMAP) amplitude in the gastrocnemius muscle and its muscle weight were larger with SKP-SCs transplantation, suggesting better muscle reinnervation. The reparative ability of SKP-SCs approached that of nerves that were immediately sutured and were not chronically denervated (Walsh et al. 2010). SKP-SCs therapy also improves behavioral recovery after acute, chronic, and nerve graft repair beyond the current standard of microsurgical nerve repair (Khuong et al. 2014).

The SKPs with neurotropic function show a full capacity of differentiating into SCs and promoting axon regeneration in vivo (Chen et al. 2012). In one study, SKPs were injected into neural scaffolds (NGCs) that had been prepared with L-lactide–trimethylene carbonate (L-lac/TMC) copolymer or type I collagen, respectively, to generate a TENG, which was then used to bridge a 16-mm sciatic nerve gap in rats. The results of the study confirmed the beneficial effects of SKPs on nerve regeneration (Marchesi et al. 2007). In another study, porcine SKPs were found to induce prominent nerve regeneration in porcine peripheral nerve injury sites after SKPs were added to a collagen/fibrin NGC for bridging a 10-mm femoral nerve gap in pigs (Park et al. 2012). More studies demonstrated further evidence for the effectiveness of using SKPs as support cells in TENGs (Walsh et al. 2009; Chen et al. 2012).

10.1.2 Application Strategy of Stem Cells for Peripheral Nerve Regeneration

10.1.2.1 Transplantation of Stem Cells Combined with Scaffolds for Repairing Peripheral Nerve Injury

It is reported in a recent study that evaluated the long-term safety of using support cells-containing TENGs to repair a 50-mm-long median nerve gap in monkeys in terms of the data from blood test, immunological and tumor marker detection, and histopathological examination of organs and glands (Hu et al. 2013). They also directly transplant MSCs combined with chitosan/PLGA scaffold to successfully repair the 50- and 60-mm sciatic nerve gap in dogs (Ding et al. 2010; Xue et al. 2012), and they developed a new design of TENGs by introducing bone marrow mesenchymal stem cells (MSCs) of rats, as support cells, into a silk fibroin (SF)-based scaffold, which was composed of an SF nerve guidance conduit and oriented SF filaments as the conduit lumen filler. The biomaterial SF had been tested to possess good biocompatibility and noncytoxicity with MSCs before the TENG was implanted to bridge a 10-mm-long gap in rat sciatic nerve. Functional and histological assessments showed that at 12 weeks after nerve grafting, TENGs yielded an improved outcome of nerve regeneration and functional recovery, which was better than that achieved by SF scaffolds and close to that achieved by autologous nerve grafts. During 1–4 weeks after nerve grafting, MSCs contained in the TENG significantly accelerated axonal growth, displaying a positive reaction to S-100 (a Schwann cell marker). During 1–3 weeks after nerve grafting, MSCs contained in the TENG led to gene expression upregulation of S100 and several growth factors (brain-derived neurotrophic factor, ciliary neurotrophic factor, and basic fibroblast growth factor). The cell behaviors and neurotrophic functions of MSCs might be responsible for their promoting effects on peripheral nerve regeneration (Yang et al. 2011).

Several papers showed the xenogeneic or allogenic acellular nerve grafts implanted with MSCs promote nerve regeneration effectively in rats or even monkeys (Wang et al. 2008; Zhang et al. 2010; Hu et al. 2007; Jia et al. 2012). So far, most studies directly implant stem cells together with biomaterials for repairing peripheral nerve injury. Other types of stem cells are also mostly implanted in the same way as MSCs mentioned above (Guerout et al. 2014; Lokanathan et al. 2014; Wang et al. 2011; Uemura et al. 2012; Walsh et al. 2009; Ikeda et al. 2014).

10.1.2.2 Joint Use of Stem Cells, Cytokines, and Scaffolds for Repairing Peripheral Nerve Injury

Upon injury to peripheral nerves, the local presence of growth factors at the injury sites plays a vital and complex role in modulating phenotypic changes of a variety of neural and nonneural cells. Although the endogenous growth factors secreted by neural cells in the distal nerve stump can support axon regeneration, the supportive action may not be sustained indefinitely due to an obvious decline with time in cellular production of growth factors, and hence, the continuous supply of growth factors is critically required, which is mainly dependent on the addition of exogenous growth factors. To date, the most commonly used growth factors belong to two classes: (1) neurotrophins, including NGF, brain-derived neurotrophic factor (BDNF), and neurotrophin-3 (NT-3); and (2) growth factors with neurotrophic actions, including glial cell line-derived neurotrophic factor (GDNF), ciliary neurotrophic factor (CNTF), and fibroblast growth factors (FGFs). The details about their application for neural tissue engineering are available in review papers (Gu et al. 2011; Jiang et al. 2010).

The cell-based delivery of growth factors has been developed. A silicone-based NGC seeded with genetically modified SCs overexpressing fibroblast growth factor-2 (FGF-2) was used to bridge a 15-mm sciatic nerve gap in adult rats. Different FGF isoforms overexpressed by implanted SCs improved both lengths and number of regenerating myelinated axons over different time periods post-grafting (Timmer et al. 2003; Haastert et al. 2008). Likewise, the design of GDNF-transduced Schwann cell grafts for enhancing regeneration of erectile nerves represents the same attempt of cell-based delivery of neurotrophic factors (May et al. 2008, 2013). In addition, a combination of iPSc-derived neurospheres and basic fibroblast growth factor (bFGF)-containing gelatin microspheres was incorporated into a neural scaffold (a synthetic polymer-based, double-layered NGC), and the constructed TENG was used to bridge 5-mm-long sciatic nerve gaps in mice, achieving regenerative outcomes to some degree (Ikeda et al. 2014). The above-mentioned are typical examples of the combined use of support cells and growth factors as biochemical cues within TENGs.

MSCs not only enhance functional outcomes of nerve repair through production of various growth factors, but also act as gene delivery vehicles of growth factors that are released to nerve injury sites for augmenting axonal growth (Lu et al. 2009). For example, it has been reported that transduction of MSCs to overexpress a certain factor, for example, brain-derived neurotrophic factor, could result in significant increase in the extent and diversity of axonal growth in the host nerves (Chen et al. 2005; Lu et al. 2005). Although gene transfer to SCs, MSCs or other stem cells, even to nerve tissues usually facilitates a continuous release of growth factors to encourage peripheral nerve regeneration, it has also found that lentiviral vector-mediated overexpression of glial cell line-derived neurotrophic factor causes trapping of regenerating axons and failure of appropriate target reinnervation during nerve regeneration in a rat sciatic nerve injury model (Tannemaat et al. 2008). It seems that the possible side effects are not negligible in gene-based therapy.

10.1.2.3 Construction of Tissue-Engineered Nerve Grafts In Vitro for Repairing Peripheral Nerve Injury

TENGs are considered as a promising alternative to autologous nerve grafts used for peripheral nerve repair. The differences between these two types of nerve grafts are mainly in the regenerative microenvironment established by them. To construct ideal TENGs, it is therefore required to develop a better way to introduce biochemical cues into a neural scaffold, as compared to single or combined use of support cells and growth factors.

Tang et al. (2012) have reported that a coculture system of dorsal root ganglia (DRGs) and SCs could give rise to an in vitro cultured nerve equivalent that was likely to mimic the native nerve microenvironment. Their TENGs consisted of an in vitro cultured nerve equivalent residing in a SF-based scaffold were used to bridge a 10-mm sciatic nerve defect in rats. At 12 weeks after nerve grafting, a series of measurements were performed to evaluate the regenerative capacity of these TENGs. The recovery in the motor function of the injured hindlimb in TENG group, as indexed by the SFI value, was close to that in autograft group without significant difference between each other, and prevailed over that in scaffold group. The restoration of electrophysiological properties for 3 grafted groups was reflected in detectable CAMP data, which represented an important measure for the conduction function of peripheral nerves. The comparison in the CMAP amplitude between 3 grafted groups provided further evidence that functional recovery in TENG group was more close to that in autograft group than that in scaffold group. Histological analysis showed that either the regenerated nerve or target gastrocnemius muscle achieved the similar reconstruction, both in qualitative and quantitative aspects, between TENG and autograft groups, and these similar results were significantly better than those in scaffold group. Their findings suggested that more axons might successfully grow through our developed TENG to reach the distal stumps for reinnervation of target muscle. In other words, the incorporation of a nerve equivalent into SF-based scaffold led to an enhanced repair capacity for peripheral nerve injuries. Successful myelination of PNS depends on induction of major protein components of myelin including PMP22, and myelin stability is also sensitive to PMP22 levels (Wrabetz et al. 2006). The high expressions of N-cadherin and PMP22 meant that in their developed nerve equivalent, as in the autologous nerve, dynamic interactions between axons and SCs contributed to the establishment of an ideal microenvironment for nerve regeneration via the increased expression of several bioactive molecules. The introduction of an in vitro cultured nerve equivalent into a scaffold might contribute to establishing a native-like microenvironment for nerve regeneration.

This culture system in vitro is also the same applies in candidate stem cells of the TENG for repairing peripheral nerve injury. The stem cells cocultured with DRGs or seeded onto the biomaterial scaffold in vitro for a period of time provide chance for cells in TENG to communicate with each other or establish a native-like microenvironment effectively.

The rotary bioreactors could influence major cellular events such as differentiation, proliferation, viability, and cell cycle. Introduced by the National Aeronautics and Space Administration (NASA) of the USA in 1987, simulate microgravity (SMG) cultures seem to be ideal for overcoming some drawbacks associated with static culturing systems (Vunjak-Novakovic et al. 1999; Goodwin et al. 1993). For instance, SMG conditions allow the cells to well proliferate in a rotary cell culture system (RCCS) of microgravity environment but at low shear stress and low turbulence environment. SMG culture conditions can provide appropriate microenvironments which have proven advantageous for intercellular communication on tissue-specific cell assembly, cell adhesion, signal transduction, glandular structures, and function (Meyers et al. 2005; Goodwin et al. 1993). In addition, when cells are maintained in a 3D growth environment, they tend to aggregate. SMG culture can enhance cell–cell interactions and supply such 3D growth microenvironment. So, the aggregating growth of cells is an important effect of SMG on some cell lines (Unsworth and Lelkes 1998). SMG promoted porcine liver cells to grow into 3D cell aggregation, which displayed that SMG culture system was suitable for long-term and expanding cell culture (Dabos et al. 2001). The dynamic flow of RCCS might improve nutrient supply and increase metabolic waste removal for the cells in the interior cellular spheres. Thus, comparison with the static culture, rotating simulated microgravity culture environment can show better cellular vitality and function for some cultivated cells. There is also specific application in the construction of TENG in vitro recently (Luo et al. 2014).

Nowadays, for the construction of neural tissue engineering, it needs to be addressed how to obtain adequate and unified standardized conditions to promote the peripheral nerve regeneration. A series of problems should be solved before wide clinical application of the RCCS in tissue engineering technology.

10.1.2.4 Construction of Acellular Tissue-Engineered Nerve Grafts In Vitro for Repairing Peripheral Nerve Injury

The field of stem cells and regenerative medicine offers considerable promise as a means of delivering new treatments for a wide range of diseases. In order to maximize the effectiveness of cell-based therapies—whether stimulating expansion of endogenous cells or transplanting cells into patients—it is essential to understand the niche signals that regulate stem cell behavior. One of those signals is from the ECM. New technologies have offered insights into how stem cells sense signals from the ECM and how they respond to these signals at the molecular level, which ultimately regulate their fate (Watt and Huck 2013).

To date, however, cellular and molecular therapies directed at peripheral nerve repair have not yet gone beyond the laboratory stage, and their translation to the clinic has been beset with numerous challenges, such as the type and quantity of cells or factors, their delivery, cell viability or factor activity, cell phenotypic stability, timing of treatment, regulatory issues, and high costs (McAllister et al.

2008; Burdick et al. 2013). Therefore, alternate approaches are being developed to substitute an inclusion of support cells or growth factors in a nerve graft.

The ECM is composed of diverse molecules, including proteins, glycoproteins, and glycosaminoglycans, produced by the resident cells in tissues or organs. The composition and structure of ECM are dependent on the phenotype of the resident cells and the function of the tissues or organs. In turn, ECM affects the phenotype and behavior of the resident cells (Bissell et al. 1982; Boudreau et al. 1995; Ingber 1991). It is clear that ECM contains sufficient biological cues to regulate cell phenotype and function in tissues or organs. Based on this knowledge, the mimicking of the native ECM of peripheral nerves within a nerve scaffold seems to be a promising strategy to replace cellular or molecular components added to the scaffold. As an early attempt, nerve scaffolds were prepared with purified individual ECM components, such as collagen, fibrin, laminin, fibronectin, and hyaluronan (Gu et al. 2011; Schmidt and Leach 2003; Deumens et al. 2010; Jiang et al. 2010; Khaing and Schmidt 2012). Some of these scaffolds (e.g., NeuraGen®, NeuroMatrix™, Neuroflex™, NeuraWrap™, and NeuroMend™) have been commercially available and approved by FDA (Kehoe et al. 2012; Meek and Coert 2008). Unfortunately, individual ECM components fail to create an extracellular environment similar to that in vivo within a scaffold in a comprehensive manner because various ECM components have different functions and any individual component cannot substitute for the complete ECM (Ravindran et al. 2012). Indeed, an ECM scaffold can be engineered by using acellular biomaterials, which have been considered a feasible alternative to cellular and/or molecular therapy (Khaing and Schmidt 2012; Burdick et al. 2013). Allogeneic and xenogeneic nerve (or nonnerve) tissues are treated with chemical or thermal decellularization to produce a tissue-derived ECM, which represents one of acellular biomaterials suitable for preparing nerve scaffolds. The resulting tissue-derived ECM scaffolds, commonly called acellular nerve grafts, have been well studied, and some of them are of considerable commercial interest with a FDA-approved product on the market (e.g., Avance®) (Kehoe et al. 2012). As compared to scaffolds prepared with individual ECM components, tissue-derived ECM scaffolds (also called acellular nerve grafts) have a better ability to retain the basic structure of native nerves and promote peripheral nerve regeneration (Whitlock et al. 2009). Mounting evidence indicates that although tissue-derived ECM scaffolds recapitulate biochemical and biophysical cues intrinsic to tissues (Gilbert et al. 2006; Lu et al. 2011; Badylak et al. 2009; Hoshiba et al. 2010; Wolchok and Tresco 2010), they may suffer from several drawbacks, including tissue scarcity, host responses, pathogen transfer, insufficient mechanical properties, and uncontrollable degradation kinetics (Badylak et al. 2009; Cheng et al. 2009; Liao et al. 2010; Skora et al. 2012). In contrast, cultured cell-derived ECM scaffolds have recently attracted attention. They are similar to or even better than tissue-derived ECM scaffolds because the former excludes pathogen transfer when cultured and expanded under pathogen-free conditions and maintains the desired geometry and flexibility when reconstituted with common biomaterials either of synthetic or natural origin (Wolchok and Tresco 2010; Lu et al. 2011; Cheng et al. 2009; Narayanan et al. 2009; Volpato et al. 2013). In consequence, cultured cell-derived ECM scaffolds have been used in some fields of tissue engineering (Wolchok and Tresco 2010; Liao et al. 2010; Lu et al. 2011; Choi et al. 2010). The application of this type of ECM scaffolds, however, has not been fully studied in nerve tissue engineering.

As discussed in detail elsewhere (Williams 2008, 2014), these paradigms move away from the search for biomaterials and structures that passively allow cells to express new ECM; instead, these materials have to be actively involved in the delivery of cues to cells. Indeed, it should be borne in mind that a tissue engineering template should replicate, as far as possible, the niche of those target cells. ECM plays a prominent role in establishing and maintaining an ideal microenvironment for tissue regeneration, and ECM scaffolds are used as a feasible alternative to cellular and molecular therapy in the fields of tissue engineering. Because of their advantages over tissue-derived ECM scaffolds, cultured cell-derived ECM scaffolds are beginning to attract attention, but they have been scarcely studied for peripheral nerve repair.

10.1.3 Outlook on Stem Cells for Peripheral Nerve Regeneration

The research of peripheral nerve regeneration dates back to many years ago, and the past century has witnessed the accelerated development in peripheral nerve repair strategies, especially a significant progress from early artificial tubular NGCs to current TENGs, but clinical applications of state-of-the-art approaches are still limited and the relevant functional outcomes are not completely satisfactory, largely depending on many factors including the size and location of injured nerves as well as the age of patients. As has been repeatedly indicated by previous literature (Fields et al. 1989), only a clear and thorough understanding of the fundamental events, which occur after nerve injury and during nerve regeneration, could result in a medical breakthrough for peripheral nerve repair. The challenges and possibilities facing surgical management of peripheral nerve injuries are likely to be summarized as two main aspects: (1) survival of the damaged neurons and establishment of the microenvironment that facilitates the neurite outgrowth; (2) accurate pathfinding that guides axons to their targets for appropriate reinnervation. Obviously, peripheral nerve regeneration is determined by the large quantity and high speed of axonal outgrowth, the remyelination of axons by SCs, and the maturity of regenerating nerve fibers. Incorporation of support cells and/or growth factors has proven effective for modulating the above cellular behaviors as evidenced by a wide range of animal experiments, but it has not come to clinical use due to multiple barriers.

Target reinnervation is another pivotal cellular event that predetermines functional recovery following peripheral nerve repair. When a nerve is transected with a damage of basal lamina tubes, axonal sprouts are not restrained to their original basal lamina tubes and axons become unable to grow faithfully along original pathways to their target regions. The mismatch of regenerated axons to their targets may contribute, at least in part, to unsatisfactory functional recovery after nerve grafting. Further elucidation of the pathfinding mechanisms of axons is necessary for developing more efficient methods that will enable nerve regeneration to better mimic natural process of neurogenesis. To conclude, nerve regeneration, even in the simpler PNS (as compared to CNS), is also a quite complex phenomenon that is, to date, still far from being fully understood. However, with the advancement of tissue engineering and regenerative medicine and with an accumulated knowledge in the neuroscience field, one can expect that TENGs with a close similarity to native nerve tissues, both in structure and function, will be eventually developed.

The heterogeneity of the stem cell population could also pose a problem, which may be exacerbated if cell manipulation and growth in culture introduce additional genomic variants, which may affect normal functioning. Culture systems can add heterogeneity to the phenotype and genotype, thus complicating selection criteria for transplantation (Bara et al. 2014). Incorrect or incomplete differentiation can also be a concern, as in the case of improperly differentiated endothelium from transplanted PSCs, which can lead to maladapted fibrosis and affect organ function (Ginsberg et al. 2012). Although these potential risks can be partly investigated and addressed by transplantation of cells into suitable immunosuppressed animals, these models may only approximate the human disease and often do not account for the intact human immune system.

Despite the gold standard for peripheral nerve gap repair, autologous nerve grafts fail to achieve an entirely satisfactory restoration of function after they are implanted. Intended to supplement and replace autologous nerve grafts, TENGs should be able to compete with or even surpass autologous nerve grafts in the outcomes of nerve regeneration and functional recovery. Therefore, although the past several decades have witnessed great advance from the earliest nerve tube to the state-of-the-art TENG, neural tissue engineering needs further significant progress toward the development of ideal TENGs and their translation to clinical applications.

It has to be recognized that many materials have been tested and used for TENGs without clear resolution of the optimal structure, a fact by itself which indicates that there is much to learn about their performance. We have delineated here the experimental use of various inorganic and indeed metallic materials, which go beyond the framework of traditional biomaterials, for neural template fabrication. We have also addressed the favorable features of some nanostructured neural scaffolds due to their topographical resemblance of natural ECM architecture. This is a key issue in light of the need for the template to replicate the niche of the target cells. It is unlikely that conventional materials, including most synthetic polymers, will meet the strict requirements of this cell niche concept; the use of decellularized natural tissues and various forms of biopolymers, including hydrogel forms of both proteins and polysaccharides, are clearly very important here. We have described the incorporation of SKPs and AMSCs (as support cells) into neural templates in the construction of TENGs and illustrate newly developed delivery systems for growth factors within them. Certainly, there are still other prospective cues that can be integrated within TENGs, such as molecular inflammatory mediators (Kiefer et al. 2001; Wang et al. 2012; Camara-Lemarroy et al. 2010; Tang et al. 2013), bioactive peptides (Schense et al. 2000; Cheng et al. 2014), and antioxidant reagents (Shen et al. 2007; Wilson et al. 2007). These extra cues are ready to demonstrate their promising applications in neural tissue engineering.

Although the promise of cell therapy for tissue regeneration is exciting, troublesome details persist, such as cell selection, delivery, viability, and phenotypic stability, in addition to timing of treatments, regulatory issues, and high costs (McAllister et al. 2008; Mummery et al. 2010). Nowadays, TENGs used in the clinic, however, are limited to those composed of a neural scaffold alone without any biochemical components due to the presence of various barriers. The construction of an effective TENG should be considered as a complex scientific and engineering problem that involves multifaceted interactions between a diverse array of physicochemical and biological cues, which have been and are still being elucidated within the constantly updated knowledge of peripheral nerve injury and regeneration. The various cues have distinctive effects on the performance of TENGs, but it is necessary to understand and implement the orchestration of the different cue-induced effects. A considerable number of comparative studies must be conducted to decipher. For example, which are more prominent cues, whether and how different cues are interrelated to and interfere with each other. Obviously, the research on these issues will benefit from an improved insight into the molecular events and mechanisms that underlie peripheral nerve injury and regeneration (Navarro et al. 2007; Raimondo et al. 2011; Napoli et al. 2012; Fricker et al. 2013). So far, not only animal models that have TENGs achieved good results, but clinical trials with TENGs to treat human patients with peripheral nerve injury have also met with a certain degree of success (Kehoe et al. 2012; Meek and Coert 2008; Lin et al. 2013; Rinkel et al. 2013; Zhang et al. 2013). Especially, many commercial available products of neural scaffolds have been used in the clinic with promising outcomes (Gu et al. 2011; Kehoe et al. 2012; Meek and Coert 2008).

Although it may be possible to promote the activity of endogenous stem cells to enhance tissue repair, it is becoming increasingly clear that as much attention needs to be paid to the environment in which the cells reside as to the nature of the cells themselves (Watt and Hogan 2000).

The physical niche for any somatic stem cell is composed of two basic components: an acellular ECM and local cellular constituents. The molecular components of the matrix have a profound effect on the biology of stem cells in regulating their quiescence, proliferation, symmetric and asymmetric divisions, and fate (Watt and Huck 2013); conversely, stem cells typically produce constituents of their own matrix. The cells found in the niche, such as differentiated cells of the tissue, interstitial mesenchymal cells, and cellular components of the vasculature, may influence stem cell functionality by direct contact or by locally secreted paracrine factors. Despite considerable success of acellular nerve grafts in supporting peripheral nerve regeneration (Khaing and Schmidt 2012; Krekoski et al. 2001; Kim et al. 2004), they are still associated with several drawbacks, which inspire the emergence of cultured cell-derived ECM scaffolds. The developed nerve scaffolds might have not these drawbacks because (1) the isolated SCs can be in vitro expanded, and even commercially available SC cell lines are also used to obtain ECM; (2) SCs can be cultured in a pathogen-free medium; and (3) the joint use of SC-derived ECM and natural biomaterials allows our developed scaffolds to have robust mechanical properties and exhibit tailored degradation profiles. The safety examination indicated that adverse effects, including adverse immune responses, were not observed in animals after bridging rat sciatic nerve gap with different scaffolds, which provided a necessary basis for the translation of our developed scaffolds to the clinic. Targeted activation of endogenous stem cells to repair the tissues in which they reside requires an understanding of the molecular pathways that normally control stem cell function and how those signals might have changed in the setting of injury or disease, perhaps rendering the stem cells less responsive to extrinsic cues.

A major challenge for the development of targeted therapeutics to enhance endogenous stem cell function is the modeling, in vitro, of the environment that sustains the stem cells in a state of reduced responsiveness in vivo. Typically, plating stem cells in culture induces them to begin dividing and differentiating, thus altering the cellular state that would be the therapeutic target (Dimmeler et al. 2014). To push the translation of stem cell application for peripheral nerve regeneration into the clinic, we anticipate that the most appropriate stem cells and other biological cues with a close proximity to the regenerative microenvironment of the PNS will be developed.

References

Abdallah BM, Kassem M. Human mesenchymal stem cells: from basic biology to clinical applications. Gene Ther. 2008;15(2):109–16.

Alessandri G, Emanueli C, Madeddu P. Genetically engineered stem cell therapy for tissue regeneration. Ann NY Acad Sci. 2004;1015:271–84.

Andrews MR, Stelzner DJ. Modification of the regenerative response of dorsal column axons by olfactory ensheathing cells or peripheral axotomy in adult rat. Exp Neurol. 2004;190(2):311–27.

Artico M, et al. Birthday of peripheral nervous system surgery: the contribution of Gabriele Ferrara (1543–1627). Neurosurgery. 1996;39(2):380–2 (discussion 382–3).

Asplund M, et al. Incidence of traumatic peripheral nerve injuries and amputations in Sweden between 1998 and 2006. Neuroepidemiology. 2009;32(3):217–28.

Au E, et al. SPARC from olfactory ensheathing cells stimulates Schwann cells to promote neurite outgrowth and enhances spinal cord repair. J Neurosci. 2007;27(27):7208–21.

Badylak SF, Freytes DO, Gilbert TW. Extracellular matrix as a biological scaffold material: structure and function. Acta Biomater. 2009;5(1):1–13.

Bara JJ, et al. Concise review: bone marrow-derived mesenchymal stem cells change phenotype following in vitro culture: implications for basic research and the clinic. Stem Cells. 2014;32(7):1713–23.

Battiston B, et al. Chapter 1: Peripheral nerve repair and regeneration research: a historical note. Int Rev Neurobiol. 2009;87:1–7.
Bianco P, et al. Bone marrow stromal stem cells: nature, biology, and potential applications. Stem Cells. 2001;19(3):180–92.
Bissell MJ, Hall HG, Parry G. How does the extracellular matrix direct gene expression? J Theor Biol. 1982;99(1):31–68.
Blazejewska EA, et al. Corneal limbal microenvironment can induce transdifferentiation of hair follicle stem cells into corneal epithelial-like cells. Stem Cells. 2009;27(3):642–52.
Borlongan CV, et al. Bone marrow grafts restore cerebral blood flow and blood brain barrier in stroke rats. Brain Res. 2004;1010(1–2):108–16.
Boudreau N, Myers C, Bissell MJ. From laminin to lamin: regulation of tissue-specific gene expression by the ECM. Trends Cell Biol. 1995;5(1):1–4.
Burdick JA, et al. Acellular biomaterials: an evolving alternative to cell-based therapies. Sci Transl Med. 2013;5(176):176ps4.
Camara-Lemarroy CR, Guzman-de la Garza FJ, Fernandez-Garza NE. Molecular inflammatory mediators in peripheral nerve degeneration and regeneration. NeuroImmunoModulation. 2010;17(5):314–24.
Cao L, et al. Olfactory ensheathing cells promote migration of Schwann cells by secreted nerve growth factor. Glia. 2007;55(9):897–904.
Caplan AI, Dennis JE. Mesenchymal stem cells as trophic mediators. J Cell Biochem. 2006;98(5):1076–84.
Carriel V, et al. Combination of fibrin-agarose hydrogels and adipose-derived mesenchymal stem cells for peripheral nerve regeneration. J Neural Eng. 2013;10(2):026022.
Chalfoun CT, Wirth GA, Evans GR. Tissue engineered nerve constructs: where do we stand? J Cell Mol Med. 2006;10(2):309–17.
Chen X, et al. Ischemic rat brain extracts induce human marrow stromal cell growth factor production. Neuropathology. 2002;22(4):275–9.
Chen Q, et al. Protective effects of bone marrow stromal cell transplantation in injured rodent brain: synthesis of neurotrophic factors. J Neurosci Res. 2005;80(5):611–9.
Chen X, et al. Study of in vivo differentiation of rat bone marrow stromal cells into schwann cell-like cells. Microsurgery. 2006;26(2):111–5.
Chen CJ, et al. Transplantation of bone marrow stromal cells for peripheral nerve repair. Exp Neurol. 2007;204(1):443–53.
Chen Z, et al. Skin-derived precursors as a source of progenitors for cutaneous nerve regeneration. Stem Cells. 2012;30(10):2261–70.
Cheng NC, et al. Chondrogenic differentiation of adipose-derived adult stem cells by a porous scaffold derived from native articular cartilage extracellular matrix. Tissue Eng Part A. 2009a;15(2):231–41.
Cheng HW, et al. Decellularization of chondrocyte-encapsulated collagen microspheres: a three-dimensional model to study the effects of acellular matrix on stem cell fate. Tissue Eng Part C Methods. 2009b;15(4):697–706.
Cheng Q, et al. Neurotrophic and neuroprotective actions of achyranthes bidentata polypeptides on cultured dorsal root ganglia of rats and on crushed common peroneal nerve of rabbits. Neurosci Lett. 2014;562:7–12.
Choi KH, et al. The chondrogenic differentiation of mesenchymal stem cells on an extracellular matrix scaffold derived from porcine chondrocytes. Biomaterials. 2010;31(20):5355–65.
Chopp M, Li Y. Treatment of neural injury with marrow stromal cells. Lancet Neurol. 2002;1(2):92–100.
Chopp M, et al. Spinal cord injury in rat: treatment with bone marrow stromal cell transplantation. NeuroReport. 2000;11(13):3001–5.
Craff MN, et al. Embryonic stem cell-derived motor neurons preserve muscle after peripheral nerve injury. Plast Reconstr Surg. 2007;119(1):235–45.
Crane JF, Trainor PA. Neural crest stem and progenitor cells. Annu Rev Cell Dev Biol. 2006;22:267–86.

Crigler L, et al. Human mesenchymal stem cell subpopulations express a variety of neuro-regulatory molecules and promote neuronal cell survival and neuritogenesis. Exp Neurol. 2006;198(1):54–64.
Cuevas P, et al. Peripheral nerve regeneration by bone marrow stromal cells. Neurol Res. 2002;24(7):634–8.
Cui L, et al. Transplantation of embryonic stem cells improves nerve repair and functional recovery after severe sciatic nerve axotomy in rats. Stem Cells. 2008;26(5):1356–65.
Dabos KJ, et al. The simulated microgravity environment maintains key metabolic functions and promotes aggregation of primary porcine hepatocytes. Biochim Biophys Acta. 2001;1526(2):119–30.
Deryugina EI, Muller-Sieburg CE. Stromal cells in long-term cultures: keys to the elucidation of hematopoietic development? Crit Rev Immunol. 1993;13(2):115–50.
Deumens R, et al. Neurite outgrowth promoting effects of enriched and mixed OEC/ONF cultures. Neurosci Lett. 2006;397(1–2):20–4.
Deumens R, et al. Repairing injured peripheral nerves: bridging the gap. Prog Neurobiol. 2010;92(3):245–76.
Dezawa M. Insights into autotransplantation: the unexpected discovery of specific induction systems in bone marrow stromal cells. Cell Mol Life Sci. 2006;63(23):2764–72.
Dezawa M, et al. Sciatic nerve regeneration in rats induced by transplantation of in vitro differentiated bone-marrow stromal cells. Eur J Neurosci. 2001;14(11):1771–6.
Dezawa M, et al. Specific induction of neuronal cells from bone marrow stromal cells and application for autologous transplantation. J Clin Invest. 2004;113(12):1701–10.
di Summa PG, et al. Long-term in vivo regeneration of peripheral nerves through bioengineered nerve grafts. Neuroscience. 2011;181:278–91.
Dimmeler S, et al. Translational strategies and challenges in regenerative medicine. Nat Med. 2014;20(8):814–21.
Ding F, et al. Use of tissue-engineered nerve grafts consisting of a chitosan/poly(lactic-co-glycolic acid)-based scaffold included with bone marrow mesenchymal cells for bridging 50-mm dog sciatic nerve gaps. Tissue Eng Part A. 2010;16(12):3779–90.
Dombrowski MA, et al. Myelination and nodal formation of regenerated peripheral nerve fibers following transplantation of acutely prepared olfactory ensheathing cells. Brain Res. 2006;1125(1):1–8.
Dupouey P, et al. Immunochemical studies of myelin basic protein in shiverer mouse devoid of major dense line of myelin. Neurosci Lett. 1979;12(1):113–8.
Erba P, et al. Regeneration potential and survival of transplanted undifferentiated adipose tissue-derived stem cells in peripheral nerve conduits. J Plast Reconstr Aesthet Surg. 2010;63(12):e811–7.
Fairless R, Barnett SC. Olfactory ensheathing cells: their role in central nervous system repair. Int J Biochem Cell Biol. 2005;37(4):693–9.
Fernandes KJ, et al. A dermal niche for multipotent adult skin-derived precursor cells. Nat Cell Biol. 2004;6(11):1082–93.
Fickert S, Fiedler J, Brenner RE. Identification, quantification and isolation of mesenchymal progenitor cells from osteoarthritic synovium by fluorescence automated cell sorting. Osteoarthritis Cartilage. 2003;11(11):790–800.
Fields RD, et al. Nerve regeneration through artificial tubular implants. Prog Neurobiol. 1989;33(2):87–134.
Franchi S, et al. Intravenous neural stem cells abolish nociceptive hypersensitivity and trigger nerve regeneration in experimental neuropathy. Pain. 2012;153(4):850–61.
Fricker FR, et al. Axonal neuregulin 1 is a rate limiting but not essential factor for nerve remyelination. Brain. 2013;136(Pt 7):2279–97.
Gilbert TW, Sellaro TL, Badylak SF. Decellularization of tissues and organs. Biomaterials. 2006;27(19):3675–83.
Ginsberg M, et al. Efficient direct reprogramming of mature amniotic cells into endothelial cells by ETS factors and TGFbeta suppression. Cell. 2012;151(3):559–75.

Gokhan S, Mehler MF. Basic and clinical neuroscience applications of embryonic stem cells. Anat Rec. 2001;265(3):142–56.
Goodwin TJ, et al. Reduced shear stress: a major component in the ability of mammalian tissues to form three-dimensional assemblies in simulated microgravity. J Cell Biochem. 1993a;51(3):301–11.
Goodwin TJ, et al. Rotating-wall vessel coculture of small intestine as a prelude to tissue modeling: aspects of simulated microgravity. Proc Soc Exp Biol Med. 1993b;202(2):181–92.
Gu Y, et al. Neurotrophic actions of bone marrow stromal cells on primary culture of dorsal root ganglion tissues and neurons. J Mol Neurosci. 2010;40(3):332–41.
Gu X, et al. Construction of tissue engineered nerve grafts and their application in peripheral nerve regeneration. Prog Neurobiol. 2011;93(2):204–30.
Gu JH, et al. Transplantation of adipose derived stem cells for peripheral nerve regeneration in sciatic nerve defects of the rat. Curr Stem Cell Res Ther. 2012;7(5):347–55.
Guerout N, et al. Transplantation of olfactory ensheathing cells promotes axonal regeneration and functional recovery of peripheral nerve lesion in rats. Muscle Nerve. 2011;43(4):543–51.
Guerout N, et al. Transplantation of olfactory ensheathing cells to evaluate functional recovery after peripheral nerve injury. J Vis Exp. 2014;84:e50590.
Guo BF, Dong MM. Application of neural stem cells in tissue-engineered artificial nerve. Otolaryngol Head Neck Surg. 2009;140(2):159–64.
Haastert K, et al. The effects of FGF-2 gene therapy combined with voluntary exercise on axonal regeneration across peripheral nerve gaps. Neurosci Lett. 2008;443(3):179–83.
Heine W, et al. Transplanted neural stem cells promote axonal regeneration through chronically denervated peripheral nerves. Exp Neurol. 2004;189(2):231–40.
Hofstetter CP, et al. Marrow stromal cells form guiding strands in the injured spinal cord and promote recovery. Proc Natl Acad Sci USA. 2002;99(4):2199–204.
Honkanen H, et al. Isolation, purification and expansion of myelination-competent, neonatal mouse Schwann cells. Eur J Neurosci. 2007;26(4):953–64.
Horwitz EM, Dominici M. How do mesenchymal stromal cells exert their therapeutic benefit? Cytotherapy. 2008;10(8):771–4.
Horwitz EM, et al. Isolated allogeneic bone marrow-derived mesenchymal cells engraft and stimulate growth in children with osteogenesis imperfecta: implications for cell therapy of bone. Proc Natl Acad Sci USA. 2002;99(13):8932–7.
Hoshiba T, et al. Decellularized matrices for tissue engineering. Expert Opin Biol Ther. 2010;10(12):1717–28.
Hsu SH, Su CH, Chiu IM. A novel approach to align adult neural stem cells on micropatterned conduits for peripheral nerve regeneration: a feasibility study. Artif Organs. 2009;33(1):26–35.
Hu J, et al. Repair of extended peripheral nerve lesions in rhesus monkeys using acellular allogenic nerve grafts implanted with autologous mesenchymal stem cells. Exp Neurol. 2007;204(2):658–66.
Hu N, et al. Long-term outcome of the repair of 50 mm long median nerve defects in rhesus monkeys with marrow mesenchymal stem cells-containing, chitosan-based tissue engineered nerve grafts. Biomaterials. 2013;34(1):100–11.
Ikeda M, et al. Acceleration of peripheral nerve regeneration using nerve conduits in combination with induced pluripotent stem cell technology and a basic fibroblast growth factor drug delivery system. J Biomed Mater Res A. 2014;102(5):1370–8.
Ingber D. Extracellular matrix and cell shape: potential control points for inhibition of angiogenesis. J Cell Biochem. 1991;47(3):236–41.
Jahoda CA, Horne KA, Oliver RF. Induction of hair growth by implantation of cultured dermal papilla cells. Nature. 1984;311(5986):560–2.
Jakob H. Stem cells and embryo-derived cell lines: tools for study of gene expression. Cell Differ. 1984;15(2–4):77–80.

Jia H, et al. Sciatic nerve repair by acellular nerve xenografts implanted with BMSCs in rats xenograft combined with BMSCs. Synapse. 2012;66(3):256–69.
Jiang X, et al. Current applications and future perspectives of artificial nerve conduits. Exp Neurol. 2010;223(1):86–101.
Jinno H, et al. Convergent genesis of an adult neural crest-like dermal stem cell from distinct developmental origins. Stem Cells. 2010;28(11):2027–40.
Johnson A, Dorshkind K. Stromal cells in myeloid and lymphoid long-term bone marrow cultures can support multiple hemopoietic lineages and modulate their production of hemopoietic growth factors. Blood. 1986;68(6):1348–54.
Johnson EO, Soucacos PN. Nerve repair: experimental and clinical evaluation of biodegradable artificial nerve guides. Injury. 2008;39(Suppl 3):S30–6.
Johnson EO, Zoubos AB, Soucacos PN. Regeneration and repair of peripheral nerves. Injury. 2005;36(Suppl 4):S24–9.
Johnson EO, Charchanti A, Soucacos PN. Nerve repair: experimental and clinical evaluation of neurotrophic factors in peripheral nerve regeneration. Injury. 2008;39(Suppl 3):S37–42.
Kehoe S, Zhang XF, Boyd D. FDA approved guidance conduits and wraps for peripheral nerve injury: a review of materials and efficacy. Injury. 2012;43(5):553–72.
Keilhoff G, et al. Peripheral nerve tissue engineering: autologous Schwann cells vs. transdifferentiated mesenchymal stem cells. Tissue Eng. 2006a;12(6):1451–65.
Keilhoff G, et al. Transdifferentiated mesenchymal stem cells as alternative therapy in supporting nerve regeneration and myelination. Cell Mol Neurobiol. 2006b;26(7–8):1235–52.
Kelsey JL, et al. Upper extremity disorders: frequency, impact, and cost. New York, NY: Churchill Livingstone; 1997.
Khaing ZZ, Schmidt CE. Advances in natural biomaterials for nerve tissue repair. Neurosci Lett. 2012;519(2):103–14.
Khuong HT, et al. Skin derived precursor Schwann cells improve behavioral recovery for acute and delayed nerve repair. Exp Neurol. 2014;254:168–79.
Kiefer R, et al. The role of macrophages in immune-mediated damage to the peripheral nervous system. Prog Neurobiol. 2001;64(2):109–27.
Kim BS, Yoo JJ, Atala A. Peripheral nerve regeneration using acellular nerve grafts. J Biomed Mater Res A. 2004;68(2):201–9.
Kingham PJ, et al. Adipose-derived stem cells differentiate into a Schwann cell phenotype and promote neurite outgrowth in vitro. Exp Neurol. 2007;207(2):267–74.
Kocsis JD, et al. Cell transplantation of peripheral-myelin-forming cells to repair the injured spinal cord. J Rehabil Res Dev. 2002;39(2):287–98.
Kopen GC, Prockop DJ, Phinney DG. Marrow stromal cells migrate throughout forebrain and cerebellum, and they differentiate into astrocytes after injection into neonatal mouse brains. Proc Natl Acad Sci USA. 1999;96(19):10711–6.
Krekoski CA, et al. Axonal regeneration into acellular nerve grafts is enhanced by degradation of chondroitin sulfate proteoglycan. J Neurosci. 2001;21(16):6206–13.
Kunter U, et al. Transplanted mesenchymal stem cells accelerate glomerular healing in experimental glomerulonephritis. J Am Soc Nephrol. 2006;17(8):2202–12.
Lee RH, et al. Multipotent stromal cells from human marrow home to and promote repair of pancreatic islets and renal glomeruli in diabetic NOD/scid mice. Proc Natl Acad Sci USA. 2006;103(46):17438–43.
Lee G, et al. Isolation and directed differentiation of neural crest stem cells derived from human embryonic stem cells. Nat Biotechnol. 2007;25(12):1468–75.
Liao J, et al. Bioactive polymer/extracellular matrix scaffolds fabricated with a flow perfusion bioreactor for cartilage tissue engineering. Biomaterials. 2010;31(34):8911–20.
Lin MY, Manzano G, Gupta R. Nerve allografts and conduits in peripheral nerve repair. Hand Clin. 2013;29(3):331–48.
Liu CH, Hwang SM. Cytokine interactions in mesenchymal stem cells from cord blood. Cytokine. 2005;32(6):270–9.

Lokanathan Y, et al. Olfactory ensheathing cells seeded muscle-stuffed vein as nerve conduit for peripheral nerve repair: a nerve conduction study. J Biosci Bioeng. 2014;118(2):231–4.

Lu P, Jones LL, Tuszynski MH. BDNF-expressing marrow stromal cells support extensive axonal growth at sites of spinal cord injury. Exp Neurol. 2005;191(2):344–60.

Lu J, et al. Adult bone marrow cells differentiate into neural phenotypes and improve functional recovery in rats following traumatic brain injury. Neurosci Lett. 2006;398(1–2):12–7.

Lu L, et al. Morphological and functional characterization of predifferentiation of myelinating glia-like cells from human bone marrow stromal cells through activation of F3/Notch signaling in mouse retina. Stem Cells. 2008;26(2):580–90.

Lu Z, et al. Overexpression of CNTF in Mesenchymal Stem Cells reduces demyelination and induces clinical recovery in experimental autoimmune encephalomyelitis mice. J Neuroimmunol. 2009;206(1–2):58–69.

Lu H, et al. Autologous extracellular matrix scaffolds for tissue engineering. Biomaterials. 2011a;32(10):2489–99.

Lu H, et al. Cultured cell-derived extracellular matrix scaffolds for tissue engineering. Biomaterials. 2011b;32(36):9658–66.

Luo H, et al. Tissue-engineered nerve constructs under a microgravity system for peripheral nerve regeneration. Tissue Eng Part A. 2014;21:267.

Mackinnon SE, Hudson AR. Clinical application of peripheral nerve transplantation. Plast Reconstr Surg. 1992;90(4):695–9.

Malin D, et al. The extracellular-matrix protein matrilin 2 participates in peripheral nerve regeneration. J Cell Sci. 2009;122(Pt 7):995–1004.

Marchesi C, et al. Skin-derived stem cells transplanted into resorbable guides provide functional nerve regeneration after sciatic nerve resection. Glia. 2007;55(4):425–38.

May F, et al. GDNF-transduced Schwann cell grafts enhance regeneration of erectile nerves. Eur Urol. 2008;54(5):1179–87.

May F, et al. Schwann cell-mediated delivery of glial cell line-derived neurotrophic factor restores erectile function after cavernous nerve injury. Int J Urol. 2013;20(3):344–8.

McAllister TN, et al. Cell-based therapeutics from an economic perspective: primed for a commercial success or a research sinkhole? Regen Med. 2008;3(6):925–37.

McKenzie IA, et al. Skin-derived precursors generate myelinating Schwann cells for the injured and dysmyelinated nervous system. J Neurosci. 2006;26(24):6651–60.

Meek MF, Coert JH. US Food and Drug Administration/Conformit Europe-approved absorbable nerve conduits for clinical repair of peripheral and cranial nerves. Ann Plast Surg. 2008;60(4):466–72.

Meyers VE, et al. RhoA and cytoskeletal disruption mediate reduced osteoblastogenesis and enhanced adipogenesis of human mesenchymal stem cells in modeled microgravity. J Bone Miner Res. 2005;20(10):1858–66.

Mimura T, et al. Peripheral nerve regeneration by transplantation of bone marrow stromal cell-derived Schwann cells in adult rats. J Neurosurg. 2004;101(5):806–12.

Minguell JJ, Erices A. Mesenchymal stem cells and the treatment of cardiac disease. Exp Biol Med (Maywood). 2006;231(1):39–49.

Mohammadi R, Azizi S, Amini K. Effects of undifferentiated cultured omental adipose-derived stem cells on peripheral nerve regeneration. J Surg Res. 2013;180(2):e91–7.

Mohanna PN, et al. A composite poly-hydroxybutyrate-glial growth factor conduit for long nerve gap repairs. J Anat. 2003;203(6):553–65.

Morrison SJ, et al. Prospective identification, isolation by flow cytometry, and in vivo self-renewal of multipotent mammalian neural crest stem cells. Cell. 1999;96(5):737–49.

Muir D. The potentiation of peripheral nerve sheaths in regeneration and repair. Exp Neurol. 2010;223(1):102–11.

Mummery CL, Davis RP, Krieger JE. Challenges in using stem cells for cardiac repair. Sci Transl Med. 2010;2(27):27ps17.

Munoz JR, et al. Human stem/progenitor cells from bone marrow promote neurogenesis of endogenous neural stem cells in the hippocampus of mice. Proc Natl Acad Sci USA. 2005;102(50):18171–6.
Munoz-Elias G, Woodbury D, Black IB. Marrow stromal cells, mitosis, and neuronal differentiation: stem cell and precursor functions. Stem Cells. 2003;21(4):437–48.
Napoli I, et al. A central role for the ERK-signaling pathway in controlling Schwann cell plasticity and peripheral nerve regeneration in vivo. Neuron. 2012;73(4):729–42.
Narayanan K, et al. Three-dimensional reconstituted extracellular matrix scaffolds for tissue engineering. Biomaterials. 2009;30(26):4309–17.
Navarro X, Vivo M, Valero-Cabre A. Neural plasticity after peripheral nerve injury and regeneration. Prog Neurobiol. 2007;82(4):163–201.
Neuhuber B, et al. Axon growth and recovery of function supported by human bone marrow stromal cells in the injured spinal cord exhibit donor variations. Brain Res. 2005;1035(1):73–85.
Noble J, et al. Analysis of upper and lower extremity peripheral nerve injuries in a population of patients with multiple injuries. J Trauma. 1998;45(1):116–22.
Orbay H, et al. Differentiated and undifferentiated adipose-derived stem cells improve function in rats with peripheral nerve gaps. J Plast Reconstr Aesthet Surg. 2012;65(5):657–64.
Ortiguela ME, Wood MB, Cahill DR. Anatomy of the sural nerve complex. J Hand Surg Am. 1987;12(6):1119–23.
Ortiz LA, et al. Mesenchymal stem cell engraftment in lung is enhanced in response to bleomycin exposure and ameliorates its fibrotic effects. Proc Natl Acad Sci USA. 2003;100(14):8407–11.
Park IH, et al. Reprogramming of human somatic cells to pluripotency with defined factors. Nature. 2008;451(7175):141–6.
Park BW, et al. Peripheral nerve regeneration using autologous porcine skin-derived mesenchymal stem cells. J Tissue Eng Regen Med. 2012;6(2):113–24.
Phinney DG, Prockop DJ. Concise review: mesenchymal stem/multipotent stromal cells: the state of transdifferentiation and modes of tissue repair–current views. Stem Cells. 2007;25(11):2896–902.
Radtke C, Kocsis JD. Peripheral nerve injuries and transplantation of olfactory ensheathing cells for axonal regeneration and remyelination: fact or fiction? Int J Mol Sci. 2012;13(10):12911–24.
Radtke C, et al. Transplantation of olfactory ensheathing cells enhances peripheral nerve regeneration after microsurgical nerve repair. Brain Res. 2009;1254:10–7.
Raimondo S, et al. Perspectives in regeneration and tissue engineering of peripheral nerves. Ann Anat. 2011;193(4):334–40.
Rajaram A, Chen XB, Schreyer DJ. Strategic design and recent fabrication techniques for bioengineered tissue scaffolds to improve peripheral nerve regeneration. Tissue Eng Part B Rev. 2012;18(6):454–67.
Rao MS, Anderson DJ. Immortalization and controlled in vitro differentiation of murine multipotent neural crest stem cells. J Neurobiol. 1997;32(7):722–46.
Ravindran S, et al. Biomimetic extracellular matrix-incorporated scaffold induces osteogenic gene expression in human marrow stromal cells. Tissue Eng Part A. 2012;18(3–4):295–309.
Rider DA, et al. Autocrine fibroblast growth factor 2 increases the multipotentiality of human adipose-derived mesenchymal stem cells. Stem Cells. 2008;26(6):1598–608.
Ringden O, et al. Mesenchymal stem cells for treatment of therapy-resistant graft-versus-host disease. Transplantation. 2006;81(10):1390–7.
Rinkel WD, et al. What is evidence based in the reconstruction of digital nerves? A systematic review. J Plast Reconstr Aesthet Surg. 2013;66(2):151–64.
Robinson LR. Traumatic injury to peripheral nerves. Muscle Nerve. 2000;23(6):863–73.
Rumsey JW, et al. Node of Ranvier formation on motoneurons in vitro. Biomaterials. 2009;30(21):3567–72.

Salzer JL, Brophy PJ, Peles E. Molecular domains of myelinated axons in the peripheral nervous system. Glia. 2008;56(14):1532–40.
Sanchez-Ramos J, et al. Adult bone marrow stromal cells differentiate into neural cells in vitro. Exp Neurol. 2000;164(2):247–56.
Sauka-Spengler T, Bronner-Fraser M. A gene regulatory network orchestrates neural crest formation. Nat Rev Mol Cell Biol. 2008;9(7):557–68.
Schense JC, et al. Enzymatic incorporation of bioactive peptides into fibrin matrices enhances neurite extension. Nat Biotechnol. 2000;18(4):415–9.
Schmidt CE, Leach JB. Neural tissue engineering: strategies for repair and regeneration. Annu Rev Biomed Eng. 2003;5:293–347.
Scholz T, et al. Neuronal differentiation of human adipose tissue-derived stem cells for peripheral nerve regeneration in vivo. Arch Surg. 2011;146(6):666–74.
Schuldiner M, et al. Effects of eight growth factors on the differentiation of cells derived from human embryonic stem cells. Proc Natl Acad Sci USA. 2000;97(21):11307–12.
Schuldiner M, et al. Induced neuronal differentiation of human embryonic stem cells. Brain Res. 2001;913(2):201–5.
Seidlits SK, Lee JY, Schmidt CE. Nanostructured scaffolds for neural applications. Nanomedicine (Lond). 2008;3(2):183–99.
Shakhbazau A, et al. Sensory recovery after cell therapy in peripheral nerve repair: effects of naive and skin precursor-derived Schwann cells. J Neurosurg. 2014;121(2):423–31.
Shen Y, et al. Neuroprotective effect of carnosine on necrotic cell death in PC12 cells. Neurosci Lett. 2007;414(2):145–9.
Shen CC, Yang YC, Liu BS. Peripheral nerve repair of transplanted undifferentiated adipose tissue-derived stem cells in a biodegradable reinforced nerve conduit. J Biomed Mater Res A. 2012;100(1):48–63.
Shi Y, et al. Transplantation of neural stem cells overexpressing glia-derived neurotrophic factor promotes facial nerve regeneration. Acta Otolaryngol. 2009;129(8):906–14.
Skora J, et al. Evaluation of the humoral and cellular immune responses after implantation of a PTFE vascular prosthesis. Postepy Hig Med Dosw (Online). 2012;66:469–74.
Su Z, et al. Olfactory ensheathing cells: the primary innate immunocytes in the olfactory pathway to engulf apoptotic olfactory nerve debris. Glia. 2013;61(4):490–503.
Suganuma S, et al. Uncultured adipose-derived regenerative cells promote peripheral nerve regeneration. J Orthop Sci. 2013;18(1):145–51.
Sun F, et al. Combined use of decellularized allogeneic artery conduits with autologous transdifferentiated adipose-derived stem cells for facial nerve regeneration in rats. Biomaterials. 2011;32(32):8118–28.
Suzuki H, et al. Neurospheres induced from bone marrow stromal cells are multipotent for differentiation into neuron, astrocyte, and oligodendrocyte phenotypes. Biochem Biophys Res Commun. 2004;322(3):918–22.
Takahashi K, et al. Induction of pluripotent stem cells from adult human fibroblasts by defined factors. Cell. 2007;131(5):861–72.
Tan CW, et al. Sciatic nerve repair with tissue engineered nerve: olfactory ensheathing cells seeded poly (lactic-co-glygolic acid) conduit in an animal model. Indian J Orthop. 2013;47(6):547–52.
Tang X, et al. Bridging peripheral nerve defects with a tissue engineered nerve graft composed of an in vitro cultured nerve equivalent and a silk fibroin-based scaffold. Biomaterials. 2012;33(15):3860–7.
Tang X, et al. Signaling pathways regulating dose-dependent dual effects of TNF-alpha on primary cultured Schwann cells. Mol Cell Biochem. 2013;378(1–2):237–46.
Tannemaat MR, et al. Differential effects of lentiviral vector-mediated overexpression of nerve growth factor and glial cell line-derived neurotrophic factor on regenerating sensory and motor axons in the transected peripheral nerve. Eur J Neurosci. 2008;28(8):1467–79.
Taylor CA, et al. The incidence of peripheral nerve injury in extremity trauma. Am J Phys Med Rehabil. 2008;87(5):381–5.

Timmer M, et al. Axonal regeneration across long gaps in silicone chambers filled with Schwann cells overexpressing high molecular weight FGF-2. Cell Transplant. 2003;12(3):265–77.
Tohill M, Terenghi G. Stem-cell plasticity and therapy for injuries of the peripheral nervous system. Biotechnol Appl Biochem. 2004;40(Pt 1):17–24.
Toma JG, et al. Isolation of multipotent adult stem cells from the dermis of mammalian skin. Nat Cell Biol. 2001;3(9):778–84.
Tomita K, et al. Differentiated adipose-derived stem cells promote myelination and enhance functional recovery in a rat model of chronic denervation. J Neurosci Res. 2012;90(7):1392–402.
Tomita K, et al. Glial differentiation of human adipose-derived stem cells: implications for cell-based transplantation therapy. Neuroscience. 2013;236:55–65.
Uemura T, et al. Transplantation of induced pluripotent stem cell-derived neurospheres for peripheral nerve repair. Biochem Biophys Res Commun. 2012;419(1):130–5.
Unsworth BR, Lelkes PI. Growing tissues in microgravity. Nat Med. 1998;4(8):901–7.
Verdu E, et al. Olfactory bulb ensheathing cells enhance peripheral nerve regeneration. NeuroReport. 1999;10(5):1097–101.
Volpato FZ, et al. Using extracellular matrix for regenerative medicine in the spinal cord. Biomaterials. 2013;34(21):4945–55.
Vunjak-Novakovic G, et al. Bioreactor cultivation conditions modulate the composition and mechanical properties of tissue-engineered cartilage. J Orthop Res. 1999;17(1):130–8.
Walsh S, et al. Supplementation of acellular nerve grafts with skin derived precursor cells promotes peripheral nerve regeneration. Neuroscience. 2009;164(3):1097–107.
Walsh SK, et al. Skin-derived precursor cells enhance peripheral nerve regeneration following chronic denervation. Exp Neurol. 2010;223(1):221–8.
Wang D, et al. Bridging small-gap peripheral nerve defects using acellular nerve allograft implanted with autologous bone marrow stromal cells in primates. Brain Res. 2008;1188:44–53.
Wang J, et al. Bone marrow mesenchymal stem cells promote cell proliferation and neurotrophic function of Schwann cells in vitro and in vivo. Brain Res. 2009;1262:7–15.
Wang A, et al. Induced pluripotent stem cells for neural tissue engineering. Biomaterials. 2011;32(22):5023–32.
Wang Y, et al. Gene network revealed involvements of Birc2, Birc3 and Tnfrsf1a in anti-apoptosis of injured peripheral nerves. PLoS ONE. 2012;7(9):e43436.
Wanner IB, et al. Role of N-cadherin in Schwann cell precursors of growing nerves. Glia. 2006;54(5):439–59.
Watt FM, Hogan BL. Out of Eden: stem cells and their niches. Science. 2000;287(5457):1427–30.
Watt FM, Huck WT. Role of the extracellular matrix in regulating stem cell fate. Nat Rev Mol Cell Biol. 2013;14(8):467–73.
Wei Y, et al. Chitosan/silk fibroin-based tissue-engineered graft seeded with adipose-derived stem cells enhances nerve regeneration in a rat model. J Mater Sci Mater Med. 2011;22(8):1947–64.
Weimann JM, et al. Contribution of transplanted bone marrow cells to Purkinje neurons in human adult brains. Proc Natl Acad Sci USA. 2003a;100(4):2088–93.
Weimann JM, et al. Stable reprogrammed heterokaryons form spontaneously in Purkinje neurons after bone marrow transplant. Nat Cell Biol. 2003b;5(11):959–66.
Whitlock EL, et al. Processed allografts and type I collagen conduits for repair of peripheral nerve gaps. Muscle Nerve. 2009;39(6):787–99.
Williams DF. On the mechanisms of biocompatibility. Biomaterials. 2008;29(20):2941–53.
Williams DF. Essential biomaterials science. Cambridge: Cambridge University Press; 2014.
Wilson AD, et al. Delayed acetyl-L-carnitine administration and its effect on sensory neuronal rescue after peripheral nerve injury. J Plast Reconstr Aesthet Surg. 2007;60(2):114–8.
Wislet-Gendebien S, et al. Plasticity of cultured mesenchymal stem cells: switch from nestin-positive to excitable neuron-like phenotype. Stem Cells. 2005;23(3):392–402.

Wolchok JC, Tresco PA. The isolation of cell derived extracellular matrix constructs using sacrificial open-cell foams. Biomaterials. 2010;31(36):9595–603.

Woodbury D, et al. Adult rat and human bone marrow stromal cells differentiate into neurons. J Neurosci Res. 2000;61(4):364–70.

Wrabetz L, et al. Different intracellular pathomechanisms produce diverse myelin protein zero neuropathies in transgenic mice. J Neurosci. 2006;26(8):2358–68.

Xiong Y, et al. Synaptic transmission of neural stem cells seeded in 3-dimensional PLGA scaffolds. Biomaterials. 2009;30(22):3711–22.

Xu Y, et al. Myelin-forming ability of Schwann cell-like cells induced from rat adipose-derived stem cells in vitro. Brain Res. 2008;1239:49–55.

Xue C, et al. Joint use of a chitosan/PLGA scaffold and MSCs to bridge an extra large gap in dog sciatic nerve. Neurorehabil Neural Repair. 2012;26(1):96–106.

Yang Y, et al. Repair of rat sciatic nerve gap by a silk fibroin-based scaffold added with bone marrow mesenchymal stem cells. Tissue Eng Part A. 2011;17(17–18):2231–44.

Yohn DC, et al. Transplanted mouse embryonic stem-cell-derived motoneurons form functional motor units and reduce muscle atrophy. J Neurosci. 2008;28(47):12409–18.

Yu J, et al. Induced pluripotent stem cell lines derived from human somatic cells. Science. 2007;318(5858):1917–20.

Zhang P, et al. Bridging small-gap peripheral nerve defects using biodegradable chitin conduits with cultured schwann and bone marrow stromal cells in rats. J Reconstr Microsurg. 2005;21(8):565–71.

Zhang Y, et al. A nerve graft constructed with xenogeneic acellular nerve matrix and autologous adipose-derived mesenchymal stem cells. Biomaterials. 2010;31(20):5312–24.

Zhang P, et al. Biodegradable conduit small gap tubulization for peripheral nerve mutilation: a substitute for traditional epineurial neurorrhaphy. Int J Med Sci. 2013;10(2):171–5.

Zochodne DW. The challenges and beauty of peripheral nerve regrowth. J Peripher Nerv Syst. 2012;17(1):1–18.

Chapter 11
Biomaterial-Assisted Stem Cell Engineering for Tissue Construction and Regeneration

Xiaojun Yan, Bingjie Wang, Chunxiao Qi, Jonathan Joseph Lowrey and Yanan Du

Abstract Stem cells offer optimal cell sources to generate in vitro cellular models for physio/pathological studies and drug screening as well as for cell-based regenerative therapy. Despite the great potential for stem-cell-based applications, challenges including difficulties in efficient in vitro expansion, controlled in vitro tissue formation, and effective in vivo regeneration are hindering the pace of their translation. To overcome these limitations, biomaterials have been used to recreate the complex three-dimensional (3D) natural niches for regulation of stem cell behaviors. This chapter aims to review the advances in biomaterial-assisted stem cell engineering in three aspects: (1) two-dimensional (2D) surfaces and 3D hydrogels/scaffolds engineered to maintain stemness and sustain self-renewal of stem cells for providing reliable cell sources; (2) 3D biomaterials employed to effectively and efficiently differentiate stem cells to desired lineages as well as to establish organotypic constructs mimicking in vivo organogenesis; and (3) decellularized scaffolds, synthetic scaffolds, and hydrogels designed as vehicles for delivering and protecting exogenous stem cells or as stimulators for activating endogenous stem cells to repair and regenerate injured or diseased tissues. Besides introducing some of the most representative works in these areas, this chapter also gives special attention to the advances made in China. Finally, future perspectives and trends in this field will be discussed to highlight the importance of biomaterials in tissue construction and regeneration.

X. Yan · B. Wang · C. Qi · Y. Du
Department of Biomedical Engineering, School of Medicine,
Tsinghua University, Beijing 100084, China

B. Wang
School of Life Sciences, Tsinghua University, Beijing 100084, People's Republic of China

J.J. Lowrey
Department of Biomedical Engineering, The Johns Hopkins University, Baltimore, MD, USA

Y. Du (✉)
Department of Biomedical Engineering, School of Medicine,
Collaborative Innovation Center for Diagnosis and Treatment of Infectious Diseases,
Tsinghua University, Beijing 100084, China
e-mail: duyanan@tsinghua.edu.cn

R.C. Zhao (ed.), *Stem Cells: Basics and Clinical Translation*, Translational Medicine Research 1, DOI 10.1007/978-94-017-7273-0_11

Keywords Stem cell · Biomaterial · Tissue engineering · Regeneration

11.1 Introduction

Stem cells play critical roles in biological understanding of organ and disease development. They have great potential in generating disease models for drug screening as well as in regenerative medicine for replacing or repairing damaged tissues (Main et al. 2014; Mimeault et al. 2007; Nirmalanandhan and Sittampalam 2009). A cell is classified as a stem cell when it is able to proliferate while maintaining its undifferentiated state and has the capacity to differentiate into various specialized cell types. Depending on their origins, stem cells are generally classified as embryonic stem cells (ESCs), induced pluripotent stem cells (iPSCs), and adult stem cells (Wu and Hothedlinger 2011; Anversa et al. 2013). ESCs are undifferentiated cells derived from the inner mass cells of a human embryo that possess the ability to differentiate into any lineage of cells (pluripotent) (Jensen et al. 2009; Keller 2005). Adult stem cells, on the other hand, are a reserve of cells in matured bodies that can multiply to repair adult tissues or organs. Such stem cells have limited self-renewal ability and more committed cell lineages. However, not all organs have been confirmed to have such a reserve of cells (Locke et al. 2011; Bianco et al. 2013). iPSCs were only discovered in the last decade and are produced by reprogramming adult cells to pluripotent stem cells with external introduction of four specific genes (Wu and Hothedlinger 2011). Due to the relative availability of supplies (such as bone marrow and adipose tissue) and lack of ethical issues, adult stem cells are currently the most commonly used cell source in regenerative medicine.

Challenges for successful translation of stem cell technologies to biomedical applications include expansion of cells in vitro without losing stemness, controlled differentiation into desired lineages with high efficiency and efficacy, and achieving desired performance after transplantation into a foreign tissue environment for effective and safe therapy (Fuchs et al. 2004). A steady and reliable source of cells retaining their pluripotency is the prerequisite for any cell-based application; hence, it is desirable for the limited stem cells obtained from various origins to be reproducibly expanded in large-scale in vitro. Complete or controlled differentiation of stem cells is another essential requirement for favorable outcomes in applications, as undesired proliferation or cell types will result in tumors, decreased therapeutic effects, or unreliable in vitro models (Toh et al. 2010). Upon transplantation into the damaged tissues for repair, stem cells or their derivatives need to be able to perform as anticipated, but without undesirable side effects, for effective cell-based therapy.

To meet these requirements, it is essential to understand how stem cells usually behave and develop in their natural microenvironment, the so-called stem cell niche. (Fuchs et al. 2004). A stem cell niche is usually a complex three-dimensional (3D) microenvironment containing extracellular matrix (ECM), soluble factors, and supporting cells. ECM is secreted and remodeled by resident cells, and some common ECMs are collagen, laminin, fibronectin, and polysaccarides [e.g., hyaluronic acid (HA)]. These components provide binding motifs, mostly specific peptide sequences, which interact with integrins on cell membranes. Recent studies have revealed that ECM is not merely a collection of adhesion sites for cell attachment and migration, but the binding is specific to different receptors on different cell membranes. ECM also provides binding domains for several growth factors such as fibroblast growth factor (FGF), vascular endothelial growth factor (VEGF), and hepatic growth factor (HGF) (Crapo et al. 2011; Gilbert et al. 2006; Freytes et al. 2004; Gilbert et al. 2005). Thus, this chapter aims to review the engineering efforts made in biomaterials to exploit such biochemical and biophysical factors to regulate stem cell behaviors, especially in in vitro expansion of stem cells, in vitro 3D tissue construction, and in vivo tissue regeneration.

Current advances in techniques to expand stem cells either involve 2D biomimetic materials or 3D hydrogels and scaffolds to reduce loss of stemness. Efficiency and efficacy of in vitro differentiation are also enhanced when 3D hydrogels and scaffolds are used for culture and lineage induction. Besides cell level phenotypic improvements, researchers have also mimicked in vivo conditions to construct 3D organ-like tissues ex vivo with the aid of suitable biomaterials. Hydrogels or scaffolds are further applied to tissue regeneration in vivo to deliver stem cells, protect them from attack by host immune system, or activate endogenous stem cells to repair and regenerate injured or diseased organs. As there has been much research in this field, this chapter will not be exhaustive in its examples but will only provide some of the most representative works, with special focus given to advances made in China. Due to its promising potential in biomedical and regenerative medical applications, the future prospects of stem cell technologies and engineering will also be discussed.

11.2 Biomaterials for Stem Cell Expansion

In order to expand stem cells without losing their stemness and harvest enough cell source, many methods have been developed to control stem cell fate. In traditional 2D cell culture in vitro, soluble factors in the culture medium are very important. Soluble factors are cell-secreted molecules recruited for stimulating cellular growth, expansion, and differentiation. Soluble factors are commonly used as supplements in basal culture medium for controlling stem cells fate. The use of feeder cells such as mouse embryonic fibroblast (MEF) for human pluripotent stem cells (hPSCs) and mouse embryonic fibroblast cell line (STO) for hepatoblast-like cells (HBCs) is also a common method for maintaining stemness in traditional biology research. Co-culturing ESCs with supporting cells recapitulates key features of the natural stem cell niche, which usually contains multiple cell types. Co-culture systems can promote the differentiation of ESCs via a paracrine signaling pathway, where inductive factors are continuously secreted by supporting cells (Duester 2008). Besides traditional methods, engineering systems are widely used in biology research in current years, and greatly assisted traditional scientific research. In this part, we will focus on engineering methods for optimizing cell niche to maintain stemness of stem cells in both 2D and 3D cell culture, thereby providing reliable cell sources.

11.2.1 Biomaterial-Assisted Stem Cell Expansion in 2D

Both biochemical and biophysical cues can be provided by biomaterials in 2D cell culture, for instance, coating with biomaterials to mimic ECM components, stiffness, and topology. It has been demonstrated that the stiffness of 2D substrate has an effect on stem cell adhesion and cytoskeleton assembly (Georges and Janmey 1985). The effect of substrate properties such as chemical constitution, ECM stiffness, and topology on stem cell behaviors has received extensive attention, especially for adult stem cells. Several research groups have reported that mesenchymal stem cells (MSCs) can sense the stiffness of substrates and differentiate into various downstream lineages, and the underlying mechanisms have been extensively explored (Engler et al. 2006; Park et al. 2011; Khetan et al. 2013; Huebsch et al. 2010).

Previous studies have shown that maintenance of undifferentiated human ESCs (hESCs) requires culture on MEF feeders. In 2001, Xu et al. first established a feeder-free system of hESC culture and published the findings in *Nature Biotechnology*. In this system, hESCs are cultured on Matrigel or laminin in medium conditioned by MEF. Through this process, undifferentiated cells can be maintained for at least 130 passages (Xu et al. 2001). The establishment of self-renewing HBCs from hPSCs would realize a stable supply of HBCs for medical applications. Takayama et al. established a system of long-term self-renewal of human ES/iPS-derived HBCs on human laminin 111-coated dishes. When cultured on the laminin 111-coated dishes, hPSC-derived HBCs were maintained for more than 3 months and had the ability to differentiate into both hepatocyte-like cells and cholangiocyte-like cells (Takayama et al. 2013). Lee et al. revealed that polyethylene terephthalate (PET) with optimal stiffness (0.345 GPa) could inhibit hESC differentiation and promote maintenance of hESC self-renewal. No expression of ectoderm, mesoderm, or endoderm markers, which indicated a loss of stemness, was observed in hESCs cultured on a PET substrate with this stiffness. In addition, Rho/ROCK signaling pathway, one of the ECM stiffness-based signaling pathways, was down-regulated when cells were cultured on such material (Lee et al. 2011). In 2010, Gilbert et al. published their work titled *Substrate Elasticity Regulates Skeletal Muscle Stem Cell Self-Renewal in Culture* on *Science* (Gilbert et al. 2010). Muscle stem cells (MuSCs) exhibit robust regenerative capacity in vivo that is rapidly lost in traditional culture. MuSCs cultured on soft hydrogel substrates, mimicking the elasticity of muscle (12 kPa), self-renew in vitro and contribute extensively to muscle regeneration when subsequently transplanted into mice (Gilbert et al. 2010).

Tissue surface is usually not smooth or flattened but covered with grooves, ridges, pits, and pores (Chai and Leong 2007), hence topographical features of the substrate are also key in regulating stem cell fate. Mian Long and his group of Chinese scientists focus on biomaterial-assisted stem cell fate control, and in 2014, they achieved differential regulation of morphology and stemness of mouse ESCs (mESCs) by controlling substrate stiffness and topography. In their work, mESCs are able to grow effectively on both polystyrene and polyacrylamide substrates in the absence of feeder cells. They found that mESCs formed relatively flattened colony on substrates with grooves or square pillars but formed spheroid colony on substrate with hexagonal projections. These cells also formed 3D structure when cultured on substrates with either grooves or hexagonal projections. Their work demonstrated that topography is a critical factor for manipulating stemness and furthered the understanding of stem cell morphology and stemness in a microenvironment that mimics physiological conditions (Lu et al. 2014).

11.2.2 Biomaterial-Assisted Stem Cell Expansion in 3D

In parallel with 2D biomaterial-assisted stem cell niches for cell expansion, numerous 3D hydrogels/scaffolds of synthetic or natural origin with designated properties such as minimized cytotoxicity, good biocompatibility, defined porosity, appropriate pore sizes, and suitable interconnectivity are under development for cell culture. These are expected to greatly assist stem cell fate regulation and some noteworthy examples are summarized below.

In 2010, Li et al. established a system of self-renewing hESCs in 3D porous scaffolds comprising of chitosan and alginate, without the use of feeder cells or conditioned medium. They assessed the pluripotency of the renewed hESCs both in vitro, by evaluation of cellular proliferation, functionality, and gene activities for 21 days; and in vivo, by implantation of the stem cell populated scaffolds in an immunodeficient mouse model to induce teratoma formation (Fig. 11.1a) (Li et al. 2010). Lee et al. achieved long-term maintenance of mESC pluripotency by manipulating integrin signaling within 3D scaffolds without activating Stat3 by the traditionally used exogenous chemical factor, leukemia inhibitory factor (LIF). It was proposed that the 3D environment formed using hydrogel scaffolds could provide specific integrin ligation that could mimic the effect of LIF in a LIF-free medium (Lee et al. 2012). In 2013, Huch et al. published their excellent work about In vitro *expansion of single Lgr5^{+} liver stem cells induced by Wnt-driven regeneration* in *Nature*. Their work used Matrigel to assist Lgr5^{+} liver stem cells to assemble into 3D spheroids while maintaining their bipotential differentiation ability (Fig. 11.1b) (Huch et al. 2013). In the same year, Chinese scientists led by Jianwu Dai published their work on maintenance of self-renewal properties of neural progenitor cells (NPCs) cultured in 3D collagen scaffolds by controlling the

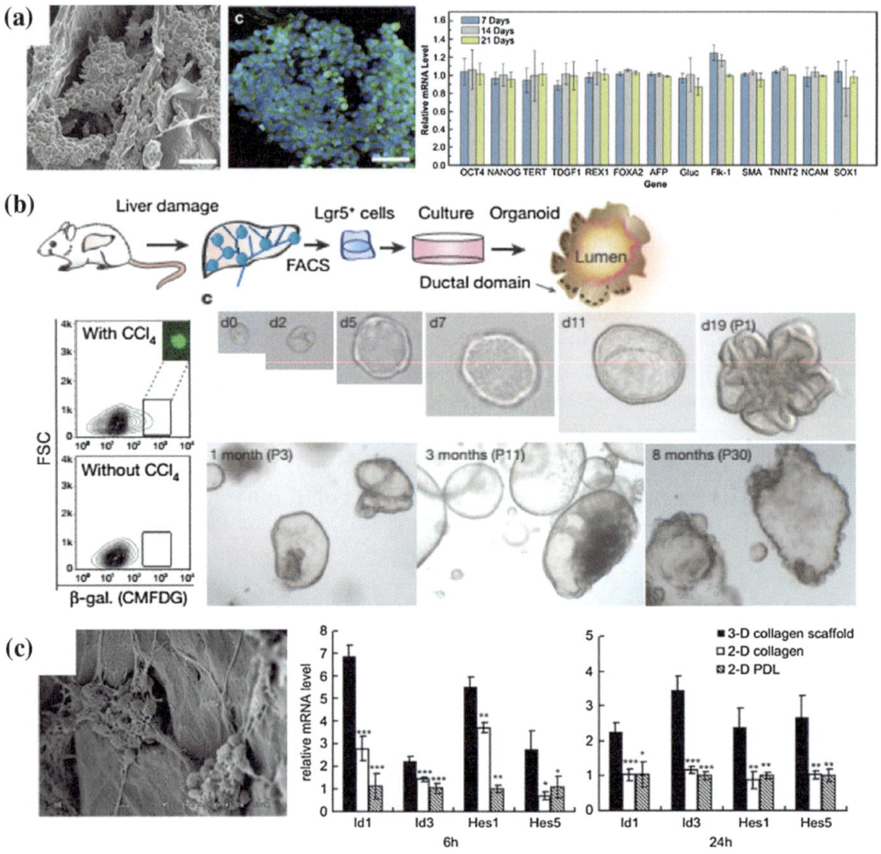

Fig. 11.1 Biomaterial-assisted stem cell engineering for cell expansion in 3D. **a** In vitro assessment of pluripotency of hESCs in chitosan–alginate (*CA*) scaffolds (with hESCs grown on human fibroblast feeder cells layers as controls). SEM image and immuneostaining of SSEA4 to illustrate cell morphology as well as gene activity of hESCs cultured in CA scaffolds for 21 days, assessed by RT-PCR (Li et al. 2010). **b** In vitro expansion of single $Lgr5^+$ cells from adult liver tissue. Lgr5-lacZ mice were injected intraperitoneal with corn oil or CCl_4. Sorted isolated Lgr5-$LacZ^+$ cells were cultured at a ratio of one single Lgr5-$LacZ^+$ cell per well (Huch et al. 2013). **c** Maintenance of the self-renewal properties of neural progenitor cells (*NPCs*) in 3D scaffolds. The inhibitory basic helix-loop-helix (bHLH, important transcription factors for NPCs cell fate) was overexpressed by NPCs in 3D collagen scaffolds. The expressions of inhibitory bHLH factors (Id1, Id3, Hes1, and Hes5) of NPCs in 3D collagen scaffolds were higher than those on 2D plates. The result is presented by fold change relative to the mRNA level of 2D PDL group. $*p < 0.05$, $**p < 0.01$, $***p < 0.001$ (Han et al. 2013). (Images reproduced with permission)

REDD1-mTOR signal pathway. In their work, sponge-like collagen scaffolds were used to assess how 3D culture would affect the differentiation and self-renewal of NPCs. Cultured in differentiation medium without growth factors, cells in 3D collagen scaffolds yielded much higher colony formation efficiency and expressed less neuron marker, TUJ1, indicating higher stemness maintenance, compared with cells cultured on 2D plates. It was found that cells cultured in 3D conditions expressed higher level of REDD1, which inhibited mTOR. mTOR inactivation was shown to support the self-renewal of NPCs in 3D culture. Knocking-down REDD1 induced the differentiation of NPCs in 3D collagen scaffolds (Fig. 11.1c) (Han et al. 2013).

11.3 Biomaterials for 3D Tissue Construction in Vitro

During embryonic development, the differentiation of embryo into the three germ layers and determined lineages is tightly regulated by cell–matrix and cell–cell interactions with high spatial and temporal precision. To mimic the 3D architecture and biological role of ECM, there is an increasing interest in developing engineering approaches that enable modulation of the behaviors of stem cells (Kraehenbuehl et al. 2011).

It is proposed that hPSCs and adult stem cells (e.g. MSCs, liver progenitor stem cells) can be regulated by controlling the 3D microenvironment. This approach is based on the premise that cellular responses to environmental factors are predictable (Burdick and Vunjak-Novakovic 2009), and that 3D culture models could permit recapitulation of development in vitro to a degree of complexity which is not achievable in 2D culture systems (Kraehenbuehl et al. 2011). There are emerging trends to utilize 3D microenvironments as stem cell niches to direct stem cell differentiation toward a particular cell type or even to utilize cell self-organization to realize organogenesis in vitro. Herein, we highlight some of the latest advances on stem cell differentiation and tissue formation in two aspects, namely (1) controlling stem cell differentiation in 3D microenvironment and (2) biomaterial-assisted organogenesis.

11.3.1 Controlling Stem Cell Differentiation in 3D Microenvironment

Hydrogels can encompass biological functional entities such as cells, tissues, organs, or entire organisms in 3D culture and can closely mimic natural tissues with their soft and rubbery consistency (Lieleg and Ribbeck 2011). Gerecht et al. synthesized photopolymerized methacrylated hyaluronic acid (HA) hydrogels for cultivation of hESCs in a 3D configuration to examine the function of HA as an engineered stem cell microenvironment. When encapsulated in HA hydrogel disks and cultured in MEF-conditioned medium, hESCs retained their undifferentiated state, normal karyotype, and pluripotency. ESC differentiation could be induced in situ within the same hydrogel simply by altering to endothelial growth medium supplemented with VEGF (Gerecht et al. 2007). Chayosumrit et al. established a 3D model to culture and induce hESC differentiation by encapsulating cells in calcium alginate microcapsules. This work showed that microcapsules could support the differentiation of hESCs into definitive endoderm in 3D and could have potential application for immune isolation and prevention of teratoma formation of hESCs during transplantation in vivo (Chayosumrit et al. 2010).

A paper in PNAS in 2014 by Dixon et al. demonstrated a novel hybrid hydrogel that could switch hPSC fate from self-renewal to differentiation. In this paper, they addressed the dual need of differentiation and renewal by developing a hydrogel-based material that uses ionic de-cross-linking to remove a self-renewal permissive hydrogel (alginate), leaving behind a differentiation-permissive microenvironment (collagen). Adjusting the timing of this switch could preferentially steer hPSC differentiation to mimic lineage commitment during gastrulation to ectoderm (early switch) or mesoderm/endoderm (late switch). They showed that directing early lineage specification using this single system could promote cardiogenesis in high-density cell populations. This work has a potential application of facilitating regenerative medicine by allowing in situ hPSC expansion to be coupled with early lineage specification within defined tissue geometries (Dixon et al. 2014).

3D scaffolds have also been widely applied in tissue engineering as a special microenvironment for cell growth and fate decision (Schussler et al. 2010). In recent years, excellent works about controlling stem cell differentiation in scaffolds have been published, and Chinese scientists contributed greatly to these fields.

A paper by Changyong Wang in China in 2014 showed the promotion of cardiac differentiation of brown adipose-derived stem cells (BADSCs) by chitosan hydrogel for repair after myocardial infarction. In their work, the group explored an injectable tissue engineering strategy to repair damaged myocardium, in which chitosan hydrogels were investigated as a carrier for BADSCs. In vitro, the effect and mechanism of chitosan components on the cardiac differentiation of BADSCs had also been investigated (Wang et al. 2014). Collaboration between Jianwu Dai's and Ruxiang Xu's groups resulted in an in vitro system for accelerated proliferation of neural stem/progenitor cells (NS/PCs) in collagen sponges immobilized with engineered basic fibroblast growth factor (bFGF) for nervous system tissue engineering. They showed that this natural biological neural scaffold consisting of collagen sponges, engineered bFGF, and NS/PCs works as an effective carrier for NS/PCs, and that neural reconstruction was not only contributed by the proliferating cells but also directly by the engineered bFGF retained in the scaffolds. The two aspects of these neural scaffolds produced synergistic effects and represent a promising candidate for nervous system repair (Ma et al. 2014). Guoqiang Chen's group in China focus on the effects of polyhydroxyalkanoates (PHA) and other related biomaterials on stem cell fate regulation. In their work on microRNA regulation-associated chondrogenesis of mouse MSCs grown on PHA, they found that PHA produced by microorganisms, including poly(3-hydroxybutyrate-co-3-hydroxyhexanoate) (PHBHHx), induced chondrogenesis of MSCs, preserved chondrocytic phenotype, and supported chondrocyte-specific ECM secretion. This was the first time that cell–matrix interaction was shown to mediate microRNA-associated regulation of stem cell differentiation (Yan et al. 2011).

11.3.2 Biomaterial-Assisted Organogenesis

From the above, it is obvious that over the past decade, scientists have developed many excellent protocols to achieve high differentiation efficiency and obtain homogeneous cell populations. These homogeneous cell populations resulting from the high differentiation efficiency can provide great cell sources for drug screening and regenerative medicine in vitro, but they lack complete functions as found in in vivo. Recently, scientists have focused on a new area of stem cell research—the spatiotemporal control of dynamic cellular interactions—to initiate organogenesis in vitro and develop functional mini-organs. Organogenesis shows great potential in tissue engineering and opens a new avenue for next-generation regenerative medicine.

Knowledge of the principles by which organ architecture develops through complex collective cell behaviors is very limited. The behavior of stem cells, when they work collectively, can be much more sophisticated than one might expect from their individual programming. Self-organization is the spontaneous formation of ordered patterns and structures from a population of elements (or individuals) that have no or minimal pattern (Babloyantz and Kaczmarek 1979; Isaeva 2012; Saetzler et al. 2011). Self-organization refers to situations in which an internal system is composed of definable elements that lack pre-patterns and can be conceptually distinguished from an external system (Sasai 2013).

In 3D self-organizing culture, progenitors induced by differentiation conditions (e.g., liver bud induction and retinal positional information) undergo multiple cellular interactions. These local interactions build up to emergent collective behaviors, leading to self-organization of complex structures (Sasai 2013). In this part, we will cover recent studies of in vitro tissue formation from stem cells using 3D culture.

Recent works have shown that the complex shapes of tissues such as the intestinal organoids, optic cups, inner ear sensory epithelia, and liver bud form by self-organization in vitro from a homogeneous population of stem cells. In this process, scientists used Matrigel as an assisting biomaterial to help stem cells self-organize into 3D functional micro-tissues.

In 2011, Spence et al. established a robust and efficient process to direct the differentiation of hPSCs into intestinal tissue in vitro using a temporal series of growth factor manipulations to mimic embryonic intestinal development. Their work used Matrigel to assist formation of 3D intestinal "organoids" consisting of a polarized, columnar epithelium that was patterned into villus-like structures and crypt-like proliferative zones that expressed intestinal stem cell markers. The epithelium contained functional enterocytes, as well as goblet, Paneth, and enteroendocrine cells (Fig. 11.2a) (Spence et al. 2011). Nakano et al. published their

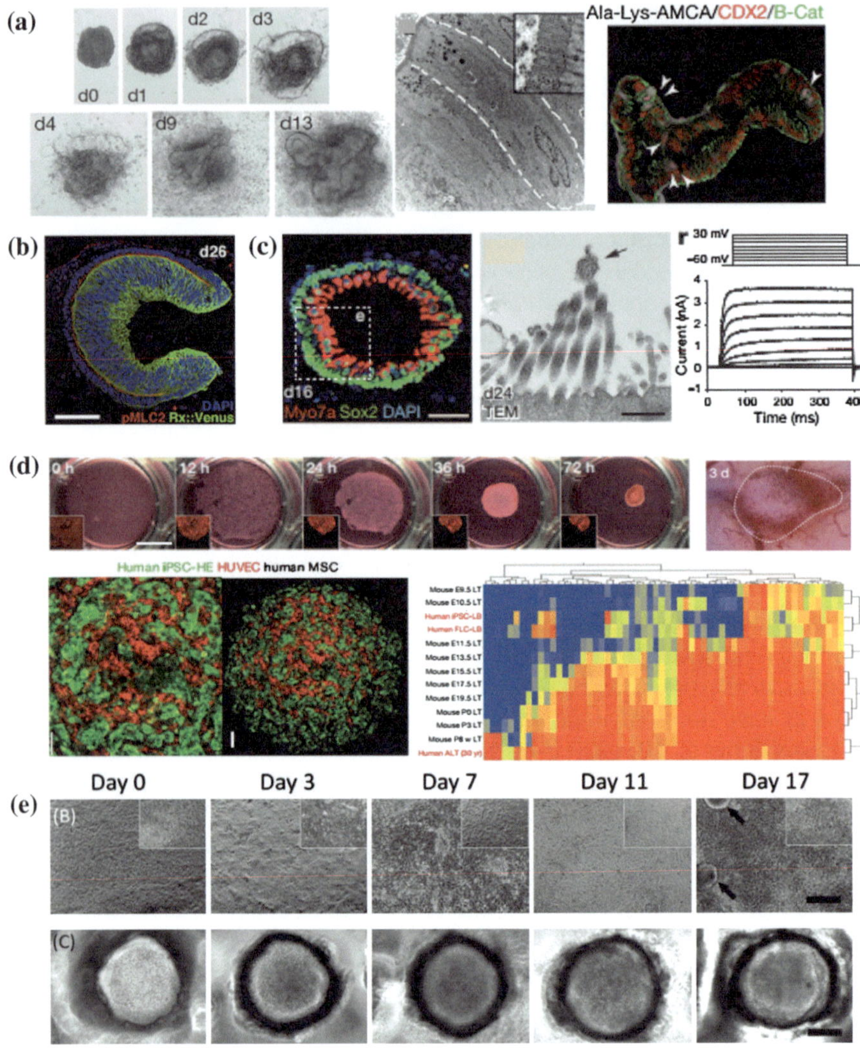

work in 2012 about the self-formation of optic cups and storable stratified neural retina from hESCs. They demonstrated that an optic cup structure could be formed by self-organization in Matrigel. hESC-derived neural retina self-assembled into multilayered tissue containing both rods and cones and they showed that an optimized vitrification method enabled *en bloc* cryopreservation of stratified neural retina of human origin (Fig. 11.2b) (Nakano et al. 2012). In 2013, Koehler et al. successfully generated inner ear sensory epithelia from PSCs in 3D culture, also using Matrigel. They reported the stepwise differentiation of inner ear sensory epithelia from mESCs in 3D culture and showed that by recapitulating in vivo development with precise temporal control of signaling pathways, ESC aggregates transform sequentially into non-neural, pre-placodal, and otic-placode-like

◀ **Fig. 11.2** Biomaterial-assisted organogenesis. **a** Human ESCs and iPSCs form 3D intestine-like organoids—a time course shows that intestinal organoids formed highly convoluted epithelial structures surrounded by mesenchyme after 13 days and electron micrograph showing an enterocyte cell with a characteristic brush border with microvilli (*inset*) and epithelial uptake of the fluorescently labeled dipeptide d-Ala-Lys-AMCA (*arrowheads*) indicating a functional peptide transport system (Spence et al. 2011). **b** Retinal epithelia in 3D hESC culture—immunostaining of the day-26 optic cup (Rx::venus+: *green*) for phospho-MLC2 (pMLC2: *red*). In the optic cup, pMLC2 was strongly accumulated on the apical side of the RPE, but not substantially observed in the NR (Nakano et al. 2012). **c** Non-neural and pre-placodal ectoderm induction in 3D culture. Whole-mount immunofluorescence for Myo7a and Sox2 and 3D reconstruction of a vesicle in a day 20 BMP/SB-FGF/LDN aggregate. Transmission electron micrograph (*TEM*) of stereocilia bundles and kinocilium (*arrow*). Representative voltage–current responses recorded from hair cells (Koehler et al. 2013). **d** Generation of human liver buds from human iPSCs. Self-organization of three-dimensional human iPSC-liver buds in co-cultures of human iPSC-hepatic endoderm cells with Human umbilical vein endothelial cells (*HUVECs*) and human MSCs; Comparison of liver developmental gene signatures among human iPSC-liver buds, human fetal cell-derived liver buds, human adult liver tissue (*ALT*), and mouse liver tissue (*LT*) of various developmental stages (from E9.5 to 8 weeks after birth) (Takebe et al. 2013). **e** Morphological change of hESCs during differentiation as 2D format on TCP (*arrows* indicating uncontrollable cellular overgrowth) and morphological change of cells in the micro-wells resulting in uniform multilayered colonies. Scale bar, 200 μm (Yao et al. 2014). (Images reproduced with permission)

epithelia. Moreover, these stem-cell-derived hair cells exhibit functional properties of native mechano-sensitive hair cells and form specialized synapses with sensory neurons that had also arisen from ESCs in the culture. They also demonstrated that the apparent otic placodes gave rise to vesicles containing pro-sensory cells which are structurally and biochemically comparable to developing vestibular end organs (Fig. 11.2c) (Koehler et al. 2013).

Liver cells derived from human iPSCs and cultured with developmentally important progenitor cells can self-organize into functional, three-dimensional liver buds, according to a research published in *Nature* in 2013 by Takebe et al. (2013). The liver buds exhibited metabolisms resembling, in some aspects, that of human livers and, when transplanted into mice, connected with the host circulatory system. Co-culture systems recapitulate key features of the natural stem cell niche, which usually contains multiple cell types. Matrigel is an appropriate biomaterial which can help the co-cultured cells self-organize into 3D liver buds. Takebe and his colleagues cultured human hepatocytes derived from iPSCs with human umbilical vein endothelial cells and human MSCs, two other developmentally important cell types, to recapitulate liver development. Within 48 h of combining these three cell types, the mass of liver cells self-organized into three-dimensional clusters and formed blood vessels (Fig. 11.2d). Immunostaining and gene-expression analyses revealed a resemblance between in vitro grown iPSC derived LBs (iPSC-LBs) and in vivo liver buds (Fig. 11.2d). Vasculatures in iPSC-LB transplants became functional by connecting to the host vessels within 48 h. The formation of functional vasculatures stimulated the maturation of iPSC-LBs into tissue resembling the adult liver. Microarray analyses of 83 genes known to be active during liver development also revealed that gene expression in the iPSC-derived liver buds was more similar to human fetal liver tissue than that of iPSC-derived hepatocytes created by other groups (Fig. 11.2d) (Takebe et al. 2013).

Recently, hepatic differentiation of hESCs in 3D configuration with better mimicry of embryonic liver development represents incremental efforts to improve differentiation efficiency and hepatic maturation. Chinese scientists in Yanan Du's group have developed off-the-shelf micro-stencil arrays which readily fit into commercial multi-well culture plates to generate adherent multilayered colonies composed of hESC-derived cells subjected to hepatic lineage differentiation. Uninterrupted cellular differentiation and proliferation was achieved within 17 days to recapitulate the continuous and multi-stage liver development. The multilayered colonies as novel 3D configuration for hepatic differentiation of hESCs represent a significant step toward efficient generation of functional hepatocytes for regenerative medicine and drug discovery. The off-the-shelf micro-stencil arrays platform is a potential system to achieve organogenesis and produce organ-on-a-chip in vitro (Fig. 11.2e) (Yao et al. 2014).

All the above examples are excellent works on in vitro organogenesis assisted by biomaterials and represent new hopes and opportunities for regenerative medicine.

11.4 Biomaterial-Assisted Tissue Regeneration

The tremendous advancements in tissue construction in vitro introduced in the above sections provide a great basis for regenerative medicine and stem-cell-based therapy (Fu 2014). Tissue engineering has been envisioned to be a promising alternative to organ transplantation that potentially offers effective method to cure many severe end-stage diseases. Organ transplantation, as commonly known, is greatly restricted by the shortage of organ source (Soto-Gutierrez et al. 2012; Matesanz 2012). Like the rest of the world, China faces the same crisis in the shortage of transplantable tissues and organs, especially liver, heart, kidney, and spinal cord. The waiting list is long and sometimes patients wait in vain and never get the necessary life-saving transplant (Fu 2014). Over the past 20 years, tremendous efforts and progress have been made in regenerative medicine and tissue engineering (Fu 2014). Cell-based therapy may be used to slowdown disease progression, giving patients more time to wait for suitable organs for transplantation (Matesanz 2012).

Cell transplantation therapy has shown excellent therapeutic effects in animal experiments, e.g., transplantation of MSCs to treat acute and chronic liver diseases (Meier et al. 2013). Types of cells selected for therapy depend on the disease to be treated. Among them, stem cells have attracted great attentions due to their inherent advantages in the ability to differentiate into different cell types and to secrete nutrients. Such stem cells could be isolated from a variety of sources, such as embryos, umbilical cord, and adult tissue (Dawson et al. 2008). Stem cells are conventionally injected into recipients directly or pre-differentiated into desired cell types in vitro prior to transplantation. However, the challenge of such therapy is that a large number of transplanted cells would die within a few days after transplantation due to lack of blood perfusion, inflammation, leakage from injection site, and anoikis. These factors result in low cell retention and engraftment at diseased sites and hence greatly impair the efficacy of stem cell therapy (Wang et al. 2010; Robey et al. 2008; Yu et al. 2010). For example, only 11 and 0.6 % of an intra-myocardially injected MSC suspension were retained in rat or porcine heart after 90 min and 24 h, respectively (Yu et al. 2010; Hou et al. 2005). Such observations were also made in clinical trials (Blocklet et al. 2006; Hofmann et al. 2005). Regardless of transplantation method or cell type, it is reported that the immediate retention of cells in saline/media is less than 10 % (Terrovitis et al. 2009). To overcome such barriers in translating cell therapy to clinical applications, various biomaterials have been investigated to improve survival rates and functions of transplanted cells in vivo, either by protecting them from being washed away by high blood flow or from immune attacks (Roche et al. 2014). Numerous biomaterials such as collagen, gelatin, silk, alginate, hyaluronic acid, fibrin, and PLGA have been fabricated into porous scaffolds or hydrogels to deliver cells via transplantation or injection with syringes to target sites or protect them from lymphocytes and immunoglobulins (Venugopal et al. 2012). Taking the advantages of stem cells, biomaterial-assisted cell therapy has been used to provide support for, and in some cases even completely substitute, the failed organs.

11.4.1 Biomaterials for Stem Cell Retention

In order to prevent transplanted cells from being washed away by blood flow, biomaterials have been developed to retain transplanted cells at target sites, thereby improving cell survival rate and treatment efficacy. Naturally derived scaffolds and synthetic hydrogels and/or scaffolds are two main types of materials used for cell transplantation.

11.4.1.1 Naturally Derived Scaffolds

Whole organ decellularization is a major technique to generate scaffolds with obvious advantages of retaining natural structures and intact vascular system (Gilbert et al. 2006). It also keeps ECM intact which can provide suitable physical (e.g., cell adhesion) and physiological (e.g., oxygen and nutrient) conditions for seeded cells (Crapo et al. 2011). Because ECM is biocompatible and biodegradable, decellularized scaffold is a highly suitable and optimal scaffold for in vivo transplantation into a multitude of organs such as urinary bladder (Freytes et al. 2004; Gilbert et al. 2005; Rosario et al. 2008), skin (Chen et al. 2004), heart (Bader et al. 1998; Booth et al. 2002; Kasimir et al. 2003), lung (Price et al. 2010; Daly et al. 2012), tendon (Cartmell and Dunn 2000), blood vessels (Conklin et al. 2002; Dahl et al. 2003; Uchimura et al. 2003), nerves (Hudson et al. 2004; Kim et al. 2004), skeletal muscle (Borschel et al. 2004), ligaments (Woods and Gratzer

2005), small intestinal submucosa (Badylak et al. 1989, 1995; Wang et al. 2003), and liver (Lin et al. 2004). Such scaffolds promote cell survival as seeded cells are in a niche similar to that in vivo, hence repopulated decellularized scaffolds can act as a functional organ graft for transplantation and a bioreactor for three-dimensional cell culture (Crapo et al. 2011). Efficient engraftment and effective therapy have been achieved in several studies based on animals, some of which include treatments for damaged bladder (Yoo et al. 1998), skin (Schechner et al. 2003), heart (Ott et al. 2008; Wainwright et al. 2010), and lung (Cortiella et al. 2010; Petersen et al. 2010).

In 2008, two groups realized whole-heart decellularization for the first time by perfusing detergents via the coronaries in rat and porcine (Fig. 11.3a) (Ott et al. 2008). These decellularized hearts were reported to retain ECM, a perfusable vascular system, and intact chamber geometry, which maintained functional contraction and electrical stimulation of the seeded cardiac or endothelial cells

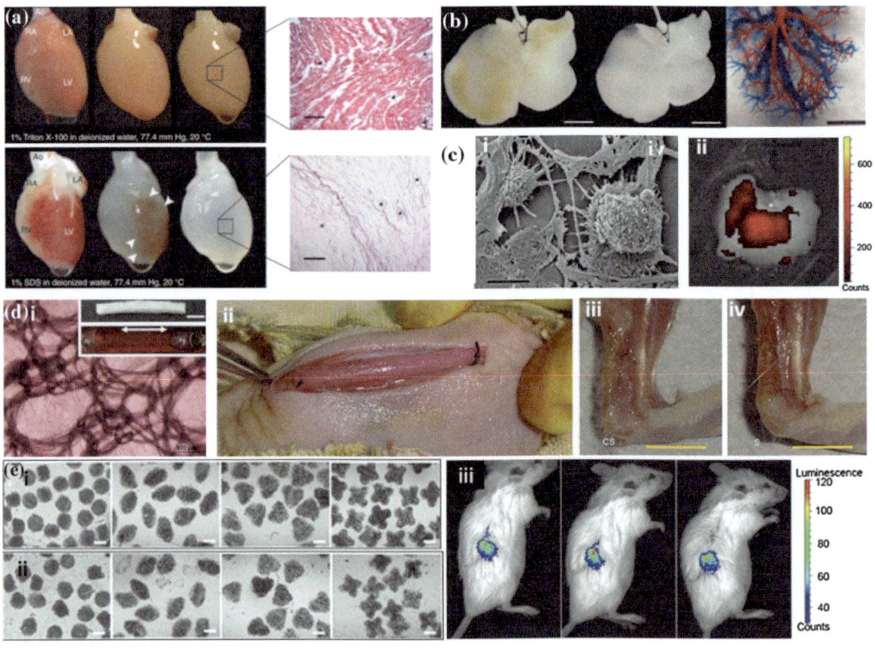

Fig. 11.3 Naturally derived and synthetic scaffolds for cell delivery. **a** Perfusion decellularization of whole rat hearts and HE staining at different stages. **b** Decellularization of ischemic rat livers and corrosion cast model of left lobe of decellularized liver matrix. **c** (i) SEM analysis of decellularized liver scaffold after recellularization on day 7, scale bar = 5 μm; (ii) live imaging of GFP-labeled MSCs in decellularized liver scaffold 6 weeks after transplantation into mouse with CCl_4 induction. **d** (i) Morphology of silk fiber; (ii) transplantation of silk fiber with cells into dorsum of nude mice; (iii and iv) morphology of repaired tendons after 4 weeks, *CS* refers to cells + scaffold; *S* refers to scaffold only. Scale bar = 1 cm. **e** Microscopic images of micro-scaled cryo-gels of different shapes (i) before and (ii) after injection. (iii) Cells in micro-cryo-gels survived till day 7 after injection subcutaneous. (Images reproduced with permission)

for 28 days after seeding (Ott et al. 2008). Taylor's group further optimized the method for cell seeding to obtain more efficient cell distribution and cellularization of the scaffold by infusion via the brachiocephalic artery or via both the inferior vena cava and brachiocephalic artery. The recellularized hearts were then transplanted into recipients, and no immune rejection occurred as demonstrated by the lack of host immune cells observed in the scaffolds 7 days later (Badylak et al. 2011). Vincentelli and colleagues recellularized porcine pulmonary artery with bone marrow mononuclear cells (BMMCs) or MSCs implanted in decellularized pulmonary artery under cardiopulmonary bypass. At 4 months of post-implantation, as shown in echocardiograph and histological results, recolonization and re-endothelialization were achieved in both groups, while significantly greater valve thickening and inflammatory cell infiltration were found in the BMMC recellularization group compared to the MSCs recellularization group. Furthermore, valves from the MSCs group showed ECM and cell disposition that were closer to those of native pulmonary valves (Vincentelli et al. 2007).

Uygun and colleagues decellularized an entire liver and seeded the decellularized scaffold with human fetal hepatocytes and endothelial cells through the intact vascular network (Fig. 11.3b) (Uygun et al. 2010). The cells in the scaffold expressed typical endothelial, hepatic, and biliary epithelial markers indicating that an in vitro liver tissue was successfully created. Ji et al. seeded MSCs into decellularized liver scaffold and cultured in the presence of growth factors in vitro to differentiate MSCs along hepatic lineage. After they observed hepatocytes' ultra-structural characteristics, they transplanted the complex into mice with CCl_4-induced liver injury. As a result, mice survival rate and liver function were increased partially via the influence of paracrine factors on liver stellate cells and native hepatocytes (Fig. 11.3c) (Ji et al. 2012).

Other tissues such as blood vessel (Conklin et al. 2002; Dahl et al. 2003; Schmidt and Baier 2000), airway (Gray et al. 2012), cornea (Ponce Marquez et al. 2009), muscle (Qing and Qin 2009), ligaments (Woods and Gratzer 2005), nerves (Hudson et al. 2004), skin (Chen et al. 2004), tendon (Cartmell and Dunn 2000), intestine (Badylak et al. 1989, 1995; Kropp et al. 1995), and kidney (Sullivan et al. 2012; Ross et al. 2009) have also been decellularized and recellularized successfully so far, all of which provide promising alternatives to organ transplantation in clinical applications.

11.4.1.2 Synthetic Scaffolds

Synthetic scaffolds are made by crosslinking polymers to form hydrogels and/or porous scaffolds. Cells seeded into such scaffolds are usually cultured in vitro for several days, during which ECM and growth factors are secreted and deposited on these scaffolds.

Roche and colleagues compared the cell delivery ability of various biomaterial vehicles [i.e., two injectable hydrogels (alginate, chitosan/ß-glycerophosphate (chitosan/ß-GP) and two epicardial patches (alginate and collagen)] in search for an improvement in the immediate retention of transplanted cells (Roche et al. 2014). They found that all four biomaterials could retain 50–60 % of the transplanted cells as compared to only 10 % when cells were transplanted as a suspension.

Such retention function can improve treatment efficacy significantly. For instance, research group headed by Hongwei Ouyang in China developed a knitted silk-collagen sponge scaffold within which human ESC-derived MSCs (hESC-MSCs) were seeded into diseased tendon. After transplantation, tenocyte-like morphology (scleraxis) and proteins (intergrins and myosin) were observed on hESC-MSCs sheets as well as other mechanosensory structures (cilia) with the expression of markers relating to tendon such as Collagen type I & III and Epha4. Transplanted hESC-MSCs modified the microenvironment at implantation site as well (Chen et al. 2010; Shen et al. 2010). Another Chinese group, Yang et al., incorporated MSCs in silk fibroin and transplanted them into animals with sciatic nerve gaps of 10 mm in rats, 50 and 60 mm in dogs, and 50 mm in rhesus monkeys. These transplanted MSCs promoted reconstruction of injured nerve trunks and restored nerve continuity and axonal functions (e.g., electrical conduction and axoplasmic transport) (Fig. 11.3d) (Yang et al. 2011; Xue et al. 2012).

The above examples utilize invasive methods to deliver cells, i.e., transplantation. However, patients at terminal stage of diseases such as liver failure and heart failure may not be able to withstand the extra trauma caused by transplantation. Furthermore, some patients may place high value on esthetics and would like to minimize scar formation during therapy. Hence, injection as a minimally invasive cell delivery method is an attractive option. Commonly used injectable biomaterials include thermal-sensitive hydrogels made from collagen, Matrigel, and other polymers (Matton et al. 1985). Cells are mixed with polymer solutions at low temperature and injected into target sites. Upon increment to physiological temperature in vivo, a hydrogel is formed which retains cells at desired locations. Other than hydrogels, microporous cryo-gels have been reported to be injectable. Unlike thermal-responsive hydrogels, which have been reported to result in a large number of cell deaths due to high pressure when passing through the needles during injection (Sheikh et al. 2007; Liu et al. 2014), injectable cryo-gels could shield cells from high injection pressures. Koshy et al. synthesized highly porous injectable gelatin scaffolds that could reduce the damage imposed on cells during injection (Koshy et al. 2014). Cell adhesion to scaffold, viability, and proliferation was minimally affected after passing through the needles when seeded in these scaffolds. Minimal host response was observed after transplantation in mice subcutaneously, except for formation of a thick fibrous capsule. Another injectable cryo-gel system developed by Liu and colleagues in China is realized at the microscale level (Liu et al. 2014). Human MSCs in such micro-cryo-gels could be shielded from mechanical stresses and necrosis caused by injection of cell suspension. After subcutaneous injection in mice, cells seeded in micro-cryo-gels have shown concentrated localization and enhanced retention at the injection site (Fig. 11.3e).

11.4.2 Biomaterials for Stem Cell Protection

When delivered into hosts, transplanted cells are inevitably in contact with the host circulatory system and hence are at risk of immunoreactions (Hernandez et al. 2010; Brun-Graeppi et al. 2011). Hosts would recognize these transplanted cells as foreign substances and trigger immune cells to attack and eliminate them (Uludag et al. 2000), thereby resulting in loss of functions of transplanted cells (Hernandez et al. 2010; Jang et al. 2004; Pukel et al. 1988).

Encapsulation of cells resolves such problems, to a certain extent, by reducing immunoreactions and thus prolonging survival of transplanted cells (Hernandez et al. 2010; Chien et al. 2012). Coating with various types of hydrogels is currently commonly used as an immune protection technique (Portero et al. 2010; Orive et al. 2004). Selection of suitable encapsulating materials should take into consideration the appropriate porosity by which nutrients, proteins, DNA, or drugs can be transported but antibodies and immune cells will be blocked, the stability provided by the mechanical property of the material, and the ease of handling such encapsulating membranes (Fig. 11.4a). Satisfying these requirements means that the pore size and thickness of entrapping membrane are restricted to the microscale ranges (Kang et al. 2014). Though there have been several entrapping methods developed, satisfaction of all the above demands remains challenging (Uludag et al. 2000).

The easiest method to encapsulate cells in hydrogels involves simple steps of suspending cells in a pre-gel solution, injecting the mixture into a container, and forming the gel via temperature changes, chemical reaction, or photocuring process (Fig. 11.4d-f) (Kang et al. 2014). There are also other reported methods of gelation such as ultrasonication-induced gelation (Wang et al. 2008). Many kinds of biomaterials have also been applied in beads and sheets formation, including natural polymers such as alginate, chitosan, collagen, and synthetic polymers such as PEG, PLGA, and PLA (Kang et al. 2014). Among these, alginate is most widely used because of its rapid gelation, good biocompatibility, and biodegradability. Microencapsulation commonly describes spherical capsules within the range of 100–1500 μm in diameter. In contrast, macro-encapsulation typically refers to larger constructs with a planar or cylindrical geometry, such as flat sheet or hollow fiber configurations (Fig. 11.4d–f) (Hernandez et al. 2010). Immunoprotection by size-based semipermeable membranes of transplanted cells or tissues allows in situ delivery of secreted proteins to treat many diseases, such as central nervous system (CNS) diseases, diabetes mellitus, hepatic diseases, amyotrophic lateral sclerosis, hemophilia, hypothyroidism, and cardiovascular diseases (Zhang et al. 2008; Grandoso et al. 2007; Colton 1995; Desai et al. 2000).

In 1991, Lacy et al. incorporated islets encapsulated in alginate into acrylic hollow fibers and transplanted them in mice with diabetes. Because islet survival time was prolonged to 60 days, mice showed more efficient diabetes reversal than mice transplanted with islets without encapsulation (Lacy et al. 1991). Ngoc et al. encapsulated MSCs derived from mouse bone marrow or human umbilical cord in an alginate membrane. These capsules were injected intraperitonealy into diabetic mice. Body weight and blood glucose were monitored as indices of treatment efficiency, and total white blood cells as that of immune reactions. Over a 30-day period, the mice's body weight recovered gradually with about a 10 g decrease compared against a control group that received no cell transplantation. The total white blood cell count was lesser in the group treated with encapsulated cells than the group injected with pure MSC suspension (Ngoc et al. 2011). Chinese scientists, Zhang et al., modified Chinese hamster ovary (CHO) cells with VEGF genes and then enveloped these modified cells within semipermeable microcapsules (Fig. 11.4b). After transplantation into myocardial infarction rat models, immune response to encapsulated cells was decreased when compared to control group without encapsulation. They found that the amount of serum anti-CHO antibodies in the encapsulation group was significantly lower than that of control group, resulting in longer implant survival and hence corresponding to better treatment efficacy and cardiac function recovery (Fig. 11.4c) (Zhang et al. 2008).

Multilayered capsulation will no doubt improve mechanical stability, selective permeability, and total entrapment of cells, which is known to block transplanted cells from immune attack and sustained cell survival more efficiently (Chia et al. 2002; Yin et al. 2003). However, immune rejection is, in fact, not completely blocked. In vivo experiments demonstrated that anti-human antibody was detectable in rat serum after 3 days of post-transplantation of encapsulated hepG2, which meant that the rat immune system responded to the entrapped cells (Kang et al. 2014).

Despite the achievements discussed above, several barriers have slowed clinical application, such as biocompatibility, maintenance of long-term implants, biodegradability, control of thickness and porosity, and also the cost of such treatment (Hernandez et al. 2010). Rapid development in micro-fabrication techniques and biomaterials may solve these difficulties and achieve improved immune isolation of transplanted cells in cell therapy (Kang et al. 2014).

11.4.3 Biomaterials for Endogenous Stem Cell Stimulation

Besides transplanting biomaterials pre-loaded with stem cells for disease treatment, another approach to reverse damaged tissue is to recruit and stimulate endogenous stem cells or progenitor cells (Lee et al. 2008, 2009) to repopulate desired locations in situ via intervention with biomaterials (Lee et al. 2008). It has been widely accepted that many tissues contain certain types of stem cells or progenitor cells. Such tissues include brain, skin, circulating blood, fat, liver, heart, and muscle (Lee et al. 2008; Zhang et al. 2003; Guillot et al. 2007; Langer and Vacanti 1993; Ossendorf et al. 2007; Shin'oka et al. 2005). When small and non-life-threatening damages occur in the body, these stem cells would undergo regeneration and differentiation to repair the injury. However, wound recovery will fail if the damage is too severe (Lee et al. 2008). To help severely damaged tissue recruit more stem cells for regeneration, biomaterials can be transplanted into injured sites (Lutolf and Hubbell 2005). Some biomaterials themselves can play the role of stimulators in situ after transplantation into damaged sites, such as collagen (Kin et al. 2007), while others can be modified with bioactive molecules to achieve such effects (Nair et al. 2011).

A myriad of bioactive factors has been incorporated in biomaterials, such as bone morphogenetic protein-2 (BMP-2) (Yamamoto et al. 2006) and FGF-2 (Kodama et al. 2009) in gelatin and VEGF and HGF in PEG scaffolds (Salimath et al. 2012). Transplanted biomaterials release bioactive factors continuously and unlock the body's own regenerative ability. Furthermore, they provide a proper microenvironment for cell survival, proliferation, and differentiation and also guide the reconstruction of 3D tissue (Davis et al. 2005). In China, Zhang et al. produced a matrix environment in situ which is beneficial to cartilage-derived MSCs (C-MSCs) and synovial membrane-derived MSCs (SM-MSCs) to promote cartilage repair spontaneously. They coated type 1 collagen (col1) scaffolds with stromal cell-derived factors-1 (SDF-1), and transplanted them into rabbits with partial-thickness cartilage defects. As results shown, C-MSCs and SM-MSCs migrated and adhered to the col1 scaffolds, achieving self-repair of cartilage defects as assessed by histological score (Zhang et al. 2013). In another research, Shi et al. modified Sca-1 (a member of the Ly-6 family, which is a common marker of adult hematopoietic stem cells) on collagen scaffolds, and then transplanted these scaffolds into mice with surgically induced heart defects. 2 days after transplantation, Sca-1 positive cells were observed on the scaffold. 4 weeks later, orderly arranged collagen fibers were found, and finally regenerated myofibers appeared at 12 weeks after transplantation (Fig. 11.4g–i) (Shi et al. 2011).

From the above examples, it is clear that scientists have improved biomaterial-assisted stem cell therapy in many aspects, such as protecting stem cells during transplantation by seeding the cells in different kinds of hydrogels or scaffolds, after transplantation by entrapping stem cells in semipermeable biomaterials, and by recruiting local stem cells to differentiate or proliferate by transplanting growth factor modified biomaterials. While there is still no optimal system at the moment, it is clear that more efforts are on the way to optimize various conditions to meet all of the therapeutic requirements of transplanted stem cells, such as surviving soundly in vivo without being attacked by the immune system, receiving an adequate supply of oxygen and nutrients, and functioning well enough to reverse or even cure diseases.

11.5 Conclusion and Perspectives

Despite the great advancements made in stem-cell-based technology, challenges still remain for their translation to regenerative medicine and actual applications, including the difficulties in efficient in vitro expansion, controlled in vitro tissue formation, and effective in vivo regeneration. Scientists and engineers have utilized biomaterials to overcome these limitations by recreating the complex 3D natural

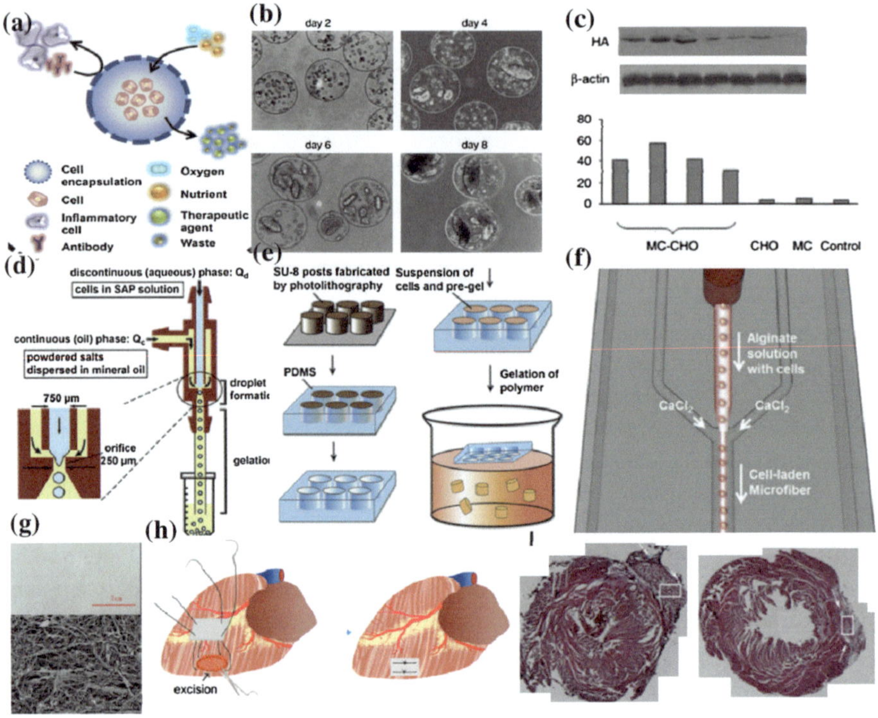

Fig. 11.4 Biomaterials for cell protection and endogenous cell stimulation. **a** Cell microencapsulation protects transplanted cells in cell therapy by enclosing cells in a semi-permeable membrane which allows diffusion of nutrients, oxygen, and waste, as well as paracrine factors secreted by transplanted cells but circumvents immune rejection. **b** Growth of microencapsulated engineered CHO cells at day 4 and day 8, viewed by a phase-contrast microscope (40X). **c** In situ expression of VEGF–hemagglutinin (VEGF—HA) by encapsulated CHO cells. The *left four lanes* were loaded with samples from injection sites of four rats in the microencapsulated CHO (*MC-CHO*) group, and the *right three lanes* were loaded with samples from each rat in the CHO, microcapsule, and control groups, respectively. **d** A mono-dispersed cell-laden microbead fabrication method. **e** Formation of cell-embedded micro-modules using a photo-lithographically fabricated PDMS template. **f** A PDMS-based microfluidic chip for three-dimensional co-culture of hybrid spheroids. **g** SEM of collagen scaffold, scale bar = 100 μm. **h** Surgical procedure of cardiac failure using cardiac patch. **i** HE staining of myocardium regeneration at 90 days after surgery. (Images reproduced with permission)

niches to regulate stem cell behaviors. We have reviewed some of the representative works, especially those advances made in China, in biomaterial-assisted stem cell engineering from three aspects: (1) two-dimensional (2D) substrates and 3D hydrogels/scaffolds engineered to maintain stemness and sustain self-renewal of stem cells for providing reliable cell sources; (2) 3D biomaterials employed to effectively and efficiently differentiate stem cells to desired lineages as well as to establish organotypic constructs mimicking in vivo organogenesis; and (3) decellularized scaffolds, synthetic scaffolds, and hydrogels designed as vehicles for delivering and protecting exogenous stem cells or as stimulators for activating endogenous stem cells to repair and regenerate injured or diseased tissues.

The works reviewed here are all excellent examples illustrating how biomaterials and engineering means could help to re-establish favorable niches for stem cell growth, differentiation, and formation or regeneration of tissue. Many proof-of-concept demonstrations of mini-organ constructions, transplantations, and stem-cell-based therapy also provide promising new approaches to study regenerative medicine. However, more efforts and research have to be done before we can achieve the long-term goal of tissue engineering, which is to synthesize or design novel types of tissue or organs de novo to truly cure and replace degenerated organs in human. We believe that in this process of technological advancement, biomaterial-assisted stem cell engineering for tissue construction and regeneration will play an important role in both in vitro organogenesis and in vivo cell therapy.

Contributions

Xiaojun Yan, Bingjie Wang, Chunxiao Qi, and Yanan Du conceived the ideas and planned the structure of this chapter together. Xiaojun Yan contributed the abstract, introduction, and conclusion, while overseeing the completion and organization, as well as editing, all parts of this chapter. Bingjie Wang contributed literal details and figures for Sects. 11.2 and 11.3 on in vitro stem cell expansion, differentiation, and organogenesis. Chunxiao Qi contributed both literal details and figures for Sect. 11.4 on in vivo tissue regeneration. Jonathan Lowrey and Yanan Du edited the completed versions of this chapter.

Acknowledgments This work was financially supported by the Natural Science Foundation of China (81171474, 51273106), the Beijing Municipal Natural Science Foundation (157142090). All authors would sincerely like to thank Professor Robert Zhao for the invitation to write this chapter. Supports from members of Du-lab are greatly appreciated.

References

Anversa P, et al. Regenerating new heart with stem cells. J Clin Invest. 2013;123(1):62–70.

Babloyantz A, Kaczmarek LK. Self-organization in biological systems with multiple cellular contacts. Bull Math Biol. 1979;41(2):193–201.

Bader A, et al. Tissue engineering of heart valves–human endothelial cell seeding of detergent acellularized porcine valves. Eur J Cardiothorac Surg. 1998;14(3):279–84.

Badylak SF, et al. Small intestinal submucosa as a large diameter vascular graft in the dog. J Surg Res. 1989;47(1):74–80.

Badylak SF, et al. The use of xenogeneic small intestinal submucosa as a biomaterial for Achilles tendon repair in a dog model. J Biomed Mater Res. 1995;29(8):977–85.

Badylak SF, Taylor D, Uygun K. Whole-organ tissue engineering: decellularization and recellularization of three-dimensional matrix scaffolds. Annu Rev Biomed Eng. 2011;13:27–53.

Bianco P, et al. The meaning, the sense and the significance: translating the science of mesenchymal stem cells into medicine. Nat Med. 2013;19(1):35–42.

Blocklet D, et al. Myocardial homing of nonmobilized peripheral-blood CD34+ cells after intracoronary injection. Stem Cells. 2006;24(2):333–6.

Booth C, et al. Tissue engineering of cardiac valve prostheses I: development and histological characterization of an acellular porcine scaffold. J Heart Valve Dis. 2002;11(4):457–62.

Borschel GH, Dennis RG, Kuzon WM Jr. Contractile skeletal muscle tissue-engineered on an acellular scaffold. Plast Reconstr Surg. 2004;113(2):595–602 (discussion 603–4).
Brun-Graeppi AK, et al. Cell microcarriers and microcapsules of stimuli-responsive polymers. J Control Release. 2011;149(3):209–24.
Burdick JA, Vunjak-Novakovic G. Engineered microenvironments for controlled stem cell differentiation. Tissue Eng Part A. 2009;15(2):205–19.
Cartmell JS, Dunn MG. Effect of chemical treatments on tendon cellularity and mechanical properties. J Biomed Mater Res. 2000;49(1):134–40.
Chai C, Leong KW. Biomaterials approach to expand and direct differentiation of stem cells. Mol Ther. 2007;15(3):467–80.
Chayosumrit M, Tuch B, Sidhu K. Alginate microcapsule for propagation and directed differentiation of hESCs to definitive endoderm. Biomaterials. 2010;31(3):505–14.
Chen RN, et al. Process development of an acellular dermal matrix (ADM) for biomedical applications. Biomaterials. 2004;25(13):2679–86.
Chen JL, et al. Efficacy of hESC-MSCs in knitted silk-collagen scaffold for tendon tissue engineering and their roles. Biomaterials. 2010;31(36):9438–51.
Chia SM, et al. Multi-layered microcapsules for cell encapsulation. Biomaterials. 2002;23(3):849–56.
Chien HW, Tsai WB, Jiang S. Direct cell encapsulation in biodegradable and functionalizable carboxybetaine hydrogels. Biomaterials. 2012;33(23):5706–12.
Colton CK. Implantable biohybrid artificial organs. Cell Transplant. 1995;4(4):415–36.
Conklin BS, et al. Development and evaluation of a novel decellularized vascular xenograft. Med Eng Phys. 2002;24(3):173–83.
Cortiella J, et al. Influence of acellular natural lung matrix on murine embryonic stem cell differentiation and tissue formation. Tissue Eng Part A. 2010;16(8):2565–80.
Crapo PM, Gilbert TW, Badylak SF. An overview of tissue and whole organ decellularization processes. Biomaterials. 2011;32(12):3233–43.
Dahl SL, et al. Decellularized native and engineered arterial scaffolds for transplantation. Cell Transplant. 2003;12(6):659–66.
Daly AB, et al. Initial binding and recellularization of decellularized mouse lung scaffolds with bone marrow-derived mesenchymal stromal cells. Tissue Eng Part A. 2012;18(1–2):1–16.
Davis ME, et al. Custom design of the cardiac microenvironment with biomaterials. Circ Res. 2005;97(1):8–15.
Dawson E, et al. Biomaterials for stem cell differentiation. Adv Drug Deliv Rev. 2008;60(2):215–28.
Desai TA, Hansford DJ, Ferrari M. Micromachined interfaces: new approaches in cell immunoisolation and biomolecular separation. Biomol Eng. 2000;17(1):23–36.
Dixon JE, et al. Combined hydrogels that switch human pluripotent stem cells from self-renewal to differentiation. Proc Natl Acad Sci USA. 2014;111(15):5580–5.
Duester G. Retinoic acid synthesis and signaling during early organogenesis. Cell. 2008;134(6):921–31.
Engler AJ, et al. Matrix elasticity directs stem cell lineage specification. Cell. 2006;126(4):677–89.
Freytes DO, et al. Biaxial strength of multilaminated extracellular matrix scaffolds. Biomaterials. 2004;25(12):2353–61.
Fu X. Regenerative medicine research in China: from basic research to clinical practice. Sci China Life Sci. 2014;57(2):155–6.
Fuchs E, Tumbar T, Guasch G. Socializing with the neighbors: stem cells and their niche. Cell. 2004;116(6):769–78.
Georges PC, Janmey PA. Cell type-specific response to growth on soft materials. J Appl Physiol. 1985;98(4):1547–53.
Gerecht S, et al. Hyaluronic acid hydrogel for controlled self-renewal and differentiation of human embryonic stem cells. Proc Natl Acad Sci USA. 2007;104(27):11298–303.

Gilbert TW, et al. Production and characterization of ECM powder: implications for tissue engineering applications. Biomaterials. 2005;26(12):1431–5.
Gilbert TW, Sellaro TL, Badylak SF. Decellularization of tissues and organs. Biomaterials. 2006;27(19):3675–83.
Gilbert PM, et al. Substrate elasticity regulates skeletal muscle stem cell self-renewal in culture. Science. 2010;329(5995):1078–81.
Grandoso L, et al. Long-term survival of encapsulated GDNF secreting cells implanted within the striatum of parkinsonized rats. Int J Pharm. 2007;343(1–2):69–78.
Gray FL, et al. Prenatal tracheal reconstruction with a hybrid amniotic mesenchymal stem cells-engineered construct derived from decellularized airway. J Pediatr Surg. 2012;47(6):1072–9.
Guillot PV, et al. Stem cell differentiation and expansion for clinical applications of tissue engineering. J Cell Mol Med. 2007;11(5):935–44.
Han J, et al. Maintenance of the self-renewal properties of neural progenitor cells cultured in three-dimensional collagen scaffolds by the REDD1-mTOR signal pathway. Biomaterials. 2013;34(8):1921–8.
Hernandez RM, et al. Microcapsules and microcarriers for in situ cell delivery. Adv Drug Deliv Rev. 2010;62(7–8):711–30.
Hofmann M, et al. Monitoring of bone marrow cell homing into the infarcted human myocardium. Circulation. 2005;111(17):2198–202.
Hou D, et al. Radiolabeled cell distribution after intramyocardial, intracoronary, and interstitial retrograde coronary venous delivery: implications for current clinical trials. Circulation. 2005;112(9 Suppl):I150–6.
Huch M, et al. In vitro expansion of single $Lgr5^+$ liver stem cells induced by Wnt-driven regeneration. Nature. 2013;494(7436):247–50.
Hudson TW, Liu SY, Schmidt CE. Engineering an improved acellular nerve graft via optimized chemical processing. Tissue Eng. 2004;10(9–10):1346–58.
Huebsch N, et al. Harnessing traction-mediated manipulation of the cell/matrix interface to control stem-cell fate. Nat Mater. 2010;9(6):518–26.
Isaeva VV. Self-organization in biological systems. Izv Akad Nauk Ser Biol. 2012;2:144–53.
Jang JY, et al. Immune reactions of lymphocytes and macrophages against PEG-grafted pancreatic islets. Biomaterials. 2004;25(17):3663–9.
Jensen J, Hyllner J, Bjorquist P. Human embryonic stem cell technologies and drug discovery. J Cell Physiol. 2009;219(3):513–9.
Ji R, et al. The differentiation of MSCs into functional hepatocyte-like cells in a liver biomatrix scaffold and their transplantation into liver-fibrotic mice. Biomaterials. 2012;33(35):8995–9008.
Kang A, et al. Cell encapsulation via microtechnologies. Biomaterials. 2014;35(9):2651–63.
Kasimir MT, et al. Comparison of different decellularization procedures of porcine heart valves. Int J Artif Organs. 2003;26(5):421–7.
Keller G. Embryonic stem cell differentiation: emergence of a new era in biology and medicine. Genes Dev. 2005;19(10):1129–55.
Khetan S, et al. Degradation-mediated cellular traction directs stem cell fate in covalently crosslinked three-dimensional hydrogels. Nat Mater. 2013;12(5):458–65.
Kim BS, Yoo JJ, Atala A. Peripheral nerve regeneration using acellular nerve grafts. J Biomed Mater Res A. 2004;68(2):201–9.
Kin S, et al. Regeneration of skeletal muscle using in situ tissue engineering on an acellular collagen sponge scaffold in a rabbit model. ASAIO J. 2007;53(4):506–13.
Kodama N, et al. A local bone anabolic effect of rhFGF2-impregnated gelatin hydrogel by promoting cell proliferation and coordinating osteoblastic differentiation. Bone. 2009;44(4):699–707.
Koehler KR, et al. Generation of inner ear sensory epithelia from pluripotent stem cells in 3D culture. Nature. 2013;500(7461):217–21.
Koshy ST, et al. Injectable, porous, and cell-responsive gelatin cryogels. Biomaterials. 2014;35(8):2477–87.

Kraehenbuehl TP, Langer R, Ferreira LS. Three-dimensional biomaterials for the study of human pluripotent stem cells. Nat Methods. 2011;8(9):731–6.
Kropp BP, et al. Experimental assessment of small intestinal submucosa as a bladder wall substitute. Urology. 1995;46(3):396–400.
Lacy PE, et al. Maintenance of normoglycemia in diabetic mice by subcutaneous xenografts of encapsulated islets. Science. 1991;254(5039):1782–4.
Langer R, Vacanti JP. Tissue engineering. Science. 1993;260(5110):920–6.
Lee SJ, et al. Host cell mobilization for in situ tissue regeneration. Rejuvenation Res. 2008;11(4):747–56.
Lee KY, et al. Electrospinning of polysaccharides for regenerative medicine. Adv Drug Deliv Rev. 2009;61(12):1020–32.
Lee S, et al. The effects of the physical properties of culture substrates on the growth and differentiation of human embryonic stem cells. Biomaterials. 2011;32(34):8816–29.
Lee ST, et al. Long-term maintenance of mouse embryonic stem cell pluripotency by manipulating integrin signaling within 3D scaffolds without active Stat3. Biomaterials. 2012;33(35):8934–42.
Li Z, et al. Feeder-free self-renewal of human embryonic stem cells in 3D porous natural polymer scaffolds. Biomaterials. 2010;31(3):404–12.
Lieleg O, Ribbeck K. Biological hydrogels as selective diffusion barriers. Trends Cell Biol. 2011;21(9):543–51.
Lin P, et al. Assessing porcine liver-derived biomatrix for hepatic tissue engineering. Tissue Eng. 2004;10(7–8):1046–53.
Liu W, et al. Microcryogels as injectable 3-D cellular microniches for site-directed and augmented cell delivery. Acta Biomater. 2014;10(5):1864–75.
Locke M, Feisst V, Dunbar R. Concise review: human adipose-derived stem cells: separating promise from clinical need. Stem Cells. 2011;29(3):404–11.
Lu D, et al. Differential regulation of morphology and stemness of mouse embryonic stem cells by substrate stiffness and topography. Biomaterials. 2014;35(13):3945–55.
Lutolf MP, Hubbell JA. Synthetic biomaterials as instructive extracellular microenvironments for morphogenesis in tissue engineering. Nat Biotechnol. 2005;23(1):47–55.
Ma F, et al. Accelerating proliferation of neural stem/progenitor cells in collagen sponges immobilized with engineered basic fibroblast growth factor for nervous system tissue engineering. Biomacromolecules. 2014;15(3):1062–8.
Main H, Munsie M, O'Connor MD. Managing the potential and pitfalls during clinical translation of emerging stem cell therapies. Clin Transl Med. 2014;3:10.
Matesanz R. The frontiers of organ transplantation and cell therapy. Adv Exp Med Biol. 2012;741:1–12.
Matton G, Anseeuw A, De Keyser F. The history of injectable biomaterials and the biology of collagen. Aesthetic Plast Surg. 1985;9(2):133–40.
Meier RP, et al. Transplantation of mesenchymal stem cells for the treatment of liver diseases, is there enough evidence? Stem Cell Res. 2013;11(3):1348–64.
Mimeault M, Hauke R, Batra SK. Stem cells: a revolution in therapeutics—recent advances in stem cell biology and their therapeutic applications in regenerative medicine and cancer therapies. Clin Pharmacol Ther. 2007;82(3):252–64.
Nair A, et al. Biomaterial implants mediate autologous stem cell recruitment in mice. Acta Biomater. 2011;7(11):3887–95.
Nakano T, et al. Self-formation of optic cups and storable stratified neural retina from human ESCs. Cell Stem Cell. 2012;10(6):771–85.
Ngoc PK, et al. Improving the efficacy of type 1 diabetes therapy by transplantation of immunoisolated insulin-producing cells. Hum Cell. 2011;24(2):86–95.
Nirmalanandhan VS, Sittampalam GS. Stem cells in drug discovery, tissue engineering, and regenerative medicine: emerging opportunities and challenges. J Biomol Screen. 2009;14(7):755–68.

Orive G, et al. History, challenges and perspectives of cell microencapsulation. Trends Biotechnol. 2004;22(2):87–92.
Ossendorf C, et al. Treatment of posttraumatic and focal osteoarthritic cartilage defects of the knee with autologous polymer-based three-dimensional chondrocyte grafts: 2-year clinical results. Arthritis Res Ther. 2007;9(2):R41.
Ott HC, et al. Perfusion-decellularized matrix: using nature's platform to engineer a bioartificial heart. Nat Med. 2008;14(2):213–21.
Park JS, et al. The effect of matrix stiffness on the differentiation of mesenchymal stem cells in response to TGF-beta. Biomaterials. 2011;32(16):3921–30.
Petersen TH, et al. Tissue-engineered lungs for in vivo implantation. Science. 2010;329(5991):538–41.
Ponce Marquez S, et al. Decellularization of bovine corneas for tissue engineering applications. Acta Biomater. 2009;5(6):1839–47.
Portero A, et al. Cell encapsulation for the treatment of central nervous system disorders. Rev Neurol. 2010;50(7):409–19.
Price AP, et al. Development of a decellularized lung bioreactor system for bioengineering the lung: the matrix reloaded. Tissue Eng Part A. 2010;16(8):2581–91.
Pukel C, Baquerizo H, Rabinovitch A. Destruction of rat islet cell monolayers by cytokines. Synergistic interactions of interferon-gamma, tumor necrosis factor, lymphotoxin, and interleukin 1. Diabetes. 1988;37(1):133–6.
Qing Q, Qin T. Optimal method for rat skeletal muscle decellularization. Zhongguo Xiu Fu Chong Jian Wai Ke Za Zhi. 2009;23(7):836–9.
Robey TE, et al. Systems approaches to preventing transplanted cell death in cardiac repair. J Mol Cell Cardiol. 2008;45(4):567–81.
Roche ET, et al. Comparison of biomaterial delivery vehicles for improving acute retention of stem cells in the infarcted heart. Biomaterials. 2014;35(25):6850–8.
Rosario DJ, et al. Decellularization and sterilization of porcine urinary bladder matrix for tissue engineering in the lower urinary tract. Regen Med. 2008;3(2):145–56.
Ross EA, et al. Embryonic stem cells proliferate and differentiate when seeded into kidney scaffolds. J Am Soc Nephrol. 2009;20(11):2338–47.
Saetzler K, Sonnenschein C, Soto AM. Systems biology beyond networks: generating order from disorder through self-organization. Semin Cancer Biol. 2011;21(3):165–74.
Salimath AS, et al. Dual delivery of hepatocyte and vascular endothelial growth factors via a protease-degradable hydrogel improves cardiac function in rats. PLoS ONE. 2012;7(11):e50980.
Sasai Y. Cytosystems dynamics in self-organization of tissue architecture. Nature. 2013a;493(7432):318–26.
Sasai Y. Next-generation regenerative medicine: organogenesis from stem cells in 3D culture. Cell Stem Cell. 2013b;12(5):520–30.
Schechner JS, et al. Engraftment of a vascularized human skin equivalent. FASEB J. 2003;17(15):2250–6.
Schmidt CE, Baier JM. Acellular vascular tissues: natural biomaterials for tissue repair and tissue engineering. Biomaterials. 2000;21(22):2215–31.
Schussler O, et al. 3-dimensional structures to enhance cell therapy and engineer contractile tissue. Asian Cardiovasc Thorac Ann. 2010;18(2):188–98.
Sheikh AY, et al. Molecular imaging of bone marrow mononuclear cell homing and engraftment in ischemic myocardium. Stem Cells. 2007;25(10):2677–84.
Shen W, et al. The effect of incorporation of exogenous stromal cell-derived factor-1 alpha within a knitted silk-collagen sponge scaffold on tendon regeneration. Biomaterials. 2010;31(28):7239–49.
Shi C, et al. Stem-cell-capturing collagen scaffold promotes cardiac tissue regeneration. Biomaterials. 2011;32(10):2508–15.
Shin'oka T, et al. Midterm clinical result of tissue-engineered vascular autografts seeded with autologous bone marrow cells. J Thorac Cardiovasc Surg. 2005;129(6):1330–8.

Soto-Gutierrez A, et al. Perspectives on whole-organ assembly: moving toward transplantation on demand. J Clin Invest. 2012;122(11):3817–23.
Spence JR, et al. Directed differentiation of human pluripotent stem cells into intestinal tissue in vitro. Nature. 2011;470(7332):105–9.
Sullivan DC, et al. Decellularization methods of porcine kidneys for whole organ engineering using a high-throughput system. Biomaterials. 2012;33(31):7756–64.
Takayama K, et al. Long-term self-renewal of human ES/iPS-derived Hepatoblast-like cells on human Laminin 111-coated dishes. Stem Cell Rep. 2013;1(4):322–35.
Takebe T, et al. Vascularized and functional human liver from an iPSC-derived organ bud transplant. Nature. 2013;499(7459):481–4.
Terrovitis J, et al. Noninvasive quantification and optimization of acute cell retention by in vivo positron emission tomography after intramyocardial cardiac-derived stem cell delivery. J Am Coll Cardiol. 2009;54(17):1619–26.
Toh YC, Blagovic K, Voldman J. Advancing stem cell research with microtechnologies: opportunities and challenges. Integr Biol. 2010;2(7–8):305–25.
Uchimura E, et al. Novel method of preparing acellular cardiovascular grafts by decellularization with poly(ethylene glycol). J Biomed Mater Res A. 2003;67(3):834–7.
Uludag H, De Vos P, Tresco PA. Technology of mammalian cell encapsulation. Adv Drug Deliv Rev. 2000;42(1–2):29–64.
Uygun BE, et al. Organ reengineering through development of a transplantable recellularized liver graft using decellularized liver matrix. Nat Med. 2010;16(7):814–20.
Venugopal JR, et al. Biomaterial strategies for alleviation of myocardial infarction. J R Soc Interface. 2012;9(66):1–19.
Vincentelli A, et al. In vivo autologous recellularization of a tissue-engineered heart valve: are bone marrow mesenchymal stem cells the best candidates? J Thorac Cardiovasc Surg. 2007;134(2):424–32.
Wainwright JM, et al. Preparation of cardiac extracellular matrix from an intact porcine heart. Tissue Eng Part C Methods. 2010;16(3):525–32.
Wang ZQ, Watanabe Y, Toki A. Experimental assessment of small intestinal submucosa as a small bowel graft in a rat model. J Pediatr Surg. 2003;38(11):1596–601.
Wang X, et al. Sonication-induced gelation of silk fibroin for cell encapsulation. Biomaterials. 2008;29(8):1054–64.
Wang H, et al. Injectable cardiac tissue engineering for the treatment of myocardial infarction. J Cell Mol Med. 2010;14(5):1044–55.
Wang H, et al. Promotion of cardiac differentiation of brown adipose derived stem cells by chitosan hydrogel for repair after myocardial infarction. Biomaterials. 2014;35(13):3986–98.
Woods T, Gratzer PF. Effectiveness of three extraction techniques in the development of a decellularized bone-anterior cruciate ligament-bone graft. Biomaterials. 2005;26(35):7339–49.
Wu SM, Hothedlinger K. Harnessing the potential of induced pluripotent stem cells for regenerative medicine. Nat Cell Biol. 2011;13(5):497–505.
Xu C, et al. Feeder-free growth of undifferentiated human embryonic stem cells. Nat Biotechnol. 2001;19(10):971–4.
Xue C, et al. Joint use of a chitosan/PLGA scaffold and MSCs to bridge an extra large gap in dog sciatic nerve. Neurorehabil Neural Repair. 2012;26(1):96–106.
Yamamoto M, Takahashi Y, Tabata Y. Enhanced bone regeneration at a segmental bone defect by controlled release of bone morphogenetic protein-2 from a biodegradable hydrogel. Tissue Eng. 2006;12(5):1305–11.
Yan C, et al. MicroRNA regulation associated chondrogenesis of mouse MSCs grown on polyhydroxyalkanoates. Biomaterials. 2011;32(27):6435–44.
Yang Y, et al. Repair of rat sciatic nerve gap by a silk fibroin-based scaffold added with bone marrow mesenchymal stem cells. Tissue Eng Part A. 2011;17(17–18):2231–44.
Yao R, et al. Hepatic differentiation of human embryonic stem cells as microscaled multilayered colonies leading to enhanced homogeneity and maturation. Small. 2014;10:4311–23.

Yin C, et al. Microcapsules with improved mechanical stability for hepatocyte culture. Biomaterials. 2003;24(10):1771–80.
Yoo JJ, et al. Bladder augmentation using allogenic bladder submucosa seeded with cells. Urology. 1998;51(2):221–5.
Yu J, et al. The use of human mesenchymal stem cells encapsulated in RGD modified alginate microspheres in the repair of myocardial infarction in the rat. Biomaterials. 2010;31(27):7012–20.
Zhang Y, Bai XF, Huang CX. Hepatic stem cells: existence and origin. World J Gastroenterol. 2003;9(2):201–4.
Zhang X, et al. A biodegradable, immunoprotective, dual nanoporous capsule for cell-based therapies. Biomaterials. 2008a;29(31):4253–9.
Zhang H, et al. Transplantation of microencapsulated genetically modified xenogeneic cells augments angiogenesis and improves heart function. Gene Ther. 2008b;15(1):40–8.
Zhang W, et al. The use of type 1 collagen scaffold containing stromal cell-derived factor-1 to create a matrix environment conducive to partial-thickness cartilage defects repair. Biomaterials. 2013;34(3):713–23.

Part II
Clinical Translation

Chapter 12
Stem Cells in the Treatment of Myocardial Infarction and Cardiomyopathy

Robert J. Henning

Abstract Cardiovascular investigators are currently investigating adult bone marrow stem cells, cardiac stem cells, and adipose stem cells as potential new regenerative cell treatments for patients with acute myocardial infarctions and cardiomyopathies. The initial ten-year experience with autologous, unfractionated bone marrow aspirates, which contain hematopoietic and mesenchymal stem cells (MSCs), suggested that patients with myocardial infarctions who receive these cells demonstrate 2–3 % increases in the left ventricular (LV) ejection fraction of the heart, 4.8 ml decreases in the left ventricular end-systolic volume (LVESV), and approximately 5 % reductions in infarction size without experiencing significant side effects from these cells. The bone marrow stem cells are thought to act by releasing biologically active factors that limit myocardial inflammation, injury, and necrosis. The LateTIME, the TIME, and the Swiss Myocardial Infarction trials have recently addressed the questions of the optimal time for autologous, unfractionated bone marrow cell administration after acute myocardial infarction and coronary angioplasty and whether these cells limit myocardial damage in comparison with the patients treated with percutaneous coronary angioplasty and current medical care without cell transplantation. In these studies, the myocardial infarction sizes and the left ventricular ejection fractions (LVEFs) were not significantly different between the cell-treated patients with standard medical care and patients treated with standard medical care without bone marrow cells (BMCs). The lack of differences between treatments may have been due to the performance of coronary angioplasty within 4–5 h of the onset of patient's symptoms of myocardial infarction, but may also have been due to heterogeneous patient bone marrow cell populations, red blood cell contamination of stem cells, heparin inhibition of stem cell

---

R.J. Henning (✉)
Center for Cardiovascular Research and the James A. Haley Hospital,
13000 Bruce B. Downs Blvd, Tampa, FL 33612, USA
e-mail: robert.henning@va.gov

R.J. Henning
The University Of South Florida College of Medicine,
12901 Bruce B. Downs Blvd, Tampa, Fl 33612, USA

R.C. Zhao (ed.), *Stem Cells: Basics and Clinical Translation*, Translational Medicine Research 1, DOI 10.1007/978-94-017-7273-0_12

migration, and expulsion of the stem cells from the contracting heart shortly after injection. Current trials of patients with myocardial infarction are examining specific bone marrow stem cells including MSCs and $CD34^+$ endothelial stem cells. In addition, cardiac stem cells isolated from human hearts are being investigated in the treatment of patients with infarcted hearts. Cardiac stem cell treatments in patients with infarcted hearts have reportedly decreased left ventricular infarct scar sizes by 11.9–15.7 g and increased left ventricular viable mass by 17.9–22.6 g. Successful cell-based therapy for patients with heart disease requires the close cooperation and interaction of basic scientists and clinicians throughout the world. In this manner, the cell-based therapy in the twenty-first century will offer new hope to the millions of patients with heart disease throughout the world who would otherwise suffer from the inexorable downward progression of heart disease, heart failure, and death.

Keywords Stem cell · Myocardial infarction · Cardiomyopathy

12.1 Introduction

Each year in the USA, more than one million people experience a myocardial infarction and approximately 400,000 people die as a result of their myocardial infarction (Go et al. 2014). Single or recurrent myocardial infarction with loss of more than 25 % of the left ventricular myocardium is associated with low cardiac output and congestive heart failure in patients. Currently, five million people in the USA have congestive heart failure and more than one million people are hospitalized with heart failure each year (Go et al. 2014). In addition, approximately 58,000 patients with heart failure die annually.

Left ventricular scar due to myocardial infarction is largely acellular and causes left ventricular mechanical dysfunction, electrical uncoupling and cardiac arrhythmias, and ultimately cardiomyopathy. The cardiomyocyte deficit in patients with myocardial infarction and cardiomyopathy is approximately one billion cardiac myocytes (Beltrami et al. 1994). Although the heart has some ability to regenerate cardiac myocytes after myocardial infarction, endogenous myocardial muscle restoration is inadequate to compensate for the cardiac myocyte loss with myocardial infarction. Consequently, cardiovascular investigators are currently exploring the use of adult bone marrow stem cells, cardiac progenitor cells, and adipose stem cells as potential new regenerative cell treatments for patients with myocardial infarctions and cardiomyopathies.

12.2 Bone Marrow Stem Cells in Cardiac Repair: The First Ten Years

Due to the limited availability of human embryonic stem cells for cardiac repair, many cardiovascular investigators turned to adult bone marrow cells (BMCs) to potentially reduce in patients the size and the fibrosis of myocardial infarctions, limit post-infarction left ventricular (LV) remodeling, and improve left ventricular wall thickening and compliance. Human bone marrow contains hematopoietic and mesenchymal stem cells (MSCs) which constitute less than 0.01 % of the BMCs. Bone marrow MSCs can generate in vitro myocytes as well as osteoblasts, chondrocytes, and adipose cells. Bone marrow hematopoietic stem cells can produce endothelial progenitor cells and also red blood cells, megakaryocytes, myeloid cells, and lymphocytes. Consequently, cardiovascular investigators theorized that bone marrow mesenchymal and hematopoietic stem cells could transdifferentiate into cardiomyocytes and vascular endothelial cells when implanted in infarcted myocardium. Subsequent investigations in research animals with myocardial infarctions demonstrated that BMCs could decrease infarct size and improve LV ejection fraction and paved the way for studies in patients with acute myocardial infarctions. Based on the studies in research animals, transdifferentiation of BMCs to cardiac myocytes and vascular endothelial cells in the infarcted LV does not appear to occur and does not explain the cardiac changes that can occur with bone marrow cell transplantation in the heart. Rather, bone marrow stem cells appear to act by paracrine mechanisms with the release of biologically active growth factors and anti-inflammatory cytokines that limit myocardial inflammation, injury, and infarct-associated decreases in heart contractility (Gnecchi et al. 2005; Kinnaird et al. 2004).

The first trials using autologous bone marrow stems cells for the treatment of patients with acute myocardial infarctions were reported in 2002. Since that time, many of the clinical trials have used autologous *unfractionated bone marrow mononuclear cells* that contain hematopoietic, MSCs, and other BMCs. The Reinfusion of Enriched Progenitor Cells and Infarct Remodeling in Acute Myocardial Infarction (REPAIR-AMI) trial was a randomized study that reported in 2006 that bone marrow cell therapy significantly increased in patients with myocardial infarctions the left ventricular ejection fractions (LVEFs) and reduced the 1-year combined clinical endpoint of death, recurrence of myocardial infarction and revascularization especially in patients with myocardial infarctions and pulmonary congestion as determined by *n*-terminal brain natriuretic peptide (NT-BNP) values >733 pg/ml (Schaechinger et al. 2006a, b). In contrast to the REPAIR-AMI study, the Bone Marrow Transfer to Enhance ST-Elevation Infarct Regeneration (BOOST) trial reported that adult bone marrow cell-treated patients with myocardial infarctions demonstrated an initial increase in LV ejection fraction at 6 months but no increase in LV ejection fraction after 18 months in comparison with patients treated with optimal medical treatment without bone marrow cells (Wollert et al. 2004; Meyer et al. 2006).

The discrepancies in these and other initial bone marrow stem cell studies in patients with myocardial infarctions are due to significant variations in bone marrow cell processing and characterization of cells, the timing of cell transplantation and the technique of injection, the number and volume of injected cells, the ability of autologous BMCs to migrate to ischemic and infarcted tissue and propagate in patients with chronic diseases, the presence of red blood cell contamination of BMCs, adjunctive medical therapy, and different observation periods after bone marrow cell treatment (Kissel et al. 2007). In addition, the methodology of quantification of patient's cardiac performance after bone marrow cell transplantation is not standardized among cardiovascular investigators, which makes comparison of studies from different research investigators extremely difficult.

Several major meta-analyses of cardiac cell transplantation have been published during the first decade of stem cell transplantation into patient's hearts that permit general conclusions regarding the effects of BMCs in the treatment of patients with myocardial infarctions. An analysis by Lipinski et al. (2007) included 10 studies of 698 patients that were treated with percutaneous coronary angioplasty after acute myocardial infarction and then allocated to treatment with either intracoronary autologous adult bone marrow cell therapy or standard medical therapy. The patients were followed for a mean of 6 (range 3–18) months. Seven of the 10 studies randomized the patients to either cell treatment or placebo controls. In this meta-analysis, patients who received BMCs showed statistically significant 3 % (range 1.9–4.1 %) increases in LV ejection fraction, decreases in end-systolic volume of 7.4 ml (range −12.2 to −2.7 ml), and reductions in infarct size of 5.6 % (−8.7 to −2.5 %) (Lipinski et al. 2007). Patients who received intracoronary bone marrow cell infusions had a significant decrease in recurrent myocardial infarctions but no difference in rehospitalization for heart failure during the study. Table 12.1 summarizes the studies analyzed by Lipinski and coworkers (Schaechinger et al. 2006a, b; Meyer et al. 2006; Strauer et al. 2002; Bartunek et al. 2005; Li et al. 2007; Janssens et al. 2006; Wollert et al. 2004; Kang et al. 2006; Lunde et al. 2006, 2008; Ge et al. 2006; Meluzín et al. 2006, 2008; Dill et al. 2008).

Martin-Rendon et al. (2008) performed a Cochrane Systematic Review of 13 randomized controlled trials with 14 different comparisons involving 811 patients with acute myocardial infractions from 9 different countries that compared percutaneous coronary intervention plus autologous BMCs with percutaneous coronary angioplasty plus saline or heparinized plasma. All patients were followed for 3–6 months and 3 trials followed patients for greater than 12 months. Autologous bone marrow cell therapy was found to be safe and increased LV ejection fraction by 2.99 % (range 1.26–4.72 %), reduced LV end-systolic volume by 4.74 ml (range −7.84 to −1.64 ml), and decreased myocardial infarction size by 3.51 % (range −5.91 to −1.11 %) (Martin-Rendon et al. 2008). Subgroup analysis indicated that there was a statistically significant increase in LV ejection fraction when cells were infused within 7 days following the acute myocardial infarction and when the cell dose administered was greater than 10^8 cells. However, the trials in this Cochrane Review were too small to demonstrate whether bone marrow cell therapy reduced patient mortality after myocardial infarction.

Abdel-Latif examined 999 patients in a meta-analysis of 18 randomized and non-randomized trials of BMCs (Abdel-Latif et al. 2008). Twelve trials included patients with acute myocardial infarction, and 6 studies included patients with acute myocardial infarction and ischemic cardiomyopathy. The cells used for

Table 12.1 Adult bone marrow studies in patients with acute myocardial infarction

Study	Patient (*N*)	Randomized	Days post-MI	Cell dose	Baseline LVEF (%)	LVEF change (%)	Duration/ months	Other findings
Strauer et al. (2002)	20	Cohort	8	$2.8 \pm 2.2 \times 10^7$	57 ± 8	+5	3	Increased regional but not global LVEF; decreased LVESV and infarct size
Bartunek et al. (2005)	35	Cohort	10	$12.6 + 2.22 \times 10^6$	45 ± 2.5	+7	4	Increased LV regional function, perfusion, and restenosis
Li et al. (2007)	70	Cohort	6	$7.3 \pm 7.3 \times 10^7$	50 ± 8.2	+7	6	Decreased LVESV, LV wall motion score
Janssen et al. (2006)	67	Yes	1	172×10^6	48.5 ± 7.2	+3.3	4	Decreased infarct size
Wollert et al. (2004)	60	Yes	4.8	24.6×10^8	50.0 ± 10.0	No change	6–18	Increased LVEF at 6 but not at 18 months
Kang et al. (2006)	96	Yes	4	$1–2 \times 10^9$	52.0 ± 9.9	+5.1 AMI	6	Decreased LVESV and infarction in acute MI; no change ESV and no change OMI
Lunde et al. (2006, 2008)	100	Yes	6	68×10^6	41.3 ± 11.0	No change	6–12	Increased LVEF in treated and controls; no change EDV and infarct size
Ge et al. (2006)	20	Yes	1	4×10^7	53.8 ± 9.2	+4.8	6	Increased LV regional wall perfusion by SPECT
Meluzin et al. (2006, 2008)	66	Yes	5–9	$10^7–10^8$	42 ± 2	+3–5	3–12	LVEF increased 3 % with 10^7 LVEF increased 5–7 % with 10^8 at 3–12 months
Schachinger et al. (2006a, b), Dill et al. (2008)	204	Yes	3–8	2.4×10^8	48.3 ± 9.2	+6–7	4–12	Increased EF when Rx >4 days post-MI and when EF <49 %; increased LV perfusion

AMI Acute myocardial infarction; *BMC* Bone marrow cell; *CPC* Circulating progenitor cell; *EF* Ejection fraction; *ESV* End-systolic volume; *G-CSF* Granulocyte colony-stimulating factor; *IC* Intracoronary injection; *ICM* Ischemic cardiomyopathy; *IM* Intramyocardial injection; *LVEF* left ventricular ejection fraction; *LV* = Left Ventricle or Left Ventricular; *OMI* Old myocardial infarction; *SPECT* Single-photon emission computer tomography

Adapted and reproduced with permission from Henning (2011)

treatment included bone marrow mononuclear cells, MSCs, or bone marrow-derived circulating blood progenitor cells. The cells were administered into the coronary arteries of patients in 15 trials and directly into the myocardium in 3 studies. Compared with control patients, BMCs significantly increased LV ejection fraction by 3.66 % (range 1.93–5.4 %), decreased LV end-systolic volume by 4.8 ml (range −8.20 to −1.41 ml), and reduced infarct size by 5.49 % (range −9.10 to 1.88 %) (Abdel-Latif et al. 2008). There were no major local or systemic complications. In this meta-analysis, there was no significant difference between groups that received less or more than 80 × 10^6 cells. However, injection of BMCs 5–20 days after myocardial infarction resulted in a greater than threefold reduction in myocardial infarction size and better reduction in the LV end-systolic volume compared with injection of BMCs within the first 5 days after acute myocardial infarction and/or percutaneous coronary intervention (Abdel-Latif et al. 2008). Table 12.2 summarizes the 18 studies analyzed by Abdel-Latif and coworkers (Ge et al. 2006; Assmus et al. 2006; Bartunek et al. 2006; Chen et al. 2004; Erbs et al. 2005; Ge et al. 2006; Hendrikx et al. 2006; Katritsis et al. 2005; Mocini et al. 2006; Perin et al. 2004; Ruan et al. 2005; Strauer et al. 2005).

None of the meta-analyses of bone marrow trials reported an increased incidence of cardiac arrhythmias with bone marrow cell transplantation into the heart muscle. The meta-analyses suggest that in patients with myocardial infarctions, modest decreases in infarct size and increases in left ventricular ejection fraction can be achieved with bone marrow cell therapy and that the therapy is safe.

Although the increases in cardiac LV ejection fraction with bone marrow stem cell therapy are modest, they are comparable to what has been reported with pharmacology therapy and with angioplasty in patients with myocardial infarctions and ischemic cardiomyopathies. In the Valsartan in Acute Myocardial Infarction (VALIANT) Study, treatment of patients with myocardial infarctions with the drug valsartan increased the LV ejection fraction 1.3 ± 6.7 %, while treatment with captopril increased the LV ejection fraction by 2.7 ± 7.2 %, and combined valsartan and captopril treatment increased the LV ejection fraction by 1.9 ± 7.3 % (Solomon et al. 2005). In the intravenous streptokinase in acute myocardial infarction (ISAM) trial, the LV ejection fractions in patients treated with thrombolytic streptokinase therapy averaged 56.8 ± 0.7 % versus 53.9 ± 0.7 % in control patients (The I.S.A.M. Study Group 1986). In a comparison of thrombolytic therapy with streptokinase versus acute coronary angioplasty in the treatment of acute myocardial infarction, the LV ejection fraction was 45 ± 12 % in patients treated with streptokinase and 51 ± 11 % in patients with acute coronary angioplasty (Zijlstra et al. 1993). Consequently, the meta-analyses suggest that the increases in cardiac function with stem cell therapy are comparable to pharmacologic therapy and angioplasty for the treatment of acute myocardial infarction.

Table 12.2 Bone marrow and circulating progenitor cells in coronary artery disease patients

Study	Patient (*N*)	Randomized	Time-post PCI and/or MI (days)	Cells dose	Injection route	Baseline LVEF (%)	LVEF change (%)	Duration (months)	Other findings
Assmus et al. (2006)	92	Yes	2348–2470	22 ± 10^6 CPC $205 \pm 110 \times 10^6$ BMC	IC	CPC 39 ± 10; BMC 41 ± 11	CPC: -0.4 BMC: +2.9	3	Patients with previous MI; increased LVEF in BMC but not CPC treatment
Bartunek et al. (2005, 2006)	35	Cohort	10	$12.6 \pm 2.2 \times 10^6$	IC	45 ± 2.5	+7	4	Increased LV regional function, increased perfusion, and restenosis
Chen et al. (2004)	69	Yes	18.4 ± 0.5	$8–10 \times 10^9$	IC	49 ± 9	+18	6	Increased LVEF by ventriculogram; increased perfusion; decreased ESV
Erbs et al. (2005)	26	Yes	225 ± 87	$69 \pm 14 \times 10^6$	IC	51.7 ± 3.7	+7.2	3	Patients with chronic CAD occlusion Rx with CPC; EF by MRI; infarct size decreased 16 %
Ge et al. (2006a, b)	20	Yes	1 day	$39 \pm 22 \times 10^6$	IC	53.8 ± 9.2	+4.8	6	Increased perfusion by SPECT
Hendrikx et al. (2006)	20	Yes	217 ± 162	$60 \pm 31 \times 10^6$	IM	42.9 ± 10.3	+5	4	CABG in patients with previous CAD; increased regional but not global LV function; 6/9 with induced VTach

(continued)

Table 12.2 (continued)

Study	Patient (*N*)	Randomized	Time-post PCI and/or MI (days)	Cells dose	Injection route	Baseline LVEF (%)	LVEF change (%)	Duration (months)	Other findings
Janssen et al. (2006)	67	Yes	1	172×10^6	IC	48.5 ± 7.2	+3.3	4	Decreased infarct size
Kang et al. (2006)	96	Yes	<14 AMI >14 OMI	$1–2 \times 10^9$	IC	52.0 ± 9.9	+5.1 AMI	6	G-CSF for 3 days; decreased ESV and infarct size in AMI; no change EF, ESV, and infarct size in OMI
Katritsis et al. (2005)	22	Cohort	224 ± 470	$2–4 \times 10^6$	IC	39.7 ± 9.3	+1.6	4	Increased regional but not global LV function
Lunde et al. (2006, 2008)	100	Yes	6 ± 1.3	68×10^6 (median) $54–130 \times 10^6$	IC	41.3 ± 11.0	No change	6–12	Increased LVEF in treated and controls; no change EDV and infarct size
Meyer et al. (2006)	60	Yes	4.8 ± 1.3	$24.6 \pm 9.4 \times 10^8$	IC	50 ± 10	+5.9	18 ± 6	Increased LVEF by MRI significantly at 6 but not at 18 months
Mocini et al. (2006)	36	Cohort	AMI < 6 months	$292 \pm 232 \times 10^6$	IM	46 ± 6	+5	3–12	CABG in all; troponin increased
Perin et al. (2004)	20	Cohort	ICM	$25.5 \pm 6.3 \times 10^6$	IM transendocardial	30 ± 6	+5.1	12	LVEF no change controls; increased LV perfusion and exercise

(continued)

Table 12.2 (continued)

Study	Patient (*N*)	Randomized	Time-post PCI and/or MI (days)	Cells dose	Injection route	Baseline LVEF (%)	LVEF change (%)	Duration (months)	Other findings
Ruan et al. (2005)	20	Yes	~1	NR	IC	53.5 ± 5.8	+5.8	6	Increased LV segmental contraction
Schachinger et al. (2006a), Dill et al. (2008)	204	Yes	3–8	2.4×10^8	IC	48.3 ± 9.2	+6–7	4–12	Increased EF when Rx >4 days post-MI and when EF $\leq$49 %; increased LV perfusion
Strauer et al. (2002)	20	Cohort	5–9	$2.8 \pm 2.2 \times 10^7$	IC	57 ± 8	+5	3	Increased regional but not global LVEF; decreased ESV and infarct size
Strauer et al. (2005)	36	Cohort	823 ± 945	9.0×10^7	IC	52 ± 9	+8	3	Decreased infarct size 30 %; increased VO_2 max
Li et al. (2007)	70	Yes	7 ± 5	$7.3 \pm 7.3 \times 10^7$	IC	50 ± 8.2	+7	6	G-CSF for 5 days; decreased LV ESV; decreased LV wall motion score

AMI Acute myocardial infarction; *BMC* Bone marrow cell; *CABG* Coronary artery bypass surgery; *CPC* Circulating progenitor cell; *EF* Ejection fraction; *ESV* End-systolic volume; *G-CSF* Granulocyte colony-stimulating factor; *IC* Intracoronary injection; *ICM* Ischemic cardiomyopathy; *IM*: Intramyocardial injection; *LVEF* left ventricular ejection fraction; *NR* Not recorded; *OMI* Old myocardial infarction; *PCI* Percutaneous coronary intervention; *Rx* Treatment; *SPECT* Single-photon emission computer tomography; *VTach* Ventricular tachycardia

Adapted and Reproduced with Permission from Henning (2011)

12.3 Bone Marrow Stem Cells in Ischemic Cardiomyopathy

In contrast to the large number of bone marrow cell trials in patients with acute myocardial infarctions, few clinical trials have investigated the use of stem cells in the treatment of patients with chronic ischemic cardiomyopathies. Moreover, the cardiomyopathy trials frequently often involved small numbers of patients and often did not randomize patients to bone marrow cell treatment or no cell treatment. Nevertheless, the trials demonstrated that BMCs could be safely administered to patients with ischemic cardiomyopathies.

In a randomized study of patients with ischemic cardiomyopathies who were at least 3 months of post-myocardial infarction, Assmus injected into the coronary arteries autologous BMCs or a placebo into the patients' coronary arteries (Assmus et al. 2006). At 3 months post-injection, LV ejection fraction, determined by LV ventriculograms with contrast material, increased significantly by 2.9 ± 3.6 % among patients receiving BMCs but declined by 1.2 ± 3.0 % among patients who did not receive cell infusions. Magnetic resonance imaging (MRI) of LV regional function in a subgroup of patients treated with BMCs demonstrated that the hypocontractile segments decreased from 10.1 ± 3.6 segments to 8.7 ± 3.6 segments and the normal contractile segments increased significantly from 3.8 ± 4.5 to 5.4 ± 4.6 segments.

Strauer reported in a non-randomized study the five-year follow-up of the intracoronary administration of BMCs in 191 patients with ischemic cardiomyopathies with LVEF <35 % and chronic heart failure (Strauer et al. 2010). All patients received dobutamine intravenously to augment contractility for 24 h after the coronary injection of BMCs. In 181 of these patients at 60 months, BMCs increased the LVEF by 6.2 ± 8.4 %, improved the New York Heart Association Classification from 3.2 ± 0.5 to 1.5 ± 0.5, and increased the survival in patients with heart failure compared to the group not treated with BMCs (Strauer et al. 2010).

In contrast to the studies by Assmus and Strauer, several trials have failed to confirm the beneficial effects of intracoronary delivery of BMCs in patients with ischemic cardiomyopathies. In a study of 20 patients with chronic ischemic heart disease, Ang et al. (2008) injected autologous BMCs either directly into myocardial scars or into a coronary artery graft supplying scarred myocardium in 10 patients. In this study, BMCs did not significantly increase myocardial regional contractile thickening and LV ejection fraction or decrease infarct scar volume when measured by dobutamine stress echocardiography and magnetic resonance imaging (MRI) 6 months after cell transplantation in comparison with patients treated with myocardial revascularization without cell transplantation.

Hendrikx et al. (2006) reported that autologous BMCs directly injected into the infarct scar borders in 10 patients with ischemic cardiomyopathy at the time of coronary artery bypass surgery improved regional myocardial wall function but not global contractile function when measured by MRI at 4 months after cell transplantation in comparison with 10 patients who received coronary artery bypass grafts without bone marrow stem cells. Monomorphic ventricular tachycardia was induced during electrophysiological studies in 5 patients, and polymorphic ventricular tachycardia was induced in 1 patient in in a total of ten patients who received BMCs (Hendrikx et al. 2006). Automatic implantable defibrillators were inserted in 3 patients, and 3 patients were treated with amiodarone.

The studies by Ang and Hendrikx suggest that stem cells when injected into myocardial scars in patients with cardiomyopathies are associated with little or no left ventricular hemodynamic improvement. In contrast, stem cell injection into *viable myocardium* in patients with ischemic cardiomyopathies, either at the time of cardiac surgery or with the use of the NOGA catheter navigation system during LV catheterization, appears to be associated with modest hemodynamic improvement.

The Cochrane Collaboration has published an evaluation of 23 randomized control trials that investigated the use of bone marrow stem cells in the treatment of patients with chronic ischemic heart disease and congestive heart failure (Fisher et al. 2014). Co-interventions, such as primary angioplasty, coronary bypass surgery, or administration of stem cell mobilizing agents, were included in the meta-analysis when the interventions were equally administered to stem cell-treated and control patients. The 23 studies included 659 patients treated with bone marrow-derived stem cells and 478 control patients. The quality of the evidence in the 23 studies was stated by the Cochrane Collaboration to be *low to moderate quality* (Fisher et al. 2014). Nevertheless based on the data available, the collaboration concluded that bone marrow stem cell treatment in patients with ischemic heart disease was associated with a reduction in left ventricular end-systolic volume (LVESV) (mean difference −14.64 ml, 95 % confidence intervals (CI) −20.88 to −8.39 ml in 153 patients), an increase in LV stroke volume index (mean difference 6.52, 95 % CI 1.51–11.54 in 62 patients), an improvement in LVEF (mean difference 2.62 %, 95 % CI 0.50–4.73 % in 6 studies), a reduced incidence of mortality (risk ratio 0.28, 95 % CI 0.14–0.53), and rehospitalization due to heart failure (relative risk 0.26, 95 % CI 0.07–0.94) over ≥12 months (Fisher et al. 2014) (see Table 12.3). Patients with LVEF less than 30 % or patients with symptomatic congestive heart failure benefited more from BMC treatment than patients with LVEF >30 % and/or patients without symptoms of heart failure. Of 19 trials in which adverse events were reported, adverse events due to either BMC treatment or related procedures occurred in only four individuals. No long-term adverse events were reported. The Cochrane Collaboration emphasized the low to moderate quality of the evidence and concluded that there is a need for large-scale, adequately powered studies with well-defined participant cohorts and long-term follow-up to confirm the modest beneficial effects of BMCs for patients with chronic ischemic heart disease and congestive heart failure (Fisher et al. 2014).

Table 12.3 Comparison of stem cell versus no stem cell treatment on all mortality

Study or subgroup	BMSC n/N	no BMSC n/N	Risk Ratio M-H, Fixed, 95% CI
Short-term follow-up (<12 months)			
Ang 2008	1/42	1/19	
Assmus 2006	0/52	1/23	
Erbs 2005	0/12	0/11	
Hendrikx 2006	1/11	1/12	
Honold 2012	0/22	0/10	
Hu 2011	0/31	1/28	
Kang 2006	1/16	0/16	
Losordo 2007	0/18	0/28	
Losordo 2011	0/111	0/55	
Patel 2005	0/10	0/10	
Perin 2011	0/20	0/10	
Perin 2012a	1/61	0/31	
Perin 2012b	0/10	0/10	

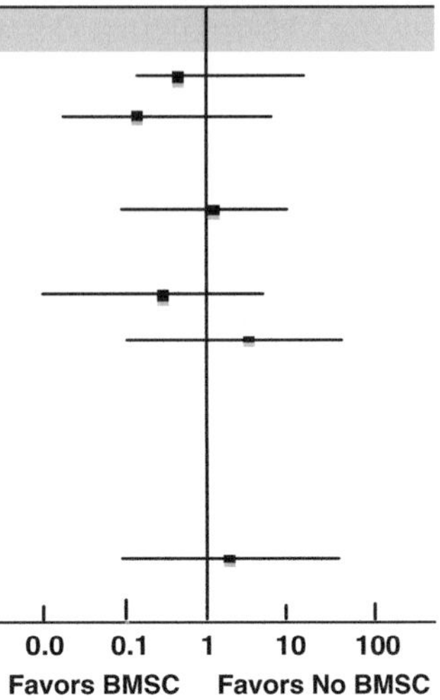

Adapted and Reproduced with permission from Fisher et al. (2014)

12.4 Bone Marrow Cells in Non-ischemic Cardiomyopathies

Patients with non-ischemic cardiomyopathies have dilated cardiac ventricles, often with non-homogenous LV tissue perfusion, which makes targeted administration of stem cells through coronary arteries challenging. However, these patients often have greater numbers of progenitor cells in the systemic circulation with better functional capacity than patients with ischemic cardiomyopathies, which suggests that bone marrow cell therapy might be beneficial in patients with non-ischemic cardiomyopathies (Vrtovec et al. 2011). See Table 12.4.

Fifty-five patients with non-ischemic cardiomyopathies in an open-labeled study received intracoronary bone marrow $CD34^+$ endothelial progenitor cells or placebo. At 5 years after treatment, stem cell therapy was associated with an increase in left ventricular ejection fraction (from 24.3 ± 6.5 to 30.0 ± 5.1 %), an increase in 6-min walk distance (from 344 ± 90 to 477 ± 130 m), and a decrease in pulmonary congestion as determined by N-terminal B-type natriuretic peptide (from 2322 ± 1234 to 1011 ± 893 pg/ml) (Vrtovec et al. 2011). The improvement in the LVEF was most significant in patients with the greatest attraction of injected $CD34^+$ cells to the myocardium. In this study, patient mortality at five years was 14 % in the stem cell treatment group and 35 % in the non-stem cell-treated patients (Vrtovec et al. 2011).

In the Autologous Bone Marrow Cells in Dilated Cardiomyopathy (ABCD) trial of 24 patients with non-ischemic cardiomyopathy who received intracoronary BMCs, the LVEF increased by an average of 5.9 % (22.5 ± 8.3 % to 28.4 ± 11.8)

Table 12.4 Prospective randomized trials of stem cell therapy in non-ischemic heart failure

Study	Patient (*N*)	Randomized	Cell dose	Injection type	Baseline LVEF	LVEF increase	Duration (months)
Bocchi et al. (2010)	22	No	GSF stimulation BMC	IC	21	8.8	15
Fischer-Rasokat et al. (2009)	33	No	$259 \pm 135 \times 10^6$ BMC	IC	30	3.4	3
Seth et al. (2010)	85	Yes	$168 \pm 96 \times 10^6$ BMC	IC	23	5.9	36
Vrtovec et al. (2011)	55	Yes	$123 \pm 23 \times 10^6$ BMC	IC	26	4.6	12
Vrtovec et al. (2013)	55	Yes	$113 \pm 26 \times 10^6$ BMC	IC	24	5.6	60

LVEF Left ventricular ejection fraction; *BMC* Bone marrow cells; *IC* Intracoronary injection
Adapted with permission from Methodist DeBakey Cardiovascular Journal 2013;iX (4):196

which was associated with an improvement in New York Heart Association functional class III and IV patients (Seth et al. 2010). The three-year follow-up showed persistent improvement in LVEF due to decreases in LVESV. However, this hemodynamic improvement was not associated with significant improvement in patient survival because 12 (24.4 %) patients died in the treated group and 14 (30 %) patients died in the control group (Seth et al. 2010).

In a third study of non-ischemic cardiomyopathy patients, Fischer-Roasokat infused BMCs into the coronary arteries of 33 patients and analyzed patient hemodynamics at 3 months (Fischer-Roasokat et al. 2009). In this study, the LVEF increased by 3.2 % from 30.2 ± 10.9 to 33.4 ± 11.5 %, and the NT-proBNP decreased at one year from 1610 ± 993 to 1473 ± 1147 pg/ml. The increase of regional contractile function and LVEF was directly related to the functionality of the infused cells as measured by their colony-forming capacity (Fischer-Rasokat et al. 2009). Based on this study and the previous two studies, the number of BMCs retained in the myocardium and the functionality of the BMCs appear to be important factors in the response of patients with cardiomyopathy to bone marrow cell therapy.

The studies of bone marrow stem cells in patients with ischemic and non-ischemic cardiomyopathies are promising. However, the effects of BMCs in patients with cardiomyopathies require much larger numbers of patients in clinical trials for longer periods of time in order to permit definitive conclusions about the effects of stem cells in these patients. The amount of hibernating myocardium in patients with ischemic cardiomyopathies varies greatly, and patients with non-ischemic cardiomyopathies frequently have myocardial fibrosis and non-homogenous myocardial perfusion. These confounding factors contribute to the heterogeneity of the treatment responses in the different studies. In addition, the type of cells injected (adult bone marrow mononuclear cells, $CD34^+$ cells, or mesenchymal cells), the location of injection (viable LV myocardial tissue or scar), the methods of measurement of LV function, and the duration of patient follow-up should be standardized in order to permit comparisons of different studies and ultimately permit reasonable conclusions regarding the amount of benefit of bone marrow cell therapy in patients with ischemic and non-ischemic cardiomyopathies.

12.5 The Second Decade: TIME Trials of Bone Marrow Stem Cells in Cardiac Repair

The initial ten-year experience with stem cells, primarily from bone marrow aspirates, suggested that patients with myocardial infarctions who received autologous unfractionated bone marrow mononuclear cells showed significant 2–3 % (range 1.9–5.4 %) increases in LVEF, decreases in LVESV of 4.8 ml (range −1.4 to −8.2 ml), and reductions in infarct size of approximately 5 % (−1.9 to −9.1 %) without experiencing significant side effects (Henning 2011, 2012, 2013). However, many of the initial bone marrow cell studies consisted of small numbers of patients and not all the studies randomized patients to treatment with bone marrow mononuclear cells or placebo.

Despite promising but modest results from the initial studies, major questions have persisted in the treatment of patients with acute myocardial infarctions with bone marrow mononuclear cells or other cells for heart repair. What is the optimal cell for treatment of patients with myocardial infarction? When is the optimal time to inject cells in patients with myocardial infarctions? What is the viability of the stem cells prior to injection into patients? What is the best technique to monitor cardiac patients after stem cell treatment? The LateTIME, the TIME, and the Swiss Myocardial Infarction trials were multicenter trials that addressed the questions whether unfractionated bone marrow mononuclear cells and standard of care limit myocardial damage in comparison with patients treated with standard of care and what is the optimal time for cell administration after acute myocardial infarctions.

12.6 LateTIME Trial: Transplantation in Myocardial Infarction Evaluation 2–3 Weeks Following Acute Myocardial Infarction (AMI)

The LateTIME trial was a randomized, double-blind, placebo-controlled trial designed to determine whether unfractionated bone marrow mononuclear cells administered to patients 2–3 weeks after AMI would be safe and effective in limiting infarct size and improving LV function (Traverse et al. 2011). All patients were successfully treated initially with primary percutaneous coronary angioplasty (PTCA) within a median time of 4 h after the onset of chest pain. More than 90 % of the patients had anterior wall myocardial infarctions.

All patients underwent bone marrow aspirations. The bone marrow aspirate was processed at each site with a closed, automated cell processing system (Sepax, Biosafe SA) to ensure a uniform cellular product for administration. The BMCs contained 2.6 % $CD34^+$ and 1.2 % $CD133^+$ hematopoietic cells, and the viability of the cells was >70 %. Fifty-eight patients were given 150×10^6 autologous bone marrow mononuclear cells into the infarct-related coronary artery, and 20 patients were given 5 % human serum albumin plus 100 μl of autologous blood as a placebo, 2–3 weeks after acute myocardial infarction (Traverse et al. 2011). All patients received heparin during the procedure as well as aspirin, clopidogrel, and American Heart Association guideline recommended post-AMI medications.

Infarct volume and global and regional LV function were measured by MRI with gadolinium prior to each intracoronary injection and at 6 months after injection. The LVEFs prior to infusion of cells or placebo averaged 48.7 % in the BMC group and 45.3 % in the placebo group. The changes between baseline and 6 months in BMC group for infarct volume, LVEF, wall motion in the infarct zone, and wall motion in the border zone of the infarction were not statistically different from the placebo group (Traverse et al. 2011). No significant improvement was observed in the recovery of LV function in the group of AMI patients with the most depressed LVEF at baseline. However, the BMC treatment group had fewer clinical adverse events than the placebo group, and the bone marrow cell infusions were felt to be safe.

The LateTIME trial investigators concluded that among patients with AMI and LV dysfunction following reperfusion with PCI, intracoronary infusion of autologous unfractionated BMCs 2–3 weeks after PCI did not improve global or regional LV function at 6 months (Traverse et al. 2011).

12.7 TIME Trial: Transplantation in Myocardial Infarction Evaluation

In 2006, the investigators in the REPAIR-AMI trial reported that delivery of BMCs to patients 5–7 days after AMI resulted in a 5.1 % absolute increase in LV ejection fraction (Schaechinger et al. 2006b). This increase in LV ejection fraction contrasted with patients treated 3–4 days after AMI in which there was no significant increase in LVEF in comparison with placebo-treated patients (Schaechinger et al. 2006b). Based on the REPAIR-AMI trial, additional studies were recommended. The TIME trial was a double-blind, placebo-controlled trial that investigated the intracoronary administration of autologous bone marrow mononuclear cells or placebo in patients 3 or 7 days after an acute myocardial infarction (Traverse et al. 2012). All patients had successful coronary reperfusion with coronary angioplasty within a median time of 3–4 h after the onset of ischemic symptoms. More than 81 % of the patients had anterior wall infarctions.

All patients had bone marrow aspirations. The mean time from PCI to bone marrow aspiration and cell processing was 3.3 days in the 3 day and 7.4 days in the 7 day group. Bone marrow mononuclear cells were isolated in each center with the Sepax system (Biosafe), and the cells or placebo was infused within 12 h of bone marrow aspiration and cell processing. The BMCs contained 2.3 % $CD34^+$ and 1.1 % $CD34^+$ plus $CD131^+$ hematopoietic cells. All patients had baseline cardiac MRIs with gadolinium at day 3 or at day 7 after AMI and at 6 months after the AMI (Traverse et al. 2012).

Forty-three patients received unfractionated BMCs on day 3, and 36 patients received unfractionated BMCs on day 7 after AMI. Each patient received approximately 147×10^6 bone marrow mononuclear cells within 12 h of aspiration and cell processing. Forty-one patients received a placebo. All patients received heparin during the procedure as well as aspirin and clopidogrel.

The differences between the BMC treatment and the placebo treatments in the 3 day group and in the 7 day group were not significant (Traverse et al. 2012). When both BMC groups were combined ($n = 75$) to include patients with MRI measurements at baseline and at 6 months and compared with the combined placebo group ($n = 37$), there was no significant increase in the LVEF for the BMC group in comparison with the placebo group. Moreover, there was no significant difference between the changes in regional wall motion in the infarct zone and the border zone between BMC and placebo groups. Infarct volumes uniformly decreased in both groups, but the differences were not statistically significant. Major coronary adverse events were rare among all treatment groups.

The TIME trial investigators concluded that among patients with ST segment elevation myocardial infarction treated with primary PCI, the administration of intracoronary autologous unfractionated BMCs at either day 3 or day 7 after AMI had no significant effect on recovery of global or regional LV function compared with placebo (Traverse et al. 2012).

12.8 Swiss Multicenter Intracoronary Stem Cell Study in Acute Myocardial Infarction Trial (Swiss AMI Trial)

The Swiss Multicenter Intracoronary Stem Cell Study in Acute Myocardial Infarction trial randomized patients with AMIs with LVEF <45 % by ventriculography or echocardiography, who had been successfully treated with PCI of the infarct-related artery within a median of 5 h of onset of chest pain, to either the intracoronary administration of 140–160 million autologous bone marrow mononuclear cells at a median of 6 days after AMI (early group $n = 58$) or at a median of 24 days after AMI (late group, $n = 49$) or to a placebo group ($n = 60$) (Sürder et al. 2013). Ninety-two percent of the patients had anterior wall infarctions. Bone marrow aspirates were performed only in patients assigned to the BMC treatment. Each 10 ml aspirate was treated with 1000 IU heparin to prevent clot formation. The bone marrow mononuclear cell fraction was isolated by density gradient centrifugation at a centralized processing facility and contained 1–1.3 % $CD34^+$ hematopoietic cells. The median percentage of mononuclear cells that exhibit migration capacity was only 29 % (Sürder et al. 2013).

Cardiac MRI with gadolinium was performed on patients at baseline prior to infusion and at 4 months after the injection of BMCs into the infarct-related coronary artery and was compared with MRIs of control patients treated with best medical care at the same times.

At 4 months after coronary infusion, there were no significant differences in infarct scar size or LV myocardial wall thickening in patients treated with BMCs at either 5–7 days or 3–4 weeks after AMI in comparison with control patients. Moreover, LV function did not significantly improve at 4 months after the intracoronary infusion of autologous BMCs in either the early or late treated groups in comparison with the placebo group. In all cell and placebo treatment groups, LV scar, determined by late gadolinium enhancement on MRI, decreased by more than 10 g with a 4–5 % decrease in the ratio of myocardial scar to myocardial mass. There were no significant differences in adverse events between BMC-treated and control patients (Sürder et al. 2013).

12.9 Critique of LateTIME, TIME, and Swiss Bone Marrow Cell Trials

The primary endpoints of the LateTIME, TIME, and Swiss Bone Marrow Cell trials were not met, and the functional benefit of autologous unfractionated bone marrow mononuclear cells remains in doubt. Nevertheless, these trials provide insights into stem cell trial designs and stem cell functions in patients with AMIs.

Patients with AMIs in the LateTIME, TIME, and Swiss Multicenter trials were treated with percutaneous coronary angioplasty within a median of approximately 4–5 h of the onset of chest pain. Thereafter, the patients were treated with American and European Heart Association guided best medical therapy. Consequently, myocardial infarction sizes and the extent of LV remodeling in the trial patients were significantly limited, and the differences between BMC-treated patients and placebo-treated patients were small. Although the initial qualifying LVEFs by echocardiography after PCI in the LateTIME and TIME trials patients were <45 %, the LVEFs by MRI at the time of BMC injection were larger than 45 %. BMCs are much less effective in patients with small myocardial infarctions with near normal LVEFs. Moreover, placebo-treated patients continue to improve with best medical therapy after myocardial infarctions as exemplified by the control patients in the BOOST trial in which the LVEFs continued to improve and equaled or exceeded the increases in the LVEFs in the BMC-treated patients at 18 months after AMI (Meyer et al. 2006). In addition, the Valsartan in Acute Myocardial infarction trial and trials of neurohormonal blockade of patients with acute myocardial infarctions have demonstrated that optimal medical therapy of patients with AMIs can increase LVEF by a mean of 2.7 % points at 20 months

(Solomon et al. 2005; Henning 2011). Consequently, much larger numbers of patients will be required in clinical trials to demonstrate statistically significant differences between BMC-treated patients and placebo-treated patients who receive PCI early after the onset of AMI and guideline directed optimal medical therapy. The BAMI trial (the effect of intracoronary reinfusion of bone marrow-derived mononuclear cells on all-cause mortality in acute myocardial infarction) is recruiting 3000 patients with LVEFs <45 % within 7 days of AMIs, who have undergone successful coronary reperfusion therapy, for randomization into treatment with either intracoronary autologous unfractionated bone marrow mononuclear cells or placebo (Mathur 2013). Perhaps, the BAMI trial will provide a definitive answer to the question not only whether autologous unfractionated BMCs can significantly decrease patient mortality due to myocardial infarction but also substantially reduce infarct size and improve LVEF in comparison with patients treated with best medical therapy over three years.

The lack of differences between bone marrow cell-treated patients and placebo-treated patients with AMIs in the LateTIME, the TIME, and the Swiss Multicenter trials may be due to important factors other than prompt coronary angioplasty after AMI and optimal medical therapy in these trials. Several important factors are discussed in the following sections.

12.9.1 Heterogeneous Bone Marrow Cell Populations

Unfractionated bone marrow mononuclear cells are a heterogeneous group of cells that contain less than 3 % $CD34^{+}$ and 1 % $CD34^{+}/CD133^{+}$ hematopoietic progenitor cells and <1 % $CD105^{+}$ MSCs when marrow cells are separated by Ficoll density gradient-based separation. However, the bone marrow aspirates in the LateTIME and TIME trials were separated by an automated cell process system (Sepax, Biosafe), which recovered only 23.6 % of the total nucleated cells (Richman et al. 2012). Consequently, the bone marrow mononuclear cells delivered in the LateTIME and TIME trials may have contained smaller numbers of $CD34^{+}$ and $CD105^{+}$ cells. In addition, 150–160 $\times$ 10^{6} unfractionated BMCs may not be the most optimal dose of BMCs for stem cell treatment of patients with AMI. In addition, bone marrow mononuclear cells from patients with advanced age and patients with chronic diseases, such as ischemic heart disease or diabetes mellitus, are often functionally impaired, propagate poorly, and have a shortened life span (Kissel et al. 2007; Fadini et al. 2010; Orlandi et al. 2010). In meta-analyses of stem cell trials of patients with myocardial infractions or ischemic cardiomyopathies, bone marrow mononuclear cells produce only a modest increase in the LVEF of approximately 2–3 % (Henning 2011, 2012). Consequently, despite well-conducted clinical trials, autologous unfractionated BMCs have a small therapeutic effect and may not be the most optimal cells for the treatment of patients with AMIs or ischemic cardiomyopathies.

12.9.2 Red Blood Cell Contamination of Stem Cells

Red blood cell contamination of bone marrow mononuclear cells can significantly decrease the migration ability and the efficacy of BMCs. Large numbers of red blood cells in the cell preparations cause reduced BMC viability and decreased colony-forming unit capacity and are associated with reduced recovery of LVEF in patients with myocardial infarctions (Assmus et al. 2010). In patients in the REPAIR-AMI trial, univariate and multivariate analysis demonstrated that red blood cell contamination of the BMCs prior to infusion into patients with myocardial infarctions independently predicted reduced recovery of LVEF (Schaechinger et al. 2006b). Moreover, the addition of red blood cells to BMCs dose-dependently decreased neovascularization in ischemic hind-limbs of research animals compared to treatment with BMCs without red blood cells (Assmus et al. 2010). The mechanism by which red blood cells interfere with bone marrow cell propagation, migration, and neovascularization involves a dose-dependent reduction of BMC mitochondrial membrane potential and a decrease in BMC mitochondrial adenosine triphosphate (ATP) production (Assmus et al. 2010). As a consequence, mitochondrial metabolism and function, stem cell self-renewal, and differentiation are decreased.

12.9.3 Heparin Decreases Stem Cell Migration

Heparin is another factor that can impact on the efficacy of BMCs in patients with AMIs. Heparin in a dose-dependent manner can inhibit stromal cell-derived factor 1 (SDF-1) induced BMC migration (Seeger et al. 2012; Heeschen et al. 2004; Murphy et al. 2007). In this regard, homing of BMCs to areas of myocardial ischemia is primarily guided by SDF-1 and its receptor termed chemokine receptor 4 (CXCR4). Heparin can bind to SDF-1 and CXCR4 and thereby block CXCR4 signaling (Seeger et al. 2012). Incubation of BMCs with 20 U/ml of heparin for 30 min abrogates SDF-1 BMC migration by 84 % in vitro and significantly reduces the homing of injected BMCs to injured and infarcted myocardium by 50 % in research animals (Seeger et al. 2012). Decreased migratory capacity of BMCs also correlates with reduced neovascularization and functional capacity in research animals with limb ischemia (Heeschen et al. 2004). The minimal dose of heparin that inhibits SDF-1-induced migration of BMC is 0.05 U/ml which is significantly less than the heparin dose used in several large trials of BMCs in AMI (Seeger et al. 2012). Heparin also interferes with activation of the cell survival factor Akt (Protein Kinase B) by SDF-1 and CXCR4 and in this manner interferes with cell survival and growth. In addition, heparin decreases the levels of vascular endothelial growth factor and in this manner limits neovascularization (Seeger et al. 2012). In contrast, the thrombin inhibitor bivalirudin does not appear to interfere with BMC homing or SDF-1/CXCR4 signaling and does not decrease vascular endothelial growth factor.

Consequently, not only the BMC isolation protocol but also the use of anticoagulants such as heparin can have major impact on the functional activity of BMCs (Seeger et al. 2007, 2012). The assessment of bone marrow cell number and viability by Trypan Blue staining does not accurately reflect the functional capacity of BMCs or other stem cells when injected into patients with AMIs. Colony-forming unit capacity is a better measure of stem cell viability than Trypan Blue staining.

12.9.4 Stem Cell Expulsion from Myocardium

An important factor that impacts on the efficacy of stem cells in patients with myocardial infarctions and ischemic cardiomyopathies is the time the stem cells actually reside in the myocardium. The majority (90–97 %) of unfractionated BMCs injected directly into the myocardium or into the coronary arteries leave the myocardium in less than 2 h (Hofmann et al. 2005; Hou et al. 2005). Most of the cells are ejected out of the myocardium through the injection sites or through the coronary veins and lymphatics into the right heart due to the massaging action of the contracting myocardium. The cells are ultimately lodged in the lungs, liver, spleen, and kidneys. In addition, approximately 12 % of cells are retained in the catheter delivery system after injection (Hou et al. 2005). With the intravenous injection of bone marrow or other cells for cardiac repair, the majority of the cells become entrapped in the lungs. Consequently, fourfold or greater numbers of stem cells are required above that required for intramyocardial or intracoronary injection for repair of myocardial infarctions (Henning 2011).

12.10 Unfractionated Bone Marrow Stem Cells: Quo Vadas? ("Whither Goest Thou?")

A meta-analysis published in 2014 concluded after reviewing 22 randomized control trials between 2002 and 2013 of unfractionated bone marrow-derived mononuclear cell therapy in patients with acute myocardial infarction that BMC therapy is safe but does not significantly enhance cardiac function based on MRI-derived parameters and does not improve 6-month patient outcome (de Jong et al. 2014). The results of the BAMI trial, which examines the effects of BMC on patient mortality over 3 years, will be important in determining the future of autologous, unfractionated bone marrow mononuclear cells in the treatment of patients with AMIs (Mathur 2013). In the interim, MSCs, adipose-derived stem cells (ADSCs) (which include MSCs), and cardiac stem cells are being investigated for cardiac repair in patients with myocardial infarctions and cardiomyopathies.

12.11 Mesenchymal Stem Cells in Cardiac Repair

In 1970, Friedenstein demonstrated that bone marrow contains not only hematopoietic stem cells but also a small population of MSCs which are also known as stromal cells (Friedenstein et al. 1968). These MSCs support the hematopoietic stem cells and the development of hematopoietic lineages but also can differentiate in vitro into osteoblasts, chondrocytes, adipocytes, and a myocyte phenotype.

The International Society for Cell Therapy has published specific criteria for identifying MSCs that include (1) the expression of cell surface proteins CD73, CD90, and CD105 in the absence of surface proteins such as CD34, CD45, HLA-DR, CD14, CD11b, CD79a, or CD19 when cells are analyzed by fluorescence-activated cell sorting; (2) cell adherence to plastic culture dishes during standard cell culture conditions; and (3) a cell capacity for differentiation in vitro into osteoblasts, adipocytes, and chondroblasts (Dominici et al. 2006).

Human MSCs express modest levels of major histocompatibility complex class I human leukocyte antigens (HLA), lack major histocompatibility complex class II expression, and do not express co-stimulatory molecules B7 and CD40 ligand (Williams and Hare 2011; Majumdar et al. 2003). Consequently, MSCs do not cause T-cell proliferation in mixed lymphocyte cultures.

Autologous MSCs have been examined in patients with acute myocardial infarctions and ischemic cardiomyopathies and do not cause cardiac arrhythmias or significant patient side effects. In this regard, 69 patients who underwent primary percutaneous coronary angioplasty within 8 ± 3.7 h after onset of acute myocardial infarction were randomized to autologous MSCs or saline 3 weeks after angioplasty. In the 34 MSC-treated patients, LVEF increased from 49 ± 9 to 67 ± 3 % at 6 months in comparison with the 35 patients in the control group in which the LVEF increased from 48 ± 10 to 54 ± 5 % (Chen et al. 2004).

In a separate study, 20 patients were treated with intravenous allogeneic MSCs from a single healthy donor 1–10 days post-myocardial infarction and were compared with 14 patients treated with placebo (Hare et al. 2009). Although the baseline LVEF was similar in both groups (47.3 ± 3.3 % vs. 45.2 ± 3.4 %), the MSC-treated patient showed a 5.2 ± 1.9 % increase in MRI-determined LVEF relative to baseline after 12 months, whereas the control patients showed an increase of only 1.8 ± 1 %.

To date, three studies have examined MSCs in patients with ischemic cardiomyopathies. In 2007, eight patients with chronic heart failure were reported that received MSCs and endothelial progenitor cells at the time of revascularization with either coronary artery bypass grafting or percutaneous coronary angioplasty (Mohyeddin-Bonab et al. 2007). The LVEF increased significantly from 38.7 ± 13 % at baseline to 48.8 ± 6.4 % at 18 months in the cell treated patients in comparison with a control group in which the LVEF only slightly increased from 41.9 ± 8.4 to 42.5 ± 8.9 %. Infarct scar size, measured by Thalium scan, decreased from 11 segments to 7.75 segments in the cell treatment group and also decreased slightly but significantly from 10.88 to 9.75 segments in the control group (Mohyeddin-Bonab et al. 2007).

In the POSEIDON trial, either autologous or allogeneic MSCs were administered to 30 patients with ischemic cardiomyopathies (Hare et al. 2012). This was an open-label study without a control group. In this study, the LVEF did not increase, but allogeneic and autologous MSCs reduced LV scar by 31.6 % (C.I. −49.24 to −13.99 %) and 34.9 % (C.I. 48.18–21.68 %), respectively, and decreased LV remodeling. The 6-min walk test increased significantly by 65.8 m at 12 months in the autologous MSC group but also increased slightly but not significantly in the allogeneic MSC group by 19.7 m (Hare et al. 2012).

In the C-CURE trial, 32 patients with ischemic cardiomyopathies were treated with MSCs that were injected with the aid of the NOGA endocardial catheter mapping system into viable LV myocardium (Bartunek et al. 2013). Before injection, the MSCs were treated with a cardiogenic cocktail (transforming growth factor, bone morphogenetic protein, activin, fibroblast growth factor, cardiotrophin, thrombin, diaminopyrimidine) that triggered expression of cardiac transcription factors. In the treated patients, LVEF, measured by echocardiography, improved with MSC therapy from 27.5 ± 1.0 to 34.5 ± 1.1 % versus patients treated with standard care in which the LVEF did not significantly change at 6 months after treatment. Cell therapy also improved the 6-min walk test in MSC-treated patients by 62 ± 18 m versus a decrease in the 6-min walk test of 15 ± 20 m in the standard care group (Bartunek et al. 2013).

The studies of MSCs in patients with myocardial infarctions and ischemic cardiomyopathies are promising, but clinical conclusions are limited due to the small numbers of studies, the small numbers of patients in each of the studies, and the relatively short-term follow-up of patients that received MSCs. Nevertheless, the fact that these cells can improve quality of life as evidence by an increase in 6-min walk test in patients with ischemic cardiomyopathy is encouraging and suggests that the otherwise dire prognosis of these patients with cardiomyopathies can be improved at least for the short term. The Safety Study of Allogeneic Mesenchymal Precursor Cell Infusion in MyoCardial Infarction (AMICI) in www.ClinicalTrials.gov and the Safety and Efficacy of Adipose-Derived Regenerative Cells (ADRCs) Delivered Via the Intracoronary Route in the Treatment of Patients With ST-elevation Acute Myocardial Infarction (ADVANCE) in www.ClinicalTrials.gov are investigating the effects of mesenchymal and mesenchymal-like cells on cardiac repair in more than 200 patients with myocardial infarctions.

12.12 Adipose-Derived Stem Cells (ADSCs)

Adipose tissue is mesodermally derived tissue that consists of adipocytes of various sizes. The adipocytes are interspersed in stroma composed of endothelial cells, preadipocytes, smooth muscle cells, fibroblasts, leukocytes, and macrophages, which are collectively termed the adipose stromal cell fraction (Lin et al. 2010). The frequency of adipose stem cells in adult human adipose tissue ranges from 1:30 to 1:100 cells per total nucleated cells and is dependent on the location of the adipose tissue and the age of the patient. ADSCs contain mesenchymal surface markers such as CD 105, CD90, CD73, CD51, CD49e, and CD29 and also express surface markers CD166, CD13, and CD44 but do not express hematopoietic cell markers. Microarray analysis and real-time polymerase chain reaction (PCR) determinations of ADSCs and BM-MSCs demonstrate that these cells exhibit virtually identical transcriptional profiles for the stem-related genes OCT4, Nanog, and Sox2 (Witkowska-Zimny and Walenko 2011).

Less than 1 % of ADSCs expresses HLA-DR protein, which makes these cells suitable for allogeneic transplantation without significant risk of rejection (Russo et al. 2014). ADSCs can suppress peripheral blood mononuclear cell proliferation in vitro and shift lymphocyte Th1 cytotoxic responses to lymphocyte Th2 anti-inflammatory responses. This anti-inflammatory ADSC effect exceeds that of bone marrow MSCs (Witkowska-Zimny and Walenko 2011; Russo et al. 2014).

Adipose cells can be readily obtained by liposuction of a patient and immediately prepared for autologous transplantation without the need for adipose cell culture and expansion. Human ADSC has significantly reduced myocardial infarction size in 8/9 animal studies, substantially improved LVEF, determined by echocardiography, in 9/10 animal studies, and significantly increased the number of blood vessels in the infarcted area in 10/10 animal studies (Naaijkens et al. 2014). The predominant mechanism of action of ADSC in AMI is through the paracrine release of antiapoptotic, immunomodulatory, and proangiogenic factors (Yang et al. 2013). These factors are thought to salvage injured cardiomyocytes and stimulate neoangiogenesis in the infarct border zone, thereby limiting scar size and left ventricular remodeling after myocardial infarction. Consequently, ADSCs currently represent an alternative to bone marrow MSCs for the treatment of patients with myocardial infarctions.

Based on the promising results of ADSCs in animal studies, several studies of adipose stem cells have been performed in patients with acute myocardial infarction and ischemic cardiomyopathy.

The APOLLO trial was a randomized, double-blind, placebo-controlled clinical trial of ADSCs, in concentrations of 17.4 ± 4.1 million cells, administered into the coronary arteries of 10 patients with acute ST-elevation myocardial infarction (Houtgraaf et al. 2012). In the ADSC-treated patients, the percentage of infarcted left ventricle was significantly reduced from 31.6 ± 5.3 to 15.3 ± 2.6 % at six months after infarction in contrast to no change in infarct size in the placebo-treated AMI patients (24.7 ± 9.2 % vs. 24.7 ± 4.1). This decrease in infarct size in ADSC-treated patients was associated with a significant decrease of the LV perfusion defect in ADSC-treated patients from 16.9 ± 2.1 to 10.9 ± 2.4 % at six-month follow-up. However, the LVEF in the ADSC-treated patients, measured by single-photon emission computerized tomography (SPECT), did not significantly differ from the placebo group. This study suggests that autologous ADSCs can be safely obtained by liposuction and administered via intracoronary infusion to patients with AMI and can produce some reduction of cardiac damage (Houtgraaf et al. 2012).

In the PRECISE trial, patients with ischemic cardiomyopathies with Canadian Cardiovascular society Class II to IV angina and/or New York Heart Association Class II to II heart failure, not amendable to coronary revascularization, were randomized to receive either the stromal vascular fraction (SVF) of autologous adipose tissue or placebo (Perin et al. 2014). Twenty-one patients received the SVF and 6 patients received placebo. The SVF or placebo was injected transendocardially into ischemic areas, which was determined by NOGA unipolar catheter LV voltages > 6.9 mV. In the SVF-treated patients, the total LV mass, determined by cardiac MRI, increased from baseline to 6 months from 128.1 ± 26 to 149.5 ± 32 g but did not significantly change in the control group (from 144.6 ± 52.7 to 152.6 ± 59.6 g). The absolute mass of the LV infarctions did not significantly change (35.1 + 20.4 to 34 + 16.5 g) in the treated patients but increased in the control group over 6 months from 29.6 ± 15.9 to 39.0 ± 15.4 g. The global LV wall motion score index improved slightly but significantly in the SVF-treated patients but did not significantly change in the control patients (Perin et al. 2014). The exercise tolerance, measured by metabolic equivalent (MET) values, was preserved over time in the ADSC-treated group (4.9 ± 0.8 to 4.9 ± 1.4) but decreased significantly (5.3 ± 2.5 to 4.2 ± 2.1) in the control group at 18 months (Perin et al. 2014).

Other trials that are examining ADSC include the ATHENA trial (Taylor and Dabkowski 2012) and the MyStromalCell trial (Qayyum et al. 2012), which are two prospective, randomized, double-blind, placebo-controlled Phase II studies that are investigating the effects of ADSCs in patients with chronic ischemic heart disease.

Although cellular therapy using adipose tissue in patients with acute myocardial infarctions and ischemic cardiomyopathies is promising, we currently do not know the most optimal adipose cell type (mesenchymal cell vs. SVF), the most optimal cell number to inject, the most optimal time to inject these cells in patients with myocardial infarction, or the most optimal technique (intracoronary or transendocardial) with which to inject these cells. Large patient randomized studies of ADSCs and SVF are necessary to answer these important questions. In addition, MSCs from adipose tissue should be compared with MSCs from bone marrow MSCs for the treatment of patients with acute myocardial infarction and ischemic cardiomyopathies.

12.13 Cardiac Stem/Progenitor Cells

Cardiovascular investigators have sought alternative stem cells to unfractionated bone marrow stem cells and adipose stem cells for cardiac repair in patients with ischemic heart disease. Cardiac stem cells are multipotent stem cells that are present in niches in the heart. These cells contribute to the physiological turnover of myocytes and vascular endothelial cells in the heart. The number of cardiac stem cells in the heart is small with an estimated one cardiac stem cell per 10,000 cardiac myocytes (Beltrami et al. 2003). Consequently, endogenous cardiac stem cells are not normally able to reverse heart damage due to myocardial infarctions. The physiologic turnover of myocytes by cardiac stem cells in the heart occurs at rates of approximately 1 % per year and is dependent on the age, sex, and the health of the individual (Kikuchi and Poss 2012).

Autologous cardiac stem cells have been isolated, cultured, propagated, and delivered to patients with injured and infarcted myocardium. Two major cardiac stem cell types have been investigated in the SCIPIO and CADUCEUS clinical trials in patients with ischemic cardiomyopathies: c-kit$^+$ lineage negative cardiac stem cells isolated from right atrial appendages and cardiosphere-derived cells (CDCs) from cardiospheres grown from right ventricular muscle biopsies.

12.14 C-Kit$^+$ Stem Cells

C-kit$^+$ cardiac stem cells have the capacity for self-renewal, clonogenicity, and multi-potency (Bearzi et al. 2007; Anversa et al. 2013). These stem cells can express the cardiac transcription factors GATA-4, Nkx2.5, and MEF2 and are reported to differentiate into myogenic, vascular endothelial, and smooth muscles cells (Bearzi et al. 2007; Kajstura et al. 2010). C-kit is a cell surface receptor for stem cell factor, and stem cell factor can chemoattract these stem cells to ischemic and injured myocardium. In research animals with myocardial infarctions, c-kit$^+$ cardiac stem cells are reported to form new myocytes in the heart (Bearzi et al. 2007). Consequently, these cells represent an important area of investigation for cardiac repair.

Autologous c-kit cardiac stem cells from right atrial appendages have recently been investigated for the treatment of patients with ischemic cardiomyopathies in the open-labeled Cardiac Stem Cell Infusion in Patients with Ischemic Cardiomyopathy (SCIPIO) trial (Bolli et al. 2011; Chugh et al. 2012; Bolli 2012). In this trial, the right atrial appendage was removed from patients during cardiopulmonary bypass for coronary artery surgery. C-kit positive stem cells were then isolated from each patient's atrial appendage and propagated in cell culture. Four months later, approximately one million autologous cardiac stem cells were injected back into each patient's saphenous vein grafts and coronary arteries supplying the infarcted myocardium.

In the SCIPIO trial, the LVEF, which was measured by three-dimensional echocardiography and also by MRI with gadolinium in patients who received cardiac stem cells, increased by 11.9 ± 2.7 % absolute units in 12 patients at 2 years after treatment (Bolli 2012). Left ventricular infarct scar in 6 patients, determined by cardiac MRI, decreased by 15.7 ± 4.7 g at 2 years. This decrease in myocardial scar was associated with an increase in viable muscle of 17.9 ± 12.1 g ($N = 6$) at 2 years (Bolli 2012). New York Heart Association Functional Class score improved in these patients by 0.9 ± 0.2 at 2 years ($N = 13$). In this study, c-kit cardiac stem cells were postulated to chemoattract the patients' native stem cells to areas of myocardial injury and also to transdifferentiate to myocytes for cardiac repair. A Phase 2 trial of safety and efficacy of c-kit cardiac stem cells in a larger group of patients with cardiomyopathy is currently being planned.

12.15 Cardiosphere-Derived Cells (CDCs)

Percutaneous endomyocardial biopsy specimens of the right ventricular septal wall in patients, when grown in culture, can yield spherical multicellular clusters termed cardiospheres. Cardiospheres are a mixture of stromal, mesenchymal, and hematopoietic progenitor cells that contain cells that express CD 105 (a transforming growth factor beta-receptor subunit commonly associated with MSCs) and partially express c-kit (Smith et al. 2007; Li et al. 2012). CDCs, when injected into the border zone of myocardial infarctions in mice, engrafted and increased the viable myocardium (Li et al. 2012). The functional benefit of CDCs is thought to be predominantly due to the secretion of growth factors and the recruitment of endogenous stem cells to injured and infarcted myocardium for myocyte generation (Li et al. 2012; Chimenti et al. 2010). In this regard, cardiospheres and CDCs can secrete the growth factors angiopoietin-2, basic fibroblastic growth factor, hepatocyte growth factor, insulin-like growth factor-1, stromal-derived factor-1, and vascular endothelial growth factor (Li et al. 2012; Chimenti et al. 2010).

Autologous CDCs have been investigated in the treatment of patients with ischemic cardiomyopathies in the open-labeled Cardiosphere-derived Autologous Stem Cells to Reverse Ventricular Dysfunction (CADUCEUS) trial (Makkar et al. 2012). In this trial, 17 patients, post-myocardial infarction with LVEFs of 25–45 %, underwent endomyocardial biopsies of the right ventricular septum. CDCs were obtained from cultures of the endomyocardial biopsies from each patient, and the cells were propagated in cell culture. A total of 12.5 and 25 million autologous CDCs were then given directly into the infarct-related coronary artery of each of the 17 patients 1.5–3 months after their myocardial infarctions in the CADUCEUS trial. The one-year follow-up of 12 of the 17 patients treated with autologous CDCs and 8 control patients have been presented (Makkar et al. 2012; Malliaras et al. 2014). Cardiac MRI with gadolinium was used for the determination of most endpoints. Left ventricular scar mass significantly decreased by a mean of 11.9 ± 6.8 g in CDC-treated patients and by 1.7 ± 7.8 g in patient controls. Left ventricular viable mass increased substantially by a mean of 22.6 ± 9.4 g in treated patients in comparison with 1.8 ± 8.7 g in patient controls. LVEFs did not significantly increase but the regional wall function of infarcted segments did increase and correlated with the decrease in LV myocardial scar size (Makkar et al. 2012; Malliaras et al. 2014). Although adverse events were slightly greater in the treated patients than in the control patients, the events were not significantly different between the two groups. The ALLSTAR trial is a Phase 2 study of CDCs currently in progress that involves *allogeneic* CDCs for the treatment of patients after myocardial infarction (www.ClinicalTrials.gov).

12.16 Critique of the Scipio and Caduceus Trials

The SCIPIO and CADUCEUS trials utilized unique cardiac stem cell populations in a highly selected patient population with myocardial infarctions. In the SCIPIO trial, 1545 patients were evaluated. Two hundred and thirteen patients had LVEFs <40 %, and 20 patients were treated with CSCs. All patients had echocardiographic determinations of LVEF, and 12 of 20 patients had MRI determinations of left ventricular function. Control patients did not have MRI determinations of left ventricular function. In the CADUCEUS trial, approximately 436 patients were evaluated and 17 patients received CDCs. Consequently, these trials report a highly selected patient population, and the results of these trials cannot be applied to all patients with myocardial infarctions and ischemic cardiomyopathies. Much larger trials of each of these cell types in patients with myocardial infarctions are necessary.

In each of these studies, LV infarction was defined by MRI of delayed enhancement of myocardium in the region of coronary artery occlusion/reperfusion due to gadolinium that leaked from myocardial capillaries and pooled in the myocardial interstitial space and intracellular spaces of infarcted myocytes. In these patients, the gadolinium volume of distribution was increased and washout from the myocardium was reduced. However, c-kit$^+$ cardiac stem cells and CDCs can incorporate into damaged blood vessels in infarcted myocardium. In addition, these stem cells can chemoattract endogenous stem cells that can form entirely new blood vessels (vasculogenesis) and can also secrete angiogenic growth factors that stimulate new blood vessels from preexisting vessels (angiogenesis). Consequently, the blood vessels in the damaged myocardium of patients treated with these stem cells were possibly less permeable to gadolinium (Murry 2012). Infarct scars can potentially appear smaller on MRI due to less gadolinium leak as well as myocardial infarction contracture, and therefore, "viable" myocardium can actually appear larger with MRI. MRI also cannot distinguish hypertrophic cardiac myocytes from myocyte hyperplasia (Murry 2012). Moreover, inter-scan variability and intra- and inter-observer variability in infarct measurements and interpreting MRI scans can account for some myocardial changes between pre- and post-stem cell infusion (Kwong and Farzaneh-Far 2011). Rebuttals to these arguments against the use of contrast-enhanced MRI in estimating infarct size and myocardial regeneration after stem cell treatment have been published by the Caduceus Investigators (Malliaras et al. 2013). The rebuttal is based on a porcine myocardial infarction study in which allogeneic CDCs decreased infarct scar size and lead to cardiomyocyte hyperplasia on MRI and also on histological examination (Malliaras et al. 2013). Nevertheless, anatomical and histological examinations of myocardial biopsies or myocardial autopsy examinations of patients treated with these stem cells are necessary to determine whether the infarct fibrosis is significantly decreased and whether the substantial generation of new myocytes occurs in patients treated with c-kit$^+$ cardiac stem cells and CDCs.

The fact that there is long-term improvement in the LVEFs by echocardiography and MRI in the SCIPIO trial and improvement in LV regional wall motion and LV thickening by MRI in the CADEUCUS trial suggests that the cardiac stem cells can reduce myocardial inflammation and scar formation, preserve injured myocytes, and chemoattract endogenous stem cells for myocardial repair. Trials of larger numbers of patients treated with CDCs and cardiac stem cells, such as the ALLSTAR trial (Marban NCT0 1458) and the proposed Phase II SCIPIO trial (Loughran et al. 2012), are warranted to determine the precise mechanisms of cardiac stem cell action and their benefit in patients over long periods.

12.17 Other Cell Types: Human Umbilical Cord Blood Stem Cells (hUCBC)

Four million births occur each year in the USA and approximately one hundred and thirty-four million births occur each year throughout the world. Human umbilical cord blood mononuclear cells (hUCBCs) are a source of hematopoietic, endothelial, and MSCs (Broxmeyer 1998; Broxmeyer et al. 1992; Bieback et al. 2004). The total content of hematopoietic progenitor cells in umbilical cord blood equals or exceeds that of bone marrow, but the highly proliferative hematopoietic stem cells are eightfold higher in hUCBC than in bone marrow and can be enriched by as much as 77–95 % (Broxmeyer et al. 1989; Piacibello et al. 1997). Human umbilical cord mesenchymal stem cells, which are present in cord blood and also umbilical cord tissue, are in the G0/G1 stage of the cell cycle but are capable of proliferating with a population-doubling time of 48 h (Bieback et al. 2004; Erices et al. 2000).

Human cord blood mononuclear cells are currently used for repopulating BMCs in patients treated for acute leukemia, chronic myeloid leukemia, myelodysplastic syndrome, neuroblastoma, and non-malignant diseases such as Fanconi's anemia and aplastic anemia (Broxmeyer 1998; Gluckman 2009). These cord blood cells contain less $CD3^+$, $CD4^+$, and $CD8^+$ immune cells than human adult blood cells and rarely express HLA class II antigens. In addition, cord blood T cells express CD45RA antigen which indicates that they are immunologically naïve (Broxmeyer 1998). This significantly reduces the risk of rejection by the host (Broxmeyer 1998; Henning et al. 2004). Moreover, hUCBC can be cryopreserved for periods of 20 or more years with the recovery of 60–100 % viable cells (Broxmeyer 1998). Consequently, hUCBC can be readily available for the treatment of damaged hearts.

Human umbilical cord mononuclear cells have been given to research animals with acute myocardial infarctions. In animal studies, hUCBCs have significantly limited the size of myocardial infarctions by $\geq$50 % and reduced LV remodeling thereby preserving LV ejection fraction and the rate of rise and fall of LV pressure (dP/dt) without requirements for host immune suppression (Henning et al. 2004, 2006, 2007, 2008, 2010). The optimal number of hUCBC for infarct size reduction in rodents with myocardial infarctions is four million cord cells when administered directly into the peri-infarct area or into the coronary arteries and 16 million cord blood cells when administered intravenously (Henning et al. 2007). The optimal time for injection of these cells in order to minimize infarct size is 2–24 h after the onset of acute myocardial infarction (Henning et al. 2006).

Human umbilical cord blood cells significantly limit the expression of inflammatory cytokines in acutely inflamed and infarcted myocardium. Within 12 h of acute infarctions in untreated research animals, the myocardial concentration of tumor necrosis factor alpha, monocyte chemoattraction protein, fractalkine, IL6 ciliary neurotrophic protein, macrophage inflammatory protein, and interferon-gamma increases two to as much as eightfold in comparison with cytokine concentrations in non-infarcted myocardium (Henning et al. 2008). In contrast, these inflammatory cytokines do not significantly change between 2 and 72 h after coronary occlusion in myocardial infarctions treated with hUCBC (Henning et al. 2008). Moreover, hUCBCs also significantly limit the myocardial infiltration of inflammatory neutrophils and lymphocytes into acute infarctions. For example, the percentage of neutrophils in untreated myocardial infarctions within 12 h of coronary occlusion significantly increases more than 130-fold from 0.04 ± 0.2 to 5.3 ± 1.2 %/50,000 ventricular myocytes in research animals (Henning et al. 2008). In contrast in the hUCBC-treated myocardial infarctions, the percentage of neutrophils is significantly less and averages only 1.3 ± 0.7 %/50,000 heart myocytes. Moreover, the percentages of neutrophils/50,000 cardiac myocytes at 24 and 72 h in hUCBC-treated infarcted hearts are not significantly different from normal controls (Henning et al. 2008). Similarly, at 24 and 72 h after coronary occlusion, the percentage of CD3 and CD4 lymphocytes in infarcted myocardium are twofold greater in untreated infarcted hearts in comparison with hUCBC-treated infarcted hearts (Henning et al. 2008). The hUCBC-induced reduction in inflammatory cells and inflammatory cytokines in these investigations is associated with left ventricular infarct sizes that are more than 40–50 % smaller in hUCBC-treated infarctions and LV ejection fractions that are more than 10 % greater at 1 and 2 months post-infarction than untreated infarctions (Henning et al. 2008).

An additional mechanism whereby hUCBC may be beneficial in ischemic/infarcted myocardium is by stimulating new blood vessel formation (Henning et al. 2004; Ma et al. 2005; Murohara et al. 2000). Endothelial progenitor cells are normal components of umbilical cord blood that can release pro-angiogenic molecules such as vascular endothelial growth factor (Ma et al. 2005; Murohara et al. 2000). These cells can also express KDR, Tie2/Tek, and VE-cadherin, which are expressed by endothelial cells during new blood vessel formation (Murohara et al. 2000; Nieda et al. 1997). In addition, $CD34^+$ hUCBCs integrate into the walls of blood vessels in the periphery of injured tissue and can increase capillary density in ischemic/infarcted muscles (Murohara et al. 2000; Pesce et al. 2003; Hirata et al. 2005).

When subjected to 1 % oxygen or hydrogen peroxide (H_2O_2)-induced free oxygen radical stress, hUCBC significantly increase the secretion of hepatocyte growth factor, insulin-like growth factor vascular endothelial cell growth factor, placental growth factor, IL-10, and angiogenin (Pesce et al. 2003; Henning et al. 2012; Jin et al. 2013; Henning et al. 2014). These hUCBC paracrine factors can significantly increase in cardiac myocytes the activation of the cell survival protein Akt (Protein kinase B) which can decrease activation of the myocyte death proteins JNK and p38 and thereby preserve myocyte viability by limiting or preventing myocyte apoptosis and necrosis (Henning et al. 2012; Jin et al. 2013; Henning et al. 2014).

Currently, hUCBCs are being processed for the development of vascular grafts and heart valves for the treatment of newborns with congenital heart defects. In addition, Phase 1 studies are being performed with the use of hUCBC in the treatment of patients with angina pectoris that is refractory to medical therapy and who are not candidates for surgical or angioplasty coronary revascularization.

12.18 A Stem Cell Perspective

Although different stem cells are available for cardiac repair, the optimal stem cell for the treatment of all patients with infarcted myocardium remains to be determined. The optimal stem cell should permit transplantation into different patients without requirements for patient immune suppression therapy. Current cell candidates that are undergoing investigations in patients include allogeneic bone marrow MSCs and CDCs. New techniques must be developed to enhance the survival and propagation of these stem cells without increasing the risks of neoplastic differentiation.

Cell banks should be established that provide readily available, undifferentiated, but accurately characterized allogeneic stem cells that have significant capacity for in vitro and in vivo propagation for the treatment of patients with heart disease. The optimal number of stem cells and the optimal timing of stem cell transplantation into patients' hearts after myocardial infarction must be systematically investigated to maximize the chemoattraction of stem cells to ischemic and infarcted myocardium and facilitate myocardial healing. In this regard, the repeated administration of stem cells to patients will probably be necessary via intracoronary or intravenous injections and should be investigated.

Investigations must determine whether intramyocardial, intracoronary, or intravenous injection is most optimal for cardiac repair. With intracoronary or intravenous injections, large numbers of MSCs can cause cell clumping and microinfarctions in the heart or lungs. Multiple intramyocardial injections can be associated with high rates of stem cell leakage from the myocardium, disruption of the extracellular matrix of the myocardium, and scar formation, thereby potentiating the formation of arrhythmogenic foci. Although some stem cells injected intravenously do reach the heart, many stem cells become lodged in the lungs. Consequently with intravenous injections, the number of stem cells required for cardiac repair can be fourfold greater than the numbers required for intramyocardial or intracoronary injection for cardiac repair (Henning et al. 2007). Pulmonary function studies and oxygen saturation levels should be monitored. Moreover, strategies must be developed to facilitate the homing to the heart of stem cells that are injected intravenously.

Enhancement of stem cell engraftment in the heart is mandatory for optimizing the therapeutic benefits of these cells. Currently, less than 10 % of the stem cells remains in the heart 1–2 h after injection into the beating heart (Hofmann et al. 2005; Hou et al. 2005). Ninety percent or more of the cells are expelled from the intramyocardial injection site or are extruded from the myocardium through the coronary veins and lymphatics due to the massaging action of the heart. The stem cells then migrate to the lungs, liver, spleen, and kidneys (Hofmann et al. 2005). Potential treatment strategies to ensure that stem cells remain in the myocardium include co-delivery of stem cells with extracellular matrix molecules, nanofibers, fibrin glues, or other drugs, or applying stem cell patches to the epicardium (Henning 2011, 2012).

Significant discrepancies exist between the paucity of stem cells that actually engraft in the heart and the improvement in heart function that can occur with stem cell therapy. This suggests that the beneficial effects of stem cells are due to the release of biologically active growth factors and anti-inflammatory cytokines that protect cardiomyocytes and vascular endothelial cells in the injured myocardium. These biologically active factors can potentially limit myocyte apoptosis, necrosis, and extracellular matrix remodeling, stimulate angiogenesis, and recruit endogenous stem cells to the damaged myocardium. Consequently, growth factors and anti-inflammatory cytokines secreted by stem cells must be isolated, identified, purified, expanded, and investigated as new pharmacologic therapies for cardiac repair.

Imaging and hemodynamic measurement endpoints must be uniformly employed to demonstrate benefit, permit comparisons of different stem cell investigations, and provide insights into stem cell mechanisms of action. In this regard, MRI should be uniformly employed to measure changes in cardiac regional wall motion, ejection fraction, ventricular end-systolic and end-diastolic volumes, and left ventricular mass.

Additional basic science, preclinical, and clinical studies are required in order to address and answer the unresolved issues discussed in this chapter regarding stem cells in cardiac repair. These studies will require the close cooperation, the interaction, and the financial support of basic scientists and clinicians throughout the world. In this way, cell-based therapy in the twenty-first century will offer new hope to the millions of patients with heart disease who would otherwise suffer from the inexorable downward progression of the heart disease and heart failure.

Acknowledgments This work was support, in part, by grants from the Florida-King Biomedical Research Program, the Muscular Dystrophy Association, the Robert O. Law Foundation, and the Cornelius Foundation and by facilities at the James A. Haley Hospital and the University of South Florida College Of Medicine. Sections of this chapter were adapted from the author's publications in Future Cardiology

References

Due to significant space limitations the author could not include all the relevant papers that have been published on stem cells and heart disease.

Abdel-Latif A, Bolli A, Zuba-Surma EK, Tleyjeh IM, Hornung CA, Dawn B. Granulocyte colony-stimulating factor therapy for cardiac repair after acute myocardial infarction: a systematic review and meta-analysis of randomized controlled trials. Am Heart J. 2008;156:216–26. doi:10.1016/j.ahj.2008.03.024.

Ang KL, Chin D, Leyva F, Foley P, Kubal C, Chalil S, Srinivasan L, Bernhardt L, Stevens S, Shenje LT, Galiñanes M. Randomized, controlled trial of intramuscular or intracoronary injection of autologous bone marrow cells into scarred myocardium during CABG versus CABG alone. Nat Clin Pract Cardiovasc Med. 2008;5:663–70. doi:10.1038/ncpcardio1321.

Anversa P, Kajstura J, Rota M, Leri A. Regenerating new heart with stem cells. J Clin Invest. 2013;123:62–70. doi:10.1172/JCI63068.

Assmus B, Honold J, Schächinger V, Britten MB, Fischer-Rasokat U, Lehmann R, Teupe C, Pistorius K, Martin H, Abolmaali ND, Tonn T, Dimmeler S, Zeiher AM. Transcoronary transplantation of progenitor cells after myocardial infarction. N Engl J Med. 2006;355:1222–32.

Assmus B, Tonn T, Seeger FH, Yoon CH, Leistner D, Klotsche J, Schächinger V, Seifried E, Zeiher AM, Dimmeler S. Red blood cell contamination of the final cell product impairs the efficacy of autologous bone marrow mononuclear cell therapy. J Am Coll Cardiol. 2010;55:1385–94. doi:10.1016/j.jacc.2009.10.059.

Bartunek J, Vanderheyden M, Vandekerckhove B, Mansour S, De Bruyne B, De Bondt P, Van Haute I, Lootens N, Heyndrickx G, Wijns W. Intracoronary injection of CD133-positive enriched bone marrow progenitor cells promotes cardiac recovery after recent myocardial infarction: feasibility and safety. Circulation 2005;112(9 Suppl):I178–83.

Bartunek J, Vanderheyden M, Vandekerckhove B, Mansour S, De Bruyne B, De Bondt P, Van Haute I, Lootens N, Heyndrickx G, Wijns W. Intracoronary injection of CD133-positive enriched bone marrow progenitor cells promotes cardiac recovery after recent myocardial infarction: feasibility and safety. Circulation. 2006;112(Suppl):I178–83.

Bartunek J, Behfar A, Dolatabadi D, Vanderheyden M, Ostojic M, Dens J, El Nakadi B, Banovic M, Beleslin B, Vrolix M, Legrand V, Vrints C, Vanoverschelde JL, Crespo-Diaz R, Homsy C, Tendera M, Waldman S, Wijns W, Terzic A. Cardiopoietic stem cell therapy in heart failure: the C-CURE (Cardiopoietic stem Cell therapy in heart failURE) multicenter randomized trial with lineage-specified biologics. J Am Coll Cardiol. 2013;61:2329–38. doi:10.1016/j.jacc.2013.02.071.

Bearzi C, Rota M, Hosoda T, Tillmanns J, Nascimbene A, De Angelis A, Yasuzawa-Amano S, Trofimova I, Siggins RW, Lecapitaine N, Cascapera S, Beltrami AP, D'Alessandro DA, Zias E, Quaini F, Urbanek K, Michler RE, Bolli R, Kajstura J, Leri A, Anversa P. Human cardiac stem cells. Proc Natl Acad Sci USA. 2007;104:14068–73.

Beltrami C, Finato N, Rocco M, Feruglio G, Puricelli C, Cigola E, Quaini F, Sonnenblick E, Olivetti G, Anversa P. Structural basis of end-stage failure in ischemic cardiomyopathy in humans. Circulation. 1994;89:151–63.

Beltrami AP, Barlucchi L, Torella D, Baker M, Limana F, Chimenti S, Kasahara H, Rota M, Musso E, Urbanek H, Leri A, Kajstura J, Nadal-Ginard B, Anversa P. Adult cardiac stem cells are multipotent and support myocardial regeneration. Cell. 2003;114:763–76.

Bieback K, Kern S, Klüter H, Eichler H. Critical parameters for the isolation of mesenchymal stem cells from umbilical cord blood. Stem Cells. 2004;22:625–34.

Bocchi EA, Bacal F, Guimarães G, Mendroni A, Mocelin A, Filho AE, Dores da Cruz F, Resende M, Chamone D. Granulocyte-colony stimulating factor or granulocyte-colony stimulating factor associated to stem cell intracoronary infusion effects in non-ischemic refractory hear failure. Int J Cardiol. 2010;138:94–97. doi:10.1016/j.ijcard.2008.06.002.

Bolli R. Effect of cardiac stem cells in patients with ischemic cardiomyopathy: interim results of the SCIPIO trial up to 2 years after therapy. In: American Heart Association Scientific Sessions. Los Angeles, CA, USA, 3–7 Nov 2012.

Bolli R, Chugh AR, D'Amario D, Loughran JH, Stoddard MF, Ikram S, Beache GM, Wagner SG, Leri A, Hosoda T, Sanada F, Elmore JB, Goichberg P, Cappetta D, Solankhi NK, Fahsah I, Rokosh DG, Slaughter MS, Kajstura J, Anversa P. Cardiac stem cells in patients with ischaemic cardiomyopathy (SCIPIO): initial results of a randomised phase 1 trial. Lancet. 2011;378:1847–57. doi:10.1016/S0140-6736(11)61590-0.

Broxmeyer HE. Cellular characteristics of cord blood and cord blood transplantation. Bethesda: AABB Press; 1998.

Broxmeyer HE, Douglas GW, Hangoc G, Cooper S, Bard J, English D, Arny M, Thomas L, Boyse ES. Human umbilical cord blood as a potential source of transplantable hematopoietic stem/progenitor cells. Proc Natl Acad Sci USA. 1989;86:3828–32.

Broxmeyer HE, Hangoc G, Cooper S, Ribeiro RC, Graves V, Yoder M, Wagner J, Vadhan-Raj S, Benninger L, Rubinstein P. Growth characteristics and expansion of human umbilical cord blood and estimation of Its potential for transplantation adult. Proc Natl AcadSci USA. 1992;89:4109–13.

Chen SL, Fang WW, Ye F, Liu YH, Qian J, Shan SJ, Zhang JJ, Hunhua RZ, Liao LM, Lin S, Sun JP. Effect on left ventricular function of intracoronary transplantation of autologous bone marrow mesenchymal stem cell in patients with acute myocardial infarction. Am J Cardiol. 2004;94:92–5.

Chimenti I, Smith RR, Li TS, Gerstenblith G, Messina E, Giacomello A, Marbán E. Relative roles of direct regeneration versus paracrine effect of human cardiosphere-derived cells transplanted into infarcted mice. Circ Res. 2010;106:971–80.doi:10.1161/CIRCRES AHA.109.210682.

Chugh AR, Beache GM, Loughran JH, Mewton N, Elmore JB, Kajstura J, Pappas P, Tatooles A, Stoddard MF, Lima JA, Slaughter MS, Anversa P, Bolli R. Administration of cardiac stem cells in patients with ischemic cardiomyopathy: the SCIPIO trial: surgical aspects and interim analysis of myocardial function and viability by magnetic resonance. Circulation. 2012;126:S54–64.

de Jong R, Houtgraaf JH, Samiei S, Boersma E, Duckers HJ. Intracoronary stem cell infusion after acute myocardial infarction: a meta-analysis and update on clinical trials. Circ Cardiovasc Interv. 2014;7:156–67 10.1161/CIRCINTERVENTIONS.113.001009.

Dill T, Schächinger V, Rolf A, Möllmann S, Thiele H, Tillmanns H, Assmus B, Dimmeler S, Zeiher AM, Hamm C. Intracoronary administration of bone marrow-derived progenitor cells improves left ventricular function in patients at risk for adverse remodeling after acute ST-segment elevation myocardial infarction: results of the reinfusion of enriched progenitor cells and infarct remodeling in acute myocardial infarction study (REPAIR-AMI) cardiac magnetic resonance imaging substudy. Am Heart J. 2008;157:541–7. doi:10.1016/j.ahj.2008.11.011.

Dominici M, Le Blanc K, Mueller I, Slaper-Cortenbach I, Marini F, Krause D, Deans R, Keating A, Prockopj Dj, Horwitz E. Minimal criteria for defining multipotent mesenchymal stromal cells. The International Society for Cellular Therapy position statement. Cytotherapy. 2006;8:315–7.

Erbs S, Linke A, Adams V, Lenk K, Thiele H, Diederich KW, Emmrich F, Kluge R, Kendziorra K, Sabri O, Schuler G, Hambrecht R. Transplantation of blood-derived progenitor cells after recanalization of chronic coronary artery occlusion: first randomized and placebo-controlled study. Circ Res. 2005;97:756–62.

Erices A, Conget P, Minguell JJ. Mesenchymal progenitor cells in human umbilical cord blood. Br J Haematol. 2000;109:235–42.

Fadini GP, Boscaro E, de Kreutzenberg S, Agostini C, Seeger F, Dimmeler S, Zeiher A, Tiengo A, Avogaro A. Time course and mechanisms of circulating progenitor cell reduction in the natural history of type 2 diabetes. Diabetes Care. 2010;33:1097–102. doi:10.2337/dc09-1999.

Fischer-Rasokat U, Assmus B, Seeger FH, Honold J, Leistner D, Fichtlscherer S, Schächinger V, Tonn T, Martin H, Dimmeler S, Zeiher AM. A pilot trial to assess potential effects of selective intracoronary bone marrow-derived progenitor cell infusion in patients with nonischemic dilated cardiomyopathy: final 1-year results of the transplantation of progenitor cells and functional regeneration enhancement pilot trial in patients with nonischemic dilated cardiomyopathy. Circ Heart Fail. 2009;2:417–23. doi:10.1161/CIRCHEARTFAILURE.109.855023.

Fisher SA, Brunskill SJ, Doree C, Mathur A, Taggart DP, Martin-Rendon, E. Stem cell therapy for chronic ischaemic heart disease and congestive heart failure. Cochrane Database Syst Rev. 2014;4:CD007888. doi:10.1002/14651858.CD007888.pub2.

Friedenstein AJ, Petrakova KV, Kurolesova AI, Frolova GP. Heterotopic of bone marrow. Analysis of precursor cells for osteogenic and hematopoietic tissues. Transplantation. 1968;6:230–47.

Ge J, Li Y, Qian J, Shi J, Wang Q, Niu Y, Fan B, Liu X, Zhang S, Sun A, Zou Y. Efficacy of emergent transcatheter transplantation of stem cells for treatment of acute myocardial infarction (TCT-STAMI). Heart. 2006;92:1764–7.

Gluckman E. Ten years of cord blood transplantation: from bench to bedside. Br J Haematol. 2009;147:192–9. doi:10.1111/j.1365-2141.2009.07780.x.

Gnecchi M, He H, Liang O, Melo L, Moello F, Mu H, Noiseux N, Zhang L, Pratt R, Ingwall J, Dzau V. Paracrine action accounts for marked protection of ischemic heart by Akt modified mesenchymal stem cells. Nat Med. 2005;11:367–8.

Go AS, Mozaffarian D, Roger V, Benjamin E, Berry J, Blaha M, Dai S, Ford E, Fox C, Franco S, Fullerton H, Gillespie C, Hailpern S, Heit J, Howard V, Huffman M, Judd S, Kissela B, Kittner S, Lackland D, Lichtman J, Lisabeth L, Mackey R, Magid D, Marcus G, Marelli A, Matchar D, McGuire D, Mohler E III, Moy C, Mussolino E, Neumar R, Nichol G, Pandey D, Paynter N, Reeves M, Sorlie P, Stein J, Towfighi A, Turan T, Virani S, Wong N, Woo D, Turner M. On behalf of the American Heart Association Statistics Committee and stroke statistics subcommittee heart disease and stroke statistics. Circulation. 2014;129:e28–292.

Hare JM, Traverse JH, Henry TD, Dib N, Strumpf RJ, Schulman SP, Gerstenblith G, DeMaria AN, Denktas AE, Gammon RS, Hermiller JB Jr, Reisman MA, Schaer GL, Sherman W. A randomized, double-blind, placebo-controlled, dose-escalation study of intravenous adult human mesenchymal stem cells (prochymal) after acute myocardial infarction. J Am Coll Cardiol. 2009;54:2277–86. doi:10.1016/j.jacc.2009.06.055.

Hare JM, Fishman JE, Gerstenblith G, DiFede Velazquez DL, Zambrano JP, Suncion VY, Tracy M, Ghersin E, Johnston PV, Brinker JA, Breton E, Davis-Sproul J, Schulman IH, Byrnes J, Mendizabal AM, Lowery MH, Rouy D, Altman P, Wong Po Foo C, Ruiz P, Amador A, Da Silva J, McNiece IK, Heldman AW, George R, Lardo A. Comparison of allogeneic vs autologous bone marrow–derived mesenchymal stem cells delivered by transendocardial injection in patients with ischemic cardiomyopathy: the POSEIDON randomized trial. JAMA. 2012;308:2369–79.

Heeschen C, Lehmann R, Honold J, Assmus B, Aicher A, Walter DH, Martin H, Zeiher AM, Dimmeler S. Profoundly reduced neovascularization capacity of bone marrow mononuclear cells derived from patients with chronic ischemic heart disease. Circulation 2004;109:1615–22.

Hendrikx M, Hensen K, Clijsters C, Jongen H, Koninckx R, Bijnens E, Ingels M, Jacobs A, Geukens R, Dendale P, Vijgen J, Dilling D, Steels P, Meesens U. Recovery of regional but not global contractile function by the direct intramyocardial autologous bone marrow transplantation: results from a randomized controlled clinical trial. Circ. 2006;114(Suppl):I101–7.

Henning RJ. Stem cells in cardiac repair. Future Cardiol. 2011;7:99–117. doi:10.2217/fca.10.109.

Henning RJ. Stem cells in cardiac repair-recent developments and future directions. Intervent Cardiol. 2012;7:10–1.

Henning RJ. Stem cells for cardiac repair: problems and possibilities. Future Cardiol. 2013;9:875–84. doi:10.2217/fca.13.78.

Henning RJ, Abu-Ali H, Balis JU, Morgan MB, Willing AE, Sanberg PR. Human umbilical cord blood mononuclear cells for the treatment of acute myocardial infarction. Cell Transplant. 2004;13:729–39.

Henning RJ, Burgos JD, Ondrovic L, Sanberg P, Balis J, Morgan MB. Human umbilical cord blood progenitor cells are attracted to infarcted myocardium and significantly reduce myocardial infarction size. Cell Transplant. 2006;15:647–58.

Henning RJ, Burgos JD, Vasko M, Alvarado F, Sanberg CD, Sanberg PR, Morgan MB. Human cord blood cells and myocardial infarction: effect of dose and route of administration on infarct size. Cell Transplant. 2007;16:907–17.

Henning RJ, Shariff M, Eadula U, Alvarado F, Vasko M, Sanberg PR, Sanberg CD, Delostia V. Human cord blood mononuclear cells decrease cytokines and inflammatory cells in acute myocardial infarction. Stem Cells Dev. 2008;17:1207–19. doi:10.1089/scd.2008.0023.

Henning RJ, Aufman J, Shariff M, Sawmiller D, DeLostia V, Sanberg P, Morgan M. Human umbilical cord blood mononuclear cells decrease fibrosis and increase cardiac function in cardiomyopathy. Regen Med. 2010;5:45–54. doi:10.2217/rme.09.71.

Henning RJ, Dennis S, Sawmiller D, Hunter L, Sanberg P, Miller L. Human umbilical cord blood mononuclear cells activate the survival protein Akt in cardiac myocytes and endothelial cells that limits apoptosis and necrosis during hypoxia. Transl Res. 2012;159:497–506. doi:10.10 16/TRSL.2012.02.004

Henning RJ, Sanberg P, Jimenez E. Human cord blood stem cell paracrine factors activate the survival protein kinase Akt and inhibit death protein kinases JNK and p38 in injured cardiomyocytes. Cytotherapy. 2014;S1465–3249(14):00444–7. doi:10.1016/j.jcyt.2014.01.415.

Hirata Y, Sata M, Motomura N, Takanashi M, Suematsu Y, Ono M, Takamoto S. Human umbilical cord blood cells improve cardiac function after myocardial infarction. Biochem Biophys Res Commun. 2005;327:609–614.

Hofmann M, Wollert KC, Meyer GP, Menke A, Arseniev L, Hertenstein B, Ganser A, Knapp WH, Drexler H. Monitoring of bone marrow cell homing into the infarcted human myocardium. Circulation. 2005;111:2198–202.

Hou D, Youssef EA, Brinton TJ, Zhang P, Rogers P, Price ET, Yeung AC, Johnstone BH, Yock PG, March KL. Radiolabeled cell distribution after intramyocardial, intracoronary, and interstitial retrograde coronary venous delivery: implications for current clinical trials. Circulation. 2005;112(9 Suppl):I150–6.

Houtgraaf JH, den Dekker WK, van Dalen BM, Springeling T, de Jong R, van Geuns RJ, Geleijnse ML, Fernandez-Aviles F, Zijlsta F, Serruys PW, Duckers HJ. First experience in humans using adipose tissue-derived regenerative cells in the treatment of patients with ST-segment elevation myocardial infarction. J Am Coll Cardiol. 2012;59:539–40. doi:10.1016/j.jacc.2011.09.065.

Janssens S, Dubois C, Bogaert J, Theunissen K, Deroose C, Desmet W, Kalantzi M, Herbots L, Sinnaeve P, Dens J, Maertens J, Rademakers F, Dymarkowski S, Gheysens O, Van Cleemput J, Bormans G, Nuyts J, Belmans A, Mortelmans L, Boogaerts M, Van de Werf F. Autologous bone marrow-derived stem-cell transfer in patients with ST-segment elevation myocardial infarction: double-blind, randomised controlled trial. Lancet. 2006;367:113–21.

Jin H, Sanberg PR, Henning RJ. Human umbilical cord blood mononuclear cell-conditioned media inhibits hypoxic-induced apoptosis in human coronary artery endothelial cells and cardiac myocytes by activation of the survival protein Akt. Cell Transplant. 2013;22:1637–50. doi:10.3727/096368912X661427.

Kajstura J, Gurusamy N, Ogórek B, Goichberg P, Clavo-Rondon C, Thosoda D D'Amario, Bardelli S, Beltrami AP, Cesselli D, del Rbussani F, Monte F Quaini, Rota M, Beltrami CA, Buchholz BA, Leri A, Anversa P. Myocyte turnover in the aging human heart. Circ Res. 2010;107:1374–86 10.1161/CIRCRESAHA.110.231498.

Kang HJ, Lee HY, Na SH, Chang SA, Park KW, Kim HK, Kim SY, Chang HJ, Lee W, Kang WJ, Koo BK, Kim YJ, Lee DS, Sohn DW, Han KS, Oh BH, Park YB, Kim HS. Differential effect of intracoronary infusion of mobilized peripheral blood stem cells by granulocyte colony-stimulating factor on left ventricular function and remodeling in patients with acute myocardial infarction versus old myocardial infarction: the MAGIC Cell-3-DES randomized, controlled trial. Circulation. 2006;114(1 Suppl):I145–51.

Katritsis DG, Sotiropoulou PA, Karvouni E, Karabinos I, Korovesis S, Perez SA, Voridis EM, Papamichail M. Transcoronary transplantation of autologous mesenchymal stem cells and endothelial progenitors into infarcted human myocardium. Catheter Cardiovasc Interv. 2005;65:321–9.

Kikuchi K, Poss KD. Cardiac regenerative capacity and mechanisms. Annu Rev Cell Dev Biol. 2012;28:719–41. doi:10.1146/annurev-cellbio-101011-155739.

Kinnaird T, Stabile E, Burnett MS, Lee CW, Barr S, Fuchs S, Epstein SE. Marrow-derived stromal cells express genes encoding a broad spectrum of arteriogenic cytokines and promote in-vitro and in-vivo arteriogenesis through paracrine mechanisms. Cir. 2004;94:678–85.

Kissel C, Lehmann R, Assmus B, Aicher A, Honold J, Fischer-Rasokat U, Heeschen C, Sptridopoulos I, Dimmeler S, Zeiher A. Selective functional exhaustion of hematopoietic progenitor cells in the bone marrow of patients with postinfarction heart failure. J Am Coll Cardiol. 2007;49:2341–9.

Kwong RY, Farzaneh-Far A. Measuring myocardial scar by cardiac magnetic resonance imaging. JACC Cardiovasc Imaging. 2011;4:157–60. doi:10.1016/j.jcmg.2010.12.004.

Li ZQ, Zhang M, Jing YZ, Zhang WW, Liu Y, Cui LJ, Yuan L, Liu XZ, Yu X, Hu TS. The clinical study of autologous peripheral blood stem cell transplantation by intracoronary infusion in patients with acute myocardial infarction (AMI). Int J Cardiol. 2007;115:52–6.

Li TS, Cheng K, Malliaras K, Smith R, Zhang Y, Sun B, Matsushita N, Blusztajn A, Terrovitis J, Kusuoka H, Marban L, Marban E. Direct comparison of different stem cell types and subpopulations reveals superior paracrine potency and myocardial repair efficacy with cardiosphere-derived cell. J Am Coll Cardiol. 2012;59:942–53.

Lin CS, Xin ZC, Deng CH, Ning H, Lin G, Lue TF. Defining adipose tissue-derived stem cells in tissue and in culture. Histol Histopathol. 2010;25:807–15.

Lipinski M, Biondi-Zoccai G, Abbate A, Khianey R, Sheiban I, Bartunek J, Vanderheyden M, Kim H-S, Kang H-J, Strauer B, Vetrovec G. Impact of intracoronary cell therapy on left ventricular function in the setting of acute myocardial infarction: a collaborative systematic review and meta-analysis of controlled clinical trials. J Am Coll Cardiol. 2007;50:1761–7.

Loughran JH, Elmore JB, Waqar M, Chugh AR, Bolli R. Cardiac stem cells in patients with ischemic cardiomyopathy: discovery, translation, and clinical investigation. Curr Atheroscler Rep. 2012;14:491–503. doi:10.1007/s11883-012-0273-9.

Lunde K, Solheim S, Aakhus S, Arnesen H, Abdelnoor M, Egeland T, Endresen K, Llebekk A, Mangschau A, Fjeld JG, Smith HJ, Taraldsrud E, Grøgaard HK, Bjørnerheim R, Brekke M, Müller C, Hopp E, Ragnarsson A, Brinchmann JE, Forfang K. Intracoronary injection of mononuclear bone marrow cells in acute myocardial infarction. N Engl J Med. 2006;355:1199–209.

Lunde K, Solheim S, Forfang K, Arnesen H, Brinch L, Bjørnerheim R, Ragnarsson A, Egeland T, Endresen K, Ilebekk A, Mangschau A, Aakhus S. Anterior myocardial infarction with acute percutaneous coronary intervention and intracoronary injection of autologous mononuclear bone marrow cells: safety, clinical outcome, and serial changes in left ventricular function during 12-months' follow-up. J Am Coll Cardiol. 2008;51:674–6 10.1016.

Ma N, Stamm C, Kaminski A, Li W, Kleine HD, Müller-Hilke B, Zhang L, Ladilov Y, Egger D, Steinhoff G. Human cord blood cells induce angiogenesis following myocardial infarction in NOD/scid-mice. Cardiovasc Res. 2005;66:45–54.

Majumdar M, Keane-Moore M, Buyaner D, Hardy W, Moorman M, McIntosh K, Monsca J. Characterization and functionality of cell surface molecules on human mesenchymal stem cells. J Biomed Sci. 2003;10:228–41.

Makkar RR, Smith RR, Cheng K, Malliaras K, Thomson LE, Berman D, Czer LS, Marbán L, Mendizabal A, Johnston PV, Russell SD, Schuleri KH, Lardo AC, Gerstenblith G, Marbán E. Intracoronary cardiosphere-derived cells for heart regeneration after myocardial infarction (CADUCEUS): a prospective, randomised phase 1 trial. Lancet. 2012;379:895–904. doi:10.1016/S0140-6736(12)60195-0.

Malliaras K, Smith RR, Kanazawa H, Yee K, Seinfeld J, Tseliou E, Dawkins JF, Kreke M, Cheng K, Luthringer D, Ho CS, Blusztajn A, Valle I, Chowdhury S, Makkar RR, Dharmakumar R,

Li D, Marbán L, Marbán E. Validation of contrast-enhanced magnetic resonance imaging to monitor regenerative efficacy after cell therapy in a porcine model of convalescent myocardial infarction. Circ. 2013;128:2764–75. doi:10.1161/CIRCULATIONAHA.113.002863.

Malliaras K, Makkar RR, Smith RR, Cheng K, Wu E, Bonow RO, Marbán L, Mendizabal A, Cingolani E, Johnston PV, Gerstenblith G, Schuleri KH, Lardo AC, Marbán E. Intracoronary cardiosphere-derived cells after myocardial infarction: evidence of therapeutic regeneration in the final 1-year results of the CADUCEUS trial (CArdiosphere-Derived aUtologous stem CElls to reverse ventricUlar dySfunction). J Am Coll Cardiol. 2014;63:110–22. doi:10.1016/j.jacc.2013.08.724.

Marban E, Clinical Investigators. Allogeneic heart stem cells to achieve myocardial regeneration (ALLSTAR). NCT01458405. www.ClinicalTrials.gov.

Martin-Rendon E, Brunskill SJ, Hyde CJ, Stanworth SJ, Mathur A, Watt SM. Autologous bone marrow stem cells to treat acute myocardial infarction: a systematic review. Eur Heart 2008;J29:1807–18. doi: 10.1093/eurheartj/ehn220.

Mathur A. The effect of intracoronary reinfusion of bone marrow-derived mononuclear cells (BM-MNC) on all cause mortality in acute myocardial infarction. NC T01569178. ClinicalTrials. Gov.

Meluzín J, Mayer J, Groch L, Janousek S, Hornácek I, Hlinomaz O, Kala P, Panovský R, Prásek J, Kamínek M, Stanícek J, Klabusay M, Korístek Z, Navrátil M, Dusek L, Vinklárková J. Autologous transplantation of mononuclear bone marrow cells in patients with acute myocardial infarction: the effect of the dose of transplanted cells on myocardial function. Am Heart J. 2006;152:975.

Meluzín J, Janousek S, Mayer J, Groch L, Hornácek I, Hlinomaz O, Kala P, Panovský R, Prásek J, Kamínek M, Stanícek J, Klabusay M, Korístek Z, Navrátil M, Dusek L, Vinklárková J. Three-, 6-, and 12-month results of autologous transplantation of mononuclear bone marrow cells in patients with acute myocardial infarction. Int J Cardiol. 2008;128:185–92.

Meyer G, Wollert K, Lotz J, Steffens J, Pippolt P, Fichtner S, Hecker H, Schaefer A, Arseniev L, Hertenstein B, Ganser A, Drexler H. Intracoronary bone marrow cell transfer after myocardial infarction: eighteen months' follow-up data from the randomized, controlled BOOST (Bone marrow transfer to enhance ST-elevation infarct regeneration) trial. Circulation. 2006;113:1287–94.

Mocini D, Staibano M, Mele L, Giannantoni P, Menichella G, Colivicchi F, Sordini P, Salera P, Tubaro M, Santini M. Autologous bone marrow mononuclear cell transplantation in patients undergoing coronary artery bypass grafting. Am Heart J. 2006;151:192–7.

Mohyeddin-Bonab M, Mohamad-Hassani MR, Alimoghaddam K, Sanatkar M, Gasemi M, Mirkhani H, Radmehr H, Salehi M, Eslami M, Farhig-Parsa A, Emami-Razavi H, Alemohammad MG, Solimani AA, Ghavamzadeh A, Nikbin B. Autologous in-vitro expanded mesenchymal stem cell therapy for human old myocardial infarction. Arch Iran Med. 2007;10:467–73.

Murohara T, Ikeda H, Duan J, Shintani S, Sasaki K, Eguchi H, Onitsuka I, Matsui K, Imaizumi T. Transplanted cord blood-derived endothelial precursor cells augment postnatal neovascularization. J Clin Invest. 2000;105:1527–36.

Murphy JW, Cho Y, Sachpatzidis A, Fan C, Hodsdon ME, Lolis E. Structural and functional basis of CXCL12 (stromal cell-derived factor-1 alpha) binding to heparin. J Biol Chem. 2007;282:10018–27.

Murry C. Critique of the CADUCEUS trial. (2012). In: American Heart Association Scientific Sessions, Los Angeles, CA, USA, 3–7 Nov 2012.

Naaijkens BA, van Dijk A, Kamp O, Krijnen PA, Niessen HW, Juffermans LJ. Therapeutic application of adipose derived stem cells in acute myocardial infarction: lessons from animal models. Stem Cell Rev. 2014;10:389–98. doi:10.1007/s12015-014-9502-7.

Nieda M, Nicol A, Denning-Kendall P, Sweetenham J, Bradley B, Hows J. Endothelial cell precursors are normal components of human umbilical cord blood. Br J Haematol. 1997;98:775–7.

Orlandi A, Chavakis E, Seeger F, Tjwa M, Zeiher AM, Dimmeler S. Long-term diabetes impairs repopulation of hematopoietic progenitor cells and dysregulates the cytokine expression in the bone marrow microenvironment in mice. Basic Res Cardiol. 2010;105:703–12. doi:10.1007/s00395-010-0109-0.

Perin EC, Dohmann HF, Borojevic R, Silva SA, Sousa AL, Silva GV, Mesquita CT, Belém L, Vaughn WK, Rangel FO, Assad JA, Carvalho AC, Branco RV, Rossi MI, Dohmann HJ, Willerson JT. Improved exercise capacity and ischemia 6 and 12 months after transendocardial injection of autologous bone marrow mononuclear cells for ischemic cardiomyopathy. Circulation 2004;110(Suppl 1):II213–8.

Perin EC, Sanz-Ruiz R, Sánchez PL, Lasso J, Pérez-Cano R, Alonso-Farto JC, Pérez-David E, Fernández-Santos ME, Serruys PW, Duckers HJ, Kastrup J, Chamuleau S, Zheng Y, Silva GV, Willerson JT, Fernández-Avilés F. Adipose-derived regenerative cells in patients with ischemic cardiomyopathy: the PRECISE Trial. Am Heart J. 2014;168:88–95.e2. doi: 10.1016/j.ahj.2014.03.022.

Pesce M, Orlandi A, Iachininoto MG, Straino S, Torella AR, Rizzuti V, Pompilio G, Bonanno G, Scambia G, Capogrossi MC. Myoendothelial differentiation of human umbilical cord blood-derived stem cells in ischemic limb tissues. Circ Res. 2003;93:e51–62.

Piacibello W, Sanavio F, Garetto L, Severino A, Bergandi D, Ferrario J, Fagioli F, Berge M, Aglietta M. Extensive amplification and self-renewal of human primitive hematopoietic stem cells from cord blood. Blood. 1997;89:2644–53.

Qayyum AA, Haack-Sørensen M, Mathiasen AB, Jørgensen S, Ekblond A, Kastrup J. Adipose-derived mesenchymal stromal cells for chronic myocardial ischemia (MyStromalCell Trial): study design. Regen Med. 2012;7:421–8. doi:10.2217/rme.12.17.

Richman S, Gee AP, McKenna DH, Traverse JH, Henry TD, Fisk D, Pepine CJ, Bloom J, Willerson JT, Prater K, Zhao D, Koç JR, Anwaruddin S, Taylor DA, Cogle CR, Moyé LA, Simari RD, Skarlatos SI. Factors affecting the turnaround time for manufacturing, testing, and release of cellular therapy products prepared at multiple sites in support of multicenter cardiovascular regenerative medicine protocols: a cardiovascular cell therapy research network (CCTRN) study. Transfusion. 2012;52:2225–33. doi:10.1111/j.1537-2995.2011.03543.x.

Ruan W, Pan CZ, Huang GQ, Li YL, Ge JB, Shu XH. Assessment of left ventricular segmental function after autologous bone marrow stem cells transplantation in patients with acute myocardial infarction by tissue tracking and strain imaging. Chin Med J (Engl). 2005;118:1175–81.

Russo V, Young S, Hamilton A, Amsden BG, Flynn LE. Mesenchymal stem cell delivery strategies to promote cardiac regeneration following ischemic injury. Biomaterials. 2014;35:3956–74. doi:10.1016/j.biomaterials.2014.01.075.

Schaechinger V, Erbs S, Elsasser A, Haberbosch W, Hambrecht R, Ischermann H, Ju Yu, Corti R, Mathey D, Hamm C, Suselbeck T, Werner N, Haase H, Neuzner J, Germing A, Mark B, Assmus B, Ton T, Dimmeler S, Zeiher A. Intracoronary bone marrow-derived progenitor cells in acute myocardial infarction. N Engl J Med. 2006a;355:1210–21.

Schaechinger V, Erbs S, Elsasser A, Haberbosch W, Hambrecht R, Ischermann H, Ju Yu, Corti R, Mathey D, Hamm C, Suselbeck T, Werner N, Haase H, Neuzner J, Germing A, Mark B, Assmus B, Ton T, Dimmeler D, Zeiher A. Improved clinical outcome after intracoronary administration of bone-marrow-derived progenitor cells in acute myocardial infarction: final 1-year results of the REPAIR-AMI trial. Eur Heart J. 2006b;27:2775–83.

Seeger FH, Tonn T, Krzossok N, Zeiher AM, Dimmeler S. Cell isolation procedures matter: a comparison of different isolation protocols of bone marrow mononuclear cells used for cell therapy in patients with acute myocardial infarction. Eur Heart J. 2007;28:766–72.

Seeger FH, Rasper T, Fischer A, Muhly-Reinholz M, Hergenreider E, Leistner DM, Sommer K, Manavski Y, Henschler R, Chavakis E, Assmus B, Zeiher AM, Dimmeler S. Heparin disrupts the CXCR4/SDF-1 axis and impairs the functional capacity of bone marrow-derived mononuclear cells used for cardiovascular repair. Circ Res. 2012;111:854–62. doi:10.1161/CIRCRESAHA.112.265678.

Seth S, Bhargava B,Narang R, Ray R, Mohanty S, Gulati G, Kumar L, Airan B, Venugopa P, AIIMS Stem Cell Study Group. The ABCD (autologous bone marrow cells in dilated cardiomyopathy) trial a long-term follow-up study. J Am Coll Cardiol. 2010;55:1643–4. doi:10.1016/j.jacc.2009.11.070.

Smith RR, Barile L, Cho HC, Leppo MK, Hare JM, Messina E, Giacomello A, Abraham MR, Marbán E. Regenerative potential of cardiosphere-derived cells expanded from percutaneous endomyocardial biopsy specimens. Circulation. 2007;115:896–908.

Solomon SD, Skali H, Anavekar NS, Bourgoun M, Barvik S, Ghali JK, Warnica JW, Khrakovskaya M, Arnold JM, Schwartz Y, Velazquez EJ, Califf RM, McMurray JV, Pfeffer MA. Changes in ventricular size and function in patients treated with valsartan, captopril, or both after myocardial infarction. Circ. 2005;111:3411–9.

Strauer BE, Brehm M, Zeus T, Köstering M, Hernandez A, Sorg RV, Kögle G, Wernet P. Repair of infarcted myocardium by autologous intracoronary mononuclear bone marrow cell transplantation in humans. Circ. 2002;106:1913–8.

Strauer BE, Brehm M, Zeus T, Bartsch T, Schannwell C, Antke C, Sorg RV, Kögler G, Wernet P, Müller HW, Köstering M. Regeneration of human infarcted heart muscle by intracoronary autologous bone marrow cell transplantation in chronic coronary artery disease: the IACT Study. J Am Coll Cardiol. 2005;46:1651–8.

Strauer BE, Yousef M, Schannwell CM. The acute and long-term effects of intracoronary stem cell transplantation in 191 patients with chronic heart failure: the STAR-heart study. Eur J Heart Fail. 2010;12:721–9. doi:10.1093/eurjhf/hfq095.

Sürder D, Manka R, Lo Cicero V, Moccetti T, Rufibach K, Soncin S, Turchetto L, Radrizzani M, Astori G, Schwitte J, Erne P, Zuber M, Auf der Maur C, Jamshidi P, Gaemperli O, Windecker S, Moschovitis A, Wahl A, Bühler I, Wyss C, Kozerke S, Landmesser U, Lüscher TF, Corti R. Intracoronary injection of bone marrow-derived mononuclear cells early or late after acute myocardial infarction: effects on global left ventricular function. Circulation. 2013;127:1968–79. doi:10.1161/CIRCULATIONAHA.112.001035.

Taylor E, Dabkowski R. Safety and feasibility trial of adipose-derived regenerative cells in the treatment of chronic myocardial ischemia (ATHENA). NCT01556022, www.ClinicalTrials.gov, 2012.

The I.S.A.M. Study Group. A prospective trial of intravenous streptokinase in acute myocardial infarction (I.S.A.M.). Mortality, morbidity, and infarct size at 21 days. N Engl J Med. 1986;314:1465–7.

Traverse JH, Henry TD, Ellis SG, Pepine CJ, Willerson JT, Zhao DX, Forder JR, Byrne BJ, Hatzopoulos AK, Penn MS, Perin EC, Baran KW, Chambers J, Lambert C, Raveendran G, Simon DI, Vaughan DE, Simpson LM, Gee AP, Taylor DA, Cogle CR, Thomas JD, Silva GV, Jorgenson BC, Olson RE, Bowman S, Francescon J, Geither C, Handberg E, Smith DX, Baraniuk S, Piller LB, Loghin C, Aguilar D, Richman S, Zierold C, BettencourtJ J, Sayre SL, Vojvodic RW, Skarlatos SI, Gordon DJ, Ebert RF, Kwak M, Moyé LA, Simari RD. Effect of intracoronary delivery of autologous bone marrow mononuclear cells 2 to 3 weeks following acute myocardial infarction on left ventricular function: the LateTIME randomized trial. JAMA. 2011;306:2110–9. doi:10.1001/jama.2011.1670.

Traverse JH, Henry TD, Pepine CJ, Willerson JT, Zhao DX, Ellis SG, Forder JR, Anderson RD, Hatzopoulos AK, Penn MS, Perin EC, Chambers J, Baran KW, Raveendran G, Lambert C, Lerman A, Simon DI, Vaughan DE, Lai D, Gee AP, Taylor DA, Cogle CR, Thomas JD, Olson RE, Bowman S, Francescon J, Geither C, Handberg E, Kappenman C, Westbrook L, Piller LB, Simpson LM, Baraniuk S, Loghin C, Aguilar D, Richman S, Zierold C, Spoon DB, Bettencourt J, Sayre SL, Vojvodic RW, Skarlatos SI, Gordon DJ, Ebert RF, Kwak M, Moyé LA, Simari RD, Cardiovascular Cell Therapy Research Network (CCTRN). Effect of the use and timing of bone marrow mononuclear cell delivery on left ventricular function after acute myocardial infarction: the TIME randomized trial. JAMA. 2012;308:2380–9.

Vrtovec B, Poglajen G, Sever M, Lezaic L, Domanovic D, Cernelc P, Haddad F, Torre-Amione G. Effects of intracoronary stem cell transplantation in patients with dilated cardiomyopathy. J Card Fail. 2011;17(272–81):2010. doi:10.1016/j.cardfail.2010.11.007.

Vrtovec B, Poglajen G, Lezaic L, Sever M, Domanovic D, Cernelc P, Socan A, Schrepfer S, Torre-Amione G, Haddad F, Wu J. Effects of intracoronary CD34+ stem cell transplantation in nonischemic dilated cardiomyopathy patients: 5-year follow-up. Circ Res. 2013;112:165–73. doi:10.1161/CIRCRESAHA.112.276519.

Williams AR, Hare JM. Mesenchymal stem cells: biology, pathophysiology, translational findings, and therapeutic implications for cardiac disease. Circ Res. 2011;109:923–40 10.1161.

Witkowska-Zimny M, Walenko K. Stem cells from adipose tissue. Cell Mol Biol Lett. 2011;16:236–57. doi: 10.2478/s11658-011-0005-0.

Wollert KC, Meyer G, Lotz J, Ringes-Lichtenberg S, Lippolt P, Breidenbach C, Fichtner S, Korte T, Hornig B, Messinger D, Arseniev L, Hertenstin B, Ganser A, Drexler H. Intracoronary autologous bone-marrow cell transfer after myocardial infarction: the BOOST randomised controlled clinical trial. Lancet. 2004;364:141–8.

Yang D, Wang W, Li L, Peng Y, Chen P, Huang H, Guo Y, Xia X, Wang Y, Wang H, Wang WE, Zeng C. The relative contribution of paracine effect versus direct differentiation on adipose-derived stem cell transplantation mediated cardiac repair. PLoS ONE. 2013;8:e59020. doi:10.1371/journal.pone.0059020.

Zijlstra F, de Boer MJ, Hoorntje JC, Reiffers S, Reiber JH, Suryapranata H. A comparison of immediate coronary angioplasty with intravenous streptokinase in acute myocardial infarction. N Engl J Med. 1993;328:680–4.

Chapter 13
Mesenchymal Stem Cell Therapy for Bladder Dysfunction

Hong J. Lee and Sung Sik Choi

Abstract Human urological diseases, such as bladder outlet obstruction, overactive bladder, and neurogenic bladder, are caused by inappropriate muscular activity in the muscles of the bladder wall. These muscles often cannot be controlled in patients with spinal cord injury (SCI), traumatic brain injury (TBI), diabetes, or dementia. Stem cell therapy for the bladder has generally been directed to areas of bladder dysfunction, and the therapeutic efficacy of stem cells is thought to be derived from their ability to differentiate into various cell types. However, stem cell therapy for bladder dysfunction has been limited to laboratory experiments, with less emphasis on the potential of stem cell therapy to contribute to bladder regeneration. Adipose-derived mesenchymal stem cells (AD-MSCs), bone marrow mesenchymal stem cells (BM-MSCs), and muscle-derived stem cells (MDSCs) have been used for transplantation to treat bladder dysfunction; reconstitution of restoration of bladder dysfunction is achieved through effects on migration, differentiation, and paracrine signaling. In this review, we discuss the characteristics of stem cells and their potential applications in human urological diseases.

Keywords Urological disease · Bladder dysfunction · Mesenchymal stem cell

13.1 Introduction

Bladder dysfunction could arise because of neurological impairment. For example, various human urological diseases, such as neurogenic bladder, bladder outlet obstruction (BOO), and overactive bladder (OAB), are caused by inappropriate muscular activity in the bladder wall, resulting from spinal cord injury (SCI), traumatic brain injury (TBI), diabetes, or dementia. Although many patients suffer

H.J. Lee (✉) · S.S. Choi
Biomedical Research Institute, Chung-Ang University College of Medicine, Seoul, Korea
e-mail: leehj71@gmail.com

R.C. Zhao (ed.), *Stem Cells: Basics and Clinical Translation*, Translational Medicine Research 1, DOI 10.1007/978-94-017-7273-0_13

from these urological diseases, there is no effective therapy for the treatment of bladder dysfunction. Moreover, bladder dysfunction is a major medical problem that has social implications due to quality of life issues (Ku 2006).

Stem cell therapy may represent a new therapeutic strategy for the treatment of bladder dysfunction. The therapeutic efficacy of stem cells is derived from their ability to differentiate into various cell types and produce therapeutic gene products. Mesenchymal stem cells (MSCs) are multipotent stromal cells that have self-renewing capacity and can differentiate into various cell types, including osteoblasts, chondrocytes, adipocytes, myocytes, and neurons (Jiang et al. 2002; Nagai et al. 2007; Pittenger et al. 1999; Prockop 1997; Sharma et al. 2006). These stem cells can distribute and integrate throughout the whole body after transplantation. Previous studies have reported that the transplantation of bone marrow-derived MSCs induces the reduction of collagen deposition in fibrosis in various tissues, such as the liver, lung, and heart (Abdel Aziz et al. 2007; Fang et al. 2004; Ohnish et al. 2007; Ortiz et al. 2003).

Adipose-derived mesenchymal stem cells (AD-MSCs), bone marrow mesenchymal stem cells (BM-MSCs), and muscle-derived stem cells (MDSCs) are used for transplantation to treat bladder dysfunction. Stem cells can reconstitute and restore the bladder tissue via migration, differentiation, and paracrine effects.

In this review, we will discuss general information and pathological aspects of bladder dysfunction. Additionally, we will describe the current status of stem cell therapy for bladder dysfunction and discuss future directions for research on this topic.

13.2 Pathology of Bladder Dysfunction

Although numerous treatments for bladder dysfunction have previously been developed, no major improvements in bladder dysfunction have been achieved. For the study of bladder dysfunction, several bladder dysfunction models have been developed; however, these models are still not optimized or validated. Moreover, studies seeking to better understand the effects and mechanisms of BOO at a cellular level are still in preliminary phases.

13.2.1 Bladder Outlet Obstruction (BOO)

BOO is a well-known and well-established model of bladder dysfunction. Medical and surgical efforts to treat and prevent BOO are ongoing.

Benign prostatic hyperplasia (BPH) is one of the most common benign diseases in men that can lead to BOO, benign prostatic enlargement (BPE), and lower urinary tract symptoms (LUTS) (Roehrborn 2005). BOO is found in about 60 % of symptomatic and 52 % of asymptomatic men with BPH; moreover, while BOO can occur in both men and women, it is more frequent in older men (Song et al. 2012). Only, pressure-flow studies are able to accurately diagnose BOO. However, because pressure-flow studies are invasive, expensive, and time-consuming, measurements of free uroflowmetry, post-void residual urine, and prostate volume are used to estimate BOO in men with BPH in the clinical setting (Oelke et al. 2007).

BOO occurs where there is some impediment to the normal smooth, complete, and rapid voiding of the bladder and has various functional or anatomical etiologies. Interestingly, BOO has been shown to be caused by collagen deposition, which occurs frequently in the bladder during various pathological processes and eventually induces bladder fibrosis, finally resulting in a flaccid bladder. Bladder fibrosis in turn leads to weakening of the smooth muscle function and problems with incontinence. Thus, BOO alters the properties of smooth muscle and collagen in the bladder wall, including bladder instability (Kim et al. 2013).

Typical obstructive symptoms of BOO include hesitancy, sensation of incomplete bladder emptying, diminished urinary stream, and post-voiding urinary dribbling. Treatment of BOO includes medications or surgery (Dmochowski 2005).

13.2.2 Overactive Bladder (OAB)

Millions of people in the USA live with OAB. OAB is a symptom syndrome, not a diagnosis, and involves urinary symptoms of urgency, with or without urge urinary incontinence, usually associated with frequency and nocturia (Tincello et al. 2014). Symptoms may also be made worse by stress, caffeine in tea, coffee, cola, and alcohol. It is estimated that 0.3–75 % of adults suffer from OAB in the presence or absence of urge urinary incontinence, with incidences varying based on sex and age (Rashid and Ockrim 2013). In Germany, the cost of OAB is around 3.57 billion Euros per year, similar to that of diabetes and dementia.

Conservative treatment for OAB involves lifestyle modifications and pelvic floor exercises. Additional and available treatments for OAB include anticholinergics and newer beta-three agonists via oral and transdermal administration and percutaneous posterior tibial nerve stimulation. OAB can be managed by surgical approaches such as intravesical botulinum toxin and sacral neuromodulation, with augmentation cystoplasty and urinary diversion occurring during end-stage disease (Tincello et al. 2014).

13.2.3 Neurogenic Bladder

Neurogenic bladder is the result of problems with nerves in the body that may control how the bladder stores or empties urine. These conditions include OAB, incontinence, and obstructive bladder, in which the flow of urine is blocked. Many women and men suffer from these bladder problems, including people with illness and nerve injury, such as multiple sclerosis (MS), Parkinson's disease (PD), stroke, or SCI. People who are born with problems of the spinal cord, such as spina bifida, may also have this type of bladder problem.

In people with neurogenic bladder, the nerves and muscles do not work together correctly, and as a result, the bladder may not fill or empty correctly. Because of damage to nerves, bladder muscles may be overactive and contract involuntarily more often than normal and before the bladder is full. In some people, the bladder muscles may be underactive and too loose, letting urine to pass. Additionally, the bladder may not contract when it is full and will not empty completely. The sphincter muscles around the urethra may also not work properly. Treatment approaches for neurogenic bladder include medicines, physical–psychological therapy, electrical stimulatory therapy, intermittent self-catheterization, and surgery.

13.3 Animal Models of Bladder Dysfunction

Research of human diseases depends on clinical research and both in vitro and in vivo model systems. Animal models for human disease are used to understand the pathogenesis of human diseases at a cellular and molecular level and to provide systems for developing and screening new therapies (Lieschke and Currie 2007).

Because of the similarities of human and animal genetics, anatomy, and physiology, animal models are used in the study of human disease. In addition, animal models are often preferable for disease research due to their unlimited supply and ease of manipulation. Rodents are the most common type of animal for use in experimental studies, and mice and rats in particular are frequently used for human disease studies, not only because their genomes are so similar to that of humans, but also because of their availability, ease of handling, high reproductive rates, short life span, and relatively low cost. Because it is important to mimic the pathological aspects of human disease in animal models, the use of appropriate animal models is required to achieve successful results.

13.3.1 Bladder Outlet Model

BOO is one of the most common problems in elderly men with collagen deposition, resulting in bladder fibrosis and flaccid bladder, which further compromises smooth muscle function and bladder elasticity (Elbadawi et al. 1993). BOO changes the properties of smooth muscle and collagen, affecting bladder stability (Steers and De Groat 1088). Compensated bladder dysfunction with OAB is expected after 6 weeks (Elbadawi et al. 1993).

An animal model of BOO can be artificially created by partial surgical obstruction of the bladder outlet. Obstruction of the bladder outlet decreases local blood flow and induces significant tissue ischemia. Increased intraluminal pressure induces vessel compression, following by fibrosis and hypertrophy, thereby resulting in bladder dysfunction (Ghafar et al. 2002; Levin et al. 2003).

The cryoinjured model also induces bladder hypertrophy with loss of smooth muscle and increased collagen, a mechanism similar to that of the BOO model (De Coppi et al. 2007).

13.3.2 OAB Model

The spontaneous hypertensive rat (SHR) is a genetic model of multifactorial hypertension. This model resembles human hypertension. SHRs have been shown to exhibit abnormal bladder function and hyperactive behavior, concurrent with reduced bladder capacity and micturition volume, increased urinary frequency, and a greater occurrence of nonvoiding contractions. This model can be used to study detrusor overactivity (DO) and OAB (McMurray et al. 2006; Jin et al. 2009).

There is correlation between LUTS and erectile dysfunction (Ponholzer et al. 2004; Boyle et al. 2003). When rats were fed a high-fat and/or high-cholesterol diet, fatty rats exhibited increased urinary frequency and showed a greater number of nonvoiding contractions during bladder filling on awake cystometry (Rahman et al. 2007; Son et al. 2007). Chronic ischemia is induced in hyperlipidemic animal models, and transforming growth factor-β1 increases in the bladder, leading to fibrosis and noncompliance (Azadzoi et al. 1999).

Bladder ischemia has been shown to be associated with detrusor contractions during bladder filling in rabbits (Azadzoi et al. 1999). The upregulation of purinergic receptors in the urothelium and bladder nerve bundles has been reported in bladder ischemia (Rahman et al. 2007). The bladder ischemia model is established using bilateral iliac artery ligation or hyperlipidemia (Chen et al. 2012; Huang et al. 2010). Artery stenosis and blood insufficiency can cause alterations in bladder structure and functionality (Azadzoi 2003). The pathological and functional changes in the ischemic model are similar to those observed in human aging detrusor (Chen et al. 2012).

Diabetes also induces bladder dysfunction in a time-dependent manner, leading to both storage and voiding problems in the bladder. In the early phase, this model exhibits DO with urinary frequency and urgency. However, the progression of diabetes, oxidative stress, and neuropathy leads to the underactive or atonic bladder (Daneshgari et al. 2009).

13.3.3 Neurogenic Bladder Model

Injuries in or diseases of the central nervous system (CNS) can disrupt voluntary control of micturition in the bladder, leading to bladder dysfunction (Andersson and Pehrson 2003). These mechanisms are highly complex and depend on the location and extent of the neurological injury.

SCI is one of the commonest neurological models and induces complete deterioration of bladder compliance, function, infection, and other lower urinary tract complications (Yoshiyama et al. 1999). In a rat SCI animal model, increased bladder wall thickness has been observed, with the bladder wall containing a higher proportion of collagen (Nagatomi et al. 2004). The goal of bladder treatment in patients with SCI is to reduce infections, preserve renal function, and improve patients' quality of life. Because it is difficult to regenerate new neurons and glial cells in the CNS, bladder functional recovery is limited following SCI.

A number of neurodegenerative disorders cause bladder dysfunction in humans, including cerebrovascular events, Alzheimer's disease (AD), PD, amyotrophic lateral sclerosis (ALS), and MS. This leads to the development of DO with an increase in micturition frequency and a reduction in bladder capacity (Yokoyama et al. 1997), and these changes persist for several months (Yokoyama et al. 1998). Some transgenic animal models of these diseases, e.g., AD and ALS, are available.

13.4 Mesenchymal Stem Cells (MSCs)

Stem cells are pluripotent cells that propagate through self-renewal and differentiate into mature cells of multiple lineages (Masters et al. 2008). Given their unique abilities of site-specific migration, plasticity, and potential for tissue repair or regeneration, stem cells and their relationship to repair injury or damage in various organ systems have attracted much interest in a variety of fields.

MSCs are self-renewing cells with pluripotent capacity to differentiate into various cell types, including osteoblasts, chondrocytes, myocytes, adipocytes, and neurons (Jiang et al. 2002). All MSCs, including BM-MSCs, skeletal MDSCs, and AD-MSCs, exhibit similar biological properties and therapeutic capabilities.

BM-MSCs are clinically attractive cells because of their availability; bone marrow samples can be obtained easily from patients at the bedside, and BM-MSCs can then be identified by the detection of cell surface markers. After expansion in culture, these autologous MSCs are transplanted back into the patient, avoiding ethical problems and immune-related rejection. The cells can then be delivered by various transplantation routes, such as direct injection into the cerebrospinal fluid (CSF) or intravenous (IV) delivery. (Bakshi et al. 2006).

MSCs augment healing through cell replacement and stimulation of cell proliferation and angiogenesis. While numerous reports have demonstrated the ability of MSCs to engraft tissues, such as the lungs, liver, heart, and brain, data are still scarce concerning the ability of MSCs to repair bladder dysfunction (Zhao et al. 2005; Sakaida et al. 2004; Li et al. 2008). However, AD-MSCs, pluripotent stem cells that are abundant and easy to collect (Zuk et al. 2001; Lin et al. 2008), have demonstrated efficacy in experimental studies of urologic conditions (Albersen et al. 2010; Huang et al. 2010). Additionally, MDSCs, which can be obtained easily and safely during surgery for autologous transplantation, are used mainly in models of artificial injury, including pelvic nerve injury. Indeed, transplantation of these cells can lead to differentiation into skeletal muscle cells, vascular cells, and peripheral nervous cells (Tamaki et al. 2005, 2007), supporting the potential application of MDSCs in human urological disorders.

13.5 Stem Cell Therapy

BM-MSCs can give rise to neuronal cells in vitro and in vivo (Bakshi et al. 2006; Mezey et al. 2000). In multiple studies, unfractionated bone marrow cells differentiated into neural cells, with remyelination observed after transplantation into the damaged area of the CNS (Kabos et al. 2002; Brazelton et al. 2000). BM-MSCs have also been shown to improve remyelination and survival of oligodendrocytes in demyelinating injury (Zhang et al. 2008). These findings suggest that BM-MSCs may have a therapeutic role in SCI and have the ability to form functional myelin (Akiyama et al. 2002). The therapeutic effects in various animal models were listed in Table 13.1.

Table 13.1 The effect of stem cells for bladder dysfunction

Animal model	Stem cell	Gene	Transplantation route	Results	References
BOO in rat	BM-MSC		Direct transplantation	Improvement of bladder function (BF)	58
BOO in rat	BM-MSC	HGF	Direct transplantation	HGF gene is helped to improve bladder function, smooth muscle differentiation	12
BOO in mice	BM-MSC		IV	Expression of chemokine CCL2, improvement of bladder function	56
BOO in mice	BMD cell		Direct transplantation	CCL2 induction increased, improvement of bladder function	57
BOO in rat	BM-MSC		Direct transplantation	Improvement of bladder function, smooth muscle differentiation	59
Chronic ischemia in rat	BM-MSC		Intra-arterial injection	Improvement of bladder function	31

(continued)

Table 13.1 (continued)

Animal model	Stem cell	Gene	Transplantation route	Results	References
Chronic ischemia in rat	AD-MSC		Direct transplantation/IV	Smooth muscle differentiation, improvement of bladder function	48
Diabetes in rat	AD-MSC		Direct transplantation/IV	Improvement of bladder function	70
Cryoinjury in rat	BM-MSC		Direct transplantation	Smooth muscle differentiation	23
Cryoinjury in mice	MDSC		Direct transplantation	Ex vivo gene transfer using β-galactosidase, improvement of bladder function, smooth muscle differentiation	64
Cryoinjury in mice	AD-MSC		Direct transplantation	Smooth muscle differentiation via TGF-β signaling	65
Pelvic nerve injured in rat	MDSC		Damaged nerve lesion	Improvement of bladder function, smooth muscle differentiation	73
Pelvic nerve injured in rat	MDSC		Damaged nerve lesion	Autograft, improvement of bladder function	66
SCI in rat	NSC		Damaged cord lesion	Improvement of bladder function	66
SCI in rat	BM-MSC		Damaged cord lesion	Improvement of bladder function	62
SCI in rat	BM-MSC		IV	Improvement of bladder function	67

Cells in the $CD34^+/CD45^-$ fraction (Sk-34 cells) and $CD3^-/CD45^-$ fraction (Sk-DN cells) are able to synchronously reconstitute nerve–muscle–blood vessel units after transplantation (Tamaki et al. 2005, 2007). Moreover, CCL2 transcript levels have been shown to be increased twofold in animals transplanted with MSCs (Woo et al. 2011). Changes in CCL2 expression have also been reported in an animal model of BOO; in this study, transplanted BM-MSCs were integrated and survived in the urothelial and stromal layers, inducing activation of epidermal growth factor receptor to facilitate survival (Tanaka et al. 2009). Thus, transplantation of MSCs may be used to improve bladder compliance.

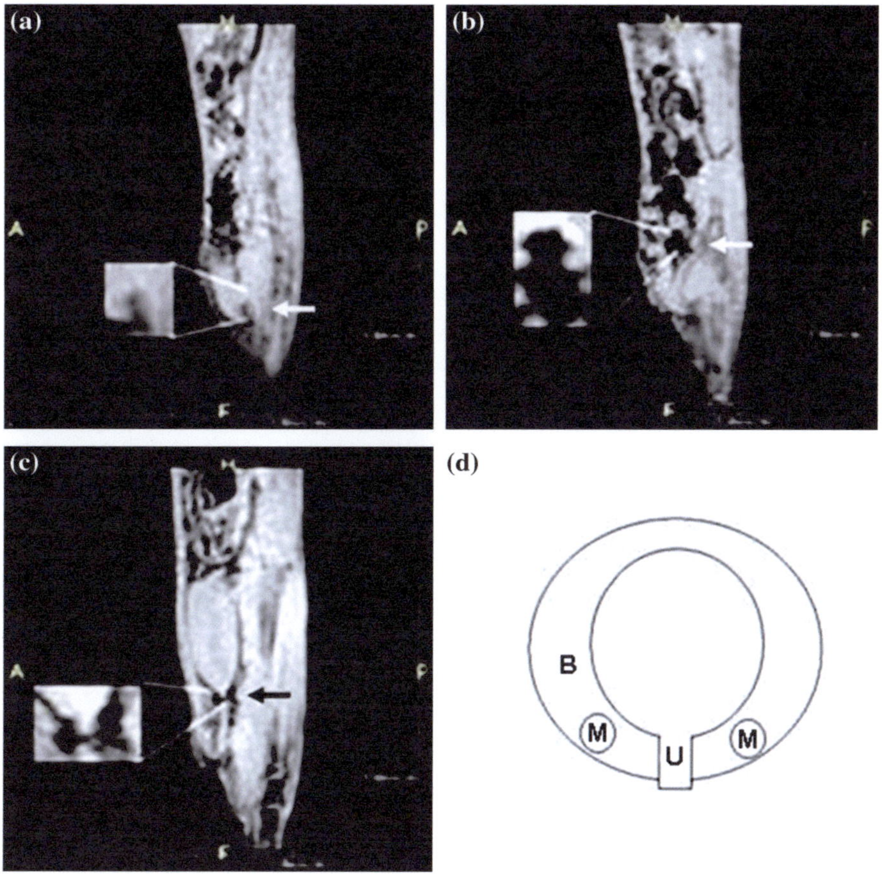

Fig. 13.1 In vivo MRI of SPION-labeled human MSCs into the bladder of BOO model of rats. MRI showed a clear hypointense signal intensity, and the areas of decreased MR signal intensity in the bladder were confined locally (*arrows*). **a** Before transplantation of SPION-labeled human MSCs and **b** immediately after transplantation of SPION-labeled human MSCs. **c** At 4 weeks after transplantation, SPION-labeled human MSCs were survived and able to detect by MRI. **d** Schematic drawing of bladder, urethra, and transplanted MSCs. *B* bladder wall; *M* injected MSCs; *U* urethra. (Reproduced from Lee et al. 2012)

Additional studies have supported this notion. Indeed, in a recent study of primary human MSCs labeled with nanoparticles containing super paramagnetic iron oxide, direct transplantation into the bladder wall of rats with BOO blocked bladder fibrosis and improved bladder function. The transplanted cells could be easily traced up to 4 weeks by MRI (Fig. 13.1) and survived and integrated in bladder wall (Fig. 13.2) (Lee et al. 2012). In another study, transplanted BMCs were

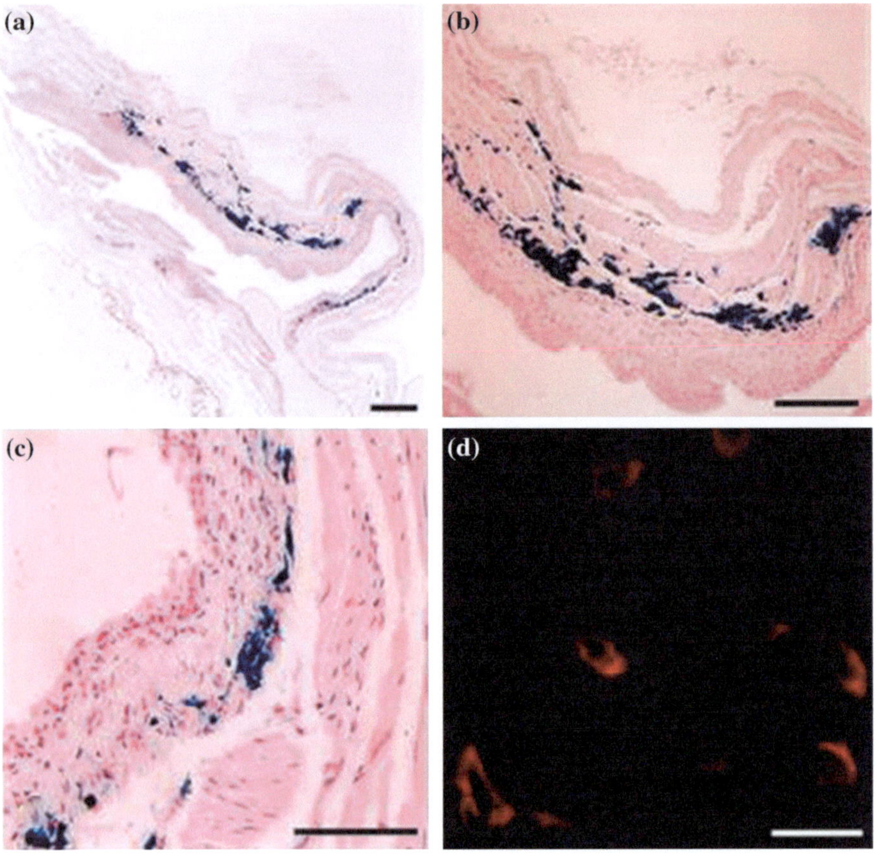

Fig. 13.2 Histologic findings of SPION-labeled human MSCs in rat bladder. Prussian blue staining at 4 weeks after MSC transplantation. (**a**–**c**) Intracytoplasmic SPION particles (*blue dots*) are clearly visible. Scale bars: 20 μm (**a**), 40 μm (**b**), and 80 μm (**c**). **d** The presence of human MSCs in rat bladder at 4 weeks after transplantation was confirmed by immunostaining with antihuman nuclear matrix antibody that is specific for human cells. Scale bar: 50 μm (**d**). *U* urothelium of bladder; *I* intraluminal side of bladder; *E* extraluminal side of bladder. (Reproduced from Lee et al. 2012)

shown to improve bladder contractility by differentiating into smooth muscle-like cells in the bladders of model rats with BOO (Nishijima et al. 2007). Injections of a stem cell suspension into the common iliac artery in a rat ischemic model led to regeneration of bladder tissue, thereby improving bladder detrusor function (Chen et al. 2012). Direct transplantation of MSCs (B10) into bladder wall was able to improve the bladder function as shown by cytometry (CMG), intercontraction interval (ICI), maximal voiding pressure (MVP), and pressure threshold (PT) test (Fig. 13.3) (Lee et al. 2014). AD-MSCs have also been shown to improve urodynamics and tissue in a rat model with hyperlipidemia by transplantation via direct injection to the bladder or intravenous injection (Huang et al. 2010). In three additional studies, transplanted neural progenitor cells were shown to promote the recovery of bladder function through regeneration of the injured site, despite difficulties in inducing neural regeneration in the CNS (Mitsui et al. 2003, 2005; Temeltas et al. 2009). Finally, therapy with smooth muscle cells and dedifferentiated fat cells has been shown to ameliorate bladder dysfunction in a cryoinjured bladder and pelvic nerve model in rats. These cells exerted important effects on the remodeling process and were recovered functionally in the injured bladder (Somogyi et al. 2002; Huard et al. 2002; Sakuma et al. 2009; Kwon et al. 2005).

In most of these studies, stem cells have been commonly administered via the direct injection method; however, they can also be transplanted via various other routes, such as intravenous injection. When BM-MSCs were transplanted intravenously, they could survive in the L3-4 region of the spinal cord and facilitated the recovery of bladder function in SCI rats (Hu et al. 2012).

Some studies have demonstrated that embryoid body-derived stem cells or BMSCs can improve the regeneration of partially cystectomized bladders after transplantation into the small intestinal submucosa (SIS) (Chung et al. 2005; Frimberger et al. 2005; Zhang et al. 2005). In addition, AD-MSC transplantation into the bladder acellular matrix (BAM) has also been reported to regenerate bladder tissue (Zhu et al. 2010). Some researchers have reported the use of scaffolds to help bladder regeneration. When BMSCs or AD-MSCs were seeded on poly (1,8-octanediol-cocitrate)-thin film (Sharma et al. 2010), poly-L-lactic acid scaffold (Nitta et al. 2010; Tian et al. 2010), or poly-lactic glycolic acid (Jack et al. 2009), these cells supported partial bladder regeneration and maintained bladder capacity and compliance. Although nonengineered MSCs have therapeutic effects on bladder dysfunction, bladder tissue engineering using MSCs might be a helpful tool for achieving better results than the results obtained from using differentiated cells. These cells may be able to achieve fast replacement of grafts and would be expected to show appropriate neural function and less fibrosis (Sakuma et al. 2009).

13.6 Gene Therapy

Stem cells have therapeutic effects associated with the regeneration and substitution of cells and tissues themselves. For the development of new medicinal drugs, it is necessary to gain a deeper understanding of the genetic factors, roles, and mechanisms of bladder dysfunction.

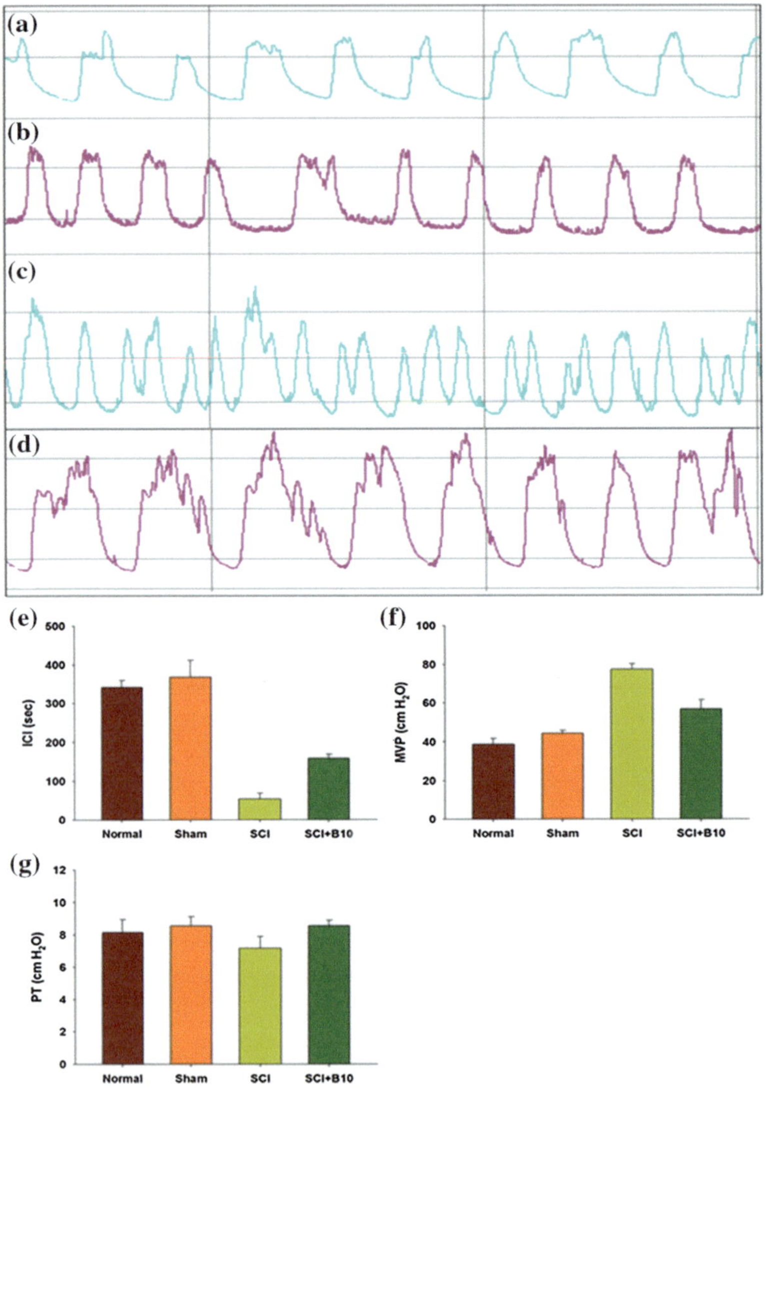
(a)
(b)
(c)
(d)
(e)
ICI (sec)
500
400
300
200
100
0
Normal
Sham
SCI
SCI+B10
(f)
MVP (cm H2O)
100
80
60
40
20
0
Normal
Sham
SCI
SCI+B10
(g)
PT (cm H2O)
12
10
8
6
4
2
0
Normal
Sham
SCI
SCI+B10

Fig. 13.3 Improvement of bladder function after B10 cell transplantation. **a–d** CMG: **a** control; **b** sham; **c** ICI decreased in SCI rats; **d** ICI recovered in SCI rats + B10 cells transplantation. **e–f** Analysis of CMG. **e** The SCI rats showed decreased ICI compared with the sham-operated group, and this decrease was reversed after transplantation of B10 hMSCs ($p < 0.05$). **f** The SCI rats showed increased MVP compared with the sham-operated group, and this increase was reversed after transplantation of B10 cells ($p < 0.05$). **g** There was no difference in PT between the groups. B10 = human mesenchymal stem cells; *CMG* cystometry; *ICI* intercontraction interval; *MVP* maximal voiding pressure; *PT* pressure threshold; Normal = control; *SCI* rats with spinal cord injury; SCI + B10 = B10 cell transplantation rats with spinal cord injury. (Reproduced from Lee et al. 2014)

While transplantation of AD-MSCs and BM-MSCs was shown to have protective effects on bladder dysfunction, stem cell transplantation in general has only a limited effect on smooth muscle regeneration (De Coppi et al. 2007). Additionally, transplantation of AD-MSCs has been shown to improve voiding function in rats, reducing apoptosis and preserving cells in the suburothelial capillary network. Although some transplanted stem cells could differentiate into smooth muscle cells, their paracrine effects may play a major role in this process as well (Zhang et al. 2012). Indeed, growth factors have been shown to be associated with bladder development and the remodeling of the bladder wall after outlet obstruction (Baskin et al. 1996).

In recent studies, MSCs overexpressing hepatic growth factor (HGF) have been shown to inhibit collagen deposition (Fig. 13.4) and improve cystometric parameters as compared to the effects of primary MSCs (Song et al. 2012). Similar studies have shown that MSCs can secrete therapeutic factors, such as HGF, and contribute to reduced fibrosis through paracrine mechanisms rather than cell incorporation (Abdel Aziz et al. 2007; Kinnaird et al. 2004; Matsuda-Hashii et al. 2004). HGF secretion by MSCs plays an essential role in the angiogenesis and regeneration of tissue and acts as a potent antifibrotic agent (Nakamura and Mizuno 2010; Schmidt et al. 1995). Therefore, a detailed understanding of the mechanisms of fibrosis is necessary to improving the therapeutic effects of different treatments for bladder dysfunction. Taken together, these results suggest that MSCs carrying therapeutic genes, such as HGF, may be promising agents for the treatment of bladder fibrosis.

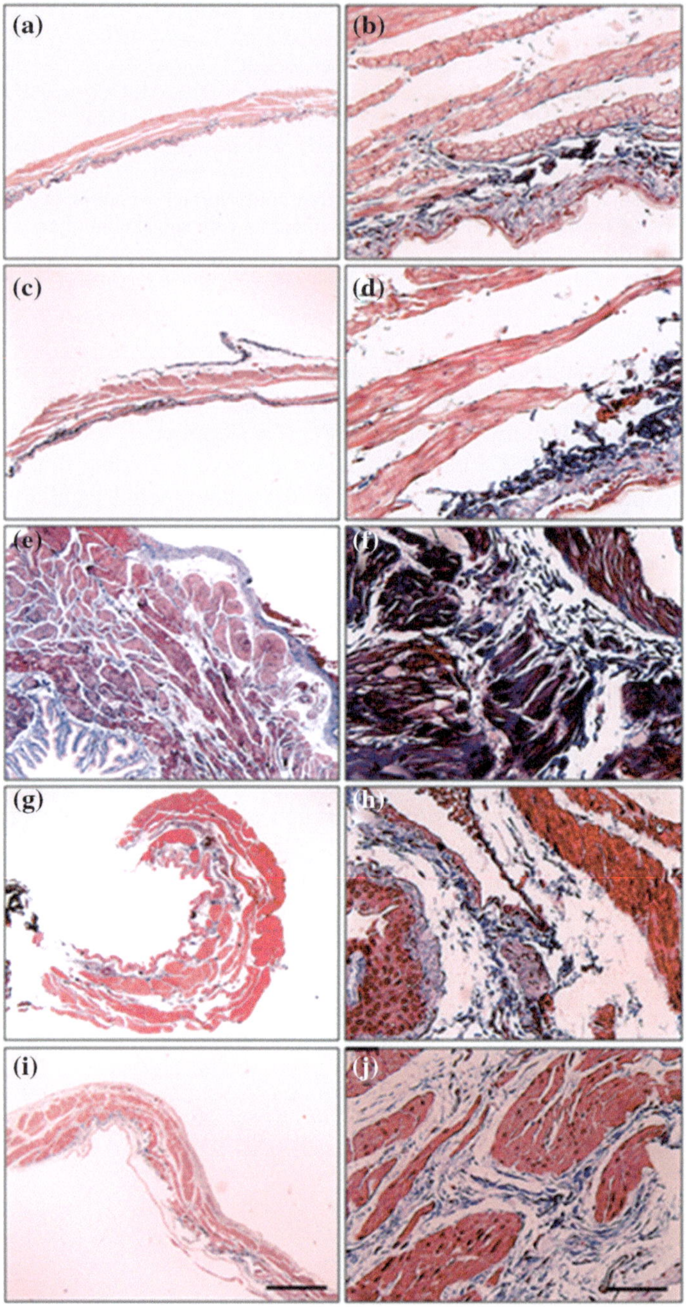
(a)
(b)
(c)
(d)
(e)
(f)
(g)
(h)
(i)
(j)

◀ **Fig. 13.4** Histological change of collagen deposition after MSC transplantation. The collagen deposition increased in the group with bladder outlet obstruction (BOO) and recovered after transplantation of B10.HGF cells. (**a**, **b**) Control. (**c**, **d**) Sham. (**e**, **f**) BOO. (**g**, **h**) BOO + B10 cells. (**i**, **j**) BOO + B10.HGF cells. Masson's trichrome staining. Sham sham operation; *BOO*, bladder outlet obstruction; BOO + B10, B10 MSC transplantation at 2 weeks after BOO; BOO + B10.HGF, B10.HGF MSC transplantation at 2 weeks after BOO. Scale bar: 100 μm. (Reproduced from Song et al. 2012)

13.7 Conclusions

At this time, no effective therapeutic tools exist for the treatment of bladder dysfunction. However, as we have outlined here, recent research has provided interesting results regarding the potential use of stem cells to treat bladder dysfunction, as highlighted by the use of MSCs in various animal models.

Although human trials have not yet been performed, MSCs have the potential to become an important therapeutic agent and major source of cells for the treatment of bladder dysfunction. However, much more work is required in order to determine the exact role of stem cells in the treatment of bladder dysfunction.

Acknowledgments This research was supported by grants from the KRIBB Research Initiative Program (KGM4611512), Republic of Korea.

References

Abdel Aziz MT, Atta HM, Mahfouz S, Fouad HH, Roshdy NK, Ahmed HH, Rashed LA, Sabry D, Hassouna AA, Hasan NM. Therapeutic potential of bone marrow-derived mesenchymal stem cells on experimental liver fibrosis. Clin Biochem. 2007;40:893–9.

Akiyama Y, Radtke C, Kocsis JD. Remyelination of the rat spinal cord by transplantation of identified bone marrow stromal cells. J Neurosci. 2002;22:6623–30.

Albersen M, Fandel TM, Lin G, Wang G, Banie L, Lin CS, Lue TF. Injections of adipose tissue-derived stem cells and stem cell lysate improve recovery of erectile function in a rat model of cavernous nerve injury. J Sex Med. 2010;7:3331–40.

Andersson KE, Pehrson R. CNS involvement in overactive bladder: pathophysiology and opportunities for pharmacological intervention. Drugs. 2003;63:2595–611.

Azadzoi KM. Effect of chronic ischemia on bladder structure and function. Adv Exp Med Biol. 2003;539:271–80.

Azadzoi KM, Tarcan T, Kozlowski R, Krane RJ, Siroky MB. Overactivity and structural changes in the chronically ischemic bladder. J Urol. 1999;162:1768–78.

Bakshi A, Barshinger AL, Swanger SA, Madhavani V, Shumsky JS, Neuhuber B, Fischer I. Lumbar puncture delivery of bone marrow stromal cells in spinal cord contusion: a novel method for minimally invasive cell transplantation. J Neurotrauma. 2006;23:55–65.

Baskin LS, Sutherland RS, Thomson AA, Hayward SW, Cunha GR. Growth factors and receptors in bladder development and obstruction. Lab Invest. 1996;75:157–66.

Boyle P, Robertson C, Mazzetta C, Keech M, Hobbs R, Fourcade R, Kiemeney L, Lee C. The association between lower urinary tract symptoms and erectile dysfunction in four centres: the UrEpik study. BJU Int. 2003;92:719–25.

Brazelton TR, Rossi FMV, Keshet GI, Blau HM. From marrow to brain: expression of neuronal phenotypes in adult mice. Science. 2000;290:1775–9.

Chen S, Zhang HY, Zhang N, Li WH, Shan H, Liu K, Yang Y. Treatment for chronic ischaemia-induced bladder detrusor dysfunction using bone marrow mesenchymal stem cells: an experimental study. Int J Mol Med. 2012;29:416–22.

Chung SY, Krivorov NP, Rausei V, Thomas L, Frantzen M, Landsittel D, Kang YM, Chon CH, Ng CS, Fuchs GJ. Bladder reconstitution with bone marrow derived stem cells seeded on small intestinal submucosa improves morphological and molecular composition. J Urol. 2005;174:353–9.

Daneshgari F, Liu G, Birder L, Hanna-Mitchell AT, Chacko S. Diabetic bladder dysfunction: current translational knowledge. J Urol. 2009;182:S18–26.

De Coppi P, Callegari A, Chiavegato A, Gasparotto L, Piccoli M, Taiani J, Pozzobon M, Boldrin L, Okabe M, Cozzi E, Atala A, Gamba P, Sartore S. Amniotic fluid and bone marrow derived mesenchymal stem cells can be converted to smooth muscle cells in the cryo-injured rat bladder and prevent compensatory hypertrophy of surviving smooth muscle cells. J Urol. 2007;177:369–76.

Dmochowski RR. Bladder outlet obstruction: etiology and evaluation. Rev Urol. 2005;7:S3–13.

Elbadawi A, Yalla SV, Resnick NM. Structural basis of geriatric voiding dysfunction IV. Bladder outlet obstruction. J Urol. 1993;150:1681–95.

Fang B, Shi M, Liao L, Yang S, Liu Y, Zhao RC. Systemic infusion of FLK1 + mesenchymal stem cells ameliorate carbon tetrachloride-induced liver fibrosis in mice. Transplantation. 2004;78:83–8.

Frimberger D, Morales N, Shamblott M, Gearhart JD, Gearhart JP, Lakshmanan Y. Human embryoid bodyderived stemcells in bladder regeneration using rodent model. Urology. 2005;65:827–32.

Ghafar MA, Anastasiadis AG, Olsson LE, Chichester P, Kaplan SA, Buttyan R, Levin RM. Hypoxia and an angiogenic response in the partially obstructed rat bladder. Lab Invest. 2002;82:903–9.

Hu Y, Liao LM, Ju YH, Fu G, Zhang HY, Wu HX. Intravenously transplanted bone marrow stromal cells promote recovery of lower urinary tract function in rats with complete spinal cord injury. Spinal Cord. 2012;50:202–7.

Huang YC, Shindel AW, Ning H, Lin G, Harraz AM, Wang G, Garcia M, Lue TF, Lin CS. Adipose derived stem cells ameliorate hyperlipidemia associated detrusor overactivity in a rat model. J Urol. 2010a;183:1232–40.

Huang YC, Ning H, Shindel AW, Fandel TM, Lin G, Harraz AM, Lue TF, Lin CS. The effect of intracavernous injection of adipose tissue-derived stem cells on hyperlipidemia-associated erectile dysfunction in a rat model. J Sex Med. 2010b;7:1391–400.

Huard J, Yokoyama T, Pruchnic R, Qu Z, Li Y, Lee JY, Somogyi GT, de Groat WC, Chancellor MB. Muscle-derived cellmediated ex vivo gene therapy for urological dysfunction. Gene Ther. 2002;9:1617–26.

Jack GS, Zhang R, Lee M, Xu Y, Wu BM, Rodr´ıguez LV. Urinary bladder smooth muscle engineered from adipose stem cells and a three dimensional synthetic composite. Biomaterials. (2009);30:3259–3270.

Jiang Y, Jahagirdar BN, Reinhardt RL, Schwartz RE, Keene CD, Ortiz-Gonzalez XR, Reyes M, Lenvik T, Lund T, Blackstad M, Du J, Aldrich S, Lisberg A, Low WC, Largaespada DA, Verfaillie CM. Pluripotency of mesenchymal stem cells derived from adult marrow. Nature. 2002;418:41–9.

Jin LH, Andersson KE, Kwon YH, Park CS, Yoon SM, Lee T. Substantial detrusor overactivity in conscious spontaneously hypertensive rats with hyperactive behaviour. Scand J Urol Nephrol. 2009;43:3–7.

Kabos P, Kabosova A, Neuman T. Neuronal injury affects expression of helix-loop-helix transcription factors. NeuroReport. 2002;13:2385–8.

Kim JH, Lee SR, Song YS, Lee HJ. Stem cell therapy in bladder dysfunction: where are we? and where do we have to go? BioMed Res Int. 2013;2013:930713.

Kinnaird T, Stabile E, Burnett MS, Shou M, Lee CW, Barr S, Fuchs S, Epstein SE. Local delivery of marrow-derived stromal cells augments collateral perfusion through paracrinemechanisms. Circulation. 2004;109:1543–9.

Ku JH. The management of neurogenic bladder and quality of life in spinal cord injury. BJU Int. 2006;98:739–45.

Kwon D, Minnery B, Kim Y, Kim JH, de Miguel F, Yoshimura N, Chancellor MB. (2005). Neurologic recovery and improved detrusor contractility using muscle-derived cells in rat model of unilateral pelvic nerve transection. Urology 65:1249–1253.

Lee HJ, Won JH, Doo SH, Kim JH, Song KY, Lee SJ, Lim I, Chang KT, Song YS, Kim SU. Inhibition of collagen deposit in obstructed rat bladder outlet by transplantation of superparamagnetic iron oxide-labeled human mesenchymal stem cells as monitored by molecular magnetic resonance imaging (MRI). Cell Transplant. 2012;21:959–70.

Lee HJ, An J, Doo SW, Kim JH, Choi SS, Lee SR, Park SW, Song YS, Kim SU. (2014). Improvement in spinal cord injury-induced bladder fibrosis using mesenchymal stem cell transplantation into the bladder wall. Cell Transplant (in press).

Levin RM, O'Connor LJ, Leggett RE, Whitbeck C, Chichester P. Focal hypoxia of the obstructed rabbit bladder wall correlates with intermediate decompensation. Neurourol Urodyn. 2003;22:156–63.

Li JH, Zhang N, Wang JA. Improvedanti-apoptotic and anti-remodeling potency of bone marrow mesenchymal stem cells by anoxic preconditioning in diabetic cardiomyopathy. J Endocrinol Invest. 2008;31:103–10.

Lieschke GJ, Currie PD. Animal models of human disease: zebrafish swim into view. Nat Rev Genetics. 2007;8:353–67.

Lin G, Garcia M, Ning H, Banie L, Guo YL, Lue TF, Lin CS. Defining stem and progenitor cells within adipose tissue. Stem Cells Dev. 2008;17:1053–63.

Masters JR, Kane C, Yamamoto H, Ahmed A. Prostate cancer stem cell therapy: hype or hope? Prostate Cancer Prostatic Dis. 2008;11:316–9.

Matsuda-Hashii Y, Takai K, Ohta H, Fujisaki H, Tokimasa S, Osugi Y, Ozono K, Matsumoto K, Nakamura T, Hara J. Hepatocyte growth factor plays roles in the induction and autocrine maintenance of bone marrow stromal cell IL-11, SDF-1, and stem cell factor. Exp Hematol. 2004;32:955–61.

McMurray G, Casey JH, Naylor AM. Animal models in urological disease and sexual dysfunction. Br J Pharmacol. 2006;147:S62–79.

Mezey E, Chandross KJ, Harta G, Maki RA, McKercher SR. Turning blood into brain: cells bearing neuronal antigens generated in vivo from bone marrow. Science. 2000;290:1779–82.

Mitsui T, Kakizaki H, Tanaka H, Shibata T, Matsuoka I, Koyanagi T. Immortalized neural stem cells transplanted into the injured spinal cord promote recovery of voiding function in the rat. J Urol. 2003;170:1421–5.

Mitsui T, Fischer I, Shumsky JS, Murray M. Transplants of fibroblasts expressing BDNF and NT-3 promote recovery of bladder and hindlimb function following spinal contusion injury in rats. Exp Neurol. 2005;194:410–31.

Nagai A, Kim WK, Lee HJ, Jeong HS, Kim KS, Hong SH, Park IH, Kim SU. Multilineage potential of stable human mesenchymal stem cell line derived from fetal marrow. PLoS ONE. 2007;2:e1272.

Nagatomi J, Gloeckner DC, Chancellor MB, DeGroat WC, Sacks MS. Changes in the biaxial viscoelastic respons of the urinary bladder following spinal cord injury. Ann Biomed Eng. 2004;32:1409–19.

Nakamura T, Mizuno S. The discovery of hepatocyte growth factor (HGF) and its significance for cell biology, life sciences and clinical medicine. Proc Jpn Acad Ser B Phys Biol Sci. 2010;86:588–610.

Nishijima S, Sugaya K, Miyazato M, Kadekawa K, Oshiro Y, Uchida A, Hokama S, Ogawa Y. Restoration of bladder contraction by bone marrow transplantation in rats with underactive bladder. Biomed Res. 2007;28:275–80.

Nitta M, Tamaki T, Tono K, Okada Y, Masuda M, Akatsuka A, Hoshi A, Usui Y, Terachi T. Reconstitution of experimental neurogenic bladder dysfunction using skeletal musclederived multipotent stem cells. Transplantation. 2010;89:1043–9.

Oelke M, Höfner K, Jonas U, de la Rosette JJ, Ubbink DT, Wijkstra H. Diagnostic accuracy of noninvasive tests to evaluate bladder outlet obstruction in men: detrusor wall thickness, uroflowmetry, postvoid residual urine, and prostate volume. Eur Urol. 2007;52:827–34.

Ohnish S, Sumiyoshi H, Kitamura S, Nagaya N. Mesenchymal stem cells attenuate cardiac fibroblast proliferation and collagen synthesis through paracrine actions. FEBS Lett. 2007;581:3961–6.

Ortiz LA, Gambelli F, McBride C, Gaupp D, Baddoo M, Kaminski N, Phinney DG. Mesenchymal stem cell engraftment in lung is enhanced in response to bleomycin exposure and ameliorates its fibrotic effects. Proc Natl Acad Sci USA. 2003;100:8407–11.

Pittenger MF, Mackay AM, Beck SC, Jaiswal RK, Douglas R, Mosca JD, Moorman MA, Simonetti DW, Craig S, Marshak DR. Multilineage potential of adult human mesenchymal stem cells. Science. 1999;284:143–7.

Ponholzer A, Temml C, Obermayr R, Madersbacher S. Association between lower urinary tract symptoms and erectile dysfunction. Urology. 2004;64:772–6.

Prockop DJ. Marrow stromal cells as stem cells for nonhematopoietic tissues. Science. 1997;276:71–4.

Rahman NU, Phonsombat S, Bochinski D, Carrion RE, Nunes L, Lue TF. An animal model to study lower urinary tract symptoms and erectile dysfunction: the hyperlipidaemic rat. BJU Int. 2007;100:658–63.

Rashid TG, Ockrim JL. Male incontinence: onabotulinum toxin A and sacral nerve stimulation. Curr Opin Urol. 2013;23:545–51.

Roehrborn CG. Benign prostatic hyperplasia: an overview. Rev Urol. 2005;7(Suppl 9):S3–14.

Sakaida I, Terai S, Yamamoto N, Aoyama K, Ishikawa T, Nishina H, Okita K. Transplantation of bone marrow cells reduces CCl4-induced liver fibrosis in mice. Hepatology. 2004;40:1304–11.

Sakuma T, Matsumoto T, Kano K, Fukuda N, Obinata D, Yamaguchi K, Yoshida T, Takahashi S, Mugishima H. Mature, adipocyte derived, dedifferentiated fat cells can differentiate into smooth muscle-like cells and contribute to bladder tissue regeneration. J Urol. 2009;182:355–65.

Schmidt C, Bladt F, Goedecke S, Brinkmann V, Zschiesche W, Sharpe M, Gherardi E, Birchmeier C. Scatter factor/hepatocyte growth factor is essential for liver development. Nature. 1995;373:699–702.

Sharma AD, Cantz T, Manns MP, Ott M. The role of stem cells in physiology, pathophysiology, and therapy of the liver. Stem Cell Rev. 2006;2:51–8.

Sharma AK, Hota PV, Matoka DJ, Fuller NJ, Jandali D, Thaker H, Ameer GA, Cheng EY. Urinary bladder smooth muscle regeneration utilizing bone marrow derive mesenchymal stem cell seeded elastomeric poly(1,8-octanediolco-citrate) based thin films. Biomaterials. 2010;31:6207–17.

Somogyi GT, Yokoyama T, Szell EA, Smith CP, de Groat WC, Huard J, Chancellor MB. Effect of cryoinjury on the contractile parameters of bladder strips: a model of impaired detrusor contractility. Brain Res Bull. 2002;59:23–8.

Son H, Lee SL, Park WH, Park K, Park S, Kang MS, Kim DY, Kim SW, Paick JS. New unstable bladder model in hypercholesterolemia rats. Urology. 2007;69:186–90.

Song YS, Lee HJ, Doo SH, Lee SJ, Lim I, Chang KT, Kim SU. Mesenchymal stem cells overexpressing hepatocyte growth factor (HGF) inhibit collagen deposit and improve bladder function in rat model of bladder outlet obstruction. Cell Transplant. 2012;21:1641–50.

Steers WD, De Groat WC. Effect of bladder outlet obstruction on micturition reflex pathways in the rat. J Urol. 1088;140:864–71.

Tamaki T, Uchiyama Y, Okada Y, Ishikawa T, Sato M, Akatsuka A, Asahara T. Functional recovery of damaged skeletal muscle through synchronized vasculogenesis, myogenesis, and neurogenesis by muscle-derived stem cells. Circulation. 2005;112:2857–66.

Tamaki T, Okada Y, Uchiyama Y, Tono K, Masuda M, Wada M, Hoshi A, Akatsuka A. Synchronized reconstitution ofmuscle fibers, peripheral nerves and blood vessels by murine skeletal muscle-derived CD34−/45− cells. Histochem Cell Biol. 2007;128:349–60.

Tanaka ST, Martinez-Ferrer M, Makari JH, Wills ML, Thomas JC, Adams MC, 3rd Brock JW, 4th Pope JC, Bhowmick NA. (2009). Recruitment of bone marrow derived cells to the bladder after bladder outlet obstruction. J Urol; 182:1769– 1774.

Temeltas G, Dagci T, Kurt F, Evren V, Tuglu I. Bladder function recovery in rats with traumatic spinal cord injury after transplantation of neuronal-glial restricted precursors or bone marrow stromal cells. J Urol. 2009;181:2774–9.

Tian H, Bharadwaj S, Liu Y, Ma H, Ma PX, Atala A, Zhang Y. Myogenic differentiation of human bone marrow mesenchymal stem cells on a 3D nano fibrous scaffold for bladder tissue engineering. Biomaterials. 2010;31:870–7.

Tincello DG, Rashid T, Revicky V. Emerging treatments for overactive bladder: clinical potential of botulinum toxins. Res Rep Urol. 2014;6:51–7.

Woo LL, Tanaka ST, Anumanthan G, 4th Pope JC, Thomas JC, Adams MC, 3rd Brock JW, Bhowmick NA. (2011) Mesenchymal stem cell recruitment and improved bladder function after bladder outlet obstruction: preliminary data. J Urol; 185:1132–1138.

Yokoyama O, Yoshiyama M, Namiki M, de Groat WC. Influence of anesthesia on bladder hyperactivity induced by middle cerebral artery occlusion in the rat. Am J Physiol. 1997;273:R1900–7.

Yokoyama O, Komatsu K, Ishiura Y, Nakamura Y, Morikawa K, Namiki M. Change in bladder contractility associated with bladder overactivity in rats with cerebral infarction. J Urol. 1998;159:577–80.

Yoshiyama M, Nezu FM, Yokoyama O, deGroat WC, Chancellor MB. Changes in micturition after spinal cord injury in conscious rats. Urology. 1999;54:929–33.

Zhang H, X Qiu, AW Shindel, H Ning, L Ferretti, X Jin, Lin, G Lin CS and TF Lue. Adipose tissue-derived stem cells ameliorate diabetic bladder dysfunction in a type II diabetic rat model. Stem Cells Dev. 2012;21:1391–1400.

Zhang Y, Lin HK, Frimberger D, Epstein RB, Kropp BP. Growth of bone marrow stromal cells on small intestinal submucosa: an alternative cell source for tissue engineered bladder. BJU Int. 2005;96:1120–5.

Zhang J, Li Y, Zheng X, Gao Q, Liu Z, Qu R, Borneman J, Elias SB, Chopp M. Bone marrow stromal cells protect oligodendrocytes from oxygen-glucose deprivation injury. J Neurosci Res. 2008;86:1501–10.

Zhao DC, Lei JX, Chen R, Yu WH, Zhang XM, Li SN, Xiang P. Bone marrow-derived mesenchymal stem cells protect against experimental liver fibrosis in rats. World J Gastroenterol. 2005;11:3431–40.

Zhu WD, Xu YM, Feng C, Fu Q, Song LJ, Cui L. Bladder reconstruction with adipose-derived stem cell-seeded bladder acellular matrix grafts improve morphology composition. World J Urol. 2010;28:493–8.

Zuk PA, Zhu M, Mizuno H, Huang J, Futrell JW, Katz AJ, Benhaim P, Lorenz HP, Hedrick MH. Multilineage cells from human adipose tissue: implications for cell-based therapies. Tissue Eng. 2001;7:211–28.

Chapter 14
Mesenchymal Stem Cells in Cancer Therapy

Nedime Serakinci and Mahmut Cerkez Ergoren

Abstract The human mesenchymal stem cells (hMSCs) are multipotent non-hematopoietic precursor cells that can generate various types of tissue cells which supports the formation of blood and fibrous connective tissue. hMSCs desirable stem cell characteristics, their ability to avoid immune rejection, and their homing ability in addition to easiness of their isolation and expansion made these cells a great therapeutic target for many diverse diseases. MSCs can be isolated from various tissues such as bone marrow, adipose tissue, dental pulp and can be expanded without greatly compromising genetic stability. Previous studies on site-directed and/or systemic administration of MSCs have revealed their ability of engraftment in a number of tissues after injury. Since the discovery that bone marrow-derived mesenchymal stem cells can be recruited to the tumor side and can home to the tumor stoma, hMSC became strong candidate for stem cell-based cancer therapy. Despite their great potential, these cells can also suffer from replicative exhaustion and acquire critically short telomeres that might increase the risk of cancer development. During transition to cancer, most often telomerase activation occurs which is highly specific to cancer and has the consequence that cancer cells maintain telomere function and thereby compensate for cell division-associated telomere attrition and facilitate crisis-bypass, which altogether promote tumor cell immortalization. The challenges and risks for cell-based therapies are multifaceted.In this current communication, the double-faced role of mesenchymal stem cells in cancer development and their therapeutic potential use in cancer therapy will be discussed. Thus, it will be focused and discussed different mechanisms that mesenchymal stem cells can show neoplastic transformation through telomere

N. Serakinci (✉) · M.C. Ergoren
Faculty of Medicine Department of Medical Genetics,
Near East University, Mersin 10, Lefkosa, Turkey
e-mail: nedimeserakinci@gmail.com

M.C. Ergoren
e-mail: mahmutcerkez@gmail.com

R.C. Zhao (ed.), *Stem Cells: Basics and Clinical Translation*, Translational
Medicine Research 1, DOI 10.1007/978-94-017-7273-0_14

pathway and again maintain telomere homeostasis and thereby the cell's ability to be expanded in vitro, and focus on a new therapeutic area that uses hMSCs as delivery vehicles as a potential new cancer treatment.

Keywords Mesenchymal stem cell · Cancer · Therapy

14.1 Introduction

Daily human body loses many types of cells such as hepatocytes, keratinocytes, and certain blood cells as part of normal homeostasis for organisms and therefore needs to replace them for survival. Stem cells are the main source for such cell replacement; moreover, they are unique cells that have the ability to renew themselves through mitotic cell division and differentiate into a diverse range of specialized cell types. Precursor cells, progenitor cells, somatic stem cells, or adult stem cells can be used as different names for cells with stem cell-like properties. In addition, stem cells, with few notable exceptions, are cell types that show telomerase activity thus actively maintain telomere length to some degree. The ability to maintain telomere length allows them to have an extended proliferative capacity compared to the somatic cells.

Embryonic, germinal, and adult (somatic) stem cells are the main types of stem cells. The differentiation potential of stem cells varies according to type from totipotency to unipotency. On account of their differentiation, ability and generating a complete organism cells such as fertilized oocytes, up to the 8-cell blastocyst, are considered to be totipotent. Embryonic stem (ES) cells derived from the inner cell mass of a blastocyst. ES cells posses all characteristics of true stem cells. In addition to self-renewal capacity, they are pluripotent, being able to produce derivates of all three germ layers (endoderm, mesoderm, and ectoderm) (Burdon et al. 2002). A high telomerase activity prevents the ES cells from undergoing crises and reaching senescence, which is an advantage for long-term culturing. Moreover, after reintroduction into the blastocyst, ES cells retain their developmental identity (Beddington and Robertson 1989). Differentiated forms of ES cells, somatic stem cells, are known as multipotent stem cells and are capable of self-renewal and are restricted to a specific tissue or organ. These cells can be isolated from the developing organism (the fetus and the postnatal organism) as well as from the adult organism. Due to their origin from a specific tissue, the offspring of the somatic stem cells is also specific for the original tissue. Unipotent cells by definition are the stem cells that can only give rise to one type of cell (Serakinci et al. 2014). Contrary to other cell types, asymmetric cell division occurs in stem cells, whereby one daughter cell remains undifferentiated, while the other is committed to differentiation (Fig. 14.1). Furthermore, by means of transdifferentiation, somatic cells are also able to produce progeny different from their tissue of origin. These cells can be isolated from developing organism in fetal and postnatal stages

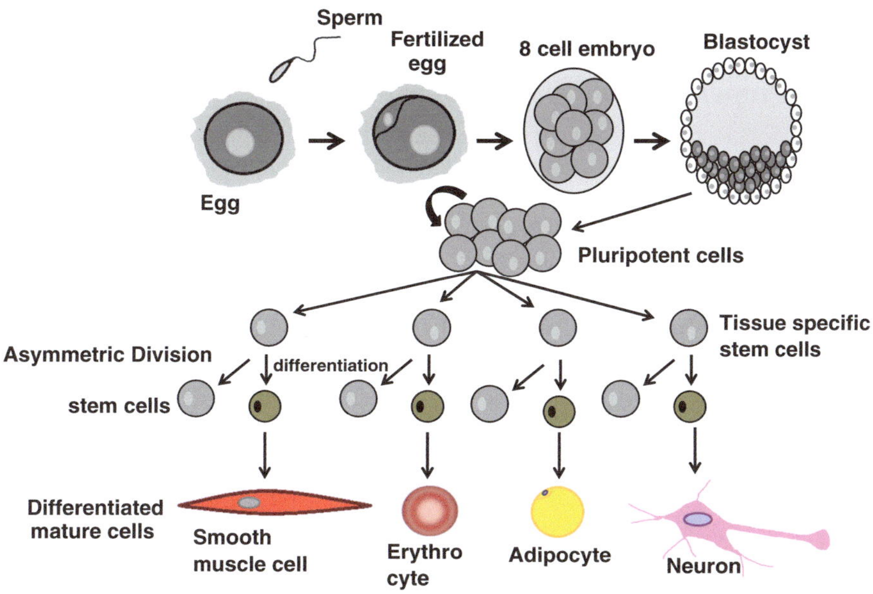

Fig. 14.1 Two types of cell division can occur in stem cells. The stem cells produce two daughter cells that are identical to the mother cells during symmetric divisions. As a result of asymmetric cell division, the mother divides to give rise to two daughter cells; one that is identical to mother cell and the other more differentiated than the mother cell. Therefore, stem cells go through more and more asymmetric divisions; they become more and more differentiated eventually becoming terminally differentiated. Stem cells have been observed at various places in human body at varying stages of differentiation. Totipotent cells are only found in the zygote; however, other stem cells exist at various stages of the organism even in adults. Tissue-specific stem cells have been identified in various tissues for instance muscle, blood, central nervous system, and adipocytes to name a few

as well as are harvested the fully developed organism. Additionally somatic cells include harvested stem cells from the brain that are capable of differentiating into the three lineages of the central nervous system (CNS) (neurons, astrocytes, and oligodendrocytes). One more example of somatic stem cells is the bone marrow stem cells that include hematopoietic and mesenchymal stem cells which are able to repopulate the blood and the bone cell systems.

In this current communication, we will focus on mesenchymal stem cells (MSCs) and their double-faced role in cancer development and their potential use in cancer therapy will be discussed. The mechanisms based on the telomerase will be discussed that MSCs can show neoplastic transformation through telomere pathway and again maintain telomere homeostasis and thereby the cell's ability to be expanded in vitro, and also focus on a new therapeutic area that uses human mesenchymal stem cells (hMSCs) as delivery vehicles as a potential new cancer treatment.

14.2 Mesenchymal Stem Cells

14.2.1 Properties and Isolation of Mesenchymal Stem Cells

Stroma is the supporting tissue that existing in bone marrow and other tissues such as thymus, ovary, and iris and thought it is a simple structural framework for the hematopoietic system. Today, it is well known that hMSCs function is very diverse. One of its most significant aspects is that it contains mesenchymal stem cells (MSCs). These cells are strongly adherent, and therefore, they can be isolated by culturing marrow on a special substrate that allows the other cells to be washed off (Dennis et al. 2004; Pittenger et al. 1999; Terskikh et al. 2006) and their properties were described in 1968 by Friedenstein (1968). MSCs undergo self-renewing divisions but also give rise to more committed progenitor cells. Therefore, during mesodermal lineage and also other embryonic lineages, they can differentiate into some cell types such as adipocytes, osteocytes, and chondrocytes (Jiang et al. 2002).

MSCs show different advantages and/or disadvantages based on the tissue that have been isolated, for example, MSCs can be isolated from adipose tissue; thus, there is evidence that cultures MSCs derived from different donors differ significantly (Kern et al. 2006). As an example of the disadvantage, MSC cultures are isolated from adult bone marrow from invasive method that is often poorly tolerated by the patients (Veryasov et al. 2014). In addition to bone marrow, MSCs have been shown to be present in a number of other adult and fetal tissues, including circulating blood (Zvaifler et al. 2000), cord blood (Weiss and Troyer 2006), placenta (Miao et al. 2006), amniotic fluid (Tsai et al. 2004), heart (Chen et al. 2008), skeletal muscle (Peault et al. 2007), adipose tissue (Zuk et al. 2001), synovial tissue (De Bari et al. 2001), pancreas (Di Rocco et al. 2008), extra embryonic tissues (Veryasov et al. 2014), skin dermis (Chen et al. 2014), and additionally, pathological tissue-like rheumatoid arthritic joints (Marinova-Mutafchieva et al. 2000). Cells with MSCs characteristics might be presented most of the postnatal organs and tissues (Chamberlain et al. 2007). Historically, the broad variety of tissue sources from which MSCs are isolated, in conjunction with disparate culture conditions such as media formulations and plating density, has led to a lack of consensus regarding the phenotype of the MSC. Nevertheless, recent reports are pointing to a strong consensus regarding the morphology of fresh MSCs, irrespective of the method by which they were isolated. The large cells with a prominent nucleoli and bleb-like projections are described as the morphology of MSCs (Jones and McGonagle 2008). These projections extend further as MSCs adhere and the morphology is different from the typically shaped spindle-shape MSCs (Jones and McGonagle 2008). Despite the historic variation in reported phenotypes of MSCs, it is widely accepted that the cultured cells, lightly their isolation and culture methods, lack expression of prototypic hematopoietic antigens including CD45, CD34, CD11b, and CD14 and express SH2 (CD105), SH3/SH4 (CD73), CD29, CD44, CD90, CD71 CD106, CD166, STRO-1, GD2, and CD146 (Pittenger et al. 1999; Simmons and Torok-Storb 1991; Shi and Gronthos 2003; Sordi et al. 2005; Martinez et al. 2007).

The methodology used in the isolation and enrichment of human mesenchymal cells is essentially based on the ability of these cells to adhere to and subsequently proliferate on tissue culture plastic with 10 % fetal calf serum. A lag phase might be experienced by the cells, as then divide rapidly. The MSCs' doubling time in vitro depends on the donor and the original plating density (Chamberlain et al. 2007). Indeed, culture selection is still widely employed as a means of MSC isolation. During negative selection, the cocktails of antibodies that deplete the bone marrow of specific cell populations are pre-enriched through cell separation strategies (Louis et al. 2001; Reyes et al. 2001) or FicollTM separation is most widely used as an initial step in MSC isolation. To date, no single and unique marker allowing for MSC isolation has been reported; thus, a range of composite cell surface phenotypes are being pre-owned. MSCs that are from enriched populations have been isolated from human bone marrow aspirates using a STRO-1 monoclonal antibody in conjunction with antibodies against VCAM-1/CD106 (Simmons and Torok-Storb 1991), CD271 (Quirici et al. 2002), D7-Fib30 and CD49a.31. In order to harvest and isolate MSCs, a needle is used to aspirate the bone marrow from the trabecular of the bone. Manipulated bone marrow can be within the laboratory to remove the red blood cells, macrophages, and other extraneous material (Boiret et al. 2005). MSCs can then be enriched by their adherence to a plastic culture dish. Using flow cytometry can sort MSCs based on the MSC surface proteins and this can be viewed under a microscope to determine that the cells look like MSCs (Quirici et al. 2002; Campagnoli et al. 2001; Li et al. 2006). MSCs have also been isolated from other species, for example, mice, rat, cats, dogs, rabbits, and baboons with varying success. However, MSCs from different species do not express the same molecules as the human cells (Javazon et al. 2004).

14.2.2 Limited Proliferation Capacity of hMSCs

Most of the human somatic cells can undergo 60–70 population doublings on average and then enter senescence (Meyerson 2000). Cellular senescence is the process by which normal cells lose the ability to divide. The "Hayflick limit" is named for this limited number of cell divisions (Hayflick 1976) (Fig. 14.2a).

In normal cells, cell divisions are preceded by replicating the DNA to form two daughter molecules each having an original strand from parent cell and one newly synthesized strand. Replication of the leading DNA strand is simple and complete, as on the lagging strand replication uses small Okazaki fragments. The purpose of using these fragments is that to ensure 5′ to 3′ addition of bases. These results in the incomplete replication of the extreme ends of the lagging strand of chromosomes, and loss of genes and it is named the "end replication problem." Two very important protein p53 and pRB that have a significant role in normal senescence and therefore normal senescence that leads initially to the arrest of cell proliferation (Campisi 2005). The senescence program operates through the Rb and p53 pathway and is activated by the telomere signal. Rb- and p53-deficient primary

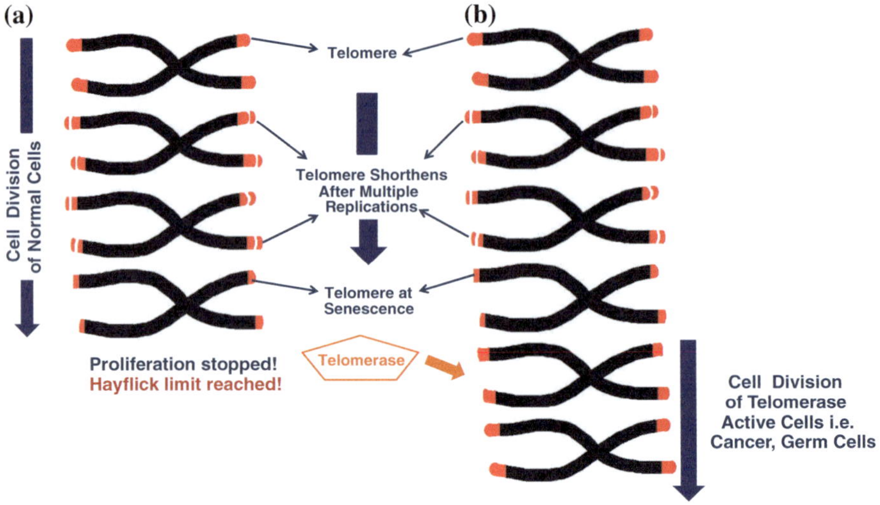

Fig. 14.2 Telomere shortening during cell divisions (**a**) and senescence at the Hayflick limit (**b**) Cells such as cancer cells or germ cells have active telomerase enzyme activity; therefore, these cells can escape from senescence

cells that continue cellular growth beyond the Hayflick limit exhibit severe telomere shortening, marked genetic instability and massive cell death—this period is referred to as crisis (Counter et al. 1992; Shay et al. 1991). It is thought that senescence plays an important role in the suppression of cancer emergence. Once the telomeres through the end replication problem reach a critical short length, the cell enters M1 stage (mortality 1) and goes into the senescence (Kim et al. 1994). If the cell escapes senescence and continues to proliferate with further shortening of telomeres, it will undergo crisis or mortality 2 (M2). If the mutational changes acquire, therefore cells overcome M1 and M2, these cells become immortal. Most cancers are the result of "immortal" cells that have evaded programmed cell death. Herein, the cells will have acquired telomerase reactivation to maintain a constant length of telomeres (Kim et al. 1994) (Fig. 14.2b). Despite their stem cell characteristics when pressed to proliferate extensively, mesenchymal 2 stem cells can also suffer from explicative senescence with critically short telomeres (Kim et al. 1994) This brings about certain limitations in therapeutic use of stem cells. One of these limitations is as mentioned above, proliferative capacity limitation, which undoubtedly can be overcome by introduction of certain genes that will enable them to continue to proliferate. It has been shown that there are numerous genes like the *TERT* gene that are capable of extending the proliferative capacity (Fig. 14.2) and immortalizing the stem cells or progenitor cells, however, with the risk that this extended life span might unmask possible malignant transformation. Therefore, telomere dynamics becomes an important issue in stem cell function especially when expanding a population of stem cells is needed. Considering the fact that stem cells in general tend to give rise to a high number of daughter cells, one would expect that stem cells would express telomerase to maintain telomere length. However, the replicative capacity in primary stem cells is limited. The telomere length pattern (also termed the telomere profile) can be monitored until the cells reach replicative arrest after approximately 10 population doublings (PD) and it has been shown that the telomere profile is conserved for this number of doublings (Graakjaer et al. 2007; Serakinci et al. 2007). The possible conservation of profile has been studied in both primary and telomerase mesenchymal stem cells where both types of cells were grown for number of passages (205 PDs) and still they maintained the profile (Graakjaer et al. 2007; Serakinci et al. 2007). Surprisingly, the long-term conservation of the telomere profile clearly suggests that also in mesenchymal stem cells there is a very low degree of random fluctuation in the telomere dynamics, as previously suggested in lymphocyte progenitor cells. Overall, there is a general agreement that adult stem cells have very low levels of telomerase, so that during life, telomeres are slowly shortened. The primary mesenchymal stem cells have also been shown to obtain a diminished mean telomere length during long-term culturing periods (Serakinci et al. 2007) have shown that the mean telomere length of the primary cells is 8 kb and the length is continuously decreasing with the population doubling level. Overall, it has been suggested that telomeres shortening may lead to genomic instability; therefore, the cells with short telomeres might be able to through neoplastic differentiation.

14.2.3 Potential Neoplastic Transformation of hMSCs During Expansion

Within the last decade, human mesenchymal stem cells have been validated as potential tools in different therapeutic approaches. Initial results are promising but there are some challenges for the use of MSCs for clinical purpose. During the isolation process from human bone marrow, only 1 in every 105 cells is MSCs and there is also the issue of low grafting efficiency and potency of MSCs. In spite of hMSCs have great potential in therapeutic use, the main rate limiting factor is that hMSCs exhibit limited mitotic potential, especially considering that, in a human system very many cells are needed for injection. Thus, it is obvious that a large scale for MSC expansion is needed (Momin et al. 2010; Teng 2010). The *TERT* gene has been highlighted for overcoming for proliferative limitation introduction. Hence, telomere dynamics plays a significant role in stem cell function, specially, while a population of stem cells expanding. Telomerase researchers succeeded to introduce a retrovirus carrying the hTERT gene has established an immortalized hMSC-telo1 cell line, which maintain their stem cells characteristics and have an expanded life span (Serakinci et al. 2007). But extending the proliferative capacity of stem cell populations through manipulation of the telomere-telomerase system brings certain risks that are associated with the possibility that stem cells may show increased susceptibility to carcinogenesis. This gene manipulation bypasses the naturally built-in controls of the cell that govern the delicate balance between cell proliferation and senescence and carcinogenesis.

Given the critical role of telomere dynamics and telomerase in tumor progression and the fact that the cancer cells rely on telomerase for its survival, it is not surprising that telomeres are rather unique structures in a given cell. Also telomerases have a significant role in protecting the chromosome ends from being recognized as DNA double-strand breaks by the DNA repair machinery. On the other hand, telomeres, when critically shortened, can lead to cellular senescence, which can be regarded as a barrier against cancer formation via the so-called telomere-mediated checkpoints. Nevertheless, the cells that are not destroyed and survived, dysfunctional telomeres can affect the genomic stability through initiating the so-called Break–Fusion–Break (BFB) cycles leading to severe genomic aberrations and ultimately to cancer development (Furlani et al. 2009; Serakinci et al. 2008). These events have been reported in literature in adipose-derived human MSCs (Rubio et al. 2005) and bone marrow-derived mouse MSCs (Miura et al. 2006; Tolar et al. 2007). Phenotypic and genotypic alterations such as chromosome instability, rapid cell proliferation, loss of contact inhibition, gradual increase in telomerase activity, and increased c-myc activity have been observed in these MSCs (Furlani et al. 2009).

However, studies in both rodent models and human mesenchymal stem cells have suggested that during long-term culturing, mesenchymal stem cells are acquiring chromosomal aberrations and subsequently exhibit a malignant transformation (Rubio et al. 2005; Zhou et al. 2006). Such studies raised the concerns that hMSCs that are forced into extensive in vitro expansion can—after being transplanted into patients—undergo spontaneous transformation. Several groups have published results showing spontaneous transformation of human mesenchymal stem cells that are expanded long term (Momin et al. 2010; Rubio et al. 2005; Serakinci et al. 2004; Wang et al. 2005). Mouse models have been studied in most studies such as Rubio et al. and these researchers have shown that adipose tissue-derived human MSC populations after long-term in vitro expansion can transform spontaneously. Rubio et al. furthermore characterized the molecular mechanisms implicated in the spontaneous transformation. Moreover, the transformation process occurred after the hMSC had bypassed senescence by upregulating c-myc and repressing p16 levels. After then, during telomerase activity acquisition, deletion of Ink4a/Arf locus and hyperphosphorylation of Rb the cells also bypassed M2 (Rubio et al. 2008).

More studies showed that cultured human MSCs derived from the bone marrow produced a subpopulation of cells and these cells would have high levels of telomerase activities, chromosomal aneuploidies, and translocations and were able to form tumors in multiple organs in NOD/SCID mice (Momin et al. 2010; Wang et al. 2005). Contrary to these studies several groups have reported no transformation of hMSC after long-term culture (Bernardo et al. 2007; Meza-Zepeda et al. 2008). Using comparative genomic hybridization, karyotyping, and subtelomeric fluorescent in situ hybridization analysis, Bernardo et al. performed extensively studies on the genetic changes in the hMSCs at different stages of long-term culture, but they did not find evidence for spontaneous transformation of hMSCs during long-term culture (Bernardo et al. 2007). Moreover, there are studies demonstrating that telomerased hMSC accumulate various genetic and epigenetic changes in spite of maintaining a normal karyotype (Serakinci et al. 2004; Burns et al. 2005). These studies are supporting the existence of neoplastic transformation of hMSCs during in vitro expansion. However, the spontaneous transformation potential of hMSCs is still a controversial issue and more evidences are needed. In addition to in vitro expansion studies, hMSCs have been observed that they might have a significant role in the carcinogenesis progression. Evidences thus suggest that exogenously administrated hMSCs can be recruited to the stroma of developing tumors when systemically infused in animal models for Karposi's sarcoma, ovarian carcinoma, glioma, colon carcinoma, and melanoma (Lazennec and Jorgensen 2008).

Furthermore, Correa et al. have demonstrated that gastric cancer may originate from hMSC (Correa and Houghton 2007). Based on these and similar studies, it can be suggested that hMSC can modulate the tumor growth, despite still remains as controversial and not fully understood. In this connection, it is relevant, as Serankici et al, have shown that telomerase-immortalized hMSCs do not necessarily show spontaneous transformation. First of all, neoplastic transformations have been observed in these cells when the cells were subjected to 2.5 Gy of gamma irradiation and after having been exposed to gamma radiation followed by further long-term culturing. Subsequently, same group investigated the stepwise acquisition of genetic changes, leading cells to acquire the malignant phenotype (Serakinci et al. 2007; Christensen et al. 2008). These investigations suggested that irradiation lead to DNA damage including telomere damage causing temporary cell cycle arrest. Subsequent culturing of the cells indicates that overall DNA damage was restored and therefore the cells returned back to cell cycle. More studies showed that there remained some degree of damaged telomere, as a result increasing detected in occurrence of anaphase bridges, suggesting uncapped telomeres that resulted by induction of F-B-F cycles. This in turn led to genomic instability (Christensen et al. 2008). Additionally, telomerase activation most often occurs at the transition to cancer which is highly specific to cancer cells. Introducing of hTERT gene in MSC did not unmask the neoplastic potential on its own but subsequent DNA damage contributed to mesenchymal tumor development. Recent findings are in agreement with observations made by a number of other groups. Thus, it appears that using telomerase to help production of large numbers of cells is effective, but it has impact on neoplastic transformation. Therefore, more knowledge about adult stem cells and how their growth is regulated is required and until then careful monitoring is crucial for clinical applications. Nevertheless, in the discussion whether or not in vitro expanded hMSCs can transform spontaneously the whole issue of cross-contamination is an important issue that definitely needs to be addressed vigorously. Thus, studies suggested that DNA fingerprinting and/or STR analysis should be used to confirm the true identity of the transformed cells (Garcia et al. 2010; Torsvik et al. 2010, 2012).

Respected amount of in vitro expansion studies showed that hMSCs could play a role in the progression of carcinogenesis. These studies will improve our knowledge for understanding cancer and according to new information applied therapeutic approaches might be changed. Lazennec et al. suggested that systemically infused exogenously administrated hMSCs can be recruited to the stroma of developing tumors in animal models (Lazennec and Jorgensen 2008). Gastric cancer might be originated from hMSC also demonstrated by Correa et al. (Correa and Houghton 2007). Despite that still remains controversial and not fully understood, based on these and similar studies, it can be suggested that hMSC can modulate the tumor growth (Dennis et al. 2004).

Studies for graft-versus-host disease (GvHD) showed encouraging results which patients suffering from this disease have received injections with MSC with no signs of tumor formation (Correa and Houghton 2007; Barkholt et al. 2013; Resnick et al. 2013). Moreover, a recent study has looked into the use of MSC and the adverse effects in Crohn's disease in a phase 2 study (Forbes et al. 2014). As a conclusion promising results obtained, however, it is clear that a larger number of studies are needed for improving registries and well-defined follow-up protocols to successfully approach the issue of tumorigenicity after MSC-based treatments (Dennis et al. 2004; Barkholt et al. 2013).

14.3 Therapeutic Potential of Mesenchymal Stem Cells

MSCs have many desirable characteristics that make them great therapeutics tools for many diverse illnesses. As we mentioned previously, MSCs can be isolated from adult donors and can easily be expanded in culture without greatly compromising genetic stability (discussion about genetic instability and neoplastic transformation is mentioned Sect. 14.2.3). MSCs are lack of immunogenicity and this allows for allogenic transplantation and their homing capacity creates from for treatment with minimal invasion (Teng 2010). In Sect. 14.2.1, isolation methods of MSCs have been well discussed. However, as previously mentioned, MSC cannot be isolated from many diverse tissues such as trabecular bone, placenta, adipose tissue.

MSCs have been used in many studies that were mainly based on the site-directed and/or systemic administration of MSCs, and both delivery methods of hMSC have shown their ability of engraftment in a number of tissue after injury (Fig. 14.3) (Barbash et al. 2003; Horwitz et al. 2002; Anversa 2001; Phinney 2003). Using stem cells as a therapeutic tool, they must be able to access the target organ to deliver their therapeutic influence. In many cases, accessing to the target organ might not be a problem, at the same time; it can be a problem if the illness is systemic in nature or if the target organ is not anatomically accessible effective delivery. MSCs provide us with a very valuable tool in the latter two scenarios as they have been shown to spread to various tissues after their intravenous administration (Devine et al. 2003). These systemically administered MSCs home (sit) to the site of injury and aid in functional recovery. Cultured MSCs have been infused into patients in early studies to support bone marrow transplant for osteogenesis imperfecta and glycogen storage disease where the therapeutic options are limited

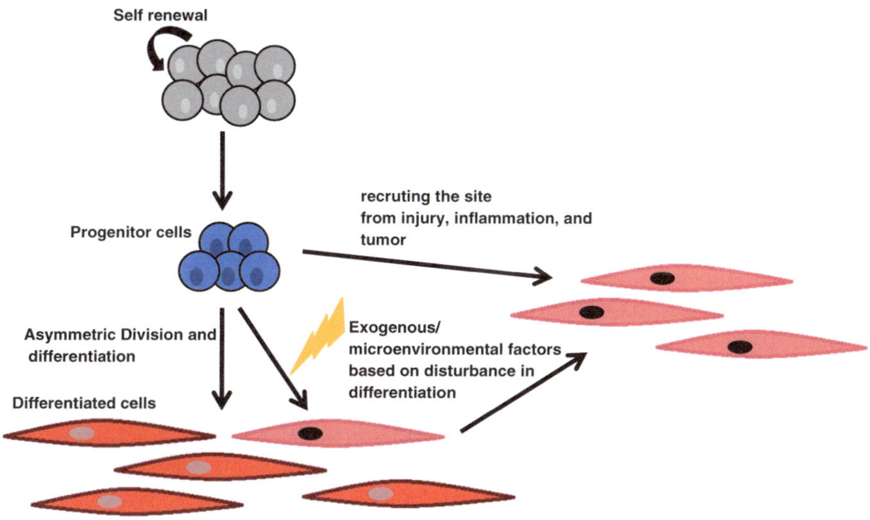

Fig. 14.3 Self-renewing potential of stem cells during symmetric divisions. Stem cells as mentioned earlier can produce differentiated daughter cell from the mother cell as a result of asymmetric. Neoplastic transformation of certain cells might occur from intrinsic and/or extrinsic factors. When injury, inflammation, or even a tumor exist in the tissue, local or distant stem cells and their progenitors at various stages of differentiation can home and cure the tissue. Isolated hMSCs can be turned and used as a therapeutic delivery vehicles

(Pittenger and Martin 2004). Additionally, their role in the treatment of Crohn's disease is being explored; therefore, they have been used in graft-versus-host disease (Mittal et al. 2009). Acute leukemia, ischemic stroke, multiple sclerosis, critical limb ischemia, articular, cartilage, and bone defects are the examples of MSCs that are also used in diverse variety.

Nevertheless, the knowledge about exact mechanism is still poor; introduction of MSCs into the infracted heart directly or through intravenous administration has resulted in improved recovery and prevented deleterious remodeling (Pittenger and Martin 2004). The ability of MSCs to repair damaged tissue is thought to be primarily from their capacity to secrete paracrine mediators such as interleukin-10, interleukin-1ra, keratinocyte growth factor, and prostaglandin (Matthay et al. 2010). However, the mechanisms underlying migration/homing of hMSC have yet to be clarified, even though there are some evidence suggesting the involvement of chemokines and their receptors (Matthay et al. 2010; Kortesidis et al. 2005; Von Luttichau et al. 2005). In addition to cellular adhesion molecules, the following chemokine/receptor pairs have been implicated in MSC migration: SDF-1/CXCR4, SCF/c-Kit, HGF/c-Met, VEGF/VEGFR, and PDGF/PDGFr (Wynn et al. 2004). However, the migration to the tumor site is believed to be non-specific since the administered MSCs have been shown to localize to lung, bone marrow, and lymphoid organs. Moreover, it was observed that whole body irradiation increases the distribution of MSCs to multiple organs (Momin et al. 2010).

Due to their stem cell properties, facilitation of engraftment of the transplanted hematopoietic stem cells and promotion of the structural and functional repair of damaged tissues are some of the first studies the scientists have focused in the MSCs field. However, because of the immunomodulatory properties of MSCs, they might have potential uses in immune-related diseases as well (Uccelli et al. 2008). Their poor immunogenicity in vitro, in preclinical trials and in human studies, is one of the advantages of using MSCs as therapeutic agents. Therefore, this would allow the use of MSCs from allogenic donors. There still remains the possibility of using autologous MSCs even in the autoimmune conditions (Bartholomew et al. 2002; Le Blanc et al. 2004; Murphy et al. 2002; Tse et al. 2003).

Uccelli have reported new findings that present the observation that MSCs derived from the bone marrow-suppressed T-cell proliferation (Uccelli et al. 2008). It has shown that there is a role of MSCs in both innate and adaptive immunity. (Table 14.1) MSCs can decrease the pro-inflammatory response potential of dendritic cells by inhibiting their production of tumor necrosis factor (TNF) in innate immunity. Their inhibitory effects on the maturation of monocytes and cord blood and CD34+ hematopoietic progenitor cells also have been shown (Jiang et al. 2005; Li et al. 2008; Nauta et al. 2006; Ramasamy et al. 2007). MSCs can also regulate an important part of innate immunity. The MSCs, through an IL-6-regulated mechanism, can dampen and delay the spontaneous apoptosis of resting and activated neutrophils (Raffaghello et al. 2008). In adaptive immunity, T cells are maintained in a state of quiescence through an MSC-mediated anti-proliferative effect that can be partially reversed through IL-2 stimulation (Zappia et al. 2005). Nevertheless, early studies mentioned also MSCs are able to modulate the intensity of immune system response through an inhibition of antigen-specific T-cell proliferation and cytotoxicity and through promotion of regulatory T-cell generation (Hwa Cho et al. 2006; Pevsner-Fischer et al. 2007; Tomchuck et al. 2008). Moreover, recent studies on mouse model to reveal the antiapoptotic characteristics of MSCs showed that bone marrow-derived MSCs decreased oxidative stress, apoptosis, and hippocampal damage in brain (Calió et al. 2014); nevertheless, there are many studies that support the antiapoptotic characteristic of MSCs (Yang et al. 2014; Bhang et al. 2012). On the other hand, recent studies showed that MSC-derived exosomes have similar functions as MSCs for instance suppressing inflammatory responses, repairing tissue damage, and also, modulating the immune system. However, it is very little known about their mechanisms and the findings still remain controversial (Yu et al. 2014).

Additionally, cellular vehicles are the one of the newer therapeutic applications for stem cells. Most scientist agree that the targeted delivery gene therapy has a promising future, particularly in cancer therapies, as most artificial vehicle systems are at present hampered by numerous problems including immune reaction, non-specific accumulation in normal tissues, and poor permeation (Bestor 2000). Cell-based carriers such as MSCs most probably will be the solution of these problems. Besides providing delivery system that immune privileged, these cells would also preferentially home to the site of a cancer and therefore serve as a delivery platform for therapeutic agents. Cells scientists are always developing

Table 14.1 The general suggested properties of MSCs are shown on the table

Multilineage differentiation potential	Isolation from various tissues	Promotion of tissue repair	Immunosuppression	Neuroprotective effect
(1) Multiple mesenchymal lineages (bone, fat, cartilage, and muscle) (2) Non-mesenchymal cell lineages (neurons, glial cells, and hepatocytes) (3) Self-renewal (4) Tissue regeneration	(1) Adhere to plastic (2) Easy in vitro expansion (3) Limitless supply	(1) Homing (2) cytokines, inflammatory mediators, extracellular matrix components, and antimicrobial proteins (3) Most important characteristic for selecting as anticancer carrier	(1) Low immunogenicity (2) No toxicity (3) Engraftment into xenogenic environment	(1) Transdifferentiation into neural cell secrete various neurotrophic and anti-inflammatory factors following transplantation (2) Conferring strong neuroprotective effects in models of amyotrophic lateral sclerosis, multiple sclerosis, Parkinson's disease, and glaucoma

creative and more effective ways of providing the gene delivery target cells corresponding to local delivery of the therapeutic transgene to neighboring cancer (Dennis et al. 2004; Bainbridge et al. 2008; Cideciyan et al. 2008; Kaplitt et al. 2007; Maguire et al. 2008; Nathwani et al. 2011).

Using telomerase-immortalized hMSC, Serakinci et al. have shown that human mesenchymal stem cells administered through a tail vein can preferentially engraft at the tumor sites and thence contribute to the population of stromal fibroblasts (Dennis et al. 2004). Their study also shows that in vivo assays were performed to investigate whether externally administered hMSCs would engraft at the tumor sites and interact with the cancer cells (Serakinci et al. 2011). Recent study from Sun et al. has investigated the use of hMCSs in U251 glioma-bearing mice as a vehicle for selective delivery of EphrinA1-PE38 to gliomas where it acts as a very specific immunotoxin against the EphA2 receptor over expressing gliomas (Sun et al. 2011). Recently, A new approach to studying gliomas in mice using the RCAS/Ntv-a system in which the mice have endogenous gliomas, where they tested the homing ability of syngenic MSCs to the gliomas has been created (Doucette et al. 2011).

14.4 Mesenchymal Stem Cells Are Double-Faced in Cancer

Recent studies have been shown that the critical role of MSCs in tumor development especially when stressed for long-term expansion. On the other hand, studies have been suggested that MSCs can be used as a cancer therapeutic agent. Studies on tumor models such as melanomas, colon adenocarcinomas, lung cancer, multiple myelomas, and glioblastomas have show that exogenously administered hMSC may contribute to tumor stroma formation in animal tumor models by promoting angiogenesis or by creating a niche to support cancer stem cell survival and most probably, the immunosuppressive ability of MSCs ought to be the reason (Momin et al. 2010; Sullivan and Hall 2009). The tumor cells may get the chance to better evade the immune surveillance, as the MSCs interact with many of the cells of the innate and acquired immune system and suppress them (Momin et al. 2010; Lazennec and Jorgensen 2008).

The development of a gastric cancer was induced by *Helicobacter felis* infection in a C57BL/6 mouse model where normal bone marrow was replaced with bone marrow cells tagged with either beta galactosidase or GFP Correa and Houghton (2007). Additionally, (Correa and Houghton 2007) results also demonstrated that cancer cells are directly derived from bone marrow stem cells. These cells have been recruited to the site, where the gastric cancer later occurred, during the helicobacter inflammation (Correa and Houghton 2007). Moreover, studies in a rat model of Barrett's metaplasia have showed similar results (Sarosi et al. 2008). In addition to these findings, in rodent models demonstrating the ability of MSCs' homing and transdifferentiation to inflammatory lesions at extra medullar sites are very valuable results (Hayashi et al. 2008; Kahler et al. 2007; Matthay 2007; Suzuki et al. 2008).

The current animal models play a significant role for understanding this existing complex relationship between MSCs and tumors. The model relies on harvesting human tumor cells and culturing them in vitro before grafting them into the animal. The downside of this approach is that during the culturing process, the cells are away from the tumor stroma and the epithelial cells and even when the cells are inside the animals, the stroma will be formed from the animal's own cells leading to a chimeric tumor that would not reproduce the interaction between the MSCs and the tumor stroma. More likely, the new generation of immunocompromised mice, for example, the NOG mouse will be able to help in the investigation of this relationship as it is better able to accept heterologous cell populations (Momin et al. 2010; Hahn and Weinberg 2002; Nakahata 2002; Kim et al. 2004; Rangarajan and Weinberg 2003; Rosen and Jordan 2009).

One of the biggest obstacles for finding a cure for cancer therapy research is no selective killing of tumor cells and therefore therapies that are more specifically directed toward cancer stem cells might result in much more durable. It has been suggested that hMSCs play a key role in supporting tumor formation, based on their homing abilities hMSC can be used as cellular vehicles for local delivery of biological agents to brain tumors. Human MSCs were transduced with a lentivirus expressing secretable TRAIL (S-TRAIL) and mCherry (red fluorescent protein) and injected into established intracranial glioma tumors in mice. The genetically modified hMSCs were able to inhibit tumor growth, resulting in significantly longer animal survival. Thus, the study demonstrated the therapeutic efficacy of hMSC S-TRAIL cells and confirmed that hMSCs can serve as a powerful cell-based delivery vehicle for the site-specific release of therapeutic proteins (Menon et al. 2009). Experiments for seeking targeted delivery, IFN-β-secreted genetically modified MSCs were able to successfully home (sit/ place beside to support like feeder layer) and engraft at melanomas growing in mice lungs and locally deliver IFN-β, while delivering locally, IFN-β was able to inhibit the growth of malignant cells both in vivo and in vitro. However, the same effect could not be achieved with systemically delivered IFN-β or IFN-β produced away from the tumor site (Studeny et al. 2002; Studeny et al. 2004). Recent study by Placencio and colleagues is shown that MSCs were recruited and incorporated into the prostate epithelium during prostate re-growth (after testosterone reintroduction). Additionally, the incorporated MSCs were used to deliver secreted frizzled-related protein-2 (SFRP2) to antagonize the Wnt-mediated cancer progression by reducing tumor growth, increasing apoptosis, and causing potential tumor necrosis (Placencio et al. 2010).

hMSCs have the potential home to the tumor stroma activities, therefore, for the tumor microenvironment, allow them to be a promising tool for the delivery of anticancer drugs. Studeny et al. have showed that this strategy worked by the observation of specific homing of intravenously administered hMSCs, engineered to produce interferon-β (IFN-β), to tumors with subsequent tumor regression in a xenogenic mouse model (Studeny et al. 2004). Their study determined that mesenchymal stem cells that expressed IFN-β could inhibit the growth of tumor cells in vivo. However, their approach required integration of the hMSC at the tumor site, as non-tumor site integrated or systemic delivery of IFN-β did not have enough tumor regressing effect. Serakinci et al. have recently confirmed these findings where they have shown that hMSC can home to tumor site and furthermore could deliver a therapeutic agent to the site (Serakinci et al. 2011).

In addition to mentioned studies, IL-12 M-expressing MSC injection directly injected into the tumor had the strongest antitumor effect compared with other injection routes such as intravenous or subcutaneous. The same study also demonstrated anti-metastatic effects for MSCs/IL-12 M when embedded in the Matrigel (Seo et al. 2011). Moreover, Correa and colleague have determined that gastric cancers might be originated from hMSCs (Burns et al. 2005). Based on these studies and our current knowledge, it can be suggested that hMSCs can modulate the tumor growth, as this issue still remains controversial and not fully understood.

Based on the previous studies and research carried on, Serakinci et al. have shown that the identification of reservoirs of multipotential stem cells within adult tissue provides exciting prospects for developing novel vehicle for stem cell-mediated gene therapy. However, this new approach seems to produce stronger and more specific anti-tumor effects, as considering continued cell growth is the one of the hallmarks of cancer development that most often correlated with activation of telomerase. Hence, the question must be raised if there is a potential cancer risk of genetically engineered cells. In the future, such approaches might be an ability to remove or inactivate the genetically engineered stem cells that homed to tumor site at the time of delivery of the targeted treatment (Aboody et al. 2008; Serakinci and Fahrioglu 2011).

Using telomerase inhibitors that may be a way to stop proliferation of these particular cells can be the one of the possible approach. Telomerase and modified structures named telomerase inhibitors are essential for the cell treatments, and they will eventually lead to cell death. However, this new attempt might face another problem namely that some of these cells might escape the cell death and this might cause genomic instability, after all causing development of a new cancer. Other possible approach, genetically modified vehicle cells are directed to commit suicide after delivering the targeted treatment. This strategy has been used in connection with the tumor-selective viruses that mediate oncolytic effects on tumors due to genetically engineered viruses that replicated in and kill targeted cancer cells. Based on the telomerase promoter sequence in case of attacking telomerase-positive cells, such viruses have been genetically modified with tumor-specific transcriptional response elements (Abdul-Ghani et al. 2000; Bilsland et al. 2005; Komata et al. 2001; Plumb et al. 2001).

The results recommended that targeted therapies can be improved by genetically modified vehicle stem cell therapies the suicide gene therapies; however, the studies also suggested that these therapies also can reduce the risk of secondary tumors. The role of stem cells in carcinogenesis still remains unclarified subject and need to be evaluated carefully with further studies.

14.5 Future Directions in HMSC-Based Cancer Therapies

In light of above-mentioned studies, it seems that to incorporate cancer-fighting genes inside stem cells grown from a patient's bone marrow mesenchymal stem cells is a promising way to go to effective cancer therapy. To achieve on effective therapy, scientist should find the way to deal with well-controlled and specific targeted vectors.

Gene and viral therapies for cancer have shown some therapeutic effects, but there has been a lack of real breakthrough. Based on others and our studies, it seems that the identification of reservoirs of multipotential stem cells within adult tissue provides exciting prospects for novel vehicle for stem cell-mediated gene therapy. This new strategy seems to produce stronger and more specific anti-tumor effects. However, considering that one of the hallmarks of cancer development is continued cell growth, which is most often correlates with activation of telomerase, the question must be raised if there is a potential cancer risk of genetically modified cells. These approaches will critically be an ability to remove or inactivate the genetically engineered stem cells which are homed to tumor site at the time of delivery of the targeted treatment.

One approach might be to use telomerase inhibitors that may be a way to stop proliferation of these particular cells. Since the telomerase is essential to the life of a cell treatment with modified structures called telomeres, inhibitors will eventually lead to cell death. But this approach might face another problem, namely that some of these cells will escape the cell death and this will lead to genomic instability particularly will be the reason of developing new cancer. After delivery of the targeted treatment another and the relatively safer approach will be focusing on to create a system that the vehicle cells to commit suicide. Such a strategy has been used in connection with the tumor-selective viruses that mediate oncolytic effects on tumors due to genetically modified viruses, which are engineered to replicate in and kill targeted cancer cells. Such viruses that have been engineered with tumor-specific transcriptional response elements based on the telomerase promoter sequence, thereby attacking telomerase-positive cells (Abdul-Ghani et al. 2000; Bilsland et al. 2005; Komata et al. 2001; Plumb et al. 2001). These studies have suggested that a combinational therapy approaches (genetically engineered vehicle stem cell therapy and the suicide gene therapy) might improve and targeted therapies and at the same time reduce the risk of secondary tumors.

In conclusion, using hMSCs still is a promising approach for the future of cancer treatment. However, It does not seem that traditional methods will substitute with few hMSCs approaches, but they might be used with conventional treatments to be more conclusive. One approach might be used combining stem cells with gene therapy in the future. Antitumoral potency and therapeutic index will be the key points of this mechanism for clinical applications. However, this theory should be investigated more and give answers of questions like "How can we ensure this mechanism will only kill the targeted cancer cells?"

References

Abdul-Ghani R, Ohana P, Matouk I, Ayesh S, Ayesh B, Laster M, Bibi O, Giladi H, Molnar-Kimber K, Sughayer MA, de Groot N, Hochberg A. Use of transcriptional regulatory sequences of telomerase (hTER and hTERT) for selective killing of cancer cells. Mol Ther. 2000;2(6):539–44.

Aboody KS, Najbauer J, Danks MK. Stem and progenitor cell mediated tumor selective gene therapy. Gene Ther. 2008;15(10):739–52.

Bainbridge JW, Smith AJ, Barker SS, Robbie S, Henderson R, Balaggan K, Viswanathan A, Holder GE, Stockman A, Tyler N, Petersen-Jones S, Bhattacharya SS, Thrasher AJ, Fitzke FW, Carter BJ, Rubin GS, Moore AT, Ali RR. Effect of gene therapy on visual function in Leber's congenital amaurosis. N Engl J Med. 2008;358(21):2231–9.

Barbash IM, Chouraqui P, Baron J, Feinberg MS, Etzion S, Tessone A, Miller L, Guetta E, Zipori D, Kedes LH, Kloner RA, Leor J. Systemic delivery of bone marrow-derived mesenchymal stem cells to the infarcted myocardium: feasibility, cell migration, and body distribution. Circulation. 2003;108(7):863–8.

Barkholt L, Flory E, Jekerle V, Lucas-Samuel S, Ahnert P, Bisset L, et al. Risk of tumorigenicity in mesenchymal stromal cell-based therapies—bridging scientific observations and regulatory viewpoints. Cytotherapy. 2013;15(7):753–9.

Bartholomew A, Sturgeon C, Siatskas M, Ferrer K, McIntosh K, Patil S, Hardy W, Devine S, Ucker D, Deans R, Moseley A, Hoffman R. Mesenchymal stem cells suppress lymphocyte proliferation in vitro and prolong skin graft survival in vivo. Exp Hematol. 2002;30(1):42–8.

Beddington RS, Robertson EJ. An assessment of the development potential of embryonic stem cells in the midgestation mouse embryo. Development. 1989;4:733–7.

Bernardo ME, Zaffaroni N, Novara F, Cometa AM, Avanzini MA, Moretta A, Montagna D, Maccario R, Villa R, Daidone MG, Zuffardi O, Locatelli F. Human bone marrow derived mesenchymal stem cells do not undergo transformation after long-term in vitro culture and do not exhibit telomere maintenance mechanisms. Cancer Res. 2007;67(19):9142–9.

Bestor TH. Gene silencing as a threat to the success of gene therapy. J Clin Invest. 2000;105(4):409–11.

Bhang SH, Lee S, Shin JY, Lee TJ, Kim BS. Transplantation of cord blood mesenchymal stem cells as spheroids enhances vascularization. Tissue Eng Part A. 2012;18(19–20):2138–47.

Bilsland AE, Fletcher-Monaghan A, Keith WN. Properties of a telomerase specific Cre/Lox switch for transcriptionally targeted cancer gene therapy. Neoplasia. 2005;7(11):1020–9.

Boiret N, Rapatel C, Veyrat-Masson R, Guillouard L, Guerin JJ, Pigeon P, Descamps S, Boisgard S, Berger MG. Characterization of nonexpanded mesenchymal progenitor cells from normal adult human bone marrow. Exp Hematol. 2005;33(2):219–25.

Burdon T, Smith A, Savaiter P. Signalling, cell cycle and pluripotency in embryonic stem cells. Trends Cell Biol. 2002;9:432–8.

Burns JS, Abdallah BM, Guldberg P, Rygaard J, Schroder HD, Kassem M. Tumorigenic heterogeneity in cancer stem cells evolved from long-term cultures of telomerase-immortalized human mesenchymal stem cells. Cancer Res. 2005;65(8):3126–35.

Calió ML, Marinho DS, Ko GM, Ribeiro RR, Carbonel AF, Oyama LM, Ormanji M, Guirao TP, Calió PL, Reis LA, Simões Mde J, Lisbôa-Nascimento T, Ferreira AT, Bertoncini CR. Transplantation of bone marrow mesenchymal stem cells decreases oxidative stress, apoptosis, and hippocampal damage in brain of a spontaneous stroke model. Free Radical Biol Med. 2014;141–54.

Campagnoli C, Roberts IA, Kumar S, Bennett PR, Bellantuono I, Fisk NM. Identification of mesenchymal stem/progenitor cells in human first-trimester fetal blood, liver, and bone marrow. Blood. 2001;98(8):2396–402.

Campisi J. Suppressing cancer: the importance of being senescent. Science. 2005;309(5736):886–7.

Chamberlain G, Fox J, Ashton B, Middleton J. Concise review: mesenchymal stem cells: their phenotype, differentiation capacity, immunological features, and potential for homing. Stem Cells. 2007;25(11):2739–49.

Chen Z, Wang Y, Shi C. Therapeutic implications of newly identified stem cell populations from the skin dermis; 2014. [Epub ahead of print].

Chen Y, Shao JZ, Xiang LX, Dong XJ, Zhang GR. Mesenchymal stem cells: a promising candidate in regenerative medicine. Int J Biochem Cell Biol. 2008;40(5):815–20.

Christensen R, Alsner J, Brandt Sorensen F, Dagnaes-Hansen F, Kolvraa S, Serakinci N. Transformation of human mesenchymal stem cells in radiation carcinogenesis: long-term effect of ionizing radiation. Regen Med. 2008;3(6):849–61.

Cideciyan AV, Hauswirth WW, Aleman TS, Kaushal S, Schwartz SB, Boye SL, Windsor EA, Conlon TJ, Sumaroka A, Pang JJ, Roman AJ, Byrne BJ, Jacobson SG. Human gene therapy for RPE65 isomerase deficiency activates the retinoid cycle of vision but with slow rod kinetics. Proc Natl Acad Sci USA. 2008;105(39):15112–7.

Correa P, Houghton J. Carcinogenesis of Helicobacter pylori. Gastroenterology. 2007a;133(2):659–72.

Correa P, Houghton J. Carcinogenesis of Helicobacter pylori. Gastroenterology. 2007b;133(2):659–72.

Counter CM, Avilion AA, LeFeuvre CE, Stewart NG, Greider CW, Harley CB, Bacchetti S. Telomere shortening associated with chromosome instability İs arrested in immortal cells which express telomerase activity. EMBO J. 1992;11(5):1921–9.

De Bari C, Dell'Accio F, Tylzanowski P, Luyten FP. Multipotent mesenchymal stem cells from adult human synovial membrane. Arthritis Rheum. 2001;44(8):1928–42.

Dennis JE, Cohen N, Goldberg VM, Caplan AI. Targeted delivery of progenitor cells for cartilage repair. J Orthop Res. 2004;22(4):735–41.

Devine SM, Cobbs C, Jennings M, Bartholomew A, Hoffman R. Mesenchymal stem cells distribute to a wide range of tissues following systemic infusion into nonhuman primates. Blood. 2003;101(8):2999–3001.

Di Rocco G, Tritarelli A, Toietta G, Gatto I, Iachininoto MG, Pagani F, Mangoni A, Straino S, Capogrossi MC. Spontaneous myogenic differentiation of Flk-1-positive cells from adult pancreas and other nonmuscle tissues. Am J Physiol Cell Physiol. 2008;294(2):C604–12.

Doucette T, Rao G, Yang Y, Gumin J, Shinojima N, Bekele BN, Qiao W, Zhang W, Lang FF. Mesenchymal stem cells display tumor-specific tropism in an RCAS/Ntv-a glioma model. Neoplasia. 2011;13(8):716–25.

Forbes GM, Sturm MJ, Leong RW, Sparrow MP, Segarajasingam D, Cummins AG, et al. A phase 2 study of allogeneic mesenchymal stromal cells for luminal Crohn's disease refractory to biologic therapy. Clin Gastroenterol Hepatol. 2014;12(1):64–71.

Friedenstein AY. Induction of bone tissue by translation epithelium. Clin Orthop Relat Res. 1968;59:21–37.

Furlani D, Li W, Pittermann E, Klopsch C, Wang L, Knopp A, Jungebluth P, Thedinga E, Havenstein C, Westien I, Ugurlucan M, Li RK, Ma N, Steinhoff G. A transformed cell population derived from cultured mesenchymal stem cells has no functional effect after transplantation into the injured heart. Cell Transplant. 2009;18(3):319–31.

Garcia S, Bernad A, Martin MC, Cigudosa JC, Garcia-Castro J, de la Fuente R. Pitfalls in spontaneous in vitro transformation of human mesenchymal stem cells. Exp Cell Res. 2010;316(9):1648–50.

Graakjaer J, Christensen R, Kolvraa S, Serakinci N. Mesenchymal stem cells with high telomerase expression do not actively restore their chromosome arm specific telomere length pattern after exposure to ionizing radiation. BMC Mol Biol. 2007;8:49.

Hahn WC, Weinberg RA. Modelling the molecular circuitry of cancer. Nat Rev Cancer. 2002;2(5):331–41.

Hayashi Y, Tsuji S, Tsujii M, Nishida T, Ishii S, Iijima H, Nakamura T, Eguchi H, Miyoshi E, Hayashi N, Kawano S. Topical transplantation of mesenchymal stem cells accelerates gastric ulcer healing in rats. Am J Physiol Gastrointest Liver Physiol. 2008;294(3):G778–86.

Hayflick L. The cell biology of human aging. N Engl J Med. 1976;295(23):1302–8.
Horwitz EM, Gordon PL, Koo WK, Marx JC, Neel MD, McNall RY, Muul L, Hofmann T. Isolated allogeneic bone marrow-derived mesenchymal cells engraft and stimulate growth in children with osteogenesis imperfecta: implications for cell therapy of bone. Proc Natl Acad Sci USA. 2002;99(13):8932–8937. ISSN 0027-8424.
Hwa Cho H, Bae YC, Jung JS. Role of toll-like receptors on human adipose derived stromal cells. Stem Cells. 2006;24(12):2744–52.
Ito M, Hiramatsu H, Kobayashi K, Suzue K, Kawahata M, Hioki K, UeyamaY, Koyanagi Y, Sugamura K, Tsuji K, Heike T, Nakahata T. NOD/SCID/gamma(c)(null) mouse: an excellent recipient mouse model for engraftment of human cells. Blood. 2002;100(9):3175–3182.
Javazon EH, Beggs KJ, Flake AW. Mesenchymal stem cells: paradoxes of passaging. Exp Hematol. 2004;32(5):414–25.
Jiang Y, Jahagirdar BN, Reinhardt RL, Schwartz RE, Keene CD, Ortiz-Gonzalez XR, Reyes M, Lenvik T, Lund T, Blackstad M, Du J, Alrich S, Lisberg A, Low WC, Largaespada DA, Verfaillie CM. Pluripotency of mesenchymal stem cells derived from adult marrow. Nature. 2002;418(6893):41–9.
Jiang XX, Zhang Y, Liu B, Zhang SX, Wu Y, Yu XD, Mao N. Human mesenchymal stem cells inhibit differentiation and function of monocyte-derived dendritic cells. Blood. 2005;105(10):4120–6.
Jones E, McGonagle D. Human bone marrow mesenchymal stem cells in vivo. Rheumatol (Oxford). 2008;47(2):126–31.
Kahler CM, Wechselberger J, Hilbe W, Gschwendtner A, Colleselli D, Niederegger H, Boneberg EM, Spizzo G, Wendel A, Gunsilius E, Patsch JR, Hamacher J. Peripheral infusion of rat bone marrow derived endothelial progenitor cells leads to homing in acute lung injury. Respir Res. 2007;8(50).
Kaplitt MG, Feigin A, Tang C, Fitzsimons HL, Mattis P, Lawlor PA, Bland RJ, Young D, Strybing K, Eidelberg D, During MJ. Safety and tolerability of gene therapy with an adenoassociated virus (AAV) borne GAD gene for Parkinson's disease: an open label, phase I trial. Lancet. 2007;369(9579):2097–105.
Kern S, Eitchler H, Stoeve J. Comparative analysis of mesenchymal stem cells from bone marrow, umbilical cord blood, or adipose tissue. Immunol Cell Biol. 2006;91(1):1294–301.
Kim NW, Piatyszek MA, Prowse KR, Harley CB, West MD, Ho PL, Coviello GM, Wright WE, Weinrich SL, Shay JW. Specific association of human telomerase activity with immortal cells and cancer. Science. 1994;266(5193):2011–5.
Kim JB, O'Hare MJ, Stein R. Models of breast cancer: is merging human and animal models the future? Breast Cancer Res. 2004;6(1):22–30.
Komata T, Kondo Y, Kanzawa T, Hirohata S, Koga S, Sumiyoshi H, Srinivasula SM, Barna BP, Germano IM, Takakura M, Inoue M, Alnemri ES, Shay JW, Kyo S, Kondo S. Treatment of malignant glioma cells with the transfer of constitutively active caspase-6 using the human telomerase catalytic subunit (human telomerase reverse transcriptase) gene promoter. Cancer Res. 2001;61(15):5796–802.
Kortesidis A, Zannettino A, Isenmann S, Shi S, Lapidot T, Gronthos S. Stromal-derived factor-1 promotes the growth, survival, and development ofhuman bone marrow stromal stem cells. Blood. 2005;105(10):3793–801.
Lazennec G, Jorgensen C. Concise review: adult multipotent stromal cells and cancer: risk or benefit? Stem Cells. 2008;26(6):1387–94.
Le Blanc K, Remberger M, Uzunel M, Mattsson J, Barkholt L, Ringden O. A comparison of nonmyeloablative and reduced-intensity conditioning for allogeneic stem-cell transplantation. Transplantation. 2004;78(7):1014–20.
Li H, Fu X, Ouyang Y, Cai C, Wang J, Sun T. Adult bone-marrow-derived mesenchymal stem cells contribute to wound healing of skin appendages. Cell Tissue Res. 2006;326(3):725–36.
Li YP, Paczesny S, Lauret E, Poirault S, Bordigoni P, Mekhloufi F, Hequet O, Bertrand Y, Ou-Yang JP, Stoltz JF, Miossec P, Eljaafari A. Human mesenchymal stem cells license adult

CD34+ hemopoietic progenitor cells to differentiate into regulatory dendritic cells through activation of the Notch pathway. J Immunol. 2008;180(3):1598–608.
Louis SA, Zapf R, Clarke E, Thomas TE, Sutherland HJ. A negative-selection strategy for depleting myeloma cells from patients' BM and/or leukapheresis blood. Cytotherapy. 2001;3(6):489–504.
Maguire AM, Simonelli F, Pierce EA, Pugh EN Jr, Mingozzi F, Bennicelli J, Banfi S, Marshall KA, Testa F, Surace EM, Rossi S, Lyubarsky A, Arruda VR, Konkle B, Stone E, Sun J, Jacobs J, Dell'Osso L, Hertle R, Ma JX, Redmond TM, Zhu X, Hauck B, Zelenaia O, Shindler KS, Maguire MG, Wright JF, Volpe NJ, McDonnell JW, Auricchio A, High KA, Bennett JIJ. Safety and efficacy of gene transfer for Leber's congenital amaurosis. N Engl J Med. 2008;358(21):2240–8.
Marinova-Mutafchieva L, Williams RO, Mauri C, Mason LJ, Walmsley MJ, Taylor PC, Feldmann M, Maini RN. A comparative study into the mechanisms of action of anti-tumor necrosis factor alpha, anti-CD4, and combined anti-tumor necrosis factor alpha/anti-CD4 treatment in early collagen-induced arthritis. Arthritis Rheum. 2000;43(3):638–44.
Martinez C, Hofmann TJ, Marino R, Dominici M, Horwitz EM. Human bone marrow mesenchymal stromal cells express the neural ganglioside GD2: a novel surface marker for the identification of MSCs. Blood. 2007;109(10):4245–8.
Matthay MA, Goolaerts A, Howard JP, Lee JW. Mesenchymal stem cells for acute lung injury: preclinical evidence. Crit Care Med. 2010;38(10):569–73.
Menon LG, Kelly K, Yang HW, Kim S-K, Black PM, Carroll RS. Human bone marrow-derived mesenchymal stromal cells expressing S-TRAIL as a cellular delivery vehicle for human glioma therapy. Stem Cells. 2009;27(9):2320–30.
Meyerson M. Role of telomerase in normal and cancer cells. J Clin Oncol. 2000;18(13):2626–34.
Meza-Zepeda LA, Noer A, Dahl JA, Micci F, Myklebost O, Collas P. High35 resolution analysis of genetic stability of human adipose tissue stem cells cultured to senescence. J Cell Mol Med. 2008;12(2):553–63.
Miao Z, Jin J, Chen L, Zhu J, Huang W, Zhao J, Qian H, Zhang X. Isolation of mesenchymal stem cells. Cell Biol Int. 2006;30(9):681–7.
Mittal S, Mifflin R, Powell DW. Cancer stem cells: the other face of Janus. Am J Med Sci. 2009;338(2):107–12.
Miura M, Miura Y, Padilla-Nash HM, Molinolo AA, Fu B, Patel V, Seo BM, Sonoyama W, Zheng JJ, Baker CC, Chen W, Ried T, Shi S. Accumulated chromosomal instability in murine bone marrow mesenchymal stem cells leads to malignant transformation. Stem Cells. 2006;24(4):1095–103.
Momin EN, Vela G, Zaidi HA, Quinones-Hinojosa A. The oncogenic potential of mesenchymal stem cells in the treatment of cancer: directions for future research. Curr Immunol Rev. 2010;6(2):137–48.
Murphy JM, Dixon K, Beck S, Fabian D, Feldman A, Barry F. Reduced chondrogenic and adipogenic activity of mesenchymal stem cells from patients with advanced osteoarthritis. Arthritis Rheum. 2002;46(3):704–13.
Nathwani AC, Tuddenham EG, Rangarajan S, Rosales C, McIntosh J, Linch DC, Chowdary P, Riddell A, Pie AJ, Harrington C, O'Beirne J, Smith K, Pasi J, Glader B, Rustagi P, Ng CY, Kay MA, Zhou J, Spence Y, Morton CL, Allay J, Coleman J, Sleep S, Cunningham JM, Srivastava D, Basner-Tschakarjan E, Mingozzi F, High KA, Gray JT, Reiss UM, Nienhuis AW, Davidoff AM. Adenovirus-associated virus vector-mediated gene transfer in hemophilia B. N Engl J Med. 2011;365(25):2357–65.
Nauta AJ, Kruisselbrink AB, Lurvink E, Willemze R, Fibbe WE. Mesenchymal stem cells inhibit generation and function of both CD34+-derived and monocyte-derived dendritic cells. J Immunol. 2006;177(4):2080–7.
Orlic D, Kajstura J, Chimenti S, Limana F, Jakoniuk I, Quaini F, Nadal-Ginard B, Bodine DM, Leri A, Anversa P. Mobilized bone marrow cells repair the infarcted heart, improving function and survival. Proc Natl Acad Sci USA. 2001;98(18):10344–10349.

Ortiz LA, Gambelli F, McBride C, Gaupp D, Baddoo M, Kaminski N, Phinney DG. Mesenchymal stem cell engraftment in lung is enhanced in response to 20 bleomycin exposure and ameliorates its fibrotic effects. Proc Natl Acad Sci USA. 2003;100(14):8407–8411.
Peault B, Rudnicki M, Torrente Y, Cossu G, Tremblay JP, Partridge T, Gussoni E, Kunkel LM, Huard J. Stem and progenitor cells in skeletal muscle development, maintenance, and therapy. Mol Ther. 2007;15(5):867–77.
Pevsner-Fischer M, Morad V, Cohen-Sfady M, Rousso-Noori L, Zanin-Zhorov A, Cohen S, Cohen IR, Zipori D. Toll-like receptors and their ligands control mesenchymal stem cell functions. Blood. 2007;109(4):1422–32.
Pittenger MF, Martin BJ. Mesenchymal stem cells and their potential as cardiac therapeutics. Circ Res. 2004;95(1):9–20.
Pittenger MF, Mackay AM, Beck SC, Jaiswal RK, Douglas R, Mosca JD, Moorman MA, Simonetti DW, Craig S, Marshak DR. Multilineage potential of adult human mesenchymal stem cells. Science. 1999;284(5411):143–7.
Placencio VR, Li X, Sherrill TP, Fritz G, Bhowmick NA. Bone marrow derived mesenchymal stem cells incorporate into the prostate during regrowth. PLoS ONE. 2010;5(9):e12920.
Plumb JA, Bilsland A, Kakani R, Zhao J, Glasspool RM, Knox RJ, Evans TR, Keith WN. Telomerase-specific suicide gene therapy vectors expressing bacterial nitroreductase sensitize human cancer cells to the pro-drug CB1954. Oncogene. 2001;20(53):7797–803.
Quirici N, Soligo D, Bossolasco P, Servida F, Lumini C, Deliliers GL. Isolation of bone marrow mesenchymal stem cells by anti-nerve growth factor receptor antibodies. Exp Hematol. 2002;30(7):783–91.
Raffaghello L, Bianchi G, Bertolotto M, Montecucco F, Busca A, Dallegri F, Ottonello L, Pistoia V. Human mesenchymal stem cells inhibit neutrophil apoptosis: a model for neutrophil preservation in the bone marrow niche. Stem Cells. 2008;26(1):151–62.
Ramasamy R, Fazekasova H, Lam EW, Soeiro I, Lombardi G, Dazzi F. Mesenchymal stem cells inhibit dendritic cell differentiation and function by preventing entry into the cell cycle. Transplantation. 2007;83(1):71–6.
Rangarajan A, Weinberg RA. Opinion: Comparative biology of mouse versus human cells: modelling human cancer in mice. Nat Rev Cancer. 2003;3(12):952–9.
Resnick IB, Barkats C, Shapira MY, Stepensky P, Bloom AI, Shimoni A, et al. Treatment of severe steroid resistant acute GVHD with mesenchymal stromal cells (MSC). Am J Blood Res. 2013;3(3):225–38.
Reyes M, Lund T, Lenvik T, Aguiar D, Koodie L, Verfaillie CM. Purification and ex vivo expansion of postnatal human marrow mesodermal progenitor cells. Blood. 2001;98(9):2615–25.
Ringden O, Uzunel M, Rasmusson I, Remberger M, Sundberg B, Lonnies H, et al. Mesenchymal stem cells for treatment of therapy-resistant graft-versus-host disease. Transplantation. 2006;81(10):1390–7.
Rosen JM, Jordan CT. The increasing complexity of the cancer stem cell paradigm. Science. 2009;324(5935):1670–3.
Rubio D, Garcia-Castro J, Martin MC, de la Fuente R, Cigudosa JC, Lloyd AC, Bernad A. Spontaneous human adult stem cell transformation. Cancer Res. 2005;65(8):3035–9.
Rubio D, Garcia S, Paz MF, De la Cueva T, Lopez-Fernandez LA, Lloyd AC, Garcia-Castro J, Bernad A. Molecular characterization of spontaneous mesenchymal stem cell transformation. PLoS One. 2008;3(1):e1398.
Sarosi G, Brown G, Jaiswal K, Feagins LA, Lee E, Crook TW, Souza RF, Zou YS, Shay JW, Spechler SJ. Bone marrow progenitor cells contribute to esophageal regeneration and metaplasia in a rat model of Barrett's esophagus. Dis Esophagus. 2008;21(1):43–50.
Seo SH, Kim KS, Park SH, Suh YS, Kim SJ, Jeun SS, Sung YC. The effects of mesenchymal stem cells injected via different routes on modified IL-12-mediated antitumor activity. Gene Ther. 2011;18(5):488–95.
Serakinci N, Fahrioglu U. Cancer stem cells—the cutting edge;2011. ISBN 978-953-307-580-8, edited by Stanley Shostak.

Serakinci N, Guldberg P, Burns JS, Abdallah B, Schrodder H, Jensen T, Kassem M. Adult human mesenchymal stem cell as a target for neoplastic transformation. Oncogene. 2004;23(29):5095–8.

Serakinci N, Christensen R, Graakjaer J, Cairney CJ, Keith WN, Alsner J, Saretzki G, Kolvraa S. Ectopically hTERT expressing adult human mesenchymal stem cells are less radiosensitive than their telomerase negative counterpart. Exp Cell Res. 2007;313(5):1056–67.

Serakinci N, Graakjaer J, Kolvraa S. Telomere stability and telomerase in mesenchymal stem cells. Biochimie. 2008;90(1):33–40.

Serakinci N, Christensen R, Fahrioglu U, Sorensen FB, Dagnæs-Hansen F, Hajek M, Jensen TH, Kolvraa S, Keith NW. Mesenchymal stem cells as therapeutic delivery vehicles targeting tumor stroma. Cancer Biother Radiopharm. 2011;26(6):767–73.

Serakinci N, Fahrioglu U, Christensen R. Mesenchymal stem cells, cancer challenges and new directions. Eur J Cancer. 2014;20(8):1522–30.

Serikov VB, Popov B, Mikhailov VM, Gupta N, Matthay MA. Evidence of temporary airway epithelial repopulation and rare clonal formation by BM-derived cells following naphthalene injury in mice. Anat Rec (Hoboken). 2007;290(9):1033–1045.

Shay JW, Pereira-Smith OM, Wright WE. A role for both RB and p53 in the regulation of human cellular senescence. Exp Cell Res. 1991;196(1):33–9.

Shi S, Gronthos S. Perivascular niche of postnatal mesenchymal stem cells in human bone marrow and dental pulp. J Bone Miner Res. 2003;18(4):696–704.

Simmons PJ, Torok-Storb B. Identification of stromal cell precursors in human bone marrow by a novel monoclonal antibody, STRO-1. Blood. 1991;78(1):55–62.

Sordi V, Malosio ML, Marchesi F, Mercalli A, Melzi R, Giordano T, Belmonte N, Ferrari G, Leone BE, Bertuzzi F, Zerbini G, Allavena P, Bonifacio E, Piemonti L. Bone marrow mesenchymal stem cells express a restricted set of functionally active chemokine receptors capable of promoting migration to pancreatic islets. Blood. 2005;106(2):419–27.

Studeny M, Marini FC, Champlin RE, Zompetta C, Fidler IJ, Andreeff M. Bone marrow-derived mesenchymal stem cells as vehicles for interferon-beta delivery into tumors. Cancer Res. 2002;62(13):3603–3608.

Studeny M, Marini FC, Dembinski JL, Zompetta C, Cabreira-Hansen M, Bekele BN, Champlin RE, Andreeff M. Mesenchymal stem cells: potential precursors for tumor stroma and targeted-delivery vehicles for anticancer agents. J Natl Cancer Inst. 2004;96(21):1593–603.

Sun XL, Xu ZM, Ke YQ, Hu CC, Wang SY, Ling GQ, Yan ZJ, Liu YJ, Song ZH, Jiang XD, Xu RX. Molecular targeting of malignant glioma cells with an EphA2-specific immunotoxin delivered by human bone marrow-derived mesenchymal stem cells. Cancer Lett. 2011;312(2):168–77.

Suzuki M, McHugh J, Tork C, Shelley B, Hayes A, Bellantuono I, Aebischer P, Svendsen CN. Direct muscle delivery of GDNF with human mesenchymal stem cells improves motor neuron survival and function in a rat model of familial ALS. Mol Ther. 2008;16(12):2002–10.

Teng MA. Mesenchymal stem cells: from bench to bedside. World J Stem Cells. 2010;2(2):13–7.

Terskikh AV, Bryant PJ, Schwartz PH. Mammalian stem cells. Pediatr Res. 2006;59(4 Pt 2):13R–20R.

Tolar J, Nauta AJ, Osborn MJ, Panoskaltsis Mortari A, McElmurry RT, Bell S, Xia L, Zhou N, Riddle M, Schroeder TM, Westendorf JJ, McIvor RS, Hogendoorn PC, Szuhai K, Oseth L, Hirsch B, Yant SR, Kay MA, Peister A, Prockop DJ, Fibbe WE, Blazar BR. Sarcoma derived from cultured mesenchymal stem cells. Stem Cells. 2007;25(2):371–9.

Tomchuck SL, Zwezdaryk KJ, Coffelt SB, Waterman RS, Danka ES, Scandurro AB. Toll-like receptors on human mesenchymal stem cells drive their migration and immunomodulating responses. Stem Cells. 2008;26(1):99–107.

Torsvik A, Rosland GV, Svendsen A, Molven A, Immervoll H, McCormack E, et al. Spontaneous malignant transformation of human mesenchymal stem cells reflects cross-contamination: putting the research field on track—letter. Cancer Res. 2010;70(15):6393–6.

Torsvik A, Rosland GV, Bjerkvig R. Spontaneous transformation of stem cells in vitro and the issue of cross-contamination. Int J Biol Sci. 2012;8(7):1051–2.

Tsai MS, Lee JL, Chang YJ, Hwang SM. Isolation of human multipotent mesenchymal stem cells from second-trimester amniotic fluid using a novel two-stage culture protocol. Hum Reprod. 2004;19(6):1450–6.

Tse WT, Pendleton JD, Beyer WM, Egalka MC, Guinan EC. Suppression of allogeneic T-cell proliferation by human marrow stromal cells: implications in transplantation. Transplantation. 2003;75(3):389–97.

Uccelli A, Moretta L, Pistoia V. Mesenchymal stem cells in health and disease. Nat Rev Immunol. 2008;8(9):726–36.

Veryasov VN, Savilova AM, Buyanovskaya OA, Chulkina MM, Pavlovich SV, Sukhikh GT. Isolation of mesenchymal stromal cells from extraembryonic tissues and their characteristics. Bull Exp Biol Med. 2014;157(1):119–24.

Von Luttichau I, Notohamiprodjo M, Wechselberger A, Peters C, Henger A, Seliger C, Djafarzadeh R, Huss R, Nelson PJ. Human adult CD34- progenitor cells functionally express the chemokine receptors CCR1, CCR4, CCR7, CXCR5, and CCR10 but not CXCR4. Stem Cells Dev. 2005;14(3):329–36.

Wang Y, Huso DL, Harrington J, Kellner J, Jeong DK, Turney J, McNiece IK. Outgrowth of a transformed cell population derived from normal human BM mesenchymal stem cell culture. Cytotherapy. 2005;7(6):509–19.

Weiss ML, Troyer DL. Stem cells in the umbilical cord. Stem Cell Rev. 2006;2(2):155–62.

Wynn RF, Hart CA, Corradi-Perini C, O'Neill L, Evans CA, Wraith JE, Fairbairn LJ, Bellantuono I. A small proportion of mesenchymal stem cells strongly expresses functionally active CXCR4 receptor capable of promoting migration to bone marrow. Blood. 2004;104(9):2643–5.

Sullivan NJ, Hall B.M. Mesenchymal stem cells in tumor stroma. In: Bagley G, Teicher BA, editors. Cancer drug discovery and development: stem cells and cancer. Humana Press. pp. 27–34.

Yang C, Lei D, Ouyang W, Ren J, Li H, Hu J, Huang S. Conditioned media from human adipose tissue-derived mesenchymal stem cells and umbilical cord-derived mesenchymal stem cells efficiently induced the apoptosis and differentiation in human glioma cell lines in vitro. Biomed Res Int. 2014;109389.

Yu B, Zhang X, Li X. Exosomes derived from mesenchymal stem cells. Int J Mol Sci. 2014;15(3):4142–57. doi:10.3390/ijms15034142.

Yusa K, Rashid ST, Strick-Marchand H, Varela I, Liu PQ, Paschon DE, Miranda E, Ordóñez A, Hannan NR, Rouhani FJ, Darche S, Alexander G, Marciniak SJ, Fusaki N, Hasegawa M, Holmes MC, Di Santo JP, Lomas DA, Bradley A, Vallier L. Targeted gene correction of alpha1-antitrypsin deficiency in induced pluripotent stem cells. Nature. 2011;478(7369):391–4.

Zappia E, Casazza S, Pedemonte E, Benvenuto F, Bonanni I, Gerdoni E, Giunti D, Ceravolo A, Cazzanti F, Frassoni F, Mancardi G, Uccelli A. Mesenchymal stem cells ameliorate experimental autoimmune encephalomyelitis inducing T-cell anergy. Blood. 2005;106(5):1755–61.

Zhou YF, Bosch-Marce M, Okuyama H, Krishnamachary B, Kimura H, Zhang L, Huso DL, Semenza GL. Spontaneous transformation of cultured mouse bone marrow-derived stromal cells. Cancer Res. 2006;66(22):10849–54.

Zuk PA, Zhu M, Mizuno H, Huang J, Futrell JW, Katz AJ, Benhaim P, Lorenz HP, Hedrick MH. Multilineage cells from human adipose tissue: implications for cell-based therapies. Tissue Eng. 2001;7(2):211–28.

Zvaifler NY, Marinova-Mutafchieva L, Adams G, Edwards CJ, Moss J, Burger JA, Maini RN. Mesencyhmal precursor cells in the blood of normal individuals. Arthritis Res. 2000;2(6):477–88.

Chapter 15
Stem Cell Therapy for GVHD

Nayoun Kim and Seok-Goo Cho

Abstract Graft-versus-host disease (GVHD) is a major complication associated with morbidity and mortality following allogeneic hematopoietic stem cell transplantation (HSCT). GVHD can be fatal, as patients frequently fail to respond to conventional immunosuppressants and no established second-line therapy is available. Stem cells have emerged as a therapeutic approach in a range of medical fields, including regenerative medicine, cancer, autoimmune diseases, and inflammatory diseases, because of their unique properties. In particular, mesenchymal stem cells (MSCs) have created increasing interest as a therapeutic agent for various immune-mediated diseases including GVHD. MSCs demonstrate therapeutic potential by exerting potent immunosuppressive effects through interactions between lymphocytes associated with both the innate and adaptive immune system. Preclinical experiments using MSCs for the treatment of GVHD have provided critical insights into the mechanism of action of MSCs and evidence for clinical translation. In the clinical setting, Ringden's pilot study using MSCs to treat a steroid-refractory GVHD pediatric patient was a milestone in the field of MSC therapy, which leads to numerous worldwide studies on steroid-refractory GVHD, de novo GVHD, chronic GVHD treatment, and GVHD prevention. In this chapter, we focus on the recent advances of MSCs for the treatment of GVHD in both preclinical and clinical settings. We discuss the current limitations and highlight the future considerations that should be made when using stem cell therapy for GVHD.

N. Kim · S.-G. Cho
Institute for Translational Research and Molecular Imaging,
The Catholic University of Korea College of Medicine, Seoul, Korea

N. Kim · S.-G. Cho
Laboratory of Immune Regulation, Convergent Research Consortium for Immunologic Disease (CRCID), The Catholic University of Korea College of Medicine, Seoul, Korea

S.-G. Cho (✉)
Department of Hematology, Catholic Blood and Marrow Transplantation Center,
Seoul St. Mary's Hospital, The Catholic University of Korea College of Medicine,
#505, Banpo-Dong, Seocho-Ku, Seoul 137-040, Korea
e-mail: chosg@catholic.ac.kr

R.C. Zhao (ed.), *Stem Cells: Basics and Clinical Translation*, Translational Medicine Research 1, DOI 10.1007/978-94-017-7273-0_15

Keywords Stem cell · Graft-versus-host disease · Mesenchymal stem cell

15.1 Introduction

Allogeneic hematopoietic stem cell transplantation (HSCT) is an effective treatment approach for a variety of hematologic and genetic disorders. Without appropriate intervention, almost all transplant recipients develop a severe inflammatory condition known as graft-versus-host disease (GHVD). Prior to HSCT, patients undergo a conditioning regimen in order to deplete remaining host stem cells and create space for donor stem cell engraftment. However, the toxicity of the regimen can cause tissue damage and inflammation releasing various pro-inflammatory cytokines that activate host antigen-presenting cells (APCs). Recipient antigens presented by host APCs, then, activate the injected donor T cells, in which a strong cytokine storm is initiated to further promote antigen presentation and activate effector T cells. Finally, the activated effector T cells attack recipient organs and tissues, such as the skin, liver, and gastrointestinal (GI) tract. The clinical outcomes of severe GVHD patients are generally poor with high morbidity and mortality rate affecting up to 40–60 % of HSCT patients (Jagasia et al. 2012). While steroids and immunosuppressive drugs have been used to improve survival rates of GVHD patients, steroid-resistant and severe cases of GVHD are not easily reversed by current treatment approaches.

Recently, mesenchymal stem cells (MSCs) have emerged as an alternative to current pharmacologic drugs in various immunological diseases due to their potent immunosuppressive properties regulating both innate and adaptive immune responses. Since Ringden's pilot study using MSCs to treat a steroid-resistant patient (Le Blanc et al. 2004), MSCs have been widely studied in both preclinical and clinical GVHD settings. While studies have provided valuable information on MSC therapy for GVHD, controversy on clinical effectiveness still remains (Kim et al. 2013a). In this chapter, we review the developments of MSC therapy for GVHD in both preclinical and clinical settings. We discuss the controversies and current limitations of MSC therapy and suggest future considerations that should be made when using stem cell therapy for GVHD.

15.2 Immunosuppressive Properties of Mesenchymal Stem Cells (MSCs)

MSCs are self-renewing multipotent progenitor cells with multilineage potential to differentiate into other cell types of mesodermal origin. The International Society for Cellular Therapy established the minimal criteria for MSCs, which defines MSCs as cells adherent to plastic, cells that express cell-surface molecules CD105, CD73, and CD90 but not CD11b, CD79a, CD19 and human leukocyte antigen

(HLA)-DR, and cells with the ability to differentiate into osteoblasts, adipocytes, and chondroblasts under standard in vitro conditions (Dominici et al. 2006). In addition to multipotent stem cell properties, MSCs possess potent immunosuppressive and anti-inflammatory effects. These unique properties of MSCs have created growing interest in the use of MSCs as a novel therapeutic approach for various immunological and inflammatory diseases (Kim and Cho 2013; Kim et al. 2013b).

15.2.1 Innate Immunity

Macrophages. MSCs are known to interact with various lymphocytes of both the innate and adaptive immune system and play a regulatory role in immune modulation (Fig. 15.1). Within the innate immune system, phagocytic cells such as macrophages and neutrophils provide a first-line defense in response to an inflammatory response. In response to an inflammatory environment, MSCs are triggered to secrete chemokine ligands including CCL3, CXCL2, and CCL12, which attracts monocytes and macrophages into inflamed tissues (Chen et al. 2008). MSCs produce TNF-stimulated gene 6 (TSG6) which may attenuate the activation of recruited "classical" M1 macrophages, which possess pro-inflammatory and antimicrobial functions (Choi et al. 2011). Furthermore, the upregulation of cyclooxygenase 2 (COX2) and indoleamine 2,3-dioxygenase (IDO) in MSCs produces increased levels of prostaglandin E2 (PGE2) and kynurenine (KYN), respectively (Nemeth et al. 2009; Maggini et al. 2010). These soluble factors create a microenvironment for both monocytes and M1 macrophages to become alternatively activated M2 macrophages. M2 macrophages are characterized by a high expression of CD206, high levels of interleukin (IL)-10 and low levels of TNF and interferon (IFN)-γ production (Kim and Hematti 2009). M2 macrophages express lower levels of co-stimulatory molecules compared to M1 macrophages and show immunosuppressive activity by secreting high levels of IL-10, which prevents neutrophil invasion and effector T-cell stimulation while promoting the generation of regulatory T cells (Tregs).

Natural Killer Cells. Natural killer (NK) cells are important effector cells of the innate immune system where its function greatly depends on the production of pro-inflammatory cytokines. Furthermore, the modulation of NK cells is of great interest in the HSCT setting as donor NK cells, in addition to donor T cells, provide therapeutic benefit by eradicating tumor cells. However, the role of NK cells in the development and pathology of GVHD is unclear. Previous murine studies have showed that NK cells are able to suppress the development of GVHD by eliminating host APCs and activated donor T cells that are susceptible to NK-mediated lysis (Olson et al. 2010; Ruggeri et al. 2002).

MSCs are known to interact and inhibit both proliferation and cytotoxicity of freshly isolated NK cells. MSCs were also able to strongly impair IL-2- and IL-15-activated NK cells in a dose-dependent manner (Sotiropoulou et al. 2006).

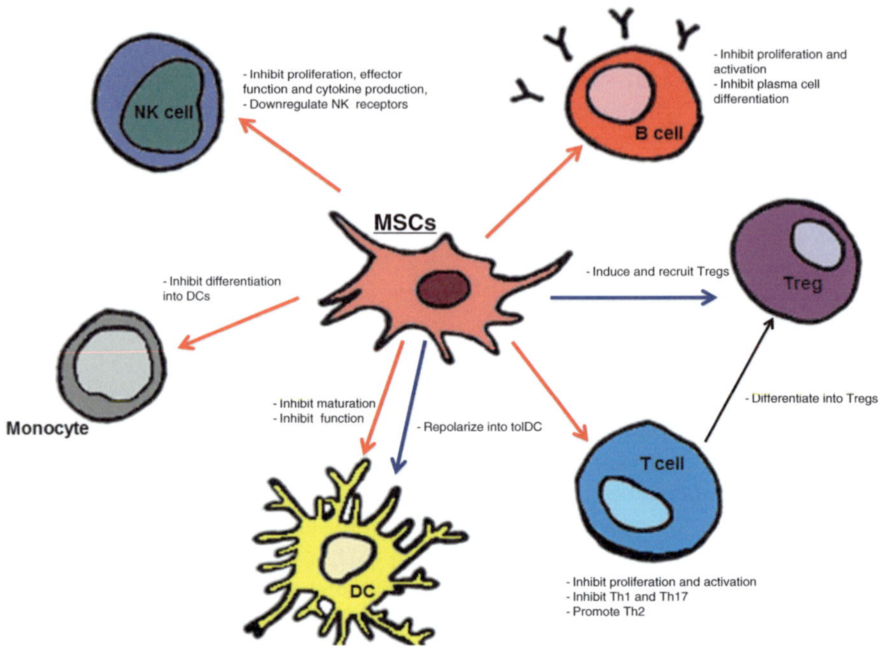

Fig. 15.1 MSCs interact with various lymphocytes of the immune system. *Red arrow* indicates inhibition. *Blue arrow* indicates promotion. *Black arrow* indicates differentiation. Abbreviations include as follows: *DC* dendritic cells; *M2* type 2 macrophages; *MSCs* mesenchymal stem cells; *NK* natural killer cell; *Th1* type 1 T helper cell; *Th2* type 2 T helper cell; *Th17* type 17 T helper cell; *tolDC* tolerogenic dendritic cells; *Treg* regulatory T cells

Furthermore, MSCs inhibit the expression of NK activating receptors such as NK44, NKp30, and NKG2D and other functional molecules that are involved in NK-cell activation and effector functions, including cytotoxic activity and cytokine production (Spaggiari et al. 2008). PGE2 and IDO are fundamental mediators of the MSC-mediated inhibition of NK-cell proliferation and effector function. Interestingly, while the inhibition of NK-cell cytotoxicity required cell-to-cell contact with MSCs, the inhibition of NK-cell proliferation did not (Sotiropoulou et al. 2006).

One limitation to the interaction between MSCs and NK cells is that NK cells are able to kill both autologous and allogeneic MSCs. MSCs are characterized by low levels of HLA class I molecules which makes them undetectable by alloreactive T cells, but susceptible to NK-cell-mediated lysis. Furthermore, MSCs express activating NK receptors including NKG2D ligands, PVR, and DNAM-1 ligands (Spaggiari et al. 2006). During the inflammatory responses, pro-inflammatory cytokines, such as IFN-γ, upregulate the surface expression of HLA class I and II on MSCs (Le Blanc et al. 2003a). However, it is important to note that only activated NK cells with high lytic potential through upregulated expression of NKp30 and NKG2D are capable of killing MSCs. Resting NK cells do not show cytotoxicity against MSCs, even at high effector to target ratio (Spaggiari et al. 2006).

Dendritic Cells. Dendritic cells (DCs) play a critical role in initiating the adaptive immune response through the priming of naïve T cells. In the GVHD setting, host DCs are critical in initiating GVHD by priming donor T cells with host antigens; however, there is a potential role of donor APCs in augmenting the graft-versus-host response as well. MSCs are able to inhibit the generation of DCs from monocytes and CD34+ precursors (Zappia et al. 2005; Nauta et al. 2006). In the presence of MSCs, monocytes did not acquire the necessary expression of CD1a and loss of the monocyte marker CD14 (Zappia et al. 2005). When the monocytes were stimulated for full DC maturation with lipopolysaccharide (LPS), the monocytes only expressed low levels of co-stimulatory molecules CD80, CD83, and CD86 suggesting that MSCs block the key features of DCs (Nauta et al. 2006; English et al. 2008). Thus, these cells that were not fully matured showed reduced ability to stimulate allogeneic T-cell proliferation in a mixed lymphocyte reaction.

In addition to the repolarization of macrophages mediated by MSCs, conventional antigen-presenting DCs can also be repolarized into anti-inflammatory tolerogenic DCs. DCs that are generated in the presence of MSCs produce high levels of anti-inflammatory cytokine IL-10 (Spaggiari et al. 2009; Zhang et al. 2009; Li et al. 2008a; Liu et al. 2012). Furthermore, tolerogenic DCs are often characterized as having a semi-mature phenotype, in which DC maturation markers are downregulated. As mentioned previously, MSCs promote DCs that express low levels of co-stimulatory molecules. Although the mechanisms on how MSCs induce tolerogenic DCs are unclear, the production of IL-6 (Nauta et al. 2006; English et al. 2008; Djouad et al. 2007) and PGE-2 (Spaggiari et al. 2009; Li et al. 2008a) with or without cell contact has been described to influence promotion of tolerogenic DCs.

15.2.2 Adaptive Immunity

T cells. Donor T cells, especially naïve T cells, play a crucial role in the pathogenesis of GVHD. In the clinical setting, many treatment approaches have aimed to target T-cell proliferation, cytokine production, and effector functions followed by activation. MSCs are able to suppress T-cell proliferation through the secretion of soluble factors including TGF-β (Keating 2008), hepatocyte growth factor (HGF) (Di Nicola et al. 2002), PGE2 (English et al. 2009), IDO (Meisel et al. 2004), nitric oxide (NO) (Ren et al. 2008), and hemoxygenase (HO) (Stagg and Galipeau 2013). In addition to production of soluble factors, MSCs can inhibit T-cell activation through cell-to-cell contact (Krampera et al. 2003).

Depending on the environment, CD4+ helper T cells (Th0) can differentiate into various subsets of effector T cells, including Th1, Th2, and Th17 subsets that have different cytokine profiles and functions. Th1 and Th17 cells produce pro-inflammatory cytokines, such as IFN-γ and IL-17, respectively, promoting GVHD, while Th2 cells produce anti-inflammatory cytokine IL-4. MSCs are able to modulate the T-cell response by orchestrating the balance between the pro-inflammatory and anti-inflammatory profiles. In an environment that consists of strong inflammatory components, such as GVHD, MSCs are able to shift the pro-inflammatory Th1 profile to an anti-inflammatory Th2 profile (Bai et al. 2009; Batten et al. 2006). MSCs also modulate Th17 cell differentiation mainly through a PGE2 mediated pathway (Duffy et al. 2011). MSCs are able to prevent the differentiation of Th17 cells from naïve Th0 cells and suppress the production of Th17 cytokines, including IL-17 and IL-22 (Ghannam et al. 2010).

Regulatory T cells. Regulatory T cells (Tregs) are a unique subpopulation of T helper cells that are characterized by the expression of forkhead box P3 (Foxp3) transcription factor and are specialized in suppressing immune responses. MSCs have been described previously to secrete soluble factors that repolarize lymphocytes toward an anti-inflammatory phenotype. Similarly, the coculture of MSCs with peripheral blood mononuclear cells (PBMCs) induces the differentiation of Foxp3+ Tregs through PGE2 and TGF-β (English et al. 2009; Maccario et al. 2005). Furthermore, purified MSCs-induced Tregs were functional with potent inhibitory effects against alloreactive T-cell proliferation in vitro (English et al. 2009; Maccario et al. 2005). Importantly, the ability of MSCs to induce and recruit Tregs to inflammatory sites has been observed in vivo in various immune-mediated models, including GVHD (Zappia et al. 2005; Joo et al. 2010; Madec et al. 2009; Choi et al. 2012; Patel et al. 2010).

B cells. While the effects of MSCs on B-cell functions remain contradictory, there is evidence that MSCs have close interactions with B cells. B cells play a critical role in the development of chronic GVHD (cGVHD), an autoimmune-like syndrome associated with autoreactive B cells and antibodies causing tissue fibrosis and antibody deposition in tissues. MSCs are able to inhibit B-cell proliferation through cell-to-cell contact (Schena et al. 2010) and by arrest of cell cycle G0/G1 (Corcione et al. 2006), but these effects are often independent of soluble factors (Schena et al. 2010). Furthermore, MSCs inhibit plasma cell differentiation induced by allostimulation (Comoli et al. 2008) and Ig production (Corcione et al. 2006; Rasmusson et al. 2007). However, studies have also suggested that while MSCs are able to suppress B cells activated by various stimuli, such as B-cell receptor activation, T-cell co-stimulation, cytokine stimulation, and toll-like receptor (TLR) activation, MSCs are unable to modulate naïve or memory B cells that do not require such signals (Traggiai et al. 2008).

15.3 First Clinical Trial Using Mesenchymal Stem Cells (MSCs) for Graft-Versus-Host Disease (GVHD)

Based on the preliminary studies that MSCs inhibit the proliferation of alloreactive T cells (Le Blanc et al. 2003b), Le Blanc et al. (2004) reported the first clinical trial using third-party haploidentical MSCs to treat severe GVHD after allogeneic stem cell transplantation. In the reported study, a 9-year-old patient with acute lymphoblastic leukemia received allogeneic HSCT from a haploidentical, unrelated, female donor. By day 11 after transplantation, the patient developed clinical symptoms of aGVHD such as maculopapular rash, diarrhea, and elevated bilirubin, and alanine aminotransferase concentrations. Despite various treatment approaches, including prednisone, psoralen, and ultraviolet-A light treatment, infliximab, daclizumab, and methylprednisolone, the patient was still unresponsive to treatment. By day 70 after transplantation, the patient developed steroid-refractory grade IV aGVHD. Patients with grade III–IV aGVHD are known to have a poor outcome where no standard treatment is available. Those with grade IV aGVHD have under 5 % long-term survival (Cahn et al. 2005).

The patient was enrolled in the study to receive third-party MSCs. MSCs express low levels of major compatibility complex and other co-stimulatory molecules, and thus act independently of MHC barriers between donor and recipient (Le Blanc et al. 2003b; Stagg et al. 2006). The patient's mother was selected as donor for MSC generation as she was readily available. MSCs were generated from bone marrow aspirate and cultured for three weeks. MSCs were confirmed for osteogenic, chondrogenic, and adipogenic differentiation as well as the expression of CD166, CD105, CD44, and CD29, and the absence of CD34 and CD56. Two million cells per kg of the patient's weight were intravenously infused on day 73 after transplantation.

There were no infusion-related toxicities observed. The patient showed improvements in clinical symptoms, such as decline in bilirubin concentration and frequency of diarrhea. During the treatment period, cyclosporine treatment was discontinued because the patient had remaining minimal residual disease in the bone marrow. The patient again showed mild GVHD symptoms and received a second infusion of MSCs (1 million cells per kg). After one week of the second infusion, he showed rapid recovery and by day 220, he was well and discharged from the hospital. Interestingly, after the first MSC infusion, female epithelial cells in the colon were detected by X and Y chromosome fluorescence in situ hybridization (FISH). These female epithelial cells may have derived from the infused MSCs of the patient's mother, as MSCs have been suggested to have healing effects on the damaged gut epithelium. However, the female epithelium may have also derived from the female donor cells following HSCT.

Ringden's study clearly demonstrates the potent immunosuppressive effects and therapeutic potential of MSCs for the treatment of GVHD patients. Furthermore, this study has provided a rationale for the use of MSCs in GVHD patients and encouraged numerous preclinical and clinical studies with MSCs for prevention and treatment of GVHD.

15.4 Preclinical Studies Using Mesenchymal Stem Cells (MSCs)

Many models have been used to investigate the potential of MSCs for the treatment of aGVHD (Application of MSCs in a preclinical cGVHD model has not yet been reported due to limited murine models.). However, unlike in vitro studies that have clearly demonstrated MSCs' immunosuppressive effects against allogeneic T-cell responses, in vivo studies have been more contradicting. Thus, many studies have focused on factors that may contribute to the therapeutic effects of MSCs in an aGVHD setting (Table 15.1).

Cell dose. MSCs are known to inhibit allogeneic T-cell responses in a dose-dependent manner in vitro; however, this dose-dependent effect has been less clear in vivo. While some studies have suggested that MSCs can effectively treat GVHD in a dose-dependent manner (Joo et al. 2010; Li et al. 2008b), others have shown that therapeutic effect could not be obtained at any dose (Badillo et al. 2008; Polchert et al. 2008; Sudres et al. 2006; Tisato et al. 2007). Li et al. (2008b), suggested that single doses of 2×10^5, 1×10^6 or 2×10^6 were able to delay the development of aGVHD and significantly prolong survival of mice. Similarly, Joo et al. (2010) demonstrated that single doses of either 1×10^6 or 2×10^6 could improve GVHD-related mortality. However, therapeutic effects were not observed at a lower dose of 5×10^5. Other studies that have used similar or higher doses of MSCs were not able to reproduce the clinical benefits. Sudres et al. (2006) used doses from 5×10^5 up to 4×10^6, but there was no effect on the incidence or severity of GVHD. A subsequent study investigated whether increasing the number of doses compared to single injection could improve treatment (Tisato et al. 2007). It is important to note that in this study, a xenogeneic GVHD model was used to test the efficacy of human umbilical cord blood (UCB)-derived MSCs. While a single dose of 3×10^6 MSCs at the time of human PBMC infusion could not alleviate GVHD symptoms, four doses of 3×10^6 MSCs given at weekly intervals improved survival and inhibited human PBMC proliferation. Surprisingly, when the same four doses of MSCs were given five weeks after human PBMC infusion, the therapeutic potential of MSC was lost. In a similar study, UCB-derived MSCs given in a single dose at the onset of GVHD could not treat GHVD, but repeated injections of MSCs could significantly increase the survival rate of mice. In contrast to the previous study, multiple doses of MSCs at the time of HSCT could not prevent GVHD (Jang et al. 2014).

Timing. In addition to cell dose, other studies began to propose that the timing of treatment might play a more critical role in MSC-mediated immune suppression. In the previously mentioned study (Tisato et al. 2007), multiple doses were more effective than a single dose at the time of allogeneic HSCT. However, the same multiple-dose treatment showed no clinical benefit five weeks after HSCT, when GVHD symptoms were evident. Whether MSCs are more suppressive when co-transplanted during HSCT is controversial. Chung et al. (2004) reported that even a single dose of MSCs during allogeneic HSCT could prevent aGVHD by immune modulation, while others demonstrated that MSCs infused during HSCT have no efficacy against aGVHD (Polchert et al. 2008; Sudres et al. 2006; Jang et al. 2014; Bruck et al. 2013). Polchert et al. (2008) investigated the effect of timing of MSC administration. MSCs were injected on day 0, 2, 20, or 30 following GVHD induction. While MSCs increased the histopathology and survival of mice when administered on days 2 and 20 after HSCT, there were no significant effects seen when administered on day 0 or 30. Thus, MSCs were ineffective

Table 15.1 Preclinical studies of MSC for GVHD

Author	Host	Donor	MSC source	Dose	Time	Results
Cell dose						
Joo et al. (2010)	BALB/c	C3H/he	Donor BM	5×10^5, 1×10^6, 2×10^6	D + 0	– Dose-dependent therapeutic effect
Li et al. (2008b)	CB6F1	C57BL/6	Donor BM	2×10^4, 2×10^5, 1×10^6, 2×10^6	D + 3	
Tisato et al. (2007)	NOD/SCID	Human PBMC	Human UCB	3×10^6 in multiple doses	D + 0, D + 0, 7, 14, 21, D + 35, 42, 49, 56	– Single dose was ineffective, but multiple doses in weekly intervals showed improvement
Jang et al. (2014)	NOD/SCID	Human PBMC	Human UCB	5×10^5	D + 0, D + 0, 3, 6, D + 0, 7, 14, D + 18, D + 18, 21, 24, D + 18, 25, 32	
Badillo et al. (2008)	CB6F1	C57BL/6	Donor BM	5×10^4, 1×10^5, 1×10^6	D + 0, D + 2, D + 0, 7, 14, D + 10 and D + 21	– No dose-dependent therapeutic effect
Polchert et al. (2008)	C57BL/6	BALB/C	Donor BM	1×10^5, 5×10^5	D + 0, D + 2, D + 20, D + 30	
Sudres et al. (2006)	BALB/c	C57/BL6	Donor BM	5×10^5, 3×10^6, 4×10^6	D + 0	
Timing						
Tisato et al. (2007)	NOD/SCID	Human PBMC	Human UCB	3×10^6 in weekly intervals	D + 0	– Multiple doses effective at the time of HSCT
Chung et al. (2004)	BALB/c	C3H/he	Donor BM	1×10^5	D + 0	– Effective at the time of HSCT
Sudres et al. (2006)	Previously described					– No therapeutic effect at the time of HSCT
Jang et al. (2014)	Previously described					
Bruck et al. (2013)	NOD/SCID or NSG	Human PBMC	Human BM	2×10^6 (I.P. injection)	D + 0, D + 0, 7, 14	
Li et al. (2008b)	Previously described					– Even a single dose was effective when infused within one week of HSCT
Polchert et al. (2008)	Previously described					

(continued)

Table 15.1 (continued)

Author	Host	Donor	MSC source	Dose	Time	Results
Pre-activation						
Polchert et al. (2008)	Previously described					– IFN-γ KO mice-induced GVHD did not respond to MSCs
						– Pretreatment of MSCs with IFN-γ was effective at time of HSCT
Trafficking						
Joo et al. (2010)	Previously described					– MSCs colocalize with donor HSCs in vivo
Sudres et al. (2006)	Previously described					– MSCs did not show therapeutic effects, but engraftment was detected
Gene transduction						
Li et al. (2014)	C57BL/6	BALB/c	Human BM and UCB (CCR7 transduced)	2×10^6	D + 0	– CCR7 transduced MSC migrated to secondary lymphoid organs
Min et al. (2007)	B6D2F1	C57BL/6	Donor BM (IL-10 transduced)	1×10^6, 2×10^6	D + 1, D + 1, 3, 5	– IL-10 transduced MSCs suppressed the development of GVHD

MSCs were injected intravenously unless otherwise noted

Abbreviations BM bone marrow; *HSCT* hematopoietic stem cell transplantation; *I.P.* intraperitoneal; *IFN-γ* interferon-gamma; *KO* knockout; *MSC* mesenchymal stem cells; *NOD/SCID* nonobese diabetic severe combined immunodeficiency mice; *NSG* NOD.Cg-$Prkdc^{scid}$ $Il2rg^{tm1Wjl}$/SzJ (also known as NOD/SCID gamma mice); *PBMC* peripheral blood mononuclear cells; *UCB* umbilical cord blood

before GVHD development or when GVHD was too severe. This study attributed the inconsistency of MSC treatment to the difference in IFN-γ levels at different time points of GVHD development. Following transplantation, circulating IFN-γ levels were initially absent on the day of transplantation followed by a spike of IFN-r by day 7. Then, the IFN-γ concentration gradually dropped from this point and absent by day 30. A single infusion at the appropriate time, usually within one week of HSCT, was effective (Li et al. 2008b; Polchert et al. 2008).

Pre-activation. MSCs are not constitutively inhibitory, but require acute inflammatory Th1 cytokines to acquire immunosuppressive effects (Marigo and Dazzi 2011). Th1 cytokine, IFN-γ, has been described to be a pivotal component in the immunomodulatory activity of MSCs in vitro (Krampera et al. 2006). Treatment of MSCs with IFN-γ induced increased secretion of chemokine receptor ligands ICAM-1, CXCL-10, and CCL-8 (Krampera et al. 2006; Dazzi and Marelli-Berg 2008), as well as increased production of immunosuppressive IDO (Hoogduijn et al. 2010). Therefore, IFN-γ treatment can enhance MSC migration to inflammatory sites by the induction of chemotactic factors, and by increased production of immunosuppressive soluble factors. The role of IFN-γ in MSC-mediated immune suppression was also demonstrated in vivo (Polchert et al. 2008). When IFN-γ knockout mice were used as effectors of GVHD, MSCs were unable to improve the survival regardless of the time of treatment. Furthermore, MSCs pretreated with IFN-γ could prevent GVHD even when administered at the time of transplantation (Polchert et al. 2008).

Recently, TLR stimulation has been proposed to induce immune-modulatory capacity. TLRs are receptors that recognize pathogen-associated molecular patterns and promote the activation of both innate and adaptive immune cells. Human MSCs express high levels of TLR3 and TLR4 (Liotta et al. 2008). Stimulation of TLR3 induces MSCs to acquire an immunosuppressive phenotype that produce increased levels of IL-10, IDO, and PGE2 (Waterman et al. 2010). However, the effectiveness of TLR3-prestimulated MSCs has not been investigated in a GVHD model.

Trafficking to inflammatory sites. MSC trafficking to inflammatory sites may increase tissue repair at the site of injury by directly providing soluble factors, transdifferentiation, and cell fusion. In a bioimaging study, donor splenocytes used to induce GVHD expressed green fluorescent protein (GFP) and MSCs used for treatment were generated from donor mice expressing red fluorescent protein (RFP) (Joo et al. 2011). MSCs were infused on the same day of transplantation and were monitored for the biodistribution of MSCs. GFP expression levels were mainly detected in the gastrointestinal (GI) tract, liver, skin, and lymph nodes after HSCT, indicating major clinical targets of GVHD. After the administration of MSCs, the RFP and GFP signals colocalized at the GVHD target sites suggesting that MSCs can home to sites of aGVHD. However, even when MSCs did not show therapeutic effects, the engraftment of MSCs was detected in recipient mice (Sudres et al. 2006). Therefore, localization of MSCs at inflammatory sites of GVHD may not directly correlate with their immunosuppressive potential, especially their ability to inhibit alloreactive T-cell proliferation. Recently, it was suggested that improving the migration of MSCs to secondary lymphoid organs by transducing the chemokine receptor 7 gene could intensify MSCs' in vivo immunomodulatory effects (Li et al. 2014).

15.5 Clinical Studies Using Mesenchymal Stem Cells (MSCs)

Treatment of aGVHD. Since Ringden's pilot study in 2004 (Le Blanc et al. 2004), MSCs have been extensively studied in steroid-refractory GVHD. In Ringden's subsequent study (Ringden et al. 2006), MSCs were administered in eight patients with grades III–IV steroid-refractory GVHD (Table 15.2) and 75 % (six out of eight patients) showed complete remission. In other words, patients treated with MSCs had a significantly better overall survival rate compared to those who did not receive MSC therapy. Furthermore, in this study, MSCs were derived from HLA-haploidentical or HLA-identical family donors and unrelated mismatched donors. No side effects were observed from unmatched MSCs, and these results reduced concerns regarding HLA disparity between the MSC donor and recipient. The European Group for Blood and Marrow Transplantation obtained similar results in a multicenter phase II study using HLA-identical, HLA-haploidentical, or third-party mismatched bone marrow-derived MSCs (Le Blanc et al. 2008). A total of 55 patients, including twenty-five patients with steroid-refractory GVHD were treated with MSCs from various donor sources and had a significantly reduced level or transplantation-related mortality.

Based on the safety and efficacy of third-party MSCs, an FDA-approved commercialized MSC product (Prochymal) was developed by Osiris Therapeutics, Inc. to promote MSCs as off-the-shelf products. Prochymal is derived from the bone marrow of healthy donors and are evaluated in other immune-mediated diseases, including Crohn's disease, and acute myocardial infarction.[1] In aGVHD, Prochymal was first evaluated in a randomized prospective study to treat GVHD directly after diagnosis (Kebriaei et al. 2009). While most clinical studies of aGVHD involved steroid-refractory patients who failed first-line treatment, this study aimed to treat de novo GVHD patients with a combination of MSCs and corticosteroids directly after diagnosis. 29 out of 32 patients (94 %) showed an initial response without any infusional toxicities or ectopic tissue formation. Prochymal was also effective in treating pediatric patients with severe steroid-refractory aGVHD (Prasad et al. 2011). Treatment of MSCs was able to significantly increase the survival of patients. Recently, Prochymal was evaluated in a phase III double-blinded, placebo-controlled clinical trial based on their encouraging results (Martin et al. 2010). Surprisingly, there was no clinical benefit seen in the MSC-treated group compared to the placebo group in either steroid-refractory

[1]Osiris Therapeutics Inc. Products. Osiris Therapeutics Inc.

or de novo GVHD patients. However, only selected patients with severe liver involvement showed improved response rates.

In most clinical trials, MSCs were derived from bone marrow; however, there are clinical studies on MSCs derived from other sources. Adipose tissue-derived MSCs were evaluated for safety and efficacy as salvage therapy for steroid-refractory aGVHD (Fang et al. 2006). Furthermore, umbilical cord blood-derived MSCs dramatically improved severe grade IV aGVHD (Wu et al. 2011). These studies showed promising results that MSCs derived from a source other than bone marrow could be used. However, direct comparison between MSCs generated from different sources has not been investigated. Future studies may shed light on the different clinical utility of various donor sources of MSCs. In addition, MSCs cultivated in platelet-lysate-containing medium have been investigated in steroid-refractory GVHD patients (Lucchini et al. 2010; von Bonin et al. 2009). Platelet lysates were used to replace fetal bovine serum (FBS) as it was previously reported in a study that anti-FBS antibodies were detected in patients that received MSCs cultivated in FBS containing medium (Sundin et al. 2007). However, the clinical significance of anti-FBS antibodies is unclear. Overall, donor source and cultivations need to define in controlled studies.

Treatment of cGVHD. The therapeutic effects of MSCs in cGVHD patients have been less clear. In one study, bone marrow-derived MSCs were used to treat patients with sclerodermatous cGVHD and all four patients showed significant improvement from MSC therapy (Zhou et al. 2010). However, most cGVHD-related studies suggest MSCs to be less effective than in aGVHD (Lucchini et al. 2010; Muller et al. 2008; Weng et al. 2010). While 14 of 19 patients (73.7 %) showed initial response, the majority of patients showed either a partial or mixed response, suggesting that MSCs are not able to fully function in the cGVHD setting (Weng et al. 2010). The number of cGVHD studies that have investigated MSC treatment is still not enough to undermine the efficacy of MSCs in cGVHD. Additional studies are needed to confirm effectiveness of MSCs and to address their limitations in the chronic setting.

Prophylaxis of GVHD. Although many preclinical studies have suggested that MSCs are not effective for GVHD prevention, clinical trials have co-transplanted MSCs with hematopoietic stem cells (HSCs) to prevent GVHD development and to facilitate engraftment (Baron et al. 2010; Kuzmina et al. 2012; Lazarus et al. 2005; Ning et al. 2008). In all studies, MSCs were well tolerated and did not show any infusion-related adverse events and other late-term complications. Culture-expanded third-party MSCs were co-transplanted with HLA-mismatched HSCs (Baron et al. 2010). The 100-day cumulative incidence of aGVHD was 45 and 56 %, and the 1-year incidence of death from GVHD was 10 and 31 % in the MSC-treated group and historic control group, respectively. MSCs were also co-transplanted with HLA-matched HSCs (Ning et al. 2008). The co-transplantation of MSCs reduced the development of grades II–IV aGVHD. However, the outcomes were not statistically significant due to the small number of subjects. In a more recent study, 37 patients were randomly divided into two groups to receive standard GVHD prophylaxis alone or GVHD prophylaxis combined with MSCs

Table 15.2 Clinical studies of MSC therapy for GVHD

Author	Phase	Patient	Grade	MSC source	Results	Comments
Acute GVHD						
Le Blanc et al. (2004)	I	1	IV	BM	CR (100 %)	– Pilot study to use third-party MSCs for GVHD
Ringden et al. (2006)	I	8	III–IV	BM	6 CR (75 %)	– HLA-haploidentical, HLA-identical, and unrelated mismatched MSCs for steroid-refractory GVHD
Le Blanc et al. (2008)	II	55	II–IV	BM	30 CR (54 %) 9 PR (16 %)	– Largest multicenter trial for steroid-refractory GVHD using HLA-haploidentical, HLA-identical, and unrelated mismatched MSCs – Pediatric patients responded better than adults
Kebriaei et al. (2009)	II	32	II–IV	BM (Prochymal)	24 CR (77 %) 5 PR (16 %)	– First study to use commercialized MSC products in de novo GVHD
Prasad et al. (2011)	I	12	III–IV	BM (Prochymal)	7 CR (58 %) 2 PR (17 %)	– Commercialized MSC products used for severe steroid-refractory GVHD in pediatric patients
Martin et al. (2010)	III			BM (Prochymal)		– Double-blinded, placebo-controlled trial: no clinical benefit in MSC-treated group compared to control group
Fang et al. (2006)	I	6	III–IV	Adipose	5 CR (83 %)	– Adipose tissue-derived MSCs were used as salvage therapy for steroid-refractory aGVHD
Wu et al. (2011)	I	2	IV	UCB	2 CR (100 %)	– UCB-derived MSCs were used to dramatically improve severe aGVHD
Lucchini et al. (2010)	I	11	I–IV	BM (platelet-lysate medium)	2 CR (23.8 %) 5 PR (47.6 %)	– Platelet lysates were used in the MSC medium to replace FBS – MSC generated in platelet-lysate medium was well tolerated and safe
Von Bonin et al. (2009)	I	13	III–IV	BM (platelet-lysate medium)	1 CR (7 %) 1 PR (7 %)	
Muller et al. (2008)	I	2	III–IV	BM	1 CR (50 %)	– First study to specifically target pediatric patients following allogeneic HSCT
Resnick et al. (2013)	II	50	II–IV	BM	17 CR (34 %) 33 CR (66 %)	– Second-largest multicenter trial involving steroid-refractory aGVHD – Most patients involved had gastrointestinal involvement

(continued)

Table 15.2 (continued)

Author	Phase	Patient	Grade	MSC source	Results	Comments
Chronic GVHD						
Zhou et al. (2010)	I	4		BM	4 improvement (100 %)	– Sclerodermatous cGVHD patients showed benefit from MSC therapy
Lucchini et al. (2010)	I	5		BM (platelet-lysate medium)	1 CR (20 %) 2 CR (40 %)	– MSC treatment efficacy was greater in aGVHD than in cGVHD
Muller et al. (2008)	I	3		BM	1 improvement (33 %)	– cGVHD patients did not respond to MSC treatment, but aGVHD patients who received MSCs did not progress to cGVHD
Weng et al. (2010)	I	19		BM	4 CR (21 %) 10 CR (52 %)	– Pilot study of MSCs combined with immunosuppressive therapies for refractory cGVHD
GVHD prophylaxis						
Baron et al. (2010)	I	20		BM	9 developed aGVHD (45 %)	– MSCs were co-transplanted in patients undergoing HLA-mismatched nonmyeloablative HSCT – MSC therapy was able to prevent aGVHD without blocking GVL effect
Kuzmina et al. (2012)	II	37		BM	1 out of 19 developed aGVHD (5.3 %)	– MSCs with standard GVHD prophylaxis treatment could significantly inhibit the development of aGVHD but not cGVHD
Lazarus et al. (2005)	I	46		BM	13 developed aGVHD (28 %)	– Co-administration of HLA-identical MSCs with HSCs were well tolerated and could prevent the development of aGVHD
Ning et al. (2008)	I	25		BM	4 out of 9 developed aGVHD (44.4 %)	– MSCs could prevent the development of aGVHD, but patients treated with MSCs had a higher relapse rate

Abbreviations aGVHD acute graft-versus-host disease; *BM* bone marrow; *cGVHD* chronic graft-versus-host disease; *CR* complete response; *FBS* fetal bovine serum; *GVHD* graft-versus-host disease; *HLA* human leukocyte antigen; *HSCT* hematopoietic stem cell transplantation; *MSC* mesenchymal stem cell; *PR* partial response; *UCB* umbilical cord blood

(Kuzmina et al. 2012). Only one of the 19 patients in the MSC-treated group developed aGVHD, where as 6 of 18 patients developed aGVHD in the non-MSC group. However, in this study, MSCs were not co-transplanted with HSCs but were infused at the time of blood count recovery, which may have affected the setting of the infused MSCs. Trials of MSCs for GVHD prevention are still lacking, and additional studies are needed to evaluate their efficacy.

Complete responders of GVHD. Due to small sample size and heterogeneous treatment protocols and MSC products, it is difficult to characterize the complete responders of MSC treatment. However, it is clear that MSC treatment is more effective in certain settings. In most trials, the response rates of patients were evaluated according to the internationally accepted criteria before and after MSC treatment (Przepiorka et al. 1994). Complete response (CR) is defined as loss of all symptoms of GVHD, while partial response is defined as the improvement of at least one grade. In many studies, MSC treatment appears to be more effective in pediatric patients. In a multicenter trial, a greater proportion of pediatric patients responded to MSCs than adults (Le Blanc et al. 2008). Subsequent trials specifically aimed to treat pediatric patients (Prasad et al. 2011; Lucchini et al. 2010; Muller et al. 2008).

Patients with skin-involved GVHD generally had a higher response rate to MSC treatment (Lucchini et al. 2010; von Bonin et al. 2009; Muller et al. 2008). In contrast, some studies have suggested that gastrointestinal GVHD may respond better. The second-largest academic multicenter trial focused on the treatment of severe gastrointestinal GVHD patients and patients with severe liver involvement (Resnick et al. 2013). MSC therapy was successful in obtaining a positive response (either CR or PR) in the majority of cases, especially patients with gastrointestinal symptoms. On the other hand, despite given intrahepatic injections, patients with liver disease showed poor response to MSCs. In the phase III trial using Prochymal as a first-line treatment in combination with steroids, majority of patients had skin involvement (Martin et al. 2010). However, these skin-involved patients responded significantly better to steroids alone even without MSC. When MSCs were added as a second-line treatment in steroid-refractory liver and GI GVHD patients, significantly improved rates were seen. However, whether this improvement is attributed to different organ involvement or the fact that MSCs were used as salvage therapy after failure of corticosteroid treatment is unclear.

Majority of studies involved patients resistant to conventional steroids and failed at least their first-line treatment (Ringden et al. 2006; Le Blanc et al. 2008; Fang et al. 2006; Wu et al. 2011; von Bonin et al. 2009). It is important to recognize, however, that there is generally a lack of studies on de novo GVHD, cGVHD, and GVHD prevention. There is some evidence that MSCs may be less effective in cGVHD (Lucchini et al. 2010) and GVHD prophylaxis (von Bonin et al. 2009). In studies that included aGVHD and cGVHD patients, the response was higher in aGVHD (Lucchini et al. 2010; Muller et al. 2008). Furthermore, the infusion of MSCs for prevention could prevent the development of aGVHD, but cGVHD was unaffected (Kuzmina et al. 2012). More specific patient recruitment and study design will allow critical analysis of the effects of MSC treatment in various GVHD settings.

15.6 Side Effects of Mesenchymal Stem Cells (MSCs) for Graft-Versus-Host Disease (GVHD)

MSC-related side effects have not yet been reported in any of the clinical trials reported. Multiple infusions of MSCs from various sources have been well tolerated in both adult and pediatric patients. However, concerns still remain regarding the use of MSCs.

A major issue for patients with hematologic malignancies is that immunosuppression induced by MSCs may abrogate the intended graft-versus-leukemia (GVL) effect of allogeneic HSCT. Preclinical studies have demonstrated that MSCs promote tumor growth and metastasis by supporting the tumor microenvironment (Zhu et al. 2006; Karnoub et al. 2007; Djouad et al. 2003). In the clinic, the impact of MSCs on GVL effect remains to be elucidated. Ning et al. (2008) observed that hematologic malignancy patients who were cotransplanted with MSCs during HSCT had higher relapse rates than in the control group. Six out of ten patients treated with MSCs experienced relapse, whereas only three out of fifteen patients in the control group relapsed. In another study, however, the co-administration of MSCs could prevent the development of GVHD while maintaining the GVL effect (Baron et al. 2010). It is important to note that the sample size of these studies is too small to draw any conclusions. Therefore, further studies are necessary to demonstrate that MSC-mediated immune suppression does not interfere with GVL effect.

In addition to the potential of promoting relapse, concerns about malignant transformation have been raised. While ex vivo expansion of human MSCs is considered safer than murine MSCs (Tolar et al. 2007; Lepperdinger et al. 2008; Bernardo et al. 2007), there have been reports that even human MSCs may be susceptible to malignant transformation during long-term cultures (Wang et al. 2005). It is, thus, important to avoid unnecessary manipulation of prolonged culture of MSCs.

15.7 Limitations of Stem Cell Therapies

Considerable progress has been made in the development of MSC treatment for GVHD. However, MSC therapy is now faced with ambiguities regarding clinical utility. Several preclinical studies have previously failed to exhibit therapeutic potential of MSCs regardless of various timing and dose (Badillo et al. 2008; Sudres et al. 2006; Prigozhina et al. 2008). MSC therapy in the clinical setting has also showed mixed results depending on age, severity, and organ involvement of GVHD patients. Recently, a phase III industry-sponsored trial (NCT00366145) examined the use of a commercial MSC product (Prochymal; Osiris Therapeutics, Inc) for the treatment of steroid-refractory GVHD. However, in 2009, it was publicly reported that Prochymal MSCs failed to achieve increased complete response rate compared to placebo controls (Martin et al. 2010). Although some studies have undoubtedly justified the clinical efficacy of MSCs, these conflicting outcomes have left unanswered questions for many investigators. Regarding the failure of Osiris' clinical trial, Galipeau (2013) presented a review article discussing the potential variables affecting MSCs. He focuses on four main variables—donor variance, epigenetic reprogramming, immunogenicity, and cryopreservation—and discusses that subtle difference during the production of MSCs, such as donor source and culture methods may result in variable potency in vivo.

We propose two additional explanations that may be associated with discrepancies in clinical outcomes. First, timing of administration of MSCs is also critical in the clinical setting. There is only a narrow window for MSC administration due to their need for licensing, as described in preclinical studies (Polchert et al. 2008). It should be noted that majority of clinical trials have focused mainly on the treatment of established aGVHD patients. However, the use of MSCs for GVHD prophylaxis and for the treatment of cGVHD is still limited. This may be due to inadequate IFN-γ concentrations needed for MSC licensing at different time points of the disease course. Based on Galipeau's perspective, different donor sources may have different IFN-γ responsiveness making some batches of MSCs more potent than others in the same environment (Galipeau 2013). Another possible explanation is that following myeloablative-conditioning regimen for allogeneic HSCT, a temporal gap until the donor-derived Tregs are fully induced may limit MSC's suppressive potential.

Second, there is emerging evidence that MSCs are limited in their ability to regulate Th17 responses. Initially, it was believed that GVHD was primarily a Th1-mediated immune response. However, studies have observed that a complex interaction between Th1 and Th17-mediated responses exist during the development of GVHD (Teshima 2011). While Th1 cytokine, IFN-γ, is known to enhance the immunosuppressive activity of MSCs (Polchert et al. 2008; Dazzi and Marelli-Berg 2008), the effects of the Th17 response on MSCs are less clear. In our study, we investigated the therapeutic potential of MSCs in Th-17-mediated collagen-induced arthritis (CIA) (Park et al. 2011); however, MSCs were ineffective in immune modulation. In fact, MSCs aggravated the disease by promoting the expansion of Th17 cells. MSCs are known producers of TGF-β and IL-6, which are key cytokines that reciprocally regulate the differentiation of naïve T cells into either Tregs or Th17 cells (Guo et al. 2009; Eljaafari et al. 2012; Chen et al. 2010). In the presence of pro-inflammatory cytokines, such as IFN-γ and TNF-α, MSCs produce significant levels of IL-6 in addition to TGF-β. This combination of TGF-β and IL-6 can induce polarization of naïve T cells into Th17 cells (Svobodova et al. 2012). In the presence of pro-inflammatory cytokines in certain GVHD settings, MSCs may aggravate GVHD by promoting Th17 cell expansion. As previously mentioned, only a few trials have attempted to treat cGVHD patients. While only limited studies exist to attribute the failure of MSCs to their inability to suppress Th17 responses, understanding the mechanism of MSCs in Th17 environment may provide valuable answers for the use of MSCs in autoimmune-like cGVHD patients.

Taken together, consideration of both MSC product development and their application in appropriate settings is needed for MSCs to achieve their full clinical potential in GVHD.

15.8 Future Considerations

Gene-transduced MSCs. MSCs' inherent homing abilities to inflammation represent an opportunity to deliver various therapeutic proteins to target sites (Karp and Leng Teo 2009; Chapel et al. 2003). Furthermore, sustained expression of therapeutic genes can enhance the potency of MSCs. In a mouse model, IL-10-transduced MSCs were used to treat aGVHD (Min et al. 2007). While untransduced MSCs were ineffective in suppressing the development of aGVHD, IL-10-transduced MSCs decreased mortality rates of mice, which correlated with decreased levels of pro-inflammatory cytokines. This clinical benefit was directly associated with MSCs ability to deliver IL-10 to GVHD target sites as systemic administration of recombinant IL-10 failed to improve GVHD. In addition, genes encoding chemokine receptors that improve the migration, such as CCR7 (Li et al. 2014), could also be considered.

Other studies have suggested genetically engineered MSCs that produce various anti-inflammatory cytokines in immune diseases. In our study, we showed that MSCs transduced with TGF-b could potently suppress CIA models compared to untransduced MSCs (Park et al. 2011). In addition, MSCs engineered to overexpress IL-4 could attenuate experimental autoimmune encephalomyelitis (Payne et al. 2012). Further studies are needed to verify the effects of different genetically engineered MSCs in the GVHD setting.

Importantly, the use of genetically engineered MSC may raise critical safety issues. In addition to manipulation of MSCs, the overexpression of anti-inflammatory cytokines may prevent the necessary GVL effect in hematologic malignancy patients. Therefore, while genetically engineered MSCs have proved to be a more powerful approach than normal MSCs, its safety in the clinical setting must be determined.

Combinatory Cell-Based Immune Modulation (CCIM). One of MSCs' major mechanisms of immune modulation is the induction of Tregs, both in vitro and in vivo (Joo et al. 2010; Burr et al. 2013). Tregs suppress effector T-cell responses themselves and are therefore considered as a source of cell therapy in various immune-mediated diseases including GVHD. In preclinical and clinical settings, the administration of Tregs was able to improve GVHD (Taylor et al. 2002; Hoffmann et al. 2002; Edinger et al. 2003; Cohen et al. 2002; Brunstein et al. 2011); however, studies have demonstrated that Tregs may lose suppressive capacity once administered and differentiate into pathogenic T cells (Rubtsov et al. 2010; Beres et al. 2011). Furthermore, MSC-mediated immune modulation is often dependent on the presence of endogenous Tregs. Early injection of MSCs may be ineffective because endogenous Tregs have not repopulated following myeloablative-conditioning regimen. In a previous study, it was reported that MSCs and Tregs do not impair each other's respective functions (Engela et al. 2013). Therefore, the combination of two immunosuppressive cell types, MSCs and Tregs, may be able to enhance the immunomodulatory effects in GVHD. We postulated that soluble factors secreted by MSCs can promote the induction of Treg differentiation, and in turn, the cytokines produced by Tregs can promote immunosuppressive potential of MSCs. In our study, we compared single cell therapy groups (MSCs or Tregs alone) with CCIM using MSCs and Tregs group in an aGVHD model (Lim et al. 2014). We observed that the CCIM approach had synergistic immunomodulatory effects in inducing long-term survival and reducing clinicopathological symptoms of GVHD. Furthermore, the CCIM group was able to effectively inhibit the IL-17 response. Thus, CCIM approach could be most effective during early post-transplant period, but also during elevated IL-17 levels of GVHD. We investigated CCIM with MSCs and Tregs in other transplantation models, including the induction of mixed chimerism following nonmyeloablative allogeneic HSCT (Im et al. 2014), and the prevention of allogeneic skin graft rejection (Lee et al. 2013), and saw similar results. Further studies will be needed to characterize the interactions between the MSCs and Tregs and the mechanisms of CCIM. Finally, future studies should address the possibility of combining other immunosuppressive cell types, such as tolerogenic DCs or IL-10-producing Type 1 Treg cells (Tr1 cells) with MSCs.

Multipotent adult progenitor cells (*MAPCs*). In addition to MSCs, MAPCs are a similar type of adult stem cells derived from bone marrow. MAPCs were first described in the rat and mouse bone marrow as stem cells with the potential to proliferate without senescence and to differentiate into cells of three germ layers (Jiang et al. 2002). While there are similarities, comparative analysis between MSCs and MAPCs concluded that the two cell types are distinct cell populations (Roobrouck et al. 2011; Jacobs et al. 2013a). Like MSCs, MAPCs exert immunomodulatory effects on T-cell proliferation through cell-to-cell contact and soluble factors. In addition to low expression of MHC class II and co-stimulatory molecules, MAPCs express low levels of MHC class I, suggesting their potential as off-the-shelf products (Jacobs et al. 2013b). However, MAPCs are smaller in size and can expand for over 70 passages in vitro (Jiang et al. 2002).

Based on these characteristics, MAPCs have been reported to inhibit aGVHD in preclinical models (Kovacsovics-Bankowski et al. 2009; Highfill et al. 2009). MAPCs are now being produced as a clinical-grade, large-scale expanded product by Athersys, Inc. (MultiStem). The efficacy of MultiStem has been confirmed in preclinical settings (Kovacsovics-Bankowski et al. 2008; Boozer et al. 2009) and is now being evaluated in a number of phase I/II clinical trials in stroke, myocardial infarction, inflammatory bowel disease, and GVHD (Vaes et al. 2012).[2] In a phase I dose-escalation study, MAPCs were assessed as a prophylaxic treatment for patients undergoing allogeneic HSCT for hematologic malignancies (Vaes et al. 2012). MAPCs were well tolerated without any adverse effects. Moreover,

[2]Athersys I MultiStem: a novel stem cell therapy. Athersys, Inc.

there was a significant reduction in the incidence of aGVHD in comparison with historical data. Further studies are needed to understand the immune-modulatory mechanisms and clinical efficacy. Nonetheless, MAPC therapy is a promising approach for GVHD patients.

Induced pluripotent stem cell-derived MSCs. Yamanaka et al., described the induction of pluripotent stem cells (also known as induced pluripotent stem cells; iPSCs) by reprogramming differentiated cells, such as fibroblasts to an embryonic-like state (Takahashi and Yamanaka 2006). The transfer of four factors, Oct3/4, Sox2, c-Myc, and KLF4 under embryonic culture conditions can convert fibroblasts to iPSCs. iPSCs exhibit characteristics of embryonic stem cells, including morphology, gene expression, differentiation potential, and teratoma formation. While iPSCs have great potential for various diseases, they have been criticized in that the insertion of genes, c-Myc in particular, may cause cancers and aberrations in reprogrammed cells (Okita et al. 2007; Gore et al. 2011). Clinical-grade iPSCs are under development, but improved reprogramming technology for safe iPSC generation is critical to avoid tumor formation caused by gene transfer.

Recently, MSCs derived from iPSCs have been suggested to overcome expansion and differentiation potential (Jung et al. 2012). iPSC-derived MSCs have been applied in ischemia and periodontitis preclinical models (Yang et al. 2014; Lian et al. 2010). These cells could be generated in clonal expansion cultures and showed the same surface marker expression as well as differentiation potentiation comparable to bone marrow-derived MSCs. Furthermore, iPSC-derived MSCs exhibited potent immunosuppressive effects in both models. Interestingly, iPSC-derived MSCs were superior to bone marrow-derived MSCs in inducing vascular and muscle regeneration in ischemic mice (Lian et al. 2010). The authors suggested that the greater potential of iPSC-derived MSCs is attributable to better survival and engraftment in vivo. However, due to safety issues and regulatory requirements of iPSCs, iPSC-derived MSCs still have a long path ahead for entry into the clinic.

15.9 Conclusion

The therapeutic potential of MSCs has provided sufficient evidence for their application in GVHD. With the appropriate dose, timing, and environment, MSCs have the capacity to alleviate clinical symptoms of GVHD. While results until today have been encouraging, inconsistencies still exist regarding their clinical utility. The accumulated observations on MSC therapy for GVHD in both preclinical and clinical settings now require critical analysis for future improvements. To achieve full clinical potential of MSCs, specific markers of MSCs, standardized protocols for expansion, dosage, and timing of MSC therapy need to be elucidated. Also, novel approaches to MSC therapy are needed to enhance their immunosuppressive effects, such as gene transduction, combined therapy, and the use of other adherent stem cell therapies. Increasing our understanding of MSCs will help facilitate the development of an improved MSC therapy for GVHD.

References

Badillo AT, Peranteau WH, Heaton TE, Quinn C, Flake AW. Murine bone marrow derived stromal progenitor cells fail to prevent or treat acute graft-versus-host disease. Br J Haematol. 2008;141:224–34.

Bai L, Lennon DP, Eaton V, Maier K, Caplan AI, Miller SD, Miller RH. Human bone marrow-derived mesenchymal stem cells induce Th2-polarized immune response and promote endogenous repair in animal models of multiple sclerosis. Glia. 2009;57:1192–203.

Baron F, Lechanteur C, Willems E, Bruck F, Baudoux E, Seidel L, Vanbellinghen JF, Hafraoui K, Lejeune M, Gothot A, Fillet G, Beguin Y. Cotransplantation of mesenchymal stem cells might prevent death from graft-versus-host disease (GVHD) without abrogating graft-versus-tumor effects after HLA-mismatched allogeneic transplantation following nonmyeloablative conditioning. Biol Blood Marrow Transplant. 2010;16:838–47.

Batten P, Sarathchandra P, Antoniw JW, Tay SS, Lowdell MW, Taylor PM, Yacoub MH. Human mesenchymal stem cells induce T cell anergy and downregulate T cell allo-responses via the TH2 pathway: relevance to tissue engineering human heart valves. Tissue Eng. 2006;12:2263–73.

Beres A, Komorowski R, Mihara M, Drobyski WR. Instability of Foxp3 expression limits the ability of induced regulatory T cells to mitigate graft versus host disease. Clin Cancer Res. 2011;17:3969–83.

Bernardo ME, Zaffaroni N, Novara F, Cometa AM, Avanzini MA, Moretta A, Montagna D, Maccario R, Villa R, Daidone MG, Zuffardi O, Locatelli F. Human bone marrow derived mesenchymal stem cells do not undergo transformation after long-term in vitro culture and do not exhibit telomere maintenance mechanisms. Cancer Res. 2007;67:9142–9.

Boozer S, Lehman N, Lakshmipathy U, Love B, Raber A, Maitra A, Deans R, Rao MS, Ting AE. Global characterization and genomic stability of human multistem, a multipotent adult progenitor cell. J Stem Cells. 2009;4:17–28.

Bruck F, Belle L, Lechanteur C, de Leval L, Hannon M, Dubois S, Castermans E, Humblet-Baron S, Rahmouni S, Beguin Y, Briquet A, Baron F. Impact of bone marrow-derived mesenchymal stromal cells on experimental xenogeneic graft-versus-host disease. Cytotherapy. 2013;15:267–79.

Brunstein CG, Miller JS, Cao Q, McKenna DH, Hippen KL, Curtsinger J, Defor T, Levine BL, June CH, Rubinstein P, McGlave PB, Blazar BR, Wagner JE. Infusion of ex vivo expanded T regulatory cells in adults transplanted with umbilical cord blood: safety profile and detection kinetics. Blood. 2011;117:1061–70.

Burr SP, Dazzi F, Garden OA. Mesenchymal stromal cells and regulatory T cells: the Yin and Yang of peripheral tolerance? Immunol Cell Biol. 2013;91:12–8.

Cahn JY, Klein JP, Lee SJ, Milpied N, Blaise D, Antin JH, Leblond V, Ifrah N, Jouet JP, Loberiza F, Ringden O, Barrett AJ, Horowitz MM, Socie G, Societe Francaise de Greffe de Moelle et Therapie Cellulaire, Dana Farber Cancer Institute, International Bone Marrow Transplant Registry. Prospective evaluation of 2 acute graft-versus-host (GVHD) grading systems: a joint Societe Francaise de Greffe de Moelle et Therapie Cellulaire (SFGM-TC), Dana Farber Cancer Institute (DFCI), and International Bone Marrow Transplant Registry (IBMTR) prospective study. Blood. 2005;106:1495–500.

Chapel A, Bertho JM, Bensidhoum M, Fouillard L, Young RG, Frick J, Demarquay C, Cuvelier F, Mathieu E, Trompier F, Dudoignon N, Germain C, Mazurier C, Aigueperse J, Borneman J, Gorin NC, Gourmelon P, Thierry D. Mesenchymal stem cells home to injured tissues when co-infused with hematopoietic cells to treat a radiation-induced multi-organ failure syndrome. J Gene Med. 2003;5:1028–38.

Chen L, Tredget EE, Wu PY, Wu Y. Paracrine factors of mesenchymal stem cells recruit macrophages and endothelial lineage cells and enhance wound healing. PLoS ONE. 2008;3:e1886.

Chen B, Hu J, Liao L, Sun Z, Han Q, Song Z, Zhao RC. Flk-1+ mesenchymal stem cells aggravate collagen-induced arthritis by up-regulating interleukin-6. Clin Exp Immunol. 2010;159:292–302.

Choi H, Lee RH, Bazhanov N, Oh JY, Prockop DJ. Anti-inflammatory protein TSG-6 secreted by activated MSCs attenuates zymosan-induced mouse peritonitis by decreasing TLR2/NF-kappaB signaling in resident macrophages. Blood. 2011;118:330–8.

Choi YS, Jeong JA, Lim DS. Mesenchymal stem cell-mediated immature dendritic cells induce regulatory T cell-based immunosuppressive effect. Immunol Invest. 2012;41:214–29.

Chung NG, Jeong DC, Park SJ, Choi BO, Cho B, Kim HK, Chun CS, Won JH, Han CW. Cotransplantation of marrow stromal cells may prevent lethal graft-versus-host disease in major histocompatibility complex mismatched murine hematopoietic stem cell transplantation. Int J Hematol. 2004;80:370–6.

Cohen JL, Trenado A, Vasey D, Klatzmann D, Salomon BL. CD4(+)CD25(+) immunoregulatory T Cells: new therapeutics for graft-versus-host disease. J Exp Med. 2002;196:401–6.

Comoli P, Ginevri F, Maccario R, Avanzini MA, Marconi M, Groff A, Cometa A, Cioni M, Porretti L, Barberi W, Frassoni F, Locatelli F. Human mesenchymal stem cells inhibit antibody production induced in vitro by allostimulation. Nephrol Dial Transplant. 2008;23:1196–202.

Corcione A, Benvenuto F, Ferretti E, Giunti D, Cappiello V, Cazzanti F, Risso M, Gualandi F, Mancardi GL, Pistoia V, Uccelli A. Human mesenchymal stem cells modulate B-cell functions. Blood. 2006;107:367–72.

Dazzi F, Marelli-Berg FM. Mesenchymal stem cells for graft-versus-host disease: close encounters with T cells. Eur J Immunol. 2008;38:1479–82.

Di Nicola M, Carlo-Stella C, Magni M, Milanesi M, Longoni PD, Matteucci P, Grisanti S, Gianni AM. Human bone marrow stromal cells suppress T-lymphocyte proliferation induced by cellular or nonspecific mitogenic stimuli. Blood. 2002;99:3838–43.

Djouad F, Plence P, Bony C, Tropel P, Apparailly F, Sany J, Noel D, Jorgensen C. Immunosuppressive effect of mesenchymal stem cells favors tumor growth in allogeneic animals. Blood. 2003;102:3837–44.

Djouad F, Charbonnier LM, Bouffi C, Louis-Plence P, Bony C, Apparailly F, Cantos C, Jorgensen C, Noel D. Mesenchymal stem cells inhibit the differentiation of dendritic cells through an interleukin-6-dependent mechanism. Stem Cells. 2007;25:2025–32.

Dominici M, Le Blanc K, Mueller I, Slaper-Cortenbach I, Marini F, Krause D, Deans R, Keating A, Prockop D, Horwitz E. Minimal criteria for defining multipotent mesenchymal stromal cells. The International Society for Cellular Therapy position statement. Cytotherapy. 2006;8:315–7.

Duffy MM, Pindjakova J, Hanley SA, McCarthy C, Weidhofer GA, Sweeney EM, English K, Shaw G, Murphy JM, Barry FP, Mahon BP, Belton O, Ceredig R, Griffin MD. Mesenchymal stem cell inhibition of T-helper 17 cell- differentiation is triggered by cell-cell contact and mediated by prostaglandin E2 via the EP4 receptor. Eur J Immunol. 2011;41:2840–51.

Edinger M, Hoffmann P, Ermann J, Drago K, Fathman CG, Strober S, Negrin RS. CD4+ CD25+ regulatory T cells preserve graft-versus-tumor activity while inhibiting graft-versus-host disease after bone marrow transplantation. Nat Med. 2003;9:1144–50.

Eljaafari A, Tartelin ML, Aissaoui H, Chevrel G, Osta B, Lavocat F, Miossec P. Bone marrow-derived and synovium-derived mesenchymal cells promote Th17 cell expansion and activation through caspase 1 activation: contribution to the chronicity of rheumatoid arthritis. Arthritis Rheum. 2012;64:2147–57.

Engela AU, Baan CC, Peeters AM, Weimar W, Hoogduijn MJ. Interaction between adipose tissue-derived mesenchymal stem cells and regulatory T-cells. Cell Transplant. 2013;22:41–54.

English K, Barry FP, Mahon BP. Murine mesenchymal stem cells suppress dendritic cell migration, maturation and antigen presentation. Immunol Lett. 2008;115:50–8.

English K, Ryan JM, Tobin L, Murphy MJ, Barry FP, Mahon BP. Cell contact, prostaglandin E(2) and transforming growth factor beta 1 play non-redundant roles in human mesenchymal stem cell induction of CD4+ CD25(High) forkhead box P3+ regulatory T cells. Clin Exp Immunol. 2009;156:149–60.

Fang B, Song YP, Liao LM, Han Q, Zhao RC. Treatment of severe therapy-resistant acute graft-versus-host disease with human adipose tissue-derived mesenchymal stem cells. Bone Marrow Transplant. 2006;38:389–90.

Galipeau J. The mesenchymal stromal cells dilemma–does a negative phase III trial of random donor mesenchymal stromal cells in steroid-resistant graft-versus-host disease represent a death knell or a bump in the road? Cytotherapy. 2013;15:2–8.

Ghannam S, Pene J, Moquet-Torcy G, Jorgensen C, Yssel H. Mesenchymal stem cells inhibit human Th17 cell differentiation and function and induce a T regulatory cell phenotype. J Immunol. 2010;185:302–12.

Gore A, Li Z, Fung HL, Young JE, Agarwal S, Antosiewicz-Bourget J, Canto I, Giorgetti A, Israel MA, Kiskinis E, Lee JH, Loh YH, Manos PD, Montserrat N, Panopoulos AD, Ruiz S, Wilbert ML, Yu J, Kirkness EF, Izpisua Belmonte JC, Rossi DJ, Thomson JA, Eggan K, Daley GQ, Goldstein LS, Zhang K. Somatic coding mutations in human induced pluripotent stem cells. Nature. 2011;471:63–7.

Guo Z, Zheng C, Chen Z, Gu D, Du W, Ge J, Han Z, Yang R. Fetal BM-derived mesenchymal stem cells promote the expansion of human Th17 cells, but inhibit the production of Th1 cells. Eur J Immunol. 2009;39:2840–9.

Highfill SL, Kelly RM, O'Shaughnessy MJ, Zhou Q, Xia L, Panoskaltsis-Mortari A, Taylor PA, Tolar J, Blazar BR. Multipotent adult progenitor cells can suppress graft-versus-host disease via prostaglandin E2 synthesis and only if localized to sites of allopriming. Blood. 2009;114:693–701.

Hoffmann P, Ermann J, Edinger M, Fathman CG, Strober S. Donor-type CD4(+)CD25(+) regulatory T cells suppress lethal acute graft-versus-host disease after allogeneic bone marrow transplantation. J Exp Med. 2002;196:389–99.

Hoogduijn MJ, Popp F, Verbeek R, Masoodi M, Nicolaou A, Baan C, Dahlke MH. The immunomodulatory properties of mesenchymal stem cells and their use for immunotherapy. Int Immunopharmacol. 2010;10:1496–500.

Im KI, Park MJ, Kim N, Lim JY, Park HS, Lee SH, Nam YS, Lee ES, Lee JH, Cho ML, Cho SG. Induction of mixed chimerism using combinatory cell-based immune modulation with mesenchymal stem cells and regulatory T cells for solid-organ transplant tolerance. Stem Cells Dev. 2014.

Jacobs SA, Pinxteren J, Roobrouck VD, Luyckx A, van't Hof W, Deans R, Verfaillie CM, Waer M, Billiau AD, Van Gool SW. Human multipotent adult progenitor cells are nonimmunogenic and exert potent immunomodulatory effects on alloreactive T-cell responses. Cell Transplant. 2013;22:1915–28.

Jacobs SA, Roobrouck VD, Verfaillie CM, Van Gool SW. Immunological characteristics of human mesenchymal stem cells and multipotent adult progenitor cells. Immunol Cell Biol. 2013;91:32–9.

Jagasia M, Arora M, Flowers ME, Chao NJ, McCarthy PL, Cutler CS, Urbano-Ispizua A, Pavletic SZ, Haagenson MD, Zhang MJ, Antin JH, Bolwell BJ, Bredeson C, Cahn JY, Cairo M, Gale RP, Gupta V, Lee SJ, Litzow M, Weisdorf DJ, Horowitz MM, Hahn T. Risk factors for acute GVHD and survival after hematopoietic cell transplantation. Blood. 2012;119:296–307.

Jang YK, Kim M, Lee YH, Oh W, Yang YS, Choi SJ. Optimization of the therapeutic efficacy of human umbilical cord blood-mesenchymal stromal cells in an NSG mouse xenograft model of graft-versus-host disease. Cytotherapy. 2014;16:298–308.

Jiang Y, Jahagirdar BN, Reinhardt RL, Schwartz RE, Keene CD, Ortiz-Gonzalez XR, Reyes M, Lenvik T, Lund T, Blackstad M, Du J, Aldrich S, Lisberg A, Low WC, Largaespada DA, Verfaillie CM. Pluripotency of mesenchymal stem cells derived from adult marrow. Nature. 2002;418:41–9.

Joo SY, Cho KA, Jung YJ, Kim HS, Park SY, Choi YB, Hong KM, Woo SY, Seoh JY, Cho SJ, Ryu KH. Mesenchymal stromal cells inhibit graft-versus-host disease of mice in a dose-dependent manner. Cytotherapy. 2010;12:361–70.

Joo SY, Cho KA, Jung YJ, Kim HS, Park SY, Choi YB, Hong KM, Woo SY, Seoh JY, Ryu KH. Bioimaging for the monitoring of the in vivo distribution of infused mesenchymal stem cells in a mouse model of the graft-versus-host reaction. Cell Biol Int. 2011;35:417–21.

Jung Y, Bauer G, Nolta JA. Concise review: Induced pluripotent stem cell-derived mesenchymal stem cells: progress toward safe clinical products. Stem Cells. 2012;30:42–7.

Karnoub AE, Dash AB, Vo AP, Sullivan A, Brooks MW, Bell GW, Richardson AL, Polyak K, Tubo R, Weinberg RA. Mesenchymal stem cells within tumour stroma promote breast cancer metastasis. Nature. 2007;449:557–63.

Karp JM, Leng Teo GS. Mesenchymal stem cell homing: the devil is in the details. Cell Stem Cell. 2009;4:206–16.

Keating A. How do mesenchymal stromal cells suppress T cells? Cell Stem Cell. 2008;2:106–8.

Kebriaei P, Isola L, Bahceci E, Holland K, Rowley S, McGuirk J, Devetten M, Jansen J, Herzig R, Schuster M, Monroy R, Uberti J. Adult human mesenchymal stem cells added to corticosteroid therapy for the treatment of acute graft-versus-host disease. Biol Blood Marrow Transplant. 2009;15:804–11.

Kim N, Cho SG. Clinical applications of mesenchymal stem cells. Korean J Intern Med. 2013;28:387–402.

Kim J, Hematti P. Mesenchymal stem cell-educated macrophages: a novel type of alternatively activated macrophages. Exp Hematol. 2009;37:1445–53.

Kim N, Im KI, Lim JY, Jeon EJ, Nam YS, Kim EJ, Cho SG. Mesenchymal stem cells for the treatment and prevention of graft-versus-host disease: experiments and practice. Ann Hematol. 2013a;92:1295–308.

Kim EJ, Kim N, Cho SG. The potential use of mesenchymal stem cells in hematopoietic stem cell transplantation. Exp Mol Med. 2013b;45:e2.

Kovacsovics-Bankowski M, Mauch K, Raber A, Streeter PR, Deans RJ, Maziarz RT, Van't Hof W. Pre-clinical safety testing supporting clinical use of allogeneic multipotent adult progenitor cells. Cytotherapy. 2008;10:730–42.

Kovacsovics-Bankowski M, Streeter PR, Mauch KA, Frey MR, Raber A, van't Hof W, Deans R, Maziarz RT. Clinical scale expanded adult pluripotent stem cells prevent graft-versus-host disease. Cell Immunol. 2009;255:55–60.

Krampera M, Glennie S, Dyson J, Scott D, Laylor R, Simpson E, Dazzi F. Bone marrow mesenchymal stem cells inhibit the response of naive and memory antigen-specific T cells to their cognate peptide. Blood. 2003;101:3722–9.

Krampera M, Cosmi L, Angeli R, Pasini A, Liotta F, Andreini A, Santarlasci V, Mazzinghi B, Pizzolo G, Vinante F, Romagnani P, Maggi E, Romagnani S, Annunziato F. Role for interferon-gamma in the immunomodulatory activity of human bone marrow mesenchymal stem cells. Stem Cells. 2006;24:386–98.

Kuzmina LA, Petinati NA, Parovichnikova EN, Lubimova LS, Gribanova EO, Gaponova TV, Shipounova IN, Zhironkina OA, Bigildeev AE, Svinareva DA, Drize NJ, Savchenko VG. Multipotent mesenchymal stromal cells for the prophylaxis of acute graft-versus-host disease-A phase II study. Stem Cells Int. 2012;2012:968213.

Lazarus HM, Koc ON, Devine SM, Curtin P, Maziarz RT, Holland HK, Shpall EJ, McCarthy P, Atkinson K, Cooper BW, Gerson SL, Laughlin MJ, Loberiza FR Jr, Moseley AB, Bacigalupo A. Cotransplantation of HLA-identical sibling culture-expanded mesenchymal stem cells and hematopoietic stem cells in hematologic malignancy patients. Biol Blood Marrow Transplant. 2005;11:389–98.

Le Blanc K, Tammik C, Rosendahl K, Zetterberg E, Ringden O. HLA expression and immunologic properties of differentiated and undifferentiated mesenchymal stem cells. Exp Hematol. 2003a;31:890–6.

Le Blanc K, Tammik L, Sundberg B, Haynesworth SE, Ringden O. Mesenchymal stem cells inhibit and stimulate mixed lymphocyte cultures and mitogenic responses independently of the major histocompatibility complex. Scand J Immunol. 2003b;57:11–20.

Le Blanc K, Rasmusson I, Sundberg B, Gotherstrom C, Hassan M, Uzunel M, Ringden O. Treatment of severe acute graft-versus-host disease with third party haploidentical mesenchymal stem cells. Lancet. 2004;363:1439–41.

Le Blanc K, Frassoni F, Ball L, Locatelli F, Roelofs H, Lewis I, Lanino E, Sundberg B, Bernardo ME, Remberger M, Dini G, Egeler RM, Bacigalupo A, Fibbe W, Ringden O, Developmental Committee of the European Group for Blood and Marrow Transplantation. Mesenchymal stem cells for treatment of steroid-resistant, severe, acute graft-versus-host disease: a phase II study. Lancet. 2008;371:1579–86.

Lee JH, Jeon EJ, Kim N, Nam YS, Im KI, Lim JY, Kim EJ, Cho ML, Han KT, Cho SG. The synergistic immunoregulatory effects of culture-expanded mesenchymal stromal cells and CD4(+)25(+)Foxp3+ regulatory T cells on skin allograft rejection. PLoS ONE. 2013;8:e70968.

Lepperdinger G, Brunauer R, Jamnig A, Laschober G, Kassem M. Controversial issue: is it safe to employ mesenchymal stem cells in cell-based therapies? Exp Gerontol. 2008;43:1018–23.

Li YP, Paczesny S, Lauret E, Poirault S, Bordigoni P, Mekhloufi F, Hequet O, Bertrand Y, Ou-Yang JP, Stoltz JF, Miossec P, Eljaafari A. Human mesenchymal stem cells license adult CD34+ hemopoietic progenitor cells to differentiate into regulatory dendritic cells through activation of the Notch pathway. J Immunol. 2008a;180:1598–608.

Li H, Guo Z, Jiang X, Zhu H, Li X, Mao N. Mesenchymal stem cells alter migratory property of T and dendritic cells to delay the development of murine lethal acute graft-versus-host disease. Stem Cells. 2008b;26:2531–41.

Li H, Jiang Y, Jiang X, Guo X, Ning H, Li Y, Liao L, Yao H, Wang X, Liu Y, Zhang Y, Chen H, Mao N. CCR7 guides migration of mesenchymal stem cell to secondary lymphoid organs: a novel approach to separate GvHD from GvL effect. Stem Cells. 2014;32:1890–903.

Lian Q, Zhang Y, Zhang J, Zhang HK, Wu X, Zhang Y, Lam FF, Kang S, Xia JC, Lai WH, Au KW, Chow YY, Siu CW, Lee CN, Tse HF. Functional mesenchymal stem cells derived from human induced pluripotent stem cells attenuate limb ischemia in mice. Circulation. 2010;121:1113–23.

Lim JY, Park MJ, Im KI, Kim N, Jeon EJ, Kim EJ, Cho ML, Cho SG. Combination cell therapy using mesenchymal stem cells and regulatory T-cells provides a synergistic immunomodulatory effect associated with reciprocal regulation of TH1/TH2 and th17/treg cells in a murine acute graft-versus-host disease model. Cell Transplant. 2014;23:703–14.

Liotta F, Angeli R, Cosmi L, Fili L, Manuelli C, Frosali F, Mazzinghi B, Maggi L, Pasini A, Lisi V, Santarlasci V, Consoloni L, Angelotti ML, Romagnani P, Parronchi P, Krampera M, Maggi E, Romagnani S, Annunziato F. Toll-like receptors 3 and 4 are expressed by human bone marrow-derived mesenchymal stem cells and can inhibit their T-cell modulatory activity by impairing Notch signaling. Stem Cells. 2008;26:279–89.

Liu X, Qu X, Chen Y, Liao L, Cheng K, Shao C, Zenke M, Keating A, Zhao RC. Mesenchymal stem/stromal cells induce the generation of novel IL-10-dependent regulatory dendritic cells by SOCS3 activation. J Immunol. 2012;189:1182–92.

Lucchini G, Introna M, Dander E, Rovelli A, Balduzzi A, Bonanomi S, Salvade A, Capelli C, Belotti D, Gaipa G, Perseghin P, Vinci P, Lanino E, Chiusolo P, Orofino MG, Marktel S, Golay J, Rambaldi A, Biondi A, D'Amico G, Biagi E. Platelet-lysate-expanded mesenchymal stromal cells as a salvage therapy for severe resistant graft-versus-host disease in a pediatric population. Biol Blood Marrow Transplant. 2010;16:1293–301.

Maccario R, Podesta M, Moretta A, Cometa A, Comoli P, Montagna D, Daudt L, Ibatici A, Piaggio G, Pozzi S, Frassoni F, Locatelli F. Interaction of human mesenchymal stem cells with cells involved in alloantigen-specific immune response favors the differentiation of CD4+ T-cell subsets expressing a regulatory/suppressive phenotype. Haematologica. 2005;90:516–25.

Madec AM, Mallone R, Afonso G, Abou Mrad E, Mesnier A, Eljaafari A, Thivolet C. Mesenchymal stem cells protect NOD mice from diabetes by inducing regulatory T cells. Diabetologia. 2009;52:1391–9.

Maggini J, Mirkin G, Bognanni I, Holmberg J, Piazzon IM, Nepomnaschy I, Costa H, Canones C, Raiden S, Vermeulen M, Geffner JR. Mouse bone marrow-derived mesenchymal stromal cells turn activated macrophages into a regulatory-like profile. PLoS ONE. 2010;5:e9252.

Marigo I, Dazzi F. The immunomodulatory properties of mesenchymal stem cells. Semin Immunopathol. 2011;33:593–602.

Martin PJ, Uberti JP, Soiffer RJ, Klingemann H, Waller EK, Daly AS, Herrmann RP, Kebriaei P. Prochymal improves response rates in patients with steroid-refractory acute graft versus host disease (SR-GVHD) involving the liver and gut: results of a randomized, placebo-controlled, multicenter phase III trial in GVHD. Biol Blood Marrow Transplant: J Am Soc Blood Marrow Transplant. 2010;16:S169–70.

Meisel R, Zibert A, Laryea M, Gobel U, Daubener W, Dilloo D. Human bone marrow stromal cells inhibit allogeneic T-cell responses by indoleamine 2,3-dioxygenase-mediated tryptophan degradation. Blood. 2004;103:4619–21.

Min CK, Kim BG, Park G, Cho B, Oh IH. IL-10-transduced bone marrow mesenchymal stem cells can attenuate the severity of acute graft-versus-host disease after experimental allogeneic stem cell transplantation. Bone Marrow Transplant. 2007;39:637–45.

Muller I, Kordowich S, Holzwarth C, Isensee G, Lang P, Neunhoeffer F, Dominici M, Greil J, Handgretinger R. Application of multipotent mesenchymal stromal cells in pediatric patients following allogeneic stem cell transplantation. Blood Cells Mol Dis. 2008;40:25–32.

Nauta AJ, Kruisselbrink AB, Lurvink E, Willemze R, Fibbe WE. Mesenchymal stem cells inhibit generation and function of both CD34+-derived and monocyte-derived dendritic cells. J Immunol. 2006;177:2080–7.

Nemeth K, Leelahavanichkul A, Yuen PS, Mayer B, Parmelee A, Doi K, Robey PG, Leelahavanichkul K, Koller BH, Brown JM, Hu X, Jelinek I, Star RA, Mezey E. Bone marrow stromal cells attenuate sepsis via prostaglandin E(2)-dependent reprogramming of host macrophages to increase their interleukin-10 production. Nat Med. 2009;15:42–9.

Ning H, Yang F, Jiang M, Hu L, Feng K, Zhang J, Yu Z, Li B, Xu C, Li Y, Wang J, Hu J, Lou X, Chen H. The correlation between cotransplantation of mesenchymal stem cells and higher recurrence rate in hematologic malignancy patients: outcome of a pilot clinical study. Leukemia. 2008;22:593–9.

Okita K, Ichisaka T, Yamanaka S. Generation of germline-competent induced pluripotent stem cells. Nature. 2007;448:313–7.

Olson JA, Leveson-Gower DB, Gill S, Baker J, Beilhack A, Negrin RS. NK cells mediate reduction of GVHD by inhibiting activated, alloreactive T cells while retaining GVT effects. Blood. 2010;115:4293–301.

Park MJ, Park HS, Cho ML, Oh HJ, Cho YG, Min SY, Chung BH, Lee JW, Kim HY, Cho SG. Transforming growth factor beta-transduced mesenchymal stem cells ameliorate experimental autoimmune arthritis through reciprocal regulation of Treg/Th17 cells and osteoclastogenesis. Arthritis Rheum. 2011;63:1668–80.

Patel SA, Meyer JR, Greco SJ, Corcoran KE, Bryan M, Rameshwar P. Mesenchymal stem cells protect breast cancer cells through regulatory T cells: role of mesenchymal stem cell-derived TGF-beta. J Immunol. 2010;184:5885–94.

Payne NL, Dantanarayana A, Sun G, Moussa L, Caine S, McDonald C, Herszfeld D, Bernard CC, Siatskas C. Early intervention with gene-modified mesenchymal stem cells overexpressing interleukin-4 enhances anti-inflammatory responses and functional recovery in experimental autoimmune demyelination. Cell Adh Migr. 2012;6:179–89.

Polchert D, Sobinsky J, Douglas G, Kidd M, Moadsiri A, Reina E, Genrich K, Mehrotra S, Setty S, Smith B, Bartholomew A. IFN-gamma activation of mesenchymal stem cells for treatment and prevention of graft versus host disease. Eur J Immunol. 2008;38:1745–55.

Prasad VK, Lucas KG, Kleiner GI, Talano JA, Jacobsohn D, Broadwater G, Monroy R, Kurtzberg J. Efficacy and safety of ex vivo cultured adult human mesenchymal stem cells (Prochymal) in pediatric patients with severe refractory acute graft-versus-host disease in a compassionate use study. Biol Blood Marrow Transplant. 2011;17:534–41.

Prigozhina TB, Khitrin S, Elkin G, Eizik O, Morecki S, Slavin S. Mesenchymal stromal cells lose their immunosuppressive potential after allotransplantation. Exp Hematol. 2008;36:1370–6.

Przepiorka D, Weisdorf D, Martin P, Klingemann HG, Beatty P, Hows J, Thomas ED. 1994 Consensus conference on acute GVHD grading. Bone Marrow Transplant. 1995;15:825–8.

Rasmusson I, Le Blanc K, Sundberg B, Ringden O. Mesenchymal stem cells stimulate antibody secretion in human B cells. Scand J Immunol. 2007;65:336–43.

Ren G, Zhang L, Zhao X, Xu G, Zhang Y, Roberts AI, Zhao RC, Shi Y. Mesenchymal stem cell-mediated immunosuppression occurs via concerted action of chemokines and nitric oxide. Cell Stem Cell. 2008;2:141–50.

Resnick IB, Barkats C, Shapira MY, Stepensky P, Bloom AI, Shimoni A, Mankuta D, Varda-Bloom N, Rheingold L, Yeshurun M, Bielorai B, Toren A, Zuckerman T, Nagler A, Or R. Treatment of severe steroid-resistant acute GVHD with mesenchymal stromal cells (MSC). Am J Blood Res. 2013;3:225–38.

Ringden O, Uzunel M, Rasmusson I, Remberger M, Sundberg B, Lonnies H, Marschall HU, Dlugosz A, Szakos A, Hassan Z, Omazic B, Aschan J, Barkholt L, Le Blanc K. Mesenchymal stem cells for treatment of therapy-resistant graft-versus-host disease. Transplantation. 2006;81:1390–7.

Roobrouck VD, Clavel C, Jacobs SA, Ulloa-Montoya F, Crippa S, Sohni A, Roberts SJ, Luyten FP, Van Gool SW, Sampaolesi M, Delforge M, Luttun A, Verfaillie CM. Differentiation potential of human postnatal mesenchymal stem cells, mesoangioblasts, and multipotent adult progenitor cells reflected in their transcriptome and partially influenced by the culture conditions. Stem Cells. 2011;29:871–82.

Rubtsov YP, Niec RE, Josefowicz S, Li L, Darce J, Mathis D, Benoist C, Rudensky AY. Stability of the regulatory T cell lineage in vivo. Science. 2010;329:1667–71.

Ruggeri L, Capanni M, Urbani E, Perruccio K, Shlomchik WD, Tosti A, Posati S, Rogaia D, Frassoni F, Aversa F, Martelli MF, Velardi A. Effectiveness of donor natural killer cell alloreactivity in mismatched hematopoietic transplants. Science. 2002;295:2097–100.

Schena F, Gambini C, Gregorio A, Mosconi M, Reverberi D, Gattorno M, Casazza S, Uccelli A, Moretta L, Martini A, Traggiai E. Interferon-gamma-dependent inhibition of B cell activation by bone marrow-derived mesenchymal stem cells in a murine model of systemic lupus erythematosus. Arthritis Rheum. 2010;62:2776–86.

Sotiropoulou PA, Perez SA, Gritzapis AD, Baxevanis CN, Papamichail M. Interactions between human mesenchymal stem cells and natural killer cells. Stem Cells. 2006;24:74–85.

Spaggiari GM, Capobianco A, Becchetti S, Mingari MC, Moretta L. Mesenchymal stem cell-natural killer cell interactions: evidence that activated NK cells are capable of killing MSCs, whereas MSCs can inhibit IL-2-induced NK-cell proliferation. Blood. 2006;107:1484–90.

Spaggiari GM, Capobianco A, Abdelrazik H, Becchetti F, Mingari MC, Moretta L. Mesenchymal stem cells inhibit natural killer-cell proliferation, cytotoxicity, and cytokine production: role of indoleamine 2,3-dioxygenase and prostaglandin E2. Blood. 2008;111:1327–33.

Spaggiari GM, Abdelrazik H, Becchetti F, Moretta L. MSCs inhibit monocyte-derived DC maturation and function by selectively interfering with the generation of immature DCs: central role of MSC-derived prostaglandin E2. Blood. 2009;113:6576–83.

Stagg J, Galipeau J. Mechanisms of immune modulation by mesenchymal stromal cells and clinical translation. Curr Mol Med. 2013;13:856–67.

Stagg J, Pommey S, Eliopoulos N, Galipeau J. Interferon-gamma-stimulated marrow stromal cells: a new type of nonhematopoietic antigen-presenting cell. Blood. 2006;107:2570–7.

Sudres M, Norol F, Trenado A, Gregoire S, Charlotte F, Levacher B, Lataillade JJ, Bourin P, Holy X, Vernant JP, Klatzmann D, Cohen JL. Bone marrow mesenchymal stem cells suppress lymphocyte proliferation in vitro but fail to prevent graft-versus-host disease in mice. J Immunol. 2006;176:7761–7.

Sundin M, Ringden O, Sundberg B, Nava S, Gotherstrom C, Le Blanc K. No alloantibodies against mesenchymal stromal cells, but presence of anti-fetal calf serum antibodies, after transplantation in allogeneic hematopoietic stem cell recipients. Haematologica. 2007;92:1208–15.

Svobodova E, Krulova M, Zajicova A, Pokorna K, Prochazkova J, Trosan P, Holan V. The role of mouse mesenchymal stem cells in differentiation of naive T-cells into anti-inflammatory regulatory T-cell or proinflammatory helper T-cell 17 population. Stem Cells Dev. 2012;21:901–10.

Takahashi K, Yamanaka S. Induction of pluripotent stem cells from mouse embryonic and adult fibroblast cultures by defined factors. Cell. 2006;126:663–76.

Taylor PA, Lees CJ, Blazar BR. The infusion of ex vivo activated and expanded CD4(+)CD25(+) immune regulatory cells inhibits graft-versus-host disease lethality. Blood. 2002;99:3493–9.

Teshima T. Th1 and Th17 join forces for acute GVHD. Blood. 2011;118:4765–7.

Tisato V, Naresh K, Girdlestone J, Navarrete C, Dazzi F. Mesenchymal stem cells of cord blood origin are effective at preventing but not treating graft-versus-host disease. Leukemia. 2007;21:1992–9.

Tolar J, Nauta AJ, Osborn MJ, Panoskaltsis Mortari A, McElmurry RT, Bell S, Xia L, Zhou N, Riddle M, Schroeder TM, Westendorf JJ, McIvor RS, Hogendoorn PC, Szuhai K, Oseth L, Hirsch B, Yant SR, Kay MA, Peister A, Prockop DJ, Fibbe WE, Blazar BR. Sarcoma derived from cultured mesenchymal stem cells. Stem Cells. 2007;25:371–9.

Traggiai E, Volpi S, Schena F, Gattorno M, Ferlito F, Moretta L, Martini A. Bone marrow-derived mesenchymal stem cells induce both polyclonal expansion and differentiation of B cells isolated from healthy donors and systemic lupus erythematosus patients. Stem Cells. 2008;26:562–9.

Vaes B, Van't Hof W, Deans R, Pinxteren J. Application of Multistem® allogeneic cells for immunomodulatory therapy: clinical progress and pre-clinical challenges in prophylaxis for graft versus host disease. Front Immunol. 2012;3:345.

von Bonin M, Stolzel F, Goedecke A, Richter K, Wuschek N, Holig K, Platzbecker U, Illmer T, Schaich M, Schetelig J, Kiani A, Ordemann R, Ehninger G, Schmitz M, Bornhauser M. Treatment of refractory acute GVHD with third-party MSC expanded in platelet lysate-containing medium. Bone Marrow Transplant. 2009;43:245–51.

Wang Y, Huso DL, Harrington J, Kellner J, Jeong DK, Turney J, McNiece IK. Outgrowth of a transformed cell population derived from normal human BM mesenchymal stem cell culture. Cytotherapy. 2005;7:509–19.

Waterman RS, Tomchuck SL, Henkle SL, Betancourt AM. A new mesenchymal stem cell (MSC) paradigm: polarization into a pro-inflammatory MSC1 or an Immunosuppressive MSC2 phenotype. PLoS ONE. 2010;5:e10088.

Weng JY, Du X, Geng SX, Peng YW, Wang Z, Lu ZS, Wu SJ, Luo CW, Guo R, Ling W, Deng CX, Liao PJ, Xiang AP. Mesenchymal stem cell as salvage treatment for refractory chronic GVHD. Bone Marrow Transplant. 2010;45:1732–40.

Wu KH, Chan CK, Tsai C, Chang YH, Sieber M, Chiu TH, Ho M, Peng CT, Wu HP, Huang JL. Effective treatment of severe steroid-resistant acute graft-versus-host disease with umbilical cord-derived mesenchymal stem cells. Transplantation. 2011;91:1412–6.

Yang H, Aprecio RM, Zhou X, Wang Q, Zhang W, Ding Y, Li Y. Therapeutic effect of TSG-6 engineered iPSC-derived MSCs on experimental periodontitis in rats: a pilot study. PLoS ONE. 2014;9:e100285.

Zappia E, Casazza S, Pedemonte E, Benvenuto F, Bonanni I, Gerdoni E, Giunti D, Ceravolo A, Cazzanti F, Frassoni F, Mancardi G, Uccelli A. Mesenchymal stem cells ameliorate experimental autoimmune encephalomyelitis inducing T-cell anergy. Blood. 2005;106:1755–61.

Zhang B, Liu R, Shi D, Liu X, Chen Y, Dou X, Zhu X, Lu C, Liang W, Liao L, Zenke M, Zhao RC. Mesenchymal stem cells induce mature dendritic cells into a novel Jagged-2-dependent regulatory dendritic cell population. Blood. 2009;113:46–57.

Zhou H, Guo M, Bian C, Sun Z, Yang Z, Zeng Y, Ai H, Zhao RC. Efficacy of bone marrow-derived mesenchymal stem cells in the treatment of sclerodermatous chronic graft-versus-host disease: clinical report. Biol Blood Marrow Transplant. 2010;16:403–12.

Zhu W, Xu W, Jiang R, Qian H, Chen M, Hu J, Cao W, Han C, Chen Y. Mesenchymal stem cells derived from bone marrow favor tumor cell growth in vivo. Exp Mol Pathol. 2006;80:267–74.

Chapter 16
Clinical Application and Molecular Mechanism of Multipotent Stem Cell Therapy for Liver Disease

Xia Zhou, Lina Cui, Yongquan Shi and Ying Han

Abstract Multipotent stem cells (MSCs) are cells that can renew themselves and transform into the major specialized cell types of tissues or organs, including hepatocyte-like cells. As MSCs provide promising potential for patients with liver disease, in this respect, we focus not only on the clinical application of MSCs but also on the mechanisms of therapy that will promote and improve treatment for liver diseases. The technology of MSCs transplantation was first used to treat patients with liver disease in 2005, indicating the supportive hepatic role of stem cell therapy in vivo. Since then, many source-derived MSCs including autologous bone marrow-derived stem cells, autologous peripheral blood-derived stem cells, autologous hematopoietic stem cells, autologous bone marrow-derived mesenchymal stem cells, allogeneic umbilical cord blood-derived stem cells, and allogeneic umbilical cord-derived mesenchymal stem cells have been used for liver disease therapy. Most of these studies have showed inspiring results of stem cell therapy for liver diseases, including the improvement of serum parameters, the recovery of hepatic function, and even the improvement of quality of life, with less adverse effects after MSCs transplantation. Indeed, after transplantation, MSCs were found to migrate to the recipient's liver and express the donor's hepatocyte markers. Recent studies have demonstrated that MSCs, especially mesenchymal stem cells, possess both a hepatic differentiation property and an immune-tolerant phenotype and immunosuppressive and immunomodulatory capacities in vitro and in vivo. The beneficial effect of stem cell therapy on liver disease may result from the supportive roles of stem cells, particularly the immunomodulation.

Keywords Mesenchymal stem cell · Immunomodulation · Liver disease

---

X. Zhou · L. Cui · Y. Shi · Y. Han (✉)
State Key Laboratory of Cancer Biology, Xijing Hospital of Digestive Diseases,
The Fourth Military Medical University, 127 West Chang-le Road,
Xi'an 710032, Shaanxi Province, China
e-mail: hanying@fmmu.edu.cn

R.C. Zhao (ed.), *Stem Cells: Basics and Clinical Translation*, Translational
Medicine Research 1, DOI 10.1007/978-94-017-7273-0_16

16.1 Introduction

The liver is the largest digestive gland in human body and the central site of material and energy metabolism. A diversity of risk elements, especially viruses, causes great damage to the liver, resulting in its dysfunction or failure associated with high levels of morbidity and mortality. In China, the HBV-carrying rate of the population has increased to 7.18 %, whereas that of HCV is 3.2 %. It is estimated that there are over 6 million patients newly diagnosed with liver cirrhosis every year, with a mortality rate of 20–30 %. According to the data, nearly 500,000 patients died from decompensated liver cirrhosis and associated complications. Currently, most of the treatment for decompensated liver cirrhosis relies on drugs or bioartificial liver systems to extend the life span of patients, with unsatisfactory results. Indeed, liver transplantation is the only effective treatment for patients with end-stage liver disease. However, the short supply of donor organs and the possible complications and immune rejection after transplantation limit its clinical application. The most promising treatment at present is the new field of regenerative medicine and in particular the use of multipotent stem cells (MSCs), which are considered to be an alternative method of organ donation to treat liver disease.

Stem cells are a class of cells that can self-renew and differentiate into multiple cell lineages of tissues or organs, including hepatocyte-like cells. The first report of stem cell transplantation that was used to treat patients with liver disease was in 2005, indicating the supportive hepatic role of stem cell therapy (Hengstler et al. 2005). Several main types of stem cells including autologous bone marrow-derived stem cells, autologous peripheral blood-derived stem cells, autologous hematopoietic stem cells, autologous bone marrow-derived mesenchymal stem cells, umbilical cord blood-derived stem cells, and umbilical cord-derived mesenchymal stem cells have been used for liver disease therapy. These stem cells possess the common stem features, yet also retain their particular advantages and disadvantages. In the following section, we discuss the preclinical and clinical application of these particular cells in liver disease and the possible molecular mechanisms by which MSCs improve hepatic function and ameliorate liver injury.

16.2 Clinical Application of Stem Cells

To date, there are many studies on multipotent stem cellular therapy for patients with liver disease, such as viral hepatitis, alcoholic liver disease, and autoimmune liver disease. Several varieties of MSCs have been considered as a source of cell therapy, a potential therapeutic treatment for organ donation, which is not always possible but the only effective method for end-stage liver disease. In liver regeneration studies, both animal and human trials have demonstrated and highlight the importance and potential of stem cell use in liver injury and repair. The aim of this study is to review these researches and compare the quality and efficacy of different stem cell treatment and to explore a possible better direction to guide future application.

16.2.1 Bone Marrow-Derived Stem Cells

Bone marrow-derived stem cells (BMSCs) are usually mononuclear cells harvested from bone marrow, mainly from the iliac crest under epidural anesthesia. There are three predominant populations of BMSCs, namely mesenchymal stem cells (MSCs), hematopoietic stem cells (HSCs), and endothelial progenitor cells (EPCs). This special cell from bone marrow has reportedly been used in hematological diseases and myocardium and blood vessel regeneration (Tateishi-Yuyama et al. 2002), and several studies have applied stem cells for radiogenic oral mucositis (Sumita et al. 2014), heart and skeletal muscle tissue repair (Souza et al. 2014), and nerve regeneration (Salomone et al. 2013). These findings suggest that bone marrow stem cells could be effective sources for liver regeneration. Recent studies have indicated that BMSCs can be used to improve liver disease, including HBV-related liver cirrhosis, alcoholic liver disease, and acute liver failure.

Indeed, animal studies focusing on the therapy of bone marrow stem cells for liver diseases have presented exciting results. Shizhu et al. (2012) suggested that systemically delivered bone marrow mononuclear cells may migrate and be retained by the injured liver, showing an overall beneficial effect in a murine model of acute liver failure. Evidence has indicated that autologous BMSC transplantation may reduce ELRLA-induced liver injury and improve survival rates in hepatic fibrosis rats (Xu et al. 2013). To confirm the potential of stem cell therapy for liver disease, the researchers traced labeled stem cells which were injected into the animal body and detected the location of the cells. One study transplanted GFP-positive BMSCs into a carbon tetrachloride (CCL_4)-induced liver injury model via the tail vein, demonstrating that BMSCs populated the damaged liver and transdifferentiated into functional mature albumin-producing hepatocytes (Terai et al. 2003). Preclinical results are promising, and studies about the clinical application of bone marrow stem cells are worthy of consideration.

Nine patients with decompensated liver cirrhosis were reported to undergo autologous bone marrow cell infusion from a peripheral vein. After transplantation, significant improvement in serum albumin levels and total protein of patients were observed during the follow-up period (Terai et al. 2006). Additionally, the infusion of BMMCs through the hepatic artery in patients with cirrhosis was found to be feasible, and improvement in liver function occurred after this intervention in some patients (Couto et al. 2011). However, the results of such studies are not always satisfactory. Laurent et al. randomly assigned 58 patients (all with cirrhosis, 81 % with alcoholic steatohepatitis) to a group receiving standard medical therapy or combined with granulocyte colony-stimulating factor (G-CSF) injections and autologous bone marrow mononuclear cell transplantation (BMMCT) into the hepatic artery (Spahr et al. 2013). However, there were no significant differences between the two groups, suggesting insufficient regenerative stimulation after BMMCT or resistance to liver regeneration. Unpredictably, when investigating the mobilization and hepatic recruitment of bone marrow (BM) stem cells in patients with alcohol liver injury and their contribution to parenchymal/non-parenchymal liver cell lineages, researchers found that BMSCs contributed to the hepatic myofibroblast population and did not promote hepatocyte repair (Dalakas et al. 2010). Thus, further studies are necessary to maximize the use of BMSCs in the therapy for liver diseases.

Presently, the only way to obtain BMSCs is by aspiration of BM from the iliac crest, which requires general anesthesia and may cause local complications. Through such studies, adverse side effects related to the transplantation of BMMCs have been reported, including hematoma, acute variceal bleeding, liver failure, and even death (Spahr et al. 2013). For optimal stem cell therapy, the presence of the injected cells in the liver should be confirmed. However, most studies have been designed to compare the therapeutic role of BMSCs with previous standard methods, ignoring the work associated with determining the positive or negative role played by BM stem cells. In addition, the culture of stem cell in vitro to increase their numbers will also increase the cost and risk of contamination before injection (Terai et al. 2006; Lyra et al. 2007). Based on past studies of BMSCs for the treatment of liver disease, it is clear that the following considerations need to be urgently addressed: the cell number, the injection route, the trace and location of cells, and the related side effects. Additionally, long-term observation and randomized control trials should make the future application of stem cells from BM more convincing.

16.2.2 Peripheral Blood-Derived Stem Cells

Peripheral blood is considered to be a good source of stem cells for basic and clinical research. Peripheral blood-derived stem cells are mononuclear cells obtained from peripheral blood; actually, these stem cells are mobilized from BM with the pretreatment of G-CSF and later collected by a special separator machine. These stem cells have the potential to differentiate into various progenitor cells, such as blood cells, endothelial cells, osteoblasts, neural cells, and hepatocytes. Furthermore, these stem cells have been reported to be associated with AML1/ETO(+) acute myeloid leukemia, multiple sclerosis (Simpson et al. 2014), and lung diseases (Bahr et al. 2013). Additionally, they show great benefit in liver diseases therapy, and their transplantation can result in tissue regeneration and repair after injury. Indeed, several distinct progenitor cell populations have been found in peripheral blood monocytes (PBMCs), including HSCs, MSCs, EPCs, and some other precursor cells, explaining their multidifferential ability in different microenvironments (Zhang and Huang 2012).

Preclinical studies play a vital role in the clinical use of stem cells. Our study group once harvested $CD14^+$ cells from PBMCs and transplanted them into the female rats with CCL4-induced liver cirrhosis, resulting in an increasing serum albumin level and a decreasing portal vein pressure (Wang et al. 2010). Later, we also illustrated that PBMCs from decompensated liver cirrhosis could differentiate into hepatocyte-like cells in vitro (Yan et al. 2007). In the ensuing research, we isolated PBMCs from one cirrhotic patient, which were mobilized by recombinant human granulocyte colony-stimulating factor for 3 consecutive days. We then transplanted these PBMCs into nude mice via tail vein after labeling them with PKH26-GL to locate the transplanted PBMCs and detect the expression of human hepatocyte markers. The results demonstrated that the PBMCs from decompensated cirrhosis liver could migrate into the liver of nude mice and express human hepatocyte markers, indicating that autologous PBMC transplantation might be an alternative therapy for decompensated liver cirrhosis (Yan et al. 2008).

Of course, animal studies are not sufficient to indicate the wide use of PBMCs to cure liver disease, and clinical research is rather important. Based on the promising preclinical data, two patients with advanced-stage alcoholic liver cirrhosis were treated with PBSCs mobilized by G-CSF (Yannaki et al. 2006). Both patients showed a lasting amelioration in the clinical course of the disease, with a safe and well-tolerated procedure, which may be considered as a bridging therapy prior to organ transplantation or to reverse decompensated cirrhosis. Our study group conducted controlled trials in hepatitis B virus-related decompensated liver cirrhosis patients, with a total of 40 subjects (31 men and 9 females) recruited into two groups. Group 1 received G-CSF mobilization, PBMC collection by leukapheresis, and PBMC transplant therapy. Group 2 only received G-CSF mobilization for 4 days. The liver function of the two groups was monitored by blood examination and ultrasonography at baseline and 6 months later. We were excited to find that compared with group 2, a significant amelioration of liver function was observed in group 1, including an elevated serum albumin level and a decreased CTP score ($P < 0.05$), and with no major adverse effects (Han et al. 2008).

In comparison with other types of stem cells, the isolation of peripheral blood is less invasive and does not require anesthesia. In addition, with autologous peripheral blood stem cells, patients do not need long-term immunosuppressive therapy and avoid the problem of ethical limitations (Bensinger et al. 1995; Hassan et al. 1996). However, as the application of G-CSF induces spleen enlargement in peripheral blood cell donors and patients with hematologic malignancies (Picardi et al. 2003), the stringent selection, especially regarding the size of spleen, may result in the low rate of eligibility. Moreover, additional and further clinical trial research is required to study the use of these cells to treat liver diseases.

16.2.3 Hematopoietic Stem Cells

HSCs, which are the source of all circulating mature blood cells, were first described in the early studies of Till and McCulloch (Till and Mc 1961). These cells are the predominant stem cells within the BM, express the important maker CD34, and are able to renew themselves and differentiate into the progenitor cells, or colony-forming units, of the hematopoietic system. However, a small portion of the cells are CD133 positive and not CD34 positive. Thus, the transplantation of autologous stem cell first requires the harvesting from BM first and then relies on antibodies and immunomagnetic beads or FACS to sort the $CD34^{+}$ and $CD133^{+}$ stem cells. HSCs have been reported to be used to alleviate kidney injury (Burst et al. 2013), acute lymphoblastic leukemia (Ribera 2011), and diabetes mellitus (Vanikar et al. 2010).

Cirrhosis represents a late stage of progressive hepatic fibrosis characterized by a distortion of the hepatic architecture and the formation of regenerative nodules. A number of scientists are working to find an alternative method to treat liver cirrhosis instead of organ transplantation, which is rarely applied due to its limitations. The most promising treatment at present may be the new field of regenerative medicine and in particular the use of stem cells. Numerous preclinical studies have demonstrated the ability of HSCs to improve liver function, alleviate liver injury and cirrhosis, and contribute to liver regeneration in animal models of liver disease (Zhan et al. 2006; Yu et al. 2012; Schmelzle et al. 2013; Tsolaki et al. 2014).

In clinical studies, five patients with liver insufficiency were injected with 1×10^6–2×10^8 autologous $CD34^{+}$ cells that were mobilized with G-CSF from peripheral blood cells (Gordon et al. 2006). In three patients, the cells were injected into the portal vein under CT scan, and in two patients, they were injected via the hepatic artery. The results of the study highlighted the dramatic improvement in liver function of patients. There were no mortalities or specific side effects, except for mild pain and discomfort at the site of $CD34^{+}$ cell infusion. In another beneficial study, six patients with end-stage liver disease were subjected to the intraportal administration of autologous BM-derived $CD133^{+}$ HSCs in comparison with mononuclear cells, with short-term (6 months) and long-term (24 months) follow-up (Nikeghbalian et al. 2011). Even though there was no significant alteration of liver serum parameters, this investigation proved the safety and practicability of this therapeutic approach to end-stage liver disease as a promising alternative for patients on waiting lists for liver transplantation. Additionally, it was reported that four patients with decompensated cirrhosis were infused with BM-derived $CD34^{+}$ HSCs through the hepatic artery in order to alleviate the damage of liver (Mohamadnejad et al. 2007). Although the liver function parameters, such as serum albumin and prothrombin time (PT), improved at first, one patient suffered from worsened total bilirubin, serum creatinine, and model for end-stage liver disease (MELD) score at the end of follow-up. Moreover, another patient developed radiocontrast nephropathy after the procedure and progressed to type 1 hepatorenal syndrome, dying of liver failure a few days later. We might conclude that the infusion of $CD34^{+}$ stem cells through the hepatic artery is not safe in decompensated cirrhosis. HSCs are also reported to be a new therapeutic option to approach drug-induced hepatitis. A case of acute liver failure related to chronic alcoholic abuse and acute drug-induced injury was also studied

(Gasbarrini et al. 2007). With $CD34^+$ cell injection into the portal vein, the patient showed a rapid improvement of synthetic liver function, with particular reference to coagulation parameters.

In addition to these beneficial results, we need to focus on the vital issues concerning the application of HSCs for liver diseases. The infusion route of HSCs is one of the important parts of research, and whether the route chosen is safe needs further study before its application. Moreover, the studies to date highlight not only the outcomes of therapy, but also the detailed methods and guarantees of safety in applying this treatment. Lastly, larger numbers of patients and more randomized control trials are needed before the wide implementation of stem cells.

16.2.4 Bone Marrow-Derived Mesenchymal Stem Cells

MSCs, originating from many mesenchymal and connective tissues (Jiang et al. 2002), possess plasticity and multidirectional differentiation potential. These cells also play a role of immunomodulation, due to their interaction with a variety of immune cells (Uccelli et al. 2008). Friedenstein et al. (1966) were the first to report culture methods for isolating BM-MSCs and testing their differentiation potential. These cells develop from "bone marrow fibroblasts," "bone marrow stromal cells" to "MSCs" or "skeletal stem cells." This group of cells can be harvested from many tissues, including BM, periosteum, trabecular bone, adipose tissue, and skeletal muscle (Barry and Murphy 2004). The MSCs have the self-renewal ability and can differentiate into multiple cells types, such as osteoblasts (Liu et al. 2013), chondrocytes (Hwang et al. 2011), hepatic cells (Pulavendran et al. 2010), and neuronlike cells (Tondreau et al. 2008). As there is no specific characteristic of MSCs, they are defined using a combination of phenotypic markers and functional properties (Kharaziha et al. 2009). Most studies report these cells to be positive for CD13, CD29, CD44, CD54, CD55, CD73, CD90, CD105, and CD166 by flow cytometry (Gotherstrom et al. 2004; Kemp et al. 2005; Le Blanc and Ringden 2005).

The ability of MSCs to differentiate into hepatocytes provides a novel method for liver diseases. Indeed, animal studies have revealed the benefits of bone marrow MSCs for the treatment of liver failure (Vaegler et al. 2014) and have shown that BM-MSC infusion can alleviate liver cirrhosis (Abdel Aziz et al. 2007). In addition, BM-MSC transplantation was found to regulate the systemic immune response and enhance recovery in liver inflammation of PBC mice (Wang et al. 2011).

Thus far, the clinical application of bone marrow MSCs has made great progress in such treatment, which indicates a bright future for the treatment of liver disease. Peng et al. (2011) studied 527 patients with liver failure caused by hepatitis B receiving the same medical treatment, and among them, 53 patients also underwent a single transplantation with autologous MMSCs. Their results demonstrated that autologous BMMSC transplantation is safe for liver failure patients caused by chronic hepatitis B, and the short-term efficacy to alleviate liver injury was favorable. Although long-term outcomes were not markedly improved, this method is preferable for patients with liver cirrhosis and may have potential for reducing their incidence of HCC and mortality. In another study, autologous MSCs obtained from the iliac crest were proliferated and injected into a peripheral vein or the portal vein of eight patients with end-stage liver disease (Kharaziha et al. 2009). After injection, the data showed that MSC injection can be used for the treatment of end-stage liver disease with satisfactory tolerability. In addition, four patients with decompensated liver cirrhosis were enrolled for MSC transplantation to determine its safety and feasibility (Mohamadnejad et al. 2007). In this phase 1 trial, there were no side effects in the patients during follow-up, and their model for end-stage liver disease scores (MELDs) and quality of life improved after the treatment. Finally, a pilot study was carried out to determine the clinical outcomes of five patients with liver failure of various causes. The patients received autologous CD34-depleted bone marrow-derived mononuclear cell (BM-MNC) transplantation, including mesenchymal stromal cells, through the hepatic artery (Park et al. 2013). Serum albumin levels, liver stiffness, liver volume, subjective healthiness, and quality of life improved in the study patients, which may suggest a promising future for autologous CD34-depleted BM-MNC transplantation as a bridge to liver transplantation in patients with liver failure, despite this being a small population.

The results of MSC therapy are inspiring, but there are still some limitations or problems that need to be resolved. First, the prolonged period of cell culture and preparation (approximately a few months) may result in the loss of critically ill patients before treatment can be initiated. Second, it is necessary to trace and localize the cells in order to further illustrate the therapeutic potential. Third, some studies on the treatment of chronic liver disease with MMSC transplantation have shortcomings, such as a small sample size, lack of controls, and the absence of evaluation on the long-term efficacy, prognosis, and safety (Levicar et al. 2008; Khan et al. 2008). Finally, the route of MMSC administration, the number of cells used for transplantation, and the homing ability that may affect the efficacy of transplantation should also be confirmed (Shim et al. 2010; Hashemi et al. 2008).

16.2.5 Umbilical Cord-Derived Mesenchymal Stem Cells

Although most of MSCs are derived from adult BM, the supply is limited and the cell numbers decrease with age. Umbilical cord mesenchymal stem cells are of particular interest because of their easy accessibility as an abundant resource from discarded umbilical cord (UC). Additionally, BM-MSCs may suffer from proliferative deficiency (Zhong et al. 2010), which also promotes the use of UC-MSCs. The UC was collected from a healthy full-term fetus with the consent of the parents, and we obtained single cells from the cord using a mechanic method and enzyme digestion; these adherent cells were then cultured on a coverslip in a 5 % CO_2 incubator at 37 °C (Cui et al. 2012).

Similar to the MSCs from BM, UC-MSCs exhibit a great potential to improve the liver function of rats with liver cirrhosis (Tsai et al. 2009). Our group elucidated specific microRNAs (miRNAs) that convert UC-MSCs into hepatocyte-like cells. We transplanted UC-MSCs and induced hepatocytes (iHep) into nude mice with CCL_4-induced liver injury and found that both of the cells can alleviate liver injury and improve liver function to some extent (Cui et al. 2012, 2013).

In another clinical research study, 45 chronic hepatitis B patients with decompensated liver cirrhosis, including 30 patients receiving UC-MSC transfusion through the peripheral vein at a concentration of 0.5×10^6/kg body weight and 15 patients receiving saline as the control, were recruited (Zhang et al. 2012). Clinical parameters during a 1-year follow-up period were detected to obtain information over time. The results showed a significant reduction in the volume of ascites and improved liver function in the patients treated with UC-MSC transfusion compared with the controls. Recently, the application of UC-MSCs achieved great improvement in patients suffering from severe autoimmune diseases, including systemic lupus erythematosus (SLE) (Sun et al. 2010), immune thrombocytopenia patients (Ma et al. 2012), and therapy-resistant rheumatoid arthritis (Liu et al. 2010), without any side effects. A single-arm survey that consisted of seven PBC patients who had an incomplete response to UDCA showed exciting results for autoimmune liver diseases therapy (Wang et al. 2013). UDCA is the only approved drug for the treatment of PBC which is a progressive autoimmune liver disease causing cholestasis, cirrhosis, liver failure, and even hepatocellular carcinoma. The patients received UC-MSCs three times at 4-week intervals and regular UDCA as well. During the follow-up period, the symptoms of fatigue were largely alleviated, and some of the patients underwent remission of pruritus. These results not only indicate a safe and feasible way to transfuse the cells, but also demonstrate the improvement of liver function and quality of life in these patients. Moreover, a study enrolling 43 acute-on-chronic liver failure (ACLF) patients for clinical research of UC-MSCs transfusion (Shi et al. 2012) showed significantly increased survival rates in ACLF patients; reduced MELDs; increased serum albumin, cholinesterase, and prothrombin activity; and increased platelet counts, indicating a novel therapeutic method for patients with ACLF.

No significant side effects and complications were observed in the above researches. However, several limitations were present in these studies. First, they did not prove that the exciting results are definitely associated with the infusion of UC-MSCs in these small study populations. Second, UC-MSCs combined with other treatment may prompt a concern that this treatment affects the function of stem cells. It is possible that more detailed follow-up times are necessary; nonetheless, the transplanted cells still need to be traced, and their numbers and optimal routines need to be optimized.

16.2.6 Umbilical Cord Blood-Derived Stem Cells

Stem cells, especially MSCs derived from umbilical cord blood (UCB), are reported to be a fascinating source for cell therapy. UCB-MSCs have the advantage of easy accessibility, low chance of viral contamination, painless procedures for donors, and less immune response (Kim et al. 2008). A human UCB sample was collected from donors after informed consent, and then, mononuclear cells, which contained MSCs, were harvested by negative separation with Ficoll-Hypaque. The UCB-derived cells are positive for mesenchymal progenitor/stem cell-related surface makers CD271, CD29, CD90, CD105, and CD73 and negative for CD31, CD79b, CD133, CD34, and CD45, indicating that these cells do not have a hematopoietic origin (Yu et al. 2012). However, $CD34^+$ cells are present in UCB and may exceed those in BM or peripheral blood. These $CD34^+$ cells can be isolated by magnetic cell sorting from UCB (Zhang et al. 2011). Stem cells from UCB tend to be more immature and with a lower immunogenicity in comparison with BMSCs.

In various animal studies, UCB-MSCs can reportedly treat intracranial glioma (Kim et al. 2008), spinal cord injury (Lim et al. 2007), lung injury (Kim et al. 2011), and bone disease (An et al. 2013). UCB-MSCs have also been reported to attenuate myocardial infarction (Kang et al. 2014), protect against cerebral ischemia (Zhu et al. 2014), and promote cutaneous wound healing (Joyce et al. 2012). A recent study showed that human HGF-overexpressing human umbilical cord blood-derived MSCs (hHGF-hUCB-MSCs) exerted a stimulatory effect on hepatocyte proliferation in vitro. When transplanted into a rat model with CCL4-induced liver fibrosis, these cells exhibited surprising results by biochemical and histological analyses compared with the control group without hHGF-hUCB-MSCs. The injected hUCB-MSCs were able to migrate to the injured liver and differentiate into hepatocyte-like cells, with the improvement of liver structure improved in experimental mice (Yu et al. 2012). Based on these studies, hUCB-MSCs could become suitable as seed cells in cellular therapy for acute liver injury. Work has been performed using hepatocyte-like cells from $CD34^+$ UCB stem cells in microcapsules in acute hepatic failure (AHF) rats to avoid possible immune rejection after transplantation (Zhang et al. 2011). Although the low immunogenicity of UCB cells makes them appropriate for heterogenic transplantation, it is still risky to use untreated cells.

Despite the inspiring results of animal studies, few clinical trials have been reported. Because confirmatory evidence in humans is currently lacking, there is a long way to go before hUCB-MSCs will be utilized clinically to benefit patients with liver diseases.

In conclusion, most of the studies to date have demonstrated exciting results of stem cell therapy for liver diseases. The transplantation of different types of stem cells could improve liver function, reduce the occurrence of complications, and prolong patient survival time. The related clinical studies are shown in Table 16.1.

Table 16.1 The clinical studies of various types of stem cells

Cell type	Liver diseases	Routine	Time (month)	Outcomes	Side effects	Reference
BMSC	Liver cirrhosis	PVI	6	Albumin levels↑; Child–Pugh scores↓; Ascites↓	Fever	Terai et al. (2006)
	Liver cirrhosis	HAI	12	Bilirubin↓; Serum albumin↑	Dissection of the hepatic artery Takotsubo syndrome	Couto et al. (2011)
	Decompensated alcoholic liver disease (ALD)	HAI	3	MELD score↓; steatosis↓	Hematoma; acute variceal bleeding; liver failure	Spahr et al. (2013)
PBSC	Alcoholic liver cirrhosis	No mention	30	CTP and MELD scores↓	Spleen enlargement	Wang et al. (2010)
	Liver insufficiency	PVI and HAI	2	Bilirubin↓; Albumin↑; Ascites↓	Mild pain and discomfort; urinary tract infection	Yan et al. (2007)
	Liver cirrhosis	PVI	6	Liver synthetic function↑	No	Yan et al. (2008)
HSC	End-stage hepatic disease	PVI	6 and 24	No significant alternations	Low-grade fevers; partial thrombosis	Nikeghbalian et al. (2011)
	Decompensated cirrhosis	HAI	6	Bilirubin↓; Albumin↑	Renal failure; type 1 hepatorenal syndrome	Mohamadnejad et al. (2007)
	Acute liver failure	PVI	2	Synthetic liver function↑	No	Gasbarrini et al. (2007)
BM-MSC	End-stage liver disease	PVI	6	MELD scores↓; Albumin↑; Bilirubin↓	No	Kharaziha et al. (2009)
	Hepatitis B liver failure	HAI	48	Albumin↑; TBIL, PT, MELD score↓	No	Peng et al. (2011)
	Liver cirrhosis	PVI	12	MELD score↓; Liver function↑	Peripheral edema	Mohamadnejad et al. (2007)

(continued)

Table 16.1 (continued)

Cell type	Liver diseases	Routine	Time (month)	Outcomes	Side effects	Reference
	Liver failure	HAI	4	Albumin levels, Liver volume↑	Transient local pain; minimal bleeding	Park et al. (2013)
UC-MSC	Hepatitis B liver cirrhosis	PVI	12	Ascites↓; Liver function↑	No	Zhang et al. (2012)
	PBC	PVI	1	Fatigue and pruritus↓	No	Wang et al. (2013)
	ACLF	Intravenously	12/18	Survival rates↑; MELD scores↓; Albumin, cholinesterase↑	Fever	Shi et al. (2012)

BMSC bone marrow-derived stem cell; *PVI* portal vein infusion; *HAI* hepatic artery infusion; *PBSC* peripheral blood-derived stem cell; *HSC* hematopoietic stem cell; *BM-MSC* bone marrow-derived mesenchymal stem cell; *UC-MSC* umbilical cord-derived mesenchymal stem cell; *PBC* primary biliary cirrhosis; *ACLF* acute and chronic liver failure; *TBIL* total bilirubin level; *PT* Prothrombin time; *MELD* Model for end-stage liver disease

16.3 Molecular Mechanism of Stem Cell Therapy

The ideal therapy for end-stage liver disease is liver transplantation. However, the scarcity of donor organs, expensive medical costs, and unpredictable complications restrict this option. Cellular therapy, especially stem cell therapy, is a new and promising field of medicine with increasingly emerging evidence for its use to treat allogeneic liver disease due to their special characteristics, including self-renewal, potential of differentiation and paracrine function, and interactions with immune cells. In previous studies, stem cells have been injected directly into the body or transplanted to induce hepatocyte-like cells. However, the fate of stem cells and induced cells once they reach the liver and the mechanisms remain uncertain. Several theories associated with cellular therapy have garnered considerable support in recent years, including cell fusion with resident hepatocytes, transdifferentiation of stem cells, and immunomodulation in the microenvironment of liver tissue. In the following section, we attempt to depict the underlying mechanisms of stem cell therapy in detail (Fig. 16.1).

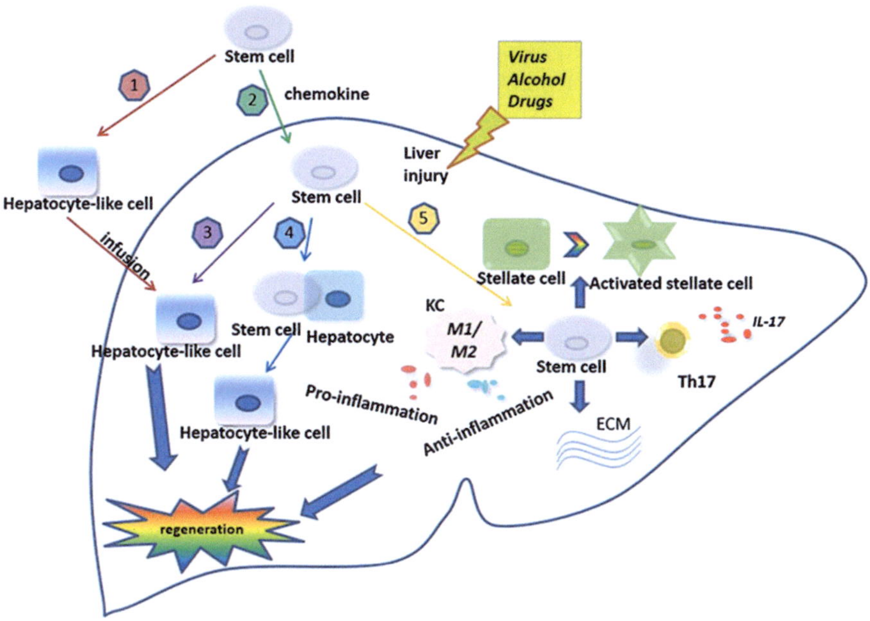

Fig. 16.1 The possible mechanisms of stem cells in liver injury therapy.① Stem cells are transdifferentiated into hepatocyte-like cells in vitro and then injected into the body. ② When stem cells are transplanted, they will migrate to the injured liver based on their chemokine receptors. ③ In the in vivo liver microenvironment, stem cells differentiate into hepatocyte-like cells to support liver function. ④ Stem cells fuse with the resident hepatocytes to generate new cells with hepatic function to repair damage. ⑤ Stem cells have a powerful immunomodulatory role when migrating to the liver: They promote the change of KCs from an M1 to an M2 phenotype; the activation of hepatic stellate cells is suppressed by stem cells which is helpful to reduce liver fibrosis; stem cells may result in a benefit to liver injury through downregulating IL-17, which is a pro-inflammatory factor mainly produced by Th17 cells; stem cells can also remodel the ECM via the regulation of matrix metalloproteinases. *Abbreviations* KCs, Kupffer cells; ECM, extracellular matrix

16.3.1 Fusion

Cell fusion is a natural process from the very beginning of life, i.e., a spermatozoid fusing with an ovum. Decades ago, cell fusion was reported to be linked with stem cells, which possess the ability of plasticity and can differentiate into different cells. Recently, the conversion of stem cells into hepatic cells has been reported in vivo and in vitro. Cell fusion is regarded as a possible mechanism related to this fate conversion with evidence from reported experiments. In a model of tyrosinemia-induced liver failure, mice with mutations in the fumarylacetoacetate hydrolase gene (Fah−/−) regained their normal liver function due to the transplantation of Fah+/+BM cells and form regenerating liver nodules with normal histology that express Fah. Moreover, the hepatic nodules contained both mutant and wild-type Fah alleles and expressed both donor and host genes, consistent with polyploidy genome formation by the fusion of host and donor cells (Vassilopoulos et al. 2003). Simultaneously, Wang et al. performed serial transplantation of bone marrow-derived hepatocytes in liver and used Southern blot analysis, showing that the repopulating cells were heterozygous for alleles unique to the donor cells. Moreover, cytogenetic analysis demonstrated that hepatocytes transplanted from female mice into male recipients were (80, XXXY) and (120, XXXXYY) hepatocyte karyotypes, indicating fusion between donor and host cells (Wang et al. 2003). Based on the low level of transdifferentiation in some animal models, cellular fusion may be a mechanism of transdifferentiation, which need further understanding and observation with regard to regeneration (Castro et al. 2002; Wagers et al. 2002). In another study, the transplantation of a single hematopoietic cell could be progenitor of both blood and functional hepatocytes. By using a Cre/lox DNA recombination-based strategy, the author showed that mature myeloid cells spontaneously fused with host hepatocytes, which indicated that the localized administration of fusogenic cells such as myeloid cells could be a novel strategy for cellular therapy of multiple tissues (Camargo et al. 2004). Myelomonocytic cells such as macrophages can produce functional epithelial cells through in vivo fusion, providing potential for cell therapy aimed at organ regeneration (Willenbring et al. 2004). In addition, umbilical cord stem cells have been tested for their ability to generate hepatocytes by cell fusion. However, it should be kept in mind that under different circumstances, stem cells are able to produce hepatocytes without the participation of cell fusion, indicating that the environment is important for the biological role of cell fusion.

Although these findings appear to favor cell fusion as the main mechanism of stem cell therapy in liver diseases, as opposed to transdifferentiation, scientists may doubt the applicability of the above animal models, such as the tyrosinemia mouse model. A number of cytogenetic abnormalities, including aberrant karyokinesis and multinucleation, have been reported in a Fah deficiency model and humans with tyrosinemia (Wilson et al. 1994; Jorquera and Tanguay 2001). Based on such reports, cell fusion, which could not occur under selective pressure, was brought into question. Thus, further studies with inherently stable animal models may be better for exploring the role of cell fusion in the repair of liver injury and hepatic regeneration after the injection of stem cells.

In conclusion, cell fusion to date has been well documented and is largely accepted. Most animal experiments show inspiring evidence for the role of cell fusion in liver repair: Fusion-derived hepatocytes tend to proliferate and even generate daughter cells with one half of the chromosomal content, which can be confirmed by maker segregation using ss-galactosidase and the Y-chromosome (Duncan et al. 2009). However, with new emerging theories, additional supportive studies and evidence are required to recognize the status of cell fusion in stem cell therapy.

16.3.2 Transdifferentiation

Transdifferentiation may be the simplest, most direct, and common mechanism of the use of stem cells for liver diseases. Such stem cells tend to generate hepatocyte-like cells with a normal function for damaged tissue to support regular metabolism. Their capacity lies in their plasticity, which refers to the ability of stem cells to produce new type of cells from different lineages (Wagers and Weissman 2004). The conversion of stem cells into hepatic cells has been observed both in vitro and in vivo. Initially, cell fusion was considered as a mechanism responsible for this conversion. However, this was observed that hepatocytes could convert from human cord blood cells when infused into NOD-SCID mice in the absence of fusion (Newsome et al. 2003).

Evidence from a number of transdifferentiation studies shows that other types of MSCs are capable to convert into hepatic cells in vitro or in vivo. Lagasse first reported liver differentiation from HSCs in vivo (Lagasse et al. 2000). Human bone marrow mesenchymal stem cells transplanted into the rat liver differentiated into hepatocytes with better outcomes compared with the MSC population (Sato et al. 2005). Stem cells obtained from UCB also have the capability to differentiate into hepatocyte-like cells and alleviate liver injury (Moon et al. 2009). Our group proved the differentiation potential of PBMCs derived from patients with HBV-related decompensated liver cirrhosis into hepatocyte-like cells. These exciting results indicate a bright future for these patients through autologous cell transplantation (Yan et al. 2007).

The transdifferentiation of hepatocytes from non-hepatic tissues, such as BM, peripheral blood, UC and other sources, has been widely studied, and different protocols of hepatic induction have been mentioned. The most frequently used medium to induce this transdifferentiation is basic medium containing epidermal growth factor (EGF), bFGF, HGF, nicotinamide, oncostatin M, dexamethasone, and ITS premix, which has an essential role in a novel two-step development process (Lee et al. 2004). Mouse embryonic and adult fibroblasts can convert into hepatocyte-like cells via the expression of a combination of transcription factors Hnf4alpha, Foxa1, Foxa2, or Foxa3; the induced cells possess some hepatocyte-specific characteristics and alleviated liver damage after transplantation (Sekiya and Suzuki 2011). In addition to transcription factors, HGF is a critical factor for induction. It has been demonstrated that HGF alone could induce characteristics of hepatocytes, including CK19, albumin, and alpha-fetoprotein (AFP) expression (Snykers et al. 2009). In our study, we directly converted human umbilical cord-derived MSCs into hepatocyte-like cells in vitro with a combination of seven microRNAs (mir-122, mor-1290, mir-148a, mir-424, mir-542-5p, mir-1246, and mir-30a) that are overexpressed during differentiation. In the nude mice model of CCL4-induced liver injury, the hepatocyte-like cells induced with these seven microRNAs resulted in a better improvement in liver function compared with the injection of MSCs (Cui et al. 2013). Therefore, stem cells transdifferentiate into hepatocytes in vitro and in vivo, acquiring the liver function needed during the process, which is the key to liver diseases therapy.

16.3.3 Immunomodulation

Some studies hold the belief that stem cell plasticity and differentiation into new hepatic cells represent the dominant mechanism for stem cell therapy, whereas many groups insist that stem cell fusion with local hepatocytes to produce new cells plays an important role in liver repair. However, the number of hepatocytes generated was low in some studies and cirrhosis is a hostile environment for the hepatic induction of stem cells. Therefore, it is possible that the supportive and regulative role of stem cells results in the beneficial effect on liver diseases.

The liver is a unique organ containing an abundance of immune cells, which contributes to oral tolerance, acceptance of allografts, and hepatic viral persistence. Kupffer cells (KCs), which are closely related to liver injury and liver repair, are special macrophages that resident in the liver and compose approximately 35 % of the total non-parenchymal cells. These cells are classified into two groups according to the function and phenotypes: pro-inflammatory M1 macrophage associated with the release of pro-inflammatory cytokines and anti-inflammatory M2 macrophages involved in tissue remodeling and the secretion of immunomodulatory mediators. These two types of cells perform diverse functions during the progression of liver disease. It has been reported that human MSCs have the ability to switch stimulated macrophages from an M1 phenotype to an M2 phenotype (Dayan et al. 2011). Lymphocytes, which mainly consist of T cells, B cells, and NK cells, participate in immune responses. In our studies, we found that serum IL-17, mainly produced from Th17 cells, in HBV-related decompensated cirrhotic patients was significantly higher than in healthy controls. However, after the transplantation of autologous stem cells, the level of IL-17 was markedly decreased. Furthermore, exogenous IL-17 can worse the liver function of CCL4-injured mice, whereas an antibody against IL-17 promotes the improvement of liver function, indicating that BMSC transplantation exerts a beneficial action on liver diseases, at least partly, by down-regulating IL-17 (Zheng et al. 2013).

Stem cells, especially MSCs, display a low immunogenicity phenotype, with the low expression of HLA-1 molecules and a lack of HLA-DR. These cells are also negative for costimulatory factors CD40, CD80, and CD86, allowing MSCs to possess an immune tolerance quality (Cui et al. 2014). Moreover, stem cells show a great immunosuppressive ability to suppress the proliferation and maturation of immune cells when treating allogeneic liver disease. Recent studies regard stem cells as candidates for cell therapy because of their paracrine effects. Stem cells can migrate to injured liver due to their chemokine receptors, such as CCR4, CCR7, and CCR10. The proliferative or anti-apoptotic cytokines released by stem cells tend to improve the proliferation of hepatocytes. With the influence of growth factors, stem cells enhance liver regeneration and promote repair in a hepatic failure model and in liver injury (Li et al. 2013). The imbalance of extracellular matrix (ECM) synthesis and degradation related to hepatic stellate cells could result in liver cirrhosis. Stem cell infusion suppresses the activation of hepatic stellate cells, key factors for the progression of liver cirrhosis, through the regulation of chemokines TGF-α and TGF-β (Tanimoto et al. 2013). Thus, the transplantation of stem cells modulates the type of KCs expressing matrix metalloproteinases, which facilitate the relief of liver injury and fibrosis (Fallowfield et al. 2007; Pellicoro et al. 2012). In autoimmune liver disease, such as primary biliary cirrhosis (PBC), the infusion of MSCs can notably alleviate symptoms in most patients who do not have a good response to traditional treatment (Wang et al. 2013). Furthermore, the changes in serum TGF-β1 and IFN-γ was observed after injection of stem cells in the PBC mouse model, indicative of MSC regulation of liver inflammation (Snykers et al. 2009).

16.4 Conclusion

Liver dysfunction caused by diseases, particularly end-stage liver cirrhosis, has posed a great challenge for therapy. Although the organ transplantation is regarded as the most effective method, it is difficult to apply transplantation to clinical treatment because of the rare source of donor organs, the expensive costs, and the possible immune rejection after surgery. Recently, regenerative medicine, especially stem cells, has shown promising results in liver disease both in preclinical and clinical studies, which could be a realistic aim for improving the function and promoting liver repair. Different types of MSCs, including BM stem cells, peripheral blood stem cells, HSCs, and MSCs from various sources, can both fuse with resident hepatocytes to generate new cells and also transdifferentiate or immunomodulate within the liver microenvironment during therapy. Indeed, studies have demonstrated that this therapeutic method can significantly support regeneration and prolong the survival times of patients via the injection of stem cells through the portal vein or hepatic artery. In addition, there have been no obvious adverse effects or complications reported that are directly related to the treatment of stem cells in clinical trials.

Despite these exciting results, the dominant mechanism by which stem cells treat liver diseases remains unclear. Thus, the best way to apply this method is uncertain, even based on the above studies. To maximize the use of stem cells, it is necessary to know the routine of injection, the cell number for therapy, the safety and efficiency of stem cells, and how to address the possible complications. Moreover, different types of stem cells display different features, while different liver diseases result from different factors. In this respect, elucidation of the special advantage of stem cell therapy in particular diseases may afford future application.

References

Abdel Aziz MT, Atta HM, Mahfouz S, Fouad HH, Roshdy NK, Ahmed HH, Rashed LA, Sabry D, Hassouna AA, Hasan NM. Therapeutic potential of bone marrow-derived mesenchymal stem cells on experimental liver fibrosis. Clin Biochem. 2007;40:893–9.

An JH, Park H, Song JA, Ki KH, Yang JY, Choi HJ, Cho SW, Kim SW, Kim SY, Yoo JJ, Baek WY, Kim JE, Choi SJ, Oh W, Shin CS. Transplantation of human umbilical cord blood-derived mesenchymal stem cells or their conditioned medium prevents bone loss in ovariectomized nude mice. Tissue Eng Part A. 2013;19:685–96.

Bahr TM, Hughes GJ, Armstrong M, Reisdorph R, Coldren CD, Edwards MG, Schnell C, Kedl R, LaFlamme DJ, Reisdorph N, Kechris KJ, Bowler RP. Peripheral blood mononuclear cell gene expression in chronic obstructive pulmonary disease. Am J Respir Cell Mol Biol. 2013;49:316–23.

Barry FP, Murphy JM. Mesenchymal stem cells: clinical applications and biological characterization. Int J Biochem Cell Biol. 2004;36:568–84.

Bensinger WI, Weaver CH, Appelbaum FR, Rowley S, Demirer T, Sanders J, Storb R, Buckner CD. Transplantation of allogeneic peripheral blood stem cells mobilized by recombinant human granulocyte colony-stimulating factor. Blood. 1995;85:1655–8.

Burst V, Putsch F, Kubacki T, Volker LA, Bartram MP, Muller RU, Gillis M, Kurschat CE, Grundmann F, Muller-Ehmsen J, Benzing T, Teschner S. Survival and distribution of injected haematopoietic stem cells in acute kidney injury. Nephrol Dial Transplant. 2013;28:1131–9.

Camargo FD, Finegold M, Goodell MA. Hematopoietic myelomonocytic cells are the major source of hepatocyte fusion partners. J Clin Inv. 2004;113:1266–70.

Castro RF, Jackson KA, Goodell MA, Robertson CS, Liu H, Shine HD. Failure of bone marrow cells to transdifferentiate into neural cells in vivo. Science. 2002;297:1299.

Couto BG, Goldenberg RC, da Fonseca LM, Thomas J, Gutfilen B, Resende CM, Azevedo F, Mercante DR, Torres AL, Coelho HS, Maiolino A, Alves AL, Dias JV, Moreira MC, Sampaio AL, Sousa MA, Kasai-Brunswick TH, Souza SA, Campos-de-Carvalho AC, Rezende GF. Bone marrow mononuclear cell therapy for patients with cirrhosis: a Phase 1 study. Liver Int. 2011;31:391–400.

Cui L, Shi Y, Han Y, Fan D. Immunological basis of stem cell therapy in liver diseases. Expert Rev Clin Immunol. 2014; 1–12.

Cui L, Zhou X, Li J, Wang L, Wang J, Li Q, Chu J, Zheng L, Wu Q, Han Z, Shi Y, Han Y, Fan D. Dynamic microRNA profiles of hepatic differentiated human umbilical cord lining-derived mesenchymal stem cells. PLoS One. 2012;7:e44737.

Cui L, Shi Y, Zhou X, Wang X, Wang J, Lan Y, Wang M, Zheng L, Li H, Wu Q, Zhang J, Fan D, Han Y. A set of microRNAs mediate direct conversion of human umbilical cord lining-derived mesenchymal stem cells into hepatocytes. Cell Death Dis. 2013;4:e918.

Dalakas E, Newsome PN, Boyle S, Brown R, Pryde A, McCall S, Hayes PC, Bickmore WA, Harrison DJ, Plevris JN. Bone marrow stem cells contribute to alcohol liver fibrosis in humans. Stem Cells Dev. 2010;19:1417–25.

Dayan V, Yannarelli G, Billia F, Filomeno P, Wang XH, Davies JE, Keating A. Mesenchymal stromal cells mediate a switch to alternatively activated monocytes/macrophages after acute myocardial infarction. Basic Res Cardiol. 2011;106:1299–310.

Duncan AW, Hickey RD, Paulk NK, Culberson AJ, Olson SB, Finegold MJ, Grompe M. Ploidy reductions in murine fusion-derived hepatocytes. PLoS Genet. 2009;5:e1000385.

Fallowfield JA, Mizuno M, Kendall TJ, Constandinou CM, Benyon RC, Duffield JS, Iredale JP. Scar-associated macrophages are a major source of hepatic matrix metalloproteinase-13 and facilitate the resolution of murine hepatic fibrosis. J Immunol. 2007;178:5288–95.

Friedenstein AJ, Piatetzky S II, Petrakova KV. Osteogenesis in transplants of bone marrow cells. J Embryol Exp Morphol. 1966;16:381–90.

Gasbarrini A, Rapaccini GL, Rutella S, Zocco MA, Tittoto P, Leone G, Pola P, Gasbarrini G, Di Campli C. Rescue therapy by portal infusion of autologous stem cells in a case of drug-induced hepatitis. Dig Liver Dis. 2007;39:878–82.

Gordon MY, Levicar N, Pai M, Bachellier P, Dimarakis I, Al-Allaf F, M'Hamdi H, Thalji T, Welsh JP, Marley SB, Davies J, Dazzi F, Marelli-Berg F, Tait P, Playford R, Jiao L, Jensen S, Nicholls JP, Ayav A, Nohandani M, Farzaneh F, Gaken J, Dodge R, Alison M, Apperley JF, Lechler R, Habib NA. Characterization and clinical application of human $CD34^+$ stem/progenitor cell populations mobilized into the blood by granulocyte colony-stimulating factor. Stem Cells. 2006;24:1822–30.

Gotherstrom C, Ringden O, Tammik C, Zetterberg E, Westgren M, Le Blanc K. Immunologic properties of human fetal mesenchymal stem cells. Am J Obstet Gynecol. 2004;190:239–45.

Han Y, Yan L, Han G, Zhou X, Hong L, Yin Z, Zhang X, Wang S, Wang J, Sun A, Liu Z, Xie H, Wu K, Ding J, Fan D. Controlled trials in hepatitis B virus-related decompensate liver cirrhosis: peripheral blood monocyte transplant versus granulocyte-colony-stimulating factor mobilization therapy. Cytotherapy. 2008;10:390–6.

Hashemi SM, Ghods S, Kolodgie FD, Parcham-Azad K, Keane M, Hamamdzic D, Young R, Rippy MK, Virmani R, Litt H, Wilensky RL. A placebo controlled, dose-ranging, safety study of allogenic mesenchymal stem cells injected by endomyocardial delivery after an acute myocardial infarction. Eur Heart J. 2008;29:251–9.

Hassan HT, Zeller W, Stockschlader M, Kruger W, Hoffknecht MM, Zander AR. Comparison between bone marrow and G-CSF-mobilized peripheral blood allografts undergoing clinical scale $CD34^+$ cell selection. Stem Cells. 1996;14:419–29.

Hengstler JG, Brulport M, Schormann W, Bauer A, Hermes M, Nussler AK, Fandrich F, Ruhnke M, Ungefroren H, Griffin L, Bockamp E, Oesch F, von Mach MA. Generation of human hepatocytes by stem cell technology: definition of the hepatocyte. Expert Opin Drug Metab Toxicol. 2005;1:61–74.

Hwang NS, Varghese S, Li H, Elisseeff J. Regulation of osteogenic and chondrogenic differentiation of mesenchymal stem cells in PEG-ECM hydrogels. Cell Tissue Res. 2011;344:499–509.

Jiang Y, Jahagirdar BN, Reinhardt RL, Schwartz RE, Keene CD, Ortiz-Gonzalez XR, Reyes M, Lenvik T, Lund T, Blackstad M, Du J, Aldrich S, Lisberg A, Low WC, Largaespada DA, Verfaillie CM. Pluripotency of mesenchymal stem cells derived from adult marrow. Nature. 2002;418:41–9.

Jorquera R, Tanguay RM. Fumarylacetoacetate, the metabolite accumulating in hereditary tyrosinemia, activates the ERK pathway and induces mitotic abnormalities and genomic instability. Hum Mol Genet. 2001;10:1741–52.

Joyce NC, Harris DL, Markov V, Zhang Z, Saitta B. Potential of human umbilical cord blood mesenchymal stem cells to heal damaged corneal endothelium. Mol Vis. 2012;18:547–64.

Kang BJ, Kim H, Lee SK, Kim J, Shen Y, Jung S, Kang KS, Im SG, Lee SY, Choi M, Hwang NS, Cho JY. Umbilical-cord-blood-derived mesenchymal stem cells seeded onto fibronectin-immobilized polycaprolactone nanofiber improve cardiac function. Acta Biomater. 2014;10:3007–17.

Kemp KC, Hows J, Donaldson C. Bone marrow-derived mesenchymal stem cells. Leuk Lymphoma. 2005;46:1531–44.

Khan AA, Parveen N, Mahaboob VS, Rajendraprasad A, Ravindraprakash HR, Venkateswarlu J, Rao SG, Narusu ML, Khaja MN, Pramila R, Habeeb A, Habibullah CM. Safety and efficacy of autologous bone marrow stem cell transplantation through hepatic artery for the treatment of chronic liver failure: a preliminary study. Transplant Proc. 2008;40:1140–4.

Kharaziha P, Hellstrom PM, Noorinayer B, Farzaneh F, Aghajani K, Jafari F, Telkabadi M, Atashi A, Honardoost M, Zali MR, Soleimani M. Improvement of liver function in liver cirrhosis patients after autologous mesenchymal stem cell injection: a phase I-II clinical trial. Eur J Gastroenterol Hepatol. 2009;21:1199–205.

Kim SM, Lim JY, Park SI, Jeong CH, Oh JH, Jeong M, Oh W, Park SH, Sung YC, Jeun SS. Gene therapy using TRAIL-secreting human umbilical cord blood-derived mesenchymal stem cells against intracranial glioma. Cancer Res. 2008;68:9614–23.

Kim ES, Chang YS, Choi SJ, Kim JK, Yoo HS, Ahn SY, Sung DK, Kim SY, Park YR, Park WS. Intratracheal transplantation of human umbilical cord blood-derived mesenchymal stem cells attenuates Escherichia coli-induced acute lung injury in mice. Respir Res. 2011;12:108.

Lagasse E, Connors H, Al-Dhalimy M, Reitsma M, Dohse M, Osborne L, Wang X, Finegold M, Weissman IL, Grompe M. Purified hematopoietic stem cells can differentiate into hepatocytes in vivo. Nat Med. 2000;6:1229–34.

Le Blanc K, Ringden O. Immunobiology of human mesenchymal stem cells and future use in hematopoietic stem cell transplantation. Biol Blood Marrow Transplant. 2005;11:321–34.

Lee KD, Kuo TK, Whang-Peng J, Chung YF, Lin CT, Chou SH, Chen JR, Chen YP, Lee OK. In vitro hepatic differentiation of human mesenchymal stem cells. Hepatology. 2004;40:1275–84.

Levicar N, Pai M, Habib NA, Tait P, Jiao LR, Marley SB, Davis J, Dazzi F, Smadja C, Jensen SL, Nicholls JP, Apperley JF, Gordon MY. Long-term clinical results of autologous infusion of mobilized adult bone marrow derived $CD34^{+}$ cells in patients with chronic liver disease. Cell Prolif. 2008;41(Suppl 1):115–25.

Li Q, Zhou X, Shi Y, Li J, Zheng L, Cui L, Zhang J, Wang L, Han Z, Han Y, Fan D. In vivo tracking and comparison of the therapeutic effects of MSCs and HSCs for liver injury. PLoS One. 2013;8:e62363.

Lim JH, Byeon YE, Ryu HH, Jeong YH, Lee YW, Kim WH, Kang KS, Kweon OK. Transplantation of canine umbilical cord blood-derived mesenchymal stem cells in experimentally induced spinal cord injured dogs. J Vet Sci. 2007;8:275–82.

Liu Y, Mu R, Wang S, Long L, Liu X, Li R, Sun J, Guo J, Zhang X, Guo J, Yu P, Li C, Liu X, Huang Z, Wang D, Li H, Gu Z, Liu B, Li Z. Therapeutic potential of human umbilical cord mesenchymal stem cells in the treatment of rheumatoid arthritis. Arthritis Res Ther. 2010;12:R210.

Liu Y, Yan F, Yang WL, Lu XF, Wang WB. Effects of zinc transporter on differentiation of bone marrow mesenchymal stem cells to osteoblasts. Biol Trace Elem Res. 2013;154:234–43.

Lyra AC, Soares MB, da Silva LF, Fortes MF, Silva AG, Mota AC, Oliveira SA, Braga EL, de Carvalho WA, Genser B, dos Santos RR, Lyra LG. Feasibility and safety of autologous bone marrow mononuclear cell transplantation in patients with advanced chronic liver disease. World J Gastroenterol. 2007;13:1067–73.

Ma L, Zhou Z, Zhang D, Yang S, Wang J, Xue F, Yang Y, Yang R. Immunosuppressive function of mesenchymal stem cells from human umbilical cord matrix in immune thrombocytopenia patients. Thromb Haemost. 2012;107:937–50.

Mohamadnejad M, Namiri M, Bagheri M, Hashemi SM, Ghanaati H, Zare Mehrjardi N, Kazemi Ashtiani S, Malekzadeh R, Baharvand H. Phase 1 human trial of autologous bone marrow-hematopoietic stem cell transplantation in patients with decompensated cirrhosis. World J Gastroenterol. 2007a;13:3359–63.

Mohamadnejad M, Alimoghaddam K, Mohyeddin-Bonab M, Bagheri M, Bashtar M, Ghanaati H, Baharvand H, Ghavamzadeh A, Malekzadeh R. Phase 1 trial of autologous bone marrow mesenchymal stem cell transplantation in patients with decompensated liver cirrhosis. Arch Iran Med. 2007b;10:459–66.

Moon YJ, Yoon HH, Lee MW, Jang IK, Lee DH, Lee JH, Lee SK, Lee KH, Kim YJ, Eom YW. Multipotent progenitor cells derived from human umbilical cord blood can differentiate into hepatocyte-like cells in a liver injury rat model. Transplant Proc. 2009;41:4357–60.

Newsome PN, Johannessen I, Boyle S, Dalakas E, McAulay KA, Samuel K, Rae F, Forrester L, Turner ML, Hayes PC, Harrison DJ, Bickmore WA, Plevris JN. Human cord blood-derived cells can differentiate into hepatocytes in the mouse liver with no evidence of cellular fusion. Gastroenterology. 2003;124:1891–900.

Nikeghbalian S, Pournasr B, Aghdami N, Rasekhi A, Geramizadeh B, Hosseini Asl SM, Ramzi M, Kakaei F, Namiri M, Malekzadeh R, Vosough Dizaj A, Malek-Hosseini SA, Baharvand H. Autologous transplantation of bone marrow-derived mononuclear and CD133$^{(+)}$ cells in patients with decompensated cirrhosis. Arch Iran Med. 2011;14:12–7.

Park CH, Bae SH, Kim HY, Kim JK, Jung ES, Chun HJ, Song MJ, Lee SE, Cho SG, Lee JW, Choi JY, Yoon SK, Han NI, Lee YS. A pilot study of autologous CD34-depleted bone marrow mononuclear cell transplantation via the hepatic artery in five patients with liver failure. Cytotherapy. 2013;15:1571–9.

Pellicoro A, Aucott RL, Ramachandran P, Robson AJ, Fallowfield JA, Snowdon VK, Hartland SN, Vernon M, Duffield JS, Benyon RC, Forbes SJ, Iredale JP. Elastin accumulation is regulated at the level of degradation by macrophage metalloelastase (MMP-12) during experimental liver fibrosis. Hepatology. 2012;55:1965–75.

Peng L, Xie DY, Lin BL, Liu J, Zhu HP, Xie C, Zheng YB, Gao ZL. Autologous bone marrow mesenchymal stem cell transplantation in liver failure patients caused by hepatitis B: short-term and long-term outcomes. Hepatology. 2011;54:820–8.

Picardi M, De Rosa G, Selleri C, Scarpato N, Soscia E, Martinelli V, Ciancia R, Rotoli B. Spleen enlargement following recombinant human granulocyte colony-stimulating factor administration for peripheral blood stem cell mobilization. Haematologica. 2003;88:794–800.

Pulavendran S, Rajam M, Rose C, Mandal AB. Hepatocyte growth factor incorporated chitosan nanoparticles differentiate murine bone marrow mesenchymal stem cell into hepatocytes in vitro. IET Nanobiotechnol. 2010;4:51–60.

Ribera JM. Acute lymphoblastic leukemia in adults. Pediatr Rep. 2011;3(Suppl 2):e1.

Salomone R, Bento RF, Costa HJ, Azzi-Nogueira D, Ovando PC, Da-Silva CF, Zanatta DB, Strauss BE, Haddad LA. Bone marrow stem cells in facial nerve regeneration from isolated stumps. Muscle Nerve. 2013;48:423–9.

Sato Y, Araki H, Kato J, Nakamura K, Kawano Y, Kobune M, Sato T, Miyanishi K, Takayama T, Takahashi M, Takimoto R, Iyama S, Matsunaga T, Ohtani S, Matsuura A, Hamada H, Niitsu Y. Human mesenchymal stem cells xenografted directly to rat liver are differentiated into human hepatocytes without fusion. Blood. 2005;106:756–63.

Schmelzle M, Duhme C, Junger W, Salhanick SD, Chen Y, Wu Y, Toxavidis V, Csizmadia E, Han L, Bian S, Furst G, Nowak M, Karp SJ, Knoefel WT, Schulte Esch J, Robson SC. CD39 modulates hematopoietic stem cell recruitment and promotes liver regeneration in mice and humans after partial hepatectomy. Ann Surg. 2013;257:693–701.

Sekiya S, Suzuki A. Direct conversion of mouse fibroblasts to hepatocyte-like cells by defined factors. Nature. 2011;475:390–3.

Shi M, Zhang Z, Xu R, Lin H, Fu J, Zou Z, Zhang A, Shi J, Chen L, Lv S, He W, Geng H, Jin L, Liu Z, Wang FS. Human mesenchymal stem cell transfusion is safe and improves liver function in acute-on-chronic liver failure patients. Stem Cells Transl Med. 2012;1:725–31.

Shim WS, Tan G, Gu Y, Qian L, Li S, Chung YY, Lim SY, Sim E, Chuah SC, Wong P. Dose-dependent systolic contribution of differentiated stem cells in post-infarct ventricular function. J Heart Lung Transplant. 2010;29:1415–26.

Shizhu J, Xiangwei M, Xun S, Mingzi H, Bingrong L, Dexia K, Xinghong W, Fenghua P. Bone marrow mononuclear cell transplant therapy in mice with CCl4-induced acute liver failure. Turk J Gastroenterol. 2012;23:344–52.

Simpson S, Jr., Stewart N, van der Mei I, Otahal P, Charlesworth J, Ponsonby AL, Blizzard L, Dwyer T, Pittas F, Gies P, Taylor B. Stimulated PBMC-produced IFN-gamma and TNF-alpha are associated with altered relapse risk in multiple sclerosis: results from a prospective cohort study. J Neurol Neurosurg Psychiatry. 2014;86(2):200 207.

Snykers S, Henkens T, De Rop E, Vinken M, Fraczek J, De Kock J, De Prins E, Geerts A, Rogiers V, Vanhaecke T. Role of epigenetics in liver-specific gene transcription, hepatocyte differentiation and stem cell reprogrammation. J Hepatol. 2009;51:187–211.

Souza BS, Azevedo CM, Lima RS, Kaneto CM, Vasconcelos JF, Guimaraes ET, Dos Santos RR, Soares MB. Bone marrow cells migrate to the heart and skeletal muscle and participate in tissue repair after Trypanosoma cruzi infection in mice. Int J Exp Pathol. 2014;95(5):321 329.

Spahr L, Chalandon Y, Terraz S, Kindler V, Rubbia-Brandt L, Frossard JL, Breguet R, Lanthier N, Farina A, Passweg J, Becker CD, Hadengue A. Autologous bone marrow mononuclear cell transplantation in patients with decompensated alcoholic liver disease: a randomized controlled trial. PLoS One. 2013;8:e53719.

I T, Sumita Y, Minamizato T, Umebayashi M, Liu Y, Tran SD, Asahina I. Bone marrow-derived cell therapy for oral mucosal repair after irradiation. J Dent Res. 2014;93(8):813–20.

Sun L, Wang D, Liang J, Zhang H, Feng X, Wang H, Hua B, Liu B, Ye S, Hu X, Xu W, Zeng X, Hou Y, Gilkeson GS, Silver RM, Lu L, Shi S. Umbilical cord mesenchymal stem cell transplantation in severe and refractory systemic lupus erythematosus. Arthritis Rheum. 2010;62:2467–75.

Tanimoto H, Terai S, Taro T, Murata Y, Fujisawa K, Yamamoto N, Sakaida I. Improvement of liver fibrosis by infusion of cultured cells derived from human bone marrow. Cell Tissue Res. 2013;354:717–28.

Tateishi-Yuyama E, Matsubara H, Murohara T, Ikeda U, Shintani S, Masaki H, Amano K, Kishimoto Y, Yoshimoto K, Akashi H, Shimada K, Iwasaka T, Imaizumi T. Therapeutic angiogenesis for patients with limb ischaemia by autologous transplantation of bone-marrow cells: a pilot study and a randomised controlled trial. Lancet. 2002;360:427–35.

Terai S, Sakaida I, Yamamoto N, Omori K, Watanabe T, Ohata S, Katada T, Miyamoto K, Shinoda K, Nishina H, Okita K. An in vivo model for monitoring trans-differentiation of bone marrow cells into functional hepatocytes. J Biochem. 2003;134:551–8.

Terai S, Ishikawa T, Omori K, Aoyama K, Marumoto Y, Urata Y, Yokoyama Y, Uchida K, Yamasaki T, Fujii Y, Okita K, Sakaida I. Improved liver function in patients with liver cirrhosis after autologous bone marrow cell infusion therapy. Stem Cells. 2006;24:2292–8.

Till JE, Mc CE. A direct measurement of the radiation sensitivity of normal mouse bone marrow cells. Radiat Res. 1961;14:213–22.

Tondreau T, Dejeneffe M, Meuleman N, Stamatopoulos B, Delforge A, Martiat P, Bron D, Lagneaux L. Gene expression pattern of functional neuronal cells derived from human bone marrow mesenchymal stromal cells. BMC Genom. 2008;9:166.

Tsai PC, Fu TW, Chen YM, Ko TL, Chen TH, Shih YH, Hung SC, Fu YS. The therapeutic potential of human umbilical mesenchymal stem cells from Wharton's jelly in the treatment of rat liver fibrosis. Liver Transpl. 2009;15:484–95.

Tsolaki E, Athanasiou E, Gounari E, Zogas N, Siotou E, Yiangou M, Anagnostopoulos A, Yannaki E. Hematopoietic stem cells and liver regeneration: differentially acting hematopoietic stem cell mobilization agents reverse induced chronic liver injury. Blood Cells Mol Dis. 2014;53(3):124–132.

Uccelli A, Moretta L, Pistoia V. Mesenchymal stem cells in health and disease. Nat Rev Immunol. 2008;8:726–36.

Vaegler M, Maerz JK, Amend B, da Silva LA, Mannheim JG, Fuchs K, Will S, Sievert KD, Stenzl A, Hart ML, Aicher WK. Labelling and tracking of human mesenchymal stromal cells in preclinical studies and large animal models of degenerative diseases. Curr Stem Cell Res Ther. 2014;9(5):444–450.

Vanikar AV, Dave SD, Thakkar UG, Trivedi HL. Cotransplantation of adipose tissue-derived insulin-secreting mesenchymal stem cells and hematopoietic stem cells: a novel therapy for insulin-dependent diabetes mellitus. Stem Cells Int. 2010;2010:582382.

Vassilopoulos G, Wang PR, Russell DW. Transplanted bone marrow regenerates liver by cell fusion. Nature. 2003;422:901–4.

Wagers AJ, Weissman IL. Plasticity of adult stem cells. Cell. 2004;116:639–48.
Wagers AJ, Sherwood RI, Christensen JL, Weissman IL. Little evidence for developmental plasticity of adult hematopoietic stem cells. Science. 2002;297:2256–9.
Wang X, Willenbring H, Akkari Y, Torimaru Y, Foster M, Al-Dhalimy M, Lagasse E, Finegold M, Olson S, Grompe M. Cell fusion is the principal source of bone-marrow-derived hepatocytes. Nature. 2003;422:897–901.
Wang JZX, Cui L, Yan L, Liang J, Cheng X, Qiao L, Shi Y, Han Z, Cao Y, Han Y, Fan D. The significance of $CD14^+$ monocytes in peripheral stem cells for the treatment of rat liver cirrhosis. Cytotherapy. 2010;12:1022–34.
Wang D, Zhang H, Liang J, Gu Z, Ma X, Huang J, Lin J, Hou Y, Lu L, Sun L. Effect of allogeneic bone marrow-derived mesenchymal stem cells transplantation in a polyI:C-induced primary biliary cirrhosis mouse model. Clin Exp Med. 2011;11:25–32.
Wang L, Li J, Liu H, Li Y, Fu J, Sun Y, Xu R, Lin H, Wang S, Lv S, Chen L, Zou Z, Li B, Shi M, Zhang Z, Wang FS. Pilot study of umbilical cord-derived mesenchymal stem cell transfusion in patients with primary biliary cirrhosis. J Gastroenterol Hepatol. 2013;28(Suppl 1):85–92.
Willenbring H, Bailey AS, Foster M, Akkari Y, Dorrell C, Olson S, Finegold M, Fleming WH, Grompe M. Myelomonocytic cells are sufficient for therapeutic cell fusion in liver. Nat Med. 2004;10:744–8.
Wilson KS, Timmons CF, Hilton DS, Weinberg AG, Tonk V. Chromosomal instability in hereditary tyrosinemia type I. Pediatr Pathol. 1994;14:1055–7.
Xu T, Wang X, Chen G, He Y, Bie P. Autologous bone marrow stem cell transplantation attenuates hepatocyte apoptosis in a rat model of ex vivo liver resection and liver autotransplantation. J Surg Res. 2013;184:1102–8.
Yan L, Han Y, Wang J, Liu J, Hong L, Fan D. Peripheral blood monocytes from patients with HBV related decompensated liver cirrhosis can differentiate into functional hepatocytes. Am J Hematol. 2007;82:949–54.
Yan L, Han Y, Wang J, Liu J, He Y, Wang H, Fan D. Peripheral blood monocytes from the decompensated liver cirrhosis could migrate into nude mouse liver with human hepatocyte-markers expression. Biochem Biophys Res Commun. 2008;371:635–8.
Yannaki E, Anagnostopoulos A, Kapetanos D, Xagorari A, Iordanidis F, Batsis I, Kaloyannidis P, Athanasiou E, Dourvas G, Kitis G, Fassas A. Lasting amelioration in the clinical course of decompensated alcoholic cirrhosis with boost infusions of mobilized peripheral blood stem cells. Exp Hematol. 2006;34:1583–7.
Yu J, Cao H, Yang J, Pan Q, Ma J, Li J, Li Y, Li J, Wang Y, Li L. In vivo hepatic differentiation of mesenchymal stem cells from human umbilical cord blood after transplantation into mice with liver injury. Biochem Biophys Res Commun. 2012;422:539–45.
Zhan Y, Wang Y, Wei L, Chen H, Cong X, Fei R, Gao Y, Liu F. Differentiation of hematopoietic stem cells into hepatocytes in liver fibrosis in rats. Transplant Proc. 2006;38:3082–5.
Zhang M, Huang B. The multi-differentiation potential of peripheral blood mononuclear cells. Stem Cell Res Ther. 2012;3:48.
Zhang FT, Wan HJ, Li MH, Ye J, Yin MJ, Huang CQ, Yu J. Transplantation of microencapsulated umbilical-cord-blood-derived hepatic-like cells for treatment of hepatic failure. World J Gastroenterol. 2011;17:938–45.
Zhang Z, Lin H, Shi M, Xu R, Fu J, Lv J, Chen L, Lv S, Li Y, Yu S, Geng H, Jin L, Lau GK, Wang FS. Human umbilical cord mesenchymal stem cells improve liver function and ascites in decompensated liver cirrhosis patients. J Gastroenterol Hepatol. 2012;27(Suppl 2):112–20.
Zheng L, Chu J, Shi Y, Zhou X, Tan L, Li Q, Cui L, Han Z, Han Y, Fan D. Bone marrow-derived stem cells ameliorate hepatic fibrosis by down-regulating interleukin-17. Cell Biosci. 2013;3:46.
Zhong YS, Lin N, Deng MH, Zhang FC, Tang ZF, Xu RY. Deficient proliferation of bone marrow-derived mesenchymal stem cells in patients with chronic hepatitis B viral infections and cirrhosis of the liver. Dig Dis Sci. 2010;55:438–45.

Zhu Y, Guan YM, Huang HL, Wang QS. Human umbilical cord blood mesenchymal stem cell transplantation suppresses inflammatory responses and neuronal apoptosis during early stage of focal cerebral ischemia in rabbits. Acta Pharmacol Sin. 2014;35:585–91.

Chapter 17
Stem Cell Therapy for Cartilage Defects

Philip S. Pastides and Wasim S. Khan

Abstract Cartilaginous defects within the articular cartilage present a treatment problem within the orthopaedic community. In cases of established osteoarthritis affecting large joints, arthroplasty is a good, well-established and predictable option. It is though a step too far for smaller and discrete lesions. Currently, surgical options include autologous chondrocyte implantation, microfracture, osteochondral autologous transplantation and even osteochondral allograft plugs. Tissue engineering techniques may prove to be the answer to this problem. There is plenty of interest in stem cell manipulation to induce chondrogenesis. The areas of research focus on the differentiation of multipotent mesenchymal stem cells but also more recently on the use of induced pluripotent stem cells. Furthermore, augmentation of repair can be facilitated by endogenous stimulants to these cells such as growth factors, gene therapy and scaffolds to maintain an optimum microenvironment. Endogenous stimulants aside, it does appear that exogenous methods of stimulation such as ultrasound and magnetic field applications can further augment and improve the reparative process. The aim of this chapter is to define the problem faced by the medical world due to the macro- and microscopic structure of cartilage and present data and reports showing the advances made in this field. The final section focuses on the current state of play surrounding the translation of these techniques to human subjects, presenting the up-to-date studies.

Keywords Cartilaginous defects · Stem cell · Therapy

P.S. Pastides (✉)
Orthopaedic Registrar, University College London Institute of Orthopaedics and Musculoskeletal Sciences, Royal National Orthopaedic Hospital, Stanmore, Middlesex HA7 4LP, UK
e-mail: ppastides@hotmail.com

W.S. Khan
Clinical Lecturer, University College London Institute of Orthopaedics and Musculoskeletal Sciences, Royal National Orthopaedic Hospital, Stanmore, Middlesex HA7 4LP, UK

R.C. Zhao (ed.), *Stem Cells: Basics and Clinical Translation*, Translational Medicine Research 1, DOI 10.1007/978-94-017-7273-0_17

17.1 Introduction

The orthopaedic community is commonly faced with problems related to cartilage defects. Osteoarthritis is a very common condition characterised by loss of articular cartilage within a joint. This leads to an alteration in joint loading and subsequent mechanics than can cause an array of symptoms from swelling and stiffness to crippling pain with associated significant restriction in movement and limitation of activities of daily living. Currently, there is no convincing method to reverse the articular cartilage loss. Treatment options are restricted to management of the symptoms, either with pain relieving intra-articular injections, analgesia and physiotherapy. In terms of surgical options, for large joints such as the hip and knee, replacement of the worn out and degenerate joint in the form of an arthroplasty procedure is an excellent option with a proven and ever growing track record. In the middle aged or elderly patient, with established cartilage loss, this is well established. For certain joints, such as the ankle joint (tibiotalar joint), the role of arthroplasty is still pending firm establishment and recognition. For smaller joints, such as the metatarsal phalangeal joints in the forefoot, arthroplasty is currently not as successful as the equivalent procedures in the hip and knee. In such cases, surgical option usually involves fusion of the joint to stop the motion within the arthritic joint and thus relieve pain.

A more significant and difficult problem to overcome is that of the younger patient presenting with a focal and discrete area of cartilage loss or even the same scenario in an older patient with an otherwise non-degenerate joint. In focal defects, the structural integrity of the articular cartilage is disrupted in circumscribed areas, for example, as a consequence of direct trauma, osteonecrosis or osteochondritis dissecans. The resulting articular cartilage defect is of a limited two-dimensional extent and characterised as being either chondral, involving only the cartilaginous zones, or osteochondral, reaching further into the subchondral bone (Orth et al. 2013). In such cases, arthroplasty is probably too morbid a procedure to consider. The ideal option would be one that would allow regeneration of the lost tissue structures within the defect, with tissue that is identical to that lost, in other words making it a perfect match. Herewith lies the challenge. The holy grail in the treatment of cartilage loss remains the establishment of methods to repair or regenerate damaged, and frail cartilage. Furthermore, it would be desirable for such techniques to as cost-effective, non-invasive, non-intrusive and as safe as possible. The recent advances in tissue engineering strategies offer a promising and potential method of treating such conditions and meeting the desired criteria.

17.2 What Is the Function of Articular Cartilage?

Articular cartilage is a highly specialised connective tissue that covers most human joints. It has a multifactorial role to aid movement of the joint. Articular cartilage is integral to the tribological properties of joints. It has a smooth, well-lubricated surface that helps to reduce the coefficient of friction. Its complex structure aids the transmission of forces across the joint and spreads the forces over a greater area in order to reduce the pressure across the joint during loading. As a material, it must resist the shear, tensile and compressive forces that occur during the joint surfaces articulation. Injury to cartilage has a well-established link to musculoskeletal pathology and patient morbidity. Whereas osteoarthritis is far more common in the older and elderly population, discrete cartilaginous defects tend to occur in younger, more active patients with an otherwise well-preserved articular surface. Cartilage lesions are common in sporting activities, with some studies showing up to 49 % of injuries associated with athletic activity (Aroen et al. 2004).

17.3 What Is Cartilage and Why Is It so Difficult to Repair?

Cartilage is an avascular and aneural structure, which is also devoid of lymphatic supply. As such, the material's own internal capacity to heal and repair is limited (Khan et al. 2009). Current treatment options for cartilage regeneration include osteochondral autologous transplantation (OATs), autologous chondrocyte implantation (ACI), microfracture and synthetic allograft osteochondral plugging of defects (Sanghvi et al. 2014). However, a recent meta-analysis of several leading databases showed that although the clinical outcomes described in the studies available appeared to be promising, the lack of large high-level evidence studies or comparable outcome measures between studies failed to show a significant advantage or disadvantage of this technique compared to other techniques such as osteochondral transplantation or microfracture (Niemeyer et al. 2012). More recently, tissue engineering techniques using stem cells have been a focal area of research. Numerous in vivo animal studies show the potential benefits of using stem cells to treat musculoskeletal diseases. However, given the complexity of the treatment and the potential ethical and economic considerations, only a few have been translated into human clinical studies (Aroen et al. 2004).

17.4 The Structure of Normal Cartilage

Articular cartilage is composed of a dense extracellular matrix (ECM), throughout which highly specialised cartilage-forming cells called chondrocytes are distributed. This ECM is primarily composed of water, collagen and proteoglycans, with other non-collagenous proteins and glycoproteins present in smaller quantities. The main role of these proteins within the ECM is to help maintain the water content and thus retain the mechanical properties required.

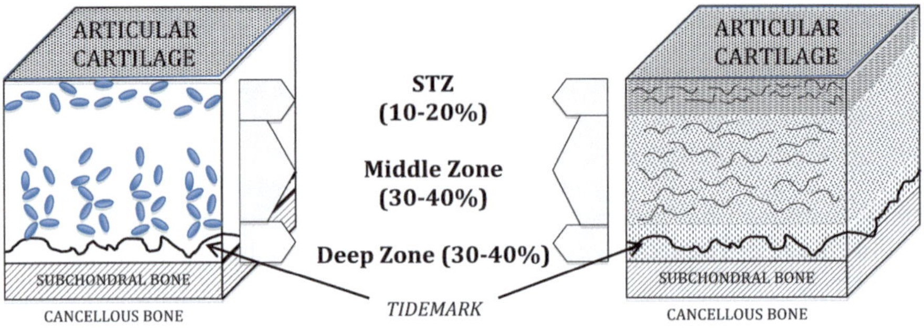

Fig. 17.1 Description of distinct zones in articular cartilage

Articular cartilage is compromised of several distinct zones; the superficial tangential zone (STZ), the middle zone, the deep zone and the calcified zone (Fig. 17.1). Each zone has a further three regions; the pericellular region, the territorial region and the interterritorial region.

The thin STZ protects the deeper layers from shear stress forces and makes up approximately 10–20 % of articular cartilage thickness. The collagen fibres of this zone are type II and IX. They are packed tightly and aligned parallel to the articular surface, hence protecting the cartilage from shear forces. It has a protective function to the deeper layers and is in contact with synovial fluid and thus responsible for most of the tensile properties of cartilage.

The middle (transitional) zone lies deep to the STZ. It is the thickest layer within the cartilage structure (40–60 % of the total volume) and provides an anatomic and functional bridge between the superficial and deep zones. It contains proteoglycans and thick collagen fibrils. In this layer, the collagen is organised obliquely, and the chondrocytes are spherical and at low density. As such, it functionally resists compressive forces.

The deep zone compromises around 30–40 % of the total cartilage volume. The collagen fibrils tend to be arranged perpendicular to the articular surface, and thus, it is responsible for providing the greatest resistance to compressive forces. The deep zone contains the highest proteoglycan content but the lowest water concentration.

The tide mark distinguishes the deep zone from the calcified cartilage. The calcified layer plays an anchor role, by securing the cartilage to underlying cancellous bone.

17.5 Subregions

The pericellular matrix is a thin layer adjacent to the cell membrane, and it completely surrounds the chondrocyte. It is thought that this matrix plays a significant role in the deformation behaviour of chondrocytes, possibly modulating cartilage development, adaptation and degeneration (Julkunen et al. 2009). The territorial matrix surrounds the pericellular matrix. This region is thicker than the pericellular matrix. The interterritorial region is the largest of the 3 matrix regions; it contributes most to the biomechanical properties of articular cartilage. It has a random arrangement and orientation of collagen fibrils, arranged parallel to the surface of the superficial zone, obliquely in the middle zone, and perpendicular to the joint surface in the deep zone (Greco et al. 1992).

17.6 Chondrocytes

The chondrocyte is the 'building' cell type in articular cartilage. Chondrocytes are specialised and metabolically active cells that help maintain, develop and repair the ECM. Chondrocytes originate from mesenchymal stem cells and constitute about 2 % of the total volume of articular cartilage. Within each of the anatomic regions of cartilage, the chondrocytes vary in shape, number and size. Within the STZ, the chondrocytes are flatter and smaller, with a greater density than that of the cells deeper in the matrix. Within each region of cartilage, the chondrocyte becomes trapped in the ECM and tends not to migrate to other areas (Alford and Cole 2005). Unfortunately, chondrocytes have a limited capacity of healing once injured. However, they respond to a variety of stimuli, including growth factors, mechanical loads, piezoelectric forces and hydrostatic pressures and for this very reason can be influenced during tissue engineering strategies (Buckwalter and Mankin 1998). This will be discussed in much more detail in later sections.

17.7 The Extracellular Matrix (ECM)

The ECM is a multicellular structure that provides structural and biochemical support to the surrounding cells. The ECM in cartilage is composed primarily of the network type II collagen and an interlocking mesh of fibrous proteins and proteoglycans (PGs), hyaluronic acid (HA) and chondroitin sulphate (CS). Cartilaginous ECM is remodelled continuously by a combination of production, degradation by matrix metalloproteinases (MMPs) and inhibition of MMPs activity by tissue inhibitors of MMPs. These fibres contain elastin and collagen. Collagen is the most abundant structural macromolecule in ECM, and it makes up about 60 % of the dry weight of cartilage. Type II collagen represents 90–95 % of the collagen in ECM and forms fibrils and fibres intertwined with proteoglycan aggregates. Collagen types I, IV, V, VI, IX and XI are also present but contribute only a minor proportion. The minor collagens help to form and stabilise the type II collagen fibril network (Plopper 2007).

17.8 Proteoglycans (PG)

Proteoglycans (PG) are heavily glycosylated protein monomers. In articular cartilage, they represent the second largest group of macromolecules in the ECM and account for 10–15 % of the total weight. Proteoglycans consist of a protein core with one or more linear glycosaminoglycan chains covalently attached. Articular cartilage contains a variety of proteoglycans that are essential for normal function, including aggrecan, decorin, biglycan and fibromodulin. The largest in size and the most abundant by weight is aggrecan, a proteoglycan that possesses more than 100 chondroitin sulphate and keratin sulphate chains (Fox et al. 2009). Aggrecan is characterised by its ability to interact with hyaluronan (HA) to form large proteoglycan aggregates via link proteins (Buckwalter and Mankin 1998).

17.9 The Goals of Tissue Engineering Techniques

As seen from the sections above, articular cartilage is a very complex structure comprised of numerous different cell types, packed and orientated in a predetermined and important manner. As a result, restoration of articular cartilage structure and hence function following pathological or traumatic damage is still considered a challenging problem in the orthopaedic and musculoskeletal community. Currently, tissue engineering-based reconstruction of articular cartilage is a feasible and continuously developing strategy to restore structure and function. Stem cells possess self-renewal capacity and exhibit long-term viability and multilineage differentiation potential. Ethical, political and religious issues surround the use of embryonic stem cells which are pluripotent in nature. In contrast, the use of autologous adult mesenchymal stem cells (MSCs), which are multipotent in nature, is generally well accepted. The exact definition of what a MSCs actually is has been somewhat debated. However, the Mesenchymal and Tissue Stem Cell Committee of the International Society for Cellular Therapy defined a minimal set of standard criteria for uniform characterisation of MSCs. They must be plastic-adherent cells when maintained in standard culture conditions and must express CD105, CD73 and CD90. In addition, they must lack surface expression of CD45, CD34, CD14 (CD11b), CD79α (CD19) and HLA-DR. Finally, and importantly, they must be capable of differentiating to cells of the mesodermal lineage (chondrocytes, osteoblasts, adipocytes) (Dominici et al. 2006).

Stem cells have been identified in a number of adult tissues, albeit in small and variable numbers. MSCs display some advantages compared with other differentiated cells that make them attractive targets for manipulation in the goal of not only cartilage regeneration but also bone regeneration. They can be easily isolated in a non-invasive and abundant manner from various tissues. In addition to bone marrow, MSCs have been identified in other tissues including bone, adipose tissue, muscle, synovium, periosteum and perichondrium, all of which can be generally isolated and subsequently expanded in culture (Malik and Khan 2011). Furthermore, MSCs are less tumourigenic than their embryonic counterpart (Raghunath et al. 2005) and provide an autologous source of cells eliminating concerns regarding rejection and disease transmission. There is also evidence to suggest that MSCs have immunosuppressive potential as co-culture with MSCs inhibits T-cell lymphocyte proliferation (Krampera et al. 2003).

The aim of this chapter is to discuss how tissue engineering treatments can be manipulated, augmented and induced to repair and regenerate articular cartilaginous defects. We aim to present the reader with the practical problems with this type of research and how the community are trying to overcome them. This chapter is divided into several sections, including the use of scaffolds, growth factors, gene therapy and exogenous methods of augmentation. The final section aims to show how these techniques are translated from the laboratory and into the human population.

17.10 Scaffolds

A key requirement for tissue engineering MSC therapy is the safe and effective delivery of the cells to the defect site, in addition to maintaining the cells in the required area and in an appropriate microenvironment to allow the reparative process to occur. Direct intra-articular injection might be possible in early stages of the disease when the defect is restricted to the cartilage layer (a chondral defect), whereas a scaffold or matrix of some kind would be required to support the MCSs in cases where the subchondral bone is exposed over large areas (an osteochondral defect). It is critical to design and fabricate a suitable scaffold for use in specific tissue regeneration, as it directly comes into contact with cells and provides structural support and guidance for subsequent tissue development. The scaffolds provide an initial mechanical support and a three-dimensional niche for transplanted cells until the regenerated tissue can stabilise the initial structure (Khamdemhosseini et al. 2009).

The ideal scaffold should be biocompatible and biodegradable upon tissue healing, highly porous so as to permit cell penetration and tissue impregnation, sufficiently permeable to allow nutrient delivery and gas exchange and adaptable to the mechanical environment. In addition, the scaffold should have a surface that is conducive to cell attachment and migration, and permits appropriate extracellular matrix formation and the transmission of signalling molecules (Susmita et al. 2012). A wide range of matrices to develop types of scaffolds have been investigated to augment tissue engineering strategies. These can broadly be divided into natural or synthetic scaffolds. Natural scaffolds include substances such as carbohydrate-based materials such as chitosan (Ye et al. 2014; García Cruz et al. 2012; Alves da Silva et al. 2011), hyaluronate (Son et al. 2013; Toh et al. 2012) or even protein-based structures such as collagen (Zhang et al. 2012, 2013a; Murphy et al. 2012), fibrin (Diederichs et al. 2012; Park et al. 2011), gelatin (Klangjorhor et al. 2012; Pruksakorn et al. 2009) and chondroitin (Chen et al. 2013; Park et al. 2010; Varghese et al. 2008). Synthetic scaffolds successfully used include polyglycolic acid, polylactic acid, poly(lactic-co-glycolic acid), polyethylene glycol and polycaprolactone (Hidalgo et al. 2013; Childs et al. 2013; Zhang et al. 2013b; Li et al. 2013).

Aside from being used in isolation with the MSCs, the application and culture of these scaffolds with other growth factors can further augment chondrogenesis, as discussed below.

17.11 Growth Factors

To help stimulate the formation of matrix and cartilage, research has focused on the use of growth factors to augment tissue engineering strategies. A growth factor tends to be a naturally occurring substance that is capable of stimulating cell growth, proliferation, healing and cellular differentiation. They typically tend to act as signal molecules between cells and thus direct cells down a certain pathway. Several growth factors in particular have been found to stimulate ECM synthesis. The majority of work so far surrounding growth factor stimulation of stem cells appears to have been performed in vitro and in animal studies. This is understandable as these sources are more readily available and associated with less ethical considerations. As the benefits begin to be more obvious, their effect in human subjects will certainly begin to be evaluated.

Freed (2001) seeded bovine calf articular chondrocytes onto biodegradable polyglycolic acid scaffolds and cultured them for four weeks. Some of the cells were augmented with IGF-1. At four weeks, the resulting engineered tissue histologically resembled cartilage and contained its major constituents: glycosaminoglycans, collagen and cells. They noted that those cells supplemented with IGF-I the independently modulated tissue morphology, growth, biochemical composition, interacted additively and produced tissue superior to that obtained by modifying these factors individually. In the same pattern of thought, Keelner et al. (2001) exposed bovine articular chondrocytes to insulin rather than IGF1. They found that the effects of insulin were similar to that of the IGF1 and thus posed an interesting proposal of employing insulin as a potent substance to improve tissue-engineered cartilage. These additive effects of IGF1 have been supported by a plethora of other researchers (Madry et al. 2013; Blunk et al. 2002; Rosa et al. 2014). Transforming growth factors, especially TGF β1 (Rosa et al. 2014; Morales 1991) and TGF β3 (Almeida et al. 2014; Wang et al. 2014), and bone morphogenetic proteins (BMP), especially BMP-2 and BMP-7 (Park et al. 2005; Kaps et al. 2002; Carreira et al. 2014), and fibroblast growth factor (FGF)-2 (Veilleux and Spector 2005) have all shown a qualitative effect on chondrogenesis. Another stimulant that has been studied is platelet-rich plasma (PRP). PRP is rich in growth factors including TGF-β, IGF, platelet-derived growth factor (PDGF), basic fibroblast growth factor (bFGF) and vascular endothelial growth factor (VEGF) (Sun et al. 2010). As a result, using adipose-derived MSCs, Mardani et al. (2013) showed has beneficial effects on MSCs differentiation to chondrocytes. In fact, there is increasing evidence that utilising several growth factors, rather than a single one, can generate a higher quality of engineered cartilage (Ertan et al. 2013; Park et al. 2009).

Furthermore, transcription factors and cytokines also have a role in up- or down-regulating cartilage formation. SOX9 is a transcription factor that has been found to be integral for chondrogenesis. It augments and enhances cartilage matrix genes, activating their expression (Lefebvre and Smits 2005). SOX9 also appears to have a role in the chondrogenic potential of chondrocytes; it is expressed highly in normal cartilage but is decreased in late osteoarthritis (Haag et al. 2008). IL-17 is a key cytokine related to chronic inflammatory conditions. IL-17 inhibited chondrogenesis of human MSCs through the suppression of PKA activity and SOX9 phosphorylation. IL-17 levels have found to be prevalent in conditions such as osteoarthritis. These results suggest that chondrogenic differentiation of MSCs can be inhibited by a mechanism triggered by IL-17 under chronic inflammation (Kondo et al. 2013).

It is clear from reading the above that we have only mention a tiny proportion of the published studies that have investigated the individual or synergistic effects of growth and transcription factors to promote articular cartilage regeneration and repair. What is needed is a large meta-analysis or systematic review to help pool all the available data together and produce a high-level study with significant findings and conclusions.

17.12 Gene Therapy

MSCs and chondrocytes are permissive to gene transfer and as such are excellent candidates for gene modification to enhance their chondrogenic phenotype and promote proliferation, avoiding detrimental cellular dedifferentiation, and senescence (Madry and Cucchiarini 2011). Gene delivery to MSCs and chondrocytes has been carried out to stimulate anabolic pathways, lost through dedifferentiation by the expression of growth factors and transcription factors, and by also inhibiting catabolic pathways to prevent degradation. The introduction of foreign DNA encoding a gene of interest directly into a living cell results in the degradation of the naked DNA and therefore requires an efficient carrier for its delivery to the cell nucleus for gene transcription and subsequent protein expression (Kay 2011). This can be achieved using viral- or non-viral-derived methods.

17.12.1 Non-Viral

Elsler et al. (2012) systematically screened a panel of 15 non-viral compounds for their ability to promote safe, efficient and durable gene expression in human bone marrow-derived MSCs (hMSCS) without impeding their commitment towards chondrogenic differentiation. Primary hMSCs were transfected with plasmid vectors carrying sequences for the *Photinus pyralis* luciferase *Escherichia coli* β-galactosidase, or human insulin-like growth factor I via 15 non-viral formulations. Transgene expression and transfection efficiencies were monitored for each component in parallel with the effects on cell viability and cytotoxicity. Upon optimisation, the most promising reagent was then evaluated for a possible influence on the chondrogenic potential of hMSCs. Among all formulations tested, GeneJammer® gave the best results for transgene expression and transfection efficacy. Notably, the application of this reagent did not affect the potential of the cells for chondrogenic differentiation when maintained in long-term (21 days) three-dimensional (aggregate) cultures.

17.12.2 Viral

Efficient transgene delivery into MSCs and articular chondrocytes has been achieved using a variety of viral-based vectors including adenovirus, recombinant adeno-associated virus (rAAV), retrovirus and lentiviral vectors (Li et al. 2004). Each of these viral vectors has inherent advantages and disadvantages. Adenoviral vectors appear to be advantageous due to their high transduction efficiency, broad cell tropism, and the reduced immunological response particularly at avascular synovial joints (Hidaka et al. 2003; Kim et al. 2011).

Among the gene candidates of value for articular cartilage repair, inhibitors of both matrix-degrading enzymes (tissue inhibitor of metalloproteinases) (Kaul et al. 2006; Kafienah et al. 2003) and of proinflammatory cytokines (IL-1Ra, the soluble receptors sIL-1R, or sTNFR) (Haupt et al. 2005; Zhang et al. 2000) as well as chondroprotective cytokines (IL-4 and IL-10) (Kim et al. 2001; Manning et al. 2010) have been applied to inhibit catabolic pathways in vitro that are potentially activated in response to cartilage damage or injury.

17.13 Embryonic and Young Stem Cells

Embryonic stem cells (ESCs) are advantageous, due to their unlimited self-renewal and pluripotency, thus representing an immortal cell source that could potentially provide an unlimited supply of chondrogenic cells for both cell- and tissue-based therapies and replacements. However, harvesting human ESCs (HESCs) is surrounded by considerable ethical, moral and religious dilemmas and barriers. There is evidence that younger cells, such as chondrocytes, are a more promising cell source for cartilage regeneration (Smeriglio et al. 2014).

A potential method of avoiding these barriers is the ability of reprogramming somatic cells into a pluripotent embryonic stem cell-like state. This was first reported by Takahashi and Yamanaka (2006). Other authors have since been able to show the feasibility of this method. Kunisato et al. (2010) showed that mouse adult bone marrow mononuclear cells are competent donor cells and can be reprogrammed into pluripotent embryonic stem cell-like cells. Sumer et al. (2010) reprogrammed mouse embryonic fibroblasts by viral induction of oct4, SOX2, c-myc and klf-4 genes. The formed induced pluripotent stem-somatic hybrids had the ability to differentiate into cell types indicative of the three germ layers and were also confirmed to be reprogrammed to a pluripotent state. The development of induced pluripotent stem cells (iPSCs) offers several advantages for potential clinical application. Since it involves the harvesting of cells from a patient to derive iPSCs, the formed cells will carry the identical genetic mutations of the patient and should also avoid immune rejection, as this is essentially a form of autologous therapy. There is limited evidence at the moment in these techniques to exploit human iPSCs and more research is needed on iPSCs technology so that it can be proved safe for therapy.

17.14 Exogenous Mechanical Stimulation

So far, we have focused on endogenous methods of manipulating and stimulating MSCs to repair and/or regenerate cartilage. There also appear to be a number of exogenous methods of mechanically stimulating cells to augment the amount of repair and improve the quality of the produced material. Kamei et al. (2013) used a novel method of magnetic-assisted delivery of magnetically labelled MSCs into full thickness patellar cartilage defects in pigs. At both 6 and 12 weeks, arthroscopic procedures and biopsies from the repair sites were performed. They found that the mean histological scores at 12 weeks were significantly better than in the control group.

Amin et al. (2014) examined the effects of moderate strength static magnetic fields (SMFs) on chondrogenic differentiation in human BMSCs in vitro, under the simultaneous influence of TGF-β3. They found that that a 0.4 T magnetic field applied for 14 days elicited a strong chondrogenic differentiation response in cultured BMSCs, so long as TGF-β3 was also present. Motoyama et al. (2010) also reached a similar conclusion, as did Mayer-Wagner et al. (2011) using a low magnetic field on MSCs under the influence of human FGF-2 and TGF-β3.

Mechanical stimulation using low intensity ultrasound, in conjunction with a variety of growth factors, scaffolds and stimulants has also shown to enhance chondrogenesis (Choi et al. 2013; Lai et al. 2010; Park et al. 2007; Lee et al. 2007). Such non-invasive and relatively low-cost methods that have been shown to enhance of potential action of MSCs may provide an additional option in the treating physician's armoury.

17.15 Human Studies

The key purpose of all this research is of course to eventually enable these techniques and materials to be used for the benefit of humans. Although there is plenty of evidence and studies showing potential benefits of stem cell treatments in in vitro studies within the confines of a laboratory and in in vivo animal studies, there is still a scarce, but slowly increasing, body of evidence investigating how such techniques translate to humans.

The authors performed a recent systematic review focusing on the clinical application of stem cell therapy to treat cartilage defects in human subjects (Pastides et al. 2013). Following an extensive review over several leading literature databases, 11 articles were included in the final review. Of these studies, there were no randomised controlled trials; they were either prospective case series, cohort and case–control studies or case reports. There were three comparative studies; a cohort study compared ACI with MSC implantation (Nejadnik et al. 2010), a case control study evaluated patients undergoing a high tibial osteotomy alone or in combination with MSC implantation (Wakitani et al. 2007), and lastly a prospective cohort study comparing open ACI procedures, arthroscopic ACI procedures and arthroscopic implantation of bone marrow-derived cells (Giannini et al. 2010).

Two studies by the same research group with 73 patients used cartilage defects arising from the talar dome in the ankle joint (Giannini et al. 2009, 2010). All of these lesions were defined as grade II lesions. The remaining nine studies with 92 patients investigated defects in the knee joint. These lesions were defined as grade III or grade IV in 78 patients. Two studies (Wakitani et al. 2007; Matsumoto et al. 2010) with 14 patients did not specify the grade of the lesion. Of the 92 knee lesions, 62 involved the femoral condyles, 17 involved the patella, four involved the trochlea, seven were multiple and two were not specified. Out of the 62 femoral condylar lesions, 29 were located at the medial condyle, nine at the lateral condyle and two at both condyles. The location of 22 femoral condylar lesions was not specified. Of the 17 patellar lesions, three were multiple.

The two studies involving the ankle joint (Giannini et al. 2009, 2010) and one study involving the knees of 20 patients (Buda et al. 2010) isolated bone marrow-derived MSCs from the posterior iliac crest, whereas all other studies used the anterior iliac crest. The implantation timing and methods also varied. The three studies using the posterior iliac crest harvest and the study with the additional AMIC technique (de Girolamo et al. 2010) implanted the bone marrow-derived cells immediately whereas the other studies culture expanded the MSCs in vitro and performed a second-stage implantation procedure at a later date. The MSCs were expanded in culture for three to four weeks in most cases; for seven patients involved in two studies (Kasemkijwattana et al. 2011; Haleem et al. 2010), the duration is not specified. Collagen was the most commonly used scaffold and was used with a periosteal flap in 54 patients and with platelet-rich fibrin in 23 patients. Hyaluronic acid was used with platelet-rich fibrin in 45 patients. In one study (Giannini et al. 2010) involving 25 patients, either a collagen or hyaluronic acid scaffold was used, but not specified. A fibrin scaffold was used for 11 patients (de Girolamo et al. 2010), and a platelet-rich fibrin with a periosteal flap for five patients (Haleem et al. 2010). One study involving two patients did not specify the scaffold used (Matsumoto et al. 2010).

Outcome measures were also extremely variable in the reported studies, thus making comparison and evaluation extremely difficult. Validated patient outcome scores were used in almost all cases, but the actual quality of the cartilage regeneration was sparsely reported. Some studies made use MRI scans to evaluate the quality of the tissue repair, whilst only one of the included studies by Haleem et al. (2010) actually performed a repeat arthroscopy and utilised the International Cartilage Research Society (ICRS) score. It is of course easy to understand why not all studies proceeded with invasive testing of their subjects.

It is clear that there is currently no agreed consensus on how to perform or follow up these early studies in humans and thus comparison and amalgamation of the outcomes is near impossible. Post-operative rehabilitation regimens, when reported, vary between individual studies. It appears that they generally involve a period of immobilisation following the procedures. However, there is no agreed consensus on details or duration of the weight-bearing status. The studies included in this review are not homogeneous and vary between each other in both study design and outcome measurements. The quality of the included studies was assessed by the authors using the Coleman Methodology Score (CMS) that assesses methodology using ten criteria, giving a total score between 0 and 100. A score of 100 indicates that the study largely avoids chance, various biases and confounding factors. The highest score was only 40 (Nejadnik et al. 2010). The remaining eight studies scored between 30 and 39.

Since the publication of our review in 2012, a phase 1, interventional, prospective and multicentric study is evaluating the efficacy of a single injection of autologous adipose-derived mesenchymal stem cells for the treatment of patients with moderate or severe osteoarthritis of the knee (http://clinicaltrials.gov/ct2/show/study/NCT01585857). The outcomes will be evaluated 365 days post-injection and patients will be assessed clinically via validated scoring systems, such as the WOMAC (Western Ontario and McMaster Universities osteoarthritis index score) and the Short Arthritis Assessment Scale (SAS), improvement in the clinical range of movement and via MRI imaging. The results are yet to be published.

Wong et al. (2013) performed a randomised trial to analyse the results of the use of intra-articular cultured autologous bone marrow-derived mesenchymal stem cell (MSC) injections in conjunction with microfracture and medial opening-wedge high tibial osteotomy (HTO). All patients underwent HTO and microfracture. The cell-recipient group received intra-articular injection of cultured MSCs with hyaluronic acid 3 weeks after surgery, whereas the control group only received hyaluronic acid. The primary outcome measure was the International Knee Documentation Committee (IKDC) score at intervals of 6 months, 1 year and 2 years post-operatively. Secondary outcome measures were Tegner and Lysholm clinical scores and 1-year post-operative Magnetic Resonance

Observation of Cartilage Repair Tissue (MOCART) scores. Adjusting for all confounding factors, the cell-recipient group showed significantly better functional scores. In addition, magnetic resonance imaging scans performed 1 year after surgical intervention showed significantly better MOCART scores for the cell-recipient group ($p < 0.001$). Lee et al. (2012) performed a prospective comparative study of combined arthroscopic microfracture and outpatient intra-articular injections of autologous bone marrow-derived MSCs and hyaluronic acid (HA) compared to implantation of the MSCs beneath a sutured periosteal patch over the defect. They found no significant difference between the two treatment options, but both groups did improve significantly. There is also some evidence to show that the intra-articular injection of MSCs alone as treatment has benefits. Orozco et al. (2013) performed a pilot study with treated twelve patients suffering with chronic knee pain secondary to osteoarthritis with autologous expanded bone marrow MSCs by intra-articular injection. At 12 months post-injection, they were found to improve symptomatically very well and subsequent MRI scans showed a highly significant decrease of poor cartilage areas with improvement of cartilage quality in 11 of the 12 patients.

As is evident, there is a slow and growing body of evidence that the effects of MSCs treatments can be successfully translated from in vitro and animal studies, to humans. It is also clear that the evidence is still sparse, diverse and non-comparable. The need for large, high-quality studies to investigate potential benefits and safety of such treatments in humans is required.

17.16 Discussion

The potential use of tissue engineering strategies and stem cell research for cartilage defects is a very attractive option with a bright future. Despite the huge amount of resources and studies being conducted around the globe, there is still a cloud of mystery surrounding the optimum strategies and treatment methods. Starting with a multipotent building block, the stem cell, the aim is to guide it down the path of chondrogenesis. There are many obstacles that need to be overcome prior to the acceptance of such techniques. These include the safe collection and isolation of the cells, optimisation of culture and expansion methods, the safe delivery of the cells with maintenance within the desired defect and detailed follow-up and rehabilitation regimens for the patients.

A major obstacle remains the use of stem cells as a form of treatment. Embryonic stem cells, which are pluripotent, are generally not accepted to be used for such research due to the moral, religious and cultural question marks surrounding them. However, the potential benefit of stem cell research to deliver new treatments for currently incurable diseases in well recognised. Most countries are open to the idea of research within this field under strict regulation. For example, in 2005, the United Kingdom established the UK Stem Cell Initiative (UKSCI) (http://webarchive.nationalarchives.gov.uk/20130107105354/http://www.dh.gov.

uk/prod_consum_dh/groups/dh_digitalassets/@dh/@en/documents/digitalasset/dh_4124088.pdf) to regulate the current and future role of the research in the country. The budget report set out the broad goals of the initiative which included a high-level review, in collaboration with public and private sector stakeholders, to formulate a ten-year vision for stem cell research in the UK, creating a platform for coordinated public and private funding of research. Induced pluripotent stem cells appear to be an attractive option, but it remains to see whether such cells show superior outcomes over MSCs.

Augmentation of MSCs using gene therapy and growth factors has shown significant benefits to cartilage formation. It is fairly well accepted that growth factors such as IGF-1 and TGF-b1 and TGF-b3 can augment chondrogenesis. Up- or downregulation of certain cytokines and transcription factors can also be used with good effect to improve outcome. Furthermore, using a combination of these techniques could cause even more synergy.

The current literature is booming with research reporting on outcomes of chondrogenesis using stem cells and tissue engineering techniques under various stimuli. It reflects the growing desire to optimise techniques with an ultimate aim of translating and applying the in vitro and animal studies into our patient population. As seen from the section on human studies, there is early interest and application of these techniques into human trials. However, there is a lack of consistency in both study design and follow-up methodology. What is required are large, mutlicentred, well-designed studies with firm and robust primary and secondary endpoints and a long follow-up.

What is certain is that stem cell techniques and tissue engineering strategies are a viable, attractive and potential cartilage regenerating method of curing the problem of articular cartilage defects. Whereas arthroplasty will almost certainly remain the mainstay treatment option for established osteoarthritis, reparative techniques for smaller, more discrete cartilaginous defects may benefit from stem cell technology. It is unclear how close we are to finding our 'holy grail' by the quest is well underway.

References

Alford JW, Cole BJ. Cartilage restoration, part I: basic science, historical perspective, patient evaluation and treatment options. Am J Sports Med. 2005;33:295–306.

Aroen A, Loken S, Heir S, Alvik E, Ekeland A, Granlund O, Engebretsen L. Articular cartilage lesions in 993 consecutive knee arthroscopies. Am J Sports Med. 2004;32:211–5.

Buckwalter JA, Mankin HJ. Articular cartilage: tissue design and chondrocytes-matrix interactions. Instr Course Lect. 1998;47:477–86.

Dominici M, Le Blanc K, Mueller I, Slaper-Cortenbach I, Marini F, Krause D, Deans R, Keating A, Prockop Dj, Horwitz E. Minimal criteria for defining multipotent mesenchymal stromal cells. The International Society for Cellular Therapy position statement. Cytotherapy 2006;8(4):315–7.

Fox AJS, Bedi A, Rodeo SA. The basic science of articular cartilage. Structure, composition and function. Sports Health. 2009;1(6):461–8.

Greco F, Specchia N, Falciglia F, Toesca A, Nori S. Ultrastructural analysis of the adaptation of articular cartilage to mechanical stimulation. Ital J Orthop Traumatol. 1992;18(3):311–21.
Julkunen P, Wilson W, Jurvelin JS, Korhonen RK. Composition of the pericellular matrix modulates the deformation behaviour of chondrocytes in articular cartilage under static loading. Med Biol Eng Comput. 2009;47(12):1281–90.
Khan WS, Malik AA, Hardingham TE. Stem cell applications and tissue engineering approaches in surgical practice. J Perioper Pract. 2009;19(4):130–5.
Krampera M, Glennie S, Dyson J, Scott D, Laylor R, Simpson E, Dazzi F. Bone marrow mesenchymal stem cells inhibit the response of naive and memory antigen-specific T cells to their cognate peptide. Blood. 2003;101(9):3722–9.
Malik A, Khan WS. Stem cell applications and tissue engineering approaches in orthopaedic surgery and musculoskeletal medicine. Curr Stem Cell Res Ther. 2011.
Niemeyer P, Salzmann G, Schmal H, Mayr H, Sudkamp NP. Autologous chondrocyte implantation for the treatment of chondral and osteochondral defects of the talus: a meta-analysis of available evidence. Knee Surg Sports Traumatol Arthrosc. 2012;20(9):1696–703.
Orth P, Cucchiarini M, Kohn D, Madry H. Alterations of the subchondral bone in osteochondral repair–translational data and clinical evidence. Eur Cell Mater. 2013;25:299–316.
Plopper G. The extracellular matrix and cell adhesion. In: Lewin B, Cassimeris L, Lingappa V, Plopper G, editors. Cells; 2007.
Raghunath J, Salacinski HJ, Sales KM, Butler PE, Seifalian AM. Advancing cartilage tissue engineering: the application of stem cell technology. Curr Opin Biotechnol. 2005;16(5):503–9.
Sanghvi D, Munshi M, Pardiwala D. Imaging of cartilage repair procedures. Indian J Radiol Imaging. 2014;24(3):249–53.

Scaffolds

Alves da Silva ML, Martins A, Costa-Pinto AR, Correlo VM, Sol P, Bhattacharya M, Faria S, Reis RL, Neves NM. Chondrogenic differentiation of human bone marrow mesenchymal stem cells in chitosan-basedscaffolds using a flow-perfusion bioreactor. J Tissue Eng Regen Med. 2011;5(9):722–32.
Susmita Bose, Roy M, Bandyopadhyay A. Recent advances in bone tissue engineering scaffolds. Trends Biotechnol. 2012;30(10):546–54.
Chen WC, Wei YH, Chu IM, Yao CL. Effect of chondroitin sulphate C on the in vitro and in vivo chondrogenesis of mesenchymal stem cells in crosslinked type II collagen scaffolds. J Tissue Eng Regen Med. 2013;7(8):665–72.
Childs A, Hemraz UD, Castro NJ, Fenniri H, Zhang LG. Novel biologically-inspired rosette nanotube PLLA scaffolds for improving human mesenchymal stem cell chondrogenic differentiation. Biomed Mater. 2013;8(6):065003.
Diederichs S, Baral K, Tanner M, Richter W. Interplay between local versus soluble transforming growth factor-beta and fibrin scaffolds: role of cells and impact on human mesenchymal stem cell chondrogenesis. Tissue Eng Part A. 2012;18(11–12):1140–50.
García Cruz DM, Salmerón-Sánchez M, Gómez-Ribelles JL. Stirred flow bioreactor modulates chondrocyte growth and extracellular matrix biosynthesis in chitosan scaffolds. J Biomed Mater Res A. 2012;100(9):2330–41.
Hidalgo IA, Sojot F, Arvelo F, Sabino MA. Functional electrospun poly (lactic acid) scaffolds for biomedical applications: experimental conditions, degradation and biocompatibility study. Mol Cell Biomech. 2013;10(2):85–105.
Khamdemhosseini A, Vacanti JP, Langer R. Progress in tissue engineering. Sci Am. 2009;300(5):64–71.
Klangjorhor J, Nimkingratana P, Settakorn J, Pruksakorn D, Leerapun T, Arpornchayanon O, Rojanasthien S, Kongtawelert P, Pothacharoen P. Hyaluronan production and chondrogenic

properties of primary human chondrocyte on gelatin based hematostatic spongostan scaffold. J Orthop Surg Res. 2012;7:40.
Li B, Yang J, Ma L, Li F, Tu Z, Gao C. Fabrication of poly(lactide-co-glycolide) scaffold filled with fibrin gel, mesenchymal stem cells, and poly(ethylene oxide)-b-poly(L-lysine)/TGF-β1 plasmid DNA complexes for cartilage restoration in vivo. J Biomed Mater Res A. 2013;101(11):3097–108.
Murphy CM, Matsiko A, Haugh MG, Gleeson JP, O'Brien FJ. Mesenchymal stem cell fate is regulated by the composition and mechanical properties of collagen-glycosaminoglycan scaffolds. J Mech Behav Biomed Mater. 2012;11:53–62.
Park JS, Yang HJ, Woo DG, Yang HN, Na K, Park KH. Chondrogenic differentiation of mesenchymal stem cells embedded in a scaffold by long-term release of TGF-beta 3 complexed with chondroitin sulfate. J Biomed Mater Res A. 2010;92(2):806–16.
Park SH, Choi BH, Park SR, Min BH. Chondrogenesis of rabbit mesenchymal stem cells in fibrin/hyaluronan composite scaffold in vitro. Tissue Eng Part A. 2011;17(9–10):1277–86.
Pruksakorn D, Khamwaen N, Pothacharoen P, Arpornchayanon O, Rojanasthien S, Kongtawelert P. Chondrogenic properties of primary human chondrocytes culture in hyaluronic acid treated gelatin scaffold. J Med Assoc Thai. 2009;92(4):483–90.
Son YJ, Yoon IS, Sung JH, Cho HJ, Chung SJ, Shim CK, Kim DD. Porous hyaluronic acid/sodium alginate composite scaffolds for human adipose-derived stem cells delivery. Int J Biol Macromol. 2013;61:175–81.
Toh WS, Lim TC, Kurisawa M, Spector M. Modulation of mesenchymal stem cell chondrogenesis in a tunable hyaluronic acid hydrogel microenvironment. Biomaterials 2012;33(15):3835–45.
Varghese S, Hwang NS, Canver AC, Theprungsirikul P, Lin DW, Elisseeff J. Chondroitin sulfate based niches for chondrogenic differentiation of mesenchymal stem cells. Matrix Biol. 2008;27(1):12–21.
Ye K, Felimban R, Traianedes K, Moulton SE, Wallace GG, Chung J, Quigley A, Choong PF, Myers DE. Chondrogenesis of infrapatellar fat pad derived adipose stem cells in 3D printed chitosan scaffold. PLoS ONE 2014;9(6).
Zhang L, Yuan T, Guo L, Zhang X. An in vitro study of collagen hydrogel to induce the chondrogenic differentiation of mesenchymal stem cells. J Biomed Mater Res A. 2012;100(10):2717–25.
Zhang S, Chen L, Jiang Y, Cai Y, Xu G, Tong T, Zhang W, Wang L, Ji J, Shi P, Ouyang HW. Bi-layer collagen/microporous electrospun nanofiber scaffold improves the osteochondral regeneration. Acta Biomater. 2013a;9(7):7236–47.
Zhang K, Zhang Y, Yan S, Gong L, Wang J, Chen X, Cui L, Yin J. Repair of an articular cartilage defect using adipose-derived stem cells loaded on a polyelectrolyte complex scaffold based on poly(l-glutamic acid) and chitosan. Acta Biomater. 2013b;9(7):7276–88.

Growth Factors

Almeida HV, Liu Y, Cunniffe GM, Mulhall KJ, Matsiko A, Buckley CT, O'Brien FJ, Kelly DJ. Controlled release of transforming growth factor-β3 from cartilage-extra-cellular-matrix-derived scaffolds to promote chondrogenesis of human-joint-tissue-derived stem cells. Acta Biomater. 2014 (pii: S1742-7061(14)00242-6).
Amin HD, Brady MA, St-Pierre JP, Stevens MM, Overby DR, Ethier CR. Stimulation of chondrogenic differentiation of adult human bone marrow-derived stromal cells by a moderate-strength static magnetic field. Tissue Eng Part A. 2014;20(11–12):1612–20.
Blunk T, Sieminski AL, Gooch KJ, Courter DL, Hollander AP, Nahir AM, Langer R, Vunjak-Novakovic G, Freed LE. Differential effects of growth factors on tissue-engineered cartilage. Tissue Eng. 2002;8(1):73–84.

Buda R, Vannini F, Cavallo M, Grigolo B, Cenacchi A, Giannini S. Osteochondral lesions of the knee: a new one-step repair technique with bone-marrow-derived cells. J Bone Joint Surg Am. 2010;92(Suppl 2):2–11.

Carreira AC, Alves GG, Zambuzzi WF, Sogayar MC, Granjeiro JM. Bone morphogenetic proteins: structure, biological function and therapeutic applications. Arch Biochem Biophys. 2014 (pii: S0003-9861(14)00254-9; Epub ahead of print).

Choi JW, Choi BH, Park SH, Pai KS, Li TZ, Min BH, Park SR. Mechanical stimulation by ultrasound enhances chondrogenic differentiation of mesenchymalstem cells in a fibrin-hyaluronic acid hydrogel. Artif Organs. 2013;37(7):648–55.

de Girolamo L, Bertolini G, Cervellin M, Sozzi G, Volpi P. Treatment of chondral defects of the knee with one step matrix-assisted technique enhanced by autologous concentrated bone marrow: in vitro characterisation of mesenchymal stem cells from iliac crest and subchondral bone. Injury. 2010;41(11):1172–7.

Elsler S, Schetting S, Schmitt G, Kohn D, Madry H, Cucchiarini M. Effective, safe nonviral gene transfer to preserve the chondrogenic differentiation potential of human mesenchymal stem cells. J Gene Med. 2012;14(7):501–11.

Ertan AB, Yılgor P, Bayyurt B, Calıkoğlu AC, Kaspar C, Kök FN, Kose GT, Hasirci V. Effect of double growth factor release on cartilage tissue engineering. J Tissue Eng Regen Med. 2013;7(2):149–60.

Freed LE. IGF-I and mechanical environment interact to modulate engineered cartilage development. Biochem Biophys Res Commun. 2001;286:909–15.

Giannini S, Buda R, Vannini F, Cavallo M, Grigolo B. One-step bone marrow-derived cell transplantation in talar osteochondral lesions. Clin Orthop Relat Res. 2009;467(12):3307–20.

Giannini S, Buda R, Cavallo M, Ruffilli A, Cenacchi A, Cavallo C, Vannini F. Cartilage repair evolution in post-traumatic osteochondral lesions of the talus: from open field autologous chondrocyte to bone-marrow-derived cells transplantation. Injury. 2010;41(11):1196–203.

Haag J, Gebhard PM, Aigner T. SOX gene expression in human osteoarthritic cartilage. Pathobiology. 2008;75:195–9.

Haleem AM, Singergy AA, Sabry D, Atta HM, Rashed LA, Chu CR, El Shewy MT, Azzam A, Abdel Aziz MT. The clinical use of human culture-expanded autologous bone marrow mesenchymal stem cells transplanted on platelet-rich fibrin glue in the treatment of articular cartilage defects: a pilot study and preliminary results. Cartilage. 2010;1(4):253–61.

Haupt JL, Frisbie DD, McIlwraith CW, Robbins PD, Ghivizzani S, Evans CH, et al. Dual transduction of insulin-like growth factor-I and interleukin-1 receptor antagonist protein controls cartilage degradation in an osteoarthritic culture model. J Orthop Res. 2005;23:118–26.

Hidaka C, Goodrich LR, Chen C-T, Warren RF, Crystal RG, Nixon AJ. Acceleration of cartilage repair by genetically modified chondrocytes over expressing bone morphogenetic protein-7. J Orthop Res. 2003;21(4):573–83.

Kafienah W, Al-Fayez F, Hollander AP, Barker MD. Inhibition of cartilage degradation: a combined tissue engineering and gene therapy approach. Arthritis Rheum. 2003;48:709–18.

Kamei G, Kobayashi T, Ohkawa S, Kongcharoensombat W, Adachi N, Takazawa K, Shibuya H, Deie M, Hattori K, Goldberg JL, Ochi M. Articular cartilage repair with magnetic mesenchymal stem cells. Am J Sports Med. 2013;41(6):1255–6.

Kaps C, Bramlage C, Smolian H, Haisch A, Ungethüm U, Burmester GR, Sittinger M, Gross G, Häupl T. Bone morphogenetic proteins promote cartilage differentiation and protect engineered artificial cartilage from fibroblast invasion and destruction. Arthritis Rheum. 2002;46(1):149–62.

Kasemkijwattana C, Hongeng S, Kesprayura S, Rungsinaporn V, Chaipinyo K, Chansiri K. Autologous bone marrow mesenchymal stem cells implantation for cartilage defects: two cases report. J Med Assoc Thai. 2011;94(3):395–400.

Kaul G, Cucchiarini M, Arntzen D, Zurakowski D, Menger MD, Kohn D, Trippel SB, Madry H. Local stimulation of articular cartilage repair by transplantation of encapsulated

chondrocytes overexpressing human fibroblast growth factor 2 (FGF-2) in vivo. J Gene Med. 2006;8:100–11.
Kay MA. State-of-the-art gene-based therapies: the road ahead. Nat Rev Genet. 2011;12(5):316–28.
Kellner K, Schulz MB, Göpferich A, Blunk T. Insulin in tissue engineering of cartilage: a potential model system for growth factor application. J Drug Target. 2001;9(6):439–48.
Kim SH, Kim S, Evans CH, Ghivizzani SC, Oligino T, Robbins PD. Effective treatment of established murine collagen-induced arthritis by systemic administration of dendritic cells genetically modified to express IL-4. J Immunol. 2001;166:3499–505.
Kim J-H, Park JS, Yang HN, Woo DG, Jeon SY, Do HJ, Lim HY, Kim JM, Park KH. The use of biodegradable PLGA nanoparticles to mediate SOX9 gene delivery in human mesenchymal stem cells (hMSCs) and induce chondrogenesis. Biomaterials. 2011;32(1):268–78.
Kondo M, Yamaoka K, Sonomoto K, Fukuyo S, Oshita K, Okada Y, Tanaka Y. IL-17 inhibits chondrogenic differentiation of human mesenchymal stem cells. PLoS ONE. 2013;8(11):e7946.
Kunisato A, Wakatsuki M, Kodama Y, Shinba H, Ishida I, Nagao K. Generation of induced pluripotent stem cells by efficient reprogramming of adult bone marrow cells. Stem Cells Dev. 2010;19(2):229–38.
Lai CH, Chen SC, Chiu LH, Yang CB, Tsai YH, Zuo CS, Chang WH, Lai WF. Effects of low-intensity pulsed ultrasound, dexamethasone/TGF-beta1 and/or BMP-2 on the transcriptional expression of genes in human mesenchymal stem cells: chondrogenic vs. osteogenic differentiation. Ultrasound Med Biol. 2010;36(6):1022–33.
Lee HJ, Choi BH, Min BH, Park SR. Low-intensity ultrasound inhibits apoptosis and enhances viability of human mesenchymal stem cells in three-dimensional alginate culture during chondrogenic differentiation. Tissue Eng. 2007;13(5):1049–57.
Lee KB, Wang VT, Chan YH, Hui JH. A novel, minimally-invasive technique of cartilage repair in the human knee using arthroscopic microfracture and injections of mesenchymal stem cells and hyaluronic acid–a prospective comparative study on safety and short-term efficacy. Ann Acad Med Singapore. 2012;41(11):511–7.
Lefebvre V, Smits P. Transcriptional control of chondrocyte fate and differentiation. Birth Defects Res C Embryo Today. 2005;75:200–12.
Li Y, Tew SR, Russell AM, Gonzalez KR, Hardingham TE, Hawkins RE. Transduction of passaged human articular chondrocytes with adenoviral, retroviral, and lentiviral vectors and the effects of enhanced expression of SOX9. Tissue Eng. 2004;10(3–4):575–84.
Madry H, Cucchiarini M. Clinical potential and challenges of using genetically modified cells for articular cartilage repair. Croatian Med J. 2011;52(3):245–61.
Madry H, Kaul G, Zurakowski D, Vunjak-Novakovic G, Cucchiarini M. Cartilage constructs engineered from chondrocytes overexpressing IGF-I improve the repair of osteochondral defects in a rabbit model. Eur Cell Mater. 2013;25:229–47.
Manning K, Rachakonda PS, Rai MF, Schmidt MF. Co-expression of insulin-like growth factor-1 and interleukin-4 in an in vitro inflammatory model. Cytokine. 2010;50:297–305.
Mardani M, Kabiri A, Esfandiari E, Esmaeili A, Pourazar A, Ansar M, Hashemibeni B. The effect of platelet rich plasma on chondrogenic differentiation of human adipose derived stem cells in transwell culture. Iran J Basic Med Sci. 2013;16(11):1163–9.
Matsumoto T, Okabe T, Ikawa T, Iida T, Yasuda H, Nakamura H, Wakitani S. Articular cartilage repair with autologous bone marrow mesenchymal cells. J Cell Physiol. 2010;225(2):291–5.
Mayer-Wagner S, Passberger A, Sievers B, Aigner J, Summer B, Schiergens TS, Jansson V, Müller PE. Effects of low frequency electromagnetic fields on the chondrogenic differentiation of human mesenchymal stem cells. Bioelectromagnetics. 2011;32(4):283–90.
Morales TI. Transforming growth factor-beta 1 stimulates synthesis of proteoglycan aggregates in calf articular cartilage organ cultures. Arch Biochem Biophys. 1991;286(1):99–106.
Motoyama M, Deie M, Kanaya A, Nishimori M, Miyamoto A, Yanada S, Adachi N, Ochi M. In vitro cartilage formation using TGF-beta-immobilized magnetic beads and

mesenchymalstem cell-magnetic bead complexes under magnetic field conditions. J Biomed Mater Res A. 2010;92(1):196–204.

Nejadnik H, Hui JH, Feng Choong EP, Tai BC, Lee EH. Autologous bone marrow-derived mesenchymal stem cells versus autologous chondrocyte implantation: an observational cohort study. Am J Sports Med. 2010;38(6):1110–6.

Orozco L, Munar A, Soler R, Alberca M, Soler F, Huguet M, Sentís J, Sánchez A, García-Sancho J. Treatment of knee osteoarthritis with autologous mesenchymal stem cells: a pilot study. Transplantation. 2013;95(12):1535–41.

Park Y, Sugimoto M, Watrin A, Chiquet M, Hunziker EB. BMP-2 induces the expression of chondrocyte-specific genes in bovine synovium-derived progenitor cells cultured in three-dimensional alginate hydrogel. Osteoarthritis Cartilage. 2005;13(6):527–36.

Park SR, Choi BH, Min BH. Low-intensity ultrasound (LIUS) as an innovative tool for chondrogenesis of mesenchymal stem cells (MSCs). Organogenesis. 2007;3(2):74–8.

Park H, Temenoff JS, Tabata Y, Caplan AI, Raphael RM, Jansen JA, Mikos AG. Effect of dual growth factor delivery on chondrogenic differentiation of rabbit marrow mesenchymal stem cellsencapsulated in injectable hydrogel composite. J Biomed Mater Res A. 2009;88(4):889–97.

Pastides P, Chimutengwende-Gordon M, Maffulli N, Khan W. Stem cell therapy for human cartilage defects: a systematic review. Osteoarthritis Cartilage. 2013;21(5):646–54.

Rosa RG, Joazeiro PP, Bianco J, Kunz M, Weber JF, Waldman SD. Growth factor stimulation improves the structure and properties of scaffold-free engineered auricular cartilage constructs. PLoS ONE. 2014;9(8):e10517.

Smeriglio P, Lai JH, Dhulipala L, Behn AW, Goodman SB, Smith RL, Maloney W, Yang F, Bhutani N. Comparative potential of juvenile and adult human articular chondrocytes for cartilage tissue formation in 3D biomimetic hydrogels. Tissue Eng Part A. 2014 (Epub ahead of print).

Sumer H, Jones KL, Liu J, Heffernan C, Tat PA, Upton KR, Verma PJ. Reprogramming of somatic cells after fusion with induced pluripotent stem cells and nuclear transfer embryonic stem cells. Stem Cells Dev. 2010;19(2):239–46.

Sun Y, Feng Y, Zhang C, Chen S, Cheng X. The regenerative effect of platelet-rich plasma on healing in large osteochondral defects. Int Orthopaed. 2010;34:589–97.

Takahashi K, Yamanaka S. Induction of pluripotent stem cells from mouse embryonic and adult fibroblast cultures by defined factors. Cell. 2006;126(4):663–76.

Veilleux N, Spector M. Effects of FGF-2 and IGF-1 on adult canine articular chondrocytes in type II collagen-glycosaminoglycan scaffolds in vitro. Osteoarthritis Cartilage. 2005;13(4):278–86.

Wakitani S, Nawata M, Tensho K, Okabe T, Machida H, Ohgushi H. Repair of articular cartilage defects in the patello-femoral joint with autologous bone marrow mesenchymal cell transplantation: three case reports involving nine defects in five knees. J Tissue Eng Regen Med. 2007;1(1):74–9.

Wang H, Li Y, Chen J, Wang X, Zhao F, Cao S. Chondrogenesis of bone marrow mesenchymal stem cells induced by transforming growth factor beta3 gene in Diannan small-ear pigs. Zhongguo Xiu Fu Chong Jian Wai Ke Za Zhi. 2014;28(2):149–54.

Wong KL, Lee KB, Tai BC, Law P, Lee EH, Hui JH. Injectable cultured bone marrow-derived mesenchymal stem cells in varus knees with cartilage defects undergoing high tibial osteotomy: a prospective, randomized controlled clinical trial with 2 years' follow-up. Arthroscopy. 2013;29(12):2020–8.

Zhang HG, Xie J, Yang P, Wang Y, Xu L, Liu D, Hsu HC, Zhou T, Edwards CK 3rd, Mountz JD. Adeno-associated virus production of soluble tumor necrosis factor receptor neutralizes tumor necrosis factor alpha and reduces arthritis. Hum Gene Ther. 2000;11:2431–42.

Chapter 18
Stem Cell Transplantation for Crohn's Disease

Jie Liang and Kaichun Wu

Abstract From the first case of bone marrow transplantation for Crohn's disease in 1993, numerous studies have shown that stem cell transplantation has potential treatment effect for inflammatory bowel disease (IBD), including ulcerative colitis (UC) and Crohn's disease (CD). The source of stem cells might include hematopoietic stem cells (HSCs) and mesenchymal stem cells (MSCs) peripheral or from the bone marrow. Further, this chapter addressed both the clinical and experimental results on the autologous or allogeneic HSC transplantation, and systemic or local MSC transplantation for CD. Ongoing or released phase I, II, and III data have all been summarized in this chapter. Later, the possible mechanisms such as mucosa healing and immune regulation were addressed. In particular, unpublished date from our group for the injection way of stem cell for the treatment of colitis has been shown. Though encouraging but inclusive from current studies, more questions like long-term, phase III, RCT studies are appealing. Best ratio of cell type and best injection way, as well as the possible issue of carcinogenic, have also been pointed out in the perspective. All in all, stem cell transplantation might be a revolution treatment for the IBD patients, but needs more confirmative, in-depth, and long-term studies from both clinical and experimental.

Keywords Stem cell · Crohn's disease · Therapy

J. Liang (✉) · K. Wu
Xijing Hospital of Digestive Diseases, 127 West ChangLe Road,
Xi'an 710032, Shannxi, People's Republic of China
e-mail: liangjie@fmmu.edu.cn

K. Wu
e-mail: kaicwu@fmmu.edu.cn

R.C. Zhao (ed.), *Stem Cells: Basics and Clinical Translation*, Translational
Medicine Research 1, DOI 10.1007/978-94-017-7273-0_18

18.1 Introduction

Stem cells are undifferentiated biological cells that can differentiate into specialized cells and can divide to produce more stem cells. In humans, there are two main types of stem cells: embryonic stem cells and adult stem cells. Embryonic stem cells are isolated from the inner cell mass of blastocysts. And adult stem cells are found in various tissues. Stem cells have the properties of self-renewal and potency definition, which has been used in the treatment of multiple diseases, such as leukemia (Craddock 2000), spinal cord injury repair (Kang et al. 2005), heart infarction (Strauer et al. 2009), wound healing (Gurtner et al. 2007), and inflammatory bowel diseases (especially Crohn's disease) (Martínez-Montiel Mdel et al. 2014a).

Inflammatory bowel disease (IBD) is a group of inflammatory conditions of the colon and small intestine, mainly including ulcerative colitis (UC) and Crohn's disease (CD) (Liang et al. 2011a). Traditional treatment contains mesalazine (Moss et al. 2014), immunosuppressant (Dulai et al. 2014), corticosteroid (Targownik et al. 2014), and newly biological treatment (Lewis 2007), such as anti-TNF-α and surgery (Levesque et al. 2014). However, all these methods might fail in certain refractory, complicated and corticosteroid-dependant or corticosteroid-resistant patients. New treatments are appealing for IBD patients (Lewis 2007).

18.2 History of Stem Cell Transplantation for IBD

There is emerging evidence that stem cell transplantation might provide long-term remission for IBD, including hematopoietic stem cells (HSCs) and mesenchymal stem cells (MSCs). The story of stem cell transplantation for the treatment of IBD came from a case of CD in bone marrow transplantation in 1993 (Drakos et al. 1993). Further, Lopez-Cubero et al. reported six leukemia patients, who underwent allogeneic marrow transplantation for mainly treatment of leukemia. Unexpectedly, five out of the six leukemia with CD were remission of CD more than 1 year after the transplantation and four of the five patients had sustained remission up to 15 years post-transplant (Lopez-Cubero et al. 1998). More practices have been approached for stem cell transplantation for IBD patients.

18.3 Hematopoietic Stem Cell Transplantation (HSCT) for IBD. Hematopoietic Stem Cell Transplantation (HSCT) for CD

Hematopoietic stem cells (HSCs) are the blood cells that give rise to all the other blood cells and are derived from mesoderm. They are located in the red bone marrow, which is contained in the core of most bones. HSCs present glycoproteins as $CD34^+$, CD 90^+, CD133, and $CD38^-$. Application of HSC transplantation (HSCT) in CD has been carried out widely. A phase I HSCT study in 12 patients with refractory CD had been carried out by Dr. Richrd K Burt's group, which showed 11 of 12 patients entered a sustained remission. After 7–37 (median 18.5) months follow-up visit, only one patient has developed a recurrence of active CD (Oyama et al. 1995). Five years later, in 2010 his group renewed the data. Eighteen of 24 patients are five or more years after transplantation, all of whom went into remission with a CDAI less than 150. The percentage of clinical relapse-free survival (defined as the percent free of restarting CD medical therapy after transplantation) is 91 % at 1 year, 63 % at 2 years, 57 % at 3 years, 39 % at 4 years, and 19 % at 5 years. The percentage of patients in remission (CDAI < 150), corticosteroid-free, or medication-free at any post-transplantation evaluation interval more than 5 years after the transplantation remained over 70, 80, and 60 %, respectively (Burt et al. 2010). Burt et al. (2011) also released HSCT for systemic sclerosis in 2011 in Lancet. In his practice, stem cells were mobilized from the peripheral blood using cyclophosphamide (2.0 g/m^2) and G-CSF (10 μg/kg/day), enriched ex vivo by $CD34^+$ selection, and re-infused after immunosuppressive conditioning with cyclophosphamide (200 mg/kg) and either equine anti-thymocyte globulin (ATG, 90 mg/kg) or rabbit ATG (6 mg/kg) (Burt et al. 2010, 2011). Later, Hasselblatt et al. (2012) reported the outcome of 12 patients with refractory CD treated with HSCT. He showed harvest following mobilization chemotherapy was successful in 11/12 patients and resulted in a clinical and endoscopic improvement in 7/12 patients. Subsequent conditioning and HSCT were performed in 9 patients with well-tolerated treatment. Among them, five patients achieved a clinical and endoscopic remission within 6 months. However, relapses occurred in 7/9 patients during 3.1-year (range 0.5–10.3 years) follow-up. The good news was that the relapsed disease could be controlled by low-dose corticosteroids and conventional immunosuppressive therapy.

Unlike Dr. Burt's group, some groups do not perform $CD34^+$ cell selection before the HSCT. Cassinotti et al. (2008) used the protocol as following: Unselected HSCs were collected after mobilization with cyclophosphamide (CTX) 1.5 g/m^2 and granulocyte colony-stimulating factor (G-CSF) 10 mg/kg. The conditioning regimen included CTX 50 mg/kg on days −5 to −2 and rabbit anti-thymocyte globulin (ATG) 2.5 mg/kg on days −4 to −2. He found no improvement or slight deterioration after mobilization. At the 3rd month, the clinical remission was achieved in all patients and complete endoscopic remission was achieved in 2/3 patients. After a median follow-up of 16.5 months, 3/4 patients maintained both clinical and endoscopic remission. After withdrawal of all drugs, complete fistula closure was observed in all affected patients. But no deaths or life-threatening infection occurred. And unexpected adverse events included a perianal abscess after mobilization in one patient, pleural and pericardial effusions in another, and BK virus-related macrohamaturia in another all rapidly resolved with conservative treatment. Generally, HSCT has been safe and effective in IBD patients whether with or without $CD34^+$ selection.

The European Crohn's and Colitis Organisation (ECCO) and the European Group for Blood and Marrow Transplantation (EMBT) have carried out the clinical trial for this issue. The phase III Autologous Stem Cell Transplantation International Crohn's Disease (ASTIC) study included the patients and randomized to two treatment arms: mobilization chemotherapy with G-CSF and autologous HSCT in 30d versus mobilization chemotherapy with G-CSF and conditioning with autologous HSCT after 13 m. ECCO released the preliminary result in 2013 that HSCT appeared to be effective in CD patients with endoscopic improvement of the disease. However, it involved the risk of adverse effects and the end results referred to the trial objective are still awaiting analysis (Hawkey et al. 2013).

18.4 Hematopoietic Stem Cell Transplantation (HSCT) for UC

Besides CD, HSCT has been used in UC patients as well. In 1995, a woman with a 7 year history of distal UC was later diagnosed of breast cancer and got HSCT. By this time, she had no symptoms relating to IBD. Till 25 months, UC relapsed and azathioprine was added. Then, the patient is asymptomatic and tumor free, on azathioprine, 36 months after relapse of UC (Martí et al. 2001).

18.5 Safety of Hematopoietic Stem Cell Transplantation (HSCT) for IBD

The overall safety of HSCT in the treatment of autoimmune diseases (including IBD) might be evaluated from the European Group for Blood and Marrow Transplantation Working Party on Autoimmune Diseases 12-year experience (Farge et al. 2010). Nine hundred patients with autoimmune diseases who underwent autologous hematopoietic stem cell transplant obtained the 5-year survival of 85 % and the progression-free survival of 43 %, although the rates varied widely according to the type of autoimmune disease. No significant influence of transplant technique was identified. Age less than 35 years ($p = 0.004$), transplantation after 2000 ($p = 0.0015$), and diagnosis ($p = 0.0007$) were associated with progression-free survival. This largest cohort studied worldwide shows that autologous HSCT can induce sustained remissions for more than 5 years in patients with severe autoimmune diseases refractory to conventional therapy. The type of autoimmune disease, rather than transplant technique, was the most relevant determinant of outcome. These data support ongoing and planned phase III trials to evaluate the autologous HSCT in the treatment strategy for severe autoimmune diseases including IBD. Accordingly, autologous HSCT might be an effective option for IBD patients. Further studies are needed to determine the best way and standard procedure for the transplantation.

18.6 Allogenic Stem Cell Transplantation (HSCT) for IBD

Allogenic stem cell transplantation (HSCT) for IBD was been reported in 2009. New England Journal of Medicine (NEJM) reported a case with mutation of the IL10RA and IL10B genes. Mutations in genes encoding the IL10R subunit proteins were found in patients with early onset colitis, involving hyperinflammatory immune responses in the intestine. Allogeneic stem cell transplantation resulted in disease remission in one patient (Glocker et al. 2009). Retrospective study from Ditschkowski et al. (2003) showed 7 CD and 4 UC patients who underwent allogeneic SCT for hematologic malignancy and myelodysplastic syndrome. After a median follow-up of 34 months post-transplantation, 10 patients are alive. None of the patients showed IBD activity after the transplantation, except one patient with mild symptoms of CD early after transplant. Colonoscopy after complete discontinuation of post-transplant immunosuppression revealed no pathologic findings. As we mentioned previously, Lopez-Cubero et al. reported six leukemia patients, who underwent allogeneic marrow transplantation for leukemia and 4/5 patients remained symptom-free during a period of 54–183 m after the transplantation (Martínez-Montiel Mdel et al. 2014b). Similar findings had been seen in a patient with CD and acute myeloid leukemia subjected to allogenic HSCT (Talbot et al. 1998). Generally, allogenic stem cell transplantation (HSCT) for IBD has not currently been recommended, except in very specific cases such as the mutation of IL-10 gene.

18.7 Mesenchymal Stem Cells (MSCs) for CD

Mesenchymal stem cells (MSCs) have a great capacity for self-renewal while maintaining their multipotency and immunosuppression (English 2013). The cultured MSCs also express on their surface CD73, CD90, and CD105 while lacking the expression of CD11b, CD14, CD19, CD34, CD45, CD79a, and HLA-DR surface markers (Dominici et al. 2006).

18.8 Systemic MSC Treatment for CD

Many clinical trials of MSC therapy for IBD have been carried out. In the first human clinical trial of systemic MSCs in CD, Onken et al. treated 10 patients who had failed the traditional medicine and had active diseases. The patients were randomized to receive allogenic bone marrow-derived MSCs 2 million cells/kg or 8 million cells/kg i.v (Onken et al. 2006). With the follow-up at day 28, 9 patients had decreased CDAI score (341 vs. 236, $p = 0.004$). The primary end point was defined as a $\geq$100-point reduction in CDAI. This response was achieved in 3 patients by day 14. And mean IBD quotient scores increased significantly from baseline to day 28 (113 vs. 146, $p = 0.008$). There had an association between mean change in IBD quotient and clinical response at day 28 ($p = 0.07$). Although not statistically significant, the mean reduction in the CDAI at day 28 was greater in the 8 million cells/kg group than that in the 2 million cells/kg group. No severe side effects had been reported. In the Digestive Disease Week in 2010, another group reported transplantation of MSC to CD patients in Russia. Clinical–morphological remission registered in 6 of 11 patients and endoscopic remission registered in 4 patients. MSC transplantation stimulates depressed synthesis of cytokines, decreases intensity of immunopathological processes in CD, and helps in decreasing GCS requirement.

A phase I study reported autologous bone marrow-derived mesenchymal stromal cell treatment for refractory luminal CD (Duijvestein et al. 2010). Ten patients with refractory CD underwent bone marrow aspiration under local anesthesia. Bone marrow MSCs were isolated and expanded ex vivo. MSCs were tested for phenotype and functionality in vitro. Nine patients received two doses of $1–2 \times 10^6$ cells/kg body weight, i.v. 7 days apart. Three patients showed clinical response (CDAI decrease ≥ 70 from baseline) 6 weeks after the treatment; conversely, three patients required surgery due to disease worsening. Phase III clinical trials have been carried out in certain companies, as Oasis in USA, Cellerix in Europe, and Anterogen in Asia (www.clinicltrials.gov). The current ongoing largest, randomized, placebo-controlled, double-blind phase III study of prochymal (allogenic marrow-derived MSCs) in CD was initiated in 2007 by Osiris company (http://clinicaltrials.gov/ct2/show/NCT00294112) (Taupin 2006). The study is planned to enroll 270 patients with active CD (CDAI 250-450) who have a history of treatment failure with or intolerance to corticosteroid, immunosuppressants, and biological agents. Patients are then randomized to receive four infusions over 2wk of either 600 million cells (low dose: two infusions of 200×10^6 hMSCs in week 1, then two infusions of 100×10^6 hMSCs in week 2) or 1200 million cells (high dose: two infusions of 400×10^6 hMSCs in week 1, then two infusions of 200×10^6 hMSCs in week 2) or placebo. The primary end point of the study is clinical remission at day 28 with secondary end points being clinical response, improved quality of life (defined as increased IBD quotient score), and decreased number of draining fistulae. In March 2009, with 207 patients enrolled, the trial was suspended because of a high placebo response. Subsequently, the United States Food and Drug Administration (US FDA) authorized the reopening of the study, though no results are available now (Mannon 2011).

Allogeneic mesenchymal stem cell transplantation had been well carried out in Nanjing GuLou Hospital in China. Liang et al. reported the results obtained in 7 patients with IBD (4 with CD and 3 with UC). In 3 cases, MSCs were obtained from the bone marrow of healthy donors, and in 4 cases, from umbilical cord. The dose administered consisted of one million cells/kg i.v. All the patients maintained the medication (corticosteroids and/or immunosuppressants) after the infusion. Five subjects showed remission. Two CD patients and one UC patient showed improved endoscopic appearance. In all three of these subjects, biopsies revealed a decrease in the extent of IBD and in intensity of lymphoid infiltrate. Side effects were mild as one facial flushing for 6 h and another insomnia during the first night following infusion (Liang et al. 2012). Recently, Forbes GM et al. published their phase II study result. In their phase II, open-label, multicenter study, they treated 16 patients who had biological agents refractory, endoscopically confirmed, active luminal CD (CDAI > 250). Subjects were given intravenous infusions of allogeneic MSCs (2×10^6 cells/kg body weight) weekly for 4 weeks. The primary end point was clinical response (decrease in CDAI > 100 points) 42 days after the first MSC administration; secondary end points were clinical remission (CDAI, <150), endoscopic improvement (a CD endoscopic index of severity value, <3 or a decrease by >5), quality of life, level of C-reactive protein, and safety. Fifteen out of sixteen patients who completed the study showed the mean CDAI score reduced from 370 (median, 327; range, 256–603) to 203 (median, 129) at day 42 ($p < 0.0001$). The mean CDAI scores decreased after each MSC infusion (370 as baseline, 269 on day 7, 240 on day 14, 209 on day 21, 182 on day 28, and 203 on day 42). Twelve patients had a clinical response (80 %; 95 % CI 72–88 %; mean reduction in CDAI, 211; range 102–367) and 8 had clinical remission (53 %; range, 43–64 %; mean CDAI at day 42, 94; range, 44–130). Seven patients had endoscopic improvement (47 %). One patient had a serious adverse event (2 dysplasia-associated lesions), but was speculated to be not caused by MSCs (Forbes et al. 2014).

For the safety of MSC treatment, human clinical trials mostly showed the safety with no toxic effects or generation of ectopic tissue. The most commonly reported side effect was transient fever (Lalu et al. 2012). MSCs may be infected with viruses (e.g., cytomegalovirus (CMV), herpes virus), and a case of infection with Epstein–Barr (EV) virus has been reported in a patient followed by the administration of MSCs due to an episode of GVHD (Ringdén et al. 2006). Meanwhile, a possible case of bacterial infection transmitted by umbilical cord hematopoietic stem cells has recently been reported (Bhatt et al. 2013). In general, systemic MSCs treatment is promising, but needs phase III clinical trials for validation.

18.9 Local MSC Treatment for CD

The first trial of local treatment of fistulae for 5 CD patients with autologous MSCs (ASCs) was published in 2005 (García-Olmo et al. 2005). In their phase I study, nine fistulas in four patients had been given the autologous MSC (ASCs) transplantation. In six fistulas, the external opening was covered with epithelium at the end of 8w, and thus, these fistulas were considered healed (around 75 %). In the other two fistulas, there was only incomplete closure of the external opening, with a decrease in output flow (not healed; 25 %). No adverse effects were observed in any patient at the end of the follow-up period (mean 22 m, 12–30 m). With this promising method, they have carried out the phase II study and released the data 4 years later. In the phase II study, fistula healing was observed in 17 of 24 patients (71 %) compared with 4 of 25 control patients (16 %) (relative risk for healing, 4.43; CI 1.74–11.27); $p < 0.001$). Quality of life scores were higher in patients who received ASCs than control. At one-year follow-up, the recurrence rate in patients treated with ASCs was 17.6 %. Both treatments were well tolerated (Garcia-Olmo et al. 2009). Another two groups published their data later. In Cho YB's dose escalation study, patients were sequentially enrolled into three dosing groups with at least three patients per group. The first three patients (group 1) were given 1×10^7 cells/ml. After 4 weeks, this dose was deemed safe, and so an additional four patients (group 2) were given 2×0^7 cells/ml. Four weeks later, after which this second dose was deemed safe, a third and final groups of three patients were given 4×10^7 cells/ml. Each patient was followed for a minimum of 8 weeks. Patients who showed complete healing at week 8 were followed up for an additional 6 months. Efficacy end point was complete healing at week 8 after injection, defined as complete closure of the fistula track and internal and external openings without drainage or signs of inflammation. There were no grade 3 or 4 severity adverse events, and there were no adverse events related to the study drug. Two patients in group 2, treated with 2×10^7 ASCs/ml, showed complete healing at week 8 after injection. Of the three patients enrolled in group 3, treated with 4×10^7 ASCs/ml, one showed complete healing. Outcome in another patient was assessed as partial healing due to incomplete closure of the external opening. All three patients with complete healing at week 8 showed a sustained effect without recurrence 8 months after injection. From this dose escalation study, it demonstrates the tolerability, safety, and potential efficacy of ASCs for the treatment of CD fistula (Cho et al. 2013).

The results of the FATTI phase III, randomized, single-blind, multicenter clinical trial were published in 2012 (Herreros et al. 2012). Two hundred subjects were randomized to three groups: 20 million ASCs (group A), 20 million ASCs with fibrin glue (group B), and fibrin glue with placebo (group C). The possibility existed of a second dose of 40 million ASCs in week 12. The primary end point was fistular sealing in week 12 and in weeks 24–26. Of the 200 randomized individuals, 183 received treatment, and 165 completed the study. After 12 wk, fistular sealing was observed in 26.5, 38.33, and 15.25 % of patients in groups A, B, and C, respectively ($p = 0.01$). A second dose was administered in 61.5 % of patients, with fistular sealing in 39.1, 43.3, and 37.3 %, respectively ($p = 0.79$). Posterior analysis stratified by center revealed far better results for patients administered ASCs on comparing the center with the greatest experience versus the rest of the participating centers: 45.55, 83.3, and 18.8 % ($P = 0.025$ for treatment) versus 35.8, 33.3, and 42.6 % for groups A, B, and C, respectively. The treatment at the center with most experience was considered to be a significant factor. It is speculated that the experience of the surgeon in using ASCs in perianal fistulas may be decisive. There were no significant differences in adverse effects among the three groups. A total of 37 serious adverse effects were recorded—three of which were related to the procedures used, but none to use of ASCs.

In that same year, the first study on donor adipose tissue expanded mesenchymal stem cells (eASCs) was published by a Spanish research group, which was open-label, single-arm clinical trial at six Spanish hospitals. Twenty-four patients were administered intralesionally with 20 million eASCs in one draining fistula tract. A subsequent administration of 40 million eASCs was performed if fistula closure was incomplete at week 12. Subjects were followed until week 24 after the initial administration and showed 69.2 % of the patients with a reduction in the number of draining fistulas at 24 weeks. Among them, 56.3 % of the patients achieved complete closure of the treated fistula achieved and 30 % of the cases presenting complete closure of all existing fistula tracts. Magnetic resonance imaging (MRI) Score of Severity also showed statistically significant differences at week 12 with a marked reduction at week 24 (de la Portilla et al. 2013). At present, a phase III study is underway involving eASCs for the treatment of complex fistulas in CD patients. This trial will contribute relevant information and may lead to future commercial use of the treatment. The safety of this method appears to have been confirmed by above different studies published to date.

18.10 Timing of MSC Transplantation for CD

Duijvestein et al. showed that immunosuppressants and anti-TNF compounds do not change the phenotype, morphology, viability, differentiation, and functional capabilities when incubated with MSCs in physiological concentrations. Conversely, the presence of MSCs did not hamper the immunosuppressive effect of those commonly used medications for the treatment of CD (Duijvestein et al. 2011). In another study examining the mechanism of various anti-TNF medications, Vos et al. (2012) found that anti-TNF antibodies differentiate blood-derived macrophage into regulatory phenotype in an FC fragment-dependent fashion. The phenotype has anti-inflammatory properties because it inhibits the proliferation of activated T cells, produces IL-10, and expresses regulatory macrophage marker CD206. Similarly, MSCs can turn macrophages into the regulatory phenotype, raising the possibility of synergism (Maggini et al. 2010). All in all, MSC treatment might be an attractive method for CD therapy and to enable more rapid tapering of immunosuppressive therapy. All the current trials have showed short-term safety and efficacy of MSC transplantation for CD. There is an apparent discrepancy in response to MSC therapy for CD based on route of administration (i.v. vs. intralesional vs. intra-arterial). The reason for this is speculative, but it might reflect the density of the injected cells at the lesion site. Ongoing trials with higher doses may help in answering this question. Many challenges remain ahead, including the best administration route, the best source of MSCs, and the density of cells needed at the lesion site to guarantee effective therapy. It will be particularly important to determine which combinations, including biologics, are more effective in the treatment of CD (Dalal et al. 2012).

18.11 Mechanisms for Stem Cell Transplantation for CD

Understanding the pathogenesis of IBD is the basis for the development of new treatments. IBD is regarded as the result of an abnormal host immune response to intraluminal antigens occurring in a genetically predisposed individual, with the production of chronic inflammation of the gastrointestinal tract, accompanied by tissue destruction. IBD is the consequence of complex interaction among genetic (Liang et al. 2011b), environmental, microbial factors (Sha et al. 2014), and some others. They produced sustained inflammation and caused the mucosal barrier and immune system defects (Sha et al. 2014). Treatment for IBD might help to repair the damage of the intestinal mucosa and regulate the immune microenvironment in the gastrointestinal system (Liang et al. 2013). Thus, transplantation for the treatment of IBD might have the following mechanisms.

18.12 Regeneration of the Damaged Gastrointestinal Track

Okamoto R et al. reported that bone marrow cells can repopulate the epithelia of the human gastrointestinal tract. From human donors and Y-FISH analysis, they found that the donor-derived epithelial cells substantially repopulated the gastrointestinal tract during epithelial regeneration after graft-versus-host disease or ulcer formation. Regeneration of gastrointestinal epithelia with donor-derived cells in humans shows a potential clinical application of bone marrow-derived cells for repairing severely damaged epithelia, not only in the gastrointestinal tract but also in other tissues (Okamoto et al. 2002). More animal experiments proved stem cells might repair the mucosal regeneration in colitis models. Brittan et al. (2005) studied on the trinitrobenzenesulfonic acid (TNBS)-induced colitis mice and showed that stem cells contribute to multiple intestinal cell lineages in colitis, with an important function in tissue regeneration and vasculogenesis after injury. The contribution of stem cells to intestinal myofibroblasts was significantly increased in regions of colitis than the noninflamed regions. Furthermore, bone marrow-derived endothelial cells, pericytes, and vascular smooth muscle cells were frequently spread throughout blood vessels to facilitate angiogenesis in tissue repair. Hayashi et al. (2007) reported the similar result in the dextran sulfate sodium (DSS)-induced colitis mice. In their study, examined by confocal microscopy and fluorescence immunohistochemistry, stem cells were frequently observed in the vimentin-positive colonic interstitial cells, which also expressed alpha-smooth muscle actin and had a spindlelike morphology, but did not express leukocyte common antigen. They also frequently transdifferentiated into subepithelial myofibroblasts and fibroblasts and often resided in the colonic subepithelia after the experimental colitis had healed. Bone marrow transplantation was observed to ameliorate the pathology in interleukin-10 (IL-10) knockout colitis mice (Bamba et al. 2006). Here, interleukin IL-$10^{-/-}$ mice were used as a model of IBD to investigate the involvement of bone marrow-derived cells in the inflamed mucosa. Body weights and histological scores showed that IL-$10^{-/-}$ mice that received wide-type mice stem cell transplantation had an improved course of colitis and decreased mucosal pro-inflammatory factors expression. Conversely, IL-$10^{-/-}$ mice receiving IL-$10^{-/-}$ bone marrow progressed to extensive colitis. The expression of MMP-7 and osteopontin was upregulated in the inflamed mucosa. In conclusion, IL-$10^{-/-}$ mice displayed ameliorated disease activity after the stem cell transplantation. Recent data exhibited that bone marrow-derived mesenchymal stem cells are a major source of interleukin-7 (IL-7) and sustain colitis by forming the niche for colitogenic CD4 memory T cells (Nemoto et al. 2013).

18.13 Mesenchymal Stem Cells (MSCs) Regulate the Immune System

MSCs provided an effective therapeutic role in inflammatory diseases by modulating inflammatory responses and tissue regeneration by their differentiation ability. Studies demonstrated the potential therapeutic use of MSCs in treating chronic DSS-induced colitis in mice (Salam et al. 2014). Mice were divided into two groups: one was treated with MSCs and the other was treated with phosphate-buffered saline (PBS) as control. Assessment of therapeutic efficacy of MSCs was carried out by measuring weight, stool score, histopathologic score, and inflammatory factors such as interleukin-23 (IL-23), tumor necrosis factor-α (TNF-α), interferon-γ (IFN-γ), and intercellular adhesion molecule-1 (ICAM-1). The results showed that MSC-treated mice showed a significant improvement in stool condition, weight gain, and normal histopathologic score compared to the control. Moreover, gene expressions of inflammatory molecules in the MSC-treated mice were also significantly lower than those of the control mice. Decreased expression of pro-inflammatory cytokines (TNF-α, IL-1β, and IL-6) and down-regulation of STAT3 phosphorylation in colon tissue were also found after MSC treatment (Chen et al. 2014). From the DSS-induced colitis mice model, it showed that MSCs might inhibit the key inflammatory molecule expression and had potential roles in the treatment for IBD. Study on the interactions between transplanted bone marrow-derived mesenchymal stem cells in the TNBS-induced colitis mice showed the similar effect (Zuo et al. 2013). The MSCs were transduced with a replication-defective recombinant lentiviral vector carrying GFP in order to be able to trace the injected cells in vivo. Prepared MSCs (1×10^6) were injected into rats with TNBS-induced colitis via the tail vein. Two weeks after the intravenous infusion, the frequency of $CD^{4+}CD^{25+}$Foxp3 cells in the peripheral blood was examined by flow cytometry. Similarly as previous DSS-induced colitis mice experiment, the systemic infusion of MSCs significantly ameliorated the clinical and histopathologic severity of TNBS-induced colitis. There was an inverse regulation of mucosal and peripheral Foxp3 expression, suggesting that the MSCs redistributed the Tregs from the mucosa to the blood. MSCs exhibit immunomodulatory functions and may be used to ameliorate or treat IBD by redistributing regulatory T cells. It showed that MSC transplantation might induce M2 macrophage polarization as the major source of TGF-beta to alleviate the colitis in DSS-induced mice model (Wang et al. 2014).

18.14 Role of Intestinal Myofibroblastive Stem Cells in Colitis

The subepithelial mesenchymal cells and their secreted basement membrane factors comprise the lamina propria, which provides a supporting network for the epithelial cells and regulates epithelial cell function. The lamina propria contains two intestinal myofibroblast populations: the interstitial cells of Cajal (ICC) and intestinal subepithelial myofibroblasts (ISEMFs). ICC are cells located in an intramuscular space between the submucosa and muscularis propria, which regulate gastrointestinal smooth muscle motility, facilitate the propagation of electrical events, and regulate neurotransmission (Sanders et al. 2002). ISEMFs, also called pericryptal fibroblasts, reside subjacent to the basement membrane of the small and large intestines (Andoh et al. 2005). ISEMFs are specialized mesenchymal cells that exhibit the features of both fibroblasts and smooth muscle cells. Previously, some controversy has been debated about the role of intestinal myofibroblastive stem cells in colitis. More data proved that intestinal myofibroblastive stem cells in colitis might also have the therapeutic effect for the colitis. Andoh et al. (2007) showed that the number of Th-17 cells was increased in the inflamed mucosa of IBD patients (Jiang et al. 2014), which could modulate the intestinal subepithelial myofibroblast to affect the colitis and bone marrow-derived stem cells in mucosal regenerative response via differentiation to ISEMF. Thus, intestinal subepithelial myofibroblasts have potential role in inflammation and regenerative response in the gut. Studies also showed that nonmyeloablative stem cell therapy enhances microcirculation and tissue regeneration in murine inflammatory bowel disease. Moderate–severe colitis in mice was induced by DSS and 2.0×10^6 immortalized $CD34^-$stem cells infused twice via the tail vein during an observation period of 35 days in a nonmyeloablative setting. It showed that nonmyeloablative stem cell therapy resulted in increased survival in severe colitis ($p < 0.0001$). Clinical activity and histologic score of the colitis severity were reduced significantly in moderate ($p = 0.0003$ or $p = 0.03$) and severe ($p < 0.0001$ or $p < 0.03$) colitis after 35 days, in addition to the DSS-induced shortening of colon length ($p = 0.002$ and $p < 0.0002$). Genetically marked stem cells were detected predominantly in the submucosa of the damaged colon epithelium. Epithelial repair in experimental IBD was mediated either by induction of improved vasculogenesis or by the differentiation of the transplanted stem cells into endothelial cells, as demonstrated by the promotion of Tie2 activity in the infused cells at the site of the damaged mucosa (Khalil et al. 2007).

Recently, Lei et al. (2014) showed intestinal subepithelial myofibroblasts support the growth of intestinal epithelial stem cells. They hypothesized that co-culture with ISEMF could enhance the growth of intestinal epithelial stem cells (ISCs) in vitro and allows for their successful in vivo implantation and engraftment. ISC-containing small intestinal crypts, FACS-sorted single ISCs, and ISEMFs were procured from C57BL/6 mice. Crypts and single ISC were grown in vitro into enteroids, in the presence or in the absence of ISEMFs. Co-culture of ISCs with supportive ISEMFs relinquished sustaining long-term growth and differentiation of ISCs. Mono- and co-cultures were implanted subcutaneously in syngeneic mice. Co-culture with ISEMFs proved necessary for the success of in vivo engraftment and proliferation of enteroids. Further study showed Wnt agonists supported ISC growth.

18.15 Best Way for MSC Transplantation for Colitis Intraperitoneal

From above data, experimental and clinical, stem cell therapy showed promising results. However, most of these studies were focused on its therapeutic effects or mechanisms. Very few reports pay attention to factors applied in the clinic, such as injection or exposure routes, which has high impact on stem cell therapeutic efficiency for IBD (Kean et al. 2013). In our group, we have performed cases of stem cell transplantation for IBD patients in clinic, which got satisfactory result. We are seeking for the best way of stem cell injection as well. So animal experiment was done for this issue (Liang and Wu, Unpublished data).

Three different MSC delivery routes, intraperitoneal injection (IP), intravenous injection (IV), and anal injection (AI), were compared on DSS-induced colitis mouse model. Disease recovery was evaluated by histological server score, gross body weight, and survival rate. MSC organ distribution and engraftment were analyzed and quantified by GFP^+ MSCs as well as near-infrared fluorescence imaging. The levels of immunomodulatory cytokines were compared by RT-PCR and ELISA. Our findings suggested that IP delivery showed higher MSCs and better experimental colitis recovery might be an ideal way for MSC therapy in IBD. Our data showed after 5 days of DSS induction, all the subject mice showed strong positive test of fecal occult blood. These mice were then randomized into 3 groups receiving different MSCs giving therapies. The three groups can decrease the mortality rate compared with PBS control; however, the IP injection showed the highest survival rate of 87.5 % ($p = 0.0021$ vs. control patient). Meanwhile, the body weight changes displayed the less weight loss and quick weight gain of IP injection group and the maximum group difference among was reached on day 3. The fecal occult blood test on the day 3 also showed nearly complete absence of occult blood in IP group. Further experiment showed that IP route promotes more MSC migration to inflamed colon. The number of therapeutic cells that can migrate and colonize at the injury site is a decisive prerequisite for the success of cytotherapy. In order to compare the difference of MSC colonization among the three delivery routes, a nontoxic NIR tracer DiR was introduced to label MSCs. DiR labeling showed no harm to the MSC viability. DiR-labeled cells were injected by three different ways, and DiR dye only was used as negative control. Twenty-four hour after injection, MSC distribution was analyzed. The DiR dye alone showed no significant different distribution among the organs. Furthermore, DiR dye showed a quick clearance, and since 24 h later, the intensity of dye was 100-fold less than the DiR-labeled cells (from the bar value 10^7 vs. 10^9). The quick clearance guarantees the less interference from the background signal that coming from free dye. The DiR-labeled cell showed very different cell distribution. In IV group, most of the MSCs were trapped in the lung, liver, and spleen, while the cell that immigrated to the colon was not too much. On the contrary, the IP and AI groups showed more engraftment cells at the inflamed colon but showed fewer trapped cells in lung, liver, and spleen (Fig. 18.1). In quantification study, the intensity of engrafted cells in IP and AI groups was significantly higher than the IV group ($p = 0.004$, $p = 0.0012$). Furthermore, the corresponding mesentery lymph nodes (MLNs) also showed relatively high MSC existence. To further confirm the MSC migration and engraftment, GFP^+ MSCs (collected from transgenic mice) were introduced and injected at the same experimental conditions. In consistent with above NIR fluorescence imaging, the GFP^+ MSCs could be found at inflamed colon in both IP and AI groups 24 h after MSC injection. Interestingly, in some IP injection mice, numerous GFP^+ cells were observed in the epithelium of the inflamed colon. In contrast, GFP^+ cells were showed in the lumen for AI group and lamina propria for IV group (Fig. 18.2) and the IV group preserved fewer cell number. In addition, from the point of cell morphological appearance, most of the MSCs kept intact in IP group, while in AI route, other than active MSCs, cell fragments and debris were also identified in the lumen (Fig. 18.2).

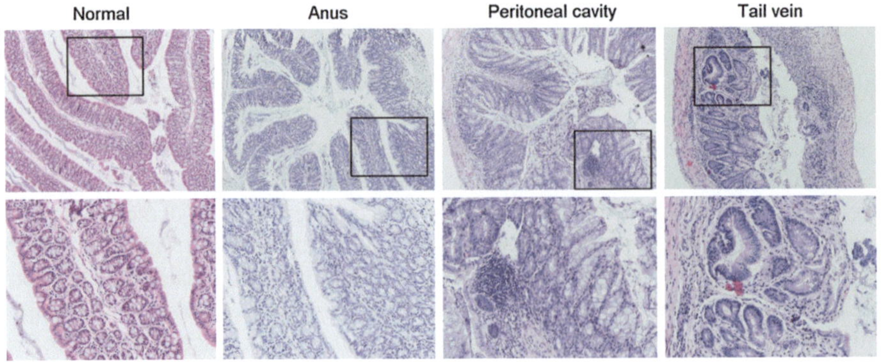

Fig. 18.1 MSC delivery routes were compared on DSS-induced colitis mouse model

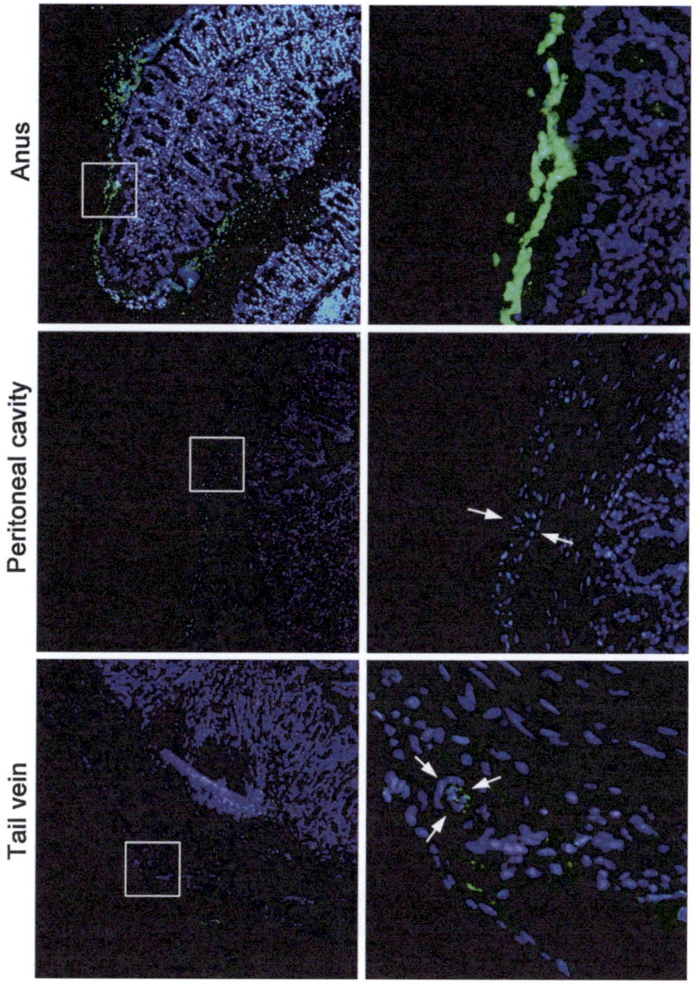

Fig. 18.2 NIR fluorescence imaging

18.16 Perspectives

Stem cell transplantation for the treatment of IBD provides a potential and effective way nowadays. Both HSCs and MSCs have obtained satisfactory results. Existing studies are encouraging but inclusive, and more open questions have been awaiting for answers. Just name a few, (1) Stem cell transplantation has been carried out clinically most in phase I and II studies. Confirmative results from phase III studies and multiple center RCTs are appealing to evaluate the effective and side effects of stem cell transplantation for IBD. (2) The follow-up time for most studies is not too long and some studies showed relapse for the diseases. Thus, long-term follow-up might be needed to address this issue. (3) For the relapse of disease, how many times can the patients get stem cell transplantation for the treatment. And how often for the routine treatment? (4) HSCs or MSCs, which is better? Or at certain ratio for both HSCs and MSCs would achieve best effect? (5) For the infusion way, systemic or local injection, as our animal experiment might have possible suggestion. But it might need clinical approval for the guidance of clinical use. The last but not the least, the possible issue of carcinogenic has got some attention recently. In particular, more requests have been carried out in China Science and Technology Division. More standardized studies have been carried out for it. The data from future studies might cause a revolution in the IBD treatment and have great effect on the IBD patients.

References

Andoh A, Yasui H, Inatomi O, Zhang Z, Deguchi Y, Hata K, Araki Y, Tsujikawa T, Kitoh K, Kim-Mitsuyama S, Takayanagi A, Shimizu N, Fujiyama Y. Interleukin-17 augments tumor necrosis factor-alpha-induced granulocyte and granulocyte/macrophage colony-stimulating factor release from human colonic myofibroblasts. J Gastroenterol. 2005;40(8):802–10.

Andoh A, Bamba S, Brittan M, Fujiyama Y, Wright NA. Role of intestinal subepithelial myofibroblasts in inflammation and regenerative response in the gut. Pharmacol Ther. 2007;114(1):94–106.

Bamba S, Lee CY, Brittan M, Preston SL, Direkze NC, Poulsom R, Alison MR, Wright NA, Otto WR. Bone marrow transplantation ameliorates pathology in interleukin-10 knockout colitic mice. J Pathol. 2006;209(2):265–73.

Bhatt AS, Freeman SS, Herrera AF, Pedamallu CS, Gevers D, Duke F, Jung J, Michaud M, Walker BJ, Young S, Earl AM, Kostic AD, Ojesina AI, Hasserjian R, Ballen KK, Chen YB, Hobbs G, Antin JH, Soiffer RJ, Baden LR, Garrett WS, Hornick JL, Marty FM, Meyerson M. Sequence-based discovery of *Bradyrhizobium enterica* in cord colitis syndrome. N Engl J Med. 2013;369:517–28.

Brittan M, Chance V, Elia G, Poulsom R, Alison MR, MacDonald TT, Wright NA. A regenerative role for bone marrow following experimental colitis: contribution to neovasculogenesis and myofibroblasts. Gastroenterology. 2005;128(7):1984–95.

Burt RK, Craig RM, Milanetti F, Quigley K, Gozdziak P, Bucha J, Testori A, Halverson A, Verda L, de Villiers WJ, Jovanovic B, Oyama Y. Autologous nonmyeloablative hematopoietic stem cell transplantation in patients with severe anti-TNF refractory Crohn disease: long-term follow-up. Blood. 2010;116(26):6123–32.

Burt RK, Shah SJ, Dill K, Grant T, Gheorghiade M, Schroeder J, Craig R, Hirano I, Marshall K, Ruderman E, Jovanovic B, Milanetti F, Jain S, Boyce K, Morgan A, Carr J, Barr W. Autologous non-myeloablative haemopoietic stem-cell transplantation compared with pulse cyclophosphamide once per month for systemic sclerosis (ASSIST): an open-label, randomised phase 2 trial. Lancet. 2011;378(9790):498–506.

Cassinotti A, Annaloro C, Ardizzone S, Onida F, Della Volpe A, Clerici M, Usardi P, Greco S, Maconi G, Porro GB, Deliliers GL. Autologous haematopoietic stem cell transplantation without CD34+ cell selection in refractory Crohn's disease. Gut. 2008;57(2):211–7.

Chen Z, He X, He X, Chen X, Lin X, Zou Y, Wu X, Lan P. Bone marrow mesenchymal stem cells ameliorate colitis-associated tumorigenesis in mice. Biochem Biophys Res Commun. 2014;450(4):1402–8.

Cho YB, Lee WY, Park KJ, Kim M, Yoo HW, Yu CS. Autologous adipose tissue-derived stem cells for the treatment of Crohn's fistula: a phase I clinical study. Cell Transplant. 2013;22(2):279–85.

Craddock C. Haemopoietic stem-cell transplantation: recent progress and future promise. Lancet Oncol. 2000;1:227–34 (Review).

Dalal J, Gandy K, Domen J. Role of mesenchymal stem cell therapy in Crohn's disease. Pediatr Res. 2012;71(4 Pt 2):445–51.

de la Portilla F, Alba F, García-Olmo D, Herrerías JM, González FX, Galindo A. Expanded allogeneic adipose-derived stem cells (eASCs) for the treatment of complex perianal fistula in Crohn's disease: results from a multicenter phase I/IIa clinical trial. Int J Colorectal Dis. 2013;28(3):313–23.

Ditschkowski M, Einsele H, Schwerdtfeger R, Bunjes D, Trenschel R, Beelen DW, Elmaagacli AH. Improvement of inflammatory bowel disease after allogeneic stem-cell transplantation. Transplantation. 2003;75(10):1745–7.

Dominici M, Le Blanc K, Mueller I, Slaper-Cortenbach I, Marini F, Krause D, Deans R, Keating A, Prockop D, Horwitz E. Minimal criteria for defining multipotent mesenchymal stromal cells. The International Society for Cellular Therapy position statement. Cytotherapy. 2006;8(4):315–7.

Drakos PE, Nagler A, Or R. Case of Crohn's disease in bone marrow transplantation. Am J Hematol. 1993;43(2):157–8.

Duijvestein M, Vos AC, Roelofs H, Wildenberg ME, Wendrich BB, Verspaget HW, Kooy-Winkelaar EM, Koning F, Zwaginga JJ, Fidder HH, Verhaar AP, Fibbe WE, van den Brink GR, Hommes DW. Autologous bone marrow-derived mesenchymal stromal cell treatment for refractory luminal Crohn's disease: results of a phase I study. Gut. 2010;59(12):1662–9.

Duijvestein M, Molendijk I, Roelofs H, Vos AC, Verhaar AP, Reinders ME, Fibbe WE, Verspaget HW, van den Brink GR, Wildenberg ME, Hommes DW. Mesenchymal stromal cell function is not affected by drugs used in the treatment of inflammatory bowel disease. Cytotherapy. 2011;13:1066–73.

Dulai PS, Siegel CA, Colombel JF, Sandborn WJ, Peyrin-Biroulet L. Systematic review: monotherapy with antitumour necrosis factor α agents versus combination therapy with an immunosuppressive for IBD. 2014. doi:10.1136/gutjnl-2014-307126 (pii: gutjnl-2014-307126; Epub ahead of print).

English K. Mechanisms of mesenchymal stromal cell immunomodulation. Immunol Cell Biol. 2013;91(1):19–26.

Farge D, Labopin M, Tyndall A, Fassas A, Mancardi GL, Van Laar J, Ouyang J, Kozak T, Moore J, Kötter I, Chesnel V, Marmont A, Gratwohl A, Saccardi R. Autologous hematopoietic stem cell transplantation for autoimmune diseases: an observational study on 12 years' experience from the European group for blood and marrow transplantation working party on autoimmune diseases. Haematologica. 2010;95(2):284–92.

Forbes GM, Sturm MJ, Leong RW, Sparrow MP, Segarajasingam D, Cummins AG, Phillips M, Herrmann RP. A phase 2 study of allogeneic mesenchymal stromal cells for luminal Crohn's disease refractory to biologic therapy. Clin Gastroenterol Hepatol. 2014;12(1):64–71.

Garcia-Olmo D, Herreros D, Pascual I, Pascual JA, Del-Valle E, Zorrilla J, De-La-Quintana P, Garcia-Arranz M, Pascual M. Expanded adipose-derived stem cells for the treatment of complex perianal fistula: a phase II clinical trial. Dis Colon Rectum. 2009;52(1):79–86.

García-Olmo D, García-Arranz M, Herreros D, Pascual I, Peiro C, Rodríguez-Montes JA. A phase I clinical trial of the treatment of Crohn's fistula by adipose mesenchymal stem cell transplantation. Dis Colon Rectum. 2005;48:1416–23.

Glocker EO, Kotlarz D, Boztug K, Gertz EM, Schäffer AA, Noyan F, Perro M, Diestelhorst J, Allroth A, Murugan D, Hätscher N, Pfeifer D, Sykora KW, Sauer M, Kreipe H, Lacher M, Nustede R, Woellner C, Baumann U, Salzer U, Koletzko S, Shah N, Segal AW, Sauerbrey A, Buderus S, Snapper SB, Grimbacher B, Klein C. Inflammatory bowel disease and mutations affecting the interleukin-10 receptor. N Engl J Med. 2009;361:2033–45.

Gurtner GC, Callaghan MJ, Longaker MT. Progress and potential for regenerative medicine. Annu Rev Med. 2007;58:299–312.

Hasselblatt P, Drognitz K, Potthoff K, Bertz H, Kruis W, Schmidt C, Stallmach A, Schmitt-Graeff A, Finke J, Kreisel W. Remission of refractory Crohn's disease by high-dose cyclophosphamide and autologous peripheral blood stem cell transplantation. Aliment Pharmacol Ther. 2012;36(8):725–35.

Hawkey C, Allez M, Ardizzone S, Clark M, Clark L, Colombel JF, Danese S, Farge-Bancel D, Labopin M, Lindsay J, Norman A, Onida F, Ricart E, Rogler G, Rovira M, Russel N, Satsangi J, Travis S, Tyndall A, Vermeire S. Clinical and endoscopic improvement following hemopoietic stem cell transplantation in the ASTIC trial. J Crohns Colitis. 2013;7:S4.

Hayashi Y, Tsuji S, Tsujii M, Nishida T, Ishii S, Nakamura T, Eguchi H, Kawano S. The transdifferentiation of bone-marrow-derived cells in colonic mucosal regeneration after dextran-sulfate-sodium-induced colitis in mice. Pharmacology. 2007;80(4):193–9.

Herreros MD, Garcia-Arranz M, Guadalajara H, De-La-Quintana P, Garcia-Olmo D. Autologous expanded adiposederived stem cells for the treatment of complex cryptoglandular perianal fistulas: a phase III randomized clinical trial (FATT 1: fistula Advanced Therapy Trial 1) and longterm evaluation. Dis Colon Rectum. 2012;55:762–72.

Jiang W, Su J, Zhang X, Cheng X, Zhou J, Shi R, Zhang H. Elevated levels of Th17 cells and Th17-related cytokines are associated with disease activity in patients with inflammatory bowel disease. Inflamm Res. 2014 (Epub ahead of print).

Kang KS, Kim SW, Oh YH, Yu JW, Kim KY, Park HK, Song CH, Han H. A 37-year-old spinal cord-injured female patient, transplanted of multipotent stem cells from human UC blood, with improved sensory perception and mobility, both functionally and morphologically: a case study. Cytotherapy. 2005;7(4):368–73.

Kean TJ, Lin P, Caplan AI, Dennis JE. MSCs: delivery routes and engraftment, cell-targeting strategies, and immune modulation. Stem Cells Int. 2013;2013:732–42.

Khalil PN, Weiler V, Nelson PJ, Khalil MN, Moosmann S, Mutschler WE, Siebeck M, Huss R. Nonmyeloablative stem cell therapy enhances microcirculation and tissue regeneration in murine inflammatory bowel disease. Gastroenterology. 2007;132(3):944–54.

Lalu MM, McIntyre L, Pugliese C, Fergusson D, Winston BW, Marshall JC, Granton J, Stewart DJ. Safety of cell therapy with mesenchymal stromal cells (SafeCell): a systematic review and meta-analysis of clinical trials. PLoS ONE. 2012;7:e47559.

Lei NY, Jabaji Z, Wang J, Joshi VS, Brinkley GJ, Khalil H, Wang F, Jaroszewicz A, Pellegrini M, Li L, Lewis M, Stelzner M, Dunn JC, Martín MG. Intestinal subepithelial myofibroblasts support the growth of intestinal epithelial stem cells. PLoS ONE. 2014;9(1):e84651.

Levesque BG, Sandborn WJ, Ruel J, Feagan BG, Sands BE, Colombel JF. Converging goals of treatment for inflammatory bowel disease, from clinical trials and practice. 2014. doi:10.1053/j.gastro.2014.08.003 (pii: S0016-5085(14)00999-8; Epub ahead of print).

Lewis JD. Anti-TNF antibodies for Crohn's disease–in pursuit of the perfect clinical trial. N Engl J Med. 2007;357(3):296–8.

Liang J, Wu KC. Unpublished data.

Liang J, Wu S, Feng B, Lei S, Luo G, Wang J, Li K, Li X, Xie H, Zhang D, Wang X, Wu K, Miao D, Fan D. Factor V Leiden and inflammatory bowel disease: a systematic review and meta-analysis. J Gastroenterol. 2011a;46(10):1158–66.

Liang J, Wu S, Feng B, Lei S, Luo G, Wang J, Li K, Li X, Xie H, Zhang D, Wang X, Wu K, Miao D, Fan D. Factor V Leiden and inflammatory bowel disease: a systematic review and meta-analysis. J Gastroenterol. 2011b;46(10):1158–66.

Liang J, Zhang H, Wang D, Feng X, Wang H, Hua B, Liu B, Sun L. Allogeneic mesenchymal stem cell transplantation in seven patients with refractory inflammatory bowel disease. Gut. 2012;61:468–9.

Liang J, Nagahashi M, Kim EY, Harikumar KB, Yamada A, Huang WC, Hait NC, Allegood JC, Price MM, Avni D, Takabe K, Kordula T, Milstien S, Spiegel S. Sphingosine-1-phosphate links persistent STAT3 activation, chronic intestinal inflammation, and development of colitis-associated cancer. Cancer Cell. 2013;23(1):107–20.

Lopez-Cubero SO, Sullivan KM, McDonald GB. Course of Crohn's disease after allogeneic marrow transplantation. Gastroenterology. 1998;114:433–40.

Maggini J, Mirkin G, Bognanni I, Holmberg J, Piazzón IM, Nepomnaschy I, Costa H, Cañones C, Raiden S, Vermeulen M, Geffner JR. Mouse bone marrow-derived mesenchymal stromal cells turn activated macrophages into a regulatory-like profile. PLoS ONE. 2010;5(2):e9252.

Mannon PJ. Remestemcel-L: human mesenchymal stem cells as an emerging therapy for Crohn's disease. Expert Opin Biol Ther. 2011;11:1249–56.

Martí JL, Mayordomo JI, Isla MD, Saenz A, Escudero P, Tres A. PBSC autotransplant for inflammatory bowel disease (IBD): a case of ulcerative colitis. Bone Marrow Transplant. 2001;28(1):109–10.

Martínez-Montiel Mdel P, Gómez-Gómez GJ, Flores AI. Therapy with stem cells in inflammatory bowel disease. World J Gastroenterol. 2014a;20(5):1211–27.

Martínez-Montiel Mdel P, Gómez-Gómez GJ, Flores AI. Therapy with stem cells in inflammatory bowel disease. World J Gastroenterol. 2014b;20(5):1211–27.

Moss AC, Lillis Y, Edwards George JB, Choudhry NK, Berg AH, Cheifetz AS, Horowitz G, Leffler DA. Attitudes to mesalamine questionnaire: a novel tool to predict mesalamine nonadherence in patients with IBD. 2014. doi:10.1038/ajg.2014.158 (Epub ahead of print).

Nemoto Y, Kanai T, Takahara M, Oshima S, Nakamura T, Okamoto R, Tsuchiya K, Watanabe M. Bone marrow-mesenchymal stem cells are a major source of interleukin-7 and sustain colitis by forming the niche for colitogenic CD4 memory T cells. Gut. 2013;62(8):1142–52.

Okamoto R, Yajima T, Yamazaki M, Kanai T, Mukai M, Okamoto S, Ikeda Y, Hibi T, Inazawa J, Watanabe M. Damaged epithelia regenerated by bone marrow-derived cells in the human gastrointestinal tract. Nat Med. 2002;8(9):1011–7.

Onken JE, et al. Successful outpatient treatment of refractory Crohn's disease using adult mesenchymal stem cells. In: American College of Gastroenterology Conference, Las Vegas, NV, Abstract 121. 2006.

Oyama Y, Craig RM, Traynor AE, Quigley K, Statkute L, Halverson A, Brush M, Verda L, Kowalska B, Krosnjar N, Kletzel M, Whitington PF, Burt RK. Autologous hematopoietic stem cell transplantation in patients with refractory Crohn's disease. Gastroenterology. 1995;128(3):552–63.

Ringdén O, Uzunel M, Rasmusson I, Remberger M, Sundberg B, Lönnies H, Marschall HU, Dlugosz A, Szakos A, Hassan Z, Omazic B, Aschan J, Barkholt L, Le Blanc K. Mesenchymal stem cells for treatment of therapy-resistant graft-versus-host disease. Transplantation. 2006;81:1390–7.

Salam AG, Ata HM, Salman TM, Rashed LA, Sabry D, Schaalan MF. Potential therapeutic utility of mesenchymal stem cells in inflammatory bowel disease in mice. Int Immunopharmacol. 2014 (pii: S1567-5769(14)00307-5).

Sanders KM, Ordög T, Ward SM. Physiology and pathophysiology of the interstitial cells of Cajal: from bench to bedside: IV. Genetic and animal models of GI motility disorders caused by loss of interstitial cells of Cajal. Am J Physiol Gastrointest Liver Physiol. 2002;282(5):G747–56.

Sha S, Liang J, Chen M, Xu B, Liang C, Wei N, Wu K. Systematic review: faecal microbiota transplantation therapy for digestive and nondigestive disorders in adults and children. Aliment Pharmacol Ther. 2014;39(10):1003–32.

Strauer BE, Schannwell CM, Brehm M. Therapeutic potentials of stem cells in cardiac diseases. Minerva Cardioangiol. 2009;57(2):249–67.

Talbot DC, Montes A, Teh WL, Nandi A, Powles RL. Remission of Crohn's disease following allogeneic bone marrow transplant for acute leukaemia. Hosp Med. 1998;59:580–1.

Targownik LE, Nugent Z, Singh H, Bernstein CN. Prevalence of and outcomes associated with corticosteroid prescription in inflammatory bowel disease. Inflamm Bowel Dis. 2014;20(4):622–30.

Taupin P. OTI-010 Osiris therapeutics/JCR pharmaceuticals. Curr Opin Investig Drugs. 2006;7:473–81.

Vos AC, Wildenberg ME, Arijs I, Duijvestein M, Verhaar AP, de Hertogh G, Vermeire S, Rutgeerts P, van den Brink GR, Hommes DW. Regulatory macrophages induced by infliximab are involved in healing in vivo and in vitro. Inflamm Bowel Dis. 2012;18(3):401–8.

Wang C, Chen J, Sun L, Liu Y. TGF-beta signaling-dependent alleviation of dextran sulfate sodium-induced colitis by mesenchymal stem cell transplantation. Mol Biol Rep. 2014;41(8):4977–83.

Zuo D, Liu X, Shou Z, Fan H, Tang Q, Duan X, Cao D, Zou Z, Zhang L. Study on the interactions between transplanted bone marrow-derived mesenchymal stem cells and regulatory T cells for the treatment of experimental colitis. Int J Mol Med. 2013;32(6):1337–44.

Chapter 19
Stem Cell Therapy for Optic Nerve Regeneration

Huasong Gao and Jianhong Zhu

Abstract Optic nerve neuropathy, such as traumatic optic nerve injury (TONI) and glaucoma, is among the leading cause of incurable vision loss across the world. What is worse, neither pharmacological nor surgical interventions are significantly effective in reversing or halting the progress. Advances in cell biology offer some hope for the victims of optic nerve damage and subsequent partial or complete visual loss. Retinal ganglion cells (RGCs) travel through optic nerve and carry all visual signals to the brain. After injury, RGCs' axons usually fail to regrow and usually die, leading to irreversible loss of vision. Various kinds of cells and factors possess the abilities that are supportive in the progress of axon regeneration for RGC. This article summarizes the latest advances in RGCs regeneration.

Keywords Stem cell · Therapy · Optic nerve neuropathy

19.1 Introduction

Optic neuropathies, including glaucoma, are characterized by progressive degeneration of retinal ganglion cells (RGCs). This characteristic process makes this group of diseases suitable candidates for cellular therapy. Stem cell-based treatment had become a promising frontier for treating neurodegenerative diseases, thus making it a potential therapeutic approach for optic neuropathies as well.

H. Gao · J. Zhu (✉)
National Key Laboratory for Medical Neurobiology, Institutes of Brain Science, Shanghai Medical College-Fudan University, Shanghai, China
e-mail: jzhu@fudan.edu.cn

H. Gao · J. Zhu
Department of Neurosurgery, Fudan University Huashan Hospital, 12# Wulumuqi Zhang Rd, Shanghai, China

R.C. Zhao (ed.), *Stem Cells: Basics and Clinical Translation*, Translational Medicine Research 1, DOI 10.1007/978-94-017-7273-0_19

Retinal ganglion cells (RGCs) start from ganglion cell layer in the retina. After traveling through the lamina cribrosa, their axons join together to form the optic nerve (Fig. 19.1), which extends within the optic canal and ends at the lateral geniculate body. RGC axons are unmyelinated while traveling through nerve fiber layer of the retina and lamina cribrosa but gain a myelin sheath after crossing the lamina cribrosa. Like neurons in the central nervous system, RGCs' axons are myelinated by oligodendrocytes, not Schwann cells. After translated by rods and cones, light signals are passed to bipolar cells, which then deliver the signals to RGCs. Later, RGCs send these signals to the brain for further processing (Moore and Goldberg 2010) (Fig. 19.1).

Various kinds of neuropathies result from the injury to the RGCs. Traumatic optic nerve injury (TONI) results in the decrease of the number of RGCs or loss of their axons. RGC damage has devastating consequences for the victim, resulting in partial or permanent vision loss. Clinical procedures like optic canal decompression therapy have received minimal success in rescuing these unlucky patients. A study (Jin et al. 2009) summarizing the therapeutic effect of optic canal decompression therapy for TONI patients included 670 cases (683 eyes). All of these patients received operation 7 h to 90 days (291 eyes within 3d, 222 eyes in 3–7d, and 170 eyes after 7d) after injury. However, only 62 % of these victims had minimal eye function improvement after 3 months to 2 years, while 38 % had no improvement. Regeneration failure of RGCs could be considered to play a key role with such visual outcomes.

Other optic neuropathies are also related to injuries to RGCs. Glaucoma is a group of chronic eye diseases that irreversibly damages RGCs and result in serious vision loss and blindness. After cataract, it is the second-leading cause of blindness worldwide and is one of the leading causes of preventable blindness (Quigley and Broman 2006). Globally, an estimated 60.5 million people (2.65 % of the global population over 40) suffered from glaucoma in 2010. Direct cost estimates for the approximately 2 million US citizens (Rein et al. 2006) and 300, 000 Australian citizens (Taylor et al. 2006) with glaucoma are $2.9 billion and AUS$144.2 million, respectively. However, these figures underestimate the true societal costs if all were to be treated, since about half of patients with glaucoma are unaware of their disease and is therefore referred to as the "silent thief of sight" (Varma et al. 2004; de Voogd et al. 2005).

Why does damage to the optic nerve and glaucoma can cause so much damage? The answer lies in RGC's incapability of regeneration after injury. Under normal conditions, mature RGCs of mammalian eyes fail to extend axons through the site of optic nerve injury. Due to a signal from amacrine cells (Fig. 19.1), RGC gradually lose its intrinsic growth ability, which plays an important part (Goldberg et al. 2002). However, this phenomenon can be reversed. In 1911, scientists discovered that RGCs can extend axons in a short distance if provided with proper substrates (Tello 1911). DeFelipe and Jones (1991) described it as "groundbreaking." After this, evidences continued to emerge supporting the view that RGCs' axons can regenerate under certain environments that were provided artificially.

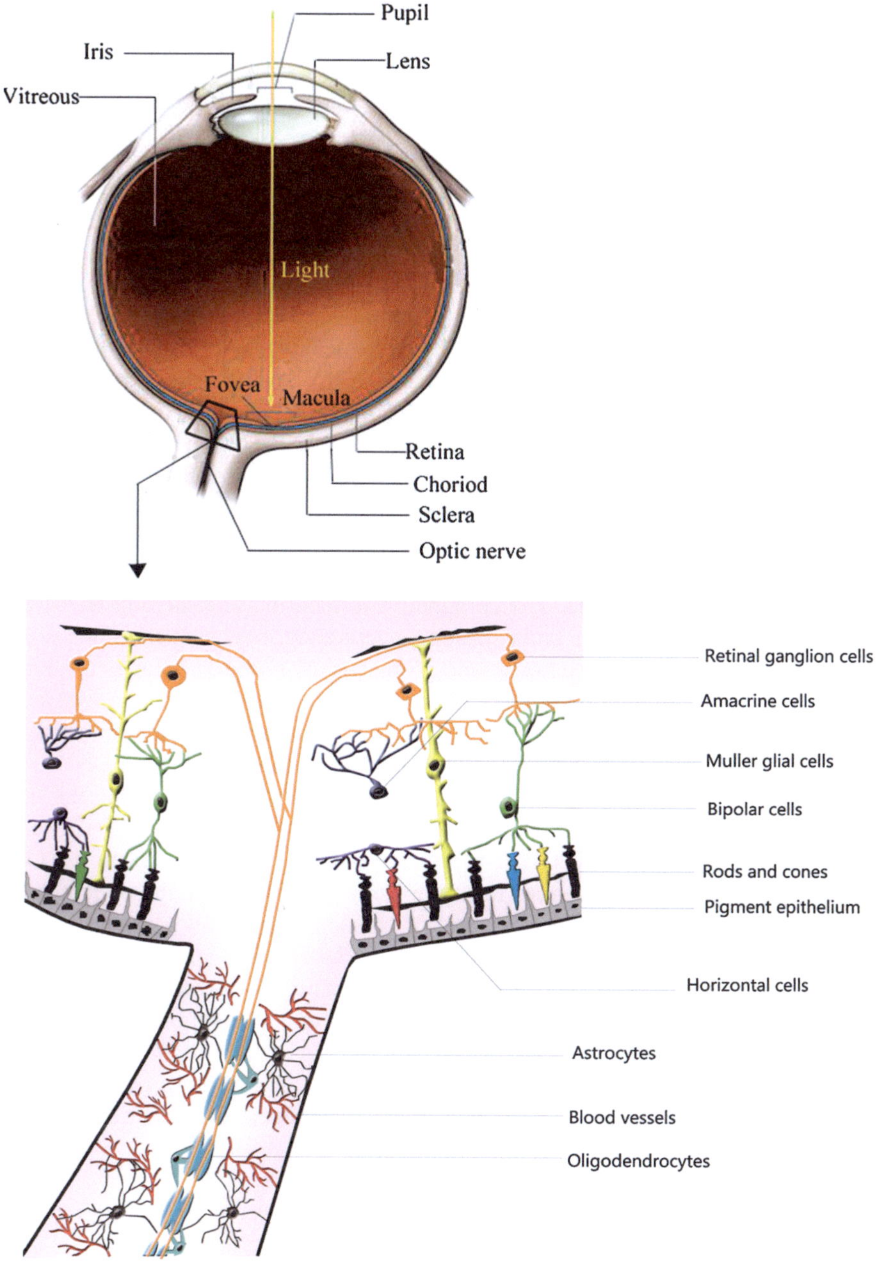

Fig. 19.1 Simplified structure of the eye and retina (close to the lamina cribrosa). After translated by rods and cones, light signals are passed to bipolar cells, which then deliver the signals to RGCs. Later, RGCs will send these signals to the brain for further progress

19.2 Animal Models

19.2.1 Optic Nerve Injury

Traumatic optic injury can be classified based on the anterior or posterior location of injury. Optic nerve head is where anterior injuries occur. Its avulsion results in intraocular hemorrhage and disruption of the optic nerve head anatomy. Posterior injuries mainly affect the foramina of the optic canal, the optic canal, and under the falciform dural fold at the edge of the optic canal that drapes the edge of the anterior clinoid process (Walsh 1966). The process of contusion necrosis that results from shearing injury to the axons and microvasculature is thought to be the basis for both kinds of injury (Walsh 1966). In order to study the regeneration process of the RGCs, we need to establish a convincing animal model. Having reviewed related articles, the following models can be found in the literature.

The optic nerve crush (ONC) injury model has the highest resemblance to the conditions which trauma exerts on the optic nerve. Thus, it is a widely used experimental model for TONI (Maeda et al. 2004; Monnier et al. 2011; Schnichels et al. 2011; Sarikcioglu et al. 2007). In this model, rats are usually anaesthetized and placed under the microscope. The conjunctiva is then incised laterally to the cornea. After this, the retractor bulbi muscle is separated, exposing the optic nerve, which is clipped 2 mm posterior to the globe for 20 s (Sarikcioglu et al. 2007), 30 s (Huang et al. 2011), or 60 s (Zhang et al. 2011) according to different methods used by researchers. The other eye (or the same eye of other rats) usually receives a sham operation with optic nerve exposure, but without the crush.

Besides the ONC model, another widely used injury model is called the optic nerve transection or axotomy model (Koeberle et al. 2010; Levkovitch-Verbin et al. 2010; Charalambous et al. 2008). The surgical procedures are similar to the ONC model, except for the treatment of the optic nerve. In this model, the optic nerve is transected partially or completely instead of crushed. It provides a suitable site for injecting solutions with specific cells, factors, etc. So, research concerning cell transplantation (Charalambous et al. 2008) at the site of injury usually favors this model.

Chinese investigators have developed a novel model using fluid percussion brain injury device (FPI) that can increase the fidelity of the force hitting the optic nerve (Ronggu et al. 2010; Yingjuan et al. 2010), which may produce a more standardized hierarchical model for the study of optic nerve injuries in the future.

19.2.2 Glaucoma

Elevated intraocular pressure (IOP) is one of the major risk factors of glaucoma and has been associated with the damage of RGCs. Thus, the main task of simulating glaucoma in animal eyes is maintaining a continuous elevation of IOP. The delicate balance between secretion and drainage of aqueous humor is one of the major ways by which IOP is regulated. Thus, if we can simulate production or block drainage of aqueous humor, a glaucomatous model eye would be created.

Among the existing methods, laser photocoagulation of the veins that carry the outflow of aqueous humor is the method favored by most scientists. Taking Purushottam Jha's research (Jha et al. 2011; Chiu et al. 2007) for example, the limbal vein and three episcleral veins are photocoagulated with an Argon laser. Approximately 80 laser spots around the limbal vein and 15 on each episcleral vein are applied. Another laser treatment of the same pattern is applied 7 days later to ensure continuous elevation of IOP. IOP can be measured with a TonoLab tonometer before and after the first laser treatment.

19.3 Signs of Optic Nerve Regeneration

In order to evaluate the effect of existing therapies for optic nerve damage, scientists have developed numerous methods. One of the most commonly used methods of identifying RGCs is the utilization of neuronal tracers. Fluorogold (FG) is the tracer favored by most laboratories (Ma et al. 2010; Xie et al. 2010; Ahmed et al. 2011) because of its high specificity (Salinas-Navarro et al. 2009). FG is usually delivered into superior colliculus region. By this technique, the surviving RGCs that still connect the retina and SC are enhanced. After counting the FG^{+} cells in the retina, researchers can compare the number of RGCs that have survived, enabling assessment of therapeutic effects. Beside FG, members of the Brn3 family can also serve as tracers of RGCs. Brn3b is being used to identify RGCs in various kinds of animals, such as mice (Buckingham et al. 2008) and rats (Bernstein et al. 2006). Brn3a has also been proved useful by Nadal-Nicola et al. (2009), Sánchez-Migallón et al. 2011) compared to that of the FG neuronal tracer. Besides, Brn3a's distribution is more sensitive to injury than FG and may replace FG in the future.

Immunodetection of specific proteins is also a common method for the identification of cells. Bex 1/2 is expressed in not only the cell body of the RGCs, but also in its axons. Thus, immunodetection of Bex 1/2 is an appropriate way for the study of RGC's morphologic changes related to the injury (Bernstein et al. 2006). Growth-associated protein 43 (*GAP-43*), a plasticity protein expressed highly during axon regeneration, is also favored by scientists who focus on the regeneration of RGC axons (Charalambous et al. 2008; Su et al. 2009).

There are still others who prefer transgenic approaches. Feng et al. (2000) and Raymond et al. (2009) have been successful in identifying RGCs, but the utilization of this technology in rats is still in its infant state.

19.4 Obstacles for Optic Nerve Regeneration

The RGCs possess features that resemble other CNS neurons, and thus, its regeneration failure may share the same inhibitory factors. The Nogo receptor (NgR) is a widely accepted neuron regeneration inhibitor receptor in the central nervous system (CNS) (Fournier et al. 2001). The Nogo gene has three isoforms: Nogo-A,

Nogo-B, and Nogo-C (Chen et al. 2000; GrandPre et al. 2000; Prinjha et al. 2000). NgR belongs to a family of three CNS-enriched glycosyl phosphatidylinositol-linked proteins (Barton et al. 2003; Lauren et al. 2003; Pignot et al. 2003). Three myelin proteins, NogoA (or Nogo66) (Chen et al. 2000; GrandPre et al. 2000), myelin-associated glycoprotein (MAG) (McKerracher et al. 1994; Mukhopadhyay et al. 1994), and oligodendrocyte-myelin glycoprotein (OMgp) (Wang et al. 2002), inhibit axon growth partly through NgR. Antagonizing NgR with function-blocking antibodies (Domeniconi et al. 2002; Li et al. 2004) has been reported to support that NgR is related to Nogo66, MAG, and Omgp inhibition. As for RGCs, blocking *NgR* is also effective for alleviating axon-regenerating inhibition. Using the ONC model, Su et al. (2009) checked GAP-43 expression in NgR knockout mice, discovering that their axon regeneration is more active than the control group. Their discovery has provided evidences that NgR plays an important role in inhibiting axonal regeneration of RGCs.

However, counteracting NgR alone has received little success in promoting RGC regeneration. Cui et al. (2004) found that application of IN-1, a NgR blocking antibody, alone failed to enhance regeneration of transected RGC axons in a peripheral nerve graft. IN-1 also failed to significantly promote crushed RGC axons to regrow into the distal part of the optic nerve (Cui et al. 2004). However, in their later studies, a combined application of IN-1 and ciliary neurotrophic factor (CNTF) had a synergistic effect that significantly enhanced RGC regeneration. This suggests that counteracting NgR, or maybe new receptors discovered in the future, will be more effective when combined with other factors such as CNTF.

19.5 Optic Nerve Regeneration Is Possible

Lower vertebrate species such as teleost fish possess the capability to regenerate new RGCs following surgical lesions even in their adulthood (Hitchcock et al. 2004; Raymond et al. 2006). The source of this regeneration are the glial cells, which are mitotically quiescent in normal state. After damaging RGCs, the glial cells reenter cell cycle and then de-differentiate to a phenotype with progenitor features (Raymond et al. 2006; Fausett and Goldman 2006; Fimbel et al. 2007; Kassen et al. 2007; Vihtelic et al. 2006). Mammalian glial cells, on the contrast, do not re-enter cell cycle after the retina is damaged (Chang et al. 2007; Close et al. 2006; Dyer and Cepko 2000; Zhao et al. 2005). Some scientists have tried growth factors on the damaged retina and detected traces of neurons such as bipolar cells and photoreceptors on some of the glial cells (Close et al. 2006; Ooto et al. 2004; Osakada et al. 2007; Wan et al. 2007, 2008). However, these traces are only cytoplasmic markers, not observed under confocal microscopic analysis. Similar studies of neurogenesis in other parts of the nervous system have caused disputes when more thorough studies were published (Dayer et al. 2005). As a result of this, scientists have turned to neuronal replacement.

19.5.1 Stem Cells as a Source for Transplantation

Stem cell transplantation is clinically attractive because it possesses the potential ability to halt or even cure degenerative and progressive conditions that are incurable such as glaucoma and TONI, respectively. Two concepts are playing dominant roles in the field of stem cell transplantation therapies: replacing malfunctioned RGCs (Minamino et al. 2005; Ramirez-Castillejo et al. 2006; Tropepe et al. 2000; Klassen et al. 2007; Mellough et al. 2004; Kinouchi et al. 2003) and implanting neurotrophic factor-associated cells (Charalambous et al. 2008; Wu et al. 2010; Zhao et al. 2011) to promote axon regeneration. Several reasons contribute to stem cells having received so much attention. They are described as follows.

The most significant potential of stem cells is their ability to generate many types of cells and to enhance tissue regeneration. Thus, stem cells may provide therapeutic hope through replacement of malfunctioned RGCs and promoting RGC's axon regeneration (Quigley and Iglesia 2004; Limb et al. 2006; Young 2005; Bull and Martin 2007; Bull and Martin 2009; Dahlmann-Noor et al. 2010). Moreover, some types of stem cells are capable of halting disease progression and enhancing survival of tissue. Ideally, RGC neuroprotection therapy would serve as adjuvants for currently used ocular hypotensive therapies and optic canal decompression surgeries (Dahlmann-Noor et al. 2010; Bull et al. 2008a). Neuroprotective approaches may be clinically utilized before cell replacement therapies, owing to far harsher requirements for the latter. However, if achieved, RGC replacement would be capable of restoring vision functionally, while neuroprotection might only preserve the remaining vision. However, clinical guideline development must carefully take type of cell, cell modification, and site for delivering into consideration. Of all the three considerations, route of delivery is the main topic.

19.5.2 Sites that Are Closely Related to RGC's Cell Body

Considering the structure of RGCs, there are four possible sites for cell transplantation: the vitreous, subretinal space, site of injury (in the optic canal), and superior colliculus region.

The vitreous is a suitable immune privilege site for cell transplantation due to the simplicity of delivery. Besides, the RGC layer lies close to the surface of the vitreous. Thus, if transplanted near the retina, stem cells may differentiate into RGCs that can grow into the retina and replace the injured RGCs. Some success has been reached with this concept. Stem cell-derived structures transplanted into the vitreous cavity of mouse eyes depleted of RGCs by NMDA injection spread on the inner retina and frequently differentiate into cells expressing RGC markers (Aoki et al. 2008). Similar results have also been reached using induced pluripotent stem cells (Parameswaran et al. 2010). Johnson et al. (2010) transplanted mesenchymal stem cell into the vitreous. The cells survived for up to 5 weeks and even migrated into the RGC layer of host retina.

Another cell transplantation target is the retina. If transplanted cells can differentiate into RGCs, extend their axons and then can establish functional connections with lateral geniculate body. Neuropathies affecting RGCs of any type can be cured including TONI. First, numerous animal experiments have supported that transplanted retinal progenitor or stem cells can migrate extensively in the retina, differentiate into neurons, elaborate the axons that reach plexiform layer, even lamina cribrosa (Young et al. 2000; Nishida et al. 2000; Klassen et al. 2004; Wojciechowski et al. 2002a, b). Second, to overcome barriers that still stand ahead before RGCs' axons reach lateral geniculate body, researchers have employed peripheral nerve grafts to break through areas of inhibition. When the sciatic nerve grafts are placed between optic nerve stump and various distant visual centers, ganglions can extend their axons and reconnect with these centers functionally (Thanos et al. 1997; Cheng et al. 1996; Carter et al. 1989; Keirstead et al. 1989). Thus, the combination of these two procedures may successfully realize the concept of mending or even replacing the malfunctioned RGCs.

As for glaucoma, graft location has been a hot topic among researchers. Some studies suggested that subretinally transplanted cells enjoy a more immune-privileged circumstance than those in vitreous cavity (Banin et al. 2006). Subretinal transplantation also ensures that transplanted cells are in close proximity to the retina. However, intravitreal introduction also holds its own advantages. For instance, it theoretically provides access to the inner retina for the grafted cells. Intravitreal transplantation is also less invasive for the reason that subretinal transplants may cause retinal detachment.

However, either of the two graft locations cannot escape potential barriers that still stand before us, like the integration of grafted cells. Future research is needed to clarify which location is more suitable for clinical utilization.

19.5.3 Sites that Are Closely Related to the RGC's Axons

For the reason that cells can be easily delivered while performing surgery, the site of injury is the mostly favored site for cell transplantation regarding optic nerve injury. If provided a "comfortable" circumstance, such as extrinsic neurotrophic factors, damaged RGCs' axons may recover with its own strength. Among the candidates that are capable of doing so, the following cells are in the lime light recently: stem or progenitor cells (Levkovitch-Verbin et al. 2010; Charalambous et al. 2008; Zhao et al. 2011), bone marrow mononuclear cells (Zaverucha-do-Valle et al. 2011) and olfactory ensheathing cells (OECs) (Wu et al. 2010). Chicken neural tube-derived stem cells (NTSCs) developed by Charalambous et al. (2008) seem to have a significant effect in stimulating RGC axon growth after optic nerve axotomy by expressing neurite growth-promoting factors like β-crystalline, γ-crystallin, and CNTF. Using GAP-43, they discovered that transected RGC axons could transcend the region of surgery and elongate within the ON distal to the site of injury. They also confirmed that RGCs even reach mid-brain region 6–8 weeks after implantation with FG. Beside stem cells, other cell types such as OECs seem to perform their effect in similar way. OEC is a kind of macroglial cells situated in the nasal olfactory mucosa and olfactory bulb. Researchers have confirmed its capability of secreting neurotrophic factors such as nerve growth factor (Roskams et al. 1996), BDNF (Lipson et al. 2003; Woodhall et al. 2001), and neurotrophin 4/5 (Boruch et al. 2001). OEC has also been proved to promote axon regeneration of RGCs (Ramon-Cueto et al. 1998; Li et al. 2003) and delayed the death of axotomized RGCs (Wu et al. 2010) through the secretion of BDNF.

While scientists have achieved significant successes about RGCs' axon regeneration after optic nerve injury, the situation of glaucoma is somewhat lagging behind. Optic nerve injury mainly damages a short length of RGCs' axons, while patients who suffer from glaucoma, especially severe glaucoma, usually lose a considerable amount of RGCs. In other words, RGCs need to be repaired in optic nerve injury but replaced in glaucoma. Because RGCs' cell body is located in the retina, researchers are mainly considering transplanting RGC precursors or stem cells in or under the retina in the hope of replacing dead RGCs. Some evidence of integration into the inner retina (Bull et al. 2008b) or migration of engrafted stem cells (Young et al. 2000; Mellough et al. 2007; Guo et al. 2003) in the retina have been reported. Axon connections that extend through the lamina cribrosa and to the brain have not yet been observed. However, incremental advances on the differentiation of stem cells to RGCs and retinal transplant integration continue to emerge. We believe that eventually successful methods will converge to completely replace dead RGCs (both cell body and axons) and grant glaucoma patient a chance of restoring sight.

There are still other possible sites that have not received much attention. For example, the superior colliculus region contains the endings of RGCs. Considering the case of how retrograde staining of FG is applied (described above), cells transplanted in this region may provide a constant supply of neurotrophic factors that can be delivered to the site of injury, enhancing the regeneration and survival of RGCs.

19.6 Achievements by Stem Cell Transplantation

The therapeutic potential of different types of stem cells for treating retinal and optic nerve diseases has already been investigated. (Bull and Martin 2009; Bull et al. 2008b) Most studies confirmed that the adult retina has limited levels of graft–host integration, whereas only few studies demonstrated successful integration of stem cells to the retina. MacLaren et al. (2006) showed that adult degenerating mammalian retinas can effectively incorporate rod photoreceptor precursor cells into the outer nuclear layer and that these cells form functional synaptic connections and even contribute to visual function. Those authors found that the success of this treatment depends mainly on the optimal ontogenetic stage of the donor cells. Qiu et al. (2005) were able to enhance photoreceptor cell differentiation and integration of retinal progenitor cells after subretinal transplantation into retinal degenerate rats by optimization of isolation, expansion, and transplantation procedures. Arnold et al. (2007) found that mesenchymal stem cells can prolong photoreceptor survival in the rhodopsin knockout mouse, also providing evidence of a therapeutic benefit in retinitis pigmentosa. The situation was not better for inner retinal diseases. Previous reports found that the inner retina, especially the inner limiting membrane, obstructs the integration of intravitreally injected cells into the retina. Bull et al. (2008a) observed minimal retinal integration (1 % of cells) when MIO-M1 stem cells or oligodendrocytes precursor cells (OPCs) were transplanted. Bull et al. (2008b) found that the Müller stem cell line MIO-M1 in glaucomatous eyes produced cells that expressed neuronal and glial cell markers but that the retina was relatively resistant to transplant integration and long-term xenograft survival was limited. Interestingly, local modulation of the retinal environment enhanced the integration of MIO-M1 cells into the glaucomatous retina (Bull et al. 2008b). On the other hand, when OPCs were used, a significant neuroprotective effect was achieved. This effect did not appear to be contact mediated or to be conferred by the myelination of naked axons; rather, it was most likely due to the release of diffusible neurotrophic factors by activated OPCs (Bull et al. 2009). All these data suggest that it may be more advantageous to use stem cells as vectors that secrete and deliver neurotrophic factors. At the same time, efforts should be expended to find ways to improve the integration of stem cells in the RGC layer, to facilitate their differentiation into RGCs, and to induce axonal sprouting to form neuronal connections within the retina and to the brain.

19.7 Future Perspectives

Not long ago, the idea that optic nerves are able to regenerate after injury was untenable. Nowadays, however, most researchers are focusing on how to improve its regeneration. The regeneration of the RGCs has received much attention since the discovery of regenerative potentiality of CNS neurons. Either blocking the inhibitory receptors or providing neuroprotective agents through various means has achieved much success in the laboratory. But there are still barriers and challenges before these procedures are utilized clinically. Effective cells and factors are put into place almost immediately after RGCs are injured in the laboratory. In the case of TONI and glaucoma, patients hardly have the chance of getting these treatments immediately or even in time. There is one study that does exist where neuroprotectants were administered 18 h after retinal ischemia and found to be effective (Chiang and Lam 2000). But the pattern in which RGCs are damaged is different from TONI and glaucoma. In a word, much is done and much remains to be done.

Acknowledgements This study was supported by grants (2010CB945500, 2012CB966300, 2009CB941100, 81271003) from the National Nature Science Foundation and Ministry of Science and Technology of China.

Reference

Ahmed Z, Kalinski H, Berry M, Almasieh M, Ashush H, Slager N, Brafman A, Spivak I, Prasad N, Mett I, Shalom E, Alpert E, Polo AD, Feinstein E, Logan A. Ocular neuroprotection by siRNA targeting caspase-2. Cell Death Dis. 2011;2:173–83.

Aoki H, Hara A, Niwa M, Motohashi T, Suzuki T, Kunisada T. Transplantation of cells from eye-like structures differentiated from embryonic stem cells in vitro and in vivo regeneration of retinal ganglion-like cells. Graefes Arch Clin Exp Ophthalmol. 2008;246:255–65.

Arnhold S, Absenger Y, Klein H, Addicks K, Schraermeyer U. Transplantation of bone marrow-derived mesenchymal stem cells rescue photoreceptor cells in the dystrophic retina of the rhodopsin knockout mouse. Graefes Arch Clin Exp Ophthalmol. 2007;245:414–22.

Banin E, Obolensky A, Idelson M, Hemo I, Reinhardtz E, Pikarsky E, Ben-Hur T, Reubinoff B. Retinal incorporation and differentiation of neural precursors derived from human embryonic stem cells. Stem Cells. 2006;24:246–57.

Barton WA, Liu BP, Tzvetkova D, Jeffrey PD, Fournier AE, Sah D, Cate R, Strittmatter SM, Nikolov DB. Structure and axon outgrowth inhibitor binding of the Nogo-66 receptor and related proteins. EMBO J. 2003;22:3291–302.

Bernstein S, Koo J, Slater B, Guo Y, Margolis F. Analysis of optic nerve stroke by retinal Bex expression. Mol Vis. 2006;12:147–55.

Boruch AV, Conners JJ, Pipitone M, Deadwyler G, Storer PD, Devries GH, Jones KJ. Neurotrophic and migratory properties of an olfactory ensheathing cell line. Glia. 2001;33:225–9.

Buckingham B, Inman D, Lambert W. Progressive ganglion cell degeneration precedes neuronal loss in a mouse model of glaucoma. J Neurosci. 2008;28:2735–44.

Bull ND, Martin KR. Optic nerve restoration: new perspectives. J Glaucoma. 2007;16:506–11.

Bull ND, Martin KR. Using stem cells to mend the retina in ocular disease. Regen Med. 2009;4:855–64.

Bull ND, Johnson TV, Martin KR. Stem cells for neuroprotection in glaucoma. Prog Brain Res. 2008a;173:511–9.

Bull ND, Limb GA, Martin KR. Human Muller stem cell (MIO-M1) transplantation in a rat model of glaucoma: survival, differentiation, and integration. Invest Ophthalmol Vis Sci. 2008b;49:3449–56.

Bull ND, Irvine KA, Franklin RJ, Martin KR. Transplanted oligodendrocyte precursor cells reduce neurodegeneration in a model of glaucoma. Invest Ophthalmol Vis Sci. 2009;50:4244–53.

Carter DA, Bray GM, Aguayo AJ. Regenerated retinal ganglion cell axons can form well-differentiated synapses in the superior colliculus of adult hamsters. J Neurosci. 1989;9:4042–50.

Chang ML, Wu CH, Jiang-Shieh YF, Shieh JY, Wen CY. Reactive changes of retinal astrocytes and Muller glial cells in kainate-induced neuroexcitotoxicity. J Anat. 2007;210:54–65.

Charalambous P, Hurst LA, Thanos S. Engrafted chicken neural tube-derived stem cells support the innate propensity for axonal regeneration within the rat optic nerve. Invest Ophthalmol Vis Sci. 2008;49:3513–24.

Chen MS, Huber AB, van der Haar ME, Frank M, Schnell L, Spillmann AA, Christ F, Schwab ME. Nogo-A is amyelin-associated neurite outgrowth inhibitor and an antigen for monoclonal antibody IN-1. Nature. 2000;403:434–439.

Cheng H, Cao Y, Olson L. Spinal cord repair in adult paraplegic rats: partial restoration of hind limb function. Science. 1996;273:510–3.

Chiang SK, Lam TT. Post-treatment at 12 or 18 hours with 3-aminobenzamide ameliorates retinal ischemia-reperfusion damage. Invest Ophthalmol Vis Sci. 2000;41:3210–4.

Chiu K, Chang R, So KF Laser-induced chronic ocular hypertension model on SD rats. J Vis Exp. 2007;(10):549.

Close JL, Liu J, Gumuscu B, Reh TA. Epidermal growth factor receptor expression regulates proliferation in the postnatal rat retina. Glia. 2006;54:94–104.

Cui Q, Cho K, So K, Yip H. Synergistic effect of Nogo-neutralizing antibody IN-1 and ciliary neurotrophic factor on axonal regeneration in adult rodent visual systems. J Neurotrauma. 2004;21:617–25.

Dahlmann-Noor A, Vijay S, Jayaram H, Limb A, Khaw PT. Current approaches and future prospects for stem cell rescue and regeneration of the retina and optic nerve. Can J Ophthalmol. 2010;45:333–41.

Dayer AG, Cleaver KM, Abouantoun T, Cameron HA. New GABAergic interneurons in the adult neocortex and striatum are generated from different precursors. J Cell Biol. 2005;168:415–27.

de Voogd S, Ikram MK, Wolfs RC, Jansonius NM, Hofman A, de Jong PT. Incidence of open-angle glaucoma in a general elderly population: the Rotterdam Study. Ophthalmology. 2005;112:1487–93.

DeFelipe J, Jones E. Cajal's degeneration and regeneration of the nervous system, history of neuroscience. Oxford 5, 1991.

Domeniconi M, Cao Z, Spencer T, Sivasankaran R, Wang K, Nikulina E, Kimura N, Cai H, Deng K, Gao Y, He Z, Filbin M. Myelin-associated glycoprotein interacts with the nogo66 receptor to inhibit neurite outgrowth. Neuron. 2002;35:283–90.

Dyer MA, Cepko CL. Control of Muller glial cell proliferation and activation following retinal injury. Nat Neurosci. 2000;3:873–80.

Fausett BV, Goldman D. A role for alpha1 tubulin-expressing Muller glia in regeneration of the injured zebrafish retina. J Neurosci. 2006;26:6303–13.

Feng G, Mellor R, Bernstein M. Imaging neuronal subsets in transgenic mice expressing multiple spectral variants of GFP. Neuron. 2000;28:41–51.

Fimbel SM, Montgomery JE, Burket CT, Hyde DR. Regeneration of inner retinal neurons after intravitreal injection of ouabain in zebrafish. J Neurosci. 2007;27:1712–24.

Fournier AE, GrandPre T, Strittmatter SM. Identification of a receptor mediating Nogo-66 inhibition of axonal regeneration. Nature. 2001;409:341–6.

Goldberg JL, Klassen MP, Hua Y, Barres BA. Amacrine-signaled loss of intrinsic axon growth ability by retinal ganglion cells. Science. 2002;296:1860–4.

GrandPre T, Nakamura F, Vartanian T, Strittmatter SM. Identification of the Nogo inhibitor of axon regeneration as a reticulon protein. Nature. 2000;403:439–44.

Guo Y, Saloupis P, Shaw SJ, Rickman DW. Engraftment of adult neural progenitor cells transplanted to rat retina injured by transient ischemia. Invest Ophthalmol Vis Sci. 2003;44:3194–201.

Hitchcock P, Ochocinska M, Sieh A, Otteson D. Persistent and injury-induced neurogenesis in the vertebrate retina. Prog Retin Eye Res. 2004;23:183–94.

Huang TL, Chang CH, Lin KH, Sheu MM, Tsai RK. Lack of protective effect of local administration of triamcinolone or systemic treatment with methylprednisolone against damages caused by optic nerve crush in rats. Exp Eye Res. 2011;92:112–9.

Jha P, Banda H, Tytarenko R, Bora PS, Bora NS. Complement mediated apoptosis leads to the loss of retinal ganglion cells in animal model of glaucoma. Mol Immunol. 2011;48:2151–8.

Jin H, Hou L, Pan C. Opportunity of optic nerve decompression in treatment of traumatic optic nerve neuropathy: review of domestic literature. Clin J Med Officer. 2009;37:6–63.

Johnson TV, Bull ND, Hunt DP, Marina N, Tomarev SI, Martin KR. Neuroprotective effects of intravitreal mesenchymal stem cell transplantation in experimental glaucoma. Invest Ophthalmol Vis Sci. 2010;51:2051–9.

Kassen SC, Ramanan V, Montgomery JE, Burket TC, Liu CG, Vihtelic TS, Hyde DR. Time course analysis of gene expression during light-induced photoreceptor cell death and regeneration in albino zebrafish. Dev Neurobiol. 2007;67:1009–1031.

Keirstead SA, Rasminsky M, Fukuda Y, Carter DA, Aguayo AJ, Vidal-Sanz M. Electrophysiologic responses in hamster superior colliculus evoked by regenerating retinal axons. Science. 1989;246:255–7.

Kinouchi R, Takeda M, Yang L, Wilhelmsson U, Lundkvist A, Pekny M, Chen DF. Robust neural integration from retinal transplants in mice deficient in GFAP and vimentin. Nat Neurosci. 2003;6:863–8.

Klassen HJ, Ng TF, Kurimoto Y, Kirov I, Shatos M, Coffey P, Young MJ. Multipotent retinal progenitors express developmental markers, differentiate into retinal neurons, and preserve light-mediated behavior. Invest Ophthalmol Vis Sci. 2004;45:4167–73.

Klassen H, Kiilgaard JF, Zahir T, Ziaeian B, Kirov I, Scherfig E, Warfvinge K, Young MJ. Progenitor cells from the porcine neural retina express photoreceptor markers after transplantation to the subretinal space of allorecipients. Stem Cells. 2007;25:1222–30.

Koeberle P, Wang Y, Schlichter L. Kv1.1 and Kv1.3 channels contribute to the degeneration of retinal ganglion cells after optic nerve transection in vivo. Cell Death Differ. 2010;17:134–44.

Lauren J, Airaksinen MS, Saarma M, Timmusk T. Two novelmammalian Nogo receptor homologs differentially expressed in the central and peripheral nervous systems. Mol Cell Neurosci. 2003;85:717–28.

Levkovitch-Verbin H, Sadan O, Vander S, Rosner M, Barhum Y, Melamed E, Offen D, Melamed S. Intravitreal injections of neurotrophic factors secreting mesenchymal stem cells are neuroprotective in rat eyes following optic nerve transection. Invest Ophthalmol Vis Sci. 2010;51:6394–9.

Li Y, Sauve Y, Li D, Lund RD, Raisman G. Transplanted olfactory ensheathing cells promote regeneration of cut adult rat optic nerve axons. J Neurosci. 2003;23:7783–8.

Li S, Liu BP, Budel S, Li M, Ji B, Walus L, Li W, Jirik A, Rabacchi S, Choi E, Worley D, Sah DW, Pepinsky B, Lee D, Relton J. Strittmatter, Blockade of Nogo-66, myelin-associated glycoprotein, and oligodendro-cyte myelin glycoprotein by soluble Nogo-66 receptor promotes axonal sprouting and recovery after spinal injury. J Neurosci. 2004;24:10511–20.

Limb GA, Daniels JT, Cambrey AD, Secker GA, Shortt AJ, Lawrence JM, Khaw PT. Current prospects for adult stem cell-based therapies in ocular repair and regeneration. Curr Eye Res. 2006;31:381–90.

Lipson AC, Widenfalk J, Lindqvist E, Ebendal T, Olson L. Neurotrophic properties of olfactory ensheathing glia. Exp Neurol. 2003;180:167–71.

Ma J, Jiang L, Zhong Y, Li Z, Xie J, Zhao C, Dong F. Neuroprotective effect on retinal ganglion cells by transpupillary laser irradiation of the optic nerve head. Neurosci Lett. 2010;476:3–8.

MacLaren RE, Pearson RA, MacNeil A, Douglas RH, Salt TE, Akimoto M, Swaroop A, Sowden JC, Ali RR. Retinal repair by transplantation of photoreceptor precursors. Nature. 2006;444:203–7.

Maeda K, Sawada A, Matsubara M, Nakai Y, Hara A, Yamamoto T. A novel neuroprotectant against retinal ganglion cell damage in a glaucoma model and an optic nerve crushmodel in the rat. Invest Ophthalmol Vis Sci. 2004;45:851–6.

McKerracher L, David S, Jackson D, Kottis V, Dunn R, Braun P. Identification of myelinassociated glycoprotein as a major myelin- derived inhibitor of neurite growth. Neuron. 1994;13:805–11.

Mellough CB, Cui Q, Spalding KL, Symons NA, Pollett MA, Snyder EY, Macklis JD, Harvey AR. Fate of multipotent neural precursor cells transplanted into mouse retina selectively depleted of retinal ganglion cells. Exp Neurol. 2004;186:6–19.

Mellough CB, Cui Q, Harvey AR. Treatment of adult neural progenitor cells prior to transplantation affects graft survival and integration in a neonatal and adult rat model of selective retinal ganglion cell depletion. Restor Neurol Neurosci. 2007;25:177–90.

Minamino K, Adachi Y, Yamada H, Higuchi A, Suzuki Y, Iwasaki M, Nakano K, Koike Y, Mukaide H, Kiriyama N, Shigematsu A, Matsumura M, Ikehara S. Long-term survival of bone marrow-derived retinal nerve cells in the retina. NeuroReport. 2005;16:1255–9.

Monnier PP, D'Onofrio PM, Magharious M, Hollander AC, Tassew N, Szydlowska K, Tymianski M, Koeberle PD. Involvement of caspase-6 and caspase-8 in neuronal apoptosis and the regenerative failure of injured retinal ganglion cells. J Neurosci. 2011;31:10494–505.

Moore DL, Goldberg JL. Four steps to optic nerve regeneration. J Neuroophthalmol. 2010;30:347–60.

Mukhopadhyay G, Doherty P, Walsh F, Crocker P, Filbin M. A novel role for myelin-associated glycoprotein as an inhibitor of axonal regeneration. Neuron. 1994;13:757–67.

Nadal-Nicola FM, Jiménez-López M, Sobrado-Calvo P, Nieto-López L, Cánovas-Martínez I, Salinas-Navarro M, Vidal-Sanz M, Agudo M. Brn3a as a marker of retinal ganglion cells: qualitative and quantitative time course studies in naïve and optic nerve–injured retinas. Invest Ophthalmol Vis Sci. 2009;50:3860–3868.

Nishida A, Takahashi M, Tanihara H, Nakano I, Takahashi JB, Mizoguchi A, Ide C, Honda Y. Incorporation and differentiation of hippocampus-derived neural stem cells transplanted in injured adult rat retina. Invest Ophthalmol Vis Sci. 2000;41:4268–74.

Ooto S, Akagi T, Kageyama R, Akita J, Mandai M, Honda Y, Takahashi M. Potential for neural regeneration after neurotoxic injury in the adult mammalian retina. Proc Natl Acad Sci USA. 2004;101:13654–9.

Osakada F, Ooto S, Akagi T, Mandai M, Akaike A, Takahashi M. Wnt signaling promotes regeneration in the retina of adult mammals. J Neurosci. 2007;27:4210–9.

Parameswaran S, Balasubramanian S, Babai N, Qiu F, Eudy JD, Thoreson WB, Ahmad I. Induced pluripotent stem cells generate both retinal ganglion cells and photoreceptors: therapeutic implications in degenerative changes in glaucoma and age-related macular degeneration. Stem Cells. 2010;28:695–703.

Pignot V, Hein AE, Barske C, Wiessner C, Walmsley AR, Kaupmann K, Mayeur H, Sommer B, Mir AK, Frentzel S. Characterization of two novel proteins, NgRH1 and NgRH2, structurally and biochemically homologous to the Nogo-66 receptor. J Neurochem. 2003;85:717–28.

Prinjha R, Moore SE, Vinson M, Blake S, Morrow R, Christie G, Michalovich D, Simmons DL, Walsh FS. Inhibitor of neurite outgrowth in humans. Nature. 2000;403:383–4.

Qiu G, Seiler MJ, Mui C, Arai S, Aramant RB, de Juan E, Sadda S. Photoreceptor differentiation and integration of retinal progenitor cells transplanted into transgenic rats. Exp Eye Res. 2005;80:515–25.

Quigley HA, Broman AT. The number of people with glaucoma worldwide in 2010 and 2020. Br J Ophthalmol. 2006;90:262–7.

Quigley HA, Iglesia DS. Stem cells to replace the optic nerve. Eye (Lond). 2004;18:1085–8.

Ramirez-Castillejo C, Sanchez-Sanchez F, Andreu-Agullo C, Ferron SR, Aroca-Aguilar JD, Sanchez P, Mira H, Escribano J, Farinas I. Pigment epithelium-derived factor is a niche signal for neural stem cell renewal. Nat Neurosci. 2006;9:331–9.

Ramon-Cueto A, Plant GW, Avila J, Bunge MB. Long-distance axonal regeneration in the transected adult rat spinal cord is promoted by olfactory ensheathing glia transplants. J Neurosci. 1998;18:3803–15.

Raymond PA, Barthel LK, Bernardos RL, Perkowski JJ. Molecular characterization of retinal stem cells and their niches in adult zebrafish. BMC Dev Biol. 2006;6:36.

Raymond I, Vila A, Huynh U, Brecha N. Cyan fluorescent protein expression in ganglion and amacrine cells in a thy1-CFP transgenic mouse retina. Mol Vis. 2009;14:1559–74.

Rein DB, Zhang P, Wirth KE, Lee PP, Hoerger TJ, McCall N, Klein R, Tielsch JM, Vijan S, Saaddine J. The economic burden of major adult visual disorders in the United States. Arch Ophthalmol. 2006;124:1754–60.

Ronggu Y, Jinguo Y, Zhansheng S, Xing W, Yingjuan L, Wei Z, Hua Y. Establishment of animal model of traumatic optic neuropathy and its function evaluation by F-VEP. Chin Ophthalmic Res. 2010;28:39–43.

Roskams AJ, Bethel MA, Hurt KJ, Ronnett GV. Sequential expression of Trks A, B, and C in the regenerating olfactory neuroepithelium. J Neurosci. 1996;16:1294–307.

Salinas-Navarro M, Mayor-Torroglosa S, Jimenez-Lopez M. A computerized analysis of the entire retinal ganglion cell population and its spatial distribution in adult rats. Vision Res. 2009;49:115–26.

Sánchez-Migallón MC, Nadal-Nicolás FM, Jiménez-López M, Sobrado-Calvo P, Vidal-Sanz M, Agudo-Barriuso M. Brain derived neurotrophic factor maintains Brn3a expression in axotomized rat retinal ganglion cells. Exp Eye Res. 2011;92:260–7.

Sarikcioglu L, Demir N, Demirtop A. A standardized method to create optic nerve crush: Yasargil aneurysm clip. Exp Eye Res. 2007;84:373–7.

Schnichels S, Heiduschka P, Julien S. Different spatial and temporal protein expressions of repulsive guidance molecule a and neogenin in the rat optic nerve after optic nerve crush with and without lens injury. J Neurosci Res. 2011;89:490–505.

Su Y, Wang F, Teng Y, Zhao S, Cui H, Pan S. Axonal regeneration of optic nerve after crush in Nogo66 receptor knockout mice. Neurosci Lett. 2009;460:223–6.

Taylor HR, Pezzullo ML, Keeffe JE. The economic impact and cost of visual impairment in Australia. Br J Ophthalmol. 2006;90:272–5.

Tello F. La influencia del neurotropismo en la regeneracio´n de los centros nerviosos. Trab Lab Invest Biol Univ Madrid. 1911;9:123–8.

Thanos S, Naskar R, Heiduschka P. Regenerating ganglion cell axons in the adult rat establish retinofugal topography and restore visual function. Exp Brain Res. 1997;114:483–91.

Tropepe V, Coles BL, Chiasson BJ, Horsford DJ, Elia AJ, McInnes RR, van der Kooy D. Retinal stem cells in the adult mammalian eye. Science. 2000;287:2032–6.

Varma R, Ying-Lai M, Francis BA, Nguyen BB, Deneen J, Wilson MR, Azen SP. Prevalence of open-angle glaucoma and ocular hypertension in Latinos: the Los Angeles Latino Eye Study. Ophthalmology. 2004;111:1439–48.

Vihtelic TS, Soverly JE, Kassen SC, Hyde DR. Retinal regional differences in photoreceptor cell death and regeneration in light-lesioned albino zebrafish. Exp Eye Res. 2006;82:558–75.

Walsh FB. Pathological-clinical correlations. I. Indirect trauma to the optic nerves and chiasm. II. Certain cerebral involvements associated with defective blood supply. Invest Ophthalmol. 1966;5:433–49.

Wan J, Zheng H, Xiao HL, She ZJ, Zhou GM. Sonic hedgehog promotes stem-cell potential of Muller glia in the mammalian retina. Biochem Biophys Res Commun. 2007;363:347–54.

Wan J, Zheng H, Chen ZL, Xiao HL, Shen ZJ, Zhou GM. Preferential regeneration of photoreceptor from Muller glia after retinal degeneration in adult rat. Vision Res. 2008;48:223–34.

Wang K, Koprivica V, Kim J, Sivasankaran R, Guo Y, Neve R, He Z. Oligodendrocyte-myelin glycoprotein is a Nogo receptor ligand that inhibits neurite outgrowth. Nature. 2002;417:941–4.

Wojciechowski AB, Englund U, Lundberg C, Warfvinge K. Long-term survival and glial differentiation of the brain-derived precursor cell line RN33B after subretinal transplantation to adult normal rats. Stem Cells. 2002a;20:163–73.

Wojciechowski AB, Englund U, Lundberg C, Wictorin K, Warfvinge K. Subretinal transplantation of brain-derived precursor cells to young RCS rats promotes photoreceptor cell survival. Exp Eye Res. 2002b;75:23–37.

Woodhall E, West AK, Chuah MI. Cultured olfactory ensheathing cells express nerve growth factor, brain-derived neurotrophic factor, glia cell line-derived neurotrophic factor and their receptors. Brain Res Mol Brain Res. 2001;88:203–13.

Wu MM, Fan DG, Tadmori I, Yang H, Furman M, Jiao XY, Young W, Sun D, You SW. Death of axotomized retinal ganglion cells delayed after intraoptic nerve transplantation of olfactory ensheathing cells in adult rats. Cell Transplant. 2010;19:159–66.

Xie J, Jiang L, Zhang T, Jin Y, Yang D, Chen F. Neuroprotective effects of Epigallocatechin-3-gallate (EGCG) in optic nerve crush model in rats. Neurosci Lett. 2010;479:26–30.

Yingjuan L, Xiulan Z, Shizhong W, Hua Y. Expression of P53, bax and caspase 3 after traumatic optic nerve injury in rats. Tianjin Med J. 2010;38:996–8.

Young MJ. Stem cells in the mammalian eye: a tool for retinal repair. APMIS. 2005;113:845–57.

Young MJ, Ray J, Whiteley SJ, Klassen H, Gage FH. Neuronal differentiation and morphological integration of hippocampal progenitor cells transplanted to the retina of immature and mature dystrophic rats. Mol Cell Neurosci. 2000;16:197–205.

Zaverucha-do-Valle C, Gubert F, Bargas-Rega M, Coronel JL, Mesentier-Louro LA, Mencalha A, Abdelhay E, Santiago MF, Mendez-Otero R. Bone marrow mononuclear cells increase retinal ganglion cell survival and axon regeneration in the adult rat. Cell Transplant. 2011;20:391–406.

Zhang R, Zhang H, Xu L, Ma K, Wallrapp C, Jonas JB. Neuroprotective effect of intravitreal cell-based glucagonlike peptide-1 production in the optic nerve crush model. Acta Ophthalmol. 2011;89:e320–6.

Zhao X, Das AV, Soto-Leon F, Ahmad I. Growth factor-responsive progenitors in the postnatal mammalian retina. Dev Dyn. 2005;232:349–58.

Zhao T, Li Y, Tang L, Fan F, Jiang B. Protective effects of human umbilical cord blood stem cell intravitreal transplantation against optic nerve injury in rats. Graefes Arch Clin Exp Ophthalmol. 2011;249:1021–8.

Chapter 20
Stem Cells and Spinal Cord Regeneration

Xinhua Zhang, Sheng Yi and Xiaosong Gu

Abstract Stem cells are characterized by self-renewal and pluripotency to become any cells in tissues/organs including the central nervous system (CNS), where they may differentiate to neurons and glial cells. The identification and characterization of stem cells have attracted great interest in their potential for treating of various diseases of different organs, as well as the CNS. The spinal cord, as a part of CNS, carries a tight bundle of neural cells and nerve pathways that connect the brain and the peripheral nervous system. Spinal cord injuries (SCI) usually begin with a sudden, mechanical trauma which results in devastating and irreversible consequences including stop of the nerve signaling and serious damage of axons and neural cell membranes beyond repair. The application of stem cells to CNS regeneration is very promising. Results from SCI models showed that transplantation of stem cells or progenitors may support spinal cord repair through the replacement of lost neural cells and the attenuation of gliosis around the rostral and dorsal terminals by the differentiated cells from the implanted stem cells. Axon regeneration-promoting and neuroprotective effects have also been credited to the transplanted stem cells. There are still issues related to stem cell transplantation that need to be resolved, including bioscaffold and ethical concerns. This chapter summarizes the latest research progress and application strategies of stem cells for SCI with the aim to push the medicine translation of stem cell application for spinal cord regeneration and implies the promising future of stem cells in SCI treatment.

Keywords Stem cell · Spinal injury · Therapy

X. Zhang · S. Yi · X. Gu (✉)
Jiangsu Key Laboratory of Neuroregeneration,
Co-innovation Center of Neuroregeneration, Nantong University,
Nantong, China
e-mail: nervegu@ntu.edu.cn

R.C. Zhao (ed.), *Stem Cells: Basics and Clinical Translation*, Translational
Medicine Research 1, DOI 10.1007/978-94-017-7273-0_20

20.1 Introduction

The spinal cord is a highly evolved and the least complex compartment of the central nervous system (CNS) with the cell bodies and dendrites of spinal neurons inside the cord and axons along the outside. It has considerable computational ability. All movements of the body below the head are controlled by the spinal cord, and injuries to it produce devastating losses of sensory, motor, and autonomic function distal to the level of trauma, arising from both the damage to the local circuitry of the spinal cord and the disruption of the ascending and descending fiber tracts. Since that, the spinal cord injury (SCI) is a devastating condition producing great personal and societal costs. Epidemiological information indicates that the global incidence of SCI is 15–40 cases per million people annually (Menezes et al. 1996; Tator 1995; Ackery et al. 2004). A newly calculated national incidence rate reported for Canada was 35 per million per year in 2006 (Rick Hansen Spinal Cord Injury Registry 2006) and an extrapolated value of 52.3 in 2010 (Farry and Baxter 2010). The estimated incidence in USA, not including those who die at the scene of the accident, is approximately 40 cases per million (National Spinal Cord Injury Statistical Center and Others 2014). Acute SCI affects 12,000 individuals annually in USA alone, 4000 of which die before reaching hospital, and another 1000 die during their hospitalization. A epidemiological research by Lee et al. (Lee et al. 2014) reported that the extrapolated regional data indicate the incidence rate is 25 per million in Central Asia and 21 per million South Asia. Li et al. (2011) and Ning et al. (2011) provided recent regional incidence data for mainland China (60.6 per million population per year in Beijing and 23.7 in Tianjin province). About 16 % SCI patients have to live with lifelong tetraplegia which is caused by high-level SCI (National Spinal Cord Injury Statistical Center and Others 2014). Current treatment includes surgery to decompress and stabilize the injury, prevention of secondary complications, management of any that do occur, and rehabilitation. Although these treatments and advances in the medical and surgical care of SCI ameliorated neurological functions, there is still no effective treatment for the neurological deficits of SCI (Tator 2006). Neurological recovery is limited, and most SCI patients still face substantial neurological dysfunction and lifelong disability. With the advances in the stem cell research, stem cell therapy is becoming a potential treatment for SCI because of their fundamental properties: high proliferative potential, self-renewal, and the ability to differentiate into multiple cell types. So far, a variety of different stem cell types, including embryonic stem cells, neural stem cell, and stem cells from non-neural tissues such as bone marrow stem cells, have been evaluated in animal models and humans after SCI, with the goal of promoting repair and recovery from the injury. The strategies for spinal cord repair using stem cell include replacement of damaged neuronal and glial cells, secretion of trophic factors, regulation of gliosis and scar formation, remyelination of spared axons, enhancement of axon elongation, restoration of neuronal circuitry, production of trophic factors, anti-inflammatory cytokines, and other molecules to promote tissue sparing and neovascularization, and a permissive environment for plasticity and axonal regeneration. In order to select the best time-point for therapeutic cell transplantation, an understanding of the timeline of secondary damage cascades is important (Su et al. 2011). The inflammatory response, glial cell activation, and the inhibitory microenvironment that exists in the acute phase after trauma largely act as a negative obstruction to any form of cellular therapy. On the other hand, the pathological alterations of the lesion site in a chronic patient may not be reversible due to the formation of a glial scar, the permanent demyelination/dysmyelination of spared axons, and the apoptosis of spared neurons. Therefore, the optimal transplantation time-window most probably lies in the subacute phase.

This chapter summarizes the latest research progress and application strategies of stem cells for SCI with the aim to push the medicine translation of stem cell application for spinal cord regeneration and implies the promising future of stem cells in SCI treatment.

20.2 Candidates of Stem Cells for Therapies in SCI

20.2.1 Endogenous Stem Cells in Central Canal Region

The region surrounding the central canal of the spinal cord is developed from the neural tube and retains a substantial degree of plasticity. The central canal region comprises several cell types, including ependymocytes, tanycytes [also referred to as radial ependymocytes (Seitz et al. 1981)], and neuronal-like cells that are located either in direct contact with the lumen or in subependymal position. The ependymocytes are the primary cell type found around the central canal. Numerous tanycytes are found lateral sides along the entire central canal region (Seitz et al. 1981), where they radiate from the ependyma into the gray matter that surrounds the central canal and send a long basal process terminating as foot processes in association with blood vessels (Horstmann 1954; Rafols and Goshgarian 1985). The cell body of tanycyte is located in either subependymal or ependymal region, but the process is in contact with the lumen (Rafols and Goshgarian 1985; Meletis et al. 2008), implying a potential link between the cerebral spinal fluid and blood. The neuronal-like cells in the central canal region are very common and well-described in lower vertebrates and mammals (Vigh et al. 2004; Hugnot and Franzen 2011). They are found sporadically distributed around the canal with a soma in an ependymal or subependymal position and a single thick dendritic-like process terminating by a large bulge in the lumen (Sabourin et al. 2009). These cells maintain some degree of immaturity through the adult because they still express PSA-NCAM, Dcx, and GAP43 (Sabourin et al. 2009; Marichal et al. 2009; Seki and Arai 1993; Stoeckel et al. 2003) that are involved in plasticity and migration. These cells are found to be produced from embryogenesis but not adult spinal cord neurogenesis (Marichal et al. 2009). The fact that these cell types are localized at specific locations and express characteristic markers involved in migration and neuronal differentiation and show potentially different functions raises the possibility that the central canal region is a reservoir of cells with capability of neurogenesis in "standby mode."

To identify the identity of the cell types in the central canal region, the GFAP-GFP and FoxJ1-GFP transgenic animals were recruited in several research groups. In total, 0.2 % of GFP^+ cells were able to generate neurospheres and most primary neurospheres derived from spinal cord of these transgenic animals contained $GFAP^+/GFP^+$ cells (Sabourin et al. 2009), suggesting that the central canal cells with astrocytic features have neural stem cell properties as observed in the SVZ. FoxJ1 expression is restricted to the central canal region (Meletis et al. 2008). The FoxJ1 positive cells from the FoxJ1-GFP transgenic mouse sorted by flow cytometric isolation formed the neurospheres in vitro, which have the multipotential to differentiate into neurons, astrocytes, and oligodendrocytes (Meletis et al. 2008). All the data support the presence of stem cells in the central canal region.

It has been demonstrated that SCI activated the ependymal cells around the central canal to proliferate and generate progeny cells and recruited them to the injured site leaving behind an intact ependymal layer (Meletis et al. 2008; Barnabe-Heider et al. 2010). In rodents, SCI causes progenitor cell proliferation in both the white matter and the ependymal zone (Horky et al. 2006; Mothe and Tator 2005), but the main fate of endogenous stem cells after SCI is to differentiate exclusively into oligodendrocytes and astrocytes (Meletis et al. 2008; Barnabe-Heider and Frisen 2008). The generation of oligodendrocytes contributes to remyelination and is likely to underlie some restoration of function. It is attractive to consider ways to facilitate this process. In contrast, the generation of astrocytes by endogenous stem/progenitor cells may contribute to glial scar formation that may potentially inhibit axonal growth. There is a need for a more focused approach to understand the molecular properties of adult endogenous stem/progenitor cells in spinal cord and develop optimal strategies for modulating their response during SCI, such as the new techniques for directing their commitment toward specific fates according to the desired function (for example, oligodendrocytes for remyelination, or new neurons for cell replacement). Research is now focusing on the manipulation of ependymal cells to produce cells of the oligodendrocyte or neuronal lineage. Ohori et al. (2006) reported that genetic manipulation by overexpressing neurogenin-2 and Mash1 in the spinal cord stem cells, together with growth factor treatment, enhanced the production and maturation of new neurons and oligodendrocytes, respectively. These results demonstrate that modulating the endogenous stem/progenitors may offer a potential therapy to replace neurons and oligodendrocytes lost to insults in the injured spinal cord.

20.2.2 Embryonic Stem Cells

Embryonic stem (ES) cells possess two properties that make them especially well suited for cell therapy. First, because embryonic stem cells are obtained from inner cell mass of developing blastocyst embryos, they retain the flexibility to differentiate into all three primary germ layers including more than 200 cell types that make up the human body (Evans and Kaufman 1981; Puri and Nagy 2012). Stem cells with such flexibility are described as "pluripotent," to indicate their high potential to differentiate into a wide variety of cell types. The second feature is their ability to remain in an undifferentiated state and to divide indefinitely. The property of "self-renewal" implies that essentially unlimited numbers of the identical, well-defined, genetically and genomically characterized stem cells can be produced in culture for medical use. Human ES cells are typically obtained from preimplantation or blastocyst-stage embryos created during in vitro fertilization procedures and can also be generated by somatic cell nuclear transfer or parthenogenetic activation of eggs. For the ES cells that will form teratomas in vivo after transplantation, ES cells must be predifferentiated prior to grafting. The rapid advances in understanding the signal or molecular cues involved in programming differentiation of stem cells develop specific protocols to differentiate ES cells into various cell types, including neural precursors (Zhang et al. 2001; Tropepe et al. 2001; Reubinoff et al. 2001) and neuronal (Wichterle et al. 2002; Li et al. 2008; Wada et al. 2009) and glial lineages (Brustle et al. 1999; Nistor et al. 2005). The refined subtypes of adult cells are being generated from ES cells, such as retinal cells (Lamba et al. 2006; Osakada et al. 2008), forebrain interneurons (Li et al. 2009; Watanabe et al. 2005), midbrain dopaminergic neurons (Kawasaki et al. 2000; Perrier et al. 2004), and cerebellar neurons (Tao et al. 2010; Erceg et al. 2010). Initial methods for differentiation of ES cells to motor neurons relied on two principal patterning signals—RA that induces neutralization and caudalization of ESCs and sonic hedgehog (SHH) that directs ventralization of the spinal neural progenitor cells (Wichterle et al. 2002). Many protocols differentiating ES cells into motor neurons are based on the two molecules (Sabourin et al. 2009; Li et al. 2008; Wu et al. 2012; Wichterle and Peljto 2008; Soundararajan et al. 2007). The differentiation efficiency of motor neurons from ES cells has been increased to 50 % or more using modified methods (Wada et al. 2009; Wu et al. 2012). After transplantation into the injured spinal cord, the predifferentiated mouse ES cells survived and differentiated into astrocytes, oligodendrocytes, and neurons and migrated forward the lesion region, and more importantly, the animals showed partial functional recovery (McDonald et al. 1999). Studies showed that the predifferentiated oligodendrocyte progenitor cells (OPCs) from ES cells remyelinated the spared axons and improved recovery when transplanted subacutely into the injured rat spinal cord (Keirstead et al. 2005; Sharp et al. 2010). Transplantation of the differentiated motor neurons from ES cells has been widely reported to be an efficient strategy in repair of SCI (Wichterle et al. 2002; Deshpande et al. 2006; Soundararajan et al. 2006; Chiba et al. 2003; Lopez-Gonzalez et al. 2009; Erceg et al. 2010). The differentiated motor neurons from ES cells promote functional recovery post-transplanted into a SCI model and improve significantly the electrophysiological and motor assessments in addition to a structural regeneration of neuronal pathways. These motor neurons formed functional synapses with denervated host muscle, which resulted in the ability to produce motor units between grafted motor neuron and muscle, and even the denervation-associated muscle atrophy was significantly reduced (Deshpande et al. 2006; Yohn et al. 2008). These results demonstrate the capacity for ES cell-derived motor neurons not only to incorporate into the adult host spinal cord and target tissue, but also to form functional motor units.

Based on promising preclinical data of human ES cell-derived OPCs transplants in rodent SCI models (Keirstead et al. 2005; Sharp et al. 2010), the US Food and Drug Administration (FDA) approved the first human ES cell trial in 2009. This phase I safety trial in SCI sponsored by Geron Corp. began in 2010 after further preclinical safety data were obtained concerning abnormal cyst formation in transplanted animals. Two million GRNOPC1 cell line (human ES cell-derived OPCs) was directly transplanted into the spinal cord of five patients with complete thoracic SCI, and the patients were followed with immunosuppression for the first two months after transplantation. However, this program finally terminated by Geron in 2011 because of funding challenges. No safety issues were reported in the five patients who received GRNOPC1 transplants, but complete results have not been published.

20.2.3 Mesenchymal Stem Cells

Mesenchymal stem cell (MSC) lineage is a kind of selfrenewing and multipotent stem cell, which was initially identified from the bone marrow (BM) (Prockop 1997). It can differentiate into bone, cartilage, adipose tissue, liver, and neural tissue (Dezawa et al. 2004). MSCs are widespread throughout a variety of tissues (Young et al. 1995), including bone marrow, Wharton's jelly of the umbilical cord, adipose tissue, adult muscle, and the dental pulp of deciduous baby teeth. Bone marrow and umbilical cord are usually the rich sources of these cells. Most of the studies in SCI use MSC derived from bone marrow and adipose tissue, but it is also possible to get MSC from a perinatal source such as umbilical cord blood, umbilical cord matrix (Karahuseyinoglu et al. 2007; Wang et al. 2004), amniotic fluid (De Coppi et al. 2007), and placenta (Yen et al. 2005; Fukuchi et al. 2004). MSC therapy has attracted the attention of scientists and clinicians around the world for the reasons: (1) ease of isolation and cryopreservation, (2) maintenance of viability and regenerative capacity after cryopreservation, (3) rapid replication with high-quality progenitor cells and high potential of multi-lineage differentiation, and (4) minimal or no immunoreactivity and graft versus host reaction of transplanted allogeneic MSCs.

MSCs from various tissues have been exploited in the effective treatment for many organic or functional diseases, including experimental autoimmune encephalomyelitis, experimental brain ischemia, and in animals undergoing brain or SCI. Basic and preclinical experimental studies have highlighted the positive effects of MSC treatment after SCI. Bone marrow MSCs (BMSCs) become the most common non-neural cell types for transplantation in SCI. Many studies have examined

BMSCs in SCI rodents, with some showing improved locomotor recovery (Hofstetter et al. 2002; Himes et al. 2006). The BMSCs labeled with fluorescence or iron-oxide nanoparticles were tracked via fluorescence or magnetic resonance after transplantation into the spinal cord. Results indicated that some of these cells migrate toward injured site and express neuronal or astrocytic markers and the animals with SCI demonstrated functional recovery (Yano et al. 2005; Sykova and Jendelova 2005; Lee et al. 2003). The above studies all showed that a very small number of MSCs differentiated into neurons and most of them were destined to astrocytes. Therefore, neuronal differentiation is unlikely to be a major factor in functional recovery after BMSC replacement for SCI. Other mechanisms involved in recovery consist of neuroprotection, formation of a favorable environment for regeneration, expression of growth factors or cytokines, and vascular effects or remyelination. These mechanisms are not mutually exclusive, and it is likely that more than one contribute to functional recovery.

MSC extracted from adipose tissue is considered an attractive source of cells due to easiness of isolation, a large amount of cells per donor, and the fact that this tissue is usually discarded after liposuctions. In SCI models, treatment with these cells have resulted in cell survival, neuroprotection, attenuation of secondary damage, axonal regeneration, decrease of gliosis, angiogenesis, and enhanced functional recovery (Kang et al. 2006; Zhang et al. 2009; Oh et al. 2012). MSCs extracted from perinatal tissues also present a therapeutic potential in spite of less investigation in SCI treatment. The umbilical cord matrix, also known as Wharton's jelly, contains a stem cell population that present some advantages in comparison with the other sources because they can proliferate more rapidly and extensively than adult MSC (Weiss et al. 2006; Troyer and Weiss 2008) and also because they are easily obtained after normal and cesarean births, with low risk of viral contamination. Other advantage is the possibility of using them for allogenic transplantation because they act by suppressing immune response and are, therefore, considered non-immunogenic cells (Weiss et al. 2008). Transplantation studies indicated that umbilical cord matrix-derived MSCs can survive and promote repair and recovery in the injured spinal cord, where they exert immunomodulatory and trophic effects through secretion of glial-derived neurotrophic factor (GDNF), BDNF, and nerve growth factor (NGF) which are known as supporters of cell survival and regeneration (Yang et al. 2008; Hu et al. 2010). The MSCs from umbilical cord blood or amniotic fluid are also demonstrated to use for recovery of SCI after transplantation into spinal cord, where they not only differentiate into neural cells, but also downregulate the fas/caspase-3 pathway and increased levels of anti-apoptotic proteins in neurons and oligodendrocytes (Dasari et al. 2007, 2008; Wu et al. 2006; Sankar and Muthusamy 2003).

Cumulative evidences suggest the therapeutic effects of MSCs are likely due to their ability to differentiate into other cell lineages, modulate inflammatory and immunomodulatory responses, reduce cell apoptosis, secrete several neurotrophic factors, respond to tissue injury, and promote angiogenesis. There are a number of completed and currently ongoing SCI clinical trials involving autologous MSC transplantation. Despite these potential benefits, there are a small number of patients treated with MSC transplants showing no adverse effects. Collectively, evidence suggests that MSC engraftment is far from being as good as those described in experimental studies. Therefore, there is an urgent need to seek for standardization of protocols in terms of source of cells, culture conditions, time of treatment after injury, and number and via of administration of cells.

20.2.4 Neural Stem/Progenitor Cells

Neural stem/progenitor cells (NSPCs) are found in both fetal and adult CNS (Gage 2000). NSPCs were first demonstrated in the subventricular zone (SVZ) of the mouse in 1989 (Temple 1989) and were isolated from the mouse striatal tissue and SVZ for the first time in 1992 (Reynolds and Weiss 1992; Stemple and Anderson 1992). These cells were capable of self-renewal and generating the main phenotypes (neurons, astrocytes and oligodendrocytes) of CNS cells in vitro and in vivo (Reubinoff et al. 2001). NSPCs can continually proliferate in vitro and maintain in a particular and unique living cluster called neurosphere. Neurosphere culture system developed by Reynolds and Weiss (1992) is the main method to obtain and preserve NSPCs, which have been extensively utilized for stem cell research. NSPCs predominantly reside within two areas of the adult mammalian brain: the SVZ lining the lateral ventricles of the forebrain (Gritti et al. 1996; Morshead et al. 1994), the subgranular zone (SGZ) of the hippocampal dentate gyrus (DG) (Palmer et al. 1997), and the central canal region of the spinal cord (Weiss et al. 1996). NSPCs in SVZ give rise to neuroblasts that migrate in chains to the olfactory bulb through the rostral migratory stream where they differentiate into granule and periglomerular neurons (Lois and Alvarez-Buylla 1994; Lois et al. 1996). In the adult DG, new neurons are born from NSPCs in the SGZ and migrate a short distance to differentiate into granule cells that project their axons to the CA3 region of the hippocampus and establish synaptic connection with local neurons (Deng et al. 2009; Markakis and Gage 1999).

Transplantation of NSPCs has been widely applied in the therapeutic study of SCI. Scientists have attempted to restore neural functions via a number of different strategies including neuronal differentiation, axon regeneration, remyelination, and nutrient secretion. The differentiated neurons and neural cells transplanted into the spinal cord can survive for the least time of 6 weeks and express neuronal and astrocytic phenotypic markers in these surviving cells, suggesting a promising survival rate and differentiation in the injured spinal cord in vivo, although most of them likely differentiate into astrocytes (Akesson et al. 2007). In most cases, in vivo directly transplanted NSPCs differentiate into glial lineages, especially astrocytes (Cao et al. 2001; Webber et al. 2007). Therefore, the direct transplantation of NSPCs is not an efficient strategy for functional recovery after SCI, the reasons for which are probably the high glial differentiation rate (40 %), low neuronal differentiation of the grafted stem cells, and the failure of axon regeneration beyond the lesion site. Transplantation of the in vitro predifferentiated NSPCs is more efficient than direct implantation, for example, hNSPCs treated with bFGF, heparin, and laminin for priming before transplantation. After transplantation into the contusion lesion of rats, these primed hNSPCs result in an optimized survival rate, neuronal and oligodendroglia differentiation, and improved trunk stability (Yan et al. 2007). The Olig2-NSPCs genetically expressing Olig2 showed high proliferative activity, increased volume of spared white matter, and reduced cavity volume. Additionally and importantly, the thickened myelin sheath was induced by the differentiated oligodendrocytes from grafted cells and the hindlimb functions got significant locomotor recovery (Hwang et al. 2009). The above studies indicated that oligodendrocyte differentiation from grafted NSPCs is vital to the functional recovery through remyelination by oligodendrocytes in the spinal cord. Grafted NSCs can also differentiate into neurons with certain pretreatments. Besides remyelination, synaptic contact reformation is also important for the reconstruction of neurofunctional circuitry. The pretreated neural progenitors of the human fetal spinal cord-derived NSPCs generated a large-scale neuronal differentiation, axon regeneration, and extensive synaptic contacts reformation with host motor neurons integrating into the host neural circuits (Yan et al. 2007). Another strategy is to transplant the NSPCs into the injured spinal cord with concomitant infusion of biomolecules or compounds, such as growth factors, which promote oligodendrocytic or neuronal differentiation and remyelination (Karimi-Abdolrezaee et al. 2006, 2010; Abematsu et al. 2010). The differentiated neurons from grafted NSPCs integrated into the injured cord and improved recovery. NSPCs have also demonstrated some immunomodulatory and pathotropic ability by homing toward damaged tissue (Ziv et al. 2006) as well as secreting various neurotrophic factors and cytokines (Yan et al. 2004; Lu et al. 2003; Hawryluk et al. 2012).

A registry Web site (http://www.clinicaltrials.gov) provides results database of publicly and privately supported clinical studies of human participants around the world where you can learn not only the history, policies, and laws, but also the progress about clinical studies. Recently, Stem Cells Inc. started a phase I/II (safety/efficacy) trial in Switzerland involving transplantation of human fetal brain stem cells into patients with thoracic SCI. Currently, this is the only human trial involving NSPCs for SCI, and these patients require immunosuppression. The drawbacks for human NSPCs include ethical concerns about fetal-derived cells, difficulties in expanding adult-derived cells to clinically sufficient numbers, and unavailability of autologous sources.

20.2.5 Induced Pluripotent Stem Cells

Mouse iPSCs were first established by Takahashi and Yamanaka in 2006 (Takahashi and Yamanaka 2006), and in recent years, the methods for their production are continuously progress. Initially, iPSCs were generated from mouse fibroblasts by retroviral introduction of the transcription factors Oct4, Sox2, c-Myc, and Klf4 (Takahashi and Yamanaka 2006). Other combinations of different factors are successively used to generate iPSCs from different somatic cells, such as Nanog/Oct4 (Okita et al. 2007), Nanog/Oct4/Sox2/lin28 (Yu et al. 2007). The actions of these transcription factors are thought to reprogram somatic cells into ES cell-like pluripotent cells through multiple stochastic epigenetic events and activation of various pluripotent genes. Compared ES cells, iPSCs share many key properties with ES cells, including morphology, pluripotency, self-renewal, and gene expression (Puri and Nagy 2012). The multipotential differentiation of iPSCs into electrophysiologically functional neurons, astrocytes, and oligodendrocytes makes them to be more useful and potential in regenerative therapies. If it were possible to perform custom cell transplantation therapy by generating iPS cells from patients themselves and transplanting them into SCI sites after the iPS cells had been induced to differentiate into neural cells, it would also be possible to avoid both the ethical problem of using human fetal tissue and the possibility of immunological rejection.

One of the main problems with generation of iPSCs is the expression of reprogramming factors associated with teratoma formation (Ben-David and Benvenisty 2011). To avoid the permanent transgene integration, several alternative delivery methods have been developed for reprogramming, such as adenovirus, the piggyBac transposon, and direct protein transduction (Puri and Nagy 2012; Gonzalez et al. 2011). These reprogramming factors are needed to initiate but not sustain somatic cell transformation into iPSCs, which is very important from a therapeutic standpoint. However, for clinical translation, it is necessary to develop reproducible protocols for iPSC differentiation to specific neural lineages. The iPSCs have been tried to generate NSPCs, motor neurons, OPCs, and olfactory ensheathing cells in vitro, which were restricted to one specific cell lineage with low risk of tumorigenesis after implantation into the spinal cord. Although the NSPCs can be derived from human iPSCs, some types of iPSC-derived neural cells have an increased likelihood of tumor formation after transplantation into the CNS. Thus, safe iPSC-derived clones need to be screened and selected for transplantation (Miura et al. 2009; Tsuji et al. 2010). Subacute transplantation of the preselected "safe" iPSC-derived neurospheres into lesion site after contusion SCI exerts positive effects on the injured spinal cord, including remyelination, axonal outgrowth of serotonergic fibers, and promotion of locomotor recovery. However, the "unsafe" iPSC-derived neurospheres resulted in robust teratoma formation and loss of locomotor function (Tsuji et al. 2010). The grafted human iPSC-derived neurospheres survived, migrated, and differentiated into the three major neural lineages (neurons, astrocytes, and oligodendrocytes) within the injured spinal cord and showed both cell-autonomous and non-cell-autonomous effects, including synapse formation, expression of neurotrophic factors, angiogenesis, axonal regrowth, and increased amounts of myelin in the injured area, resulting in significantly functional recovery (Nori et al. 2011). No tumor formation occurred in the grafted mice with the preselected clones. Motor neurons are recently generated using a high-efficient method (Karumbayaram et al. 2009; Zeng et al. 2010) that utilizes adherent cultures by forming neural rosettes that were treated with RA and SHH and supplemented with GDNF, BDNF, and CNTF or IGF. Differentiated motor neurons under such conditions were electrophysiological active because they presented repetitive firing of action potentials in response to current injection. To date, there is no available information on motor neuron grafting in non-human primate models of SCI. To our knowledge, no clinical protocols involving grafting of motor neuron derived from iPSCs have been set up. iPS cells can be derived from tissues of the same patient, enabling the possibility of autologous transplantation after in vitro differentiation. Additionally, iPS cell derivation from motor neuron disease patients allows the generation of cellular models that would improve our knowledge about mechanisms and design new preventive strategies to avoid motor neuron death.

20.3 Strategies of Stem Cells for Spinal Cord Regeneration

SCI researchers have taken great efforts for decades to bridge the injured spinal cord and repopulate the area of injury with cells that might restore axonal continuity to promote axonal growth back to its distal targets. An emerging strategy for replacing and/or regenerating damaged tissue is the implantation of stem cells and/or artificial biomaterials such as scaffolds combined with cells to form tissue bridges between damaged spinal cord stumps.

20.3.1 Transplantation of Stem Cells

Cell transplantation represents a potentially powerful treatment method for SCI via the possible means of (1) parasecreting permissive neurotrophic molecules at the lesion site to enhance the regenerative capacity; (2) providing a scaffold for the regeneration of axons; and (3) replacing lost neurons and neural cells. Direct transplantation of the single stem cells was performed in the early progress of cell replacement therapy, which requires the knowledge of characteristics of the candidate stem cells.

20.3.1.1 Transplantation of Embryonic Stem Cells

The fate of stem cells or progenitor cells that were derived from embryonic CNS or human umbilical cord blood has been studied in many laboratories after transplanted into the injured adult rodent spinal cord (McDonald et al. 1999; Han et al. 2002, 2004; Hill et al. 2004; Cao et al. 2002; Ogawa et al. 2002; Teng et al. 2002). The potential of human fetal stem cells is currently being investigated in treating animal models of SCI. Neural progenitors derived from human fetuses have been transplanted into immunosuppressed rodent and non-human primates after contusion (Cummings et al. 2005; Iwanami et al. 2005). The transplanted cells survived and differentiated into cells with characteristics of oligodendrocytes and neurons and were associated with locomotor improvements. Pre-differentiation treatment on the embryonic CNS-derived stem/progenitor cells is recently a successful approach for transplantation. Mitsui et al. (2005) transplanted neuron- and glia-restricted precursors into rat spinal cord after contusion injury. The grafted cells survived, filled the lesion site, and differentiated into cells with some characteristics of neurons and glia, resulting in sparing/sprouting of descending pathways and improved bladder and motor function. The oligodendrocyte-restricted progenitor cells were usually used to evaluate their potential in SCI because of remyelination function. Indeed, the pre-differentiated OPCs from ES cells survived, migrated over short distances and differentiated into oligodendrocytes, and finally remyelinated the spared axons and improved recovery when transplanted subacutely into the injured rat spinal cord (Keirstead et al. 2005; Sharp et al. 2010).

20.3.1.2 Transplantation of Adult Stem Cells

In the field of cell therapy, adult stem cells are emerging as a clear alternative. Adult stem cells have been isolated from a variety of different organs throughout the human body using a variety of techniques. Since that, adult stem cells have low ethical concerns. Various adult progenitor cells have been implanted in rodent models of SCI, ranging from olfactory ensheathing cells, cultured spinal cord stem cells, bone marrow derived stem cells, dermis derived stem cells, and a few others (Schultz 2005). Transplantation of MSCs from bone marrow significantly improves function recovery after SCI in mice and rats (Hofstetter et al. 2002; Himes et al. 2006; Yano et al. 2005; Sykova and Jendelova 2005; Lee et al. 2003; Wu et al. 2003). The potential mechanisms mediating BMSCs efficiency are proposed as neurotrophy, axonal elongation, formation of a favorable environment for regeneration, expression of growth factors or cytokines, and vascular effects or remyelination. A human trial was conducted in which autologous BMSCs were intravenously delivered to nine patients with SCI (Reier 2004). The improvements observed appeared to fall within an expected range of spontaneous recovery, and one participant advanced from ASIA category B to D. Nevertheless, without controls or some indication of cell viability within those lesions, this preclinical trial can demonstrate a measure of procedural safety. After transplantation of adult NSPCs into the intact and injured murine spinal cord, a small number of neuronal differentiation is observed (Akesson et al. 2007), but most grafted cells differentiated into astrocytes or oligodendrocytes (Cao et al. 2001; Vroemen et al. 2003). Recently, combination of cell transplantation with growth factors that selective increases oligodendrocytic or neuronal differentiation, which is potential on functional recovery of the injured spinal cord. The pre-treated NSPCs transplanted 2 weeks post-injury survived, migrated, integrated in the injured spinal cord tissue, generated mature oligodendrocytes that remyelinated the injured axons, and promoted some functional recovery. However, NPCs transplanted 8 weeks post-injury did not survive and failed to exert similar effects (Karimi-Abdolrezaee et al. 2006). Therefore, there is a need to find and neutralize the inhibitory obstacles present in chronic SCI that interfere with NPC survival after transplantation. The data from the same group combined the sustained infusion of chondroitinase ABC to target chondroitin sulfate proteoglycans (CSPG) with subsequent transplants of NPCs and transient infusion of growth factors, EGF, bFGF, and PDGF-AA (Karimi-Abdolrezaee et al. 2010). Results demonstrate that perturbing CSPGs dramatically optimizes NPC transplantation in chronic SCI. Engrafted NPCs successfully integrate and extensively migrate within the host spinal cord and principally differentiate into oligodendrocytes. Additionally, the combined strategy promoted the axonal integrity and plasticity of the corticospinal tract and enhanced the plasticity of descending serotonergic pathways. These neuroanatomical changes were also associated with significantly improved neurobehavioral recovery after chronic SCI. To regulate the survival number of grafted cells in spinal cord and fate to promote recovery, it will be necessary to determine which molecules are involved in governing neural stem cell proliferation, migration, and differentiation.

20.3.1.3 Transplantation of Engineered Stem Cells

The microenvironment in the injured adult spinal cord is poor for cell survival, neuronal differentiation, and maturation. Therefore, it is necessary to enhance the capacity of stem cells in CNS repair. Researchers are recently trying to genetically modify the stem cells to potentiate them better survival, and desired differentiation and maturation properties. Neural stem cells transducted with neurogenin-2 that is necessary for granule progenitor amplification greatly suppressed astrocytic differentiation of engrafted cells in spinal cord and improved the positive effects of engrafted stem cells, including increased amounts of myelin in the injured area, recovery of hindlimb locomotor function, and hindlimb sensory responses (Hofstetter et al. 2005). In order to increase the survival of transplanted rat ES cells, BCL2, an anti-apoptotic protein, was forced to be overexpressed in ES cells. This led to tumor-like growth of cells, accompanied by increased morbidity and mortality (Howard et al. 2005). More promisingly, when transplanted in the compressed mouse spinal cord, engineered mouse ES cells expressing the cell adhesion molecule L1, that is, able to enhance neuronal survival and neurite outgrowth, survived longer and migrated rostrally and caudally from the lesion. Corticospinal tract axons showed interdigitation with L1-transfected ES cells and extended into and, in some cases, beyond the lesion site (Chen et al. 2005). The embryonic spinal cord-derived glial-restricted precursor cells expressing multineurotrophin D15A with both BDNF and NT3 activities significantly increased the oligodendrocytic differentiation and formed morphologically normal-appearing myelin sheaths around the axons in the ventrolateral funiculus of spinal cord, and facilitate functional recovery after traumatic SCI when transplanted them into the contused adult spinal cord (Cao et al. 2005). MASH1 is both necessary and sufficient for the maintenance of neural precursors and neurogenesis in the ventral spinal cord (Parras et al. 2002). Hamada et al. (2006) introduced MASH1 gene into ES cells which yielded purified spinal motor neuron precursors after transplantation into injured spinal cord. The transplanted cells showed extensive outgrowth of axons and connected to each other, suggesting reconstitution of neuronal pathways. Interestingly, MASH1 transfection into ES cells downmodulated the expression of Nogo receptor that mediates axonal growth (Stoeckel et al. 2003; GrandPre et al. 2002; Li et al. 2004). This may be one of the most important mechanisms for neuronal regeneration in spinal cord.

It is apparent that transplantation of genetically modified stem/progenitor cells after SCI will not lead to more optimal recovery than the stem/progenitor cells alone; combination strategies will be necessary for optimum return of function. Advances in molecular biology have facilitated the manipulation of these cells to express molecules of interest. These types of combination strategy are promising but need further development and careful animal testing, individually and jointly, before any clinical trial can be started.

20.3.2 Biomaterial-Based Interventions

In acute cases with SCI, transplanted cells may either replace dead or dying cells or provide bioactive factors that improve endogenous regeneration and prevent apoptosis and cavity formation. However, in chronic SCI, the cystic cavity and glial scar are already developed, and probably, cell transplantation alone is not sufficient to promote injured spinal cord regeneration. Therefore, tissue repair in these cases requires "bridging" the lesion with a matrix that provides a permissive environment, fills the tissue gap, and, concomitantly, provides structural support for axonal regrowth and functional reconnection. An optimal candidate as bridge is the biomaterial that has biocompatibility supporting cell attachment, growth and differentiation, mechanical properties matching the neural tissue, and, in addition, porosity and permeability.

The recently used biomaterials in spinal cord tissue engineering include biodegradable, either natural or synthetic, and nonbiodegradable polymers. Biodegradable polymers hold the potential for the ultimate restoration of function and full regeneration of the tissue. To achieve this goal, the material degradation should match the tissue regeneration and maturation rates in the implanted site. Among synthetic biodegradable polymers, aliphatic polyesters, such as poly (lactide), poly (glycolide) and their copolymers, and poly (ecaprolactone) are the most explored (De Laporte et al. 2009; Wong et al. 2008; Pego et al. 2001). Nonbiodegradable synthetic materials, including cross-linked synthetic polymers based on methacrylate hydrogels, such as poly (2-hydroxyethyl methacrylate) (PHEMA) and poly [N-(2-hydroxypropyl) methacrylamide] (PHPMA) (Hejcl et al. 2008; Tsai et al. 2004; Woerly et al. 2008), are also being studied. Their application in a clinical scenario requires the establishment of their safety in terms of foreign body reaction upon implantation. Generally, natural polymers are biocompatible and suitable for adhesion and can minimize the occurrence of cytotoxic effects, making natural polymers advantageous materials for nerve tissue engineering (Ciardelli and Chiono 2006), despite the fact that naturally harvested materials have higher batch-to-batch variability and can, in some cases, induce immunogenic reactions. Naturally derived polymers comprise collagen, fibrin, hyaluronic acid, agarose, alginate, chitosan, fibroin, or poly (b-hydroxybutyrate). Many reports have demonstrated these promising scaffolding materials for the treatment of spinal cord lesions. In terms of low mechanical strength of nature biomaterial, the natural and synthetic composites can combine the biocompatible properties of natural materials and the mechanical strength and tunable degradation rates of synthetic materials (Bhattarai et al 2009). However, as most natural and synthetic polymers do not have cell adhesion property, an additional surface modification is needed to promote cell–surface interactions. Modifying the surface of PHEMA-based hydrogels with different surface charges showed that, after implantation into the hemisected spinal cord, hydrogels with positively charged functional groups promoted connective tissue infiltration and extended axonal ingrowth into the hydrogel scaffold (Hejcl et al. 2009). It is reported that biomaterial surfaces are modified by the pre-coating or immobilization of full-length ECM proteins or their functional protein sequences for integrin receptor binding sites, such as those from fibronectin, laminin, and collagens (Gunn et al. 2005). PHEMA hydrogels modified with the laminin-derived peptide (Ile-Lys-Val-Ala-Val, IKVAV) resulted in improved cell attachment and spread, as well as the improved differentiation of neural fetal precursor cells (Kubinova et al. 2010). Recently, Stupp and coworkers have designed and synthesized a broad range of peptide amphiphiles to create new self-assembling biomaterials with structural features of amphiphilic surfactants and the functions of bioactive peptides (Cui et al. 2010). One of these systems incorporates the sequence IKVAV and has been used to prepare in situ forming hydrogels to promote SCI regeneration (Tysseling-Mattiace et al. 2008).

The scaffolds should be highly porous with interconnected pores to allow fluid and nutrient exchange as well as neovascularization of the implant, thus creating a permissive environment for axonal growth. Cells can be either adhered to the pores of the scaffolds or incorporated as a cell suspension in a hydrogel that can also be combined with a more rigid scaffold that is supplemented with therapeutic agents, such as neurotrophins. Such matrices may enhance cell survival after transplantation and promote differentiation into desired phenotypes. Sakiyama-Elbert and coworkers report that combination of growth factors and fibrin scaffold enhanced cell survival of embryonic stem cell-derived neural progenitors present and increased neuronal differentiation after transplantation in the treated spinal cords and exhibited an increase in behavioral function (Johnson et al. 2010). Macroporous PHPMA hydrogels modified with the RGD peptidic sequence (Arg-Gly-Asp) are shown to have ability to promote tissue regeneration, axonal ingrowth, and angiogenesis when implanted into SCI (Woerly et al. 2001). Hejcl et al. (2010) seeded PHPMA-RGD hydrogels with MSCs and implanted into rat chronic spinal cord lesions. The hydrogels successfully bridged the spinal cord cavity and provided a scaffold for tissue regeneration. Behavioral analysis showed a statistically significant improvement of motor and sensory scores.

Biomaterials are receiving increased attention, either as vehicles for the cells or as vectors of therapeutic agents for the spinal cord regenerative process. In addition, the advantages of using a biomaterial-based scaffold to promote axonal regeneration are becoming more apparent in the treatment of chronic spinal cord lesions, when the cystic cavity is already developed and cell transplantation alone is not sufficient to promote tissue regeneration.

20.4 Mechanisms Underlying Cellular Therapy

Transplanted stem cells can exert plastic changes by replacing the missing cells, remyelinating denuded axons, scaffolding for axons, promoting neurite outgrowth, and secreting trophic factors which help reducing cell death and axonal dieback.

20.4.1 Cell Replacement

The fact that pluripotent-derived stem cells can differentiate into both neurons and oligodendrocytes in vitro and in vivo demonstrates the mechanism of cell replacement strategies for spinal cord repair [as reviewed in (Ruff and Fehlings 2010)]. NSPCs or the committed precursors can home to areas of damage, proliferate, and differentiate into the missing cell type resulting in cell replacement (Yan et al. 2007; Hwang et al. 2009). OPCs is most easily isolated and grown in large quantities and shares the characteristics of stem cells. Replacement of damaged oligodendrocytes by new OPCs has been shown to have benefit for remyelinating denuded axons in SCI (McTigue and Tripathi 2008). It should be noted that, although neuronal cell replacement is a feasible option for other conditions (e.g., secretory cell replacement), long descending motor tracts that commingle with pain fibers limit the clinical application of stem cell-derived neuronal replacement strategies in SCI.

20.4.2 Neurotrophic Factor

Perhaps, the most important role of stem and associated cell transplantation for SCI lies not with physical support, but in the considerable capacity to become trophic mediators in the neuronal regenerative responses. The transplanted stem cells secrete key intermediates that can enhance neuronal survival, axonal sparing, plasticity, and regeneration (Teng et al. 2006). Neurotrophic modulation is purportedly the primary mechanism of BMSCs, which are known to secrete BDNF, NGF, VEGF, TGF-β, IGF1, BNP, and SCF1 (Parr et al. 2007; Mahmood et al. 2004; Hu et al. 2005; Crigler et al. 2006). Evidences suggest that NGF may induce hyperalgesia by mediating plasticity of nociceptive circuitry (Ruiz and Banos 2009). In addition, cytokine growth factors and TGF-β family members, including GDNF, can support axonal outgrowth in SCI models, each playing a distinct but overlapping role (Jones et al. 2001; Glazova et al. 2009; Peng et al. 2003). FGF isoforms have also been linked into survival and neurite outgrowth in certain neuronal subtypes (Pataky et al. 2000). Ciliary neurotrophic factor (CNTF), secreted by glial cells or BMSCs, can promote neurite outgrowth and oligodendrocyte differentiation and OPC/neuronal survival (Tebar et al. 2008; Sleeman et al. 2000; Cao et al. 2010). BMSCs induced to become Schwann-like cells showed additional upregulation of hepatocyte growth factor (HGF), VEGF, BDNF, and NGF, promoting neuritogenesis and neuron preservation in cocultures and ex vivo preparations (Park et al. 2010). As with other biological instruments, stem cells possess unique characteristics that act in a context-specific manner as boon and burden, with myelination capacity, chemotrophic potential, and endogenous presence varying between cells. Therefore, in vitro mechanisms of cellular actions must be evaluated by in vivo modeling systems.

20.4.3 Axonal Regeneration

Several exogenous cell types can greatly increase axonal regeneration and fiber density in the injured spinal cord, such as schwann cell and olfactory ensheathing cells (Ramon-Cueto et al. 1998; Guest et al. 1997). However, neither cell type is able to significantly affect corticospinal fiber regeneration, or crossing of lesion sites, especially in the chronic stage. Transplanted NSPCs demonstrate moderate increases in axonal sprouting and regeneration, including the corticospinal tract (Karimi-Abdolrezaee et al. 2010; Hofstetter et al. 2005). Injection of BMSCs may increase axon density in the lesion site (Lu et al. 2007). In some cases, increased axonal plasticity does not translate to functional recovery of locomotion. This is likely due to tissue sparing as transplanted cells do not integrate well with host tissue.

20.4.4 Environmental Modification

After SCI, the inhibitory microenvironment, including the inflammatory response, glial cell activation, and gliosis appear and largely act as a negative obstruction to any form of cellular therapy. Environmental modification and tissue sparing are potentially more efficacious approaches to increase plastic repair. Gliosis and astrocyte reactivity has been shown to be reduced with transplantation of

BMSCs in acute and chronic SCI (Lu et al. 2007). Nearly, all cell transplantation approaches have demonstrated the tissue sparing and increases in gray and white matter. The most pronounced increases in white matter are with NSPCs and glial-restricted progenitors (Karimi-Abdolrezaee et al. 2010; Mitsui et al. 2005). NSPCs and OPCs can efficiently remyelinate host axons with organized, compact myelin in subacutic or chronic stage lesion, but only if the glial scar is degraded (Keirstead et al. 2005; Karimi-Abdolrezaee et al. 2010; Eftekharpour et al. 2007).

20.5 Clinical Translation

With progress in vivo studies, scientists and surgeons have tried to conduct clinical trials to explore the therapeutic effects of cell transplantation on spinal cord patients using various cell types and strategies on different kinds of SCI patients. Premature clinical trials of stem cell therapy carry the risk of creating more harm than benefit. Furthermore, any negative results or serious adverse effects could potentially damage the long-term development of the field. Nowadays, many uncontrolled and scientifically dubious stem cell therapies have been offered worldwide for SCI with little or no rationale in many cases (Blight et al. 2009). Therefore, so far, no safe and effective cell therapies for SCI patients have been achieved and little evidence has been obtained of clinically significant benefits. The clinical trials and challenges of stem cell-based transplantation strategies for the treatment of SCI are briefly described as followings.

20.5.1 Current Status

In 2009, Geron Corporation was approved by US FDA to run the first clinical trial of stem cell therapy for SCI. In the next year, Geron Corporation initiated the clinical trial (Phase I) to test the safety of human ES cell-derived OPCs, GRNOPC1, within patients who were suffering from complete thoracic-level paraplegia with the loss of motor and sensory function. Two million GRNOPC1 was administered into the lesion site within 14 days of injury. To date, there are no serious adverse events reported in the long-term follow-up. Furthermore, they plan to test the safety in patients with a higher cell concentration with 20 million cells in the next step. Unfortunately, this SCI stem cell research program was announced to be ended in 2011 because of financial reasons. In the Phase I clinical trial, So far, no further safety issues and therapeutic improvements were reported, although Geron was looking mainly at the safety profile at this stage. Bretzner et al. (2011) raised a comment to argue the target population selection in the clinical trial of GRNOPC1 and suggest a more detailed criteria for selecting patients for different study purposes: (1) chronic complete SCI patients for a safety trial, (2) subacute incomplete SCI patients for an efficacy trial, and (3) perhaps primary progressive multiple sclerosis patients for a combined safety and efficacy trial. They proposed that the chronic completed SCI patients may be a more preferable target population than subacute complete SCI patients in the phase I clinical trial, because simultaneous recovery may occur in some subacute complete SCI patients, which may confound results. In addition, the chronic complete lesion site may ensure a stable microenvironment after cell transplantation in which to assess the safety of transplanted cells.

Geffner et al. (2008) reported their investigation about the administration of autologous BMSCs into eight patients with SCI (four acute and four chronic) in 2008. Some improvements in bladder function and changes in spinal cord were observed during the 2-year follow-up. The safety of transplantation of BMSCs in patients with SCI have been tested by other groups, and all indicate that administration of these cells does not cause any adverse effects (Sykova et al. 2006; Pal et al. 2009). In addition, Yoon et al. (2007) studied the effects of autologous human bone marrow cell transplantation in combination with the administration of granulocyte-macrophage colony-stimulating factor to 35 patients with complete SCI. No serious complications were reported (Yoon et al. 2007).

In 2012, Park et al. (2012) and Karamouzian et al. (2012) reported two clinical trials for SCI by using MSCs transplantation. Although some improvements were noticed in some patients, the therapeutic effects of MSCs transplantation have not been established in human SCI patients. In the study of Park et al., MSCs were administered three times during the course of the study on 10 traumatic cervical SCI patients with severe paralysis. The first 8×10^6 autologous MSCs were directly injected into the intradural space. After 4 and 8 weeks, another 5×10^6 MSCs, each time, were injected into the spinal cord above the lesion cavity and the cavity, respectively. Finally, 3 patient exhibited improvements of daily living activities, increased motor power of the upper extremities, shrinked lesion cavity size, and electrophysiological improvement. Various partial improvements were observed in the rest of 7 patients. They also claimed that there are no permanent complications associated with MSCs transplantation observed. Karamouzian et al. (2012) transplanted autologous MSCs into the cerebrospinal fluid via lumbar puncture for 7 SCI patients with complete thoracic injuries. As they observed, 5 of 11 patients in the MSCs transplantation group and 3 of 20 patients in the control group showed marked function recovery; however, the differences between the two groups were not significant. On the other hand, no adverse reaction and complications in both groups were experienced by patients, which may indicate the safety of intrathecal administration of MSCs in human patients

A large number of clinical trials are underway in China involving transplantation of mononuclear cells derived from umbilical cord blood, which includes a population of stem cells (see http://www.stemcellschina.com/). And, so far, no convinced results from these trials have yet been published. The SCI Network USA (see http://www.scinetusa.org/) is also planning a study of umbilical cord blood mononuclear cell transplants combined with lithium treatment in people with chronic traumatic SCI.

20.5.2 Challenges for Clinical Translation of Stem Cell Therapy for SCI

Clinical translation of stem cell therapy for SCI still faces enormous challenges, although much has been learned from previous SCI and other trials. Most clinical trials are at phase I conducted with small numbers of patients without controls, and thus, assessment of efficacy is not possible. Enrolling sufficient numbers of SCI patients for clinical trials is very difficult because of differing severity and level of injury, age of patient, and associated injuries. Generally, for cell replacement trials, the target SCI population is ASIA A patients to avoid causing further damage, but these patients have minimal ability to recover, and demonstration of effectiveness is impaired due to insensitive outcome measures. Some of the obstacles that the Geron trial encountered were the need to screen large numbers of patients, the need of a large number of cells (2 million cells), and a relatively long waiting time to evaluate clinical efficacy (more than 6 months). The process of creating clinically acceptable ES cell- or iPS-derived cells is costly. Recent progress in direct conversion methods indicates great potential for clinical stem cell therapy, but more work is needed. The complexities of attenuating the tissue damage and secondary complications due to trauma and reconstructing the cytoarchitecture of the injured spinal cord are very challenging, but hopefully. The rapid advances being made in stem cell biology will result in effective experimental and clinical trials of stem cell therapy for SCI.

20.6 Conclusions and Perspectives

Current cell-based approaches are aimed at (1) regeneration of new neurons that die within the first minutes to days after trauma; (2) providing a source of cells to promote remyelination; and (3) delivering trophic molecules that can promote cellular protection and plasticity. Many of the strategies that achieve functional recovery take a multi-component approach. These strategies can be successful because one cell line can provide secreted factors and trophic support for the other cells. Protein therapeutic strategies, such as the use of growth factors, enzymes, and neutralizing antibodies, help promote regeneration after SCI. Damaged axons may regenerate through lesions if intrinsic growth mechanisms are bolstered, and extrinsic barriers are diminished. For example, Chondroitinase ABC is employed to degrade the glial scar to facilitate lesion entry. Cells genetically modified to secrete growth factors have the advantage of providing continuous influx of proteins that facilitate local niche suitable for cell survival, axonal extending, and specific cell-type differentiation.

Overall, cell therapy combined with other regeneration promoting strategies holds the most promise for restoring function from SCI. Many of these strategies have demonstrated efficacy of stem cell-based therapy in preclinical trials.

For successful translation to the clinic, these strategies should be tested in chronic models of SCI and functional recovery should be assessed using a variety of tests. Additionally, the development of embryonic and adult stem cell technology will require further research to produce patient-specific stem cells and culture methods necessary for clinical trials. It will be benefit to make the candidate stem cells easily accessible for hospitals, and this ideally includes the ability to bank cells and to be able to distribute them as an off-the-shelf product. The more simplified the cell handling techniques required by the clinic, the greater potential it has to become a widely used therapy. On the other hand, strategies to affect endogenous neural stem cells in the adult spinal cord appear to be a more distant scenario. Efforts must be put onto acquiring a better knowledge of this stem cell reservoir. However, this is an exciting line of research that ultimately may result in pharmacological therapies circumventing the need for invasive and allogeneic strategies.

References

Abematsu M, et al. Neurons derived from transplanted neural stem cells restore disrupted neuronal circuitry in a mouse model of spinal cord injury. J Clin Invest. 2010;120(9):3255–66.

Ackery A, Tator C, Krassioukov A. A global perspective on spinal cord injury epidemiology. J Neurotrauma. 2004;21(10):1355–70.

Akesson E, et al. Long-term culture and neuronal survival after intraspinal transplantation of human spinal cord-derived neurospheres. Physiol Behav. 2007;92(1–2):60–6.

Barnabe-Heider F, Frisen J. Stem cells for spinal cord repair. Cell Stem Cell. 2008;3(1):16–24.

Barnabe-Heider F, et al. Origin of new glial cells in intact and injured adult spinal cord. Cell Stem Cell. 2010;7(4):470–82.

Ben-David U, Benvenisty N. The tumorigenicity of human embryonic and induced pluripotent stem cells. Nat Rev Cancer. 2011;11(4):268–77.

Bhattarai N, et al. Natural-synthetic polyblend nanofibers for biomedical applications. Adv Mater. 2009;21(27):2792–7.

Blight A, et al. Position statement on the sale of unproven cellular therapies for spinal cord injury: the international campaign for cures of spinal cord injury paralysis. Spinal Cord. 2009;47(9):713–4.

Bretzner F, et al. Target populations for first-in-human embryonic stem cell research in spinal cord injury. Cell Stem Cell. 2011;8(5):468–75.

Brustle O, et al. Embryonic stem cell-derived glial precursors: a source of myelinating transplants. Science. 1999;285(5428):754–6.

Cao QL, et al. Pluripotent stem cells engrafted into the normal or lesioned adult rat spinal cord are restricted to a glial lineage. Exp Neurol. 2001;167(1):48–58.

Cao QL, et al. Differentiation of engrafted neuronal-restricted precursor cells is inhibited in the traumatically injured spinal cord. Exp Neurol. 2002;177(2):349–59.

Cao Q, et al. Functional recovery in traumatic spinal cord injury after transplantation of multineurotrophin-expressing glial-restricted precursor cells. J Neurosci. 2005;25(30):6947–57.

Cao Q, et al. Transplantation of ciliary neurotrophic factor-expressing adult oligodendrocyte precursor cells promotes remyelination and functional recovery after spinal cord injury. J Neurosci. 2010;30(8):2989–3001.

Chen J, et al. Cell adhesion molecule l1-transfected embryonic stem cells with enhanced survival support regrowth of corticospinal tract axons in mice after spinal cord injury. J Neurotrauma. 2005;22(8):896–906.

Chiba S, et al. Transplantation of motoneuron-enriched neural cells derived from mouse embryonic stem cells improves motor function of hemiplegic mice. Cell Transplant. 2003;12(5):457–68.

Ciardelli G, Chiono V. Materials for peripheral nerve regeneration. Macromol Biosci. 2006;6(1):13–26.

Crigler L, et al. Human mesenchymal stem cell subpopulations express a variety of neuro-regulatory molecules and promote neuronal cell survival and neuritogenesis. Exp Neurol. 2006;198(1):54–64.

Cui H, Webber MJ, Stupp SI. Self-assembly of peptide amphiphiles: from molecules to nano-structures to biomaterials. Biopolymers. 2010;94(1):1–18.

Cummings BJ, et al. Human neural stem cells differentiate and promote locomotor recovery in spinal cord-injured mice. Proc Natl Acad Sci USA. 2005;102(39):14069–74.

Dasari VR, et al. Axonal remyelination by cord blood stem cells after spinal cord injury. J Neurotrauma. 2007;24(2):391–410.

Dasari VR, et al. Umbilical cord blood stem cell mediated downregulation of fas improves functional recovery of rats after spinal cord injury. Neurochem Res. 2008;33(1):134–49.

De Coppi P, et al. Isolation of amniotic stem cell lines with potential for therapy. Nat Biotechnol. 2007;25(1):100–6.

De Laporte L, Yan AL, Shea LD. Local gene delivery from ECM-coated poly(lactide-co-glycolide) multiple channel bridges after spinal cord injury. Biomaterials. 2009;30(12):2361–8.

Deng W, et al. Adult-born hippocampal dentate granule cells undergoing maturation modulate learning and memory in the brain. J Neurosci. 2009;29(43):13532–42.

Deshpande DM, et al. Recovery from paralysis in adult rats using embryonic stem cells. Ann Neurol. 2006;60(1):32–44.

Dezawa M, et al. Specific induction of neuronal cells from bone marrow stromal cells and application for autologous transplantation. J Clin Invest. 2004;113(12):1701–10.

Eftekharpour E, et al. Myelination of congenitally dysmyelinated spinal cord axons by adult neural precursor cells results in formation of nodes of Ranvier and improved axonal conduction. J Neurosci. 2007;27(13):3416–28.

Erceg S, et al. Efficient differentiation of human embryonic stem cells into functional cerebellar-like cells. Stem Cells Dev. 2010a;19(11):1745–56.

Erceg S, et al. Transplanted oligodendrocytes and motoneuron progenitors generated from human embryonic stem cells promote locomotor recovery after spinal cord transection. Stem Cells. 2010b;28(9):1541–9.

Evans MJ, Kaufman MH. Establishment in culture of pluripotential cells from mouse embryos. Nature. 1981;292(5819):154–6.

Farry A, Baxter D, editors. The incidence and prevalence of spinal cord injury in Canada: overview and estimates based on current evidence. Canada: Rick Hansen Institute and Urban Futures Institute; 2010. p. 1–49.

Fukuchi Y, et al. Human placenta-derived cells have mesenchymal stem/progenitor cell potential. Stem Cells. 2004;22(5):649–58.

Gage FH. Mammalian neural stem cells. Science. 2000;287(5457):1433–8.

Geffner LF, et al. Administration of autologous bone marrow stem cells into spinal cord injury patients via multiple routes is safe and improves their quality of life: comprehensive case studies. Cell Transplant. 2008;17(12):1277–93.

Glazova M, et al. Pre-differentiated embryonic stem cells promote neuronal regeneration by cross-coupling of BDNF and IL-6 signaling pathways in the host tissue. J Neurotrauma. 2009;26(7):1029–42.

Gonzalez F, Boue S, Izpisua Belmonte JC. Methods for making induced pluripotent stem cells: reprogramming a la carte. Nat Rev Genet. 2011;12(4):231–42.

GrandPre T, Li S, Strittmatter SM. Nogo-66 receptor antagonist peptide promotes axonal regeneration. Nature. 2002;417(6888):547–51.

Gritti A, et al. Multipotential stem cells from the adult mouse brain proliferate and self-renew in response to basic fibroblast growth factor. J Neurosci. 1996;16(3):1091–100.

Guest JD, et al. The ability of human Schwann cell grafts to promote regeneration in the transected nude rat spinal cord. Exp Neurol. 1997;148(2):502–22.
Gunn JW, Turner SD, Mann BK. Adhesive and mechanical properties of hydrogels influence neurite extension. J Biomed Mater Res A. 2005;72(1):91–7.
Hamada M, et al. Introduction of the MASH1 gene into mouse embryonic stem cells leads to differentiation of motoneuron precursors lacking Nogo receptor expression that can be applicable for transplantation to spinal cord injury. Neurobiol Dis. 2006;22(3):509–22.
Han SS, et al. Grafted lineage-restricted precursors differentiate exclusively into neurons in the adult spinal cord. Exp Neurol. 2002;177(2):360–75.
Han SS, et al. Transplantation of glial-restricted precursor cells into the adult spinal cord: survival, glial-specific differentiation, and preferential migration in white matter. Glia. 2004;45(1):1–16.
Hawryluk GW, et al. In vitro characterization of trophic factor expression in neural precursor cells. Stem Cells Dev. 2012;21(3):432–47.
Hejcl A, et al. Biocompatible hydrogels in spinal cord injury repair. Physiol Res. 2008;57(Suppl 3):S121–32.
Hejcl A, et al. Macroporous hydrogels based on 2-hydroxyethyl methacrylate. Part 6: 3D hydrogels with positive and negative surface charges and polyelectrolyte complexes in spinal cord injury repair. J Mater Sci Mater Med. 2009;20(7):1571–7.
Hejcl A, et al. HPMA-RGD hydrogels seeded with mesenchymal stem cells improve functional outcome in chronic spinal cord injury. Stem Cells Dev. 2010;19(10):1535–46.
Hill CE, et al. Acute transplantation of glial-restricted precursor cells into spinal cord contusion injuries: survival, differentiation, and effects on lesion environment and axonal regeneration. Exp Neurol. 2004;190(2):289–310.
Himes BT, et al. Recovery of function following grafting of human bone marrow-derived stromal cells into the injured spinal cord. Neurorehabil Neural Repair. 2006;20(2):278–96.
Hofstetter CP, et al. Marrow stromal cells form guiding strands in the injured spinal cord and promote recovery. Proc Natl Acad Sci USA. 2002;99(4):2199–204.
Hofstetter CP, et al. Allodynia limits the usefulness of intraspinal neural stem cell grafts; directed differentiation improves outcome. Nat Neurosci. 2005;8(3):346–53.
Horky LL, et al. Fate of endogenous stem/progenitor cells following spinal cord injury. J Comp Neurol. 2006;498(4):525–38.
Horstmann E. The fiber glia of selacean brain. Z Zellforsch Mikrosk Anat. 1954;39(6):588–617.
Howard MJ, et al. Transplantation of apoptosis-resistant embryonic stem cells into the injured rat spinal cord. Somatosens Mot Res. 2005;22(1–2):37–44.
Hu DZ, et al. Upregulated gene expression of local brain-derived neurotrophic factor and nerve growth factor after intracisternal administration of marrow stromal cells in rats with traumatic brain injury. Chin J Traumatol. 2005;8(1):23–6.
Hu SL, et al. Functional recovery in acute traumatic spinal cord injury after transplantation of human umbilical cord mesenchymal stem cells. Crit Care Med. 2010;38(11):2181–9.
Hugnot JP, Franzen R. The spinal cord ependymal region: a stem cell niche in the caudal central nervous system. Front Biosci (Landmark Ed). 2011;16:1044–59.
Hwang DH, et al. Transplantation of human neural stem cells transduced with Olig2 transcription factor improves locomotor recovery and enhances myelination in the white matter of rat spinal cord following contusive injury. BMC Neurosci. 2009;10:117.
Iwanami A, et al. Transplantation of human neural stem cells for spinal cord injury in primates. J Neurosci Res. 2005;80(2):182–90.
Johnson PJ, et al. Tissue-engineered fibrin scaffolds containing neural progenitors enhance functional recovery in a subacute model of SCI. Soft Matter. 2010;6(20):5127–37.
Jones LL, et al. Neurotrophic factors, cellular bridges and gene therapy for spinal cord injury. J Physiol. 2001;533(Pt 1):83–9.
Kang SK, et al. Autologous adipose tissue-derived stromal cells for treatment of spinal cord injury. Stem Cells Dev. 2006;15(4):583–94.

Karahuseyinoglu S, et al. Biology of stem cells in human umbilical cord stroma: in situ and in vitro surveys. Stem Cells. 2007;25(2):319–31.

Karamouzian S, et al. Clinical safety and primary efficacy of bone marrow mesenchymal cell transplantation in subacute spinal cord injured patients. Clin Neurol Neurosurg. 2012;114(7):935–9.

Karimi-Abdolrezaee S, et al. Delayed transplantation of adult neural precursor cells promotes remyelination and functional neurological recovery after spinal cord injury. J Neurosci. 2006;26(13):3377–89.

Karimi-Abdolrezaee S, et al. Synergistic effects of transplanted adult neural stem/progenitor cells, chondroitinase, and growth factors promote functional repair and plasticity of the chronically injured spinal cord. J Neurosci. 2010;30(5):1657–76.

Karumbayaram S, et al. Directed differentiation of human-induced pluripotent stem cells generates active motor neurons. Stem Cells. 2009;27(4):806–11.

Kawasaki H, et al. Induction of midbrain dopaminergic neurons from ES cells by stromal cell-derived inducing activity. Neuron. 2000;28(1):31–40.

Keirstead HS, et al. Human embryonic stem cell-derived oligodendrocyte progenitor cell transplants remyelinate and restore locomotion after spinal cord injury. J Neurosci. 2005;25(19):4694–705.

Kubinova S, et al. The use of superporous Ac-CGGASIKVAVS-OH-modified PHEMA scaffolds to promote cell adhesion and the differentiation of human fetal neural precursors. Biomaterials. 2010;31(23):5966–75.

Lamba DA, et al. Efficient generation of retinal progenitor cells from human embryonic stem cells. Proc Natl Acad Sci USA. 2006;103(34):12769–74.

Lee J, et al. Migration and differentiation of nuclear fluorescence-labeled bone marrow stromal cells after transplantation into cerebral infarct and spinal cord injury in mice. Neuropathology. 2003;23(3):169–80.

Lee BB, et al. The global map for traumatic spinal cord injury epidemiology: update 2011, global incidence rate. Spinal Cord. 2014;52(2):110–6.

Li S, et al. Blockade of Nogo-66, myelin-associated glycoprotein, and oligodendrocyte myelin glycoprotein by soluble Nogo-66 receptor promotes axonal sprouting and recovery after spinal injury. J Neurosci. 2004;24(46):10511–20.

Li XJ, et al. Directed differentiation of ventral spinal progenitors and motor neurons from human embryonic stem cells by small molecules. Stem Cells. 2008;26(4):886–93.

Li XJ, et al. Coordination of sonic hedgehog and Wnt signaling determines ventral and dorsal telencephalic neuron types from human embryonic stem cells. Development. 2009;136(23):4055–63.

Li J, et al. The epidemiological survey of acute traumatic spinal cord injury (ATSCI) of 2002 in Beijing municipality. Spinal Cord. 2011;49(7):777–82.

Lois C, Alvarez-Buylla A. Long-distance neuronal migration in the adult mammalian brain. Science. 1994;264(5162):1145–8.

Lois C, Garcia-Verdugo JM, Alvarez-Buylla A. Chain migration of neuronal precursors. Science. 1996;271(5251):978–81.

Lopez-Gonzalez R, Kunckles P, Velasco I. Transient recovery in a rat model of familial amyotrophic lateral sclerosis after transplantation of motor neurons derived from mouse embryonic stem cells. Cell Transplant. 2009;18(10):1171–81.

Lu P, et al. Neural stem cells constitutively secrete neurotrophic factors and promote extensive host axonal growth after spinal cord injury. Exp Neurol. 2003;181(2):115–29.

Lu P, Jones LL, Tuszynski MH. Axon regeneration through scars and into sites of chronic spinal cord injury. Exp Neurol. 2007;203(1):8–21.

Mahmood A, Lu D, Chopp M. Intravenous administration of marrow stromal cells (MSCs) increases the expression of growth factors in rat brain after traumatic brain injury. J Neurotrauma. 2004;21(1):33–9.

Marichal N, et al. Enigmatic central canal contacting cells: immature neurons in "standby mode"? J Neurosci. 2009;29(32):10010–24.

Markakis EA, Gage FH. Adult-generated neurons in the dentate gyrus send axonal projections to field CA3 and are surrounded by synaptic vesicles. J Comp Neurol. 1999;406(4):449–60.

McDonald JW, et al. Transplanted embryonic stem cells survive, differentiate and promote recovery in injured rat spinal cord. Nat Med. 1999;5(12):1410–2.

McTigue DM, Tripathi RB. The life, death, and replacement of oligodendrocytes in the adult CNS. J Neurochem. 2008;107(1):1–19.

Meletis K, et al. Spinal cord injury reveals multilineage differentiation of ependymal cells. PLoS Biol. 2008;6(7):e182.

Menezes AH, et al, editors. Principles of spinal surgery. New York: McGraw Hill Healthcare; 1996.

Mitsui T, et al. Transplantation of neuronal and glial restricted precursors into contused spinal cord improves bladder and motor functions, decreases thermal hypersensitivity, and modifies intraspinal circuitry. J Neurosci. 2005;25(42):9624–36.

Miura K, et al. Variation in the safety of induced pluripotent stem cell lines. Nat Biotechnol. 2009;27(8):743–5.

Morshead CM, et al. Neural stem cells in the adult mammalian forebrain: a relatively quiescent subpopulation of subependymal cells. Neuron. 1994;13(5):1071–82.

Mothe AJ, Tator CH. Proliferation, migration, and differentiation of endogenous ependymal region stem/progenitor cells following minimal spinal cord injury in the adult rat. Neuroscience. 2005;131(1):177–87.

National Spinal Cord Injury Statistical Center and Others. Spinal cord injury facts and figures at a glance. J Spinal Cord Med. 2014;37(2):243–4.

Ning GZ, et al. Epidemiology of traumatic spinal cord injury in Tianjin, China. Spinal Cord. 2011;49(3):386–90.

Nistor GI, et al. Human embryonic stem cells differentiate into oligodendrocytes in high purity and myelinate after spinal cord transplantation. Glia. 2005;49(3):385–96.

Nori S, et al. Grafted human-induced pluripotent stem-cell-derived neurospheres promote motor functional recovery after spinal cord injury in mice. Proc Natl Acad Sci USA. 2011;108(40):16825–30.

Ogawa Y, et al. Transplantation of in vitro-expanded fetal neural progenitor cells results in neurogenesis and functional recovery after spinal cord contusion injury in adult rats. J Neurosci Res. 2002;69(6):925–33.

Oh JS, et al. Transplantation of an adipose stem cell cluster in a spinal cord injury. NeuroReport. 2012;23(5):277–82.

Ohori Y, et al. Growth factor treatment and genetic manipulation stimulate neurogenesis and oligodendrogenesis by endogenous neural progenitors in the injured adult spinal cord. J Neurosci. 2006;26(46):11948–60.

Okita K, Ichisaka T, Yamanaka S. Generation of germline-competent induced pluripotent stem cells. Nature. 2007;448(7151):313–7.

Osakada F, et al. Toward the generation of rod and cone photoreceptors from mouse, monkey and human embryonic stem cells. Nat Biotechnol. 2008;26(2):215–24.

Pal R, et al. Ex vivo-expanded autologous bone marrow-derived mesenchymal stromal cells in human spinal cord injury/paraplegia: a pilot clinical study. Cytotherapy. 2009;11(7):897–911.

Palmer TD, Takahashi J, Gage FH. The adult rat hippocampus contains primordial neural stem cells. Mol Cell Neurosci. 1997;8(6):389–404.

Park JH, et al. Long-term results of spinal cord injury therapy using mesenchymal stem cells derived from bone marrow in humans. Neurosurgery. 2012;70(5):1238–47; discussion 1247.

Park HW, et al. Human mesenchymal stem cell-derived Schwann cell-like cells exhibit neurotrophic effects, via distinct growth factor production, in a model of spinal cord injury. Glia. 2010;58(9):1118–32.

Parr AM, Tator CH, Keating A. Bone marrow-derived mesenchymal stromal cells for the repair of central nervous system injury. Bone Marrow Transplant. 2007;40(7):609–19.

Parras CM, et al. Divergent functions of the proneural genes Mash1 and Ngn2 in the specification of neuronal subtype identity. Genes Dev. 2002;16(3):324–38.
Pataky DM, et al. Fibroblast growth factor treatment produces differential effects on survival and neurite outgrowth from identified bulbospinal neurons in vitro. Exp Neurol. 2000;163(2):357–72.
Pego AP, et al. Copolymers of trimethylene carbonate and epsilon-caprolactone for porous nerve guides: synthesis and properties. J Biomater Sci Polym Ed. 2001;12(1):35–53.
Peng HB, et al. Differential effects of neurotrophins and schwann cell-derived signals on neuronal survival/growth and synaptogenesis. J Neurosci. 2003;23(12):5050–60.
Perrier AL, et al. Derivation of midbrain dopamine neurons from human embryonic stem cells. Proc Natl Acad Sci USA. 2004;101(34):12543–8.
Prockop DJ. Marrow stromal cells as stem cells for nonhematopoietic tissues. Science. 1997;276(5309):71–4.
Puri MC, Nagy A. Concise review: embryonic stem cells versus induced pluripotent stem cells: the game is on. Stem Cells. 2012;30(1):10–4.
Rafols JA, Goshgarian HG. Spinal tanycytes in the adult rat: a correlative Golgi gold-toning study. Anat Rec. 1985;211(1):75–86.
Ramon-Cueto A, et al. Long-distance axonal regeneration in the transected adult rat spinal cord is promoted by olfactory ensheathing glia transplants. J Neurosci. 1998;18(10):3803–15.
Reier PJ. Cellular transplantation strategies for spinal cord injury and translational neurobiology. NeuroRx. 2004;1(4):424–51.
Reubinoff BE, et al. Neural progenitors from human embryonic stem cells. Nat Biotechnol. 2001;19(12):1134–40.
Reynolds BA, Weiss S. Generation of neurons and astrocytes from isolated cells of the adult mammalian central nervous system. Science. 1992;255(5052):1707–10.
Rick Hansen Spinal Cord Injury Registry. Spinal Cord Injury Facts and Statistics, Vancouver, British Columbia, Canada. 2006. p. 1–11.
Ruff CA, Fehlings MG. Neural stem cells in regenerative medicine: bridging the gap. Panminerva Med. 2010;52(2):125–47.
Ruiz G, Banos JE. Heat hyperalgesia induced by endoneurial nerve growth factor and the expression of substance P in primary sensory neurons. Int J Neurosci. 2009;119(2):185–203.
Sabourin JC, et al. A mesenchymal-like ZEB1(+) niche harbors dorsal radial glial fibrillary acidic protein-positive stem cells in the spinal cord. Stem Cells. 2009;27(11):2722–33.
Sankar V, Muthusamy R. Role of human amniotic epithelial cell transplantation in spinal cord injury repair research. Neuroscience. 2003;118(1):11–7.
Schultz SS. Adult stem cell application in spinal cord injury. Curr Drug Targets. 2005;6(1):63–73.
Seitz R, Lohler J, Schwendemann G. Ependyma and meninges of the spinal cord of the mouse. A light-and electron-microscopic study. Cell Tissue Res. 1981;220(1):61–72.
Seki T, Arai Y. Highly polysialylated NCAM expression in the developing and adult rat spinal cord. Brain Res Dev Brain Res. 1993;73(1):141–5.
Sharp J, et al. Human embryonic stem cell-derived oligodendrocyte progenitor cell transplants improve recovery after cervical spinal cord injury. Stem Cells. 2010;28(1):152–63.
Sleeman MW, et al. The ciliary neurotrophic factor and its receptor, CNTFR alpha. Pharm Acta Helv. 2000;74(2–3):265–72.
Soundararajan P, et al. Motoneurons derived from embryonic stem cells express transcription factors and develop phenotypes characteristic of medial motor column neurons. J Neurosci. 2006;26(12):3256–68.
Soundararajan P, et al. Easy and rapid differentiation of embryonic stem cells into functional motoneurons using sonic hedgehog-producing cells. Stem Cells. 2007;25(7):1697–706.
Stemple DL, Anderson DJ. Isolation of a stem cell for neurons and glia from the mammalian neural crest. Cell. 1992;71(6):973–85.

Stoeckel ME, et al. Cerebrospinal fluid-contacting neurons in the rat spinal cord, a gamma-aminobutyric acidergic system expressing the P2X2 subunit of purinergic receptors, PSA-NCAM, and GAP-43 immunoreactivities: light and electron microscopic study. J Comp Neurol. 2003;457(2):159–74.
Su H, et al. Optimal time point for neuronal generation of transplanted neural progenitor cells in injured spinal cord following root avulsion. Cell Transplant. 2011;20(2):167–76.
Sykova E, Jendelova P. Magnetic resonance tracking of implanted adult and embryonic stem cells in injured brain and spinal cord. Ann NY Acad Sci. 2005;1049:146–60.
Sykova E, et al. Autologous bone marrow transplantation in patients with subacute and chronic spinal cord injury. Cell Transplant. 2006;15(8–9):675–87.
Takahashi K, Yamanaka S. Induction of pluripotent stem cells from mouse embryonic and adult fibroblast cultures by defined factors. Cell. 2006;126(4):663–76.
Tao O, et al. Efficient generation of mature cerebellar Purkinje cells from mouse embryonic stem cells. J Neurosci Res. 2010;88(2):234–47.
Tator CH. Update on the pathophysiology and pathology of acute spinal cord injury. Brain Pathol. 1995;5(4):407–13.
Tator CH. Review of treatment trials in human spinal cord injury: issues, difficulties, and recommendations. Neurosurgery. 2006;59(5):957–82; discussion 982–7.
Tebar LA, et al. Deletion of the mouse RegIIIbeta (Reg2) gene disrupts ciliary neurotrophic factor signaling and delays myelination of mouse cranial motor neurons. Proc Natl Acad Sci USA. 2008;105(32):11400–5.
Temple S. Division and differentiation of isolated CNS blast cells in microculture. Nature. 1989;340(6233):471–3.
Teng YD, et al. Functional recovery following traumatic spinal cord injury mediated by a unique polymer scaffold seeded with neural stem cells. Proc Natl Acad Sci USA. 2002;99(5):3024–9.
Teng YD, et al. Physical activity-mediated functional recovery after spinal cord injury: potential roles of neural stem cells. Regen Med. 2006;1(6):763–76.
Tropepe V, et al. Direct neural fate specification from embryonic stem cells: a primitive mammalian neural stem cell stage acquired through a default mechanism. Neuron. 2001;30(1):65–78.
Troyer DL, Weiss ML. Wharton's jelly-derived cells are a primitive stromal cell population. Stem Cells. 2008;26(3):591–9.
Tsai EC, et al. Synthetic hydrogel guidance channels facilitate regeneration of adult rat brainstem motor axons after complete spinal cord transection. J Neurotrauma. 2004;21(6):789–804.
Tsuji O, et al. Therapeutic potential of appropriately evaluated safe-induced pluripotent stem cells for spinal cord injury. Proc Natl Acad Sci USA. 2010;107(28):12704–9.
Tysseling-Mattiace VM, et al. Self-assembling nanofibers inhibit glial scar formation and promote axon elongation after spinal cord injury. J Neurosci. 2008;28(14):3814–23.
Vigh B, et al. The system of cerebrospinal fluid-contacting neurons. Its supposed role in the nonsynaptic signal transmission of the brain. Histol Histopathol. 2004;19(2):607–28.
Vroemen M, et al. Adult neural progenitor cell grafts survive after acute spinal cord injury and integrate along axonal pathways. Eur J Neurosci. 2003;18(4):743–51.
Wada T, et al. Highly efficient differentiation and enrichment of spinal motor neurons derived from human and monkey embryonic stem cells. PLoS ONE. 2009;4(8):e6722.
Wang HS, et al. Mesenchymal stem cells in the Wharton's jelly of the human umbilical cord. Stem Cells. 2004;22(7):1330–7.
Watanabe K, et al. Directed differentiation of telencephalic precursors from embryonic stem cells. Nat Neurosci. 2005;8(3):288–96.
Webber DJ, et al. Transplanted neural progenitor cells survive and differentiate but achieve limited functional recovery in the lesioned adult rat spinal cord. Regen Med. 2007;2(6):929–45.
Weiss S, et al. Multipotent CNS stem cells are present in the adult mammalian spinal cord and ventricular neuroaxis. J Neurosci. 1996;16(23):7599–609.

Weiss ML, et al. Human umbilical cord matrix stem cells: preliminary characterization and effect of transplantation in a rodent model of Parkinson's disease. Stem Cells. 2006;24(3):781–92.

Weiss ML, et al. Immune properties of human umbilical cord Wharton's jelly-derived cells. Stem Cells. 2008;26(11):2865–74.

Wichterle H, Peljto M. Differentiation of mouse embryonic stem cells to spinal motor neurons. Curr Protoc Stem Cell Biol. 2008. Chapter 1:p. Unit 1H 1 1–1H 1 9.

Wichterle H, et al. Directed differentiation of embryonic stem cells into motor neurons. Cell. 2002;110(3):385–97.

Woerly S, et al. Spinal cord repair with PHPMA hydrogel containing RGD peptides (NeuroGel). Biomaterials. 2001;22(10):1095–111.

Woerly S, et al. Development of a sialic acid-containing hydrogel of poly[N-(2-hydroxypropyl) methacrylamide]: characterization and implantation study. Biomacromolecules. 2008;9(9):2329–37.

Wong DY, et al. Macro-architectures in spinal cord scaffold implants influence regeneration. J Neurotrauma. 2008;25(8):1027–37.

Wu S, et al. Bone marrow stromal cells enhance differentiation of cocultured neurosphere cells and promote regeneration of injured spinal cord. J Neurosci Res. 2003;72(3):343–51.

Wu ZY, et al. Transplantation of human amniotic epithelial cells improves hindlimb function in rats with spinal cord injury. Chin Med J (Engl). 2006;119(24):2101–7.

Wu CY, et al. Efficient differentiation of mouse embryonic stem cells into motor neurons. J Vis Exp. 2012;64:e3813.

Yan J, et al. Differentiation and tropic/trophic effects of exogenous neural precursors in the adult spinal cord. J Comp Neurol. 2004;480(1):101–14.

Yan J, et al. Extensive neuronal differentiation of human neural stem cell grafts in adult rat spinal cord. PLoS Med. 2007;4(2):e39.

Yang CC, et al. Transplantation of human umbilical mesenchymal stem cells from Wharton's jelly after complete transection of the rat spinal cord. PLoS ONE. 2008;3(10):e3336.

Yano S, et al. In vivo fluorescence tracking of bone marrow stromal cells transplanted into a pneumatic injury model of rat spinal cord. J Neurotrauma. 2005;22(8):907–18.

Yen BL, et al. Isolation of multipotent cells from human term placenta. Stem Cells. 2005;23(1):3–9.

Yohn DC, et al. Transplanted mouse embryonic stem-cell-derived motoneurons form functional motor units and reduce muscle atrophy. J Neurosci. 2008;28(47):12409–18.

Yoon SH, et al. Complete spinal cord injury treatment using autologous bone marrow cell transplantation and bone marrow stimulation with granulocyte macrophage-colony stimulating factor: Phase I/II clinical trial. Stem Cells. 2007;25(8):2066–73.

Young HE, et al. Mesenchymal stem cells reside within the connective tissues of many organs. Dev Dyn. 1995;202(2):137–44.

Yu J, et al. Induced pluripotent stem cell lines derived from human somatic cells. Science. 2007;318(5858):1917–20.

Zeng H, et al. Specification of region-specific neurons including forebrain glutamatergic neurons from human induced pluripotent stem cells. PLoS ONE. 2010;5(7):e11853.

Zhang SC, et al. In vitro differentiation of transplantable neural precursors from human embryonic stem cells. Nat Biotechnol. 2001;19(12):1129–33.

Zhang HT, et al. Effects of differentiated versus undifferentiated adipose tissue-derived stromal cell grafts on functional recovery after spinal cord contusion. Cell Mol Neurobiol. 2009;29(8):1283–92.

Ziv Y, et al. Synergy between immune cells and adult neural stem/progenitor cells promotes functional recovery from spinal cord injury. Proc Natl Acad Sci USA. 2006;103(35):13174–9.

Chapter 21
Cell Transplantation Therapy for Diabetes Mellitus: From Embryonic Stem Cells to Transdifferentiation of Adult Cells

Xinhua Xiao and Yijing Liu

Abstract Pancreatic transplant and islet cell transplantation are alternative procedure to “cure” diabetes. The last one has already become an accepted practice to stabilize frequent hypoglycemia or severe glycemic lability in highly selected subjects with poor glycemic control. Advancements during the last decade in the fields of regenerative medicine, tissue engineering, immunomodulatory therapy, and gene therapy have drawn us a step closer to making the application of stem cell therapy a feasible reality in the cure of diabetes. However, a combinatorial approach that can combine safe and effective stem cell strategies with reliable existing therapies such as islet transplantation, as well as the latest immunosuppressive and immunomodulatory drug regimens and/or novel bioengineering techniques, would ensure an optimistic scenario for successful translation of stem cell therapy in the cure of diabetes. In short, the application of stem cell therapy in the cure for diabetes appears extremely promising, with bona fide hope for a permanent cure.

Keywords Stem cell · Diabetes · Therapy

Diabetes is a major source of morbidity and mortality due to progressive chronic micro- and macrovascular complications, it is associated with a reduction and dysfunction of ß–pancreatic cells, and the type 1 diabetes mellitus (T1DM) is associated with significant morbidity and mortality in the affected population. Currently, insulin replacement via exogenously administered insulin remains the mainstay

The two authors contributed equally to this work.

X. Xiao (✉)
Department of Endocrinology, Key Laboratory of Endocrinology, Ministry of Health, Peking Union Medical College Hospital, Diabetes Research Center of Chinese Academy of Medical Sciences and Peking Union Medical College, Beijing 100730, China
e-mail: xiaoxinhua@medmail.com.cn

Y. Liu
Department of Gastroenterology, China-Japan Friendship Hospital, Beijing 100029, China

R.C. Zhao (ed.), *Stem Cells: Basics and Clinical Translation*, Translational Medicine Research 1, DOI 10.1007/978-94-017-7273-0_21

of T1DM treatment. Unfortunately, insulin therapy requires either multiple daily injections of insulin or the use of insulin pumps as well as major lifestyle modifications (Silverstein et al. 2005). There has been a search for permanent diabetes "cure" that replaces ß-pancreatic cells, freeing patients from insulin requirements. The radical solution is whole pancreas transplantation, a surgical procedure developed dating back to 1966 (Kelly et al. 1967).

The pancreatic transplant was first introduced by Kelly. Currently, there are 3 types of solid pancreatic transplant and pancreatic islet cell transplants used, with the latter being a recent development. These are simultaneous pancreas kidney transplant, pancreas after kidney transplant for those with a renal transplant at an earlier stage for end-stage renal failure, and pancreas transplant alone for those with severe unawareness of hypoglycemic episodes with no renal disease (Gruessner 2011). Pancreatic transplant has progressed significantly in the past 25 years, and it has success rates comparable to those of other organ transplantations: 80 % of patients achieved and maintained insulin independence for 6–8 years (Gruessner 2011). However, pancreas transplantation requires a significant immunosuppressive therapy to prevent organ rejection, (Lo Monte et al. 1999; Ciancio et al. 2000) and therefore, is performed only in combination with kidney transplantation in diabetic dialysis patients with end-stage renal disease, but unfortunately, this intervention has many complications (Ireland 2011; Ciancio et al. 2000). Recent review studies of the 24-year practice in pancreatic surgery have shown improved outcomes and survival, both in the graft and the patient, and better outcomes including reversal of microvascular complications of diabetes (ophthalmic and neurovascular). Patient survival has now reached more than 95 % at 1 year after transplant and more than 83 % at 5 years after transplant (Gruessner 2011). A recent study by Lindahl and associates, (Lindahl et al. 2013) in Norway, showed that simultaneous kidney and pancreatic transplant patients had a better long-term survival from all causes than either live-donor kidney or deceased-donor kidney recipients. Their results are attributed to better glycemic control, better blood pressure, and lipid control (which lowers cardiovascular risk), as more than 50 % of deaths in this population are attributed to cardiovascular disease (Humar et al. 2000).

Advances in pancreatic transplant have improved patient outcomes, survival, and quality of life. As islet cell transplant is growing and more research studies are coming to light, and implantation is a much simpler method technically than transplant, in the last few years, islet transplantation has proffered an alternative promising therapy, and recent results for islet transplantation demonstrate major improvement in outcomes. While Shcarp et al's work established the liver as an ideal site for islet transplantation, (Scharp et al. 1973) further work by Najarian et al. (1977) in 1977 reported the first successful clinical islet transplant paired with the administration of azathioprine and corticosteroids. Over the past 10 years, there has been significant progress in the development of the islet transplantation procedure. There are five types of isolated pancreatic islet transplantation: autotransplantation in patients after pancreatectomy, allotransplantation in patients after total pancreatomy, allotransplantation in patients with type 1 diabetes, fetal allotransplantation or xenotransplantation in type 1 diabetes, and allotransplantation of islets in type 2 diabetes (Poradzka et al. 2013). The successful transplants of islets have been noted from donors after cardiac death. The selection of donors is strict and includes, e.g., age, cold and warm ischemia times, BMI, cause of death, serum amylase or lipase levels, glycemia, incidence of hypertension, alcoholism, or smoking. In order to achieve normal glycemic level, most recipients are required to undergo 2–3 infusions. It is estimated that to achieve full insulin independence up to 10,000 (5000–10,000) islet equivalents per kilogram are necessary. It is impossible to get that amount from a single donor (Poradzka et al. 2013). Achieving post-transplant islet function and even insulin independence with single donors would help to transition islet transplantation from therapy for few to a cure for all with T1DM. It would also avoid exposing the recipient to polyhuman leukocyte-associated (HLA) antigens that may preclude future transplants for the recipient. Single-donor islet transplant success has been achieved in certain centers, most notably at the University of Minnesota and University of California, San Francisco (UCSF) (Fig. 21.1).

Islet cell transplantation is an attractive alternative therapy to conventional insulin treatment or vascularized whole pancreas transplantation for type 1 diabetic patients by islet cells' isolation from donor pancreata and embolization into the recipient liver via the portal vein (Saidi 2012). Islet-alone transplantation has recently become an accepted practice to stabilize frequent hypoglycemia or severe glycemic lability in highly selected subjects with poor glycemic control Ryan et al. (2006). The Collaborative Islet Transplant Registry Report (CITR) from 2009 reported that 70 % of recipients achieved insulin independence for at least a two-week period, and after one- and two-year post-transplant, 49 and 39 %, respectively, of the recipients remained insulin independent (Hirshberg 2009; Group CR 2009). Moreover, marked improvements in clinical islet transplantation have been observed from 2007 to 2010 as evidenced by retained C-peptide levels, reduction of HbA1c, and reduced islet reinfusion rates (Barton et al. 2012). As the most advanced preparations often result in recovery of only 20–50 % of the potential islet mass (Nanji and Shapiro 2006), most patients require multiple islet

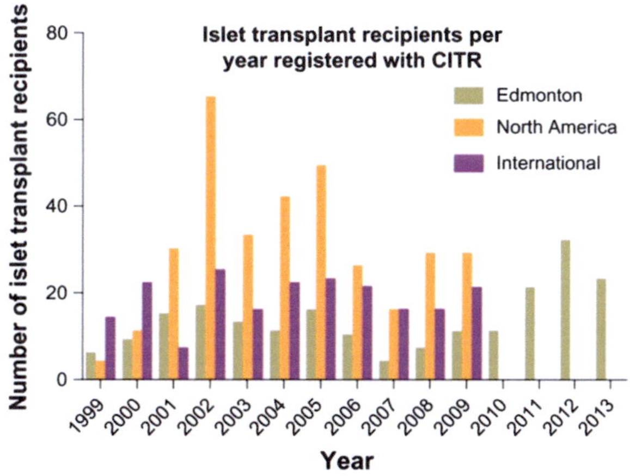

Fig. 21.1 Number of islet transplant recipients from 1999 to 2013 in Edmonton, North America, and International Islet Transplant Centers (Bruni et al. 2014)

infusions to achieve euglycemia and insulin independence (de Kort et al. 2011), and immunosuppressant treatment is necessary to achieve the greatest level of engraftment. Patients must then adhere to a lifelong immunosuppressant regimen, which carries its own adverse effects such as increased risk of infection and cancer (Robertson 2010).

Today, intrahepatic islet infusion via the portal vein accounts for virtually all clinical islet transplants conducted worldwide. A significant amount of intraportal islet mass is lost immediately post-transplant due to innate immune pathways involving platelet and complement activation described in the "Islet engraftment section." Numerous sites have been proposed and tested, both experimentally and in some cases clinically, including the liver, kidney subcapsule, spleen, pancreas, omentum, gastrointestinal wall, immune privileged sites, and subcutaneous spaces. Pepper et al. (2013) and Veriter et al. (2013) have summarized the utility of alternative transplant sites in experimental models and their prospective applicability to the clinical setting. While some alternative sites may be advantageous in experimental models, their feasibility and translation into clinical settings are limited to date. They concluded that the gold standards for islet transplantation in the clinical and experimental settings remain the intrahepatic portal infusion and kidney capsule, respectively.

21.1 Variable Stem Cells

While both pancreas transplantation and islet transplantation are limited by donor shortages, patient-derived induced pluripotent stem (iPS) cells may provide an unlimited supply of transplantable cells for ß-cell replacement therapy in diabetic patients.

21.2 Embryonic Stem Cells

There are two broad types of stem cells, embryonic stem cells (ESCs) and adult stem cells. ESCs are isolated from inner cell mass of blastocyst and can trans-differentiate into cells of all three germ layers. Islet neogenesis is the generation of new islets by locating of pancreatic stem or progenitor cells at the postnatal pancreatic duct. Unfortunately, adult stem cells are rare and difficult to expand in culture. The use of human ESC-derived insulin-producing cells in clinical trials is likely to remain a "nonstarter" because of a number of ethical and scientific considerations. Recent studies have demonstrated that ESCs (Lumelsky et al. 2001; D'Amour et al. 2006; Jiang et al. 2007) iPS cells (Tateishi et al. 2008; Maehr et al. 2009), and adult stem cells from bone marrow (BM) (Xie et al. 2009), pancreas (Noguchi et al. 2010; Seaberg et al. 2004), liver (Yang et al. 2002), umbilical cord blood (Sun et al. 2007), Wharton's jelly (Chao et al. 2008), or placenta (Chang et al. 2007), can differentiate into insulin-producing elements. Derived from ESCs, researchers found "insulin-producing cell clusters" (IPCCs), which does not represent surrogate ß-cells but rather more closely resembles an α and ß cell hybrid. The ability of ESC-derived islet-like cell clusters to survive and function in vivo was tested by grafting clusters subcutaneously into the shoulders of streptozotocin diabetic mice (Lumelsky et al. 2001). IPCC insulin released in response to minimal glucose stimulation (3.3 mmol/L glucose) but not significantly by greater glucose stimulation (25 mmol/L glucose), which indicates IPCCs are defective in glucose-sensing capacity in vivo. But currently, ESCs' use is under debate for ethical and legal issues as well as the risks of teratoma formation (Lee et al. 2009).

21.3 Adipose-Derived Stem Cells

An alternative to ESCs could be adipose-derived stem cells (ADSCs), which could differentiate into insulin-, somatostatin-, and glucagon-expressing or C-peptide positive cells (Timper et al. 2006; Lee et al. 2008). Chandra et al. (2011) reported their ability to restore normoglycemic conditions in streptozotocin-induced diabetic mice, and the study also found that undifferentiated ADSCs exerted moderate control of blood glucose levels, leading to the speculation that the autocrine and paracrine factors of a regenerating pancreas and an hyperglycemic local diabetic micro-environment can contribute to ADSC differentiation. The similar phenomenon also had been shown for bone marrow-derived stem cells (Phadnis et al. 2011). ADSCs have been reported to possess inherent regenerative angiogenic potential and anti-apoptotic capability through their secretion of trophic factors (Moon et al. 2006). ADSCs also have anti-inflammatory and immunomodulatory properties, including suppression of T-cell proliferation (Constantin et al. 2009). Therefore, ADSCs can potentially allow improved engraftment of transplanted islets with enhanced vascularization and suppression of inflammation.

It is now undeniable that the utility of ADSCs in the treatment of diabetes is extremely promising. The abundance of available source tissue, high frequency and multipotency of adipose-derived mesenchymal stem cells, and its trophic and regenerative capabilities all serve as valuable solutions to the ever-increasing diabetic population and associated health crises observed around the world. The exact mechanism of generating insulin-producing cells from ADSCs as well as further maturation of those cells into functional pancreatic islets still needs to be further explored.

21.4 Mesenchymal Stromal Cells

Unlike ESCs, adult stem cells show restricted proliferation and lineage differentiation. Adult stem cells that undergo mesodermal lineage-specific differentiation to osteocytes, adipocytes, and chondrocytes are named as mesenchymal stem cells. MSCs are characterized by their adherence to tissue culture-treated plates and the absence of hematopoietic marker expression. Upon exposure to differentiation media, MSCs undergo differentiation into osteocytes, chondrocytes, and adipocytes. The differentiation potential of MSCs to connective tissues has been exploited for tissue engineering.

Islet-like cells have also been reported from human placenta-derived mesenchymal stem cells, which is able to restore normoglycemia when transplanted under the kidney capsules of streptozotocin-induced diabetic mice (Kadam and Bhonde 2010). Mesenchymal stromal cells (MSCs) are fibroblast-like group of cells from the BM capable of differentiating into specific tissues, representing only 0.001–0.01 % of nucleated marrow cells (Figliuzzi et al. 2014). In recent years, MSCs have been highlighted as a new emerging regenerative therapy due to their multipotency but also due to their paracrine secretion of angiogenic factors, cytokines, and immunomodulatory substances. Although originally potential of MSCs to transdifferentiate into other types of cells such as β cells was considered to be their major regenerative potential, more recent evidence indicates that MSCs carry out tissue repair processes mainly by preferentially migrating to the site of tissue injury (Sordi et al. 2005; Fox et al. 2007) and participating in the repair processes (Caplan and Dennis 2006; Caplan 2009). In this regard, Mathew et al. (2004) provided evidence that bone marrow-derived endothelial cells, transplanted after a pancreatic injury, migrate to the site of damage helping injured ß-cells to recover even if they did not differentiate into insulin-producing cells. Hematopoietic stem cells and endothelial progenitor cells have also been reported in the similar efficacy (Asahara et al. 1997; Di Santo et al. 2009; Park et al. 2010). In other studies, MSCs could support the in vivo survival of the islet through secretion of trophic factors and by creating an appropriate microenvironmental niche, and what's more, MSCs would even promote proper blood circulation in the area around the islet due to their angiogenic capacities. Considering these properties, co-transplantation of MSCs with pancreatic islets holds great potential for improving long-term islet allograft survival and function (Hematti et al. 2013).

An attractive feature of MSCs in the treatment of diabetic microvascular complications is the reported ability of this cell type to benefit microvascular complications through systemic effects. Thus, it is conceivable that MSCs can have a trophic effect on the underlying diabetic microvascular complications. A variety of pre-clinical and initial clinical studies have indicated that MSCs have potential as a regenerative medicine in diabetes-associated microvascular and secondary diabetic complications (Davey et al. 2014).

MSCs have differential effects on the proliferation and cytokine secretion profile of a subpopulation of immune cells. MSCs inhibit T-cell activation, which is independent of the MHC status, allowing the administration of third-party MSCs for immunomodulation. What's more, Interaction between B cells and MSCs produces different results depending upon the culture conditions and species involved. Immunomodulatory properties of MSCs along with clinical success in GVHD have sparked interest in MSC-based immune therapy in allogeneic cell and organ transplantation.

21.5 Transdifferentiation of Adult Cells

Since original stem cells are not easy to get, researchers have been trying to find another way beyond stem cell-based therapy, like transdifferentiation of adult cells. Although the mechanisms responsible for the development of these endocrine cell types are not fully understood, several pancreatic transcription factors have been identified: Sox9, Pdx1, Ngn3, IA1, Pax4, Arx, Nkx2.2, Nkx6.1, Nkx6.2, Pax6, and MafA (Kordowich et al. 2010). Kaneto et al. (2005) showed that concomitant adenoviral application of 2 factors, Pdx1 and Ngn3 or NeuroD, in the livers of mice produced transdifferentiation of hepatic cells into insulin-producing elements associated with significant amelioration of glucose tolerance. Interestingly, exocrine pancreatic cells have the capacity to generate their endocrine counterparts when exposed in vitro to a particular microenvironment, consisting of epidermal growth factor and leukemia inhibitory factor, agonists of the JAK2/STAT3 signaling pathway (Baeyens and Bouwens 2008).

21.6 Apoptosis in Islet Transplantation

However, in the week following pancreatic islet transplantation, up to 50 % of transplanted islets are lost due to apoptotic cell death. Thus, inhibition of islet apoptosis is an attractive and potentially effective therapeutic strategy to prevent the loss of functional islet mass post-transplantation and improve clinical islet transplant outcomes. The key enzymes mediating the progression of apoptosis within a cell are the cysteine aspartate protease family of enzymes called caspases. Caspases reside within a cell as inactive pro-caspases (zymogens) until they are activated in response to pro-apoptotic stimuli. Once activated, caspases proceed to activate other caspases in a hierarchical manner, leading to the amplification of the apoptotic signaling cascade and cell death. In transplantation, rejection of islet grafts involves the activation of the adaptive immune system; in addition, perforin and granzyme are primary mediators of islet cell death following transplantation. In the early post-transplant period, multiple mechanisms are at play that negatively impact β-cell function and lead to islet cell apoptosis. Of these mechanisms, pro-inflammatory cytokines such as IL1β, IL6, IFNγ, tumor necrosis factor-α (TNF-α), and cyclooxygenase-2 mount a fierce inflammatory attack against the newly transplanted islet graft, triggering islet cell death. The release of pro-inflammatory cytokines has been observed in rats, and pro-inflammatory cytokines mediate their inflammatory effects under the control of the nuclear factor κB and MAPK cell signaling pathways (Fig. 21.2).

Engineering of β-cell lines that can protect against pro-inflammatory cytokine-mediated damage represents an interesting alternative to isolated islets for transplantation. Chen et al. (2000) developed a cytokine-resistant rat insulinoma INS-1 cell line capable of protecting against IL1β- and IFNγ-mediated apoptosis more

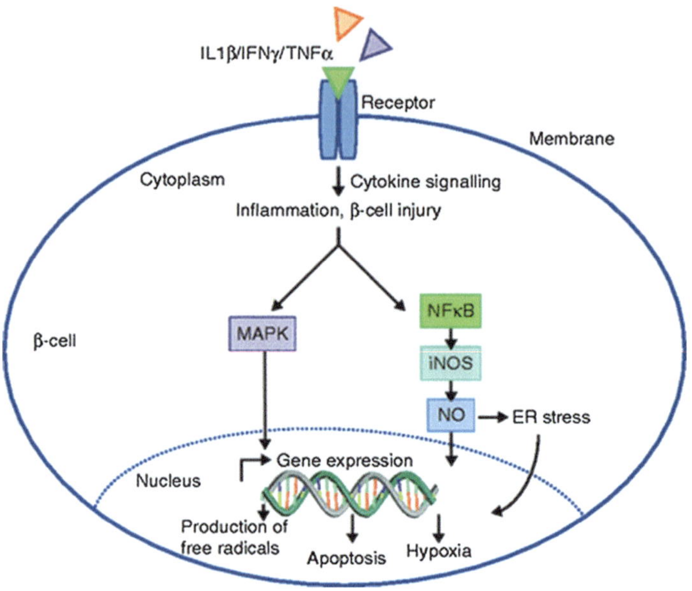

Fig. 21.2 Hughes et al. (2014) cell signaling pathways activated in the pancreatic islets following pro-inflammatory cytokine exposure. Brain death is associated with the endogenous production of pro-inflammatory cytokines such as TNFα, IL1β, and IFNγ (*colored triangles*). The release of these cytokines is associated with inflammation and their presence triggers the activation of the NFκB (*green text box*) and MAPK stress response pathways (*purple text box*), and renders islets non-functional, following induction of inducible nitric oxide synthase (iNOS; *light blue box*), the formation of NO (*blue text box*), and the induction of endoplasmic reticulum (ER) stress. As a result, islets experience subsequent stresses including the production of free radicals, apoptosis, and hypoxia, among others

efficiently than cells stably overexpressing the anti-apoptotic gene *Bcl2*. There was an enhanced anti-apoptotic effect when the cytokine selection strategy was applied to the *Bcl2* overexpressing cells. Importantly, the cells displayed no loss of glucose responsiveness, a critical function that ordinarily disappears very early during apoptosis. Jourdan et al. (2011) provided another islet cell line combining cytokine-resistant selection strategy by overexpressing IGF2.

21.7 Conclusion

Several uncertain factors in stem cell-based cell therapy for diabetes still remain: (1) the absence of gold-standard, reproducible differentiation protocol for generating insulin-producing cells from adult stem cells; (2) an exact dosage of stem cell-derived β-cells to reverse diabetic conditions and feasibility of producing such dosage in vitro; (3) proliferative capacity and maintenance of differentiated insulin-producing cells; (4) sensitivity to counter-regulatory hormones; (5) potential adverse effects of undifferentiated adult stem cells; and (6) potential in vivo migration of differentiated cells following implantation. A number of challenges need to be overcome before based on insulin-producing cell therapy will be utilized for diabetes treatment. One of the major problems to overcome is full maturation of IPCs in vitro, and another is long-term survival of IPCs grafts. Although these problems still remain, we have every reason to believe that islet cell replacement therapy and regenerative medicine will eventually be widely used in clinical practice and that these methods will result in new clinical efficacy or possibly even a cure for diabetes.

References

Asahara T, Murohara T, Sullivan A, Silver M, van der Zee R, Li T, Witzenbichler B, Schatteman G, Isner JM. Isolation of putative progenitor endothelial cells for angiogenesis. Science. 1997;275:964–7.

Baeyens L, Bouwens L. Can beta-cells be derived from exocrine pancreas? Diab Obes Metab. 2008;10(Suppl 4):170–8.

Barton FB, Rickels MR, Alejandro R, Hering BJ, Wease S, Naziruddin B, Oberholzer J, Odorico JS, Garfinkel MR, Levy M, Pattou F, Berney T, Secchi A, Messinger S, Senior PA, Maffi P, Posselt A, Stock PG, Kaufman DB, Luo X, Kandeel F, Cagliero E, Turgeon NA, Witkowski P, Naji A, O'Connell PJ, Greenbaum C, Kudva YC, Brayman KL, Aull MJ, Larsen C, Kay TW, Fernandez LA, Vantyghem MC, Bellin M, Shapiro AM. Improvement in outcomes of clinical islet transplantation: 1999–2010. Diab Care. 2012;35:1436–45.

Bruni A, Gala-Lopez B, Pepper AR, Abualhassan NS, Shapiro AJ. Islet cell transplantation for the treatment of type 1 diabetes: recent advances and future challenges. Diabet Metab Syndr Obes Targets Ther. 2014;7:211–23.

Caplan AI. Why are MSCs therapeutic? New data: new insight. J Pathol. 2009;217:318–24.

Caplan AI, Dennis JE. Mesenchymal stem cells as trophic mediators. J Cell Biochem. 2006;98:1076–84.

Chandra V, Swetha G, Muthyala S, Jaiswal AK, Bellare JR, Nair PD, Bhonde RR. Islet-like cell aggregates generated from human adipose tissue derived stem cells ameliorate experimental diabetes in mice. PLoS One. 2011;6:e20615.

Chang CM, Kao CL, Chang YL, Yang MJ, Chen YC, Sung BL, Tsai TH, Chao KC, Chiou SH, Ku HH. Placenta-derived multipotent stem cells induced to differentiate into insulin-positive cells. Biochem Biophys Res Commun. 2007;357:414–20.

Chao KC, Chao KF, Fu YS, Liu SH. Islet-like clusters derived from mesenchymal stem cells in Wharton's Jelly of the human umbilical cord for transplantation to control type 1 diabetes. PLoS One. 2008;3:e1451.

Chen G, Hohmeier HE, Gasa R, Tran VV, Newgard CB. Selection of insulinoma cell lines with resistance to interleukin-1beta- and gamma-interferon-induced cytotoxicity. Diabetes. 2000;49:562–70.

Ciancio G, Lo Monte A, Buscemi G, Miller J, Burke GW. Use of tacrolimus and mycophenolate mofetil as induction and maintenance in simultaneous pancreas-kidney transplantation. Transpl Int Official J Eur Soc Organ Transplant. 2000a;13(Suppl 1):S191–4.

Ciancio G, Lo Monte A, Julian JF, Romano M, Miller J, Burke GW. Vascular complications following bladder drained, simultaneous pancreas-kidney transplantation: the University of Miami experience. Transpl Int Official J Eur Soc Organ Transplant. 2000b;13(Suppl 1):S187–90.

Constantin G, Marconi S, Rossi B, Angiari S, Calderan L, Anghileri E, Gini B, Bach SD, Martinello M, Bifari F, Galie M, Turano E, Budui S, Sbarbati A, Krampera M, Bonetti B. Adipose-derived mesenchymal stem cells ameliorate chronic experimental autoimmune encephalomyelitis. Stem Cells. 2009;27:2624–35.

D'Amour KA, Bang AG, Eliazer S, Kelly OG, Agulnick AD, Smart NG, Moorman MA, Kroon E, Carpenter MK, Baetge EE. Production of pancreatic hormone-expressing endocrine cells from human embryonic stem cells. Nat Biotechnol. 2006;24:1392–401.

Davey GC, Patil SB, O'Loughlin A, O'Brien T. Mesenchymal stem cell-based treatment for microvascular and secondary complications of diabetes mellitus. Front Endocrinol. 2014;5:86.

de Kort H, de Koning EJ, Rabelink TJ, Bruijn JA, Bajema IM. Islet transplantation in type 1 diabetes. BMJ. 2011;342:d217.

Di Santo S, Yang Z, Wyler von Ballmoos M, Voelzmann J, Diehm N, Baumgartner I, Kalka C. Novel cell-free strategy for therapeutic angiogenesis: in vitro generated conditioned medium can replace progenitor cell transplantation. PLoS One. 2009;4:e5643.

Figliuzzi M, Bonandrini B, Silvani S, Remuzzi A. Mesenchymal stem cells help pancreatic islet transplantation to control type 1 diabetes. World J Stem Cells. 2014;6:163–72.

Fox JM, Chamberlain G, Ashton BA, Middleton J. Recent advances into the understanding of mesenchymal stem cell trafficking. Br J Haematol. 2007;137:491–502.

Group CR. 2007 update on allogeneic islet transplantation from the Collaborative Islet Transplant Registry (CITR). Cell Transpl. 2009;18:753–67.

Gruessner AC. 2011 update on pancreas transplantation: comprehensive trend analysis of 25,000 cases followed up over the course of twenty-four years at the International Pancreas Transplant Registry (IPTR). The review of diabetic studies : RDS. 2011;8:6–16.

Hematti P, Kim J, Stein AP, Kaufman D. Potential role of mesenchymal stromal cells in pancreatic islet transplantation. Transplant Rev. 2013;27:21–9.

Hirshberg B. Implications of the 2008 update from the collaborative islet transplant registry. Curr Diab Rep. 2009;9:258–9.

Hughes A, Rojas-Canales D, Drogemuller C, Voelcker NH, Grey ST, Coates PT. IGF2: an endocrine hormone to improve islet transplant survival. J Endocrinol. 2014;221:R41–8.

Humar A, Kandaswamy R, Granger D, Gruessner RW, Gruessner AC, Sutherland DE. Decreased surgical risks of pancreas transplantation in the modern era. Ann Surg. 2000;231:269–75.

Ireland R. Transplantation: effect of pancreas-after-kidney transplant on renal graft survival. Nat Rev Nephrol. 2011;7:486.

Jiang J, Au M, Lu K, Eshpeter A, Korbutt G, Fisk G, Majumdar AS. Generation of insulin-producing islet-like clusters from human embryonic stem cells. Stem Cells. 2007;25:1940–53.

Jourdan G, Dusseault J, Benhamou PY, Rosenberg L, Halle JP. Co-encapsulation of bioengineered IGF-II-producing cells and pancreatic islets: effect on beta-cell survival. Gene Ther. 2011;18:539–45.

Kadam SS, Bhonde RR. Islet neogenesis from the constitutively nestin expressing human umbilical cord matrix derived mesenchymal stem cells. Islets. 2010;2:112–20.

Kaneto H, Nakatani Y, Miyatsuka T, Matsuoka TA, Matsuhisa M, Hori M, Yamasaki Y. PDX-1/VP16 fusion protein, together with NeuroD or Ngn3, markedly induces insulin gene transcription and ameliorates glucose tolerance. Diabetes. 2005;54:1009–22.

Kelly WD, Lillehei RC, Merkel FK, Idezuki Y, Goetz FC. Allotransplantation of the pancreas and duodenum along with the kidney in diabetic nephropathy. Surgery. 1967;61:827–37.

Kordowich S, Mansouri A, Collombat P. Reprogramming into pancreatic endocrine cells based on developmental cues. Mol Cell Endocrinol. 2010;323:62–9.

Lee J, Han DJ, Kim SC. In vitro differentiation of human adipose tissue-derived stem cells into cells with pancreatic phenotype by regenerating pancreas extract. Biochem Biophys Res Commun. 2008;375:547–51.

Lee AS, Tang C, Cao F, Xie X, van der Bogt K, Hwang A, Connolly AJ, Robbins RC, Wu JC. Effects of cell number on teratoma formation by human embryonic stem cells. Cell Cycle. 2009;8:2608–12.

Lindahl JP, Hartmann A, Horneland R, Holdaas H, Reisaeter AV, Midtvedt K, Leivestad T, Oyen O, Jenssen T. Improved patient survival with simultaneous pancreas and kidney transplantation in recipients with diabetic end-stage renal disease. Diabetologia. 2013;56:1364–71.

Lo Monte AI, Ciancio G, Burke G, Roth D, Miller J. The new immunosuppressive era. Ann Ital Chir. 1999;70:1–19.

Lumelsky N, Blondel O, Laeng P, Velasco I, Ravin R, McKay R. Differentiation of embryonic stem cells to insulin-secreting structures similar to pancreatic islets. Science. 2001;292:1389–94.

Maehr R, Chen S, Snitow M, Ludwig T, Yagasaki L, Goland R, Leibel RL, Melton DA. Generation of pluripotent stem cells from patients with type 1 diabetes. Proc Natl Acad Sci USA. 2009;106:15768–73.

Mathews V, Hanson PT, Ford E, Fujita J, Polonsky KS, Graubert TA. Recruitment of bone marrow-derived endothelial cells to sites of pancreatic beta-cell injury. Diabetes. 2004;53:91–8.

Moon MH, Kim SY, Kim YJ, Kim SJ, Lee JB, Bae YC, Sung SM, Jung JS. Human adipose tissue-derived mesenchymal stem cells improve postnatal neovascularization in a mouse model of hindlimb ischemia. Cell Physiol Biochem Int J Exp Cell Physiol Biochem Pharmacol. 2006;17:279–90.

Najarian JS, Sutherland DE, Matas AJ, Steffes MW, Simmons RL, Goetz FC. Human islet transplantation: a preliminary report. Transpl Proc. 1977;9:233–6.

Nanji SA, Shapiro AM. Advances in pancreatic islet transplantation in humans. Diab Obes Metab. 2006;8:15–25.

Noguchi H, Naziruddin B, Shimoda M, Fujita Y, Chujo D, Takita M, Peng H, Sugimoto K, Itoh T, Tamura Y, Olsen GS, Kobayashi N, Onaca N, Hayashi S, Levy MF, Matsumoto S. Induction of insulin-producing cells from human pancreatic progenitor cells. Transpl Proc. 2010;42:2081–3.

Park KS, Kim YS, Kim JH, Choi B, Kim SH, Tan AH, Lee MS, Lee MK, Kwon CH, Joh JW, Kim SJ, Kim KW. Trophic molecules derived from human mesenchymal stem cells enhance survival, function, and angiogenesis of isolated islets after transplantation. Transplantation. 2010;89:509–17.

Pepper AR, Gala-Lopez B, Ziff O, Shapiro AM. Revascularization of transplanted pancreatic islets and role of the transplantation site. Clin Dev Immunol. 2013;2013:352315.

Phadnis SM, Joglekar MV, Dalvi MP, Muthyala S, Nair PD, Ghaskadbi SM, Bhonde RR, Hardikar AA. Human bone marrow-derived mesenchymal cells differentiate and mature into endocrine pancreatic lineage in vivo. Cytotherapy. 2011;13:279–93.

Poradzka A, Wronski J, Jasik M, Karnafel W, Fiedor P. Insulin replacement therapy in patients with type 1 diabetes by isolated pancreatic islet transplantation. Acta Pol Pharm. 2013;70:943–50.

Robertson RP. Islet transplantation a decade later and strategies for filling a half-full glass. Diabetes. 2010;59:1285–91.

Ryan EA, Bigam D, Shapiro AM. Current indications for pancreas or islet transplant. Diab Obes Metab. 2006;8:1–7.

Saidi RF. Current status of pancreas and islet cell transplantation. Int J Organ Transplant Med. 2012;3:54–60.

Scharp DW, Kemp CB, Knight MJ, Ballinger WF, Lacy PE. The use of ficoll in the preparation of viable islets of langerhans from the rat pancreas. Transplantation. 1973;16:686–9.

Seaberg RM, Smukler SR, Kieffer TJ, Enikolopov G, Asghar Z, Wheeler MB, Korbutt G, van der Kooy D. Clonal identification of multipotent precursors from adult mouse pancreas that generate neural and pancreatic lineages. Nat Biotechnol. 2004;22:1115–24.

Silverstein J, Klingensmith G, Copeland K, Plotnick L, Kaufman F, Laffel L, Deeb L, Grey M, Anderson B, Holzmeister LA, Clark N, American Diabetes A. Care of children and adolescents with type 1 diabetes: a statement of the American Diabetes Association. Diab Care. 2005;28:186–212.

Sordi V, Malosio ML, Marchesi F, Mercalli A, Melzi R, Giordano T, Belmonte N, Ferrari G, Leone BE, Bertuzzi F, Zerbini G, Allavena P, Bonifacio E, Piemonti L. Bone marrow mesenchymal stem cells express a restricted set of functionally active chemokine receptors capable of promoting migration to pancreatic islets. Blood. 2005;106:419–27.

Sun B, Roh KH, Lee SR, Lee YS, Kang KS. Induction of human umbilical cord blood-derived stem cells with embryonic stem cell phenotypes into insulin producing islet-like structure. Biochem Biophys Res Commun. 2007;354:919–23.

Tateishi K, He J, Taranova O, Liang G, D'Alessio AC, Zhang Y. Generation of insulin-secreting islet-like clusters from human skin fibroblasts. J Biol Chem. 2008;283:31601–7.

Timper K, Seboek D, Eberhardt M, Linscheid P, Christ-Crain M, Keller U, Muller B, Zulewski H. Human adipose tissue-derived mesenchymal stem cells differentiate into insulin, somatostatin, and glucagon expressing cells. Biochem Biophys Res Commun. 2006;341:1135–40.

Veriter S, Gianello P, Dufrane D. Bioengineered sites for islet cell transplantation. Curr Diab Rep. 2013;13:745–55.

Xie QP, Huang H, Xu B, Dong X, Gao SL, Zhang B, Wu YL. Human bone marrow mesenchymal stem cells differentiate into insulin-producing cells upon microenvironmental manipulation in vitro. Differ Res Biol Div. 2009;77:483–91.

Yang L, Li S, Hatch H, Ahrens K, Cornelius JG, Petersen BE, Peck AB. In vitro trans-differentiation of adult hepatic stem cells into pancreatic endocrine hormone-producing cells. Proc Natl Acad Sci USA. 2002;99:8078–83.

MIX
Papier aus verantwortungsvollen Quellen
Paper from responsible sources
FSC® C105338
FSC
www.fsc.org

If you have any concerns about our products, you can contact us on **ProductSafety@springernature.com**

In case Publisher is established outside the EU, the EU authorized representative is:
Springer Nature Customer Service Center GmbH
Europaplatz 3, 69115 Heidelberg, Germany

Printed by Libri Plureos GmbH
in Hamburg, Germany